CW00822389

INTRODUCTION TO

Signal Processing

PRENTICE HALL SIGNAL PROCESSING SERIES

Alan V. Oppenheim, Editor

INTRODUCTION TO

Signal Processing

Sophocles J. Orfanidis

Rutgers University

Prentice Hall, Upper Saddle River, New Jersey 07458

Library of Congress Cataloging-in-Publication Date

ORFANIDIS, SOPHOCLES J.
    Introduction to signal processing / Sophocles J. Orfanidis.
      p.   cm.
    Includes bibliographical references and index.
    ISBN 0-13-209172-0
    1. Signal processing—Digital techniques.   I. Title.
TK5102.9.073   1996
621.382'2—dc20   95- 10970
                                                 CIP

Acquisitions editor: *Tom Robbins*
Editorial production supervision: *Barbara Marttine Cappuccio*
Cover design: *Sophocles J. Orfanidis*
Copyeditor: *Martha Williams*
Manufacturing buyer: *Donna Sullivan*
Editorial assistant: *Phyllis Morgan*

 ©1996 by Prentice-Hall, Inc.

Upper Saddle River, New Jersey 07458

Printed in the United States of America

10   9   8

ISBN   0-13-209172-0

Prentice-Hall International (UK) Limited, *London*
Prentice-Hall of Australia Pty. Limited, *Sydney*
Prentice-Hall Canada Inc., *Toronto*
Prentice-Hall Hispanoamericana, S.A., *Mexico*
Prentice-Hall of India Private Limited, *New Delhi*
Prentice-Hall of Japan, Inc., *Tokyo*
Prentice-Hall Asia Pte. Ltd., *Singapore*
Editora Prentice-Hall do Brasil, Ltda., *Rio de Janerio*

To my lifelong friend George Lazos

Contents

13 *Appendices* *728*

Preface

This book provides an applications-oriented introduction to digital signal processing written primarily for electrical engineering undergraduates. Practicing engineers and graduate students may also find it useful as a first text on the subject.

Digital signal processing is everywhere. Today's college students hear "DSP"
all the time in their everyday life—from their CD players, to their electronic music
synthesizers, to the sound cards in their PCs. They hear all about "DSP chips",
"oversampling digital filters", "1-bit A/D and D/A converters", "wavetable sound
synthesis", "audio effects processors", "all-digital audio studios". By the time they
reach their junior year, they are already very eager to learn more about DSP.

Approach

The learning of DSP can be made into a rewarding, interesting, and fun experience
for the student by weaving into the material several applications, such as the above,
that serve as vehicles for teaching the basic DSP concepts, while generating and
maintaining student interest. This has been the guiding philosophy and objective
in writing this text. As a result, the book's emphasis is more on signal processing
than discrete-time system theory, although the basic principles of the latter are
adequately covered.

The book teaches by example and takes a hands-on practical approach that emphasizes the *algorithmic, computational, and programming* aspects of DSP. It contains a large number of worked examples, computer simulations and applications,
and several C and MATLAB functions for implementing various DSP operations. The
practical slant of the book makes the concepts more concrete.

Use

The book may be used at the *junior or senior* level. It is based on a junior-level
DSP course that I have taught at Rutgers since 1988. The assumed background is
only a first course on linear systems. Sections marked with an asterisk (*) are more
appropriate for a second or senior elective course on DSP. The rest can be covered
at the junior level. The included computer experiments can form the basis of an
accompanying DSP lab course, as is done at Rutgers.

A solutions manual, which also contains the results of the computer experiments, is available from the publisher. The C and MATLAB functions may be ob-

tained via anonymous FTP from the Internet site `ece.rutgers.edu` in the directory `/pub/sjo` or by pointing a Web browser to the book's WWW home page at the URL `ftp://ece.rutgers.edu/pub/sjo/intro2sp.html`, or, `http://www.ece.rut-gers.edu/~orfanidi/intro2sp.html`.

Contents and Highlights

Chapters 1 and 2 contain a discussion of the two key DSP concepts of *sampling and quantization.* The first part of Chapter 1 covers the basic issues of sampling, aliasing, and *analog reconstruction* at a level appropriate for juniors. The second part is more advanced and discusses the practical issues of choosing and defining specifications for *antialiasing prefilters and anti-image postfilters.*

Chapter 2 discusses the *quantization process* and some practical implementations of A/D and D/A converters, such as the conversion algorithm for bipolar two's complement successive approximation converters. The standard model of quantization noise is presented, as well as the techniques of *oversampling, noise shaping, and dithering.* The tradeoff between oversampling ratio and savings in bits is derived. This material is continued in Section 12.7 where the implementation and operation of delta-sigma noise shaping quantizers is considered.

Chapter 3 serves as a review of basic *discrete-time systems* concepts, such as linearity, time-invariance, impulse response, convolution, FIR and IIR filters, causality, and stability. It can be covered quickly as most of this material is assumed known from a prerequisite linear systems course.

Chapter 4 focuses on FIR filters and its purpose is to introduce two basic signal processing methods: *block-by-block* processing and *sample-by-sample* processing. In the block processing part, we discuss various approaches to convolution, transient and steady-state behavior of filters, and real-time processing on a block-by-block basis using the overlap-add method and its software implementation. This is further discussed in Section 9.9 using the FFT.

In the sample processing part, we introduce the basic building blocks of filters: adders, multipliers, and delays. We discuss *block diagrams* for FIR filters and their time-domain operation on a sample-by-sample basis. We put a lot of emphasis on the concept of *sample processing algorithm*, which is the repetitive series of computations that must be carried out on each input sample.

We discuss the concept of *circular buffers* and their use in implementing delays and FIR filters. We present a systematic treatment of the subject and carry it on to the remainder of the book. The use of circular delay-line buffers is old, dating back at least 25 years with its application to computer music. However, it has not been treated systematically in DSP texts. It has acquired a new relevance because all modern DSP chips use it to minimize the number of hardware instructions.

Chapter 5 covers the basics of *z*-transforms. We emphasize the *z*-domain view of causality, stability, and frequency spectrum. Much of this material may be known from an earlier linear system course.

Chapter 6 shows the equivalence of various ways of characterizing a linear filter and illustrates their use by example. It also discusses topics such as sinusoidal and steady-state responses, time constants of filters, simple pole/zero designs of first-

and second-order filters as well as comb and notch filters. The issues of inverse filtering and causality are also considered.

Chapter 7 develops the standard *filter realizations* of canonical, direct, and cascade forms, and their implementation with *linear and circular* buffers. Quantization effects are briefly discussed.

Chapter 8 presents three DSP application areas. The first is on digital *waveform generation*, with particular emphasis on *wavetable generators*. The second is on *digital audio effects*, such as flanging, chorusing, reverberation, multitap delays, and dynamics processors, such as compressors, limiters, expanders, and gates. These areas were chosen for their appeal to undergraduates and because they provide concrete illustrations of the use of delays, circular buffers, and filtering concepts in the context of audio signal processing.

The third area is on *noise reduction/signal enhancement*, which is one of the most important applications of DSP and is of interest to practicing engineers and scientists who remove noise from data on a routine basis. Here, we develop the basic principles for designing noise reduction and signal enhancement filters both in the frequency and time domains. We discuss the design and circular buffer implementation of *notch and comb* filters for removing periodic interference, enhancing periodic signals, signal averaging, and separating the luminance and chrominance components in digital color TV systems. We also discuss *Savitzky-Golay* filters for data smoothing and differentiation.

Chapter 9 covers *DFT/FFT algorithms*. The first part emphasizes the issues of spectral analysis, frequency resolution, windowing, and leakage. The second part discusses the computational aspects of the DFT and some of its pitfalls, the difference between physical and computational frequency resolution, the FFT, and fast convolution.

Chapter 10 covers *FIR filter design* using the window method, with particular emphasis on the *Kaiser window*. We also discuss the use of the Kaiser window in spectral analysis.

Chapter 11 discusses *IIR filter design* using the bilinear transformation based on Butterworth and Chebyshev filters. By way of introducing the bilinear transformation, we show how to design practical second-order digital audio *parametric equalizer* filters having prescribed widths, center frequencies, and gains. We also discuss the design of periodic notch and comb filters with prescribed widths.

In the two filter design chapters, we have chosen to present only a few design methods that are simple enough for our intended level of presentation and effective enough to be of practical use.

Chapter 12 discusses *interpolation, decimation, oversampling DSP systems, sample rate converters, and delta-sigma quantizers*. We discuss the use of oversampling for alleviating the need for high quality analog prefilters and postfilters. We present several practical design examples of interpolation filters, including *polyphase and multistage* designs. We consider the design of sample rate converters and study the operation of oversampled delta-sigma quantizers by simulation. This material is too advanced for juniors but not seniors. All undergraduates, however, have a strong interest in it because of its use in digital audio systems such as CD and DAT players.

The Appendix has four parts: (a) a review section on *random signals*; (b) a discussion of random number generators, including uniform, Gaussian, low frequency, and $1/f$ noise generators; (c) C functions for performing the complex arithmetic in the DFT routines; (d) listings of MATLAB functions.

Paths

Several course paths are possible through the text depending on the desired level of presentation. For example, in the 14-week junior course at Rutgers we cover Sections 1.1-1.4, 2.1-2.4, Chapters 3-7, Sections 8.1-8.2, Chapter 9, and Sections 10.1-10.2 and 11.1-11.4. One may omit certain of these sections and/or add others depending on the available time and student interest and background. In a second DSP course at the senior year, one may add Sections 1.5-1.7, 2.5, 8.3, 11.5-11.6, and Chapter 12. In a graduate course, the entire text can be covered comfortably in one semester.

Acknowledgments

I am indebted to the many generations of students who tried earlier versions of the book and helped me refine it. In particular, I would like to thank Mr. Cem Saraydar for his thorough proofreading of the manuscript. I would like to thank my colleagues Drs. Zoran Gajic, Mark Kahrs, James Kaiser, Dino Lelic, Tom Marshall, Peter Meer, and Nader Moayeri for their feedback and encouragement. I am especially indebted to Dr. James Kaiser for enriching my classes over the past eight years with his inspiring yearly lectures on the Kaiser window. I would like to thank the book's reviewers Drs. A. V. Oppenheim, J. A. Fleming, Y.-C. Jenq, W. B. Mikhael, S. J. Reeves, A. Sekey, and J. Weitzen, whose comments helped improve the book. And I would like to thank Rutgers for providing me with a sabbatical leave to finish up the project. I welcome any feedback from readers—it may be sent to orfanidi@ece.rutgers.edu.

Finally, I would like to thank my wife Monica and son John for their love, patience, encouragement, and support.

Sophocles J. Orfanidis

Sampling and Reconstruction

1.1 Introduction

Digital processing of analog signals proceeds in three stages:

1. The analog signal is *digitized*, that is, it is *sampled* and each sample *quantized* to a finite number of bits. This process is called A/D conversion.

2. The digitized samples are processed by a *digital signal processor*.

3. The resulting output samples may be converted back into *analog* form by an analog reconstructor (D/A conversion).

A typical digital signal processing system is shown below.

The digital signal processor can be programmed to perform a variety of signal processing operations, such as filtering, spectrum estimation, and other DSP algorithms. Depending on the speed and computational requirements of the application, the digital signal processor may be realized by a general purpose computer, minicomputer, special purpose DSP chip, or other digital hardware dedicated to performing a particular signal processing task.

The design and implementation of DSP algorithms will be considered in the rest of this text. In the first two chapters we discuss the two key concepts of *sampling* and *quantization*, which are prerequisites to every DSP operation.

1.2 Review of Analog Signals

We begin by reviewing some pertinent topics from analog system theory. An *analog* signal is described by a function of time, say, $x(t)$. The *Fourier transform* $X(\Omega)$ of

$x(t)$ is the *frequency spectrum* of the signal:

$$X(\Omega) = \int_{-\infty}^{\infty} x(t)\, e^{-j\Omega t}\, dt \tag{1.2.1}$$

where Ω is the radian frequency[†] in [radians/second]. The ordinary frequency f in [Hertz] or [cycles/sec] is related to Ω by

$$\Omega = 2\pi f \tag{1.2.2}$$

The physical meaning of $X(\Omega)$ is brought out by the *inverse* Fourier transform, which expresses the arbitrary signal $x(t)$ as a linear superposition of *sinusoids* of different frequencies:

$$x(t) = \int_{-\infty}^{\infty} X(\Omega)\, e^{j\Omega t}\, \frac{d\Omega}{2\pi} \tag{1.2.3}$$

The relative importance of each sinusoidal component is given by the quantity $X(\Omega)$. The *Laplace transform* is defined by

$$X(s) = \int_{-\infty}^{\infty} x(t)\, e^{-st}\, dt$$

It reduces to the Fourier transform, Eq. (1.2.1), under the substitution $s = j\Omega$. The s-plane pole/zero properties of transforms provide additional insight into the nature of signals. For example, a typical exponentially decaying sinusoid of the form

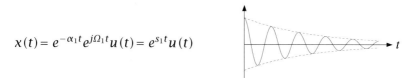

$$x(t) = e^{-\alpha_1 t} e^{j\Omega_1 t} u(t) = e^{s_1 t} u(t)$$

where $s_1 = -\alpha_1 + j\Omega_1$, has Laplace transform

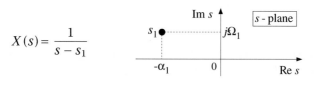

$$X(s) = \frac{1}{s - s_1}$$

with a pole at $s = s_1$, which lies in the left-hand s-plane. Next, consider the response of a *linear system* to an input signal $x(t)$:

$$x(t) \xrightarrow{\text{input}} \boxed{\begin{array}{c} \text{linear} \\ \text{system} \\ h(t) \end{array}} \xrightarrow{\text{output}} y(t)$$

[†]We use the notation Ω to denote the physical frequency in units of [radians/sec], and reserve the notation ω to denote digital frequency in [radians/sample].

The system is characterized completely by the *impulse response* function $h(t)$. The output $y(t)$ is obtained in the time domain by *convolution*:

$$y(t) = \int_{-\infty}^{\infty} h(t - t') x(t') \, dt'$$

or, in the frequency domain by *multiplication*:

$$Y(\Omega) = H(\Omega) X(\Omega) \tag{1.2.4}$$

where $H(\Omega)$ is the *frequency response* of the system, defined as the Fourier transform of the impulse response $h(t)$:

$$H(\Omega) = \int_{-\infty}^{\infty} h(t) e^{-j\Omega t} \, dt \tag{1.2.5}$$

The *steady-state* sinusoidal response of the filter, defined as its response to sinusoidal inputs, is summarized below:

$$x(t) = e^{j\Omega t} \quad \boxed{\begin{array}{c} \text{linear} \\ \text{system} \\ H(\Omega) \end{array}} \quad y(t) = H(\Omega) e^{j\Omega t}$$

sinusoid in   sinusoid out

This figure illustrates the *filtering* action of linear filters, that is, a given frequency component Ω is attenuated (or, magnified) by an amount $H(\Omega)$ by the filter. More precisely, an input sinusoid of frequency Ω will reappear at the output modified in *magnitude* by a factor $|H(\Omega)|$ and shifted in *phase* by an amount $\arg H(\Omega)$:

$$x(t) = e^{j\Omega t} \quad \Rightarrow \quad y(t) = H(\Omega) e^{j\Omega t} = |H(\Omega)| e^{j\Omega t + j\arg H(\Omega)}$$

By linear superposition, if the input consists of the sum of two sinusoids of frequencies Ω_1 and Ω_2 and relative amplitudes A_1 and A_2,

$$x(t) = A_1 e^{j\Omega_1 t} + A_2 e^{j\Omega_2 t}$$

then, after filtering, the steady-state output will be

$$y(t) = A_1 H(\Omega_1) e^{j\Omega_1 t} + A_2 H(\Omega_2) e^{j\Omega_2 t}$$

Notice how the filter changes the *relative* amplitudes of the sinusoids, but not their frequencies. The filtering effect may also be seen in the frequency domain using Eq. (1.2.4), as shown below:

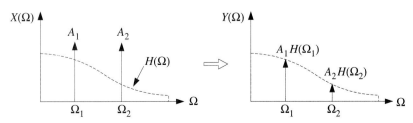

The input spectrum $X(\Omega)$ consists of two sharp spectral lines at frequencies Ω_1 and Ω_2, as can be seen by taking the Fourier transform of $x(t)$:

$$X(\Omega) = 2\pi A_1 \delta(\Omega - \Omega_1) + 2\pi A_2 \delta(\Omega - \Omega_2)$$

The corresponding output spectrum $Y(\Omega)$ is obtained from Eq. (1.2.4):

$$Y(\Omega) = H(\Omega)X(\Omega) = H(\Omega)\left(2\pi A_1 \delta(\Omega - \Omega_1) + 2\pi A_2 \delta(\Omega - \Omega_2)\right)$$
$$= 2\pi A_1 H(\Omega_1)\delta(\Omega - \Omega_1) + 2\pi A_2 H(\Omega_2)\delta(\Omega - \Omega_2)$$

What makes the subject of linear filtering useful is that the designer has complete *control* over the shape of the frequency response $H(\Omega)$ of the filter. For example, if the sinusoidal component Ω_1 represents a desired signal and Ω_2 an unwanted interference, then a filter may be designed that lets Ω_1 pass through, while at the same time it filters out the Ω_2 component. Such a filter must have $H(\Omega_1) = 1$ and $H(\Omega_2) = 0$.

1.3 Sampling Theorem

Next, we study the sampling process, illustrated in Fig. 1.3.1, where the analog signal $x(t)$ is *periodically measured* every T seconds. Thus, time is discretized in units of the *sampling interval T*:

$$t = nT, \qquad n = 0, 1, 2, \ldots$$

Considering the resulting stream of samples as an *analog* signal, we observe that the sampling process represents a very drastic chopping operation on the original signal $x(t)$, and therefore, it will introduce a lot of spurious *high-frequency* components into the frequency spectrum. Thus, for system design purposes, two questions must be answered:

1. What is the effect of sampling on the original frequency spectrum?

2. How should one choose the sampling interval T?

We will try to answer these questions intuitively, and then more formally using Fourier transforms. We will see that although the sampling process generates high frequency components, these components appear in a very regular fashion, that is, *every* frequency component of the original signal is *periodically replicated* over the entire frequency axis, with period given by the *sampling rate*:

$$\boxed{f_s = \frac{1}{T}} \tag{1.3.1}$$

This replication property will be justified first for simple sinusoidal signals and then for arbitrary signals. Consider, for example, a single sinusoid $x(t) = e^{2\pi jft}$ of frequency f. Before sampling, its spectrum consists of a single sharp spectral line

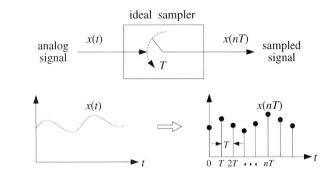

Fig. 1.3.1   Ideal sampler.

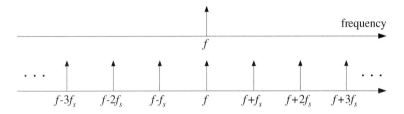

Fig. 1.3.2   Spectrum replication caused by sampling.

at f. But after sampling, the spectrum of the sampled sinusoid $x(nT) = e^{2\pi j f nT}$ will be the periodic replication of the original spectral line at intervals of f_s, as shown in Fig. 1.3.2.

Note also that starting with the replicated spectrum of the sampled signal, one cannot tell uniquely what the original frequency was. It could be *any* one of the replicated frequencies, namely, $f' = f + mf_s$, $m = 0, \pm1, \pm2, \ldots$. That is so because any one of them has the same periodic replication when sampled. This potential confusion of the original frequency with another is known as *aliasing* and can be avoided if one satisfies the conditions of the sampling theorem.

The sampling theorem provides a quantitative answer to the question of how to choose the sampling time interval T. Clearly, T must be *small enough* so that signal variations that occur between samples are not lost. But how small is small enough? It would be very impractical to choose T too small because then there would be too many samples to be processed. This is illustrated in Fig. 1.3.3, where T is small enough to resolve the details of signal 1, but is unnecessarily small for signal 2.

Another way to say the same thing is in terms of the sampling rate f_s, which is measured in units of [samples/sec] or [Hertz] and represents the "density" of samples per unit time. Thus, a rapidly varying signal must be sampled at a high sampling rate f_s, whereas a slowly varying signal may be sampled at a lower rate.

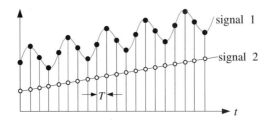

Fig. 1.3.3 Signal 2 is oversampled.

1.3.1 Sampling Theorem

A more quantitative criterion is provided by the *sampling theorem* which states that for *accurate* representation of a signal $x(t)$ by its time samples $x(nT)$, two conditions must be met:

1. The signal $x(t)$ must be *bandlimited*, that is, its frequency spectrum must be limited to contain frequencies up to some maximum frequency, say f_{max}, and no frequencies beyond that. A typical bandlimited spectrum is shown in Fig. 1.3.4.

2. The sampling rate f_s must be chosen to be at least *twice* the maximum frequency f_{max}, that is,

$$\boxed{f_s \geq 2f_{max}}$$ (1.3.2)

or, in terms of the sampling time interval: $T \leq \dfrac{1}{2f_{max}}$.

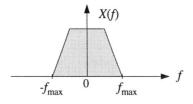

Fig. 1.3.4 Typical bandlimited spectrum.

The *minimum* sampling rate allowed by the sampling theorem, that is, $f_s = 2f_{max}$, is called the *Nyquist rate.* For arbitrary values of f_s, the quantity $f_s/2$ is called the *Nyquist frequency* or folding frequency. It defines the endpoints of the *Nyquist frequency interval*:

$$\boxed{[-\frac{f_s}{2}, \frac{f_s}{2}] = \text{Nyquist Interval}}$$

The Nyquist frequency $f_s/2$ also defines the *cutoff frequencies* of the lowpass analog prefilters and postfilters that are required in DSP operations. The values of f_{max} and f_s depend on the application. Typical sampling rates for some common DSP applications are shown in the following table.

application	f_{max}	f_s
geophysical	500 Hz	1 kHz
biomedical	1 kHz	2 kHz
mechanical	2 kHz	4 kHz
speech	4 kHz	8 kHz
audio	20 kHz	40 kHz
video	4 MHz	8 MHz

1.3.2 Antialiasing Prefilters

The practical implications of the sampling theorem are quite important. Since most signals are not bandlimited, they must be made so by lowpass filtering *before* sampling.

In order to sample a signal at a desired rate f_s and satisfy the conditions of the sampling theorem, the signal must be *prefiltered* by a lowpass *analog* filter, known as an *antialiasing* prefilter. The cutoff frequency of the prefilter, f_{max}, must be taken to be at most equal to the Nyquist frequency $f_s/2$, that is, $f_{max} \leq f_s/2$. This operation is shown in Fig. 1.3.5.

The output of the analog prefilter will then be *bandlimited* to maximum frequency f_{max} and may be sampled properly at the desired rate f_s. The spectrum replication caused by the sampling process can also be seen in Fig. 1.3.5. It will be discussed in detail in Section 1.5.

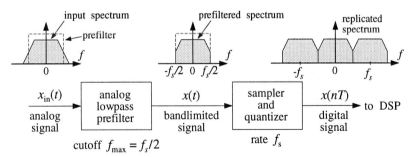

Fig. 1.3.5 Antialiasing prefilter.

It should be emphasized that the rate f_s must be chosen to be high enough so that, after the prefiltering operation, the surviving signal spectrum within the Nyquist interval $[-f_s/2, f_s/2]$ contains all the *significant* frequency components for the application at hand.

Example 1.3.1: In a hi-fi digital audio application, we wish to digitize a music piece using a sampling rate of 40 kHz. Thus, the piece must be prefiltered to contain frequencies up to 20 kHz. After the prefiltering operation, the resulting spectrum of frequencies is more than adequate for this application because the human ear can hear frequencies only up to 20 kHz. □

Example 1.3.2: Similarly, the spectrum of speech prefiltered to about 4 kHz results in very intelligible speech. Therefore, in digital speech applications it is adequate to use sampling rates of about 8 kHz and prefilter the speech waveform to about 4 kHz. □

What happens if we do not sample in accordance with the sampling theorem? If we undersample, we may be missing important time variations between sampling instants and may arrive at the erroneous conclusion that the samples represent a signal which is smoother than it actually is. In other words, we will be confusing the true frequency content of the signal with a lower frequency content. Such confusion of signals is called *aliasing* and is depicted in Fig. 1.3.6.

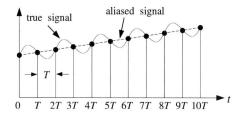

Fig. 1.3.6 Aliasing in the time domain.

1.3.3 Hardware Limits

Next, we consider the restrictions imposed on the choice of the sampling rate f_s by the hardware. The sampling theorem provides a *lower* bound on the allowed values of f_s. The hardware used in the application imposes an *upper* bound.

In real-time applications, each input sample must be acquired, quantized, and processed by the DSP, and the output sample converted back into analog format. Many of these operations can be pipelined to reduce the total processing time. For example, as the DSP is processing the present sample, the D/A may be converting the previous output sample, while the A/D may be acquiring the next input sample.

In any case, there is a total processing or computation time, say T_{proc} seconds, required for each sample. The time interval T between input samples must be greater than T_{proc}; otherwise, the processor would not be able to keep up with the incoming samples. Thus,

$$\boxed{T \geq T_{\text{proc}}}$$

or, expressed in terms of the computation or processing rate, $f_{\text{proc}} = 1/T_{\text{proc}}$, we obtain the upper bound $f_s \leq f_{\text{proc}}$, which combined with Eq. (1.3.2) restricts the choice of f_s to the range:

$$\boxed{2f_{\max} \leq f_s \leq f_{\text{proc}}}$$

In succeeding sections we will discuss the phenomenon of aliasing in more detail, provide a quantitative proof of the sampling theorem, discuss the spectrum replication property, and consider the issues of practical sampling and reconstruction and their effect on the overall quality of a digital signal processing system. Quantization will be considered later on.

1.4 *Sampling of Sinusoids*

The two conditions of the sampling theorem, namely, that $x(t)$ be bandlimited and the requirement $f_s \geq 2f_{max}$, can be derived intuitively by considering the sampling of sinusoidal signals only. Figure 1.4.1 shows a sinusoid of frequency f,

$$x(t) = \cos(2\pi f t)$$

that has been sampled at the three rates: $f_s = 8f$, $f_s = 4f$, and $f_s = 2f$. These rates correspond to taking 8, 4, and 2 samples in each cycle of the sinusoid.

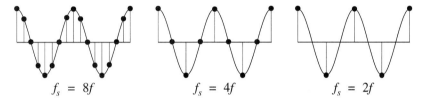

Fig. 1.4.1 Sinusoid sampled at rates $f_s = 8f, 4f, 2f$.

Simple inspection of these figures leads to the conclusion that the *minimum* acceptable number of samples per cycle is *two*. The representation of a sinusoid by two samples per cycle is hardly adequate,[†] but at least it does incorporate the basic *up-down* nature of the sinusoid. The number of samples per cycle is given by the quantity f_s/f:

$$\frac{f_s}{f} = \frac{\text{samples/sec}}{\text{cycles/sec}} = \frac{\text{samples}}{\text{cycle}}$$

Thus, to sample a single sinusoid properly, we must require

$$\frac{f_s}{f} \geq 2 \text{ samples/cycle} \qquad \Rightarrow \qquad f_s \geq 2f \qquad (1.4.1)$$

Next, consider the case of an arbitrary signal $x(t)$. According to the inverse Fourier transform of Eq. (1.2.3), $x(t)$ can be expressed as a linear combination of sinusoids. Proper sampling of $x(t)$ will be achieved only if *every* sinusoidal component of $x(t)$ is properly sampled.

[†]It also depends on the phase of the sinusoid. For example, sampling at the zero crossings instead of at the peaks, would result in zero values for the samples.

This requires that the signal $x(t)$ be bandlimited. Otherwise, it would contain sinusoidal components of arbitrarily high frequency f, and to sample those accurately, we would need, by Eq. (1.4.1), arbitrarily high rates f_s. If the signal is bandlimited to some maximum frequency f_{max}, then by choosing $f_s \geq 2f_{max}$, we are accurately sampling the *fastest-varying* component of $x(t)$, and thus a fortiori, all the slower ones. As an example, consider the special case:

$$x(t) = A_1 \cos(2\pi f_1 t) + A_2 \cos(2\pi f_2 t) + \cdots + A_{max} \cos(2\pi f_{max} t)$$

where f_i are listed in increasing order. Then, the conditions

$$2f_1 \leq 2f_2 \leq \cdots \leq 2f_{max} \leq f_s$$

imply that every component of $x(t)$, and hence $x(t)$ itself, is properly sampled.

1.4.1 Analog Reconstruction and Aliasing

Next, we discuss the aliasing effects that result if one violates the sampling theorem conditions (1.3.2) or (1.4.1). Consider the complex version of a sinusoid:

$$x(t) = e^{j\Omega t} = e^{2\pi j f t}$$

and its sampled version obtained by setting $t = nT$,

$$x(nT) = e^{j\Omega T n} = e^{2\pi j f T n}$$

Define also the following family of sinusoids, for $m = 0, \pm 1, \pm 2, \ldots$,

$$x_m(t) = e^{2\pi j(f + mf_s)t}$$

and their sampled versions,

$$x_m(nT) = e^{2\pi j(f + mf_s)Tn}$$

Using the property $f_s T = 1$ and the trigonometric identity,

$$e^{2\pi j m f_s T n} = e^{2\pi j m n} = 1$$

we find that, although the signals $x_m(t)$ are different from each other, their *sampled* values are the *same*; indeed,

$$x_m(nT) = e^{2\pi j(f + mf_s)Tn} = e^{2\pi j f T n} e^{2\pi j m f_s T n} = e^{2\pi j f T n} = x(nT)$$

In terms of their sampled values, the signals $x_m(t)$ are indistinguishable, or aliased. Knowledge of the sample values $x(nT) = x_m(nT)$ is not enough to determine which among them was the original signal that was sampled. It could have been any one of the $x_m(t)$. In other words, the set of frequencies,

$$f, \ f \pm f_s, \ f \pm 2f_s, \ \ldots, \ f \pm mf_s, \ \ldots \qquad (1.4.2)$$

are equivalent to each other. The effect of sampling was to replace the original frequency f with the replicated set (1.4.2). This is the intuitive explanation of the spectrum replication property depicted in Fig. 1.3.2. A more mathematical explanation will be given later using Fourier transforms.

Given that the sample values $x(nT)$ do not uniquely determine the analog signal they came from, the question arises: What analog signal would result if these samples were fed into an analog reconstructor, as shown in Fig. 1.4.2?

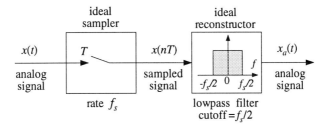

Fig. 1.4.2 Ideal reconstructor as a lowpass filter.

We will see later that an *ideal* analog reconstructor *extracts* from a sampled signal all the frequency components that lie *within* the Nyquist interval $[-f_s/2, f_s/2]$ and *removes* all frequencies outside that interval. In other words, an ideal reconstructor acts as a *lowpass filter* with cutoff frequency equal to the Nyquist frequency $f_s/2$.

Among the frequencies in the replicated set (1.4.2), there is a *unique* one that lies *within* the Nyquist interval.[†] It is obtained by reducing the original f modulo-f_s, that is, adding to or subtracting from f enough *multiples* of f_s until it lies within the *symmetric* Nyquist interval $[-f_s/2, f_s/2]$. We denote this operation by[‡]

$$\boxed{f_a = f \bmod (f_s)} \tag{1.4.3}$$

This is the frequency, in the replicated set (1.4.2), that will be extracted by the analog reconstructor. Therefore, the reconstructed sinusoid will be:

$$x_a(t) = e^{2\pi j f_a t}$$

It is easy to see that $f_a = f$ only if f lies within the Nyquist interval, that is, only if $|f| \le f_s/2$, which is equivalent to the sampling theorem requirement. If f lies outside the Nyquist interval, that is, $|f| > f_s/2$, violating the sampling theorem condition, then the "aliased" frequency f_a will be different from f and the reconstructed analog signal $x_a(t)$ will be different from $x(t)$, even though the two agree at the sampling times, $x_a(nT) = x(nT)$.

It is instructive also to plot in Fig. 1.4.3 the aliased frequency $f_a = f \bmod (f_s)$ versus the true frequency f. Observe how the straight line $f_{\text{true}} = f$ is brought down in segments by parallel translation of the Nyquist periods by multiples of f_s.

[†] The only exception is when it falls exactly on the left or right edge of the interval, $f = \pm f_s/2$.

[‡] This differs slightly from a true modulo operation; the latter would bring f into the right-sided Nyquist interval $[0, f_s]$.

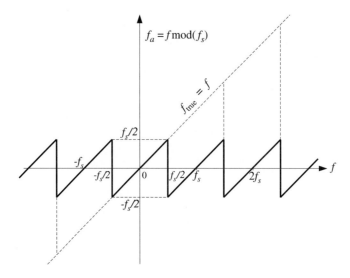

Fig. 1.4.3 $f \bmod(f_s)$ versus f.

In summary, potential aliasing effects that can arise at the reconstruction phase of DSP operations can be avoided if one makes sure that *all* frequency components of the signal to be sampled satisfy the sampling theorem condition, $|f| \leq f_s/2$, that is, all frequency components lie *within* the Nyquist interval. This is ensured by the lowpass antialiasing prefilter, which removes all frequencies beyond the Nyquist frequency $f_s/2$, as shown in Fig. 1.3.5.

Example 1.4.1: Consider a sinusoid of frequency $f = 10$ Hz sampled at a rate of $f_s = 12$ Hz. The sampled signal will contain all the replicated frequencies $10 + m12$ Hz, $m = 0, \pm 1, \pm 2, \ldots$, or,

$$\ldots, -26, \ -14, \ -2, \ 10, \ 22, \ 34, \ 46, \ldots$$

and among these only $f_a = 10 \bmod(12) = 10 - 12 = -2$ Hz lies within the Nyquist interval $[-6, 6]$ Hz. This sinusoid will appear at the output of a reconstructor as a -2 Hz sinusoid instead of a 10 Hz one.

On the other hand, had we sampled at a proper rate, that is, greater than $2f = 20$ Hz, say at $f_s = 22$ Hz, then no aliasing would result because the given frequency of 10 Hz already lies within the corresponding Nyquist interval of $[-11, 11]$ Hz. □

Example 1.4.2: Suppose a music piece is sampled at rate of 40 kHz without using a prefilter with cutoff of 20 kHz. Then, inaudible components having frequencies greater than 20 kHz can be aliased into the Nyquist interval $[-20, 20]$ distorting the true frequency components in that interval. For example, all components in the inaudible frequency range $20 \leq f \leq 60$ kHz will be aliased with $-20 = 20 - 40 \leq f - f_s \leq 60 - 40 = 20$ kHz, which are audible. □

Example 1.4.3: The following five signals, where t is in seconds, are sampled at a rate of 4 Hz:

$$-\sin(14\pi t), \quad -\sin(6\pi t), \quad \sin(2\pi t), \quad \sin(10\pi t), \quad \sin(18\pi t)$$

Show that they are all aliased with each other in the sense that their sampled values are the same.

Solution: The frequencies of the five sinusoids are:

$$-7, \quad -3, \quad 1, \quad 5, \quad 9 \quad \text{Hz}$$

They differ from each other by multiples of $f_s = 4$ Hz. Their sampled spectra will be indistinguishable from each other because each of these frequencies has the *same* periodic replication in multiples of 4 Hz.

Writing the five frequencies compactly:

$$f_m = 1 + 4m, \qquad m = -2, -1, 0, 1, 2$$

we can express the five sinusoids as:

$$x_m(t) = \sin(2\pi f_m t) = \sin(2\pi(1 + 4m)t), \qquad m = -2, -1, 0, 1, 2$$

Replacing $t = nT = n/f_s = n/4$ sec, we obtain the sampled signals:

$$x_m(nT) = \sin(2\pi(1 + 4m)nT) = \sin(2\pi(1 + 4m)n/4)$$

$$= \sin(2\pi n/4 + 2\pi mn) = \sin(2\pi n/4)$$

which are the same, independently of m. The following figure shows the five sinusoids over the interval $0 \le t \le 1$ sec.

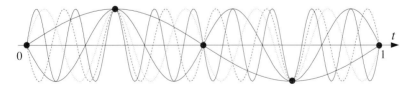

They all intersect at the sampling time instants $t = nT = n/4$ sec. We will reconsider this example in terms of rotating wheels in Section 1.4.2.   □

Example 1.4.4: Let $x(t)$ be the sum of sinusoidal signals

$$x(t) = 4 + 3\cos(\pi t) + 2\cos(2\pi t) + \cos(3\pi t)$$

where t is in milliseconds. Determine the minimum sampling rate that will not cause any aliasing effects, that is, the Nyquist rate. To observe such aliasing effects, suppose this signal is sampled at half its Nyquist rate. Determine the signal $x_a(t)$ that would be aliased with $x(t)$.

Solution: The frequencies of the four terms are: $f_1 = 0$, $f_2 = 0.5$ kHz, $f_3 = 1$ kHz, and $f_4 = 1.5$ kHz (they are in kHz because t is in msec). Thus, $f_{max} = f_4 = 1.5$ kHz and the Nyquist rate will be $2f_{max} = 3$ kHz. If $x(t)$ is now sampled at half this rate, that is, at $f_s = 1.5$ kHz, then aliasing will occur. The corresponding Nyquist interval is $[-0.75, 0.75]$ kHz. The frequencies f_1 and f_2 are already in it, and hence they are not aliased, in the sense that $f_{1a} = f_1$ and $f_{2a} = f_2$. But f_3 and f_4 lie outside the Nyquist interval and they will be aliased with

$$f_{3a} = f_3 \bmod (f_s) = 1 \bmod (1.5) = 1 - 1.5 = -0.5 \text{ kHz}$$

$$f_{4a} = f_4 \bmod (f_s) = 1.5 \bmod (1.5) = 1.5 - 1.5 = 0 \text{ kHz}$$

The aliased signal $x_a(t)$ is obtained from $x(t)$ by replacing f_1, f_2, f_3, f_4 by $f_{1a}, f_{2a}, f_{3a}, f_{4a}$. Thus, the signal

$$x(t) = 4 \cos (2\pi f_1 t) + 3 \cos (2\pi f_2 t) + 2 \cos (2\pi f_3 t) + \cos (2\pi f_4 t)$$

will be aliased with

$$x_a(t) = 4 \cos (2\pi f_{1a} t) + 3 \cos (2\pi f_{2a} t) + 2 \cos (2\pi f_{3a} t) + \cos (2\pi f_{4a} t)$$

$$= 4 + 3 \cos (\pi t) + 2 \cos (-\pi t) + \cos (0)$$

$$= 5 + 5 \cos (\pi t)$$

The signals $x(t)$ and $x_a(t)$ are shown below. Note that they agree only at their sampled values, that is, $x_a(nT) = x(nT)$. The aliased signal $x_a(t)$ is smoother, that is, it has lower frequency content than $x(t)$ because its spectrum lies entirely within the Nyquist interval, as shown below:

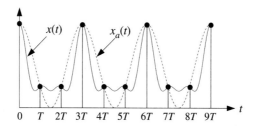

The form of $x_a(t)$ can also be derived in the frequency domain by replicating the spectrum of $x(t)$ at intervals of $f_s = 1.5$ kHz, and then extracting whatever part of the spectrum lies within the Nyquist interval. The following figure shows this procedure.

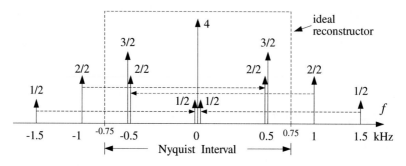

Each spectral line of $x(t)$ is replicated in the fashion of Fig. 1.3.2. The two spectral lines of strength $1/2$ at $f_3 = \pm 1.5$ kHz replicate onto $f = 0$ and the amplitudes add up to give a total amplitude of $(4 + 1/2 + 1/2) = 5$. Similarly, the two spectral lines of strength $2/2$ at $f_3 = \pm 1$ kHz replicate onto $f = \mp 0.5$ kHz and the amplitudes add to give $(3/2 + 2/2) = 2.5$ at $f = \pm 0.5$ kHz. Thus, the ideal reconstructor will extract $f_1 = 0$ of strength 5 and $f_2 = \pm 0.5$ of equal strengths 2.5, which recombine to give:

$$5 + 2.5e^{2\pi j 0.5 t} + 2.5e^{-2\pi j 0.5 t} = 5 + 5\cos(\pi t)$$

This example shows how aliasing can distort irreversibly the amplitudes of the original frequency components within the Nyquist interval. □

Example 1.4.5: The signal

$$x(t) = \sin(\pi t) + 4\sin(3\pi t)\cos(2\pi t)$$

where t is in msec, is sampled at a rate of 3 kHz. Determine the signal $x_a(t)$ aliased with $x(t)$. Then, determine two other signals $x_1(t)$ and $x_2(t)$ that are aliased with the same $x_a(t)$, that is, such that $x_1(nT) = x_2(nT) = x_a(nT)$.

Solution: To determine the frequency content of $x(t)$, we must express it as a sum of sinusoids. Using the trigonometric identity $2\sin a\cos b = \sin(a + b) + \sin(a - b)$, we find:

$$x(t) = \sin(\pi t) + 2[\sin(3\pi t + 2\pi t) + \sin(3\pi t - 2\pi t)] = 3\sin(\pi t) + 2\sin(5\pi t)$$

Thus, the frequencies present in $x(t)$ are $f_1 = 0.5$ kHz and $f_2 = 2.5$ kHz. The first already lies in the Nyquist interval $[-1.5, 1, 5]$ kHz so that $f_{1a} = f_1$. The second lies outside and can be reduced mod f_s to give $f_{2a} = f_2 \bmod(f_s) = 2.5 \bmod(3) = 2.5 - 3 = -0.5$. Thus, the given signal will "appear" as:

$$\begin{aligned} x_a(t) &= 3\sin(2\pi f_{1a} t) + 2\sin(2\pi f_{2a} t) \\ &= 3\sin(\pi t) + 2\sin(-\pi t) = 3\sin(\pi t) - 2\sin(\pi t) \\ &= \sin(\pi t) \end{aligned}$$

To find two other signals that are aliased with $x_a(t)$, we may shift the original frequencies f_1, f_2 by multiples of f_s. For example,

$$\begin{aligned} x_1(t) &= 3\sin(7\pi t) + 2\sin(5\pi t) \\ x_2(t) &= 3\sin(13\pi t) + 2\sin(11\pi t) \end{aligned}$$

where we replaced $\{f_1, f_2\}$ by $\{f_1 + f_s, f_2\} = \{3.5, 2.5\}$ for $x_1(t)$, and by $\{f_1 + 2f_s, f_2 + f_s\} = \{6.5, 5.5\}$ for $x_2(t)$. □

Example 1.4.6: Consider a periodic square wave with period $T_0 = 1$ sec, defined within its basic period $0 \le t \le 1$ by

$$x(t) = \begin{cases} 1, & \text{for} \quad 0 < t < 0.5 \\ -1, & \text{for} \quad 0.5 < t < 1 \\ 0, & \text{for} \quad t = 0, 0.5, 1 \end{cases}$$

where t is in seconds. The square wave is sampled at rate f_s and the resulting samples are reconstructed by an *ideal* reconstructor as in Fig. 1.4.2. Determine the signal $x_a(t)$ that will appear at the output of the reconstructor for the two cases $f_s = 4$ Hz and $f_s = 8$ Hz. Verify that $x_a(t)$ and $x(t)$ agree at the sampling times $t = nT$.

Solution: The Fourier series expansion of the square wave contains odd harmonics at frequencies $f_m = m/T_0 = m$ Hz, $m = 1, 3, 5, 7, \ldots$. It is given by

$$x(t) = \sum_{m=1,3,5,\ldots} b_m \sin(2\pi m t) =$$

$$= b_1 \sin(2\pi t) + b_3 \sin(6\pi t) + b_5 \sin(10\pi t) + \cdots \qquad (1.4.4)$$

where $b_m = 4/(\pi m)$, $m = 1, 3, 5, \ldots$. Because of the presence of an infinite number of harmonics, the square wave is not bandlimited and, thus, cannot be sampled properly at any rate. For the rate $f_s = 4$ Hz, only the $f_1 = 1$ harmonic lies within the Nyquist interval $[-2, 2]$ Hz. For the rate $f_s = 8$ Hz, only $f_1 = 1$ and $f_3 = 3$ Hz lie in $[-4, 4]$ Hz. The following table shows the true frequencies and the corresponding aliased frequencies in the two cases:

f_s	f	1	3	5	7	9	11	13	15	$\cdots$
4 Hz	$f \bmod(4)$	1	-1	1	-1	1	-1	1	-1	$\cdots$
8 Hz	$f \bmod(8)$	1	3	-3	-1	1	3	-3	-1	$\cdots$

Note the repeated patterns of aliased frequencies in the two cases. If a harmonic is aliased with $\pm f_1 = \pm 1$, then the corresponding term in Eq. (1.4.4) will appear (at the output of the reconstructor) as $\sin(\pm 2\pi f_1 t) = \pm \sin(2\pi t)$. And, if it is aliased with $\pm f_3 = \pm 3$, the term will appear as $\sin(\pm 2\pi f_3 t) = \pm \sin(6\pi t)$. Thus, for $f_s = 4$, the aliased signal will be

$$x_a(t) = b_1 \sin(2\pi t) - b_3 \sin(2\pi t) + b_5 \sin(2\pi t) - b_7 \sin(2\pi t) + \cdots$$

$$= (b_1 - b_3 + b_5 - b_7 + b_9 - b_{11} + \cdots) \sin(2\pi t)$$

$$= A \sin(2\pi t)$$

where

$$A = \sum_{k=0}^{\infty} (b_{1+4k} - b_{3+4k}) = \frac{4}{\pi} \sum_{k=0}^{\infty} \left[\frac{1}{1 + 4k} - \frac{1}{3 + 4k} \right] \qquad (1.4.5)$$

Similarly, for $f_s = 8$, grouping together the 1 and 3 Hz terms, we find the aliased signal

$$x_a(t) = (b_1 - b_7 + b_9 - b_{15} + \cdots)\sin(2\pi t) +$$

$$+ (b_3 - b_5 + b_{11} - b_{13} + \cdots)\sin(6\pi t)$$

$$= B\sin(2\pi t) + C\sin(6\pi t)$$

where

$$B = \sum_{k=0}^{\infty}(b_{1+8k} - b_{7+8k}) = \frac{4}{\pi}\sum_{k=0}^{\infty}\left[\frac{1}{1+8k} - \frac{1}{7+8k}\right]$$

$$C = \sum_{k=0}^{\infty}(b_{3+8k} - b_{5+8k}) = \frac{4}{\pi}\sum_{k=0}^{\infty}\left[\frac{1}{3+8k} - \frac{1}{5+8k}\right]$$

(1.4.6)

There are two ways to determine the aliased coefficients A, B, C. One is to demand that the *sampled* signals $x_a(nT)$ and $x(nT)$ agree. For example, in the first case we have $T = 1/f_s = 1/4$, and therefore, $x_a(nT) = A\sin(2\pi n/4) = A\sin(\pi n/2)$. The condition $x_a(nT) = x(nT)$ evaluated at $n = 1$ implies $A = 1$. The following figure shows $x(t), x_a(t)$, and their samples:

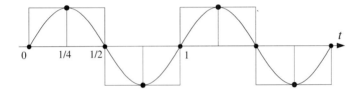

Similarly, in the second case we have $T = 1/f_s = 1/8$, resulting in the sampled aliased signal $x_a(nT) = B\sin(\pi n/4) + C\sin(3\pi n/4)$. Demanding the condition $x_a(nT) = x(nT)$ at $n = 1, 2$ gives the two equations

$$B\sin(\pi/4) + C\sin(3\pi/4) = 1 \qquad\qquad B + C = \sqrt{2}$$
$$\Rightarrow$$
$$B\sin(\pi/2) + C\sin(3\pi/2) = 1 \qquad\qquad B - C = 1$$

which can be solved to give $B = (\sqrt{2}+1)/2$ and $C = (\sqrt{2}-1)/2$. The following figure shows $x(t), x_a(t)$, and their samples:

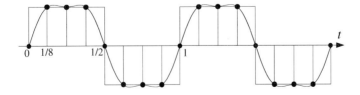

The second way of determining A, B, C is by evaluating the infinite sums of Eqs. (1.4.5) and (1.4.6). All three are special cases of the more general sum:

$$b(m, M) \equiv \frac{4}{\pi} \sum_{k=0}^{\infty} \left[\frac{1}{m + Mk} - \frac{1}{M - m + Mk} \right]$$

with $M > m > 0$. It can be computed as follows. Write

$$\frac{1}{m + Mk} - \frac{1}{M - m + Mk} = \int_0^{\infty} \left(e^{-mx} - e^{-(M-m)x} \right) e^{-Mkx} \, dx$$

then, interchange summation and integration and use the geometric series sum (for $x > 0$)

$$\sum_{k=0}^{\infty} e^{-Mkx} = \frac{1}{1 - e^{-Mx}}$$

to get

$$b(m, M) = \frac{4}{\pi} \int_0^{\infty} \frac{e^{-mx} - e^{-(M-m)x}}{1 - e^{-Mx}} \, dx$$

Looking this integral up in a table of integrals [30], we find:

$$b(m, M) = \frac{4}{M} \cot \left(\frac{m\pi}{M} \right)$$

The desired coefficients A, B, C are then:

$$A = b(1, 4) = \cot \left(\frac{\pi}{4} \right) = 1$$

$$B = b(1, 8) = \frac{1}{2} \cot \left(\frac{\pi}{8} \right) = \frac{\sqrt{2} + 1}{2}$$

$$C = b(3, 8) = \frac{1}{2} \cot \left(\frac{3\pi}{8} \right) = \frac{\sqrt{2} - 1}{2}$$

The above results generalize to any sampling rate $f_s = M$ Hz, where M is a multiple of 4. For example, if $f_s = 12$, we obtain

$$x_a(t) = b(1, 12) \sin(2\pi t) + b(3, 12) \sin(6\pi t) + b(5, 12) \sin(10\pi t)$$

and more generally

$$x_a(t) = \sum_{m=1,3,\ldots,(M/2)-1} b(m, M) \sin(2\pi m t)$$

The coefficients $b(m, M)$ tend to the original Fourier series coefficients b_m in the continuous-time limit, $M \to \infty$. Indeed, using the approximation $\cot(x) \approx 1/x$, valid for small x, we obtain the limit

$$\lim_{M\to\infty} b(m,M) = \frac{4}{M} \cdot \frac{1}{\pi m/M} = \frac{4}{\pi m} = b_m$$

The table below shows the successive improvement of the values of the aliased harmonic coefficients as the sampling rate increases:

coefficients	4 Hz	8 Hz	12 Hz	16 Hz	∞
b_1	1	1.207	1.244	1.257	1.273
b_3	–	0.207	0.333	0.374	0.424
b_5	–	–	0.089	0.167	0.255
b_7	–	–	–	0.050	0.182

In this example, the sampling rates of 4 and 8 Hz, and any multiple of 4, were chosen so that all the harmonics outside the Nyquist intervals got aliased onto *harmonics* within the intervals. For other values of f_s, such as $f_s = 13$ Hz, it is possible for the aliased harmonics to fall on non-harmonic frequencies within the Nyquist interval; thus, changing not only the relative balance of the Nyquist interval harmonics, but also the frequency values. □

When we develop DFT algorithms, we will see that the aliased Fourier series coefficients for the above type of problem can be obtained by performing a DFT, provided that the *periodic analog signal* remains a *periodic discrete-time signal* after sampling.

This requires that the sampling frequency f_s be an integral multiple of the fundamental harmonic of the given signal, that is, $f_s = Nf_1$. In such a case, the aliased coefficients can be obtained by an N-point DFT of the first N time samples $x(nT)$, $n = 0, 1, \ldots, N-1$ of the analog signal. See Section 9.7.

Example 1.4.7: A sound wave has the form:

$$x(t) = 2A\cos(10\pi t) + 2B\cos(30\pi t)$$

$$+ 2C\cos(50\pi t) + 2D\cos(60\pi t) + 2E\cos(90\pi t) + 2F\cos(125\pi t)$$

where t is in milliseconds. What is the frequency content of this signal? Which parts of it are audible and why?

This signal is prefiltered by an analog prefilter $H(f)$. Then, the output $y(t)$ of the prefilter is sampled at a rate of 40 kHz and immediately reconstructed by an ideal analog reconstructor, resulting into the final analog output $y_a(t)$, as shown below:

Determine the output signals $y(t)$ and $y_a(t)$ in the following cases:

(a) When there is no prefilter, that is, $H(f) = 1$ for all f.

(b) When $H(f)$ is the ideal prefilter with cutoff $f_s/2 = 20$ kHz.

(c) When $H(f)$ is a practical prefilter with specifications as shown below:

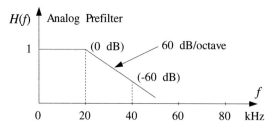

That is, it has a flat passband over the 20 kHz audio range and drops monotonically at a rate of 60 dB per octave beyond 20 kHz. Thus, at 40 kHz, which is an
octave away, the filter's response will be down by 60 dB.

For the purposes of this problem, the filter's *phase response* may be ignored
in determining the output $y(t)$. Does this filter help in removing the aliased
components?

What happens if the filter's attenuation rate is reduced to 30 dB/octave?

Solution: The six terms of $x(t)$ have frequencies:

$$f_A = 5 \text{ kHz} \qquad f_C = 25 \text{ kHz} \qquad f_E = 45 \text{ kHz}$$
$$f_B = 15 \text{ kHz} \qquad f_D = 30 \text{ kHz} \qquad f_F = 62.5 \text{ kHz}$$

Only f_A and f_B are audible; the rest are inaudible. Our ears filter out all frequencies
beyond 20 kHz, and we hear $x(t)$ as though it were the signal:

$$x_1(t) = 2A \cos(10\pi t) + 2B \cos(30\pi t)$$

Each term of $x(t)$ is represented in the frequency domain by two peaks at positive
and negative frequencies, for example, the A-term has spectrum:

$$2A \cos(2\pi f_A t) = A e^{2\pi j f_A t} + A e^{-2\pi j f_A t} \quad \longrightarrow \quad A\,\delta(f - f_A) + A\,\delta(f + f_A)$$

Therefore, the spectrum of the input $x(t)$ will be as shown below:

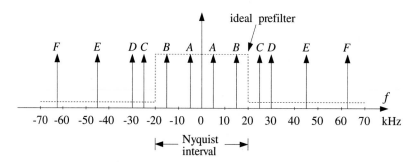

The sampling process will replicate each of these peaks at multiples of f_s = 40 kHz. The four terms C, D, E, F lie outside the $[-20, 20]$ kHz Nyquist interval and therefore will be aliased with the following frequencies inside the interval:

$$f_C = 25 \quad \Rightarrow \quad f_{C,a} = f_C \bmod (f_s) = f_C - f_s = 25 - 40 = -15$$

$$f_D = 30 \quad \Rightarrow \quad f_{D,a} = f_D \bmod (f_s) = f_D - f_s = 30 - 40 = -10$$

$$f_E = 45 \quad \Rightarrow \quad f_{E,a} = f_E \bmod (f_s) = f_E - f_s = 45 - 40 = 5$$

$$f_F = 62.5 \quad \Rightarrow \quad f_{F,a} = f_F \bmod (f_s) = f_F - 2f_s = 62.5 - 2 \times 40 = -17.5$$

In case (a), if we do not use any prefilter at all, we will have $y(t) = x(t)$ and the reconstructed signal will be:

$$y_a(t) = 2A \cos(10\pi t) + 2B \cos(30\pi t)$$
$$+ 2C \cos(-2\pi 15t) + 2D \cos(-2\pi 10t)$$
$$+ 2E \cos(2\pi 5t) + 2F \cos(-2\pi 17.5t)$$
$$= 2(A + E) \cos(10\pi t) + 2(B + C) \cos(30\pi t)$$
$$+ 2D \cos(20\pi t) + 2F \cos(35\pi t)$$

where we replaced each out-of-band frequency with its aliased self, for example,

$$2C \cos(2\pi f_C t) \rightarrow 2C \cos(2\pi f_{C,a} t)$$

The *relative* amplitudes of the 5 and 15 kHz audible components have changed and, in addition, two *new* audible components at 10 and 17.5 kHz have been introduced. Thus, $y_a(t)$ will sound very different from $x(t)$.

In case (b), if an ideal prefilter with cutoff $f_s/2$ = 20 kHz is used, then its output will be the same as the audible part of $x(t)$, that is, $y(t) = x_1(t)$. The filter's effect on the input spectrum is to remove completely all components beyond the 20 kHz Nyquist frequency, as shown below:

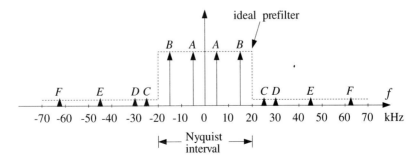

Because the prefilter's output contains no frequencies beyond the Nyquist frequency, there will be no aliasing and after reconstruction the output would sound the same as the input, $y_a(t) = y(t) = x_1(t)$.

In case (c), if the practical prefilter $H(f)$ is used, then its output $y(t)$ will be:

$$y(t) = 2A|H(f_A)|\cos(10\pi t) + 2B|H(f_B)|\cos(30\pi t)$$
$$+ 2C|H(f_C)|\cos(50\pi t) + 2D|H(f_D)|\cos(60\pi t) \qquad (1.4.7)$$
$$+ 2E|H(f_E)|\cos(90\pi t) + 2F|H(f_F)|\cos(125\pi t)$$

This follows from the steady-state sinusoidal response of a filter applied to the individual sinusoidal terms of $x(t)$, for example, the effect of $H(f)$ on A is:

$$2A\cos(2\pi f_A t) \xrightarrow{\ H\ } 2A|H(f_A)|\cos(2\pi f_A t + \theta(f_A))$$

where in Eq. (1.4.7) we ignored the phase response $\theta(f_A) = \arg H(f_A)$. The basic conclusions of this example are not affected by this simplification.

Note that Eq. (1.4.7) applies also to cases (a) and (b). In case (a), we can replace:

$$|H(f_A)| = |H(f_B)| = |H(f_C)| = |H(f_D)| = |H(f_E)| = |H(f_F)| = 1$$

and in case (b):

$$|H(f_A)| = |H(f_B)| = 1, \quad |H(f_C)| = |H(f_D)| = |H(f_E)| = |H(f_F)| = 0$$

In case (c), because f_A and f_B are in the filter's passband, we still have

$$|H(f_A)| = |H(f_B)| = 1$$

To determine $|H(f_C)|, |H(f_D)|, |H(f_E)|, |H(f_F)|$, we must find how many octaves[†] away the frequencies f_C, f_D, f_E, f_F are from the $f_s/2 = 20$ kHz edge of the passband. These are given by:

$$\log_2\left(\frac{f_C}{f_s/2}\right) = \log_2\left(\frac{25}{20}\right) = 0.322$$

$$\log_2\left(\frac{f_D}{f_s/2}\right) = \log_2\left(\frac{30}{20}\right) = 0.585$$

$$\log_2\left(\frac{f_E}{f_s/2}\right) = \log_2\left(\frac{45}{20}\right) = 1.170$$

$$\log_2\left(\frac{f_F}{f_s/2}\right) = \log_2\left(\frac{62.5}{20}\right) = 1.644$$

and therefore, the corresponding filter attenuations will be:

$$\text{at } f_C: \quad 60 \text{ dB/octave} \times 0.322 \text{ octaves} = 19.3 \text{ dB}$$
$$\text{at } f_D: \quad 60 \text{ dB/octave} \times 0.585 \text{ octaves} = 35.1 \text{ dB}$$
$$\text{at } f_E: \quad 60 \text{ dB/octave} \times 1.170 \text{ octaves} = 70.1 \text{ dB}$$
$$\text{at } f_F: \quad 60 \text{ dB/octave} \times 1.644 \text{ octaves} = 98.6 \text{ dB}$$

[†] The number of octaves is the number of powers of two, that is, if $f_2 = 2^\nu f_1 \Rightarrow \nu = \log_2(f_2/f_1)$.

By definition, an amount of A dB attenuation corresponds to reducing $|H(f)|$ by a factor $10^{-A/20}$. For example, the relative drop of $|H(f)|$ with respect to the edge of the passband $|H(f_s/2)|$ is A dB if:

$$\frac{|H(f)|}{|H(f_s/2)|} = 10^{-A/20}$$

Assuming that the passband has 0 dB normalization, $|H(f_s/2)| = 1$, we find the following values for the filter responses:

$$|H(f_C)| = 10^{-19.3/20} = \frac{1}{9}$$

$$|H(f_D)| = 10^{-35.1/20} = \frac{1}{57}$$

$$|H(f_E)| = 10^{-70.1/20} = \frac{1}{3234}$$

$$|H(f_F)| = 10^{-98.6/20} = \frac{1}{85114}$$

It follows from Eq. (1.4.7) that the output $y(t)$ of the prefilter will be:

$$y(t) = 2A\cos(10\pi t) + 2B\cos(30\pi t)$$

$$+ \frac{2C}{9}\cos(50\pi t) + \frac{2D}{57}\cos(60\pi t) \qquad (1.4.8)$$

$$+ \frac{2E}{3234}\cos(90\pi t) + \frac{2F}{85114}\cos(125\pi t)$$

Its spectrum is shown below:

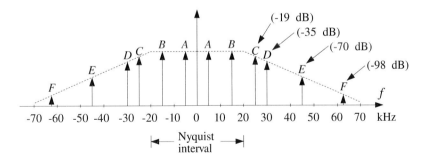

Notice how the inaudible out-of-band components have been attenuated by the pre-filter, so that when they get aliased back into the Nyquist interval because of sampling, their distorting effect will be *much less*. The wrapping of frequencies into the Nyquist interval is the same as in case (a). Therefore, after sampling and reconstruction we will get:

$$y_a(t) = 2\left(A + \frac{E}{3234}\right)\cos(10\pi t) + 2\left(B + \frac{C}{9}\right)\cos(30\pi t)$$

$$+ \frac{2D}{57}\cos(20\pi t) + \frac{2F}{85114}\cos(35\pi t)$$

Now, all aliased components have been reduced in magnitude. The component closest to the Nyquist frequency, namely f_C, causes the most distortion because it does not get attenuated much by the filter.

We will see in Section 1.5.3 that the prefilter's rate of attenuation in dB/octave is related to the filter's order N by $\alpha = 6N$ so that $\alpha = 60$ dB/octave corresponds to $60 = 6N$ or $N = 10$. Therefore, the given filter is already a fairly complex analog filter. Decreasing the filter's complexity to $\alpha = 30$ dB/octave, corresponding to filter order $N = 5$, would reduce all the attenuations by half, that is,

$$\text{at } f_C: \quad 30 \text{ dB/octave} \times 0.322 \text{ octaves} = 9.7 \text{ dB}$$

$$\text{at } f_D: \quad 30 \text{ dB/octave} \times 0.585 \text{ octaves} = 17.6 \text{ dB}$$

$$\text{at } f_E: \quad 30 \text{ dB/octave} \times 1.170 \text{ octaves} = 35.1 \text{ dB}$$

$$\text{at } f_F: \quad 30 \text{ dB/octave} \times 1.644 \text{ octaves} = 49.3 \text{ dB}$$

and, in absolute units:

$$|H(f_C)| = 10^{-9.7/20} = \frac{1}{3}$$

$$|H(f_D)| = 10^{-17.6/20} = \frac{1}{7.5}$$

$$|H(f_E)| = 10^{-35.1/20} = \frac{1}{57}$$

$$|H(f_F)| = 10^{-49.3/20} = \frac{1}{292}$$

Therefore, the resulting signal after reconstruction would be:

$$y_a(t) = 2\left(A + \frac{E}{57}\right)\cos(10\pi t) + 2\left(B + \frac{C}{3}\right)\cos(30\pi t)$$

$$+ \frac{2D}{7.5}\cos(20\pi t) + \frac{2F}{292}\cos(35\pi t) \tag{1.4.9}$$

Now the C and D terms are not as small and aliasing would still be significant. The situation can be remedied by oversampling, as discussed in the next example.   □

Example 1.4.8: *Oversampling* can be used to reduce the attenuation requirements of the prefilter, and thus its order. Oversampling increases the gap between spectral replicas reducing aliasing and allowing less sharp cutoffs for the prefilter.

For the previous example, if we oversample by a factor of 2, $f_s = 2 \times 40 = 80$ kHz, the new Nyquist interval will be $[-40, 40]$ kHz. Only the $f_E = 45$ kHz and $f_F = 62.5$ kHz components lie outside this interval, and they will be aliased with

$$f_{E,a} = f_E - f_s = 45 - 80 = -35 \text{ kHz}$$

$$f_{F,a} = f_F - f_s = 62.5 - 80 = -17.5 \text{ kHz}$$

Only $f_{F,a}$ lies in the audio band and will cause distortions, unless we attenuate f_F using a prefilter before it gets wrapped into the audio band. Without a prefilter, the reconstructed signal will be:

$$y_a(t) = 2A\cos(10\pi t) + 2B\cos(30\pi t)$$
$$+ 2C\cos(50\pi t) + 2D\cos(60\pi t)$$
$$+ 2E\cos(-2\pi 35t) + 2F\cos(-2\pi 17.5t)$$
$$= 2A\cos(10\pi t) + 2B\cos(30\pi t)$$
$$+ 2C\cos(50\pi t) + 2D\cos(60\pi t) + 2E\cos(70\pi t) + 2F\cos(35\pi t)$$

The audible components in $y_a(t)$ are:

$$y_1(t) = 2A\cos(10\pi t) + 2B\cos(30\pi t) + 2F\cos(35\pi t)$$

Thus, oversampling eliminated almost all the aliasing from the desired audio band. Note that two types of aliasing took place here, namely, the aliasing of the E component which remained *outside* the relevant audio band, and the aliasing of the F component which does represent distortion in the audio band.

Of course, one would not want to feed the signal $y_a(t)$ into an amplifier/speaker system because the high frequencies beyond the audio band might damage the system or cause nonlinearities. (But even if they were filtered out, the F component would still be there.) □

Example 1.4.9: *Oversampling and Decimation.* Example 1.4.8 assumed that sampling at 80 kHz could be maintained throughout the digital processing stages up to reconstruction. There are applications however, where the sampling rate must eventually be dropped down to its original value. This is the case, for example, in digital audio, where the rate must be reduced eventually to the standardized value of 44.1 kHz (for CDs) or 48 kHz (for DATs).

When the sampling rate is dropped, one must make sure that aliasing will not be reintroduced. In our example, if the rate is reduced back to 40 kHz, the C and D components, which were *inside* the $[-40, 40]$ kHz Nyquist interval with respect to the 80 kHz rate, would find themselves *outside* the $[-20, 20]$ kHz Nyquist interval with respect to the 40 kHz rate, and therefore would be aliased inside that interval, as in Example 1.4.7.

To prevent C and D, as well as E, from getting aliased into the audio band, one must *remove them* by a lowpass *digital* filter *before* the sampling rate is dropped to 40 kHz. Such a filter is called a digital *decimation* filter. The overall system is shown below.

The downsampler in this diagram reduces the sampling rate from 80 down to 40 kHz by throwing away every other sample, thus, keeping only half the samples. This is equivalent to sampling at a 40 kHz rate.

The input to the digital filter is the sampled spectrum of $y(t)$, which is replicated at multiples of 80 kHz as shown below.

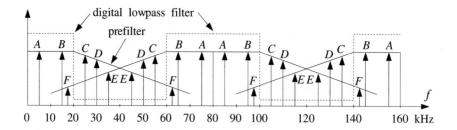

We have also assumed that the 30 dB/octave prefilter is present. The output of the digital filter will have spectrum as shown below.

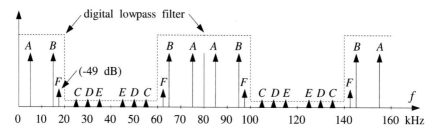

The digital filter operates at the *oversampled rate* of 80 kHz and acts as a lowpass filter within the $[-40, 40]$ kHz Nyquist interval, with a cutoff of 20 kHz. Thus, it will remove the C, D, and E components, as well as any other component that lies between $20 \le |f| \le 60$ kHz.

However, because the digital filter is periodic in f with period $f_s = 80$ kHz, it *cannot* remove any components from the interval $60 \le f \le 100$. Any components of the analog input $y(t)$ that lie in that interval would be aliased into the interval $60 - 80 \le f - f_s \le 100 - 80$, which is the desired audio band $-20 \le f - f_s \le 20$. This is what happened to the F component, as can be seen in the above figure.

The frequency components of $y(t)$ in $60 \le |f| \le 100$ can be removed *only by a pre-filter*, prior to sampling and replicating the spectrum. For example, our low-complexity 30 dB/octave prefilter would provide 47.6 dB attenuation at 60 kHz. Indeed, the number of octaves from 20 to 60 kHz is $\log_2(60/20) = 1.585$ and the attenuation there will be 30 dB/octave $\times$ 1.584 octaves = 47.6 dB.

The prefilter, being monotonic beyond 60 kHz, would suppress all potential aliased components beyond 60 kHz by more than 47.6 dB. At 100 kHz, it would provide $30 \times \log_2(100/20) = 69.7$ dB attenuation. At $f_F = 62.5$ kHz, it provides 49.3 dB suppression, as was calculated in Example 1.4.7, that is, $|H(f_F)| = 10^{-49.3/20} = 1/292$.

Therefore, assuming that the digital filter has *already removed* the C, D, and E components, and that the aliased F component has been *sufficiently attenuated* by the prefilter, we can now drop the sampling rate down to 40 kHz.

At the reduced 40 kHz rate, if we use an ideal reconstructor, it would extract only the components within the $[-20, 20]$ kHz band and the resulting reconstructed output will be:

$$y_a(t) = 2A \cos(10\pi t) + 2B \cos(30\pi t) + \frac{2F}{292} \cos(35\pi t)$$

which has a much attenuated aliased component F. This is to be compared with Eq. (1.4.9), which used the *same* prefilter but no oversampling. Oversampling in conjunction with digital decimation helped eliminate the most severe aliased components, C and D.

In summary, with oversampling, the complexity of the analog prefilter can be reduced and *traded off* for the complexity of a digital filter which is much easier to design and cheaper to implement with programmable DSPs. As we will see in Chapter 2, another benefit of oversampling is to reduce the *number of bits* representing each quantized sample. The connection between sampling rate and the savings in bits is discussed in Section 2.2. The subject of oversampling, decimation, interpolation, and the design and implementation of digital decimation and interpolation filters will be discussed in detail in Chapter 12. ☐

1.4.2 Rotational Motion

A more intuitive way to understand the sampling properties of sinusoids is to consider a representation of the complex sinusoid $x(t) = e^{2\pi jft}$ as a wheel rotating with a frequency of f revolutions per second. The wheel is seen in a dark room by means of a strobe light flashing at a rate of f_s flashes per second. The rotational frequency in [radians/sec] is $\Omega = 2\pi f$. During the time interval T between flashes, the wheel turns by an angle:

$$\omega = \Omega T = 2\pi f T = \frac{2\pi f}{f_s} \qquad (1.4.10)$$

This quantity is called the *digital frequency* and is measured in units of [radians/sample]. It represents a convenient normalization of the physical frequency f. In terms of ω, the sampled sinusoid reads simply

$$x(nT) = e^{2\pi jfTn} = e^{j\omega n}$$

In units of ω, the Nyquist frequency $f = f_s/2$ becomes $\omega = \pi$ and the Nyquist interval becomes $[-\pi, \pi]$. The replicated set $f + mf_s$ becomes

$$\frac{2\pi(f + mf_s)}{f_s} = \frac{2\pi f}{f_s} + 2\pi m = \omega + 2\pi m$$

Because the frequency $f = f_s$ corresponds to $\omega = 2\pi$, the aliased frequency given in Eq. (1.4.3) becomes in units of ω:

$$\omega_a = \omega \bmod(2\pi)$$

The quantity $f/f_s = fT$ is also called the *digital frequency* and is measured in units of [cycles/sample]. It represents another convenient normalization of the physical frequency axis, with the Nyquist interval corresponding to $[-0.5, 0.5]$.

In terms of the rotating wheel, fT represents the number of revolutions turned during the flashing interval T. If the wheel were actually turning at the higher frequency $f + mf_s$, then during time T it would turn by $(f + mf_s)T = fT + mf_sT = fT + m$ revolutions, that is, it would cover m whole additional revolutions. An observer would miss these extra m revolutions completely. The *perceived* rotational speed for an observer is always given by $f_a = f \bmod(f_s)$. The next two examples illustrate these remarks.

Example 1.4.10: Consider two wheels turning clockwise, one at $f_1 = 1$ Hz and the other at $f_2 = 5$ Hz, as shown below. Both are sampled with a strobe light flashing at $f_s = 4$ Hz. Note that the second one is turning at $f_2 = f_1 + f_s$.

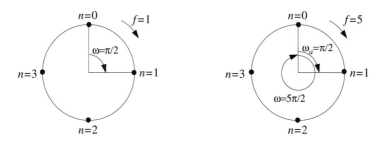

The first wheel covers $f_1 T = f_1/f_s = 1/4$ of a revolution during $T = 1/4$ second. Its angle of rotation during that time interval is $\omega_1 = 2\pi f_1/f_s = 2\pi/4 = \pi/2$ radians. During the sampled motion, an observer would observe the sequence of points $n = 0, 1, 2, 3, \ldots$ and would conclude that the wheel is turning at a speed of $1/4$ of a revolution in $1/4$ second, or,

$$\frac{1/4 \text{ cycles}}{1/4 \text{ sec}} = 1 \text{ Hz}$$

Thus, the observer would perceive the correct speed and sense of rotation. The second wheel, on the other hand, is actually turning by $f_2 T = f_2/f_s = 5/4$ revolutions in $1/4$ second, with an angle of rotation $\omega_2 = 5\pi/2$. Thus, it covers one whole extra revolution compared to the first one. However, the observer would still observe the same sequence of points $n = 0, 1, 2, 3, \ldots$, and would conclude again that the wheel is turning at $1/4$ revolution in $1/4$ second, or, 1 Hz. This result can be obtained quickly using Eq. (1.4.3):

$$f_{2a} = f_2 \bmod(f_s) = 5 \bmod(4) = 5 - 4 = 1$$

Thus, in this case the perceived speed is wrong, but the sense of rotation is still correct.

In the next figure, we see two more wheels, one turning clockwise at $f_3 = 9$ Hz and the other counterclockwise at $f_4 = -3$ Hz.

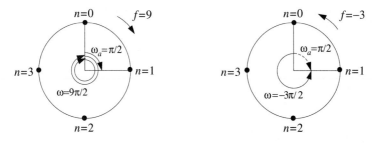

The negative sign signifies here the sense of rotation. During $T = 1/4$ sec, the third wheel covers $f_3 T = 9/4$ revolutions, that is, two whole extra revolutions over the f_1 wheel. An observer would again see the sequence of points $n = 0, 1, 2, 3, \ldots$, and would conclude that f_3 is turning at 1 Hz. Again, we can quickly compute, $f_{3a} = f_3 \bmod (f_s) = 9 \bmod (4) = 9 - 2 \cdot 4 = 1$ Hz.

The fourth wheel is more interesting. It covers $f_4 T = -3/4$ of a revolution in the counterclockwise direction. An observer captures the motion every 3/4 of a counterclockwise revolution. Thus, she will see the sequence of points $n = 0, 1, 2, 3, \ldots$, arriving at the conclusion that the wheel is turning at 1 Hz in the clockwise direction. In this case, both the perceived speed and sense of rotation are wrong. Again, the same conclusion can be reached quickly using $f_{4a} = f_4 \bmod (f_s) = (-3) \bmod (4) = -3 + 4 = 1$ Hz. Here, we added one f_s in order to bring f_4 within the Nyquist interval $[-2, 2]$. □

Example 1.4.11: The following figure shows four wheels rotating clockwise at $f = 1.5, 2, 2.5, 4$ Hz and sampled at $f_s = 4$ Hz by a strobe light.

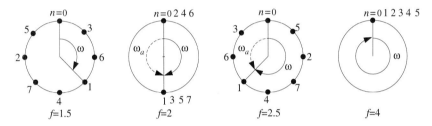

This example is meant to show that if a wheel is turning by less than *half* of a revolution between sampling instants, that is, $fT < 1/2$ or $\omega = 2\pi fT < \pi$, then the motion is perceived correctly and there is no aliasing. The conditions $fT < 1/2$ or $\omega < \pi$ are equivalent to the sampling theorem condition $f_s > 2f$. But if the wheel is turning by more than half of a revolution, it will be perceived as turning in the opposite direction and aliasing will occur.

The first wheel turns by $fT = 3/8$ of a revolution every T seconds. Thus, an observer would see the sequence of points $n = 0, 1, 2, 3, \ldots$ and perceive the right motion.

The second wheel is turning by exactly half of a revolution $fT = 1/2$ or angle $\omega = 2\pi fT = \pi$ radians. An observer would perceive an up-down motion and lose sense of direction, not being able to tell which way the wheel is turning.

The third wheel turns by more than half of a revolution, $fT = 5/8$. An observer would see the sequence of points $n = 0, 1, 2, 3, \ldots$, corresponding to successive rotations by $\omega = 5\pi/4$ radians. An observer always perceives the motion in terms of the lesser angle of rotation, and therefore will think that the wheel is turning the other way by an angle $\omega_a = \omega \bmod (2\pi) = (5\pi/4) \bmod (2\pi) = 5\pi/4 - 2\pi = -3\pi/4$ or frequency $f_a = -(3/8 \text{ cycle}) / (1/4 \text{ sec}) = -1.5$ Hz.

The fourth wheel will appear to be stationary because $f = f_s = 4$ and the motion is sampled once every revolution, $\omega = 2\pi$. The perceived frequency will be $f_a = f \bmod (f_s) = 4 \bmod (4) = 4 - 4 = 0$. □

1.4.3 DSP Frequency Units

Figure 1.4.4 compares the various frequency scales that are commonly used in DSP, and the corresponding Nyquist intervals. A sampled sinusoid takes the form in these units:

$$e^{2\pi j f T n} = e^{2\pi j (f/f_s) n} = e^{j\Omega T n} = e^{j\omega n}$$

being expressed more simply in terms of ω. Sometimes f is normalized with respect to the Nyquist frequency $f_N = f_s/2$, that is, in units of f/f_N. In this case, the Nyquist interval becomes $[-1, 1]$. In multirate applications, where successive digital processing stages operate at different sampling rates, the most convenient set of units is simply in terms of f. In fixed-rate applications, the units of ω or f/f_s are the most convenient.

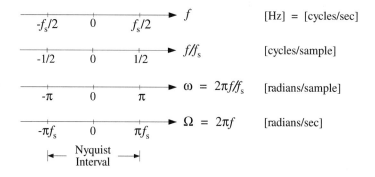

Fig. 1.4.4 Commonly used frequency units.

1.5   Spectra of Sampled Signals*

Next, we discuss the effects of sampling using Fourier transforms. Figure 1.3.1 shows an ideal sampler that *instantaneously* measures the analog signal $x(t)$ at the sampling instants $t = nT$. The output of the sampler can be considered to be an *analog* signal consisting of the linear superposition of impulses occurring at the sampling times, with each impulse weighted by the corresponding sample value. Thus, the *sampled signal* is

$$\hat{x}(t) = \sum_{n=-\infty}^{\infty} x(nT)\,\delta(t - nT) \tag{1.5.1}$$

In practical sampling, each sample must be held constant for a short period of time, say τ seconds, in order for the A/D converter to accurately convert the sample to digital format. This holding operation may be achieved by a sample/hold circuit. In this case, the sampled signal will be:

$$x_{\text{flat}}(t) = \sum_{n=-\infty}^{\infty} x(nT)p(t-nT) \tag{1.5.2}$$

where $p(t)$ is a flat-top pulse of duration of τ seconds such that $\tau \ll T$. Ideal sampling corresponds to the limit $\tau \to 0$. Figure 1.5.1 illustrates the ideal and practical cases.

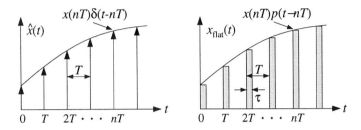

Fig. 1.5.1   Ideal and practical sampling.

We will consider only the ideal case, Eq. (1.5.1), because it captures all the essential features of the sampling process. Our objective is to determine the spectrum of the sampled signal $\hat{x}(t)$ and compare it with the spectrum of the original signal $x(t)$. Problem 1.21 explores practical sampling.

Our main result will be to express the spectrum of $\hat{x}(t)$ in two ways. The first relates the sampled spectrum to the *discrete-time samples* $x(nT)$ and leads to the discrete-time Fourier transform. The second relates it to the *original spectrum* and implies the spectrum replication property that was mentioned earlier.

1.5.1 Discrete-Time Fourier Transform

The spectrum of the sampled signal $\hat{x}(t)$ is the Fourier transform:

$$\hat{X}(f) = \int_{-\infty}^{\infty} \hat{x}(t)e^{-2\pi jft}\,dt \tag{1.5.3}$$

Inserting Eq. (1.5.1) into Eq. (1.5.3) and interchanging integration and summation, we obtain:

$$\hat{X}(f) = \int_{-\infty}^{\infty} \sum_{n=-\infty}^{\infty} x(nT)\delta(t-nT)e^{-2\pi jft}\,dt$$

$$= \sum_{n=-\infty}^{\infty} x(nT)\int_{-\infty}^{\infty} \delta(t-nT)e^{-2\pi jft}\,dt \qquad \text{or,}$$

$$\boxed{\hat{X}(f) = \sum_{n=-\infty}^{\infty} x(nT)e^{-2\pi jfTn}} \tag{1.5.4}$$

This is the first way of expressing $\hat{X}(f)$. Several remarks are in order:

1. *DTFT.* Eq. (1.5.4) is known as the *Discrete-Time Fourier Transform* (DTFT)[†] of the sequence of samples $x(nT)$. $\hat{X}(f)$ is computable only from the knowledge of the sample values $x(nT)$.

2. *Periodicity.* $\hat{X}(f)$ is a *periodic* function of f with period f_s, hence, $\hat{X}(f + f_s) = \hat{X}(f)$. This follows from the fact that $e^{-2\pi j f T n}$ is periodic in f. Because of this periodicity, one may restrict the frequency interval to just one period, namely, the Nyquist interval, $[-f_s/2, f_s/2]$.

   The periodicity in f implies that $\hat{X}(f)$ will extend over the entire frequency axis, in accordance with our expectation that the sampling process introduces high frequencies into the original spectrum. Although not obvious yet, the periodicity in f is related to the periodic replication of the original spectrum.

3. *Fourier Series.* Mathematically, Eq. (1.5.4) may be thought of as the *Fourier series* expansion of the periodic function $\hat{X}(f)$, with the samples $x(nT)$ being the corresponding Fourier series *coefficients.* Thus, $x(nT)$ may be recovered from $\hat{X}(f)$ by the inverse Fourier series:

$$\boxed{\; x(nT) = \frac{1}{f_s} \int_{-f_s/2}^{f_s/2} \hat{X}(f)\, e^{2\pi j f T n}\, df = \int_{-\pi}^{\pi} \hat{X}(\omega)\, e^{j\omega n}\, \frac{d\omega}{2\pi} \;} \qquad (1.5.5)$$

   where in the second equation we changed variables from f to $\omega = 2\pi f/f_s$.[‡] Eq. (1.5.5) is the *inverse DTFT* and expresses the discrete-time signal $x(nT)$ as a superposition of discrete-time sinusoids $e^{j\omega n}$.

4. *Numerical Approximation.* Eq. (1.5.4) may be thought of as a *numerical approximation* to the frequency spectrum of the original analog signal $x(t)$. Indeed, using the definition of integrals, we may write approximately,

$$X(f) = \int_{-\infty}^{\infty} x(t)\, e^{-2\pi j f t}\, dt \simeq \sum_{n=-\infty}^{\infty} x(nT)\, e^{-2\pi j f n T} \cdot T \qquad \text{or,}$$

$$X(f) \simeq T\hat{X}(f) \qquad (1.5.6)$$

   This approximation becomes exact in the continuous-time limit:

$$X(f) = \lim_{T \to 0} T\hat{X}(f) \qquad (1.5.7)$$

   It is precisely this limiting result and the approximation of Eq. (1.5.6) that *justify* the use of discrete Fourier transforms to compute actual spectra of analog signals.

5. *Practical Approximations.* In an actual spectrum computation, two additional approximations must be made before anything can be computed:

[†] Not to be confused with the *Discrete Fourier Transform* (DFT), which is a special case of the DTFT.
[‡] Abusing the notation slightly, we wrote $\hat{X}(\omega)$ for $\hat{X}(f)$.

(a) We must keep only a *finite* number of time samples $x(nT)$, say L samples, $n = 0, 1, 2, \ldots, L - 1$, so that Eq. (1.5.4) is computed approximately by the truncated sum:

$$\hat{X}(f) \simeq \hat{X}_L(f) = \sum_{n=0}^{L-1} x(nT) e^{-2\pi j f T n} \qquad (1.5.8)$$

This approximation leads to the concept of a *time window* and the related effects of *smearing* and *leakage* of the spectrum. These concepts are central in the area of spectral analysis and will be discussed in Chapter 9.

(b) We must decide on a *finite* set of frequencies f at which to evaluate $\hat{X}(f)$. Proper choice of this set allows the development of various efficient computational algorithms for the DFT, such as the Fast Fourier Transform (FFT), presented also in Chapter 9.

6. *z-transform*. Finally, we note that Eq. (1.5.4) leads to the concept of the *z-transform*, much like the ordinary Fourier transform leads to the Laplace transform. Setting $z = e^{j\omega} = e^{2\pi j f T}$, we may write Eq. (1.5.4) as the *z*-transform[†]

$$\hat{X}(z) = \sum_{n=-\infty}^{\infty} x(nT) z^{-n}$$

1.5.2 Spectrum Replication

Next, we show the spectrum replication property by deriving the precise relationship between the spectrum $\hat{X}(f)$ of the sampled signal $\hat{x}(t)$ and the original spectrum $X(f)$ of the analog signal $x(t)$.

The nth term $x(nT)\delta(t - nT)$ in Eq. (1.5.1) may be replaced by $x(t)\delta(t - nT)$ because the term is nonzero only at $t = nT$. Then, $x(t)$ can be factored out of the sum in Eq. (1.5.1) as a common factor:

$$\hat{x}(t) = x(t) \sum_{n=-\infty}^{\infty} \delta(t - nT) \equiv x(t) s(t) \qquad (1.5.9)$$

Thinking of this as the *modulation* of the "carrier" $s(t)$ by the "baseband" signal $x(t)$, we expect to get frequency translations of the original spectrum, much like the AM modulation of a sinusoidal carrier. The frequency translation effect may be seen by expanding the (periodic in time) sampling function $s(t)$ into its Fourier series representation as a linear combination of harmonics. It is easily shown that

$$s(t) = \sum_{n=-\infty}^{\infty} \delta(t - nT) = \frac{1}{T} \sum_{m=-\infty}^{\infty} e^{2\pi j m f_s t} \qquad (1.5.10)$$

[†]Again, abusing the notation, we wrote $\hat{X}(z)$ for $\hat{X}(f)$.

which expresses the sampling function $s(t)$ as a linear combination of sinusoidal carriers, each causing its own frequency shift. Writing Eq. (1.5.9) as

$$\hat{x}(t) = x(t)s(t) = \frac{1}{T} \sum_{m=-\infty}^{\infty} x(t) e^{2\pi jmf_s t}$$

and using the modulation property of Fourier transforms, which states that if $X(f)$ is the transform of $x(t)$ then $X(f - f_c)$ is the transform of $x(t) e^{2\pi jf_c t}$, we obtain by taking Fourier transforms of both sides,

$$\boxed{\hat{X}(f) = \frac{1}{T} \sum_{m=-\infty}^{\infty} X(f - mf_s)} \tag{1.5.11}$$

This represents the *periodic replication* of the original spectrum $X(f)$ at intervals of the sampling rate f_s. Fig. 1.5.2 shows $T\hat{X}(f)$ as the sum of the periodic replicas of $X(f)$.

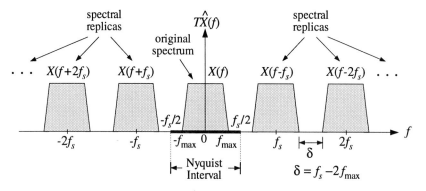

Fig. 1.5.2 Spectrum replication caused by sampling.

Another way to prove Eq. (1.5.11) is as follows. Because $\hat{x}(t)$ is the product of $x(t)$ and $s(t)$, its Fourier transform will be the convolution of the corresponding transforms, that is,

$$\hat{X}(f) = \int_{-\infty}^{\infty} X(f - f') S(f') \, df' \tag{1.5.12}$$

On the other hand, it follows from Eq. (1.5.10) that the Fourier transform of $s(t)$ will be the sum of the transforms of the individual harmonics:

$$S(f) = \frac{1}{T} \sum_{m=-\infty}^{\infty} \delta(f - mf_s) \tag{1.5.13}$$

Inserting this into Eq. (1.5.12) and interchanging the summation over m with the integration over f', we obtain

$$\hat{X}(f) = \frac{1}{T} \sum_{m=-\infty}^{\infty} \int_{-\infty}^{\infty} X(f - f') \delta(f' - mf_s) \, df' = \frac{1}{T} \sum_{m=-\infty}^{\infty} X(f - mf_s)$$

Combining Eqs. (1.5.4) and (1.5.11), we obtain the two alternative expressions for the spectrum $\hat{X}(f)$

$$\hat{X}(f) = \sum_{n=-\infty}^{\infty} x(nT)\,e^{-2\pi jfTn} = \frac{1}{T}\sum_{m=-\infty}^{\infty} X(f - mf_s) \qquad (1.5.14)$$

This is known as the *Poisson summation* formula. We also see from Fig. 1.5.2 that as we let $T \to 0$, or equivalently, $f_s \to \infty$, the replicas move out to infinity leaving behind only the original spectrum $X(f)$. Therefore, Eq. (1.5.7) follows.

We emphasize that Eq. (1.5.14) holds for *arbitrary* signals $x(t)$, not necessarily bandlimited ones. In the special case when $x(t)$ is bandlimited to some maximum frequency f_{max}, as suggested by Fig. 1.5.2, we immediately obtain the sampling theorem condition, Eq. (1.3.2).

It is seen in Fig. 1.5.2 that the replicas are separated from each other by a distance $\delta = f_s - 2f_{max}$, known as the *guard band*. It follows that the replicas will *not* overlap if $\delta \geq 0$, or equivalently, $f_s \geq 2f_{max}$. But they will overlap if $f_s < 2f_{max}$ or $\delta < 0$ and aliasing of frequencies will take place as the tails of the replicas enter into the Nyquist interval and add to the original spectrum, distorting it. This case is shown in Fig. 1.5.3.

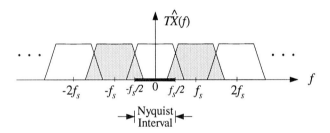

Fig. 1.5.3 Aliasing caused by overlapping spectral replicas.

It is evident by inspecting Fig. 1.5.2 that if the signal is bandlimited and f_s is large enough so that the replicas do not overlap, then the portion of the sampled signal spectrum $\hat{X}(f)$ that lies *within* the Nyquist interval $[-f_s/2, f_s/2]$ will be identical to the original spectrum $X(f)$, that is,

$$T\hat{X}(f) = X(f), \qquad \text{for} \quad -\frac{f_s}{2} \leq f \leq \frac{f_s}{2} \qquad (1.5.15)$$

This is an important result for DSP. Not only does it make possible the analog reconstruction of the sampled signal, but it also guarantees that any *subsequent* digital processing of the sampled signal will be applied to the original spectrum $X(f)$ and not to some aliased and distorted version thereof.

For example, a subsequent digital filtering operation will transform the input samples $x(nT)$ into a sequence of output samples $y(nT)$. Just like analog filtering, digital filtering is equivalent to spectral shaping in the frequency domain. If the digital filter has frequency response $H_{\text{DSP}}(f)$, the spectrum $\hat{X}(f)$ of the input sequence will be reshaped into the output spectrum

$$\hat{Y}(f) = H_{\text{DSP}}(f)\hat{X}(f)$$

If Eq. (1.5.15) holds, then the digital filter will reshape the original spectrum $X(f)$. Note that because all digital filters have periodic frequency responses, the periodicity of the sampled spectrum is preserved by the digital filtering operation. Therefore, the output samples could be recovered from Eq. (1.5.5)

$$y(nT) = \int_{-\pi}^{\pi} \hat{Y}(\omega)e^{j\omega n}\frac{d\omega}{2\pi} = \int_{-\pi}^{\pi} H_{\text{DSP}}(\omega)\hat{X}(\omega)e^{j\omega n}\frac{d\omega}{2\pi}$$

If the spectrum $X(f)$ is not bandlimited, or, if it is bandlimited but the sampling rate f_s is so low that the replicas overlap, then Eq. (1.5.15) does not hold. Any subsequent filtering will reshape the *wrong* spectrum. Therefore, it is essential to use a *lowpass antialiasing prefilter*, as shown in Fig. 1.3.5, to bandlimit the input spectrum to within the Nyquist interval, so that the resulting replicas after sampling will not overlap.

Example 1.5.1: Consider a pure sinusoid of frequency f_0, $x(t) = e^{2\pi j f_0 t}$. Its Fourier transform is the spectral line $X(f) = \delta(f - f_0)$. It follows from Eq. (1.5.11) that the sampled sinusoid

$$\hat{x}(t) = \sum_{n=-\infty}^{\infty} x(nT)\delta(t - nT) = \sum_{n=-\infty}^{\infty} e^{2\pi j f_0 Tn}\delta(t - nT)$$

will have Fourier spectrum

$$\hat{X}(f) = \frac{1}{T}\sum_{m=-\infty}^{\infty} \delta(f - f_0 - mf_s)$$

Thus, the spectrum of the sampled sinusoid consists of all the frequencies in the replicated set $\{f_0 + mf_s, m = 0, \pm 1, \pm 2, \dots\}$ in accordance with Fig. 1.3.2 and our remarks in Sections 1.4 and 1.3.   □

Example 1.5.2: This example illustrates the effect of sampling on a non-bandlimited signal and the degree to which the portion of the spectrum $\hat{X}(f)$ within the Nyquist interval approximates the original spectrum $X(f)$. Consider the exponentially decaying signal and its spectrum:

$$x(t) = e^{-at}u(t)$$

$$X(f) = \frac{1}{a + 2\pi j f}$$

The frequency spectrum of the sampled signal $\hat{x}(t)$ may be obtained in two ways. Using Eq. (1.5.4)

$$\hat{X}(f) = \sum_{n=-\infty}^{\infty} x(nT)\, e^{-2\pi jfTn} = \sum_{n=0}^{\infty} e^{-aTn}\, e^{-2\pi jfTn}$$

and summing the geometric series, we get

$$\hat{X}(f) = \frac{1}{1 - e^{-aT}\, e^{-2\pi jfT}} = \frac{1}{1 - e^{-aT}\, e^{-j\omega}}$$

Its magnitude square is

$$|\hat{X}(f)|^2 = \frac{1}{1 - 2e^{-aT}\cos(2\pi fT) + e^{-2aT}}$$

The periodicity in f is evident because the dependence on f comes through the periodic cosine function. Alternatively, we may use Eq. (1.5.11) and sum the replicas of the original spectrum to get

$$\hat{X}(f) = \frac{1}{T} \sum_{m=-\infty}^{\infty} X(f - mf_s) = \frac{1}{T} \sum_{m=-\infty}^{\infty} \frac{1}{a + 2\pi j(f - mf_s)}$$

Combining the two expression for $\hat{X}(f)$, we obtain the not-so-obvious identity in the parameters a, f, T:

$$\frac{1}{T} \sum_{m=-\infty}^{\infty} \frac{1}{a + 2\pi j(f - mf_s)} = \frac{1}{1 - e^{-aT}\, e^{-2\pi jfT}}$$

The left graph in Fig. 1.5.4 compares the periodic spectrum $|T\hat{X}(f)|^2$ with the original analog spectrum $|X(f)|^2 = 1/(a^2 + (2\pi f)^2)$. The spectra are shown in decibels, that is, $20\log_{10}|X(f)|$. The parameter a was $a = 0.2$ sec^{-1}. Two values of the sampling rate $f_s = 1/T$ are shown, $f_s = 1$ Hz and $f_s = 2$ Hz. The two Nyquist intervals are $[-0.5, 0.5]$ Hz and $[-1, 1]$ Hz, respectively. Outside these intervals, the sampled spectra repeat periodically.

Notice that even with the scale factor T taken into account, the two spectra $X(f)$ and $T\hat{X}(f)$ are very different from each other. However, within the central Nyquist interval $[-f_s/2, f_s/2]$, they agree approximately, especially at low frequencies. This approximation gets better as f_s increases.

The limit as $T \to 0$ or $f_s \to \infty$ can be seen explicitly in this example. Using the approximation $e^{-x} \simeq 1 - x$, valid for small x, or L'Hospital's rule, we obtain

$$\lim_{T \to 0} T\hat{X}(f) = \lim_{T \to 0} \frac{T}{1 - e^{-aT}\, e^{-2\pi jfT}} = \frac{1}{a + 2\pi jf} = X(f)$$

In the right graph of Fig. 1.5.4, we show the effect of using a length-L *time window* and approximating the spectrum by Eq. (1.5.8). The parameter values were $a = 0.2$, $f_s = 2$, and $L = 10$ samples.

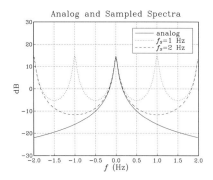 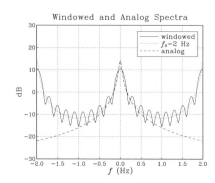

Fig. 1.5.4 Spectra of analog, sampled, and windowed signals.

That figure compares what we would *like* to compute, that is, $|X(f)|^2$, with what we can *at best* hope to compute based on our sampled signal, $|T\hat{X}(f)|^2$, and with what we can *actually* compute based on a finite record of samples, $|T\hat{X}_L(f)|^2$.

The windowed spectrum $|T\hat{X}_L(f)|^2$ can be improved by taking longer L and using a non-rectangular window, such as a Hamming window. At best, however, it will approach the sampled spectrum $|T\hat{X}(f)|^2$ and not $|X(f)|^2$. The approximation of $X(f)$ by $T\hat{X}(f)$ can be improved only by increasing the sampling rate f_s.

The quantity $\hat{X}_L(f)$ can be computed by sending the L samples $x(nT) = e^{-anT}$, $n = 0, 1, \ldots, L-1$ into a general DFT routine. In this particular example, $\hat{X}_L(f)$ can also be computed in closed form. Using the finite geometric series:

$$\sum_{n=0}^{L-1} x^n = \frac{1 - x^L}{1 - x}$$

we obtain:

$$\hat{X}_L(f) = \sum_{n=0}^{L-1} e^{-aTn} e^{-2\pi j f T n} = \frac{1 - e^{-aTL} e^{-2\pi j f TL}}{1 - e^{-aT} e^{-2\pi j f T}}$$

It is evident that $\hat{X}_L(f) \to \hat{X}(f)$ as $L \to \infty$.   □

1.5.3 *Practical Antialiasing Prefilters*

An ideal analog prefilter is shown in Fig. 1.5.5. It acts as an ideal lowpass filter removing all frequency components of the analog input signal that lie beyond the Nyquist frequency $f_s/2$.

The antialiasing prefilters used in practice are not ideal and do not completely remove all the frequency components outside the Nyquist interval. Thus, some aliasing will take place. However, by proper design the prefilters may be made as good as desired and the amount of aliasing reduced to tolerable levels. A practical antialiasing lowpass filter is shown in Fig. 1.5.6. Its passband $[-f_{\text{pass}}, f_{\text{pass}}]$ is

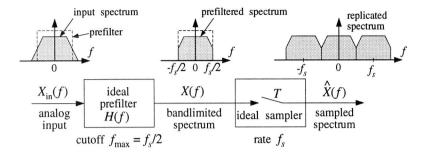

Fig. 1.5.5 Ideal antialiasing prefilter.

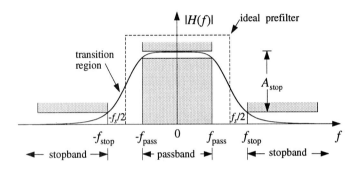

Fig. 1.5.6 Practical antialiasing lowpass prefilter.

usually taken to be the *frequency range of interest* for the application at hand and must lie entirely within the Nyquist interval.

The prefilter must be essentially *flat* over this passband in order not to distort the frequencies of interest. Even if it is not completely flat over the passband, it can be "equalized" *digitally* at a subsequent processing stage by a digital filter, say $H_{EQ}(f)$, whose frequency response is the *inverse* of the response of the prefilter *over* the passband range:

$$\boxed{H_{EQ}(f) = \frac{1}{H(f)}}, \qquad \text{for} \quad -f_{pass} \le f \le f_{pass}$$

The digital filter $H_{EQ}(f)$, being periodic with period f_s, cannot be the inverse of the prefilter over the entire frequency axis, but it can be the inverse over the passband.

The *stopband* frequency f_{stop} of the prefilter and the *minimum stopband attenuation* A_{stop} in dB must be chosen appropriately to minimize aliasing effects. It will become evident from the examples below that f_{stop} must be chosen as

$$\boxed{f_{stop} = f_s - f_{pass}} \tag{1.5.16}$$

or, equivalently,

$$f_s = f_{\text{pass}} + f_{\text{stop}}$$

This places the Nyquist frequency $f_s/2$ exactly in the middle of the transition region of the prefilter, as shown in Fig. 1.5.6. The attenuation of the filter in *decibels* is defined in terms of its magnitude response by:

$$A(f) = -20\log_{10}\left|\frac{H(f)}{H(f_0)}\right| \qquad \text{(attenuation in dB)}$$

where f_0 is a convenient reference frequency, typically taken to be at DC for a lowpass filter. Therefore, the stopband specification of the filter, depicted in this figure, is $A(f) \ge A_{\text{stop}}$, for $|f| \ge f_{\text{stop}}$.

Transfer functions of analog filters typically drop like a power $H(s) \sim 1/s^N$ for large s, where N is the filter order. Thus, their magnitude response drops like $|H(f)| \sim 1/f^N$ for large f, and their attenuation will be, up to an *additive* constant,

$$A(f) = -20\log_{10}\left|1/f^N\right| = \alpha_{10}\log_{10}f, \qquad \text{(for large f)} \qquad (1.5.17)$$

where α_{10} is the attenuation in *dB per decade* defined by:

$$\alpha_{10} = 20N \qquad \text{(dB per decade)}$$

It represents the increase in attenuation when f is changed by a factor of ten, that is, $A(10f) - A(f) = \alpha_{10}$. Engineers also like to measure attenuation in *dB per octave*, that is, the amount of change per *doubling* of f. This is obtained by using logs in base two, that is, writing Eq. (1.5.17) in the form:

$$A(f) = \alpha_2 \log_2 f = \alpha_{10}\log_{10}f$$

where α_2 is in dB/octave and is related to α_{10} by:

$$\alpha_2 = \alpha_{10}\log_{10}2 = 6N \qquad \text{(dB per octave)}$$

Figure 1.5.5 shows the effect on the input spectrum $X_{\text{in}}(f)$ of an ideal prefilter with a sharp cutoff. For a practical prefilter, the output spectrum is given by:

$$X(f) = H(f)X_{\text{in}}(f)$$

or, in terms of attenuations in dB:

$$A_X(f) = A(f) + A_{X_{\text{in}}}(f) \qquad (1.5.18)$$

where $A_X(f) = -20\log_{10}|X(f)/X(f_0)|$ and similarly for $A_{X_{\text{in}}}(f)$. Thus, attenuations are *additive*. The spectrum $X(f)$ will be replicated by the subsequent sampling operation and therefore, the amount of attenuation in $A_X(f)$ will determine the degree of overlapping of the spectral replicas, that is, the degree of aliasing.

The specifications of the prefilter can be adjusted so that its attenuation $A(f)$, in *combination* with the attenuation $A_{X_{\text{in}}}(f)$ of the input spectrum, will result in sufficient attenuation of $X(f)$ to reduce the amount of aliasing within the desired frequency band. The next few examples illustrate these remarks.

Example 1.5.3: The frequency *range of interest* of an analog signal extends to 4 kHz. Beyond 4 kHz, the spectrum attenuates at a rate of 15 dB per octave. Ideally, we would sample at a rate of 8 kHz provided the sampling operation is preceded by a perfect lowpass antialiasing prefilter with cutoff of 4 kHz. As a practical alternative to designing a perfect prefilter, we decide to sample at the higher rate of 12 kHz.

    (a) If we do not use any prefilter at all, determine the amount of aliasing that will be introduced by the sampling process into the frequency range of interest, that is, into the 4 kHz range.

    (b) We wish to suppress the aliased components *within* the frequency range of interest by more than 50 dB. Determine the least stringent specifications of the lowpass antialiasing prefilter that must be used.

Solution: Both parts are answered with the help of the figure below, which shows the original spectrum and its first replicas centered at $\pm f_s = \pm 12$ kHz.

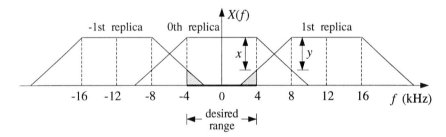

By the even symmetry of the spectra, it follows that the left tail of the 1st replica will be the same as the right tail of the 0th replica. Thus, the indicated attenuations x and y at frequencies 4 and 8 kHz will be equal, $x = y$.

If we do not use any prefilter, the attenuation at 8 kHz will be $y = 15$ dB because the 0th replica attenuates by 15 dB per octave starting at 4 kHz. The aliased components within the desired 4 kHz range correspond to the shaded portion of the left side of the 1st replica that has entered into the 4 kHz interval. They are suppressed by more than x dB. Thus, $x = y = 15$ dB. This probably represents too much aliasing to be tolerable.

If we use a prefilter, its passband must extend over the desired 4 kHz range. Therefore, $f_{\text{pass}} = 4$ kHz and $f_{\text{stop}} = f_s - f_{\text{pass}} = 12 - 4 = 8$ kHz. Because attenuations are additive in dB, the total attenuation y at 8 kHz will now be the sum of the attenuation due to the signal, that is, 15 dB, and the attenuation due to the prefilter, say A_{stop} dB. The equality $y = x$ and the requirement that $x \geq 50$ dB lead to

$$y = 15 + A_{\text{stop}} = x \geq 50 \qquad \Rightarrow \qquad A_{\text{stop}} \geq 50 - 15 = 35 \text{ dB}$$

Thus, the specifications of the prefilter are a fairly flat passband over the ± 4 kHz range and a stopband starting at 8 kHz with minimum attenuation of 35 dB.   □

Example 1.5.4: The significant frequency range of a signal extends to f_{max}. Beyond f_{max}, the spectrum attenuates by α dB/octave. We have available an off-the-shelf antialiasing prefilter that has a flat passband up to f_{max} and attenuates by β dB/octave beyond that. It is required that within the $f_{\max}$ range of interest, the aliased components due to sampling be suppressed by more than A dB. Show that the *minimum* sampling rate that we should use is given by

$$f_s = f_{\max} + 2^{A/\gamma} f_{\max}$$

where $\gamma = \alpha + \beta$.

Solution: We refer to the following figure, which shows the 0th and $\pm$1st replicas.

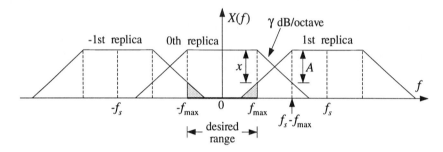

The passband edge is at $f_{\text{pass}} = f_{\max}$ and the stopband edge at $f_{\text{stop}} = f_s - f_{\max}$. Beyond the desired $f_{\max}$ range, the total attenuation (in dB) of the 0th replica will be the sum of the attenuations of the signal and the prefilter. In the notation of Eq. (1.5.18), it will be given as function of frequency by

$$A_X(f) = \alpha \log_2 \left(\frac{f}{f_{\max}} \right) + \beta \log_2 \left(\frac{f}{f_{\max}} \right) = \gamma \log_2 \left(\frac{f}{f_{\max}} \right)$$

where we have normalized the attenuation to 0 dB at $f = f_{\max}$. This is the mathematical expression of the statement that the total attenuation will be γ dB per octave.

By the even symmetry of the spectra, we have $x = A_X(f_{\text{stop}}) = A_X(f_s - f_{\max})$. Thus, the requirement that $x \geq A$ gives the condition

$$A_X(f_s - f_{\max}) \geq A \qquad \Rightarrow \qquad \gamma \log_2 \left(\frac{f_s - f_{\max}}{f_{\max}} \right) \geq A$$

Solving this as an equality gives the minimum acceptable rate f_s. If α and β had been given in dB/decade instead of dB/octave, the above condition would be valid with $\log_{10}$ instead of $\log_2$ resulting in $f_s = f_{\max} + 10^{A/\gamma} f_{\max}$. Note that the previous example corresponds to the case $A = \gamma$ giving $f_s = f_{\max} + 2 f_{\max} = 3 f_{\max}$.   □

The above examples show that to accommodate practical specifications for antialiasing prefilters, the sampling rates must be somewhat *higher* than the minimum Nyquist rate. The higher the rate, the less complex the prefilter. This idea is carried further in the method of *oversampling*, whereby the input is sampled at rates that are many times higher than the Nyquist rate. The replicas become very far separated, allowing the use of low quality, inexpensive analog prefilters. Oversampling methods will be discussed in Chapter 12.

1.6 *Analog Reconstructors**

We saw in Section 1.4.1 that an ideal reconstructor is an ideal lowpass filter with cutoff the Nyquist frequency $f_s/2$. Here, we derive this result and also consider practical reconstructors.

Analog reconstruction represents some sort of *lowpass* filtering of the sampled signal. This can be seen in Fig. 1.6.1, where practical reconstruction has been accomplished by filling the gaps between samples by holding the current sample value constant till the next sample. This is the staircase or sample/hold reconstructor.

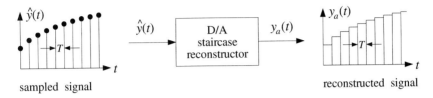

Fig. 1.6.1 Staircase reconstructor.

It must be clear from this figure that *any* reasonable way of filling the gaps between samples will result in some sort of reconstruction. Filling the gaps results in a *smoother* signal than the sampled signal. In frequency-domain language, the higher frequencies in the sampled signal are removed, that is, the sampled signal is lowpass filtered. Thus, any reconstructor may be viewed as an analog *lowpass* filter, as shown in Fig. 1.6.2.

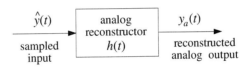

Fig. 1.6.2 Analog reconstructor as a lowpass filter.

We will determine the form of the impulse response $h(t)$ of the reconstructor both for ideal and practical reconstruction. The relationship of the reconstructed output $y_a(t)$ to the input samples $y(nT)$ can be found by inserting the sampled input signal

$$\hat{y}(t) = \sum_{n=-\infty}^{\infty} y(nT)\delta(t-nT)$$

into the convolutional equation of the reconstructor

$$y_a(t) = \int_{-\infty}^{\infty} h(t-t')\hat{y}(t')\,dt'$$

It then follows that:

$$y_a(t) = \sum_{n=-\infty}^{\infty} y(nT)h(t - nT) \qquad\qquad (1.6.1)$$

It states that the way to fill the gaps between samples is to start at the current sample $y(nT)$ and interpolate from it following the shape of $h(t)$ until the next sample. More precisely, a copy of $h(t)$ must be attached at each sample $y(nT)$, and all such contributions must be summed over—the resulting curve being the reconstructed analog signal. In the frequency domain, Eq. (1.6.1) becomes

$$Y_a(f) = H(f)\hat{Y}(f) \qquad\qquad (1.6.2)$$

where $\hat{Y}(f)$ is the replicated spectrum given by Eq. (1.5.11)

$$\hat{Y}(f) = \frac{1}{T} \sum_{m=-\infty}^{\infty} Y(f - mf_s)$$

1.6.1 Ideal Reconstructors

For *perfect* or ideal reconstruction one must require that $Y_a(f)$ be identical to the original analog spectrum $Y(f)$. If the spectrum $Y(f)$ is bandlimited and its replicas do not overlap, then within the Nyquist interval, $T\hat{Y}(f)$ will agree with $Y(f)$ in accordance with Eq. (1.5.15), that is,

$$\hat{Y}(f) = \frac{1}{T}Y(f), \qquad \text{for} \quad -\frac{f_s}{2} \le f \le \frac{f_s}{2} \qquad\qquad (1.6.3)$$

The ideal reconstruction filter $H(f)$ is an ideal lowpass filter with cutoff $f_s/2$, defined as follows:

$$H(f) = \begin{cases} T, & \text{if } |f| \le f_s/2 \\ 0, & \text{otherwise} \end{cases}$$

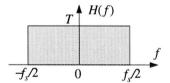

The value T for the passband gain is justified below. As shown in Fig. 1.6.3, such a filter will extract the central replica and remove all other replicas. Using Eq. (1.6.3), we have within the Nyquist interval:

$$Y_a(f) = H(f)\hat{Y}(f) = T \cdot \frac{1}{T} Y(f) = Y(f)$$

where the filter's gain factor T canceled the $1/T$ factor in the spectrum.

The same relationship also holds trivially ($0 \equiv 0$) outside the Nyquist interval. Thus, we have $Y_a(f) = Y(f)$, for all f, which implies that the reconstructed analog signal $y_a(t)$ will be identical to the original signal that was sampled, $y_a(t) = y(t)$. Combining this with Eq. (1.6.1), we obtain the *Shannon sampling theorem* [35-39] expressing the bandlimited signal $y(t)$ in terms of its samples $y(nT)$:

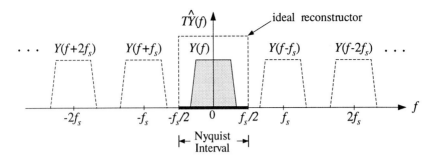

Fig. 1.6.3 Ideal reconstructor in frequency domain.

$$y(t) = \sum_{n=-\infty}^{\infty} y(nT)h(t - nT) \qquad\qquad (1.6.4)$$

The impulse response of the ideal reconstructor can be obtained from the inverse Fourier transform of $H(f)$:

$$h(t) = \int_{-\infty}^{\infty} H(f)e^{2\pi j f t}\, df = \int_{-f_s/2}^{f_s/2} Te^{2\pi j f T}\, df, \qquad \text{or,}$$

$$h(t) = \frac{\sin(\pi t/T)}{\pi t/T} = \frac{\sin(\pi f_s t)}{\pi f_s t} \qquad \text{(ideal reconstructor)} \qquad (1.6.5)$$

It is shown in Fig. 1.6.4. Unfortunately, the ideal reconstructor is not realizable. Its impulse response is not causal, having an infinite anticausal part. Therefore, alternative reconstructors, such as the staircase one, are used in practice.

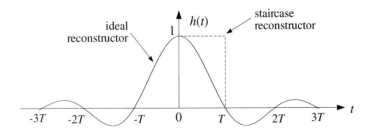

Fig. 1.6.4 Impulse response of ideal reconstructor.

An *approximation* to the ideal reconstructor, obtained by truncating it to finite length, is used in the design of digital FIR interpolation filters for oversampling and sample rate conversion applications. We will discuss it in Chapter 12.

1.6.2 Staircase Reconstructors

The staircase reconstructor shown in Fig. 1.6.1 is the simplest and most widely used reconstructor in practice. It generates a staircase approximation to the original signal. Note the similarity of this operation to practical sampling, where $h(t)$ is a sampling pulse $p(t)$ having a very narrow width $\tau \ll T$. By contrast, the impulse response of the staircase reconstructor must have duration of T seconds in order to fill the entire gap between samples. Thus, $h(t)$ is given by:

$$h(t) = u(t) - u(t - T) = \begin{cases} 1, & \text{if } 0 \le t \le T \\ 0, & \text{otherwise} \end{cases}$$

where $u(t)$ is the unit step. The staircase output, although smoother than the sampled input, still contains spurious high-frequency components arising from the sudden jumps in the staircase levels from sample to sample. This spurious frequency content may be seen by computing the frequency response of the reconstructor. The Laplace transform of $h(t) = u(t) - u(t - T)$ is

$$H(s) = \frac{1}{s} - \frac{1}{s} e^{-sT}$$

from which we obtain the Fourier transform by setting $s = 2\pi j f$:

$$H(f) = \frac{1}{2\pi j f}\left(1 - e^{-2\pi j f T}\right) = T \frac{\sin(\pi f T)}{\pi f T} e^{-\pi j f T} \tag{1.6.6}$$

It is shown in Fig. 1.6.5 in comparison to the ideal reconstructor. Notice that it vanishes at integral *multiples* of f_s — exactly where the replicas caused by sampling are centered. The spurious high frequencies mentioned above are those beyond the Nyquist frequency $f_s/2$.

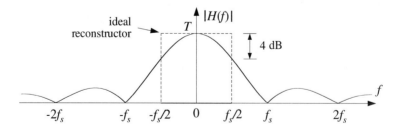

Fig. 1.6.5 Frequency response of staircase reconstructor.

Thus, the reconstructor does not completely eliminate the replicated spectral images as the ideal reconstructor does. Figure 1.6.6 compares the spectra before and after the staircase reconstructor, that is, the effect of the multiplication $Y_a(f) = H(f)\hat{Y}(f)$.

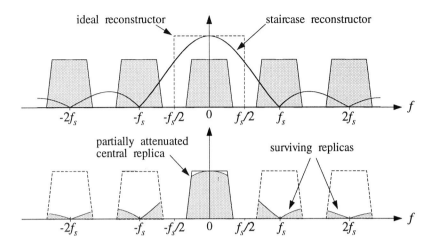

Fig. 1.6.6 Frequency response of staircase reconstructor.

1.6.3 Anti-Image Postfilters

The surviving spectral replicas may be removed by an additional *lowpass postfilter*, called an *anti-image* postfilter, whose cutoff is the Nyquist frequency $f_s/2$. This operation is shown in Fig. 1.6.7.

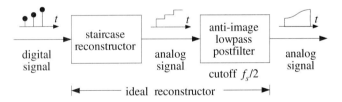

Fig. 1.6.7 Analog anti-image postfilter.

In the time domain, the postfilter has the effect of rounding off the corners of the staircase output making it smoother. In the frequency domain, the combined effect of the staircase reconstructor followed by the anti-image postfilter is to remove the spectral replicas as much as possible, that is, to emulate the ideal reconstructor. The final reconstructed spectrum at the output of the postfilter is shown in Fig. 1.6.8.

The reason for using this two-stage reconstruction procedure is the *simplicity* of implementation of the staircase reconstruction part. A typical D/A converter will act as such a reconstructor. The digital code for each sample is applied to the DAC for T seconds generating an analog output that remains constant during T.

The specifications of the postfilter are similar to those of an antialiasing prefilter, namely, a flat passband and cutoff frequency equal to the Nyquist frequency $f_s/2$. High-quality DSP applications, such as digital audio, require the use of postfilters (and prefilters) with very stringent specifications. In deciding the specifications of a postfilter, one must take into account the effect of the staircase D/A which does

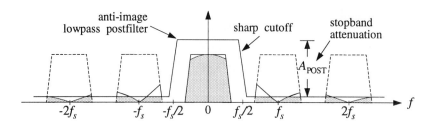

Fig. 1.6.8 Spectrum after postfilter.

part of the reconstruction.

The main function of the postfilter is to remove the remnants of the spectral images that survived the staircase D/A reconstructor. It can also be used to equalize the rolloff of the staircase response within the Nyquist interval. As shown in Fig. 1.6.5, the staircase reconstructor is not flat within the Nyquist interval, tending to attenuate more near the Nyquist frequency $f_s/2$. The maximum attenuation suffered at $f_s/2$ is about 4 dB. This can be seen as follows:

$$-20\log_{10}\left|\frac{H(f_s/2)}{H(0)}\right| = -20\log_{10}\left|\frac{\sin(\pi/2)}{\pi/2}\right| = 3.9 \text{ dB}$$

This attenuation can be compensated by proper design of the passband of the anti-image postfilter. But more conveniently, it can be compensated *digitally* before analog reconstruction, by designing an equalizing digital filter whose response matches the *inverse* of $H(f)$ *over* the Nyquist interval.

Similar techniques were mentioned in Section 1.5.3 for equalizing the imperfect passband of the antialiasing prefilter. The use of high-quality digital filters to perform these equalizations improves the overall quality of the digital processing system. By contrast, analog compensation techniques would be more cumbersome and expensive. The combined equalizer, DAC, and postfilter are shown in Fig. 1.6.9. The frequency response of the equalizer is defined as the inverse of the DAC, as given by Eq. (1.6.6):

$$H_{\text{EQ}}(f) = \frac{T}{H(f)} = \frac{\pi f T}{\sin(\pi f T)}\, e^{\pi j f T}, \qquad \text{for} \quad -\frac{f_s}{2} \le f \le \frac{f_s}{2} \qquad (1.6.7)$$

It is shown in Fig. 1.6.10. As a digital filter, $H_{\text{EQ}}(f)$ is periodic outside the Nyquist interval with period f_s. We will design such inverse filters later using the frequency sampling design method of Section 10.3. Some designs are presented in Chapter 12.

The equalizer filter transforms the sequence $y(nT)$ into the "equalized" sequence $y_{\text{EQ}}(nT)$, which is then fed into the DAC and postfilter. The frequency spectrum of $y_{\text{EQ}}(nT)$ is $\hat{Y}_{\text{EQ}}(f) = H_{\text{EQ}}(f)\hat{Y}(f)$. The spectrum of the staircase output of the DAC will be $Y_a(f) = H(f)\hat{Y}_{\text{EQ}}(f)$. Therefore, the final reconstructed spectrum at the output of the postfilter will be

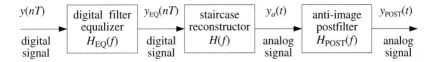

Fig. 1.6.9 Digital equalization filter for D/A conversion.

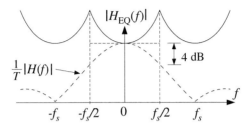

Fig. 1.6.10 Frequency response of DAC equalizer.

$$Y_{\text{POST}}(f) = H_{\text{POST}}(f)\, Y_a(f)$$

$$= H_{\text{POST}}(f)\, H(f)\, \hat{Y}_{\text{EQ}}(f)$$

$$= H_{\text{POST}}(f)\, H(f)\, H_{\text{EQ}}(f)\, \hat{Y}(f)$$

Within the Nyquist interval, using Eqs. (1.6.7) and (1.5.15) and assuming a flat postfilter there, $H_{\text{POST}}(f) \simeq 1$, we have

$$Y_{\text{POST}}(f) = H_{\text{POST}}(f)\, H(f)\, H_{\text{EQ}}(f)\, \hat{Y}(f) = 1 \cdot T \cdot \frac{1}{T} Y(f) = Y(f)$$

Outside the Nyquist interval, assuming $H_{\text{POST}}(f) \simeq 0$, we have $Y_{\text{POST}}(f) = 0$. Thus, the combination of equalizer, DAC, and postfilter acts like an ideal reconstructor.

Example 1.6.1: The signal of Example 1.5.3 that was sampled at $f_s = 12$ kHz is filtered by a digital filter designed to act as an ideal lowpass filter with cutoff frequency of $f_c = 2$ kHz. The filtered digital signal is then fed into a staircase D/A and then into a lowpass anti-image postfilter. The overall reconstructor is required to suppress the spectral images caused by sampling by more than $A = 80$ dB. Determine the *least stringent* specifications for the analog postfilter that will satisfy this requirement.

Solution: The digital lowpass filter is, by construction, periodic in f with period f_s. Thus, the spectrum of the signal after the digital filter will look as follows:

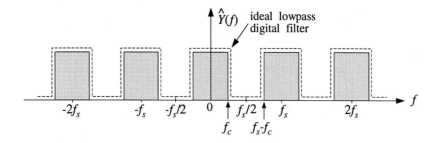

The spectral images are separated now by a distance $f_s - 2f_c = 12 - 2 \cdot 2 = 8$ kHz. After passing through the staircase reconstructor, the spectrum will be as shown below:

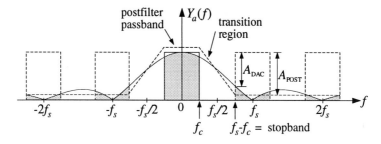

The postfilter must have a flat passband over $[-f_c, f_c]$. Its stopband must begin at $f_{stop} = f_s - f_c = 12 - 2 = 10$ kHz because the first replica is largest there. The wide transition region between f_c and $f_s - f_c$ allows the use of a less stringent postfilter.

The required stopband attenuation of the postfilter can be determined as follows. The total attenuation caused by the cascade of the DAC and postfilter is the sum of the corresponding attenuations:

$$A(f) = A_{DAC}(f) + A_{POST}(f)$$

where

$$A_{DAC}(f) = -20 \log_{10} \left| \frac{H(f)}{H(0)} \right| = -20 \log_{10} \left| \frac{\sin(\pi f / f_s)}{\pi f / f_s} \right|$$

At $f = f_{stop} = f_s - f_c$, the total attenuation must be greater than A

$$A_{DAC} + A_{POST} \geq A \qquad \Rightarrow \qquad A_{POST} \geq A - A_{DAC}$$

Numerically, we find at $f_{stop} = 10$ kHz

$$A_{DAC} = -20 \log_{10} \left| \frac{\sin(\pi 10/12)}{\pi 10/12} \right| = 14.4$$

resulting in $A_{POST} \geq 80 - 14.4 = 65.6$ dB.   □

The key idea in this example was to use the separation between spectral replicas as the *transition region* of the postfilter. The wider this separation, the less stringent the postfilter. Oversampling and digital interpolation techniques exploit this idea to its fullest and can be used to alleviate the need for expensive high-quality analog postfilters. Such oversampling systems are routinely used in CD and DAT players. They will be discussed in Chapter 12.

Example 1.6.2: A sinusoid of frequency f_0 is sampled at a rate f_s, such that $|f_0| \leq f_s/2$. The resulting sampled sinusoid is then reconstructed by an arbitrary reconstructor $H(f)$. Determine the analog signal at the output of the reconstructor when $H(f)$ is: (a) the ideal reconstructor, (b) the staircase reconstructor, (c) the staircase reconstructor followed by a very good anti-image postfilter, and (d) a staircase reconstructor equalized by the digital filter defined in Eq. (1.6.7).

Solution: Let $y(t) = e^{2\pi j f_0 t}$. Its spectrum is $Y(f) = \delta(f - f_0)$ and the spectrum of the sampled sinusoid will be the replication of $Y(f)$, as in Example 1.5.1:

$$\hat{Y}(f) = \frac{1}{T} \sum_{m=-\infty}^{\infty} \delta(f - f_0 - mf_s)$$

The spectrum of the reconstructed signal will be:

$$Y_a(f) = H(f)\hat{Y}(f) = \frac{1}{T} \sum_{m=-\infty}^{\infty} H(f)\delta(f - f_0 - mf_s)$$

$$= \frac{1}{T} \sum_{m=-\infty}^{\infty} H(f_0 + mf_s)\delta(f - f_0 - mf_s)$$

Taking inverse Fourier transforms, we obtain:

$$y_a(t) = \frac{1}{T} \sum_{m=-\infty}^{\infty} H(f_m) e^{2\pi j f_m t} \tag{1.6.8}$$

where $f_m = f_0 + mf_s$. If $H(f)$ is the ideal reconstructor, then $H(f_m)$ will be zero if f_m does not lie in the Nyquist interval. Because f_0 was assumed to lie in the interval, only the $m = 0$ term will survive the sum giving:

$$y_a(t) = \frac{1}{T} H(f_0) e^{2\pi j f_0 t} = \frac{1}{T} \cdot T e^{2\pi j f_0 t} = e^{2\pi j f_0 t}$$

thus, the sinusoid is reconstructed perfectly. If f_0 lies outside the interval, $|f_0| > f_s/2$, then there exists a unique integer m_0 such that $|f_0 + m_0 f_s| < f_s/2$, where m_0 is negative if $f_0 > 0$. In this case, only the $m = m_0$ term will survive the sum giving:

$$y_a(t) = \frac{1}{T} H(f_{m_0}) e^{2\pi j f_{m_0} t} = e^{2\pi j f_{m_0} t}$$

where $f_{m_0} = f_0 + m_0 f_s = f_0 \bmod(f_s)$. The sinusoid f_0 will be confused with the sinusoid f_{m_0}, as we discussed qualitatively in Section 1.4.1.

For the staircase reconstructor of Eq. (1.6.6), the reconstructed signal will be given by Eq. (1.6.8), which should sum up to generate the staircase approximation to the sinusoid. This is demonstrated in Example 1.6.3.

In case (c), a good postfilter will remove all frequencies outside the Nyquist interval, that is, only the $m = 0$ term will survive in Eq. (1.6.8), giving:

$$y_a(t) = \frac{1}{T} H(f_0) e^{2\pi j f_0 t}$$

where we assumed that the postfilter has unity gain over the Nyquist interval. Using Eq. (1.6.6) evaluated at $f = f_0$, we get:

$$y_a(t) = \frac{\sin(\pi f_0 T)}{\pi f_0 T} e^{-\pi j f_0 T} e^{2\pi j f_0 t}$$

Thus, there is amplitude attenuation and phase shift, which both become worse as f_0 increases towards the Nyquist frequency $f_s/2$. A digital filter that equalizes the staircase response, would anticipate this attenuation and phase shift and undo them. Indeed, in case (d), the effective reconstructor is $H_{EQ}(f)H(f)$. Therefore, Eq. (1.6.8) becomes:

$$y_a(t) = \frac{1}{T} \sum_{m=-\infty}^{\infty} H_{EQ}(f_m) H(f_m) e^{2\pi j f_m t}$$

But because of the periodicity of $H_{EQ}(f)$, we can replace $H_{EQ}(f_m) = H_{EQ}(f_0) = T/H(f_0)$, giving:

$$y_a(t) = \sum_{m=-\infty}^{\infty} \frac{H(f_m)}{H(f_0)} e^{2\pi j f_m t} \tag{1.6.9}$$

A good postfilter, extracting the $m = 0$ term, would result in the final reconstructed output $y_{post}(t) = \frac{H(f_0)}{H(f_0)} e^{2\pi j f_0 t} = e^{2\pi j f_0 t}$.   □

Example 1.6.3: The cosinusoid $y(t) = \cos(2\pi f_0 t)$ is sampled at a rate f_s and the samples are reconstructed by a staircase reconstructor $H(f)$. The reconstructed signal will be:

$$y_a(t) = \sum_{m=-\infty}^{\infty} G(f_m) \cos(2\pi f_m t + \phi(f_m)) \tag{1.6.10}$$

where $G(f)$ and $\phi(f)$ are defined as

$$G(f) = \frac{\sin(\pi f T)}{\pi f T}, \quad \phi(f) = -\pi f T \quad \Rightarrow \quad H(f) = T G(f) e^{j\phi(f)}$$

Note that $TG(f)$ and $\phi(f)$ are not quite the magnitude and phase responses of $H(f)$; those are $|H(f)| = T|G(f)|$ and $\arg H(f) = \phi(f) + \pi(1 - \text{sign } G(f))/2$. Eq. (1.6.10) is obtained by substituting $H(f) = TG(f)e^{j\phi(f)}$ into Eq. (1.6.8) and taking real parts. A computable approximation of Eq. (1.6.10) is obtained by truncating the sum to $2M + 1$ terms, that is, keeping the terms $-M \leq m \leq M$:

$$y_a(t) = \sum_{m=-M}^{M} w(m) G(f_m) \cos(2\pi f_m t + \phi(f_m)) \qquad (1.6.11)$$

where we have also introduced appropriate weights $w(m)$ to reduce the Gibbs ripples resulting from the truncation of the sum. For example, the *Hamming weights* are:

$$w(m) = 0.54 + 0.46 \cos\left(\frac{\pi m}{M}\right), \qquad -M \le m \le M$$

whereas the *rectangular* weights are $w(m) = 1$.

For the numerical values $f_0 = 0.125$ kHz, $f_s = 1$ kHz, $M = 15$, we have computed the original analog signal $y(t)$ and the reconstructed signal $y_a(t)$ as given by the approximation of Eq. (1.6.11), over the time interval $0 \le t \le 16$ msec, that is, over 16 sampling instants.

If the signal $y_a(t)$ were postfiltered by a good postfilter, only the $m = 0$ term would survive the sum, and the resulting signal would be the original f_0 sinusoid with some attenuation and phase shift:

$$y_{\text{post}}(t) = G(f_0) \cos(2\pi f_0 t + \phi(f_0))$$

The following figure compares the three signals $y(t)$, $y_a(t)$, and $y_{\text{post}}(t)$ in the two cases of using rectangular weights and Hamming weights $w(m)$.

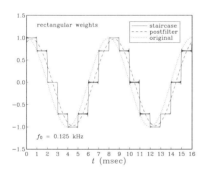
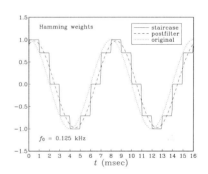

Notice how the postfilter output $y_{\text{post}}(t)$ is essentially an averaged or smoothed version of the staircase output $y_a(t)$. To see the dependence on the value of f_0 of the attenuation and phase shift of $y_{\text{post}}(t)$, the next two graphs show the cases $f_0 = 0.25$ and $f_0 = 0.5$ kHz.

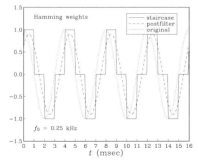
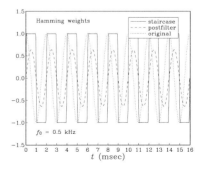

The case $f_0 = 0.5$ kHz corresponds to the Nyquist frequency $f_s/2$, having the maximum amplitude and phase distortion. In all cases, however, $y_{\text{post}}(t)$ is a smoothed version of the staircase output.

If the staircase reconstructor is preceded by an equalizer filter, as shown in Fig. 1.6.9, then the staircase output will be given by the real part of Eq. (1.6.9). We have:

$$\frac{H(f_m)}{H(f_0)} = \frac{\sin(\pi f_m T)}{\pi f_m T} \frac{\pi f_0 T}{\sin(\pi f_0 T)} e^{-j\pi(f_m - f_0)}$$

$$= \frac{\sin(\pi f_0 T + \pi m)}{\sin(\pi f_0 T)} \frac{f_0}{f_m} e^{-j\pi m}$$

$$= \frac{(-1)^m \sin(\pi f_0 T)}{\sin(\pi f_0 T)} \frac{f_0}{f_m} (-1)^m = \frac{f_0}{f_m}$$

where we used the property $\cos(x + \pi m) = (-1)^m \cos x$. Thus,

$$y_a(t) = \sum_{m=-M}^{M} w(m) \frac{f_0}{f_m} \cos(2\pi f_m t) \tag{1.6.12}$$

This signal is shown below for the case $f_0 = 0.125$ kHz.

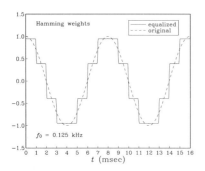 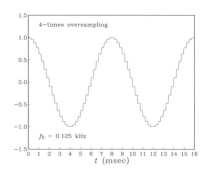

It is superimposed on the original sinusoid, corresponding to the $m = 0$ term, which is what would be extracted by a good postfilter. Notice again the smoothing effect of the postfilter. In order to remove completely all the $m \neq 0$ terms, the postfilter must be a high-quality lowpass filter with sharp cutoff at the Nyquist frequency.

To illustrate the beneficial effect of oversampling on such a reconstructor, we have also plotted the digitally equalized staircase output in the case of 4-times oversampling, as given by Eq. (1.6.12) with $f_s = 4$ kHz. Now there are four times as many staircase levels. They follow the original sinusoid more closely and can be smoothed more easily. Therefore, a lower quality, less expensive lowpass postfilter would be sufficient in this case.   □

1.7 Basic Components of DSP Systems

It follows from the discussion in Sections 1.5 and 1.6 that the minimum number of necessary components in a typical digital signal processing system must be:

1. A lowpass analog antialiasing prefilter that bandlimits the signal to be sampled to within the Nyquist interval.

2. An A/D converter (sampler and quantizer).

3. A digital signal processor.

4. A D/A converter (staircase reconstructor), possibly preceded by an equalizing digital filter.

5. A lowpass analog anti-image postfilter to complete the job of the staircase reconstructor and further remove the spectral images introduced by the sampling process.

These are shown in Fig. 1.7.1. Here, we review briefly the function of each stage and its impact on the quality of the overall digital processing system. With the exception of the sampling stage, every stage in this system may be thought of as a filtering operation by an appropriate transfer function. The sampling stage, through its spectrum replication, is a spectrum expansion operation.

The function of the antialiasing prefilter $H_{\mathrm{PRE}}(f)$ is to bandlimit the overall analog input signal $x_a(t)$ to within the Nyquist interval $[f_s/2, f_s/2]$. The output of the prefilter is now a bandlimited signal $x(t)$ and may be sampled at a rate of f_s samples per second. By design, the spectral replicas generated by the sampling process will not overlap. The sampling rate must be high enough so that the surviving input spectrum after the prefiltering operation, that is, the spectrum of $x(t)$, contains all the frequencies of interest for the application at hand.

The quality of the prefilter affects critically the quality of the overall system, that is, the degree of overlap of the spectral replicas depends on the rolloff characteristics of the prefilter.

The sampled (and quantized) signal $\hat{x}(t)$ or $x(nT)$ is then processed by a digital signal processor whose effect is to *reshape* the spectrum by means of a transfer function, say $H_{\mathrm{DSP}}(f)$, so that $\hat{Y}(f) = H_{\mathrm{DSP}}(f)\hat{X}(f)$.

The resulting output samples $\hat{y}(t)$ or $y(nT)$ are then reconstructed by the DAC into the staircase analog signal $y(t)$. Finally, the signal $y(t)$ is smoothed further by the postfilter, resulting in the overall analog output signal $y_a(t)$. Separating in Eq. (1.5.11) the central replica from the other replicas, we write

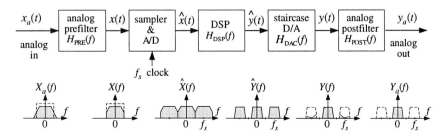

Fig. 1.7.1 Components of typical DSP system.

$$\hat{X}(f) = \frac{1}{T} \sum_{m=-\infty}^{\infty} X(f - mf_s) = \frac{1}{T} [X(f) + \text{replicas}]$$

and following backwards all the transfer function relationships, we find for the spectrum of $y_a(t)$:

$$Y_a(f) = H_{\text{POST}}(f) Y(f) = H_{\text{POST}}(f) H_{\text{DAC}}(f) \hat{Y}(f)$$

$$= H_{\text{POST}}(f) H_{\text{DAC}}(f) H_{\text{DSP}}(f) \hat{X}(f)$$

$$= H_{\text{POST}}(f) H_{\text{DAC}}(f) H_{\text{DSP}}(f) \frac{1}{T} [X(f) + \text{replicas}]$$

$$= H_{\text{POST}}(f) H_{\text{DAC}}(f) H_{\text{DSP}}(f) \frac{1}{T} [H_{\text{PRE}}(f) X_a(f) + \text{replicas}]$$

In a well-designed system, the product of the staircase DAC and the postfilter transfer functions should be effectively equal to the ideal reconstructor. Therefore, for frequencies outside the Nyquist interval the output spectrum will vanish, that is, the spectral images will be removed. But for frequencies within the Nyquist interval, that product should be equal to the gain T canceling the $1/T$ factor.

Furthermore, because the prefilter $H_{\text{PRE}}(f)$ ensures that the replicas do not overlap, the terms labeled "replicas" will vanish for all frequencies within the Nyquist interval. Finally, because the prefilter approximates an ideal lowpass filter, its passband gain will be approximately one. The upshot of all this is that *within* the Nyquist interval one has approximately

$$H_{\text{POST}}(f) H_{\text{DAC}}(f) \simeq T$$

$$\text{replicas} \simeq 0$$

$$H_{\text{PRE}}(f) \simeq 1$$

To the extent that these approximations are good—and this determines the quality of the overall system—we finally find

$$Y_a(f) = T \cdot H_{\text{DSP}}(f) \frac{1}{T} [1 \cdot X_a(f) + 0], \qquad \text{or,}$$

$$\boxed{Y_a(f) = H_{\text{DSP}}(f) X_a(f)}, \qquad \text{for} \quad |f| \leq \frac{f_s}{2} \qquad (1.7.1)$$

Thus, the above arrangement works exactly as expected, that is, it is equivalent to *linear filtering* of the analog input, with an effective transfer function $H_{\text{DSP}}(f)$ defined by the digital signal processor. This is, of course, the ultimate goal of the DSP system. The primary reasons for using digital signal processing are the *programmability, reliability, accuracy, availability, and cost* of the digital hardware.

1.8 Problems

1.1 A wheel, rotating at 6 Hz, is seen in a dark room by means of a strobe light flashing at a rate of 8 Hz. Determine the apparent rotational speed and sense of rotation of the wheel. Repeat the question if the flashes occur at 12 Hz, 16 Hz, or 24 Hz.

1.2 The analog signal $x(t) = 10 \sin(2\pi t) + 10 \sin(8\pi t) + 5 \sin(12\pi t)$, where t is in seconds, is sampled at a rate of $f_s = 5$ Hz. Determine the signal $x_a(t)$ aliased with $x(t)$. Show that the two signals have the *same* sample values, that is, show that $x(nT) = x_a(nT)$. Repeat the above questions if the sampling rate is $f_s = 10$ Hz.

1.3 The signal $x(t) = \cos(5\pi t) + 4 \sin(2\pi t) \sin(3\pi t)$, where t is in milliseconds, is sampled at a rate of 3 kHz. Determine the signal $x_a(t)$ aliased with $x(t)$.

Determine two other signals $x_1(t)$ and $x_2(t)$ that are different from each other and from $x(t)$, yet they are aliased with the same $x_a(t)$ that you found.

1.4 Let $x(t) = \cos(8\pi t) + 2 \cos(4\pi t) \cos(6\pi t)$, where t is in seconds. Determine the signal $x_a(t)$ aliased with $x(t)$, if the sampling rate is 5 Hz. Repeat for a sampling rate of 9 Hz.

1.5 The analog signal $x(t) = \sin(6\pi t)[1 + 2 \cos(4\pi t)]$, where t is in milliseconds, is sampled at a rate of 4 kHz. The resulting samples are immediately reconstructed by an ideal reconstructor. Determine the analog signal $x_a(t)$ at the output of the reconstructor.

1.6 The analog signal $x(t) = 4 \cos(2\pi t) \cos(8\pi t) \cos(12\pi t)$, where t is in seconds, is sampled at a rate of $f_s = 10$ Hz. Determine the signal $x_a(t)$ aliased with $x(t)$. Show that the two signals have the *same* sample values, that is, show that $x(nT) = x_a(nT)$. Repeat the above questions if the sampling rate is $f_s = 12$ Hz. [*Hint*: Express $x(t)$ as a *sum* of sines and cosines.]

1.7 Consider the periodic triangular waveform with period $T_0 = 1$ sec shown in Fig. 1.8.1. The waveform is sampled at rate $f_s = 8$ Hz and the resulting samples are reconstructed by an *ideal* reconstructor. Show that the signal $x_{\text{rec}}(t)$ that will appear at the output of the reconstructor will have the form:

$$x_{\text{rec}}(t) = A \sin(2\pi f_1 t) + B \sin(2\pi f_2 t)$$

and determine the numerical values of the frequencies f_1, f_2 and amplitudes A, B.

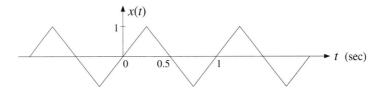

Fig. 1.8.1 Triangular waveform of Problem 1.7.

1.8 *Computer Experiment: Aliasing.* Consider an analog signal $x(t)$ consisting of three sinusoids of frequencies $f_1 = 1$ kHz, $f_2 = 4$ kHz, and $f_3 = 6$ kHz, where t is in milliseconds:

$$x(t) = 2 \sin(2\pi f_1 t) + 2 \sin(2\pi f_2 t) + \sin(2\pi f_3 t)$$

 a. The signal is sampled at a rate of 5 kHz. Determine the signal $x_a(t)$ that would be aliased with $x(t)$. On the same graph, plot the two signals $x(t)$ and $x_a(t)$ versus

t in the range $0 \leq t \leq 2$ msec. Show both analytically and graphically that the two signals have the *same* sampled values, which occur at intervals of $T = 1/f_s = 0.2$ msec.

b. Repeat with a sampling rate of $f_s = 10$ kHz.

c. On the same graph, plot the signals $x(t)$ and $x_a(t)$ of Problem 1.7, over the range $0 \leq t \leq 2$ sec, and verify that they intersect at the sampling instants at multiples of $T = 1/f_s = 0.125$ sec. In plotting, $x(t)$, you need to define it as a triangular function of t.

Repeat this part, but with sampling rate $f_s = 4$ Hz. What is $x_a(t)$ now?

1.9 Consider the following sound wave, where t is in milliseconds:

$$x(t) = \sin(10\pi t) + \sin(20\pi t) + \sin(60\pi t) + \sin(90\pi t)$$

This signal is prefiltered by an analog antialiasing prefilter $H(f)$ and then sampled at an audio rate of 40 kHz. The resulting samples are immediately reconstructed using an ideal reconstructor. Determine the output $y_a(t)$ of the reconstructor in the following cases and compare it with the *audible part* of $x(t)$:

a. When there is no prefilter, that is, $H(f) \equiv 1$.

b. When $H(f)$ is an *ideal* prefilter with cutoff of 20 kHz.

c. When $H(f)$ is a *practical* prefilter that has a flat passband up to 20 kHz and attenuates at a rate of 48 dB/octave beyond 20 kHz. (You may ignore the effects of the phase response of the filter.)

1.10 Prove the Fourier series expansion of the ideal sampling function $s(t)$ given in Eq. (1.5.10). Then, prove its Fourier transform expression (1.5.13).

1.11 Given Eq. (1.5.4), prove the inverse DTFT property (1.5.5), that is,

$$\hat{X}(f) = \sum_{n=-\infty}^{\infty} x(nT) e^{-2\pi j f T n} \quad \Rightarrow \quad x(nT) = \frac{1}{f_s} \int_{-f_s/2}^{f_s/2} \hat{X}(f) e^{2\pi j f T n} \, df$$

1.12 Consider a pure sinusoid of frequency f_0, $x(t) = \cos(2\pi f_0 t)$. Show that the spectrum of the sampled sinusoid $x(nT)$ is:

$$\hat{X}(f) = \frac{1}{2T} \sum_{m=-\infty}^{\infty} \left[\delta(f - f_0 - mf_s) + \delta(f + f_0 + mf_s) \right]$$

1.13 *Computer Experiment: Sampling of Non-Bandlimited Signals.* Consider the exponentially decaying sinusoid $x(t) = e^{-at} \cos(2\pi f_0 t)$ sampled at a rate $f_s = 1/T$. For convenience, replace it by its complex-valued version: $x(t) = e^{-at} e^{2\pi j f_0 t}$. Let $x(nT) = e^{-aTn} e^{2\pi j f_0 T n}$ be its samples, and let $x_L(nT) = x(nT)$, $n = 0, 1, \ldots, L-1$ be its windowed version to length L. Show that the magnitude spectra of the analog, sampled, and windowed signals are given by:

$$|X(f)|^2 = \frac{1}{a^2 + (2\pi(f - f_0))^2}$$

$$|\hat{X}(f)|^2 = \frac{1}{1 - 2e^{-aT} \cos(2\pi(f - f_0)) + e^{-2aT}}$$

$$|\hat{X}_L(f)|^2 = \frac{1 - 2e^{-aTL} \cos(2\pi(f - f_0)L) + e^{-2aTL}}{1 - 2e^{-aT} \cos(2\pi(f - f_0)) + e^{-2aT}}$$

Show the limits:

$$\lim_{L \to \infty} \hat{X}_L(f) = \hat{X}(f), \qquad \lim_{f_s \to \infty} T\hat{X}(f) = X(f)$$

For the numerical values $a = 0.2 \text{ sec}^{-1}$, $f_0 = 0.5$ Hz, and the two rates $f_s = 1$ Hz and $f_s = 2$ Hz, plot on the same graph the analog spectrum $|X(f)|^2$ and the sampled spectrum $|T\hat{X}(f)|^2$, over the frequency range $0 \le f \le 3$ Hz.

For $f_s = 2$, plot on another graph, the three spectra $|X(f)|^2$, $|T\hat{X}(f)|^2$, $|T\hat{X}_L(f)|^2$, over the range $0 \le f \le 3$ Hz.

What conclusions do you draw from these graphs? What are the implications of the above limits? What are the essential differences if we work with the real-valued signal?

1.14 The frequency range of interest of a signal extends to f_{max}. Beyond f_{max}, the spectrum attenuates by α dB per decade. We have available an off-the-shelf antialiasing prefilter that has a flat passband up to f_{max} and attenuates by β dB per decade beyond that. It is required that within the f_{max} range of interest, the aliased components due to sampling be suppressed by more than A dB. Show that the *minimum* sampling rate that we should use is given by

$$f_s = f_{max} + 10^{A/\gamma} f_{max}, \qquad \text{where } \gamma = \alpha + \beta$$

1.15 An analog input signal to a DSP system has spectrum:

$$|X_{in}(f)| = \frac{1}{\sqrt{1 + (0.1f)^8}}$$

where f is in kHz. The highest frequency of interest is 20 kHz. The signal is to be sampled at a rate f_s. It is required that the aliased spectral components within the frequency range of interest be suppressed by more than 60 dB *relative* to the signal components, that is, they must be at least 60 dB below the value of the signal components throughout the 20 kHz range of interest.

    a. Determine the minimum sampling rate f_s, if no antialiasing prefilter is used.

    b. Suppose a simple third-order Butterworth antialiasing prefilter is used having magnitude response

$$|H(f)| = \frac{1}{\sqrt{1 + (f/f_0)^6}}$$

    It is required that the prefilter's attenuation within the 20 kHz band of interest remain less than 1 dB. What is the value of the normalization frequency f_0 in this case? What is the minimum value of f_s that may be used? Compare your exact calculation of f_s with the approximate one using the method of Problem 1.14.

1.16 For the above example, suppose we are constrained to use a particular sampling rate, which is less than the minimum we determined above (and greater than $2f_{max}$), such as $f_s = 70$ kHz. In order to achieve the required 60 dB suppression of the aliased replicas, we must now use a more complex prefilter—one that has a steeper transition width, such as a higher-order Butterworth. An Nth order Butterworth filter has magnitude response

$$|H(f)|^2 = \frac{1}{1 + (f/f_0)^{2N}}$$

Given f_s, determine the minimum filter order N in order for the filter to attenuate less than $A_{\text{pass}} = 1$ dB in the passband and the total suppression of the spectral images to be greater than $A = 60$ dB.

1.17 *Computer Experiment: Butterworth Prefilter Design.* Using the methods of the previous problem, derive a "design curve" for the prefilter, that is, an expression for the Butterworth filter order N as a function of the sampling rate f_s and stopband attenuation A. Assume $f_{\text{max}} = 20$ kHz and $A_{\text{pass}} = 1$ dB for the passband attenuation.

For each of the attenuation values $A = 40, 50, 60, 70, 80$ dB, plot the filter order N versus f_s in the range $50 \le f_s \le 120$ kHz. Identify on these graphs the design points of the Problems 1.15 and 1.16.

1.18 The significant frequency range of an analog signal extends to 10 kHz. Beyond 10 kHz, the signal spectrum attenuates at a rate of 80 dB per decade.

The signal is to be sampled at a rate of 30 kHz. The aliased frequency components introduced into the 10 kHz range of interest must be kept below 60 dB, as compared to the signal components.

Suppose we use an antialiasing prefilter whose passband is flat over the 10 kHz interval. Beyond 10 kHz, it attenuates at a certain rate that must be steep enough to satisfy the above sampling requirements. What is this attenuation rate in *dB per decade*? Explain your reasoning. What is the minimum *filter order* that we must use?

What is the prefilter's attenuation rate if we increase the sampling rate to 50 kHz? What is the filter order in this case?

1.19 An analog input signal to a DSP system has spectrum:

$$|X_{\text{in}}(f)| = \frac{1}{\sqrt{1 + (f/f_a)^{2N_a}}}$$

where f_a and N_a are given. The highest frequency of interest is $f_{\text{max}} = 2f_a$. The signal is to be sampled at a rate f_s. It is required that the aliased spectral components within the frequency range of interest be suppressed by more than A dB *relative* to the signal components, that is, they must be at least A dB below the value of the signal components throughout the $0 \le f \le f_{\text{max}}$ range of interest.

  a. Assuming that no antialiasing prefilter is used, set up and solve an equation for the *minimum* sampling rate f_s, in terms of the quantities f_a, N_a, A.

  b. Next, suppose that an Nth order Butterworth analog prefilter is to be used to aid the sampling process. Let f_0 be the filter's 3-dB normalization frequency. It is required that the prefilter's attenuation within the $0 \le f \le f_{\text{max}}$ band of interest remain less than B dB.

  Set up an equation for f_0 that would guarantee this condition.

  Then, set up an equation for the minimum f_s that would guarantee the desired A dB suppression of the spectral images.

  c. Show that f_s is given approximately by

$$f_s = f_{\text{max}} \left[1 + 10^{A/20(N+N_a)} \right]$$

When is this approximation valid? Show that this expression also covers part (a) if you set $N = 0$. Discuss the meaning of the limit $N \rightarrow \infty$ in terms of the sampling theorem.

1.20 In Problem 1.19, we implicitly assumed that the prefilter's order N was given, and we determined f_0 and f_s. Here, we assume that f_s is given and is equal to some value above $2f_{\text{max}}$. Show that the *minimum* prefilter order that must be used to guarantee A dB suppression of the spectral images is approximately linearly related to A via an equation of the form:

$$N = aA + b$$

Determine expressions for a and b in terms of the given quantities.

1.21 The operation of flat-top practical sampling depicted in Fig. 1.5.1 may be thought of as filtering the ideally sampled signal $\hat{x}(t)$ through an analog linear filter whose impulse response is the sampling pulse $p(t)$, as shown in Fig. 1.8.2. Show that Eq. (1.5.2) can be written as the I/O convolutional equation of such a filter:

$$x_{\text{flat}}(t) = \int_{-\infty}^{\infty} p(t - t')\hat{x}(t')\, dt' = \sum_{n=-\infty}^{\infty} x(nT)p(t - nT)$$

where $\hat{x}(t)$ is given by Eq. (1.5.1). In the frequency domain, this translates to $X_{\text{flat}}(f) = P(f)\hat{X}(f)$, where $P(f)$ is the spectrum of sampling pulse $p(t)$.

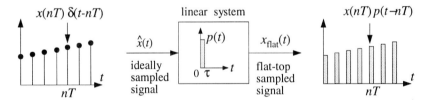

Fig. 1.8.2 Flat-top sampling as filtering.

Determine $P(f)$ for a flat pulse $p(t)$ of duration τ seconds. For the case $\tau = T/5$, make a sketch of $X_{\text{flat}}(f)$ over the range $-6f_s \leq f \leq 6f_s$.

1.22 After having been properly prefiltered by an antialiasing filter, an analog signal is sampled at a rate of 6 kHz. The digital signal is then filtered by a digital filter designed to act as an ideal lowpass filter with cutoff frequency of 1 kHz. The filtered digital signal is then fed into a staircase D/A reconstructor and then into a lowpass anti-image postfilter.

The overall reconstructor is required to suppress the spectral images caused by sampling by more than $A = 40$ dB. Determine the *least stringent* specifications for the analog postfilter that will satisfy this requirement.

1.23 Consider an arbitrary D/A reconstructing filter with impulse response $h(t)$ and corresponding frequency response $H(f)$. The analog signal at the output of the reconstructor is related to the incoming time samples $x(nT)$ by

$$x_a(t) = \sum_n x(nT)h(t - nT)$$

Show this result in two ways:

a. Using convolution in the time domain.

b. Starting with $X_a(f) = H(f)\hat{X}(f)$ and taking inverse Fourier transforms.

1.24 The sinusoidal signal $x(t) = \sin(2\pi f_0 t)$ is sampled at a rate f_s and the resulting samples are then reconstructed by an *arbitrary* analog reconstructing filter $H(f)$. Show that the analog signal at the output of the reconstructor will have the form:

$$x_{\text{rec}}(t) = \sum_{m=-\infty}^{\infty} A_m \sin(2\pi f_m t + \theta_m)$$

What are the frequencies f_m? How are the quantities A_m and θ_m related to the frequency response $H(f)$? Determine the quantities A_m and θ_m for the two cases of a staircase reconstructor and an ideal reconstructor.

1.25 The sum of sinusoids

$$y(t) = A_1 e^{2\pi j f_1 t} + A_2 e^{2\pi j f_2 t}$$

is sampled at a rate f_s such that $f_s > 2|f_1|$ and $f_s > 2|f_2|$. The resulting samples are then filtered digitally by a *staircase-equalizing* digital filter and then reconstructed by a staircase reconstructor, as shown in Fig. 1.6.9. If a final postfilter is not used, show that the resulting analog signal at the output of the reconstructor will be

$$y_a(t) = \sum_{m=-\infty}^{\infty} \left[A_{1m} e^{2\pi j f_{1m} t} + A_{2m} e^{2\pi j f_{2m} t} \right]$$

where $A_{1m} = A_1 f_1 / f_{1m}$, $A_{2m} = A_2 f_2 / f_{2m}$, and $f_{1m} = f_1 + m f_s$, $f_{2m} = f_2 + m f_s$. What would a final postfilter do to each of these terms?

2

Quantization

2.1 *Quantization Process*

Sampling and quantization are the necessary prerequisites for any digital signal processing operation on analog signals. A sampler and quantizer are shown in Fig. 2.1.1 [40–45]. The hold capacitor in the sampler holds each measured sample $x(nT)$ for at most T seconds during which time the A/D converter must convert it to a quantized sample, $x_Q(nT)$, which is representable by a finite number of bits, say B bits. The B-bit word is then shipped over to the digital signal processor.

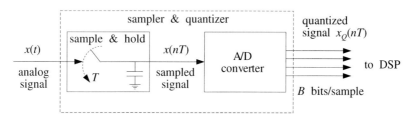

Fig. 2.1.1   Analog to digital conversion.

After digital processing, the resulting B-bit word is applied to a D/A converter which converts it back to analog format generating a staircase output. In practice, the sample/hold and ADC may be separate modules or may reside on board the same chip.

The quantized sample $x_Q(nT)$, being represented by B bits, can take only one of 2^B possible values. An A/D converter is characterized by a *full-scale range R*, which is divided equally (for a uniform quantizer) into 2^B *quantization levels*, as shown in Fig. 2.1.2. The spacing between levels, called the *quantization width* or the quantizer resolution, is given by:

$$Q = \frac{R}{2^B}$$ (2.1.1)

This equation can also be written in the form:

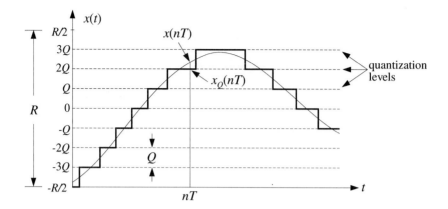

Fig. 2.1.2 Signal quantization.

$$\frac{R}{Q} = 2^B \qquad (2.1.2)$$

which gives the number of quantization levels. Typical values of R in practice are between 1-10 volts. Figure 2.1.2 shows the case of $B = 3$ or $2^B = 8$ levels, and assumes a *bipolar* ADC for which the possible quantized values lie within the symmetric range:

$$-\frac{R}{2} \leq x_Q(nT) < \frac{R}{2}$$

For a *unipolar* ADC, we would have instead $0 \leq x_Q(nT) < R$. In practice, the input signal $x(t)$ must be preconditioned by analog means to lie within the full-scale range of the quantizer, that is, $-R/2 \leq x(t) < R/2$, before it is sent to the sampler and quantizer. The upper end, $R/2$, of the full-scale range is not realized as one of the levels; rather, the maximum level is $R/2 - Q$.

In Fig. 2.1.2, quantization of $x(t)$ was done by *rounding*, that is, replacing each value $x(t)$ by the value of the *nearest* quantization level. Quantization can also be done by *truncation* whereby each value is replaced by the value of the level below it. Rounding is preferred in practice because it produces a less biased quantized representation of the analog signal.

The *quantization error* is the error that results from using the quantized signal $x_Q(nT)$ instead of the true signal $x(nT)$, that is,[†]

$$e(nT) = x_Q(nT) - x(nT) \qquad (2.1.3)$$

In general, the error in quantizing a number x that lies in $[-R/2, R/2)$ is:

$$e = x_Q - x$$

[†]A more natural definition would have been $e(nT) = x(nT) - x_Q(nT)$. The choice (2.1.3) is more convenient for making quantizer models.

where x_Q is the quantized value. If x lies between two levels, it will be rounded up or down depending on which is the closest level. If x lies in the upper (lower) half between the two levels, it will be rounded up (down). Thus, the error e can only take the values[†]

$$-\frac{Q}{2} \le e \le \frac{Q}{2} \qquad (2.1.4)$$

Therefore, the maximum error is $e_{\max} = Q/2$ in magnitude. This is an overestimate for the typical error that occurs. To obtain a more representative value for the average error, we consider the *mean* and *mean-square* values of e defined by:

$$\bar{e} = \frac{1}{Q} \int_{-Q/2}^{Q/2} e\, de = 0, \quad \text{and} \quad \overline{e^2} = \frac{1}{Q} \int_{-Q/2}^{Q/2} e^2\, de = \frac{Q^2}{12} \qquad (2.1.5)$$

The result $\bar{e} = 0$ states that on the average half of the values are rounded up and half down. Thus, $\bar{e}$ cannot be used as a representative error. A more typical value is the *root-mean-square* (rms) error defined by:

$$\boxed{e_{\text{rms}} = \sqrt{\overline{e^2}} = \frac{Q}{\sqrt{12}}} \qquad (2.1.6)$$

Equations (2.1.5) can be given a probabilistic interpretation by assuming that the quantization error e is a *random variable* which is distributed *uniformly* over the range (2.1.4), that is, having probability density:

$$p(e) = \begin{cases} \dfrac{1}{Q} & \text{if } -\dfrac{Q}{2} \le e \le \dfrac{Q}{2} \\ 0 & \text{otherwise} \end{cases}$$

The normalization $1/Q$ is needed to guarantee:

$$\int_{-Q/2}^{Q/2} p(e)\, de = 1$$

It follows that Eqs. (2.1.5) represent the *statistical expectations*:

$$E[e] = \int_{-Q/2}^{Q/2} e\, p(e)\, de \quad \text{and} \quad E[e^2] = \int_{-Q/2}^{Q/2} e^2 p(e)\, de$$

Thinking of R and Q as the ranges of the signal and quantization noise, the ratio in Eq. (2.1.2) is a *signal-to-noise ratio* (SNR). It can be expressed in dB:

$$20 \log_{10}\left(\frac{R}{Q}\right) = 20 \log_{10}(2^B) = B \cdot 20 \log_{10} 2, \quad \text{or,}$$

[†]If the midpoint between levels is always rounded up, then we should have more strictly $-Q/2 < e \le Q/2$.

$$SNR = 20\log_{10}\left(\frac{R}{Q}\right) = 6B \ \text{dB} \qquad\qquad (2.1.7)$$

which is referred to as the 6 *dB per bit* rule. Eq. (2.1.7) is called the *dynamic range* of the quantizer. Equations (2.1.1) and (2.1.6) can be used to determine the wordlength B if the full-scale range and desired rms error are given.

Example 2.1.1: In a digital audio application, the signal is sampled at a rate of 44 kHz and each sample quantized using an A/D converter having a full-scale range of 10 volts. Determine the number of bits B if the rms quantization error must be kept below 50 microvolts. Then, determine the actual rms error and the bit rate in bits per second.

Solution: Write Eq. (2.1.6) in terms of B, $e_{\text{rms}} = Q/\sqrt{12} = R2^{-B}/\sqrt{12}$ and solve for B:

$$B = \log_2\left[\frac{R}{e_{\text{rms}}\sqrt{12}}\right] = \log_2\left[\frac{10}{50\cdot 10^{-6}\sqrt{12}}\right] = 15.82$$

which is rounded to $B = 16$ bits, corresponding to $2^B = 65536$ quantization levels. With this value of B, we find $e_{\text{rms}} = R2^{-B}/\sqrt{12} = 44$ microvolts. The bit rate will be $Bf_s = 16\cdot 44 = 704$ kbits/sec. This is a typical bit rate for CD players.

The dynamic range of the quantizer is $6B = 6\cdot 16 = 96$ dB. Note that the dynamic range of the human ear is about 100 dB. Therefore, the quantization noise from 16-bit quantizers is about at the threshold of hearing. This is the reason why "CD quality" digital audio requires at least 16-bit quantization.   □

Example 2.1.2: By comparison, in digital speech processing the typical sampling rate is 8 kHz and the quantizer's wordlength 8 bits, corresponding to 256 levels. An 8-bit ADC with full-scale range of 10 volts, would generate an rms quantization noise $e_{\text{rms}} = R2^{-B}/\sqrt{12} = 11$ millivolts. The bit rate in this case is $Bf_s = 8\cdot 8 = 64$ kbits/sec.   □

The probabilistic interpretation of the quantization noise is very useful for determining the effects of quantization as they propagate through the rest of the digital processing system. Writing Eq. (2.1.3) in the form[†]

$$x_Q(n) = x(n) + e(n) \qquad\qquad (2.1.8)$$

we may think of the quantized signal $x_Q(n)$ as a noisy version of the original unquantized signal $x(n)$ to which a noise component $e(n)$ has been added. Such an additive noise model of a quantizer is shown in Fig. 2.1.3.

In general, the statistical properties of the noise sequence $e(n)$ are very complicated [46–51,54]. However, for so-called *wide-amplitude wide-band* signals, that is, signals that vary through the entire full-scale range R crossing often all the quantization levels, the sequence $e(n)$ may be assumed to be a *stationary* zero-mean *white noise* sequence with *uniform* probability density over the range $[-Q/2, Q/2]$. Moreover, $e(n)$ is assumed to be *uncorrelated* with the signal $x(n)$. The average *power* or variance of $e(n)$ has already been computed above:

[†]For simplicity, we denoted $x(nT)$ by $x(n)$, etc.

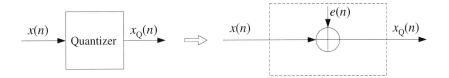

Fig. 2.1.3 Additive noise model of a quantizer.

$$\sigma_e^2 = E[e^2(n)] = \frac{Q^2}{12} \qquad (2.1.9)$$

The assumption that $e(n)$ is white noise means that it has a delta-function auto-correlation (see Appendix A.1):

$$R_{ee}(k) = E[e(n+k)e(n)] = \sigma_e^2 \delta(k) \qquad (2.1.10)$$

for all lags k. Similarly, that it is uncorrelated with $x(n)$ means that it has zero cross-correlation:

$$R_{ex}(k) = E[e(n+k)x(n)] = 0 \qquad (2.1.11)$$

for all k. Later on we will illustrate this statistical model for $e(n)$ with a simulation example and verify equations (2.1.9)-(2.1.11), as well as the uniform distribution for the density $p(e)$.

The model is not accurate for low-amplitude slowly varying signals. For example, a sinusoid that happens to lie exactly in the middle between two levels and has amplitude less than $Q/2$ will be quantized to be a square wave, with all the upper humps of the sinusoid being rounded up and all the lower ones rounded down. The resulting error $e(n)$ will be highly periodic, that is, correlated from sample to sample, and not resembling random white noise. It will also be highly correlated with input sinusoid $x(n)$.

In digital audio, quantization distortions arising from low-level signals are referred to as *granulation* noise and correspond to unpleasant sounds. They can be virtually eliminated by the use of *dither*, which is low-level noise added to the signal before quantization.

The beneficial effect of dithering is to make the overall quantization error behave as a white noise signal, which is *perceptually* much more preferable and acceptable than the gross granulation distortions of the undithered signal [52-69]. On the negative side, dithering reduces the signal-to-noise ratio somewhat—between 3 to 6 dB depending on the type of dither used. Dither will be discussed in Section 2.5.

2.2 Oversampling and Noise Shaping*

In the frequency domain, the assumption that $e(n)$ is a white noise sequence means that it has a flat spectrum. More precisely, the total average power σ_e^2 of $e(n)$ is distributed *equally* over the Nyquist interval $[-f_s/2, f_s/2]$, as shown in Fig. 2.2.1.

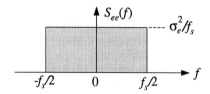

Fig. 2.2.1 Power spectrum of white quantization noise.

Thus, the power per unit frequency interval or *power spectral density* of $e(n)$ will be[†]

$$S_{ee}(f) = \frac{\sigma_e^2}{f_s}, \qquad \text{for} \quad -\frac{f_s}{2} \le f \le \frac{f_s}{2} \tag{2.2.1}$$

and it is periodic outside the interval with period f_s. The noise power within any Nyquist subinterval $[f_a, f_b]$ of width $\Delta f = f_b - f_a$ is given by

$$S_{ee}(f)\,\Delta f = \sigma_e^2\,\frac{\Delta f}{f_s} = \sigma_e^2\,\frac{f_b - f_a}{f_s}$$

As expected, the total power over the entire interval $\Delta f = f_s$ will be

$$\frac{\sigma_e^2}{f_s}\,f_s = \sigma_e^2$$

Noise shaping quantizers reshape the spectrum of the quantization noise into a more convenient shape. This is accomplished by filtering the white noise sequence $e(n)$ by a noise shaping filter $H_{NS}(f)$. The *equivalent noise model* for the quantization process is shown in Fig. 2.2.2. The corresponding quantization equation, replacing Eq. (2.1.8), becomes:

$$\boxed{x_Q(n) = x(n) + \varepsilon(n)} \tag{2.2.2}$$

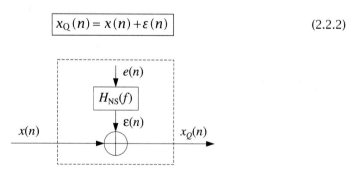

Fig. 2.2.2 Model of noise shaping quantizer.

where $\varepsilon(n)$ denotes the filtered noise. The sequence $\varepsilon(n)$ is no longer white. Its power spectral density is not flat, but acquires the shape of the filter $H_{NS}(f)$:

[†]In units of digital frequency $\omega = 2\pi f / f_s$, it is $S_{ee}(\omega) = \sigma_e^2 / 2\pi$.

$$S_{\varepsilon\varepsilon}(f) = |H_{NS}(f)|^2 S_{ee}(f) = \frac{\sigma_e^2}{f_s} |H_{NS}(f)|^2 \qquad (2.2.3)$$

The noise power within a given subinterval $[f_a, f_b]$ is obtained by integrating $S_{\varepsilon\varepsilon}(f)$ over that subinterval:

$$\text{Power in } [f_a, f_b] = \int_{f_a}^{f_b} S_{\varepsilon\varepsilon}(f)\, df = \frac{\sigma_e^2}{f_s} \int_{f_a}^{f_b} |H_{NS}(f)|^2\, df \qquad (2.2.4)$$

Noise shaping quantizers are implemented by the so-called *delta-sigma* A/D converters [276], which are increasingly being used in commercial products such as digital audio sampling systems for hard disk or digital tape recording. We will discuss implementation details in Chapter 12. Here, we give a broad overview of the advantages of such quantizers.

The concepts of sampling and quantization are independent of each other. The first corresponds to the quantization of the time axis and the second to the quantization of the amplitude axis. Nevertheless, it is possible to trade off one for the other. Oversampling was mentioned earlier as a technique to alleviate the need for high quality prefilters and postfilters. It can also be used to trade off bits for samples. In other words, if we sample at a higher rate, we can use a coarser quantizer. Each sample will be less accurate, but there will be many more of them and their effect will average out to recover the lost accuracy.

The idea is similar to performing multiple measurements of a quantity, say x. Let σ_x^2 be the mean-square error in a single measurement. If L independent measurements of x are made, it follows from the law of large numbers that the measurement error will be reduced to σ_x^2/L, improving the accuracy of measurement. Similarly, if σ_x^2 is increased, making each individual measurement worse, one can maintain the *same* level of quality as long as the number of measurements L is also increased commensurately to keep the ratio σ_x^2/L constant.

Consider two cases, one with sampling rate f_s and B bits per sample, and the other with higher sampling rate f_s' and B' bits per sample. The quantity:

$$\boxed{L = \frac{f_s'}{f_s}}$$

is called the *oversampling ratio* and is usually an integer. We would like to show that B' can be less than B and still maintain the same level of quality. Assuming the same full-scale range R for the two quantizers, we have the following quantization widths:

$$Q = R2^{-B}, \qquad Q' = R2^{-B'}$$

and quantization noise powers:

$$\sigma_e^2 = \frac{Q^2}{12}, \qquad \sigma_e'^2 = \frac{Q'^2}{12}$$

To maintain the same quality in the two cases, we require that the power spectral densities remain the same, that is, using Eq. (2.2.1):

$$\frac{\sigma_e^2}{f_s} = \frac{\sigma_e'^2}{f_s'}$$

which can be rewritten as

$$\sigma_e^2 = f_s \frac{\sigma_e'^2}{f_s'} = \frac{\sigma_e'^2}{L} \tag{2.2.5}$$

Thus, the total quantization power σ_e^2 is less than $\sigma_e'^2$ by a factor of L, making B greater than B'. The meaning of this result is shown pictorially in Fig. 2.2.3. If sampling is done at the higher rate f_s', then the total power $\sigma_e'^2$ of the quantization noise is spread evenly over the f_s' Nyquist interval.

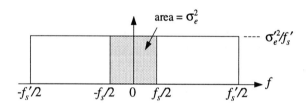

Fig. 2.2.3 Oversampled quantization noise power, without noise shaping.

The shaded area in Fig. 2.2.3 gives the *proportion* of the $\sigma_e'^2$ power that lies within the smaller f_s interval. Solving Eq. (2.2.5) for L and expressing it in terms of the difference $\Delta B = B - B'$, we find:

$$L = \frac{\sigma_e'^2}{\sigma_e^2} = 2^{2(B-B')} = 2^{2\Delta B}$$

or, equivalently

$$\boxed{\Delta B = 0.5 \log_2 L} \tag{2.2.6}$$

that is, a saving of half a bit per doubling of L. This is too small to be useful. For example, in order to reduce a 16-bit quantizer for digital audio to a 1-bit quantizer, that is, $\Delta B = 15$, one would need the unreasonable oversampling ratio of $L = 2^{30}$.

A *noise shaping* quantizer operating at the higher rate f_s' can reshape the flat noise spectrum so that most of the power is squeezed out of the f_s Nyquist interval and moved into the outside of that interval. Fig. 2.2.4 shows the power spectrum of such a quantizer.

The total quantization noise power that resides within the original f_s Nyquist interval is the shaded area in this figure. It can be calculated by integrating Eq. (2.2.4) over $[-f_s/2, f_s/2]$:

$$\sigma_e^2 = \frac{\sigma_e'^2}{f_s'} \int_{-f_s/2}^{f_s/2} |H_{\mathrm{NS}}(f)|^2 \, df \tag{2.2.7}$$

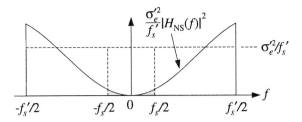

Fig. 2.2.4   Spectrum of oversampling noise shaping quantizer.

Note that it reduces to Eq. (2.2.5) if there is no noise shaping, that is, $H_{NS}(f) = 1$. We will see in Section 12.7 that a typical pth order noise shaping filter operating at the high rate f_s' has magnitude response:

$$|H_{NS}(f)|^2 = \left| 2 \sin\left(\frac{\pi f}{f_s'}\right) \right|^{2p}, \qquad \text{for} \quad -\frac{f_s'}{2} \leq f \leq \frac{f_s'}{2} \qquad (2.2.8)$$

For small frequencies f, we may use the approximation, $\sin x \simeq x$, to obtain:

$$|H_{NS}(f)|^2 = \left(\frac{2\pi f}{f_s'}\right)^{2p}, \qquad \text{for } |f| \ll f_s'/2 \qquad (2.2.9)$$

Assuming a large oversampling ratio L, we will have $f_s \ll f_s'$, and therefore, we can use the approximation (2.2.9) in the integrand of Eq. (2.2.7). This gives:

$$\sigma_e^2 = \frac{\sigma_e'^2}{f_s'} \int_{-f_s/2}^{f_s/2} \left(\frac{2\pi f}{f_s'}\right)^{2p} df = \sigma_e'^2 \frac{\pi^{2p}}{2p+1} \left(\frac{f_s}{f_s'}\right)^{2p+1}$$

$$= \sigma_e'^2 \frac{\pi^{2p}}{2p+1} \frac{1}{L^{2p+1}}$$

Using $\sigma_e^2 / \sigma_e'^2 = 2^{-2(B-B')} = 2^{-2\Delta B}$, we obtain:

$$2^{-2\Delta B} = \frac{\pi^{2p}}{2p+1} \frac{1}{L^{2p+1}}$$

Solving for ΔB, we find the gain in bits:

$$\boxed{\Delta B = (p + 0.5)\log_2 L - 0.5\log_2\left(\frac{\pi^{2p}}{2p+1}\right)} \qquad (2.2.10)$$

Now, the savings are $(p + 0.5)$ bits per doubling of L. Note that Eq. (2.2.10) reduces to Eq. (2.2.6) if there is no noise shaping, that is, $p = 0$. Practical values for the order p are at present $p = 1, 2, 3$, with $p = 4, 5$ becoming available. Table 2.2.1 compares the gain in bits ΔB versus oversampling ratio L for various quantizer orders.

The first CD player built by Philips used a first-order noise shaper with 4-times oversampling, that is, $p = 1, L = 4$, which according to the table, achieves a savings

p	L	4	8	16	32	64	128
0	$\Delta B = 0.5\log_2 L$	1.0	1.5	2.0	2.5	3.0	3.5
1	$\Delta B = 1.5\log_2 L - 0.86$	2.1	3.6	5.1	6.6	8.1	9.6
2	$\Delta B = 2.5\log_2 L - 2.14$	2.9	5.4	7.9	10.4	12.9	15.4
3	$\Delta B = 3.5\log_2 L - 3.55$	3.5	7.0	10.5	14.0	17.5	21.0
4	$\Delta B = 4.5\log_2 L - 5.02$	4.0	8.5	13.0	17.5	22.0	26.5
5	$\Delta B = 5.5\log_2 L - 6.53$	4.5	10.0	15.5	21.0	26.5	32.0

Table 2.2.1 Performance of oversampling noise shaping quantizers.

of $\Delta B = 2.1$ bits. Because of that, the Philips CD player used a 14-bit, instead of a 16-bit, D/A converter at the analog reconstructing stage [279].

We also see from the table that to achieve 16-bit CD-quality resolution using 1-bit quantizers, that is, $\Delta B = 15$, we may use a second-order 128-times oversampling quantizer. For digital audio rates $f_s = 44.1$ kHz, this would imply oversampling at $f_s' = Lf_s = 5.6$ MHz, which is feasible with the present state of the art. Alternatively, we may use third-order noise shaping with 64-times oversampling.

An overall DSP system that uses oversampling quantizers with noise shaping is shown in Fig. 2.2.5. Sampling and reconstruction are done at the fast rate f_s' and at the reduced resolution of B' bits. Intermediate processing by the DSP is done at the low rate f_s and increased resolution of B bits. The overall quality remains the same through all the processing stages. Such a system replaces the traditional DSP system, shown in Fig. 1.7.1.

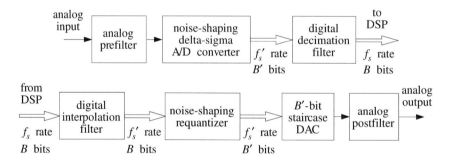

Fig. 2.2.5 Oversampling DSP system.

The faster sampling rate f_s' also allows the use of a less expensive, lower quality, antialiasing prefilter. The digital decimation filter converts the fast rate f_s' back to the desired low rate f_s at the higher resolution of B bits and removes the out-of-band quantization noise that was introduced by the noise shaping quantizer into the outside of the f_s Nyquist interval.

After digital processing by the DSP, the interpolation filter increases the sampling rate digitally back up to the fast rate f_s'. The noise shaping requantizer rounds the B-bit samples to B' bits, without reducing quality. Finally, an ordinary B'-bit staircase A/D converter reconstructs the samples to analog format and the postfilter smooths out the final output. Again, the fast rate f_s' allows the use of a low-quality postfilter.

Oversampling DSP systems are used in a variety of applications, such as digital transmission and coding of speech, the playback systems of CD players, and the sampling/playback systems of DATs. We will discuss the design of oversampling digital interpolation and decimation filters and the structure of noise shaping quantizers and $\Delta\Sigma$ converters in Chapter 12.

2.3 D/A Converters

Next, we discuss some coding details for standard A/D and D/A converters, such as binary representations of quantized samples and the successive approximation method of A/D conversion. We begin with D/A converters, because they are used as the building blocks of successive approximation ADCs. We take a functional view of such converters without getting into the electrical details of their construction.

Consider a B-bit DAC with full-scale range R, as shown in Fig. 2.3.1. Given B input bits of zeros and ones, $\mathbf{b} = [b_1, b_2, \ldots, b_B]$, the converter outputs an analog value x_Q, that lies on one of the 2^B quantization levels within the range R. If the converter is unipolar, the output x_Q falls in the range $[0, R)$. If it is bipolar, it falls in $[-R/2, R/2)$.

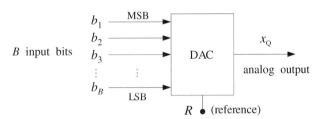

Fig. 2.3.1 B-bit D/A converter.

The manner in which the B bits $[b_1, b_2, \ldots, b_B]$ are associated with the analog value x_Q depends on the type of converter and the coding convention used. We will discuss the three widely used types: (a) unipolar natural binary, (b) bipolar offset binary, and (c) bipolar two's complement converters.

The *unipolar natural binary* converter is the simplest. Its output x_Q is computed in terms of the B bits by:

$$x_Q = R\left(b_1 2^{-1} + b_2 2^{-2} + \cdots + b_B 2^{-B}\right) \tag{2.3.1}$$

The minimum level is $x_Q = 0$ and is reached when all the bits are zero, $\mathbf{b} = [0, 0, \ldots, 0]$. The smallest nonzero level is $x_Q = Q = R2^{-B}$ and corresponds to the least significant bit (LSB) pattern $\mathbf{b} = [0, 0, \ldots, 0, 1]$. The most significant bit (MSB) pattern is $\mathbf{b} = [1, 0, 0, \ldots, 0]$ and corresponds to the output value $x_Q = R/2$.

The maximum level is reached when all bits are one, that is, $\mathbf{b} = [1, 1, \ldots, 1]$ and corresponds to the analog output:

$$x_Q = R\left(2^{-1} + 2^{-2} + \cdots + 2^{-B}\right) = R\left(1 - 2^{-B}\right) = R - Q$$

where we used the geometric series

$$2^{-1} + 2^{-2} + \cdots + 2^{-B} = 2^{-1}\left(1 + 2^{-1} + 2^{-2} + \cdots + 2^{-(B-1)}\right)$$

$$= 2^{-1}\left(\frac{1 - 2^{-B}}{1 - 2^{-1}}\right) = 1 - 2^{-B}$$

Eq. (2.3.1) can be written also in terms of the quantization width Q, as follows:

$$x_Q = R2^{-B}\left(b_1 2^{B-1} + b_2 2^{B-2} + \cdots + b_{B-1} 2^1 + b_B\right) \qquad \text{or,}$$

$$\boxed{x_Q = Qm} \tag{2.3.2}$$

where m is the integer whose binary representation is $(b_1 b_2 \cdots b_B)$, that is,

$$m = b_1 2^{B-1} + b_2 2^{B-2} + \cdots + b_{B-1} 2^1 + b_B$$

As the integer m takes on the 2^B consecutive values $m = 0, 1, 2, \ldots, 2^B - 1$, the analog output x_Q runs through the quantizer's consecutive levels. The *bipolar offset binary* converter is obtained by shifting Eq. (2.3.1) down by half-scale, $R/2$, giving the rule:

$$\boxed{x_Q = R\left(b_1 2^{-1} + b_2 2^{-2} + \cdots + b_B 2^{-B} - 0.5\right)} \tag{2.3.3}$$

The minimum and maximum attainable levels are obtained by shifting the corresponding natural binary values by $R/2$:

$$x_Q = 0 - \frac{R}{2} = -\frac{R}{2} \qquad \text{and} \qquad x_Q = (R - Q) - \frac{R}{2} = \frac{R}{2} - Q$$

The analog value x_Q can also be expressed in terms of Q, as in Eq. (2.3.2). In this case we have:

$$\boxed{x_Q = Qm'} \tag{2.3.4}$$

where m' is the integer m shifted by half the maximum scale, that is,

$$m' = m - \frac{1}{2}2^B = m - 2^{B-1}$$

It takes on the sequence of 2^B values

$$m' = -2^{B-1}, \ldots, -2, -1, 0, 1, 2, \ldots, 2^{B-1} - 1$$

One unnatural property of the offset binary code is that the level $x_Q = 0$ is represented by the nonzero bit pattern $\mathbf{b} = [1, 0, \ldots, 0]$. This is remedied by the *two's complement* code, which is the most commonly used code. It is obtained from the offset binary code by *complementing* the most significant bit, that is, replacing b_1 by $\bar{b}_1 = 1 - b_1$, so that

$$x_Q = R\,(\bar{b}_1 2^{-1} + b_2 2^{-2} + \cdots + b_B 2^{-B} - 0.5) \qquad (2.3.5)$$

Table 2.3.1 summarizes the three converter types and their input/output coding conventions and Table 2.3.2 compares the three coding conventions for the case $B = 4$ and $R = 10$ volts. The level spacing is $Q = R/2^B = 10/2^4 = 0.625$ volts. The codes $[b_1, b_2, b_3, b_4]$ in the first column, apply to both the natural and offset binary cases, but the quantized analog values that they represent are different.

Converter type	I/O relationship
natural binary	$x_Q = R\,(b_1 2^{-1} + b_2 2^{-2} + \cdots + b_B 2^{-B})$
offset binary	$x_Q = R\,(b_1 2^{-1} + b_2 2^{-2} + \cdots + b_B 2^{-B} - 0.5)$
two's complement	$x_Q = R\,(\bar{b}_1 2^{-1} + b_2 2^{-2} + \cdots + b_B 2^{-B} - 0.5)$

Table 2.3.1 Converter types.

	natural binary		offset binary		2's C
$b_1 b_2 b_3 b_4$	m	$x_Q = Qm$	m'	$x_Q = Qm'$	$b_1 b_2 b_3 b_4$
—	16	10.000	8	5.000	—
1 1 1 1	15	9.375	7	4.375	0 1 1 1
1 1 1 0	14	8.750	6	3.750	0 1 1 0
1 1 0 1	13	8.125	5	3.125	0 1 0 1
1 1 0 0	12	7.500	4	2.500	0 1 0 0
1 0 1 1	11	6.875	3	1.875	0 0 1 1
1 0 1 0	10	6.250	2	1.250	0 0 1 0
1 0 0 1	9	5.625	1	0.625	0 0 0 1
1 0 0 0	8	5.000	0	0.000	0 0 0 0
0 1 1 1	7	4.375	−1	−0.625	1 1 1 1
0 1 1 0	6	3.750	−2	−1.250	1 1 1 0
0 1 0 1	5	3.125	−3	−1.875	1 1 0 1
0 1 0 0	4	2.500	−4	−2.500	1 1 0 0
0 0 1 1	3	1.875	−5	−3.125	1 0 1 1
0 0 1 0	2	1.250	−6	−3.750	1 0 1 0
0 0 0 1	1	0.625	−7	−4.375	1 0 0 1
0 0 0 0	0	0.000	−8	−5.000	1 0 0 0

Table 2.3.2 Converter codes for $B = 4$ bits, $R = 10$ volts.

For the natural binary case, the values x_Q are positive, spanning the range $[0, 10)$ volts, with the maximum value being $R - Q = 10 - 0.625 = 9.375$. For offset binary, the level values are offset by half scale, $R/2 = 5$ volts, and span the range $[-5, 5)$ volts, with the maximum being $R/2 - Q = 5 - 0.625 = 4.375$ volts. Note that the upper ends of the full-scale range, $R = 10$ and $R/2 = 5$ volts, are shown in the table for reference and do not represent a level.

The last column shows the two's complement codes. They are obtained from the first column by complementing the MSB, b_1. The quantized values x_Q represented by these codes are the *same* as in the offset binary case, that is, given in the fifth column of the table.

The two's complement code can be understood by wrapping the linear natural binary code around in a circle, as shown in Fig. 2.3.2. This figure shows the natural binary integers m and their negative values in the lower half of the circle. The negative of any positive m in the upper semicircle can be obtained by the usual rule of complementing all its bits and adding one, that is, $m_{2c} = \overline{m} + 1$.

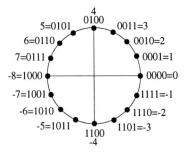

Fig. 2.3.2   Two's complement code.

Example 2.3.1: In Table 2.3.2 or Fig. 2.3.2, the level $m = 3$ corresponds to the natural binary quantized value of $x_Q = 1.875$ volts. The two's complement of m is obtained by the rule

$$m_{2c} = \overline{m} + 1 = \overline{(0011)} + (0001) = (1100) + (0001) = (1101) = -3$$

which, according to the fifth column of the table, corresponds to the two's complement quantized value $x_Q = -1.875$ volts.   □

The following C routine dac.c simulates the operation of the *bipolar two's complement* converter. Its inputs are the B bits $[b_1, b_2, \ldots, b_B]$, and the full-scale range R, and its output is the analog value x_Q computed by Eq. (2.3.5).

```
/* dac.c - bipolar two's complement D/A converter */

double dac(b, B, R)
int *b, B;                          bits are dimensioned as b[0], b[1], ... , b[B − 1]
double R;
{
        int i;
        double dac = 0;
```

```
        b[0] = 1 - b[0];                        complement MSB

        for (i = B-1; i >= 0; i--)              Hörner's rule
            dac = 0.5 * (dac + b[i]);

        dac = R * (dac - 0.5);                  shift and scale

        b[0] = 1 - b[0];                        restore MSB

        return dac;
    }
```

Its usage is:

```
    xQ = dac(b, B, R);
```

Because of the default indexing of arrays in C, the B-dimensional bit vector $b[i]$ is indexed for $i = 0, 1, \ldots, B - 1$. The declaration and dimensioning of $b[i]$ should be done in the main program. For example, if $B = 4$, the main program must include a line:

```
    int b[4];
```

The array $b[i]$ can also be allocated dynamically for any desired value of B using `calloc`. The main program must include the lines:

```
    int *b;                                 b is a pointer to int
    B = 4;                                  B can also be read from stdin
    b = (int *) calloc(B, sizeof(int));     allocates B int slots
```

The internal for-loop in `dac.c` implements a variant of Hörner's rule for evaluating a polynomial. The result is the binary sum $\bar{b}_1 2^{-1} + b_2 2^{-2} + \cdots + b_B 2^{-B}$ which is then shifted by 0.5 and scaled by R. We leave the details of Hörner's algorithm for Problems 2.10-2.13. This algorithm will be used again later for the evaluation of z-transforms and DTFTs. (See the MATLAB function `dtft.m` in Appendix D.)

The routine `dac` may be modified easily to implement the natural binary and offset binary converter types, given by Eqs. (2.3.1) and (2.3.3).

2.4 A/D Converters

A/D converters quantize an analog value x so that it is represented by B bits $[b_1, b_2, \ldots, b_B]$, as shown in Fig. 2.4.1. ADCs come in many varieties, depending on how the conversion process is implemented. One of the most popular ones is the *successive approximation* A/D converter whose main building block is a D/A converter in a feedback loop. It is shown in Fig. 2.4.2.

The conversion algorithm is as follows. Initially all B bits are cleared to zero, $\mathbf{b} = [0, 0, \ldots, 0]$, in the successive approximation register (SAR). Then, starting with the MSB b_1, each bit is turned *on* in sequence and a test is performed to determine whether that bit should be left *on* or turned *off*.

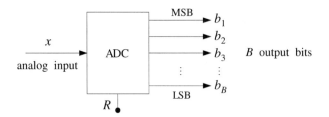

Fig. 2.4.1 *B*-bit A/D converter.

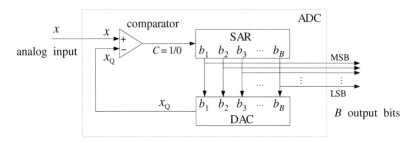

Fig. 2.4.2   Successive approximation A/D converter.

The control logic puts the correct value of that bit in the right slot in the SAR register. Then, leaving all the tested bits set at their correct values, the next bit is turned *on* in the SAR and the process repeated. After B tests, the SAR will hold the correct bit vector $\mathbf{b} = [b_1, b_2, \ldots, b_B]$, which can be sent to the output.

At each test, the SAR bit vector $\mathbf{b}$ is applied to the DAC which produces the analog quantized value x_Q. When a given bit is turned *on*, the output x_Q of the DAC is compared with the analog input x to the ADC. If $x \geq x_Q$, that bit is kept *on*; else, it is turned *off*. The output C of the comparator is the correct value of the bit being tested. The algorithm is summarized below:

> *for each x to be converted, do:*
> $\quad$*initialize* $\mathbf{b} = [0, 0, \ldots, 0]$
> $\quad$*for* $i = 1, 2, \ldots, B$ $\;$*do:*
> $\qquad b_i = 1$
> $\qquad x_Q = \text{dac}(\mathbf{b}, B, R)$
> $\qquad$*if* $(x \geq x_Q)$
> $\qquad\quad C = 1$
> $\qquad$*else*
> $\qquad\qquad C = 0$
> $\qquad b_i = C$

Therefore, C becomes a *serial representation* of the bit vector $\mathbf{b}$. The algorithm imitates the operations shown in Fig. 2.4.2. It can be written more compactly as follows:

$$\boxed{\begin{aligned}
&\textit{for each } x \textit{ to be converted, do:}\\
&\quad \textit{initialize } \mathbf{b} = [0, 0, \ldots, 0]\\
&\quad \textit{for } i = 1, 2, \ldots, B \ \textit{do:}\\
&\qquad b_i = 1\\
&\qquad x_Q = \mathrm{dac}\,(\mathbf{b}, B, R)\\
&\qquad b_i = u\,(x - x_Q)
\end{aligned}}$$

where $u\,(x)$ is the unit-step function, defined by:

$$u\,(x) = \begin{cases} 1 & \text{if } x \geq 0 \\ 0 & \text{if } x < 0 \end{cases}$$

As stated above, the algorithm applies to the natural and offset binary cases (with corresponding versions of dac). It implements *truncation* of x to the quantization level just below, instead of rounding to the nearest level.

The algorithm converges to the right quantization level by performing a *binary search* through the quantization levels. This can be seen with the help of the first column of Table 2.3.2. The test $b_1 = 1$ or 0 determines whether x lies in the upper or lower *half* of the levels. Then, the test $b_2 = 1/0$ determines whether x lies in the upper/lower half of the first half, and so on. Some examples will make this clear.

Example 2.4.1: Convert the analog values $x = 3.5$ and $x = -1.5$ volts to their offset binary representation, assuming $B = 4$ bits and $R = 10$ volts, as in Table 2.3.2.

Solution: The following table shows the successive tests of the bits, the corresponding DAC output x_Q at each test, and the comparator output $C = u\,(x - x_Q)$.

test	$b_1 b_2 b_3 b_4$	x_Q	$C = u\,(x - x_Q)$
b_1	1 0 0 0	0.000	1
b_2	1 1 0 0	2.500	1
b_3	1 1 1 0	3.750	0
b_4	1 1 0 1	3.125	1
	1 1 0 1	3.125	

For each bit pattern, the DAC inputs/outputs were looked up in the first/fifth columns of Table 2.3.2, instead of computing them via Eq. (2.3.3). When b_1 is tested, the DAC output is $x_Q = 0$ which is less than x; therefore, b_1 passes the test. Similarly, b_2 passes the test and stays on. On the other hand bit b_3 fails the test because $x < x_Q = 3.75$; thus, b_3 is turned off. Finally, b_4 passes.

The last row gives the final content of the SAR register and the corresponding quantized value $x_Q = 3.125$. Even though $x = 3.5$ lies in the upper half between the two levels $x_Q = 3.75$ and $x_Q = 3.125$, it gets truncated down to the lower level. The C column is a serial representation of the final answer.

Note also the binary searching taking place: $b_1 = 1$ selects the upper half of the levels, $b_2 = 1$ selects the upper half of the upper half, and of these, $b_3 = 0$ selects the lower half, and of those, $b_4 = 1$ selects the upper half. For the case $x = -1.5$ we have the testing table

test	$b_1 b_2 b_3 b_4$	x_Q	$C = u(x - x_Q)$
b_1	1 0 0 0	0.000	0
b_2	0 1 0 0	-2.500	1
b_3	0 1 1 0	-1.250	0
b_4	0 1 0 1	-1.875	1
	0 1 0 1	-1.875	

Bit b_1 fails the test because $x < x_Q = 0$, and therefore, $b_1 = 0$, and so on. Again, the final quantized value $x_Q = -1.875$ is that obtained by truncating $x = -1.5$ to the level below it, even though x lies nearer the level above it.   □

In order to quantize by *rounding* to the nearest level, we must shift x by half the spacing between levels, that is, use:

$$y = x + \frac{1}{2}Q$$

in place of x and perform *truncation* on y. If x is already in the upper half between two levels, then y will be brought above the upper level and will be truncated down to that level. The conversion algorithm for rounding is:

> *for each x to be converted, do:*
> $y = x + Q/2$
>   *initialize* $\mathbf{b} = [0, 0, \ldots, 0]$
>   *for* $i = 1, 2, \ldots, B$ *do:*
> $b_i = 1$
> $x_Q = \text{dac}(\mathbf{b}, B, R)$
> $b_i = u(y - x_Q)$

Example 2.4.2: To quantize the value $x = 3.5$ by rounding, we shift it to $y = x + Q/2 = 3.5 + 0.625/2 = 3.8125$. The corresponding test table will be

test	$b_1 b_2 b_3 b_4$	x_Q	$C = u(y - x_Q)$
b_1	1 0 0 0	0.000	1
b_2	1 1 0 0	2.500	1
b_3	1 1 1 0	3.750	1
b_4	1 1 1 1	4.375	0
	1 1 1 0	3.750	

Only b_4 fails the test because with it on, the DAC output $x_Q = 4.375$ exceeds y. The final value $x_Q = 3.750$ is the rounded up version of $x = 3.5$. For the case $x = -1.5$, we have $y = -1.5 + 0.625/2 = -1.1875$. The corresponding test table is

test	$b_1 b_2 b_3 b_4$	x_Q	$C = u(y - x_Q)$
b_1	1 0 0 0	0.000	0
b_2	0 1 0 0	−2.500	1
b_3	0 1 1 0	−1.250	1
b_4	0 1 1 1	−0.625	0
	0 1 1 0	−1.250	

The value $x_Q = -1.250$ is the rounded up version of $x = -1.5$.   □

The successive approximation algorithm for the *two's complement* case is slightly different. Because the MSB is complemented, it must be treated separately from the other bits. As seen in the last column of Table 2.3.2, the bit b_1 determines whether the number x is positive or negative. If $x \geq 0$ then, we must have $b_1 = 0$; else $b_1 = 1$. We can express this result by $b_1 = 1 - u(x)$, or, $b_1 = 1 - u(y)$ if we are quantizing by rounding. The remaining bits, $\{b_2, b_3, \ldots, b_B\}$, are tested in the usual manner. This leads to the following two's complement conversion algorithm with rounding:

> *for each x to be converted, do:*
> $\quad y = x + Q/2$
> $\quad$ *initialize* $\mathbf{b} = [0, 0, \ldots, 0]$
> $\quad b_1 = 1 - u(y)$
> $\quad$ *for* $i = 2, 3, \ldots, B$ *do:*
> $\quad\quad b_i = 1$
> $\quad\quad x_Q = \text{dac}(\mathbf{b}, B, R)$
> $\quad\quad b_i = u(y - x_Q)$

Example 2.4.3: The two's complement rounded 4-bit representations of $x = 3.5$ and $x = -1.5$ are:

$$x = 3.5 \quad \Rightarrow \quad x_Q = 3.750 \quad \Rightarrow \quad \mathbf{b} = [0, 1, 1, 0]$$
$$x = -1.5 \quad \Rightarrow \quad x_Q = -1.250 \quad \Rightarrow \quad \mathbf{b} = [1, 1, 1, 0]$$

They are obtained from the offset binary by complementing the MSB. The quantized values x_Q are the same as in the offset binary case — only the binary codes change.   □

Example 2.4.4: Consider the sampled sinusoid $x(n) = A \cos(2\pi f n)$, where $A = 3$ volts and $f = 0.04$ cycles/sample. The sinusoid is evaluated at the ten sampling times $n = 0, 1, \ldots, 9$ and $x(n)$ is quantized using a 4-bit successive approximation ADC with full-scale range $R = 10$ volts. The following table shows the *sampled and quantized* values $x_Q(n)$ and the offset binary and two's complement binary codes representing them.

n	$x(n)$	$x_Q(n)$	2's C	offset
0	3.000	3.125	0101	1101
1	2.906	3.125	0101	1101
2	2.629	2.500	0100	1100
3	2.187	1.875	0011	1011
4	1.607	1.875	0011	1011
5	0.927	0.625	0001	1001
6	0.188	0.000	0000	1000
7	−0.562	−0.625	1111	0111
8	−1.277	−1.250	1110	0110
9	−1.912	−1.875	1101	0101

The 2's complement and offset binary codes differ only in their MSB. The quantized values they represent are the same.   □

The following routine `adc.c` simulates the operation of a *bipolar two's complement* successive approximation ADC. It makes successive calls to `dac.c` to determine each bit.

```
/* adc.c - successive approximation A/D converter */

#include <math.h>

double dac();
int u();

void adc(x, b, B, R)
double x, R;
int *b, B;
{
       int i;
       double y, xQ, Q;

       Q = R / pow(2, B);              quantization width Q = R/2^B
       y = x + Q/2;                    rounding

       for (i = 0; i < B; i++)         initialize bit vector
              b[i] = 0;

       b[0] = 1 - u(y);                determine MSB

       for (i = 1; i < B; i++) {       loop starts with i = 1
              b[i] = 1;                turn ith bit ON
              xQ = dac(b, B, R);       compute DAC output
              b[i] = u(y-xQ);          test and correct bit
              }
}
```

The inputs to the routine are the analog value x to be converted and the full-scale range R. The outputs are the B bits $\mathbf{b} = [b_1, b_2, \ldots, b_B]$ representing x in the two's complement representation. The unit-step function $u(x)$ is implemented by the routine:

```
/* u.c - unit step function */

int u(x)
double x;
{
        if (x >= 0)
                return 1;
        else
                return 0;
}
```

Example 2.4.5: This example illustrates the usage of the routines adc and dac. Consider $L = 50$ samples of a sinusoid $x(n) = A\cos(2\pi fn)$ of digital frequency $f = 0.02$ cycles/sample and amplitude $A = 4$. The signal $x(n)$ is quantized using a successive approximation two's complement converter with rounding, as implemented by the routine adc. The following for-loop was used in the main program for calculating $x_Q(n)$:

```
for (n=0; n<L; n++) {
        x[n]  = A * cos(2 * pi * f * n);
        adc(x[n], b, B, R);
        xQ[n] = dac(b, B, R);
        }
```

where each call to adc determines the bit vector **b**, which is then passed to dac to calculate the quantized value.

The following figure shows the sampled and quantized signal $x_Q(n)$ plotted together with the exact values $x(n)$ for the two cases of a 3-bit and a 4-bit converter. The full-scale range was $R = 16$.

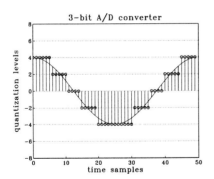 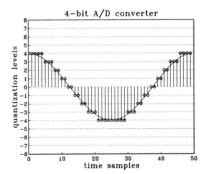

Example 2.4.6: This example illustrates the statistical properties of the quantization error. Consider L samples of the noisy sinusoidal signal:

$$x(n) = A\cos(2\pi f_0 n + \phi) + v(n), \qquad n = 0, 1, \dots, L - 1 \qquad (2.4.1)$$

where ϕ is a random phase distributed uniformly in the interval $[0, 2\pi]$ and $v(n)$ is white noise of variance σ_v^2. Using a B-bit two's complement converter with full-scale range R, these samples are quantized to give $x_Q(n)$ and the quantization error is computed:

$$e(n) = x_Q(n) - x(n) \,, \qquad n = 0, 1, \ldots, L - 1$$

According to the standard statistical model discussed in Section 2.1, the quantization noise samples $e(n)$ should be distributed uniformly over the interval $-Q/2 \le e \le Q/2$. This can be tested by computing the histogram of the L values of $e(n)$.

The theoretical statistical quantities given in Eqs. (2.1.9–2.1.11) can be calculated experimentally by the *time-average* approximations:

$$\sigma_e^2 = \frac{1}{L} \sum_{n=0}^{L-1} e^2(n) \tag{2.4.2}$$

$$R_{ee}(k) = \frac{1}{L} \sum_{n=0}^{L-1-k} e(n+k)e(n) \tag{2.4.3}$$

$$R_{ex}(k) = \frac{1}{L} \sum_{n=0}^{L-1-k} e(n+k)x(n) \tag{2.4.4}$$

We can also compute the autocorrelation of $x(n)$ itself:

$$R_{xx}(k) = \frac{1}{L} \sum_{n=0}^{L-1-k} x(n+k)x(n) \tag{2.4.5}$$

where in the last three equations, k ranges over a few lags $0 \le k \le M$, with M typically being much less than $L-1$. Note also that $\sigma_e^2 = R_{ee}(0)$. All four of the above expressions are special cases of the cross correlation, Eq. (2.4.4), which is implemented by the correlation routine `corr.c`, given in Appendix A.1.

For this experiment, we chose the following numerical values:

$B = 10$ bits
$R = 1024$ volts, so that $Q = 1$ volt
$L = 1000$ samples
$M = 50$ lags
$f_0 = 1/\sqrt{131} \simeq 0.08737$ cycles/sample
$A = R/4 = 256$ volts and $\phi = 0$

The white noise $v(n)$ was distributed uniformly over the interval $[-R/4, R/4]$. Such numbers can be generated by:

$$v = 0.5R(u - 0.5) \tag{2.4.6}$$

where u is a *uniform* random number in the standardized interval $[0, 1]$. The quantity $(u - 0.5)$ is uniform over $[-0.5, 0.5]$, making v uniform over $[-R/4, R/4]$. We used the routine `ran` of Appendix B.1 to generate the u's, but any other uniform random number generator could have been used. The samples $v(n)$ were generated by a for-loop of the form:

```
for (n=0; n<L; n++)
        v[n] = 0.5 * R * (ran(&iseed) - 0.5);
```

where the initial seed[†] was picked arbitrarily. With these choices, the sinusoidal and noise terms in $x(n)$ vary over half of the full-scale range, so that their sum varies over the full range $[-R/2, R/2]$, as is necessary for the model.

With $Q = 1$, the theoretical value of the noise variance is $\sigma_e = Q/\sqrt{12} = 1/\sqrt{12} = 0.289$. The experimental value computed using Eq. (2.4.2) was $\sigma_e = 0.287$.

The histogram of the computed $e(n)$ values was computed by dividing the interval $[-Q/2, Q/2] = [-0.5, 0.5]$ into 10 bins. It is shown below. Theoretically, for a uniform distribution, 1000 samples would distribute themselves evenly over the 10 bins giving $1000/10 = 100$ samples per bin.

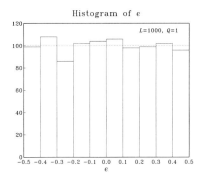

The next two figures show the standard *normalized* correlation functions:

$$\rho_{ee}(k) = \frac{R_{ee}(k)}{R_{ee}(0)}, \quad \rho_{ex}(k) = \frac{R_{ex}(k)}{\sqrt{R_{ee}(0)R_{xx}(0)}}, \quad \rho_{xx}(k) = \frac{R_{xx}(k)}{R_{xx}(0)}$$

computed for lags $k = 0, 1, \ldots, M$ using Eqs. (2.4.3–2.4.5).

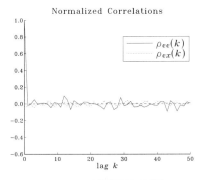

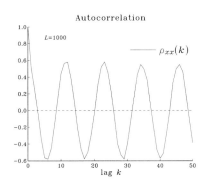

[†]Note that iseed is passed by address in ran(&iseed).

Theoretically, $\rho_{ee}(k)$ should be $\delta(k)$ and $\rho_{ex}(k)$ should be zero. Using the results of Problem 2.20, the theoretical expression for $\rho_{xx}(k)$ will be, for the particular numerical values of the parameters:

$$\rho_{xx}(k) = 0.6\cos(2\pi f_0 k) + 0.4\,\delta(k)$$

Thus, although $x(n)$ itself is highly self-correlated, the quantization noise $e(n)$ is not. The above figures show the closeness of the experimental quantities to the theoretical ones, confirming the reasonableness of the standard statistical model.   □

Successive approximation A/D converters are used routinely in applications with sampling rates of 1 MHz or less. Slower converters also exist, the so-called counter or integrating type. They convert by searching through the quantization levels in a *linear* fashion, comparing each level with the value x to be converted. Therefore, they may require up to 2^B tests to perform a conversion. This is to be compared with the B binary searching steps of the successive approximation type.

For higher rates, *parallel* or flash A/D converters must be used. They determine all the bits simultaneously, in parallel, but they are electrically complex requiring $2^B - 1$ comparators internally. For this reason they are limited at present to $B \le$ 12 bits, achieving conversion rates of 500 MHz with 8 bits, or 50 MHz with 10 bits [41]. As discussed in Problem 2.21, two or more flash A/D converters can be cascaded together, in a so-called subranging configuration, to increase the effective quantization resolution.

2.5 *Analog and Digital Dither**

Dither is a low-level white noise signal added to the input *before* quantization for the purpose of eliminating granulation or quantization distortions and making the total quantization error behave like white noise [52–69].

Analog dither can be added to the analog input signal before the A/D converter, but perhaps after the sample/hold operation. It is depicted in Fig. 2.5.1. In many applications, such as digital audio recordings, the inherent analog system noise of the microphones or mixers may already provide some degree of dithering and therefore artificial dithering may not be necessary.

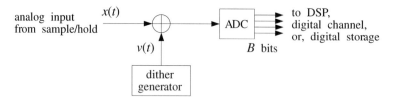

Fig. 2.5.1 Analog dither.

Digital dither can be added to a digital signal prior to a *requantization* operation that *reduces* the number of bits representing the signal.

This circumstance arises, for example, when an audio signal has been sampled and quantized with 20 bits for the purpose of high-quality digital mixing and processing, which then must be reduced to 16 bits for storing it on a CD. Another example is the noise shaping requantization required in oversampling D/A converters used in the playback systems of CD players and DAT decks.

Figure 2.5.2 shows a general model of the analog or digital dither process followed by the quantization operation. It represents a type of dither known as *non-subtractive*.

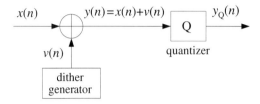

Fig. 2.5.2   Nonsubtractive dither process and quantization.

The input to the quantizer is the sum of the input signal $x(n)$ to be quantized or requantized and the dither noise $v(n)$, that is,

$$y(n) = x(n) + v(n)$$

The output of the quantizer is $y_Q(n)$, the quantized version of $y(n)$. The quantization error is

$$e(n) = y_Q(n) - y(n)$$

Thus, the *total* error resulting from dithering *and* quantization will be:

$$\epsilon(n) = y_Q(n) - x(n) \tag{2.5.1}$$

which can be written as

$$\epsilon(n) = \big(y(n) + e(n)\big) - x(n) = x(n) + v(n) + e(n) - x(n)$$

or,

$$\epsilon(n) = y_Q(n) - x(n) = e(n) + v(n) \tag{2.5.2}$$

that is, the sum of the dither noise plus the quantization error. Proper choice of the dither process $v(n)$ can guarantee that $e(n)$ and $v(n)$ will be uncorrelated from each other, and therefore the total error noise power will be

$$\boxed{\sigma_\epsilon^2 = \sigma_e^2 + \sigma_v^2 = \frac{1}{12}Q^2 + \sigma_v^2} \tag{2.5.3}$$

The statistical properties of the dither signal $v(n)$, such as its probability density function (pdf), can affect drastically the nature of the total error (2.5.2). In practice, there are three commonly used types of dithers, namely, those with Gaussian, rectangular, or triangular pdf's. The pdf's of the rectangular and triangular cases are shown in Fig. 2.5.3. Note that the areas under the curves are equal to unity. In the Gaussian case, the zero-mean pdf is:

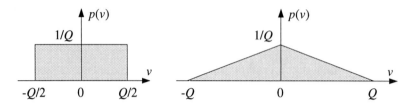

Fig. 2.5.3 Rectangular and triangular dither probability densities.

$$p(v) = \frac{1}{\sqrt{2\pi\sigma_v^2}} e^{-v^2/2\sigma_v^2} \tag{2.5.4}$$

with the recommended value for the variance:

$$\sigma_v^2 = \frac{1}{4}Q^2 \tag{2.5.5}$$

which corresponds to the rms value $v_{\text{rms}} = Q/2$, or half-LSB. It follows from Eq. (2.5.3) that the total error variance will be:

$$\sigma_\epsilon^2 = \frac{1}{12}Q^2 + \frac{1}{4}Q^2 = 4 \cdot \frac{1}{12}Q^2 = \frac{1}{3}Q^2$$

In the rectangular case, the pdf is taken to have width Q, that is, 1-LSB. Therefore, the dither signal can only take values in the interval:

$$-\frac{1}{2}Q \le v \le \frac{1}{2}Q$$

The corresponding pdf and variance are in this case:

$$p(v) = \begin{cases} \dfrac{1}{Q}, & \text{if } -\dfrac{1}{2}Q \le v \le \dfrac{1}{2}Q \\ 0, & \text{otherwise} \end{cases} \quad \text{and} \quad \sigma_v^2 = \frac{1}{12}Q^2 \tag{2.5.6}$$

Therefore, the total error variance will be:

$$\sigma_\epsilon^2 = \frac{1}{12}Q^2 + \frac{1}{12}Q^2 = 2 \cdot \frac{1}{12}Q^2 = \frac{1}{6}Q^2$$

Similarly, the width of the triangular dither pdf is taken to be $2Q$, that is, 2-LSB, and therefore, the corresponding pdf and variance are:

$$p(v) = \begin{cases} \dfrac{Q - |v|}{Q^2}, & \text{if } -Q \le v \le Q \\ 0, & \text{otherwise} \end{cases} \quad \text{and} \quad \sigma_v^2 = \frac{1}{6}Q^2 \tag{2.5.7}$$

and, the total error variance will be:

$$\sigma_\epsilon^2 = \frac{1}{12}Q^2 + \frac{1}{6}Q^2 = 3 \cdot \frac{1}{12}Q^2 = \frac{1}{4}Q^2$$

In summary, the total error variance in the three cases and the undithered case ($v = 0$) will be:

$$\sigma_\epsilon^2 = \begin{cases} Q^2/12, & \text{undithered} \\ 2Q^2/12, & \text{rectangular dither} \\ 3Q^2/12, & \text{triangular dither} \\ 4Q^2/12, & \text{Gaussian dither} \end{cases} \tag{2.5.8}$$

Thus, the noise penalty in using dither is to double, triple, or quadruple the noise of the undithered case. This corresponds to a decrease of the SNR by:

$$10\log_{10} 2 = 3 \text{ dB}$$

$$10\log_{10} 3 = 4.8 \text{ dB}$$

$$10\log_{10} 4 = 6 \text{ dB}$$

which is quite acceptable in digital audio systems that have total SNRs of the order of 95 dB.

It has been shown that the *triangular* dither is the best (of the nonsubtractive types) in the sense that it accomplishes the main objective of the dithering process, namely, to eliminate the quantization distortions of the undithered case and to render the total error (2.5.2) equivalent to white noise [56].

Rectangular, uniformly distributed dither can be generated very simply by using a uniform random number generator such as ran. For example,

$$v = Q(u - 0.5)$$

where u is the random number returned by ran, that is, uniformly distributed over the interval $[0, 1)$. The shifting and scaling of u imply that v will be uniformly distributed within $-Q/2 \le v < Q/2$.

Triangular dither can be generated just as simply by noting that the triangular pdf is the convolution of two rectangular ones and therefore v can be obtained as the *sum* of two independent rectangular random numbers, that is,

$$v = v_1 + v_2 \tag{2.5.9}$$

where v_1, v_2 are generated from two independent uniform u_1, u_2, by

$$v_1 = Q(u_1 - 0.5) \quad \text{and} \quad v_2 = Q(u_2 - 0.5)$$

Example 2.5.1: This simulation example illustrates the impact of dither on the quantization of a low-amplitude sinusoid and the removal of quantization distortions. Consider the following dithered sinusoid:

$$y(n) = x(n) + v(n) = A\cos(2\pi f_0 n) + v(n)$$

where A is taken to be below 1-LSB; for example, $A = 0.75Q$. The frequency f_0 is taken to be $f_0 = 0.025$ cycles/sample, which corresponds to $1/f_0 = 40$ samples per cycle. At an audio rate of $f_s = 40$ kHz, this frequency would correspond to a 1 kHz sinusoid.

The signal $y(n)$ is quantized using a 3-bit A/D converter, $(B = 3)$, with full-scale range of $R = 8$ volts. Therefore, $Q = R/2^B = 8/2^3 = 1$, and $A = 0.75Q = 0.75$. Triangular dither was used, generated by Eq. (2.5.9). The dithered signal $y(n)$ and its quantized version $y_Q(n)$ were generated by the following loop:

```
for (n=0; n<Ntot; n++) {
        v1 = Q * (ran(&iseed) - 0.5);
        v2 = Q * (ran(&iseed) - 0.5);
        v = v1 + v2;
        y[n] = A * cos(2 * pi * f0 * n) + v;
        adc(y[n], b, B, R);
        yQ[n] = dac(b, B, R);
        }
```

Note that v1 and v2 are independent of each other because each call to ran updates the seed to a new value.

The following graphs show the undithered sinusoid $x(n)$ and its quantized version $x_Q(n)$, together with its Fourier spectrum $|X_Q(f)|$ plotted over the right half of the Nyquist interval, that is, $0 \leq f \leq 0.5$, in units of cycles/sample.

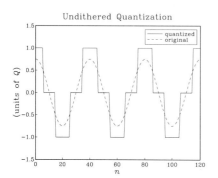
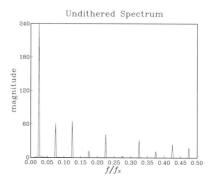

The spectrum of $x_Q(n)$ has peaks at f_0 and the odd harmonics $3f_0$, $5f_0$, and so forth. These harmonics were not present in $x(n)$. They are the artifacts of the quantization process which replaced the sinusoid by a square-wave-like signal.

The next two graphs show the dithered signal $y(n)$ and its quantized version $y_Q(n)$, together with its spectrum $|Y_Q(f)|$.

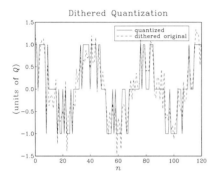
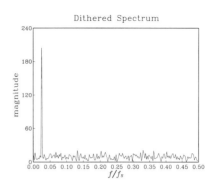

The main peak at f_0 is still there, but the odd harmonics have been eliminated by the dithering process and replaced by a typical featureless background noise spectrum. For digital audio, this noise is perceptually far more acceptable than the artificial harmonics introduced by the quantizer.

The above spectra were computed in the following way: The sequences $x_Q(n)$ and $y_Q(n)$ were generated for $0 \le n \le N_{\text{tot}} - 1$, with $N_{\text{tot}} = 1000$. Then, they were windowed using a length-N_{tot} Hamming window, that is,

$$y'_Q(n) = w(n)\, y_Q(n), \qquad n = 0, 1, \ldots, N_{\text{tot}} - 1$$

where,

$$w(n) = 0.54 - 0.46 \cos\left(\frac{2\pi n}{N_{\text{tot}} - 1}\right), \qquad n = 0, 1, \ldots, N_{\text{tot}} - 1$$

And, their DTFT

$$Y_Q(f) = \sum_{n=0}^{N_{\text{tot}}-1} y'_Q(n)\, e^{-2\pi j f n}$$

was evaluated at 200 equally spaced frequencies f over the interval $0 \le f \le 0.5$ [cycles/sample], that is, at $f_i = 0.5i/200$, $i = 0, 1, \ldots, 199$.

This example is somewhat special in that the undithered spectrum $X_Q(f)$ contained only odd harmonics of the fundamental frequency f_0. This is what one would expect if the quantized square-wave-like signal $x_Q(n)$ were an unsampled, analog, signal.

In general, the sampling process will cause all the odd harmonics that lie outside the Nyquist interval to be aliased back into the interval, onto frequencies that may or may not be odd harmonics. In the above example, because the sampling rate is an even multiple of f_0, that is, $f_s = 40 f_0$, one can show that any odd harmonic of f_0 that lies outside the Nyquist interval will be wrapped onto one of the odd harmonics inside the interval.

But, for other values of f_0, the out-of-band odd harmonics may be aliased onto in-band non-harmonic frequencies. For example, the following graphs show the undithered and dithered spectra in the case of $f_0 = 0.03$ cycles/sample.

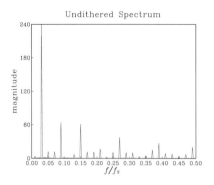
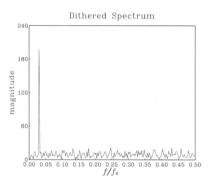

In addition to the odd harmonics at $3f_0 = 0.09$, $5f_0 = 0.15$, and so forth, one sees non-harmonic peaks at:

$$f = 0.01, 0.05, 0.07, 0.13, 0.17, 0.19, 0.23, 0.25, \ldots$$

which are the aliased versions of the following out-of-band odd harmonics:

$$33f_0, 35f_0, 31f_0, 29f_0, 39f_0, 27f_0, 41f_0, 25f_0, \ldots$$

The beneficial effect of dithering works, of course, for any value of f_0.   □

An alternative dithering strategy is to use the so-called *subtractive* dither, as shown in Fig. 2.5.4. Here, the dither noise $v(n)$ that was added during the recording or transmission phase prior to quantization is subtracted at the playback or receiving end.

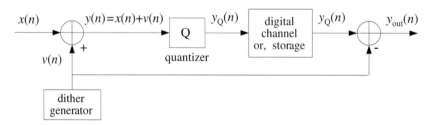

Fig. 2.5.4 Subtractive dither.

The total error in this case can be determined as follows:

$$\epsilon(n) = y_{\text{out}}(n) - x(n) = \big(y_Q(n) - v(n)\big) - x(n) = y_Q(n) - \big(x(n) + v(n)\big)$$

or,

$$\epsilon(n) = y_Q(n) - y(n) = e(n)$$

that is, only the quantizer error. Therefore, its variance remains the same as the undithered case, $\sigma_\epsilon^2 = Q^2/12$, and the SNR remains the same.

It can be shown [56] that the *best* type of dither is subtractive rectangularly distributed dither with 1-LSB width, in the sense that it not only removes the quantization distortions but also renders the total error completely *independent* of the input signal. However, its practical implementation in digital audio and other applications is difficult because it requires a copy of the dither signal at the playback or receiving end.

By contrast, the triangular nonsubtractive dither that we considered earlier does not make the total error independent of the input—it only makes the power spectrum of the error independent of the input. In digital audio, this whitening of the total error appears to be enough perceptually. Therefore, triangular nonsubtractive dither is the best choice for practical use [56].

In summary, triangular nonsubtractive dither improves the quality of a digital processing system by removing the artifacts of the quantization process with only a modest decrease in the signal-to-noise ratio. It may be applied at any intermediate processing stage that involves reduction in the number of bits and, therefore, potential quantization distortions.

2.6 Problems

2.1 Consider a 3-bit successive approximation two's complement bipolar A/D converter with full scale range of $R = 16$ volts. Using the successive approximation algorithm, determine the quantized value as well as the corresponding 3-bit representation of the following analog input values: $x = 2.9, 3.1, 3.7, 4, -2.9, -3.1, -3.7, -4$.

Repeat using an offset binary converter.

2.2 Consider the signal $x(n) = 5 \sin(2\pi f n)$, where $f = 0.04$ [cycles/sample]. This signal is to be quantized using a 4-bit successive approximation bipolar ADC whose full-scale range is $R = 16$ volts. For $n = 0, 1, \ldots, 19$, compute the numerical value of $x(n)$ and its quantized version $x_Q(n)$ as well as the corresponding bit representation at the output of the converter. Do this both for an *offset binary* converter and a *two's complement* converter.

2.3 It is desired to pick an A/D converter for a DSP application that meets the following specifications: The full-scale range of the converter should be 10 volts and the rms quantization error should be kept below 1 millivolt. How many bits should the converter have? What is the actual rms error of the converter? What is the dynamic range in dB of the converter?

2.4 Hard disk recording systems for digital audio are becoming widely available. It is often quoted that to record 1 minute of "CD quality" digital audio in *stereo*, one needs about 10 Megabytes of hard disk space. Please, derive this result, explaining your reasoning.

2.5 A digital audio mixing system uses 16 separate recording channels, each sampling at a 48 kHz rate and quantizing each sample with 20 bits. The digitized samples are saved on a hard disk for further processing.

    a. How many megabytes of hard disk space are required to record a 3-minute song for a 16-channel recording?

    b. Each channel requires about 35 multiplier/accumulation (MAC) instructions to perform the processing of each input sample. (This corresponds to about 7 second-order parametric EQ filters covering the audio band.)

In how many nanoseconds should each MAC instruction be executed for: (i) each channel? (ii) all 16 channels, assuming they are handled by a single processor? Is this within the capability of present day DSP chips?

2.6 If the quantized value x_Q is obtained by *truncation* of x instead of rounding, show that the truncation error $e = x_Q - x$ will be in the interval $-Q < e \le 0$. Assume a uniform probability density $p(e)$ over this interval, that is,

$$p(e) = \begin{cases} \dfrac{1}{Q} & \text{if } -Q < e \le 0 \\ 0 & \text{otherwise} \end{cases}$$

Determine the mean $m_e = E[e]$ and variance $\sigma_e^2 = E[(e - m_e)^2]$ in terms of Q.

2.7 Using Eq. (2.2.10), determine the value of the oversampling ratio L to achieve 16-bit resolution using 1-bit quantizers for the cases of first-, second-, and third-order noise shaping quantizers. What would be the corresponding oversampled rate Lf_s for digital audio?

2.8 In a speech codec, it is desired to maintain quality of 8-bit resolution at 8 kHz sampling rates using a 1-bit oversampled noise shaping quantizer. For quantizer orders $p = 1, 2, 3$, determine the corresponding oversampling ratio L and oversampling rate Lf_s in Hz.

2.9 Show that the two's complement expression defined in Eq. (2.3.5) can be written in the alternative form:

$$x_Q = R\left(-b_1 2^{-1} + b_2 2^{-2} + \cdots + b_B 2^{-B}\right)$$

2.10 Hörner's rule is an efficient algorithm for polynomial evaluation. Consider a polynomial of degree M

$$B(z) = b_0 + b_1 z + b_2 z^2 + \cdots + b_M z^M$$

Hörner's algorithm for evaluating $B(z)$ at some value of z, say $z = a$, can be stated as follows:

> *initialize* $p = 0$
> *for* $i = M, M-1, \ldots, 0$ *do:*
> $p = ap + b_i$

Verify that upon exit, p will be the value of the polynomial at $z = a$, that is, $p = B(a)$.

2.11 *Computer Experiment: Hörner's Rule.* Write a polynomial evaluation C routine pol.c that implements the algorithm of Problem 2.10. The routine should return the value $B(a)$ of the polynomial and should be dimensioned as follows:

```
double pol(M, b, a)
int M;                          order of polynomial
double *b, a;                   b is (M+1)-dimensional
```

2.12 Consider the following variation of Hörner's algorithm:

$$initialize \ q_{M-1} = b_M$$
$$for \ i = M-1, M-2, \ldots, 1 \ do:$$
$$q_{i-1} = aq_i + b_i$$
$$p = aq_0 + b_0$$

where the final computation yields $p = B(a)$. Show that it is equivalent to the algorithm of Problem 2.10. This version is equivalent to "synthetic division" in the following sense.

The computed coefficients $\{q_0, q_1, \ldots, q_{M-1}\}$ define a polynomial of degree $(M-1)$, namely, $Q(z) = q_0 + q_1 z + \cdots + q_{M-1} z^{M-1}$.

Show that $Q(z)$ is the *quotient* polynomial of the division of $B(z)$ by the monomial $z - a$. Moreover, the *last* computed $p = B(a)$ is the *remainder* of that division. That is, show as an identity in z

$$B(z) = (z - a) Q(z) + p$$

2.13 In the dac routines, the polynomial to be evaluated is of the form

$$B(z) = b_1 z + b_2 z^2 + \cdots + b_M z^M$$

The dac routines evaluate it at the specific value $z = 2^{-1}$. Show that the polynomial $B(z)$ can be evaluated at $z = a$ by the following modified version of Hörner's rule:

$$initialize \ p = 0$$
$$for \ i = M, M-1, \ldots, 1 \ do:$$
$$p = a(p + b_i)$$

2.14 Consider a 4-bit successive approximation A/D converter with full-scale range of 8 volts. Using the successive approximation algorithm (with rounding), determine the 4-bit codes of the voltage values $x = 1.2, 5.2, -3.2$ volts, for the following types of converters:

a. Natural binary.

a. Bipolar two's complement.

In each case, show all the steps of the successive approximation algorithm. Explain what happens if the analog voltage to be converted lies *outside* the full-scale range of the converter. This happens for $x = 5.2$ in two's complement, and $x = -3.2$ in natural binary representations.

2.15 Carry out, by hand, the successive approximation conversion of all the signal values shown in the table of Example 2.4.4, for both the offset binary and two's complement cases.

2.16 *Computer Experiment: DAC and ADC Routines.* Write C versions of the routines dac and adc for the natural binary, offset binary, and two's complement cases that implement *truncation*.

For the natural and offset binary cases, write another set of such routines that implement *rounding*.

2.17 *Computer Experiment: Simulating DAC and ADC Operations.* Generate $L = 50$ samples of a sinusoidal signal $x(n) = A \cos(2\pi f n)$, $n = 0, 1, \ldots, L - 1$ of frequency $f = 0.02$ [cycles/sample] and amplitude $A = 8$.

a. Using a 3-bit ($B = 3$) bipolar two's complement successive approximation A/D converter, as implemented by the routine adc, with full-scale range $R = 32$, quantize $x(n)$ and denote the quantized signal by $x_Q(n)$.

For $n = 0, 1, \ldots, L - 1$, print in three parallel columns the true analog value $x(n)$, the quantized value $x_Q(n)$, and the corresponding two's complement bit vector **b**.

On the same graph, plot the two signals $x(n)$ and $x_Q(n)$ versus n. Scale the vertical scales from $[-16, 16]$ and use 8 y-grid lines to indicate the 8 quantization levels.

b. Repeat part (a) using a $B = 4$ bit A/D converter. In plotting $x(n)$ and $x_Q(n)$, use the *same* vertical scales as above, namely, from $[-16, 16]$, but use 16 y-grid lines to show the 16 quantization levels.

c. What happens if the analog signal to be quantized has amplitude that *exceeds* the full-scale range of the quantizer? Most D/A converters will *saturate* to their largest (positive or negative) levels. To see this, repeat part (a) by taking the amplitude of the sinusoid to be $A = 20$.

d. What happens if we use truncation instead of rounding? Repeat part (a) using the two's complement truncation routines adc and dac that you developed in the previous problem.

2.18 *Computer Experiment: Quantization Noise Model.* Reproduce the results of Example 2.4.6.

2.19 Show that the mean and variance of the random variable v defined by Eq. (2.4.6) of Example 2.4.6 are $m_v = 0$ and $\sigma_v^2 = R^2/48$.

2.20 Show that the normalized autocorrelation function $\rho_{xx}(k)$ of the signal $x(n)$ given by Eq. (2.4.1) in Example 2.4.6, is given by

$$\rho_{xx}(k) = \frac{R_{xx}(k)}{R_{xx}(0)} = a \cos(2\pi f_0 k) + (1 - a) \delta(k), \quad \text{where} \quad a = \frac{SNR}{SNR + 1}$$

where $R_{xx}(k)$ defined as the statistical expectation value

$$R_{xx}(k) = E[x(n + k)x(n)]$$

Assume that phase of the sinusoid ϕ is not correlated with $v(n)$. The quantity *SNR* is the signal-to-noise ratio $SNR = A^2/(2\sigma_v^2)$. For the numerical values of Example 2.4.6, show $a = 0.6$.

2.21 *Computer Experiment: Subranging Converters.* It was mentioned that parallel A/D converters are at present limited in their bits. However, it is possible to use two of them in cascade. For example, using two identical 6-bit flash ADCs, one can achieve effective resolution of 12 bits at conversion rates of 10 MHz.

Consider a B-bit ADC and write B as the sum of two integers $B = B_1 + B_2$. The conversion of an analog value x to its B-bit representation can be done in two stages: First, convert x into its B_1-bit representation. This is equivalent to keeping the first B_1 most significant bits of its B-bit representation. Let x_1 be the quantized B_1-bit value. Then, form the difference $x_2 = x - x_1$ and quantize it to B_2 bits. These operations are shown in the following figure:

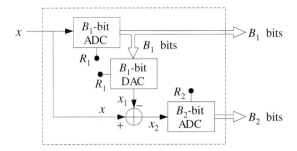

The B_1-bit word from the first ADC is sent to the output and also to a B_1-bit DAC whose output is x_1. The analog subtracter forms x_2, which is sent to the B_2-bit ADC producing the remaining B_2 bits.

a. What should be the full-scale ranges R_1 and R_2 of the two ADCs in order for this arrangement to be *equivalent* to a single $(B_1 + B_2)$-bit ADC with full-scale range R? What is the relationship of R_1 and R_2 in terms of R?

b. Using the routines adc and dac as building blocks, write a routine that implements this block diagram. Test your routine on the signal:

$$x(n) = A\cos(2\pi f_0 n), \qquad n = 0, 1, \ldots, L - 1$$

where $A = 4$, $f_0 = 0.02$, and $L = 50$. Take $B_1 = 5$, $B_2 = 5$, $B = B_1 + B_2 = 10$, and $R = 16$. Compare the results of your routine with the results of an equivalent single B-bit ADC with full-scale range R.

c. How does the block diagram generalize in the case of cascading three such converters, such that $B = B_1 + B_2 + B_3$?

2.22 *Computer Experiment: Triangular Dither.* Reproduce the results and graphs of Example 2.5.1.

3

Discrete-Time Systems

In this and the next chapter, we discuss discrete-time systems and, in particular, linear time-invariant (LTI) systems. The input/output (I/O) relationship of LTI systems is given by the discrete-time convolution of the system's impulse response with the input signal.

LTI systems can be classified into *finite impulse response* (FIR) or *infinite impulse response* (IIR) types depending on whether their impulse response has finite or infinite duration. Our main objective in these two chapters is to develop practical *computational algorithms* for the FIR case. The IIR case is considered in Chapter 7, although we do present a few simple IIR examples here.

Depending on the application and hardware, an FIR digital filtering operation can be organized to operate either on a *block* basis or a *sample-by-sample* basis.

In the *block processing* case, the input signal is considered to be a single block of signal samples. The block is filtered by *convolving* it with the filter, generating the output signal as another block of samples.

If the input signal is very long or infinite in duration, this method requires modification—for example, breaking up the input into multiple blocks of manageable size, filtering the blocks one at a time, and piecing together the resulting output blocks to form the overall output. The filtering of each block can be implemented in various ways, such as by ordinary convolution, or fast convolution via the FFT.

In the *sample processing* case, the input samples are processed *one at a time* as they arrive at the input. The filter operates as a state machine; that is, each input sample is used in conjunction with the current *internal state* of the filter to compute the current output sample and also to *update* the internal state of the filter in preparation for processing the *next* input sample.

This approach is useful in *real-time applications* involving very long input signals. It is also useful in adaptive filtering applications where the filter itself changes after processing each sample. Moreover, it is efficiently implemented with present day DSP chip families, such as the Texas Instruments TMS320, the Bell Labs AT&T DSP16/32, the Motorola DSP56K/96K, and the Analog Devices ADSP2101 families. The architectures and instruction sets of these chips are optimized for such sample-by-sample processing operations.

3.1 Input/Output Rules

A *discrete-time system*, shown in Fig. 3.1.1, is a processor that transforms an input sequence of discrete-time samples $x(n)$ into an output sequence of samples $y(n)$, according to some *input/output rule* that specifies how to compute the output sequence $y(n)$ from the knowledge of the input sequence $x(n)$. In sample-by-sample processing methods, we may think of the I/O rule as processing the input samples one at a time:[†]

$$\{x_0, x_1, x_2, \ldots, x_n, \ldots\} \xrightarrow{H} \{y_0, y_1, y_2, \ldots, y_n, \ldots\}$$

that is, $x_0 \xrightarrow{H} y_0$, $x_1 \xrightarrow{H} y_1$, $x_2 \xrightarrow{H} y_2$, and so on. In block processing methods, we think of the input sequence as a block or vector of signal samples being processed as a whole by the system, producing the corresponding output block:

$$\mathbf{x} = \begin{bmatrix} x_0 \\ x_1 \\ x_2 \\ \vdots \end{bmatrix} \xrightarrow{H} \begin{bmatrix} y_0 \\ y_1 \\ y_2 \\ \vdots \end{bmatrix} = \mathbf{y}$$

Thus, the I/O rule maps the input vector $\mathbf{x}$ into the output vector $\mathbf{y}$ according to some functional mapping:

$$\mathbf{y} = H[\mathbf{x}] \tag{3.1.1}$$

For linear systems, this mapping becomes a linear transformation by a matrix H, $\mathbf{y} = H\mathbf{x}$. For linear and time-invariant systems, the matrix H has a special structure being built in terms of the impulse response of the system.

Some examples of discrete-time systems illustrating the wide variety of possible I/O rules are given below.

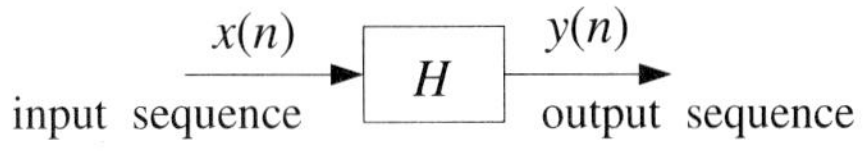

Fig. 3.1.1 Discrete-time system.

Example 3.1.1: $y(n) = 2x(n)$. It corresponds to simple scaling of the input:

$$\{x_0, x_1, x_2, x_3, x_4, \ldots\} \xrightarrow{H} \{2x_0, 2x_1, 2x_2, 2x_3, 2x_4, \ldots\}$$

Example 3.1.2: $y(n) = 2x(n) + 3x(n-1) + 4x(n-2)$. A weighted average of three successive input samples. At each time instant n, the system must remember the *previous* input samples $x(n-1)$ and $x(n-2)$ in order to use them.

[†]For brevity, we denoted $\{x(0), x(1), x(2), \ldots\}$ by subscripts $\{x_0, x_1, x_2, \ldots\}$.

Example 3.1.3: Here, the I/O rule is specified as a block processing operation by a linear transformation, transforming a length-4 input block $\{x_0, x_1, x_2, x_3\}$ into a length-6 output block:

$$\mathbf{y} = \begin{bmatrix} y_0 \\ y_1 \\ y_2 \\ y_3 \\ y_4 \\ y_5 \end{bmatrix} = \begin{bmatrix} 2 & 0 & 0 & 0 \\ 3 & 2 & 0 & 0 \\ 4 & 3 & 2 & 0 \\ 0 & 4 & 3 & 2 \\ 0 & 0 & 4 & 3 \\ 0 & 0 & 0 & 4 \end{bmatrix} \begin{bmatrix} x_0 \\ x_1 \\ x_2 \\ x_3 \end{bmatrix} = H\mathbf{x}$$

It is equivalent to the convolutional form of Example 3.1.2. The output block is longer than the input block by two samples because this filter has memory two — the last two outputs being the input-off transients generated after the input is turned off. If we had to filter length-5 input blocks $\{x_0, x_1, x_2, x_3, x_4\}$, the linear transformation would have one more column and row:

$$\mathbf{y} = \begin{bmatrix} y_0 \\ y_1 \\ y_2 \\ y_3 \\ y_4 \\ y_5 \\ y_6 \end{bmatrix} = \begin{bmatrix} 2 & 0 & 0 & 0 & 0 \\ 3 & 2 & 0 & 0 & 0 \\ 4 & 3 & 2 & 0 & 0 \\ 0 & 4 & 3 & 2 & 0 \\ 0 & 0 & 4 & 3 & 2 \\ 0 & 0 & 0 & 4 & 3 \\ 0 & 0 & 0 & 0 & 4 \end{bmatrix} \begin{bmatrix} x_0 \\ x_1 \\ x_2 \\ x_3 \\ x_4 \end{bmatrix} = H\mathbf{x}$$

Example 3.1.4: Example 3.1.2 can also be cast in an equivalent sample-by-sample processing form described by the following system of three equations:

$$y(n) = 2x(n) + 3w_1(n) + 4w_2(n)$$

$$w_2(n+1) = w_1(n)$$

$$w_1(n+1) = x(n)$$

The auxiliary signals $w_1(n)$ and $w_2(n)$ can be thought of as the internal states of the system. The present input sample $x(n)$ together with the knowledge of the present internal states $\{w_1(n), w_2(n)\}$ is sufficient to compute the present output $y(n)$. The next output $y(n+1)$ due to the next input $x(n+1)$ requires knowledge of the updated states $\{w_1(n+1), w_2(n+1)\}$, but these are already available from the nth time step; thus, at time $n+1$ we have:

$$y(n+1) = 2x(n+1) + 3w_1(n+1) + 4w_2(n+1)$$

$$w_2(n+2) = w_1(n+1)$$

$$w_1(n+2) = x(n+1)$$

The computations are repetitive from one time instant to the next and can be summarized by the following I/O *sample-by-sample processing algorithm* which tells how to process each arriving input sample x producing the corresponding output sample y and updating the internal states:[†]

[†]The symbol := denotes *assignment* not equation, that is, $a := b$ means "a takes on the value b."

> *for each new input sample x do:*
> $y := 2x + 3w_1 + 4w_2$
> $w_2 := w_1$
> $w_1 := x$

Once the current values of the internal states $\{w_1, w_2\}$ are used in the computation of the output y, they may be updated by the last two assignment equations to the values they must have for processing the *next* input sample. Therefore, $\{w_1, w_2\}$ must be saved from call to call of the algorithm. The order in which $\{w_1, w_2\}$ are updated is important, that is, w_2 is updated first and w_1 second, to prevent overwriting of the correct values.

This and the previous two examples represent equivalent formulations of the same discrete-time system. Deciding which form to use depends on the nature of the application—that is, whether the input signals are finite or infinite sequences and the samples must be processed one at a time as they arrive.

This example is a special case of more general *state-space* representations of discrete-time systems described by the following I/O sample processing algorithm:

$$y(n) = g(x(n), \mathbf{s}(n)) \qquad \text{(output equation)}$$

$$\mathbf{s}(n + 1) = \mathbf{f}(x(n), \mathbf{s}(n)) \qquad \text{(state updating equation)}$$

where $\mathbf{s}(n)$ is an internal state vector of appropriate dimension, like $\mathbf{s}(n) = \begin{bmatrix} w_1(n) \\ w_2(n) \end{bmatrix}$
of the previous example. The I/O algorithm calculates both the output $y(n)$ and the next state $\mathbf{s}(n + 1)$ from the knowledge of the present input $x(n)$ and the present state $\mathbf{s}(n)$. It can be rephrased in the repetitive algorithmic form:

> *for each new input sample x do:*
> $y := g(x, \mathbf{s})$
> $\mathbf{s} := \mathbf{f}(x, \mathbf{s})$

State-space realizations of LTI systems are described by functions $\mathbf{f}$ and g that are *linear functions* of their arguments, that is, $\mathbf{f}(x, \mathbf{s}) = A\mathbf{s} + Bx$, $g(x, \mathbf{s}) = C\mathbf{s} + Dx$, where A, B, C, D have appropriate dimensions. In particular, for the above example we have

$$y := 2x + 3w_1 + 4w_2 = [3, 4] \begin{bmatrix} w_1 \\ w_2 \end{bmatrix} + 2x = [3, 4]\mathbf{s} + 2x \equiv g(x, \mathbf{s})$$

$$\mathbf{s} = \begin{bmatrix} w_1 \\ w_2 \end{bmatrix} := \begin{bmatrix} x \\ w_1 \end{bmatrix} = \begin{bmatrix} 0 & 0 \\ 1 & 0 \end{bmatrix} \begin{bmatrix} w_1 \\ w_2 \end{bmatrix} + \begin{bmatrix} 1 \\ 0 \end{bmatrix} x = \begin{bmatrix} 0 & 0 \\ 1 & 0 \end{bmatrix} \mathbf{s} + \begin{bmatrix} 1 \\ 0 \end{bmatrix} x \equiv \mathbf{f}(x, \mathbf{s})$$

Example 3.1.5: $y(n) = 0.5y(n - 1) + 2x(n) + 3x(n - 1)$. The output is computed recursively by a constant-coefficient difference equation. At each time instant n, the system must remember the previous input *and* output samples $x(n - 1), y(n - 1)$.

Example 3.1.6: Example 3.1.5 can also be described by stating its I/O rule as a sample-by-sample processing algorithm:

$$\textit{for each new input sample x do:}$$
$$y := 0.5w_1 + 2x + 3v_1$$
$$w_1 := y$$
$$v_1 := x$$

It corresponds to the so-called *direct form* realization of the difference equation and requires the computation and updating of the auxiliary quantities $\{w_1, v_1\}$. Its equivalence to Example 3.1.5 will be seen later.

An alternative I/O computational rule for Example 3.1.5, corresponding to the so-called *canonical* realization of the system, is as follows:

$$\textit{for each new input sample x do:}$$
$$w_0 := x + 0.5w_1$$
$$y := 2w_0 + 3w_1$$
$$w_1 := w_0$$

It uses the auxiliary quantities $\{w_0, w_1\}$.

Example 3.1.7: $y(n) = \dfrac{1}{5}[x(n+2)+x(n+1)+x(n)+x(n-1)+x(n-2)]$. Smoother or averager of five successive samples. The operation is slightly non-causal, because at each time n, the system must know the next samples $x(n+1)$ and $x(n+2)$. We will see later how to handle such cases in real time.

Example 3.1.8: $y(n) = 2x(n)+3$. Scaling and shifting of the input.

Example 3.1.9: $y(n) = x^2(n)$. Squaring the input.

Example 3.1.10: $y(n) = 2x(n)+3x(n-1)+x(n)x(n-1)$. It contains a nonlinear cross-product term $x(n)x(n-1)$.

Example 3.1.11: $y(n) = \text{med}[x(n+1),x(n),x(n-1)]$. A simple *median filter*, where the operation $\text{med}[a,b,c]$ represents the median of the three numbers a,b,c obtained by sorting the three numbers in increasing order and picking the middle one.

Example 3.1.12: $y(n) = nx(n)$. It has a time-varying coefficient.

Example 3.1.13: $y(n) = \dfrac{1}{n}[x(0)+x(1)+\cdots+x(n-1)]$. Cumulative average of n numbers. It can also be expressed recursively as in the following example.

Example 3.1.14: $y(n+1) = a_n y(n)+b_n x(n)$, where $a_n = n/(n+1)$, $b_n = 1-a_n = 1/(n+1)$. It corresponds to a first-order difference equation with time-varying coefficients a_n, b_n.

Example 3.1.15: $y(n) = x(2n)$. It acts as a rate compressor or *downsampler*, keeping every other sample of $x(n)$, thus, resulting in half of the input samples. That is, the input and output sequences are:

$$\{x_0, x_1, x_2, x_3, x_4, x_5, x_6, \ldots\} \xrightarrow{H} \{x_0, x_2, x_4, x_6, \ldots\}$$

Example 3.1.16: $y(n) = \begin{cases} x(n/2), & \text{if } n \text{ is even} \\ 0, & \text{if } n \text{ is odd} \end{cases}$. It acts as a rate expander or *upsam-*

pler, inserting a zero sample between the samples of $x(n)$, thus, doubling the number of input samples. That is, the input and output sequences are:

$$\{x_0, x_1, x_2, x_3, x_4, \ldots\} \xrightarrow{H} \{x_0, 0, x_1, 0, x_2, 0, x_3, 0, x_4, \ldots\}$$

Examples 3.1.1–3.1.7 represent LTI systems; Examples 3.1.8–3.1.11 are nonlinear but time-invariant; and Examples 3.1.12–3.1.16 are linear but time-varying systems.

Although most applications of DSP use linear time-invariant systems, nonlinear and time-varying systems are used increasingly in a variety of applications. For example, median filters are used successfully in image processing because of their excellent edge-preserving and noise removal properties; time-varying filters are used in adaptive filtering applications, such as channel equalization and echo cancellation in digital data or voice channels; downsamplers and upsamplers are part of multirate signal processing systems, such as those used for interpolation, decimation, oversampling, and sample rate conversion.

3.2 Linearity and Time Invariance

A *linear system* has the property that the output signal due to a linear combination of two or more input signals can be obtained by forming the same linear combination of the individual outputs. That is, if $y_1(n)$ and $y_2(n)$ are the outputs due to the inputs $x_1(n)$ and $x_2(n)$, then the output due to the linear combination of inputs

$$x(n) = a_1 x_1(n) + a_2 x_2(n) \tag{3.2.1}$$

is given by the linear combination of outputs

$$y(n) = a_1 y_1(n) + a_2 y_2(n) \tag{3.2.2}$$

To test linearity one must determine separately the three outputs $y(n)$, $y_1(n)$, and $y_2(n)$ and then show that they satisfy Eq. (3.2.2). The required operations are shown in Fig. 3.2.1.

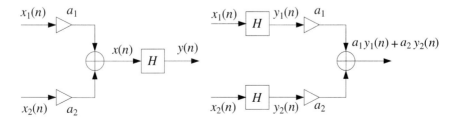

Fig. 3.2.1 Testing linearity.

Example 3.2.1: Test the linearity of the discrete-time systems defined in Examples 3.1.8 and 3.1.9, that is, defined by $y(n) = 2x(n) + 3$ and $y(n) = x^2(n)$.

Solution: The I/O equation $y(n) = 2x(n) + 3$ is a linear equation, but it does not represent a linear *system*. Indeed, the output due to the linear combination of Eq. (3.2.1) will be

$$y(n) = 2x(n) + 3 = 2[a_1 x_1(n) + a_2 x_2(n)] + 3$$

and is not equal to the linear combination in the right-hand side of Eq. (3.2.2), namely,

$$a_1 y_1(n) + a_2 y_2(n) = a_1(2x_1(n) + 3) + a_2(2x_2(n) + 3)$$

Similarly, for the quadratic system $y(n) = x^2(n)$ of Example 3.1.9, we have

$$a_1 x_1^2(n) + a_2 x_2^2(n) \neq (a_1 x_1(n) + a_2 x_2(n))^2$$

More simply, if a system is nonlinear, one can use a counterexample to show violation of linearity. For example, if the above quadratic system were linear, doubling of the input would cause doubling of the output. But in this case, doubling of the input quadruples the output. □

A *time-invariant* system is a system that remains unchanged over time. This implies that if an input is applied to the system today causing a certain output to be produced, then the same output will also be produced tomorrow if the same input is applied. The operation of waiting or delaying a signal by a time delay of, say, D units of time is shown in Fig. 3.2.2. It represents the *right translation* of $x(n)$

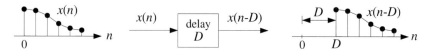

Fig. 3.2.2 Time delay by D samples.

as a whole by D samples. A *time advance* would have negative D and correspond to the left translation of $x(n)$.

The mathematical formulation of time invariance can be stated with the aid of Fig. 3.2.3. The upper diagram in this figure shows an input $x(n)$ being applied to

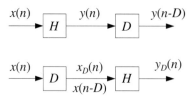

Fig. 3.2.3 Testing time invariance.

the system producing the output $y(n)$. The lower diagram shows the same input delayed by D units of time, that is, the signal:

$$x_D(n) = x(n - D) \tag{3.2.3}$$

which is then applied to the system producing an output, say, $y_D(n)$.

To test whether the system will produce the same output later as the one it is producing now, we must take the output $y(n)$ produced now and save it until later, that is, delay it by D time units, as shown in the upper diagram of Fig. 3.2.3. Then, it can be compared with the output $y_D(n)$ that will be produced later. Thus, if

$$y_D(n) = y(n - D) \tag{3.2.4}$$

the system will be time-invariant. In other words, delaying the input, Eq. (3.2.3), causes the output to be delayed by the same amount, Eq. (3.2.4). Equivalently, in terms of the samples, if

$$\{x_0, x_1, x_2, \dots\} \xrightarrow{H} \{y_0, y_1, y_2, \dots\}$$

then

$$\underbrace{\{0, 0, \dots, 0, x_0, x_1, x_2, \dots\}}_{D \text{ zeros}} \xrightarrow{H} \underbrace{\{0, 0, \dots, 0, y_0, y_1, y_2, \dots\}}_{D \text{ zeros}}$$

Example 3.2.2: Test the time invariance of the discrete-time systems defined in Examples (3.1.12) and (3.1.15), that is, defined by $y(n) = nx(n)$ and $y(n) = x(2n)$.

Solution: Because the system $y(n) = nx(n)$ has a time-varying coefficient, we expect it not to be time-invariant. According to the given I/O rule, the signal $x_D(n)$ applied to the system will cause the output $y_D(n) = nx_D(n)$. But $x_D(n)$ is the delayed version of $x(n)$, $x_D(n) = x(n - D)$. Therefore,

$$y_D(n) = nx_D(n) = nx(n - D)$$

On the other hand, delaying the output signal $y(n) = nx(n)$ by D units gives, replacing n by $n - D$:

$$y(n - D) = (n - D)x(n - D) \neq nx(n - D) = y_D(n)$$

Thus, the system is not time-invariant. Example 3.1.15 described by $y(n) = x(2n)$ is a little more subtle. According to the I/O rule, the signal $x_D(n)$ will cause the output $y_D(n) = x_D(2n)$. But, $x_D(n) = x(n - D)$ and therefore, replacing the argument n by $2n$ gives $x_D(2n) = x(2n - D)$, or,

$$y_D(n) = x_D(2n) = x(2n - D)$$

On the other hand, replacing n by $n - D$ in $y(n) = x(2n)$ gives

$$y(n - D) = x(2(n - D)) = x(2n - 2D) \neq x(2n - D) = y_D(n)$$

Thus, the downsampler is not time-invariant. This can also be seen more intuitively by considering the effect of the system on the original input sequence and its delay by one time unit. Noting that the output sequence is obtained by dropping every other input sample, we find:

$$\{x_0, x_1, x_2, x_3, x_4, x_5, x_6, \dots\} \quad \xrightarrow{H} \quad \{x_0, x_2, x_4, x_6, \dots\}$$
$$\{0, x_0, x_1, x_2, x_3, x_4, x_5, x_6, \dots\} \quad \xrightarrow{H} \quad \{0, x_1, x_3, x_5, \dots\}$$

We see that the lower output is not the upper output delayed by one time unit.   □

3.3 *Impulse Response*

Linear time-invariant systems are characterized uniquely by their *impulse response* sequence $h(n)$, which is defined as the response of the system to a *unit impulse* $\delta(n)$, as shown in Fig. 3.3.1. The unit impulse is the discrete-time analog of the Dirac delta function $\delta(t)$ and is defined as

$$\delta(n) = \begin{cases} 1 & \text{if } n = 0 \\ 0 & \text{if } n \neq 0 \end{cases}$$

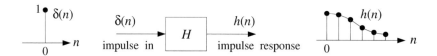

Fig. 3.3.1 Impulse response of an LTI system.

Thus, we have by definition,

$$\delta(n) \xrightarrow{H} h(n)$$

or, in terms of the sample values:[†]

$$\{1, 0, 0, 0, \dots\} \xrightarrow{H} \{h_0, h_1, h_2, h_3, \dots\}$$

Time invariance implies that if the unit impulse is delayed or time shifted by a certain amount, say D units of time, then it will cause the impulse response to be delayed by the same amount, that is, $h(n - D)$. Thus,

$$\delta(n - D) \xrightarrow{H} h(n - D)$$

for any positive or negative delay D. Figure Fig. 3.3.2 shows this property for $D = 0, 1, 2$. On the other hand, linearity implies that any linear combination of inputs causes the same linear combination of outputs, so that, for example, the sum of the three impulses of Fig. 3.3.2 will cause the sum of the three outputs, that is,

$$\delta(n) + \delta(n - 1) + \delta(n - 2) \xrightarrow{H} h(n) + h(n - 1) + h(n - 2)$$

[†]Again, we denote $\{h(0), h(1), h(2), \dots\}$ by $\{h_0, h_1, h_2, \dots\}$.

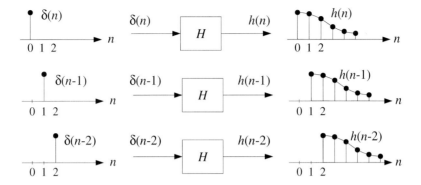

Fig. 3.3.2 Delayed impulse responses of an LTI system.

or, more generally the weighted linear combination of the three inputs:

$$x(0)\,\delta\,(n) + x(1)\,\delta\,(n-1) + x(2)\,\delta\,(n-2)$$

will cause the same weighted combination of the three outputs:

$$x(0)\,h\,(n) + x(1)\,h\,(n-1) + x(2)\,h\,(n-2)$$

as shown in Fig. 3.3.3. In general, an arbitrary input sequence $\{x(0), x(1), x(2), \dots\}$ can be thought of as the linear combination of shifted and weighted unit impulses:

$$x(n) = x(0)\,\delta\,(n) + x(1)\,\delta\,(n-1) + x(2)\,\delta\,(n-2) + x(3)\,\delta\,(n-3) + \cdots$$

This follows because each term of the right-hand side is nonzero only at the corresponding delay time, for example, at $n = 0$ only the first term is nonzero, at $n = 1$ only the second term is nonzero, and so on. Linearity and time invariance imply then that the corresponding output sequence will be obtained by replacing each delayed unit impulse by the corresponding delayed impulse response, that is,

$$y(n) = x(0)\,h\,(n) + x(1)\,h\,(n-1) + x(2)\,h\,(n-2) + x(3)\,h\,(n-3) + \cdots \qquad (3.3.1)$$

or written more compactly:

$$\boxed{y(n) = \sum_{m} x(m)\,h(n-m)} \qquad \text{(LTI form)} \qquad (3.3.2)$$

This is the discrete-time *convolution* of the input sequence $x(n)$ with the filter sequence $h(n)$. Thus, LTI systems are convolvers.

In general, the summation could extend also over negative values of m, depending on the input signal. Because it was derived using the LTI properties of the system, Eq. (3.3.2) will be referred to as the *LTI form* of convolution. Changing the index of summation, it can also be written in the alternative form:

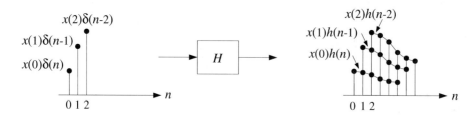

Fig. 3.3.3 Response to linear combination of inputs.

$$\boxed{y(n) = \sum_m h(m)x(n-m)}\qquad\text{(direct form)}\qquad\qquad(3.3.3)$$

For reasons that will become clear later, Eq. (3.3.3) will be referred to as the *direct form* of convolution. The computational aspects of Eqs. (3.3.2) and (3.3.3) and their realization in terms of block or sample processing methods will be discussed in detail in the next chapter.

3.4 FIR and IIR Filters

Discrete-time LTI systems can be classified into FIR or IIR systems, that is, having finite or infinite impulse response $h(n)$, as depicted in Fig. 3.4.1.

An FIR filter has impulse response $h(n)$ that extends only over a finite time interval, say $0 \le n \le M$, and is identically zero beyond that:

$$\{h_0, h_1, h_2, \ldots, h_M, 0, 0, 0, \ldots\}$$

M is referred to as the *filter order*. The *length* of the impulse response vector $\mathbf{h} = [h_0, h_1, \ldots, h_M]$ is:

$$\boxed{L_h = M + 1}$$

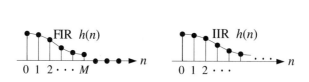

Fig. 3.4.1 FIR and IIR impulse responses.

The impulse response coefficients $\{h_0, h_1, \ldots, h_M\}$ are referred to by various names, such as *filter coefficients, filter weights,* or *filter taps,* depending on the context. In the direct form of convolution of Eq. (3.3.3), all the terms for $m > M$ and $m < 0$ will be absent because by definition $h(m)$ vanishes for these values of m; only the terms $0 \le m \le M$ are present. Therefore, Eq. (3.3.3) is simplified to the finite-sum form:

$$\boxed{y(n) = \sum_{m=0}^{M} h(m)x(n-m)}$$ \qquad (FIR filtering equation) \qquad (3.4.1)

or, explicitly

$$y(n) = h_0 x(n) + h_1 x(n-1) + h_2 x(n-2) + \cdots + h_M x(n-M) \qquad (3.4.2)$$

Thus, the I/O equation is obtained as a weighted sum of the *present* input sample $x(n)$ and the *past M* samples $x(n-1), x(n-2), \ldots, x(n-M)$.

Example 3.4.1: Second-order FIR filters are characterized by three impulse response co-efficients $\mathbf{h} = [h_0, h_1, h_2]$ and have I/O equation:

$$y(n) = h_0 x(n) + h_1 x(n-1) + h_2 x(n-2)$$

Such was the case of Example 3.1.2, which had $\mathbf{h} = [2, 3, 4]$.

Example 3.4.2: Similarly, third-order FIR filters are characterized by four weights $\mathbf{h} = [h_0, h_1, h_2, h_3]$ and have I/O equation:

$$y(n) = h_0 x(n) + h_1 x(n-1) + h_2 x(n-2) + h_3 x(n-3)$$

Example 3.4.3: Determine the impulse response $\mathbf{h}$ of the following FIR filters:

  (a) $y(n) = 2x(n) + 3x(n-1) + 5x(n-2) + 2x(n-3)$
  (b) $y(n) = x(n) - x(n-4)$

Solution: Comparing the given I/O equations with Eq. (3.4.2), we identify the impulse response coefficients:

  (a) $\mathbf{h} = [h_0, h_1, h_2, h_3] = [2, 3, 5, 2]$
  (b) $\mathbf{h} = [h_0, h_1, h_2, h_3, h_4] = [1, 0, 0, 0, -1]$

Alternatively, sending a unit impulse as input, $x(n) = \delta(n)$, will produce the impulse response sequence as output, $y(n) = h(n)$:

  (a) $h(n) = 2\delta(n) + 3\delta(n-1) + 5\delta(n-2) + 2\delta(n-3)$
  (b) $h(n) = \delta(n) - \delta(n-4)$

The expressions for $h(n)$ and $\mathbf{h}$ are equivalent. \hfill $\square$

An IIR filter, on the other hand, has an impulse response $h(n)$ of infinite duration, defined over the infinite interval $0 \le n < \infty$. Eq. (3.3.3) now has an infinite number of terms:

$$\boxed{y(n) = \sum_{m=0}^{\infty} h(m)x(n-m)}$$ \qquad (IIR filtering equation) \qquad (3.4.3)

This I/O equation is not *computationally* feasible because we cannot deal with an infinite number of terms. Therefore, we must restrict our attention to a subclass of IIR filters, namely, those for which the infinite number of filter coefficients $\{h_0, h_1, h_2, \dots\}$ are not chosen arbitrarily, but rather they are coupled to each other through *constant-coefficient linear difference equations*.

For this subclass of IIR filters, Eq. (3.4.3) can be *rearranged* as a *difference equation* allowing the efficient *recursive* computation of the output $y(n)$. Some examples will make this point clear.

Example 3.4.4: Determine the I/O difference equation of an IIR filter whose impulse response coefficients $h(n)$ are coupled to each other by the difference equation:

$$h(n) = h(n-1) + \delta(n)$$

Solution: Setting $n = 0$, we have $h(0) = h(-1) + \delta(0) = h(-1) + 1$. Assuming causal initial conditions, $h(-1) = 0$, we find $h(0) = 1$. For $n > 0$, the delta function vanishes, $\delta(n) = 0$, and therefore, the difference equation reads $h(n) = h(n-1)$. In particular $h(1) = h(0) = 1$, $h(2) = h(1) = 1$, and so on. Thus, all of the samples $h(n)$ are equal to each other. In summary, we have the (causal) solution:

$$h(n) = u(n) = \begin{cases} 1 & \text{if } n \geq 0 \\ 0 & \text{if } n \leq -1 \end{cases}$$

where $u(n)$ is the discrete-time unit-step function. Now, putting this solution into the convolutional I/O equation (3.4.3), we have

$$y(n) = \sum_{m=0}^{\infty} h(m)x(n-m) = \sum_{m=0}^{\infty} x(n-m)$$

where we set $h(m) = 1$. Writing it explicitly we have

$$y(n) = x(n) + x(n-1) + x(n-2) + x(n-3) + \cdots$$

Replacing n by $n-1$ gives the previous output

$$y(n-1) = x(n-1) + x(n-2) + x(n-3) + \cdots$$

Subtracting it from $y(n)$, we have

$$y(n) - y(n-1) = x(n)$$

Therefore, the I/O convolutional equation is equivalent to the recursive difference equation

$$y(n) = y(n-1) + x(n)$$

It represents an *accumulator*, or discrete-time integrator. Note that this is the *same* difference equation as that of $h(n)$, because by definition $h(n)$ is the output when the input is an impulse; that is, $y(n) = h(n)$ if $x(n) = \delta(n)$.   □

Example 3.4.5: Suppose the filter coefficients $h(n)$ satisfy the difference equation

$$h(n) = ah(n-1) + \delta(n)$$

where a is a constant. Determine the I/O difference equation relating a general input signal $x(n)$ to the corresponding output $y(n)$.

Solution: Arguing as in the previous example, we have

$$h(0) = ah(-1) + \delta(0) = a \cdot 0 + 1 = 1$$
$$h(1) = ah(0) + \delta(1) = a \cdot 1 + 0 = a$$
$$h(2) = ah(1) + \delta(2) = a \cdot a + 0 = a^2$$
$$h(3) = ah(2) + \delta(3) = a \cdot a^2 + 0 = a^3$$

and so on. Therefore, we find the solution

$$h(n) = a^n u(n) = \begin{cases} a^n, & \text{if } n \geq 0 \\ 0, & \text{if } n \leq -1 \end{cases}$$

Inserting this solution into Eq. (3.4.3), we have

$$y(n) = x(n) + ax(n-1) + a^2 x(n-2) + a^3 x(n-3) + \cdots$$
$$= x(n) + a[x(n-1) + ax(n-2) + a^2 x(n-3) + \cdots]$$

The sum in the brackets is recognized now as the previous output $y(n-1)$. Therefore, we obtain the I/O difference equation:

$$y(n) = ay(n-1) + x(n)$$

As expected, it is the same as the difference equation satisfied by $h(n)$. $\square$

Example 3.4.6: Determine the convolutional form and the (causal) impulse response of the IIR filter described by the following difference equation:

$$y(n) = -0.8y(n-1) + x(n)$$

Solution: This is the same example as above, with $a = -0.8$. Setting $x(n) = \delta(n)$ and $y(n) = h(n)$, we obtain the difference equation for $h(n)$:

$$h(n) = -0.8h(n-1) + \delta(n)$$

Assuming causal initial conditions, $h(-1) = 0$, and iterating a few values of n as we did in the previous example, we find the solution:

$$h(n) = (-0.8)^n u(n) = \begin{cases} (-0.8)^n, & \text{if } n \geq 0 \\ 0, & \text{if } n \leq -1 \end{cases}$$

Inserting the values for $h(n)$ into the convolutional equation (3.4.3), we find

$$y(n) = x(n) + (-0.8)x(n-1) + (-0.8)^2 x(n-2) + (-0.8)^3 x(n-3) + \cdots$$

which, in general, has an infinite number of terms. $\square$

Example 3.4.7: In this example, start with an expression for $h(n)$ and work backwards to obtain the I/O difference equation satisfied by $y(n)$ in terms of $x(n)$, and also determine the difference equation satisfied by $h(n)$. Assume the IIR filter has a causal $h(n)$ defined by

$$h(n) = \begin{cases} 2, & \text{for } n = 0 \\ 4(0.5)^{n-1}, & \text{for } n \geq 1 \end{cases}$$

Solution: The first two values $h(0)$ and $h(1)$ are chosen arbitrarily, but for $n \geq 2$ the values are recursively related to one another; for example, starting with $h(1) = 4$, we have $h(2) = 0.5h(1)$, $h(3) = 0.5h(2)$, $h(4) = 0.5h(3)$, and so on. Therefore, we expect that these recursions can be used to reassemble the I/O convolutional equation into a difference equation for $y(n)$.

Inserting the numerical values of $h(n)$ into Eq. (3.4.3), we find for the I/O equation

$$y_n = h_0 x_n + h_1 x_{n-1} + h_2 x_{n-2} + h_3 x_{n-3} + h_4 x_{n-4} + \cdots$$

$$= 2x_n + 4x_{n-1} + 2\left[x_{n-2} + 0.5x_{n-3} + 0.5^2 x_{n-4} + \cdots\right]$$

and for the previous output

$$y_{n-1} = 2x_{n-1} + 4x_{n-2} + 2\left[x_{n-3} + 0.5x_{n-4} + \cdots\right]$$

Multiplying by 0.5, we have

$$0.5y_{n-1} = x_{n-1} + 2\left[x_{n-2} + 0.5x_{n-3} + 0.5^2 x_{n-4} + \cdots\right]$$

Subtracting it from y_n, we find the I/O difference equation

$$y_n - 0.5y_{n-1} = 2x_n + 3x_{n-1}$$

and solving for $y(n)$

$$y(n) = 0.5y(n-1) + 2x(n) + 3x(n-1)$$

which is recognized as the difference equation of Example 3.1.5. Setting $x(n) = \delta(n)$ and $y(n) = h(n)$ gives the difference equation for $h(n)$:

$$h(n) = 0.5h(n-1) + 2\delta(n) + 3\delta(n-1)$$

Starting with the initial value $h(-1) = 0$ and iterating for a few values of n, one can easily verify that this difference equation generates the sequence $h(n)$ we started out with.   □

Example 3.4.8: Determine the convolutional form and the (causal) impulse response of the IIR filter described by the following difference equation:

$$y(n) = 0.25y(n-2) + x(n)$$

Solution: The impulse response $h(n)$ will satisfy the difference equation:

$$h(n) = 0.25h(n-2) + \delta(n)$$

to be iterated with zero initial conditions: $h(-1) = h(-2) = 0$. A few iterations give:

$$h(0) = 0.25h(-2) + \delta(0) = 0.25 \cdot 0 + 1 = 1$$
$$h(1) = 0.25h(-1) + \delta(1) = 0.25 \cdot 0 + 0 = 0$$
$$h(2) = 0.25h(0) + \delta(2) = 0.25 \cdot 1 + 0 = 0.25 = (0.5)^2$$
$$h(3) = 0.25h(1) + \delta(3) = 0.25 \cdot 0 + 0 = 0$$
$$h(4) = 0.25h(2) + \delta(4) = 0.25 \cdot 0.25 + 0 = (0.25)^2 = (0.5)^4$$

And, in general, for $n \geq 0$

$$h(n) = \begin{cases} (0.5)^n, & \text{if } n = \text{even} \\ 0, & \text{if } n = \text{odd} \end{cases}$$

Equivalently, we can write:

$$\mathbf{h} = [1, \ 0, \ (0.5)^2, \ 0, \ (0.5)^4, \ 0, \ (0.5)^6, \ 0, \ (0.5)^8, \ 0, \ \dots \ ]$$

And, Eq. (3.4.3) becomes:

$$y_n = x_n + 0.5^2 x_{n-2} + 0.5^4 x_{n-4} + 0.5^6 x_{n-6} + \cdots$$

which is the solution of the given difference equation in terms of $x(n)$. $\qquad \square$

Example 3.4.9: Determine the I/O difference equation of the IIR filter that has the following causal periodic impulse response:

$$h(n) = \{2, 3, 4, 5, 2, 3, 4, 5, 2, 3, 4, 5, \dots\}$$

where the dots denote the periodic repetition of the four samples $\{2, 3, 4, 5\}$.

Solution: If we delay the given response by one period, that is, 4 samples, we get

$$h(n-4) = \{0, 0, 0, 0, 2, 3, 4, 5, 2, 3, 4, 5, \dots\}$$

Subtracting it from $h(n)$, we get

$$h(n) - h(n-4) = \{2, 3, 4, 5, 0, 0, 0, 0, 0, 0, 0, 0, \dots\}$$

with all samples beyond $n = 4$ canceling to zero. These operations are depicted below.

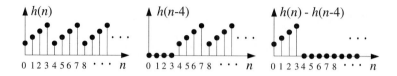

Thus, the right-hand side is nonzero only for $n = 0, 1, 2, 3$, and we can rewrite it as the difference equation

$$h(n) - h(n - 4) = 2\delta(n) + 3\delta(n - 1) + 4\delta(n - 2) + 5\delta(n - 3)$$

or, solving for $h(n)$

$$h(n) = h(n - 4) + 2\delta(n) + 3\delta(n - 1) + 4\delta(n - 2) + 5\delta(n - 3)$$

Using the method of the previous example, we can show that $y(n)$ satisfies the same difference equation:

$$y_n = y_{n-4} + 2x_n + 3x_{n-1} + 4x_{n-2} + 5x_{n-3}$$

This example shows how to construct a *digital periodic waveform generator*: Think of the waveform to be generated as the impulse response of an LTI system, determine the difference equation for that system, and then hit it with an impulse, and it will generate its impulse response, that is, the desired waveform. See Section 8.1.2.　□

More generally, the IIR filters that we will be concerned with have impulse responses $h(n)$ that satisfy constant-coefficient difference equations of the general type:

$$h(n) = \sum_{i=1}^{M} a_i h(n - i) + \sum_{i=0}^{L} b_i \delta(n - i)$$

or, written explicitly

$$h_n = a_1 h_{n-1} + a_2 h_{n-2} + \cdots + a_M h_{n-M} + b_0 \delta_n + b_1 \delta_{n-1} + \cdots + b_L \delta_{n-L}$$

Using the methods of Example 3.4.7, it can be shown that the corresponding convolutional equation (3.4.3) can be reassembled into the same difference equation for $y(n)$ in terms of $x(n)$, that is,

$$y(n) = \sum_{i=1}^{M} a_i y(n - i) + \sum_{i=0}^{L} b_i x(n - i)$$

or, explicitly

$$y_n = a_1 y_{n-1} + a_2 y_{n-2} + \cdots + a_M y_{n-M} + b_0 x_n + b_1 x_{n-1} + \cdots + b_L x_{n-L}$$

We will explore the properties of IIR filters after we discuss z-transforms. Note also that FIR filters can be thought of as special cases of the IIR difference equations when the recursive terms are absent, that is, when the recursive coefficients are zero, $a_1 = a_2 = \cdots = a_M = 0$.

Eventually, we aim to develop several mathematically *equivalent* descriptions of FIR and IIR filters, such as,

- I/O difference equation
- Convolutional equation
- Impulse response $h(n)$
- Transfer function $H(z)$
- Frequency response $H(\omega)$
- Pole/zero pattern
- Block diagram realization and sample processing algorithm

The above examples show how to go back and forth between the first three of these descriptions—from the *difference equation* to the corresponding *impulse response* to the *convolutional form* of the filtering equation. We will see later that all of the tedious *time-domain* manipulations of the above examples can be avoided by working with z-transforms.

Each description serves a different *purpose* and provides a different *insight* into the properties of the filter. For example, in a typical application, we would provide desired frequency-domain specifications for the filter, that is, specify the desired shape of $H(\omega)$. Using a filter design technique, we would design a filter whose frequency response closely approximates the desired response. The output of the filter design technique is typically the transfer function $H(z)$ for IIR filters or the impulse response $h(n)$ for FIR filters. From $H(z)$ or $h(n)$, we would obtain an appropriate block diagram realization that can be used to implement the filter in real time.

3.5 Causality and Stability

Like analog signals, discrete-time signals can be classified into causal, anticausal, or mixed signals, as shown in Fig. 3.5.1.

A *causal* or right-sided signal $x(n)$ exists only for $n \geq 0$ and vanishes for all negative times $n \leq -1$. Causal signals are the most commonly encountered signals, because they are the type that we generate in our labs — for example, when we turn on a signal generator or signal source.

An *anticausal* or left-sided signal exists only for $n \leq -1$ and vanishes for all $n \geq 0$. A *mixed* or double-sided signal has both a left-sided and a right-sided part.

The placement of the time origin, $n = 0$, along the time axis is entirely a matter of convention. Typically, it is taken to be the time when we turn on our signal generators or the time when we begin our processing operations. Therefore, a signal that is double-sided with respect to a chosen time origin is simply a signal that *has already been in existence* when we start our processing.

LTI systems can also be classified in terms of their causality properties depending on whether their impulse response $h(n)$ is causal, anticausal, or mixed. For a

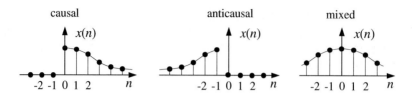

Fig. 3.5.1 Causal, anticausal, and mixed signals.

general double-sided $h(n)$, which can extend over $-\infty < n < \infty$, the I/O convolutional equation becomes

$$y(n) = \sum_{m=-\infty}^{\infty} h(m)x(n-m) \qquad (3.5.1)$$

Such systems cannot be implemented in real time, as can be seen by writing a few of the positive and negative m terms:

$$y_n = \cdots + h_{-2}x_{n+2} + h_{-1}x_{n+1} + h_0 x_n + h_1 x_{n-1} + h_2 x_{n-2} + \cdots$$

which shows that to compute the output $y(n)$ at the current time n, one needs to know the *future* input samples $x(n+1), x(n+2), \ldots$, which are not yet available for processing.

Anticausal and double-sided systems are very counter-intuitive, violating our sense of causality. For example, in response to a unit impulse $\delta(n)$, which is applied to the system at $n = 0$, the system will generate its impulse response output $h(n)$. But if $h(-1) \neq 0$, this means that the system had already produced an output sample at time $n = -1$, even before the input impulse was applied at $n = 0$!

Should we, therefore, be concerned with non-causal filters? Are they relevant, useful, or necessary in DSP? The answer to all of these questions is yes. Some typical applications where double-sided filters crop up are the design of FIR *smoothing* filters, the design of FIR *interpolation* filters used in oversampling DSP systems, and the design of *inverse* filters.

FIR smoothing and interpolation filters belong to a class of double-sided filters that are only finitely anticausal, that is, their anticausal part has *finite duration*, say over the period $-D \leq n \leq -1$. Such filters are shown in Fig. 3.5.2. In general, the causal part of $h(n)$ may be finite or infinite. The I/O equation (3.5.1) becomes for this class of filters:

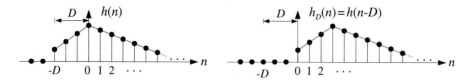

Fig. 3.5.2 Finitely anticausal filter and its causal version.

$$y(n) = \sum_{m=-D}^{\infty} h(m)x(n-m) \tag{3.5.2}$$

A standard technique for dealing with such filters is to make them causal by replacing $h(n)$ with its *delayed* version by D time units, that is,

$$h_D(n) = h(n-D)$$

As shown in Fig. 3.5.2, this operation translates $h(n)$ to the right by D units, making it causal. The I/O filtering equation for the causal filter $h_D(n)$ will be

$$y_D(n) = \sum_{m=0}^{\infty} h_D(m)x(n-m) \tag{3.5.3}$$

and will be implementable in real time. It is easily seen that the resulting sequence $y_D(n)$ is simply the delayed version of $y(n)$ as computed by Eq. (3.5.2):

$$y_D(n) = y(n-D)$$

Thus, the output samples are computed correctly, but come out with a time delay.

Example 3.5.1: Consider the typical 5-tap smoothing filter of Example 3.1.7 having filter coefficients $h(n) = 1/5$ for $-2 \le n \le 2$. The corresponding I/O convolutional equation (3.5.2) becomes

$$\begin{aligned}
y(n) &= \sum_{m=-2}^{2} h(m)x(n-m) = \frac{1}{5} \sum_{m=-2}^{2} x(n-m) \\
&= \frac{1}{5}\left[x(n+2)+x(n+1)+x(n)+x(n-1)+x(n-2)\right]
\end{aligned}$$

It is called a smoother or averager because at each n it replaces the current sample $x(n)$ by its average with the two samples ahead and two samples behind it, and therefore, it tends to average out rapid fluctuations from sample to sample.

Its anticausal part has duration $D = 2$ and can be made causal with a time delay of two units, resulting in

$$y_2(n) = y(n-2) = \frac{1}{5}\left[x(n)+x(n-1)+x(n-2)+x(n-3)+x(n-4)\right]$$

This filtering equation must be thought of as smoothing the middle sample $x(n-2)$ and not the current sample $x(n)$. □

When real-time processing is not an issue, as in block processing methods, and the input data to be processed have already been collected and saved as a block of samples on some medium such as memory or tape, one may use the non-causal form of Eq. (3.5.2) directly. This is one of the advantages of DSP that does not have a parallel in analog signal processing. An example of this may be the processing of a still picture, where all the pixel information has been gathered into a block of samples.

In addition to their causality properties, LTI systems can be classified in terms of their stability properties. A *stable* LTI system is one whose impulse response $h(n)$ goes to zero sufficiently fast as $n \to \pm\infty$, so that the output of the system $y(n)$ never diverges, that is, it remains *bounded* by some bound $|y(n)| \leq B$ if its input is *bounded*, say $|x(n)| \leq A$. That is, a system is stable if bounded inputs always generate bounded outputs.

It can be shown that a *necessary and sufficient* condition for an LTI system to be stable in the above bounded-input/bounded-output sense is that its impulse response $h(n)$ be *absolutely summable*:

$$\boxed{\sum_{n=-\infty}^{\infty} |h(n)| < \infty} \qquad \text{(stability condition)} \qquad (3.5.4)$$

Example 3.5.2: Consider the following four examples of $h(n)$:

$$h(n) = (0.5)^n u(n) \qquad \text{(stable and causal)}$$

$$h(n) = -(0.5)^n u(-n-1) \qquad \text{(unstable and anticausal)}$$

$$h(n) = 2^n u(n) \qquad \text{(unstable and causal)}$$

$$h(n) = -2^n u(-n-1) \qquad \text{(stable and anticausal)}$$

In the two causal cases, the presence of the unit step $u(n)$ causes $h(n)$ to be nonzero only for $n \geq 0$, whereas in the anticausal cases, the presence of $u(-n-1)$ makes $h(n)$ nonzero only for $-n-1 \geq 0$ or $n+1 \leq 0$ or $n \leq -1$. The first example tends to zero exponentially for $n \to \infty$; the second diverges as $n \to -\infty$; indeed, because n is negative, one can write $n = -|n|$ and

$$h(n) = -(0.5)^n u(-n-1) = -(0.5)^{-|n|} u(-n-1) = -2^{|n|} u(-n-1)$$

and therefore it blows up exponentially for large negative n. The third example blows up for $n \to \infty$ and the fourth tends to zero exponentially for $n \to -\infty$, as can be seen from

$$h(n) = -2^n u(-n-1) = -2^{-|n|} u(-n-1) = -(0.5)^{|n|} u(-n-1)$$

Thus, cases one and four are stable and cases two and three unstable. The same conclusion can also be reached by the stability criterion of Eq. (3.5.4). We have in the four cases:

$$\sum_{n=-\infty}^{\infty} |h(n)| = \sum_{n=0}^{\infty} (0.5)^n = \frac{1}{1-0.5} < \infty$$

$$\sum_{n=-\infty}^{\infty} |h(n)| = \sum_{n=-1}^{-\infty} (0.5)^n = \sum_{m=1}^{\infty} 2^m = \infty$$

$$\sum_{n=-\infty}^{\infty} |h(n)| = \sum_{n=0}^{\infty} 2^n = \infty$$

$$\sum_{n=-\infty}^{\infty} |h(n)| = \sum_{n=-1}^{-\infty} 2^n = \sum_{m=1}^{\infty} (0.5)^m = \frac{0.5}{1-0.5} < \infty$$

where in the first and fourth cases, we used the infinite geometric series formulas:

$$\sum_{m=0}^{\infty} x^m = \frac{1}{1-x} \quad \text{and} \quad \sum_{m=1}^{\infty} x^m = \frac{x}{1-x}$$

valid for $|x| < 1$. For the second and third cases the geometric series have $x > 1$ and diverge. We will see later that cases one and two have the *same* transfer function, namely, $H(z) = \dfrac{1}{1-0.5z^{-1}}$, and therefore, cannot be distinguished on the basis of just the transfer function. Similarly, cases three and four have the common transfer function $H(z) = \dfrac{1}{1-2z^{-1}}$. □

Stability is absolutely essential in hardware or software implementations of LTI systems because it guarantees that the *numerical* operations required for computing the I/O convolution sums or the equivalent difference equations remain well behaved and never grow beyond bounds. In hardware implementations, such instabilities would quickly saturate the hardware registers and in software implementations they would exceed the numerical ranges of most computers resulting in numerical nonsense.

The concepts of stability and causality are logically independent, but are *not always compatible* with each other, that is, it may not be possible to satisfy simultaneously the conditions of stability and causality, as was the case of the last two systems of Example 3.5.2. However, because of the practical numerical considerations mentioned above, we must *always* prefer stability over causality.

If the anticausal part of a stable system $h(n)$ has finite duration, then it can be handled as above, making it causal by a time delay. If, on the other hand, the anticausal part is infinite, then $h(n)$ can only be handled *approximately* by the following procedure. Because $h(n)$ is stable, it will tend to zero for large negative n. Therefore, one may pick a *sufficiently large* negative integer $n = -D$ and *clip* the left tail of $h(n)$ for $n < -D$. That is, one can replace the true $h(n)$ by its clipped approximation:

$$\tilde{h}(n) = \begin{cases} h(n), & \text{for } n \geq -D \\ 0, & \text{for } n < -D \end{cases} \qquad (3.5.5)$$

This clipped response will be of the finitely anticausal type shown in Fig. 3.5.2, and therefore, it can be made causal by a delay of D units of time, that is, $\tilde{h}_D(n) = \tilde{h}(n - D)$. The *approximation error* can be made as small as desired by increasing the value of D. To see this, let $\tilde{y}(n)$ be the output of the approximate system $\tilde{h}(n)$ for a bounded input, $|x(n)| \le A$, and let $y(n)$ be the output of the exact system $h(n)$. It is easily shown that the error in the output is bounded from above by

$$|y(n) - \tilde{y}(n)| \le A \sum_{m=-\infty}^{-D-1} |h(m)| \tag{3.5.6}$$

for all n. The above sum, being a partial sum of Eq. (3.5.4), is finite and tends to zero as D increases. For example, in case four of Example 3.5.2 we find

$$\sum_{m=-\infty}^{-D-1} |h(m)| = \sum_{m=D+1}^{\infty} (0.5)^m = (0.5)^{D+1} \frac{1}{1 - 0.5} = (0.5)^D$$

which can be made as small as desired by increasing D.

This type of stable but non-causal filter can often arise in the design of *inverse filters*. The inverse of a filter with transfer function $H(z)$ has transfer function

$$H_{\text{inv}}(z) = \frac{1}{H(z)}$$

Such inverse filters are used in various equalization applications, such as *channel equalization* for digital data transmission where $H(z)$ may represent the channel's transfer function, or the equalizer filters that we discussed in Chapter 1.

The corresponding impulse response of the inverse filter $h_{\text{inv}}(n)$ must be chosen to be stable. But, then it may be not be causal.[†] Therefore, in this case we must work with the approximate *clipped/delayed* inverse filter response

$$\tilde{h}_{\text{inv},D}(n) = \tilde{h}_{\text{inv}}(n - D)$$

We will consider examples of such designs later, after we discuss z-transforms.

3.6 Problems

3.1 Determine whether the discrete-time systems described by the following I/O equations are linear and/or time-invariant:

    a. $y(n) = 3x(n) + 5$

    b. $y(n) = x^2(n - 1) + x(2n)$,

    c. $y(n) = e^{x(n)}$

    d. $y(n) = nx(n - 3) + 3x(n)$.

    e. $y(n) = n + 3x(n)$

[†]We will see later that this circumstance can arise if some of the *zeros* of the transfer function $H(z)$ lie outside the unit circle in the z-plane.

3.2 Determine the causal impulse response $h(n)$ for $n \geq 0$ of the LTI systems described by the following I/O difference equations:

    a. $y(n) = 3x(n) - 2x(n-1) + 4x(n-3)$

    b. $y(n) = 4x(n) + x(n-1) - 3x(n-3)$

    c. $y(n) = x(n) - x(n-3)$

3.3 Determine the causal impulse response $h(n)$ for $n \geq 0$ of the LTI systems described by the following I/O difference equations:

    a. $y(n) = -0.9y(n-1) + x(n)$

    b. $y(n) = 0.9y(n-1) + x(n)$

    c. $y(n) = 0.64y(n-2) + x(n)$

    d. $y(n) = -0.81y(n-2) + x(n)$

    e. $y(n) = 0.5y(n-1) + 4x(n) + x(n-1)$

3.4 Determine the I/O difference equations relating $x(n)$ and $y(n)$ for the LTI systems having the following impulse responses:

    a. $h(n) = (0.9)^n u(n)$

    b. $h(n) = (-0.6)^n u(n)$

    c. $h(n) = (0.9)^n u(n) + (-0.9)^n u(n)$

    d. $h(n) = (0.9j)^n u(n) + (-0.9j)^n u(n)$

3.5 A causal IIR filter has impulse response $h(n) = 4\delta(n) + 3(0.5)^{n-1} u(n)$. Working with the convolutional equation $y(n) = \sum_m h(m)x(n-m)$, derive the *difference equation* satisfied by $y(n)$.

3.6 A causal IIR filter has impulse response:

$$h(n) = \begin{cases} 5, & \text{if } n = 0 \\ 6(0.8)^{n-1}, & \text{if } n \geq 1 \end{cases}$$

Working with the convolutional filtering equation, derive the *difference equation* satisfied by $y(n)$.

3.7 To understand the role played the first two values $h(0)$ and $h(1)$, redo Problem 3.6 starting with the more general expression for $h(n)$:

$$h(n) = \begin{cases} c_0 & \text{for } n = 0 \\ c_1 a^{n-1} & \text{for } n \geq 1 \end{cases}$$

which has $h(0) = c_0$ and $h(1) = c_1$. First, determine the difference equation satisfied by $h(n)$ for all $n \geq 0$. Then, using the I/O convolutional equation (3.3.3), determine the difference equation relating $y(n)$ to $x(n)$. How are $\{c_0, c_1\}$ related to the coefficients of the difference equation?

3.8 A causal linear time-invariant filter has impulse response:

$$h_n = [C_1 p_1^n + C_2 p_2^n + \cdots + C_M p_M^n] u(n)$$

Without using any z-transforms and working entirely in the time domain, show that h_n satisfies the order-M difference equation:

$$h_n + a_1 h_{n-1} + a_2 h_{n-2} + \cdots + a_M h_{n-M} = 0, \qquad \text{for } n \geq M$$

where $\{1, a_1, a_2, \ldots, a_M\}$ are the coefficients of the polynomial whose roots are the (complex) numbers $\{p_1, p_2, \ldots, p_M\}$, that is,

$$1 + a_1 z^{-1} + a_2 z^{-2} + \cdots + a_M z^{-M} = (1 - p_1 z^{-1})(1 - p_2 z^{-1}) \cdots (1 - p_M z^{-1})$$

Note that C_i are arbitrary and the restriction $n \geq M$ necessary.

3.9 A causal linear time-invariant filter has impulse response:

$$h_n = C_0 \delta(n) + C_1 p_1^n + C_2 p_2^n + \cdots + C_M p_M^n, \qquad n \geq 0$$

Show that it satisfies the same difference equation as in the previous problem, but with the restriction $n \geq M + 1$.

3.10 A causal linear time-invariant filter has impulse response:

$$h_n = C_1 p_1^n + C_2 p_2^n, \qquad n \geq 0$$

Working in the time domain, show that the difference equation satisfied by h_n for all $n \geq 0$ and the difference equation relating the input and output signals are of the form:

$$h_n + a_1 h_{n-1} + a_2 h_{n-2} = b_0 \delta(n) + b_1 \delta(n-1)$$

$$y_n + a_1 y_{n-1} + a_2 y_{n-2} = b_0 x_n + b_1 x_{n-1}$$

Determine $\{a_1, a_2, b_0, b_1\}$ in terms of $\{C_1, C_2, p_1, p_2\}$.

3.11 A causal linear time-invariant filter has impulse response:

$$h_n = C_0 \delta(n) + C_1 p_1^n + C_2 p_2^n, \qquad n \geq 0$$

Show that the difference equation satisfied by h_n for all $n \geq 0$ and the difference equation relating the input and output signals are of the form:

$$h_n + a_1 h_{n-1} + a_2 h_{n-2} = b_0 \delta(n) + b_1 \delta(n-1) + b_2 \delta(n-2)$$

$$y_n + a_1 y_{n-1} + a_2 y_{n-2} = b_0 x_n + b_1 x_{n-1} + b_2 x_{n-2}$$

Determine $\{a_1, a_2, b_0, b_1, b_2\}$ in terms of $\{C_0, C_1, C_2, p_1, p_2\}$.

3.12 A causal linear time-invariant filter has impulse response:

$$h_n = C_1 p_1^n + C_2 p_2^n + C_3 p_3^n, \qquad n \geq 0$$

Working in the time domain, show that the difference equation satisfied by h_n for all $n \geq 0$ and the difference equation relating the input and output signals are of the form:

$$h_n + a_1 h_{n-1} + a_2 h_{n-2} + a_3 h_{n-3} = b_0 \delta(n) + b_1 \delta(n-1) + b_2 \delta(n-2)$$

$$y_n + a_1 y_{n-1} + a_2 y_{n-2} + a_3 y_{n-3} = b_0 x_n + b_1 x_{n-1} + b_2 x_{n-2}$$

Determine $\{a_1, a_2, a_3, b_0, b_1\}$ in terms of $\{C_1, C_2, C_3, p_1, p_2, p_3\}$.

3.13 Using the results of the previous two problems, determine and verify the difference equations satisfied by the impulse responses:

 a. $h(n) = 5(0.5)^n u(n) + 4(0.8)^n u(n)$.
 b. $h(n) = (0.5j)^n u(n) + (-0.5j)^n u(n)$.
 c. $h(n) = [3(0.4)^n + 4(0.5)^n - 7(-0.5)^n] u(n)$.

3.14 The condition of Eq. (3.5.4) is *sufficient* for *bounded-input/bounded-output* (BIBO) stability. Assume $A = \sum_m |h(m)| < \infty$. Show that if the input is bounded, $|x(n)| \leq B$, then the output is bounded by $|y(n)| \leq AB$.

3.15 The condition of Eq. (3.5.4) is also *necessary* for BIBO stability. Assume that every bounded input results in a bounded output and consider the particular bounded input $x(n) = \text{sign}(h(-n))$ defined to be the algebraic sign of $h(-n)$. Then, the corresponding output $y(n)$ will be bounded. By considering the particular output sample $y(0)$, prove that Eq. (3.5.4) must hold. What happens if when $h(-n) = 0$ for some n?

3.16 The standard method for making an anticausal (but stable) system into a causal system is to clip off its anticausal tail at some large negative time $n = -D$ and then delay the impulse response by D time units to make it causal, that is, $h_D(n) = h(n - D)$. Let $y(n)$ be the output of $h(n)$ with input $x(n)$, and let $y_D(n)$ be the output of the delayed system $h_D(n)$ also with input $x(n)$. Working in the time domain, show that $y_D(n)$ is the delayed version of $y(n)$, that is, $y_D(n) = y(n - D)$.

3.17 In certain applications, such as data smoothing and FIR interpolation, the desired output $y(n)$ must be computed from a partly anticausal filter, that is, a filter $h(n)$ with anticausal duration of D time units. This filter can be made causal by a delay D, but this would cause the output to be delayed as we saw in the previous problem.

In order to get the correct *undelayed* output from the delayed causal filter $h_D(n)$, the input must be *time-advanced* by D time units, that is, $x_A(n) = x(n + D)$. Using time-domain convolution, show that $y(n)$ can be computed in the following two ways:

$$y(n) = \sum_m h(m)x(n - m) = \sum_m h_D(m)x_A(n - m)$$

In the z-domain, this is obvious:

$$Y(z) = H(z)X(z) = [z^{-D}H(z)][z^D X(z)]$$

In Sections 8.3.5 and 12.2.2, we show how to implement this idea by proper initialization of the filter's internal states.

3.18 Prove the inequality (3.5.6). Is the right-hand side finite? Does it get smaller as D gets larger?

4

FIR Filtering and Convolution

Practical DSP methods fall in two basic classes:

- Block processing methods.
- Sample processing methods.

In block processing methods, the data are collected and processed in blocks. Some typical applications include, FIR filtering of finite-duration signals by convolution, fast convolution of long signals which are broken up in short segments, DFT/FFT spectrum computations, speech analysis and synthesis, and image processing.

In sample processing methods, the data are processed one at a time—with each input sample being subjected to a *DSP algorithm* which transforms it into an output sample. Sample processing methods are used primarily in *real-time* applications, such as real-time filtering of long signals, digital audio effects processing, digital control systems, and adaptive signal processing. Sample processing algorithms are essentially the *state-space* realizations of LTI filters.

In this chapter, we consider block processing and sample processing methods for FIR filtering applications. We discuss the *computational* aspects of the convolution equations (3.3.2) or (3.3.3) as they apply to FIR filters and finite-duration inputs and present various *equivalent* forms of convolution, namely,

- Direct form
- Convolution table
- LTI form
- Matrix form
- Flip-and-slide form
- Overlap-add block convolution form.

Each form has its own distinct advantages. For example, the LTI form is of fundamental importance because it incorporates the consequences of linearity and time invariance; the direct form leads directly to block diagram realizations of the filter and the corresponding sample-by-sample processing algorithms; the convolution table is convenient for quick computations by hand; the flip-and-slide form shows clearly the input-on and input-off transient and steady-state behavior of a filter; the matrix form provides a compact vectorial representation of the filtering operation and is widely used in some applications such as image processing; and the overlap-add form is used whenever the input is extremely long or infinite in duration.

Then, we go on to discuss sample-by-sample processing methods for FIR filtering and discuss block diagram realizations which provide a *mechanization* of the sample processing algorithms. We develop the so-called *direct form* realizations of FIR filters and discuss some hardware issues for DSP chips. We develop also the concept of *circular addressing*, which is the "modern" way to implement delay lines, FIR, and IIR filters in both hardware and software.

4.1 Block Processing Methods

4.1.1 Convolution

In many practical applications, we sample our analog input signal (in accordance with the sampling theorem requirements) and collect a *finite* set of samples, say L samples, representing a finite time record of the input signal. The *duration* of the data record in seconds will be:[†]

$$\boxed{T_L = LT} \qquad (4.1.1)$$

where T is the sampling time interval, related to the sampling rate by $f_s = 1/T$. Conversely, we can solve for the number of time samples L contained in a record of duration T_L seconds:

$$\boxed{L = T_L f_s} \qquad (4.1.2)$$

The L collected signal samples, say $x(n)$, $n = 0, 1, \ldots, L - 1$, can be thought of as a block:

$$\mathbf{x} = [x_0, x_1, \ldots, x_{L-1}] \qquad (4.1.3)$$

which may then be processed further by a digital filter. The direct and LTI forms of convolution given by Eqs. (3.3.3) and (3.3.2)

$$y(n) = \sum_m h(m) x(n - m) = \sum_m x(m) h(n - m) \qquad (4.1.4)$$

describe the filtering equation of an LTI system in general. An alternative way of writing these equations, called the *convolution table* form, is obtained by noting that the sum of the indices of $h(m)$ and $x(n - m)$ is $m + (n - m) = n$. Therefore, Eqs. (4.1.4) can be written in the form:

[†]More correctly, $T_L = (L - 1)T$, but for large L Eq. (4.1.1) is simpler. See also Section 9.1.

$$y(n) = \sum_{\substack{i,j \\ i+j=n}} h(i)x(j) \qquad \text{(convolution table form)} \qquad (4.1.5)$$

That is, the sum of all possible products $h(i)x(j)$ with $i + j = n$. The precise range of summation with respect to m in Eqs. (4.1.4) or i, j in Eq. (4.1.5) depends on the *particular* nature of the filter and input sequences, $h(n)$ and $x(n)$.

4.1.2 Direct Form

Consider a causal FIR filter of order M with impulse response $h(n), n = 0, 1, \ldots, M$. It may be represented as a block:

$$\mathbf{h} = [h_0, h_1, \ldots, h_M] \qquad (4.1.6)$$

Its length (i.e., the number of filter coefficients) is one more than its order:

$$\boxed{L_h = M + 1} \qquad (4.1.7)$$

The convolution of the length-L input $\mathbf{x}$ of Eq. (4.1.3) with the order-M filter $\mathbf{h}$ will result in an output sequence $y(n)$. We must determine: (i) the range of values of the output index n, and (ii) the precise range of summation in m. For the direct form, we have

$$y(n) = \sum_m h(m)x(n-m)$$

The index of $h(m)$ must be within the range of indices in Eq. (4.1.6), that is, it must be restricted to the interval:

$$0 \le m \le M \qquad (4.1.8)$$

Similarly, the index of $x(n-m)$ must lie within the legal range of indices in Eq. (4.1.3), that is,

$$0 \le n - m \le L - 1 \qquad (4.1.9)$$

To determine the range of values of the output index n, we rewrite (4.1.9) in the form:

$$m \le n \le L - 1 + m$$

and use (4.1.8) to extend the limits to:

$$0 \le m \le n \le L - 1 + m \le L - 1 + M, \qquad \text{or,}$$

$$\boxed{0 \le n \le L - 1 + M} \qquad (4.1.10)$$

This is the index range of the output sequence $y(n)$. Therefore, it is represented by a block

$$\mathbf{y} = [y_0, y_1, \ldots, y_{L-1+M}] \tag{4.1.11}$$

with length

$$\boxed{L_y = L + M} \tag{4.1.12}$$

Thus, $\mathbf{y}$ is longer than the input $\mathbf{x}$ by M samples. As we will see later, this property follows from the fact that a filter of order M has *memory* M and keeps each input sample inside it for M time units. Setting $L_x = L$, and $L_h = M + 1$, we can rewrite Eq. (4.1.12) in the more familiar form:

$$\boxed{L_y = L_x + L_h - 1} \tag{4.1.13}$$

The relative block lengths are shown in Fig. 4.1.1. For any value of the output index n in the range (4.1.10), we must determine the summation range over m in the convolution equation. For fixed n, the inequalities (4.1.8) and (4.1.9) must be satisfied simultaneously by m. Changing the sign of (4.1.9), we obtain

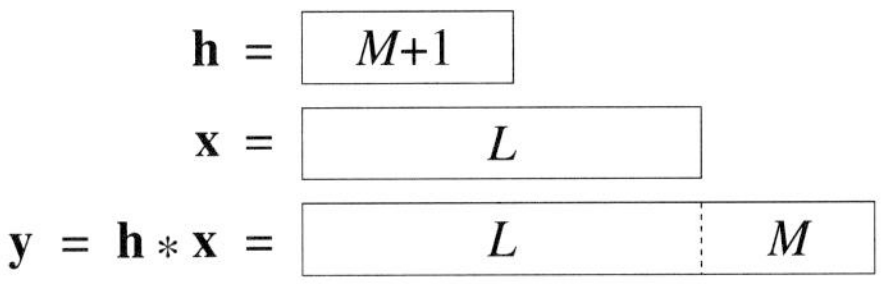

$$\mathbf{h} = \boxed{M+1}$$

$$\mathbf{x} = \boxed{\qquad L \qquad}$$

$$\mathbf{y} = \mathbf{h} * \mathbf{x} = \boxed{\qquad L \qquad}\ \boxed{M}$$

Fig. 4.1.1 Relative lengths of filter, input, and output blocks.

$$-(L - 1) \le m - n \le 0$$

and adding n to all sides

$$n - L + 1 \le m \le n \tag{4.1.14}$$

Thus, m must satisfy simultaneously the inequalities:

$$0 \le m \le M$$

$$n - L + 1 \le m \le n$$

It follows that m must be greater than the maximum of the two left-hand sides and less than the minimum of the two right-hand sides, that is,

$$\boxed{\max(0, n - L + 1) \le m \le \min(n, M)} \tag{4.1.15}$$

Therefore, in the case of an order-M FIR filter and a length-L input, the direct form of convolution is given as follows:

$$\boxed{y(n) = \sum_{m=\max(0, n-L+1)}^{\min(n,M)} h(m)x(n-m)} \qquad \text{(direct form)} \tag{4.1.16}$$

for $n = 0, 1, \ldots, L + M - 1$. Sometimes, we will indicate this convolutional operation by the compact notation:

$$\mathbf{y} = \mathbf{h} * \mathbf{x}$$

As an example, consider the case of an order-3 filter and a length-5 input signal. The filter, input, and output blocks are

$$\mathbf{h} = [h_0, h_1, h_2, h_3]$$

$$\mathbf{x} = [x_0, x_1, x_2, x_3, x_4]$$

$$\mathbf{y} = \mathbf{h} * \mathbf{x} = [y_0, y_1, y_2, y_3, y_4, y_5, y_6, y_7]$$

The output block has length $L_y = L + M = 5 + 3 = 8$ and is indexed as $0 \le n \le 7$. The convolutional equation (4.1.16) becomes:

$$y_n = \sum_{m=\max(0,n-4)}^{\min(n,3)} h_m x_{n-m}, \qquad n = 0, 1, \ldots, 7$$

For $n = 0, 1, 2, \ldots, 7$, the summation index m takes on the values:

$$
\begin{aligned}
\max(0, 0 - 4) &\le m \le \min(0, 3) &\Rightarrow&\quad m = 0 \\
\max(0, 1 - 4) &\le m \le \min(1, 3) &\Rightarrow&\quad m = 0, 1 \\
\max(0, 2 - 4) &\le m \le \min(2, 3) &\Rightarrow&\quad m = 0, 1, 2 \\
\max(0, 3 - 4) &\le m \le \min(3, 3) &\Rightarrow&\quad m = 0, 1, 2, 3 \\
\max(0, 4 - 4) &\le m \le \min(4, 3) &\Rightarrow&\quad m = 0, 1, 2, 3 \\
\max(0, 5 - 4) &\le m \le \min(5, 3) &\Rightarrow&\quad m = 1, 2, 3 \\
\max(0, 6 - 4) &\le m \le \min(6, 3) &\Rightarrow&\quad m = 2, 3 \\
\max(0, 7 - 4) &\le m \le \min(7, 3) &\Rightarrow&\quad m = 3
\end{aligned}
\tag{4.1.17}
$$

So, for example, at $n = 5$ the output y_5 will be given by

$$y_5 = \sum_{m=1,2,3} h_m x_{5-m} = h_1 x_4 + h_2 x_3 + h_3 x_2$$

Using the values in Eq. (4.1.17), we find all the output samples:

$$y_0 = h_0 x_0$$

$$y_1 = h_0 x_1 + h_1 x_0$$

$$y_2 = h_0 x_2 + h_1 x_1 + h_2 x_0$$

$$y_3 = h_0 x_3 + h_1 x_2 + h_2 x_1 + h_3 x_0$$

$$y_4 = h_0 x_4 + h_1 x_3 + h_2 x_2 + h_3 x_1 \qquad (4.1.18)$$

$$y_5 = h_1 x_4 + h_2 x_3 + h_3 x_2$$

$$y_6 = h_2 x_4 + h_3 x_3$$

$$y_7 = h_3 x_4$$

4.1.3 Convolution Table

Note how each output y_n in Eq. (4.1.18) is the sum of all possible products $h_i x_j$ with $i + j = n$. This leads directly to the convolution table of Eq. (4.1.5). For example, y_5 is obtained as

$$y_5 = \sum_{\substack{i,j \\ i+j=5}} h_i x_j = h_1 x_4 + h_2 x_3 + h_3 x_2$$

The required computations can be arranged in a table [24] as shown in Fig. 4.1.2 with the filter **h** written vertically and the input block **x** horizontally.[†]

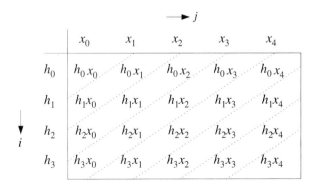

Fig. 4.1.2 Convolution table.

The nth row of the table is filled by multiplying the **x** samples by the corresponding h_n sample for that row. Then, the table is "folded" along its antidiagonal lines. In the ij-plane, the condition $i + j = n$ represents the nth antidiagonal straight line. Therefore, the entries within each antidiagonal strip are *summed* together to form the corresponding output value. There are as many antidiagonal strips as output

---

[†]It can also be arranged the other way, with the filter horizontally and the input vertically.

samples y_n. For example, the $n = 0$ strip contains only $h_0 x_0$, which is y_0; the $n = 1$ strip contains $h_0 x_1$ and $h_1 x_0$, whose sum is y_1, and so on; finally the $n = 7$ strip contains only $h_3 x_4$, which is y_7.

The convolution table is convenient for quick calculation by hand because it displays all required operations compactly.

Example 4.1.1: Calculate the convolution of the following filter and input signals:

$$\mathbf{h} = [1, 2, -1, 1], \qquad \mathbf{x} = [1, 1, 2, 1, 2, 2, 1, 1]$$

Solution: The convolution table, with $\mathbf{h}$ arranged vertically and $\mathbf{x}$ horizontally, is

h\x	1	1	2	1	2	2	1	1
1	1	1	2	1	2	2	1	1
2	2	2	4	2	4	4	2	2
-1	-1	-1	-2	-1	-2	-2	-1	-1
1	1	1	2	1	2	2	1	1

Folding the table, we get

$$\mathbf{y} = [1, 3, 3, 5, 3, 7, 4, 3, 3, 0, 1]$$

Note that there are $L_y = L + M = 8 + 3 = 11$ output samples.   □

4.1.4 LTI Form

Next, we discuss the LTI form of convolution. A more intuitive way to understand it is in terms of the linearity and time-invariance properties of the filter. Consider again the filter $\mathbf{h} = [h_0, h_1, h_2, h_3]$ in the example of Eq. (4.1.18). The input signal

$$\mathbf{x} = [x_0, x_1, x_2, x_3, x_4]$$

can be written as a linear combination of delayed impulses:

$$\mathbf{x} = x_0 [1, 0, 0, 0, 0]$$
$$+ x_1 [0, 1, 0, 0, 0]$$
$$+ x_2 [0, 0, 1, 0, 0]$$
$$+ x_3 [0, 0, 0, 1, 0]$$
$$+ x_4 [0, 0, 0, 0, 1]$$

It can also be written analytically for all n as a sum of delta functions:

$$x(n) = x_0 \delta(n) + x_1 \delta(n-1) + x_2 \delta(n-2) + x_3 \delta(n-3) + x_4 \delta(n-4)$$

The effect of the filter is to replace each delayed impulse by the corresponding delayed impulse response, that is,

$$y(n) = x_0 h(n) + x_1 h(n-1) + x_2 h(n-2) + x_3 h(n-3) + x_4 h(n-4)$$

We can represent the input and output signals as blocks:

$$\mathbf{x} = x_0 [1,0,0,0,0] \qquad\qquad \mathbf{y} = x_0 [h_0, h_1, h_2, h_3, 0, 0, 0, 0]$$
$$+ x_1 [0,1,0,0,0] \qquad\qquad\quad + x_1 [0, h_0, h_1, h_2, h_3, 0, 0, 0]$$
$$+ x_2 [0,0,1,0,0] \quad\xrightarrow{H}\quad + x_2 [0, 0, h_0, h_1, h_2, h_3, 0, 0]$$
$$+ x_3 [0,0,0,1,0] \qquad\qquad\quad + x_3 [0, 0, 0, h_0, h_1, h_2, h_3, 0]$$
$$+ x_4 [0,0,0,0,1] \qquad\qquad\quad + x_4 [0, 0, 0, 0, h_0, h_1, h_2, h_3]$$

The result is the same as Eq. (4.1.18). Indeed, the indicated linear combinations in the right-hand side give:

$$\mathbf{y} = [h_0 x_0, \; x_0 h_1 + x_1 h_0, \; x_0 h_2 + x_1 h_1 + x_2 h_0, \; \ldots, \; x_4 h_3]$$
$$= [y_0, y_1, y_2, \ldots, y_7]$$

For computational purposes, the LTI form can be represented pictorially in a table form, as shown in Fig. 4.1.3. The impulse response $\mathbf{h}$ is written horizontally, and the input $\mathbf{x}$ vertically.

	h_0	h_1	h_2	h_3	0	0	0	0
x_0	$x_0 h_0$	$x_0 h_1$	$x_0 h_2$	$x_0 h_3$	0	0	0	0
x_1	0	$x_1 h_0$	$x_1 h_1$	$x_1 h_2$	$x_1 h_3$	0	0	0
x_2	0	0	$x_2 h_0$	$x_2 h_1$	$x_2 h_2$	$x_2 h_3$	0	0
x_3	0	0	0	$x_3 h_0$	$x_3 h_1$	$x_3 h_2$	$x_3 h_3$	0
x_4	0	0	0	0	$x_4 h_0$	$x_4 h_1$	$x_4 h_2$	$x_4 h_3$
	y_0	y_1	y_2	y_3	y_4	y_5	y_6	y_7

Fig. 4.1.3 LTI form of convolution.

The rows of the table correspond to the successive delays (right shifts) of the $\mathbf{h}$ sequence—the mth row corresponds to delay by m units. Each row is scaled by the corresponding input sample, that is, the mth row represents the term $x_m h_{n-m}$ in the LTI form. After the table is filled, the table entries are summed *column-wise* to obtain the output samples $y(n)$, which is equivalent to forming the sum:

$$y(n) = \sum_m x(m) h(n-m)$$

Example 4.1.2: Calculate the convolution of Example 4.1.1 using the LTI form.

Solution: The corresponding LTI table is in this case:

n	0	1	2	3	4	5	6	7	8	9	10	
$x \backslash \mathbf{h}$	1	2	-1	1								partial output
1	1	2	-1	1								$x_0 h_n$
1		1	2	-1	1							$x_1 h_{n-1}$
2			2	4	-2	2						$x_2 h_{n-2}$
1				1	2	-1	1					$x_3 h_{n-3}$
2					2	4	-2	2				$x_4 h_{n-4}$
2						2	4	-2	2			$x_5 h_{n-5}$
1							1	2	-1	1		$x_6 h_{n-6}$
1								1	2	-1	1	$x_7 h_{n-7}$
y_n	1	3	3	5	3	7	4	3	3	0	1	$\sum_m x_m h_{n-m}$

The output samples are obtained by summing the entries in each column. The result agrees with Example 4.1.1.   □

The LTI form can also be written in a form similar to Eq. (4.1.16) by determining the proper limits of summation. Arguing as in the direct form, or interchanging the roles of $h(n)$ and $x(n)$ and correspondingly, the length quantities M and $L-1$, we obtain the following form:

$$y(n) = \sum_{m=\max(0,n-M)}^{\min(n,L-1)} x(m)h(n-m) \qquad \text{(LTI form)} \qquad (4.1.19)$$

for $n = 0, 1, \ldots, L + M - 1$.

4.1.5 Matrix Form

The convolutional equations (4.1.16) or (4.1.19) can also be written in the linear matrix form:

$$\boxed{\mathbf{y} = H\mathbf{x}} \qquad (4.1.20)$$

where H is built out of the filter's impulse response $\mathbf{h}$. Because the output vector $\mathbf{y}$ has length $L+M$ and the input vector $\mathbf{x}$ length L, the filter matrix H must be rectangular with dimensions

$$L_y \times L_x = (L + M) \times L$$

To see this, consider again the example of Eq. (4.1.18). The output samples can be arranged in the matrix form

$$
\mathbf{y} =
\begin{bmatrix}
y_0 \\
y_1 \\
y_2 \\
y_3 \\
y_4 \\
y_5 \\
y_6 \\
y_7
\end{bmatrix}
=
\begin{bmatrix}
h_0 & 0 & 0 & 0 & 0 \\
h_1 & h_0 & 0 & 0 & 0 \\
h_2 & h_1 & h_0 & 0 & 0 \\
h_3 & h_2 & h_1 & h_0 & 0 \\
0 & h_3 & h_2 & h_1 & h_0 \\
0 & 0 & h_3 & h_2 & h_1 \\
0 & 0 & 0 & h_3 & h_2 \\
0 & 0 & 0 & 0 & h_3
\end{bmatrix}
\begin{bmatrix}
x_0 \\
x_1 \\
x_2 \\
x_3 \\
x_4
\end{bmatrix}
= H\mathbf{x}
$$

Note that the *columns* of H are the successively delayed replicas of the impulse response vector $\mathbf{h}$. There are as many columns as input samples. Note also that H is a so-called *Toeplitz matrix*, in the sense that it has the same entry along each diagonal. The Toeplitz property is a direct consequence of the time invariance of the filter. Note also that Eq. (4.1.20) is equivalent to the LTI table, transposed column-wise instead of row-wise.

Example 4.1.3: Calculate the convolution of Example 4.1.1 using the matrix form.

Solution: Because $L_y = 11$ and $L_x = 8$, the filter matrix will be 11×8 dimensional. We have,

$$
H\mathbf{x} =
\begin{bmatrix}
1 & 0 & 0 & 0 & 0 & 0 & 0 & 0 \\
2 & 1 & 0 & 0 & 0 & 0 & 0 & 0 \\
-1 & 2 & 1 & 0 & 0 & 0 & 0 & 0 \\
1 & -1 & 2 & 1 & 0 & 0 & 0 & 0 \\
0 & 1 & -1 & 2 & 1 & 0 & 0 & 0 \\
0 & 0 & 1 & -1 & 2 & 1 & 0 & 0 \\
0 & 0 & 0 & 1 & -1 & 2 & 1 & 0 \\
0 & 0 & 0 & 0 & 1 & -1 & 2 & 1 \\
0 & 0 & 0 & 0 & 0 & 1 & -1 & 2 \\
0 & 0 & 0 & 0 & 0 & 0 & 1 & -1 \\
0 & 0 & 0 & 0 & 0 & 0 & 0 & 1
\end{bmatrix}
\begin{bmatrix}
1 \\
1 \\
1 \\
2 \\
1 \\
2 \\
2 \\
1
\end{bmatrix}
=
\begin{bmatrix}
1 \\
3 \\
3 \\
5 \\
3 \\
7 \\
4 \\
3 \\
3 \\
0 \\
1
\end{bmatrix}
$$

which agrees with Example 4.1.1. □

There is also an alternative matrix form written as follows:

$$
\begin{bmatrix}
y_0 \\
y_1 \\
y_2 \\
y_3 \\
y_4 \\
y_5 \\
y_6 \\
y_7
\end{bmatrix}
=
\begin{bmatrix}
x_0 & 0 & 0 & 0 \\
x_1 & x_0 & 0 & 0 \\
x_2 & x_1 & x_0 & 0 \\
x_3 & x_2 & x_1 & x_0 \\
x_4 & x_3 & x_2 & x_1 \\
0 & x_4 & x_3 & x_2 \\
0 & 0 & x_4 & x_3 \\
0 & 0 & 0 & x_4
\end{bmatrix}
\begin{bmatrix}
h_0 \\
h_1 \\
h_2 \\
h_3
\end{bmatrix}
\tag{4.1.21}
$$

Instead of a filter matrix H acting on the input data vector $\mathbf{x}$, it has a data matrix acting on the filter vector $\mathbf{h}$. It can be written compactly in the form:

$$\boxed{\mathbf{y} = X\mathbf{h}}\tag{4.1.22}$$

where the data matrix X has dimension:

$$L_y \times L_h = (L + M) \times (M + 1)$$

The *first* column of X is the given input, padded with M zeros at the end to account for the input-off transients. The remaining columns are the successively delayed (down-shifted) versions of the first one. We will see in Section 4.2.2 that this form is essentially equivalent to the sample-by-sample processing algorithm of the direct form realization of the filter—with the nth row of X representing the filter's internal states at time n.

Example 4.1.4: Calculate the convolution of Example 4.1.1 using the matrix form (4.1.22).

Solution: The X matrix will have dimension $L_y \times L_h = 11 \times 4$. Its first column is the input signal padded with 3 zeros at the end:

$$X\mathbf{h} = \begin{bmatrix} 1 & 0 & 0 & 0 \\ 1 & 1 & 0 & 0 \\ 2 & 1 & 1 & 0 \\ 1 & 2 & 1 & 1 \\ 2 & 1 & 2 & 1 \\ 2 & 2 & 1 & 2 \\ 1 & 2 & 2 & 1 \\ 1 & 1 & 2 & 2 \\ 0 & 1 & 1 & 2 \\ 0 & 0 & 1 & 1 \\ 0 & 0 & 0 & 1 \end{bmatrix} \begin{bmatrix} 1 \\ 2 \\ -1 \\ 1 \end{bmatrix} = \begin{bmatrix} 1 \\ 3 \\ 3 \\ 5 \\ 3 \\ 7 \\ 4 \\ 3 \\ 3 \\ 0 \\ 1 \end{bmatrix}$$

which agrees with Example 4.1.1.   □

Matrix representations of convolution are very useful in some applications, such as image processing, and in more advanced DSP methods such as parametric spectrum estimation and adaptive filtering.

4.1.6 Flip-and-Slide Form

The LTI form is also closely related to the popular flip-and-slide form of convolution, in which the filter $h(n)$ is flipped around or reversed and then slid over the input data sequence. At each time instant, the output sample is obtained by computing the *dot* product of the flipped filter vector $\mathbf{h}$ with the $M+1$ input samples aligned below it, as shown in Fig. 4.1.4.

The input sequence is assumed to have been extended by padding M zeros to its left and to its right. At time $n = 0$, the only nonzero contribution to the dot product comes from h_0 and x_0 which are time aligned. It takes the filter M time units before it is completely over the nonzero portion of the input sequence. The first M outputs correspond to the *input-on transient* behavior of the filter. Then,

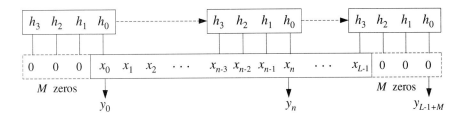

Fig. 4.1.4 Flip-and-slide form of convolution.

for a period of time $M \le n \le L - 1$, the filter remains completely over the nonzero portion of the input data, and the outputs are given by the form

$$y_n = h_0 x_n + h_1 x_{n-1} + \cdots + h_M x_{n-M}$$

This period corresponds to the *steady-state* behavior of the filter. Finally, the *last M* outputs beyond the end of the input data are the *input-off transients*, that is, they are the outputs after the input has been turned off. They correspond to the time period $L \le n \le L - 1 + M$. During this period the filter slides over the last M zeros padded at the end of the input. The very last output is obtained when h_M is aligned over x_{L-1}, which gives $y_{L-1+M} = h_M x_{L-1}$.

One can also think of the filter block **h** as being stationary and the input block **x** sliding underneath it in the opposite direction. This view leads to the sample-by-sample processing algorithms for FIR filtering.

4.1.7 Transient and Steady-State Behavior

The transient and steady-state behavior of an FIR filter can also be understood using the direct form of convolution, Eq. (4.1.16). For a length-L input and order-M filter, the output time index n will be in the range:

$$0 \le n \le L - 1 + M$$

It can be divided into *three subranges*, depicted in Fig. 4.1.5, corresponding to the input-on transients, steady state, and input-off transients:

$0 \le n < M$	(input-on transients)
$M \le n \le L - 1$	(steady state)
$L - 1 < n \le L - 1 + M$	(input-off transients)

These subranges affect differently the limits of the convolution summation equation (4.1.16). As implied in Fig. 4.1.5, we assumed that the filter length is much shorter than the length of the input, that is, $M + 1 < L$ or $M < L - 1$, otherwise the steady-state range defined above does not exist—the input is too short to exhibit any steady behavior.

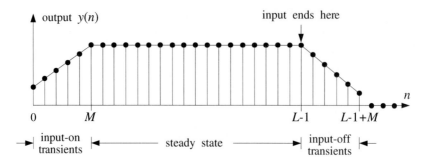

Fig. 4.1.5 Transient and steady-state filter outputs.

For the input-on transients, the restriction $0 \leq n < M < L - 1$ implies the following summation limits:

$$\max(0, n - L + 1) = 0, \qquad \min(n, M) = n$$

For the steady-state range, $M \leq n \leq L - 1$, we have:

$$\max(0, n - L + 1) = 0, \qquad \min(n, M) = M$$

And, for the input-off transients, $M < L - 1 < n \leq L - 1 + M$, we have:

$$\max(0, n - L + 1) = n - L + 1, \qquad \min(n, M) = M$$

Therefore, Eq. (4.1.16) takes the following different forms depending on the value of the output index n:

$$
y_n = \begin{cases}
\displaystyle\sum_{m=0}^{n} h_m x_{n-m}, & \text{if } 0 \leq n < M, & \text{(input-on)} \\[2em]
\displaystyle\sum_{m=0}^{M} h_m x_{n-m}, & \text{if } M \leq n \leq L - 1, & \text{(steady state)} \\[2em]
\displaystyle\sum_{m=n-L+1}^{M} h_m x_{n-m}, & \text{if } L - 1 < n \leq L - 1 + M, & \text{(input-off)}
\end{cases}
$$

During the input-on transient period, the number of terms in the sum is $n+1$ and is increasing. During the steady-state period, the number of terms is equal to the number of filter weights, $M+1$, and remains fixed. And during the input-off period, the number of terms in the sum keeps decreasing down to one because the lower summation limit is increasing.

In Eq. (4.1.18), the first three outputs $\{y_0, y_1, y_2\}$ are the input-on transients and the number of terms keeps increasing to 4. The next two outputs $\{y_3, y_4\}$ are the steady-state outputs, and the last three outputs $\{y_5, y_6, y_7\}$ are the input-off transients having a decreasing number of terms.

The I/O equation in the steady state was also discussed earlier in Eq. (3.4.1). It has a fixed number of terms:

$$y(n) = \sum_{m=0}^{M} h(m)x(n-m) \qquad \text{(steady state)} \qquad (4.1.23)$$

In a certain sense, it also incorporates the input-on and input-off transients and is quoted often as the generic I/O equation for FIR filters. To understand why, consider again the example of Eq. (4.1.18). In this case, we write

$$y_n = h_0 x_n + h_1 x_{n-1} + h_2 x_{n-2} + h_3 x_{n-3}$$

If n is in the input-on range, that is, $0 \le n \le 2$, not every term in the sum will contribute, because x_n is assumed causal. For example, if $n = 1$, we have

$$y_1 = h_0 x_1 + h_1 x_{1-1} + h_2 x_{1-2} + h_3 x_{1-3} = h_0 x_1 + h_1 x_0$$

When n is in the steady range, then all terms are contributing. And, when n is in the input-off range, not all terms contribute. For example, if $n = 6$ we have

$$y_6 = h_0 x_6 + h_1 x_{6-1} + h_2 x_{6-2} + h_3 x_{6-3} = h_2 x_4 + h_3 x_3$$

because x_n was assumed to have length 5, and therefore $x_6 = x_5 = 0$. With these caveats in mind, it is correct to write Eq. (4.1.23) as the generic I/O filtering equation of an order-M FIR filter. For programming purposes, one must of course work with Eq. (4.1.16) which does not let the indices exceed the array bounds.

4.1.8 Convolution of Infinite Sequences

By taking appropriate limits of the direct form of convolution

$$y_n = \sum_{m=\max(0,n-L+1)}^{\min(n,M)} h_m x_{n-m}$$

we can obtain the correct summation limits for the following three cases:

1. Infinite filter, finite input; i.e., $M = \infty$, $L < \infty$.
2. Finite filter, infinite input; i.e., $M < \infty$, $L = \infty$.
3. Infinite filter, infinite input; i.e., $M = \infty$, $L = \infty$.

In all three cases, the range of the output index Eq. (4.1.10) is infinite, $0 \le n < \infty$, that is, the output $y(n)$ has infinite duration. When $M \to \infty$, the upper limit in the convolution sum becomes

$$\min(n, M) = n$$

and when $L \to \infty$, the lower limit becomes

$$\max(0, n - L + 1) = 0$$

Therefore, we find in the three cases:

$$y_n = \sum_{m=\max(0,n-L+1)}^{n} h_m x_{n-m}, \qquad \text{if } M = \infty, L < \infty$$

$$y_n = \sum_{m=0}^{\min(n,M)} h_m x_{n-m}, \qquad \text{if } M < \infty, L = \infty$$

$$y_n = \sum_{m=0}^{n} h_m x_{n-m}, \qquad \text{if } M = \infty, L = \infty$$

When the filter is infinite, we define steady state as the limit of $y(n)$ for large n.

Example 4.1.5: An IIR filter has impulse response $h(n) = (0.75)^n u(n)$. Using convolution, derive closed-form expressions for the output signal $y(n)$ when the input is:

(a) A unit step, $x(n) = u(n)$.

(b) An alternating step, $x(n) = (-1)^n u(n)$.

(c) A square pulse of duration $L = 25$ samples, $x(n) = u(n) - u(n - 25)$.

In each case, determine the steady-state response of the filter.

Solution: In case (a), because both the input and filter are causal and have infinite duration, we use the formula:

$$y(n) = \sum_{m=0}^{n} h(m) x(n - m) = \sum_{m=0}^{n} (0.75)^m u(m) u(n - m)$$

or, using the finite geometric series:

$$y(n) = \sum_{m=0}^{n} (0.75)^m = \frac{1 - (0.75)^{n+1}}{1 - 0.75} = 4 - 3(0.75)^n$$

The steady-state response is the large-n limit of this formula, that is, as $n \to \infty$

$$y(n) \to \frac{1}{1 - 0.75} = 4$$

For case (b), we have

$$y(n) = \sum_{m=0}^{n} (0.75)^m (-1)^{n-m} = (-1)^n \sum_{m=0}^{n} (-0.75)^m$$

$$= (-1)^n \frac{1 - (-0.75)^{n+1}}{1 + 0.75} = \frac{4}{7} (-1)^n + \frac{3}{7} (0.75)^n$$

where in the second term, we wrote $(-1)^n(-0.75)^n = (0.75)^n$. In the large-$n$ limit, $n \to \infty$, we have

$$y(n) \to (-1)^n \frac{1}{1+0.75} = \frac{4}{7}(-1)^n$$

We will see later that these steady-state responses correspond to special cases of the sinusoidal response of the filter (at frequencies $\omega = 0$ and $\omega = \pi$), and can be obtained very simply in terms of the transfer function $H(z)$ of the filter evaluated at $z = 1$ for part (a), and $z = -1$ for part (b), that is,

$$y(n) \to H(1) \quad \text{and} \quad y(n) \to (-1)^n H(-1)$$

where in this example,

$$H(z) = \frac{1}{1 - 0.75z^{-1}} \quad \Rightarrow \quad H(1) = \frac{1}{1 - 0.75} = 4, \quad H(-1) = \frac{1}{1 + 0.75} = \frac{4}{7}$$

In part (c), the input is finite with length $L = 25$. Therefore,

$$y_n = \sum_{m=\max(0,n-L+1)}^{n} h_m x_{n-m} = \sum_{m=\max(0,n-24)}^{n} (0.75)^m$$

We must distinguish two subranges in the output index: For $0 \le n \le 24$, we have

$$y_n = \sum_{m=0}^{n} (0.75)^m = \frac{1 - (0.75)^{n+1}}{1 - 0.75} = 4 - 3(0.75)^n$$

and for $25 \le n < \infty$,

$$y_n = \sum_{m=n-24}^{n} (0.75)^m = (0.75)^{n-24} \frac{1 - (0.75)^{n-(n-24)+1}}{1 - 0.75}$$

$$= (0.75)^{n-24} \frac{1 - (0.75)^{25}}{1 - 0.75}$$

which corresponds to the input-off transient behavior. Because of the exponentially decaying nature of the impulse response, this filter acts like an RC-type integrator. During the input-on period $0 \le n \le 24$, the output "charges" up and during the input-off period $n \ge 25$, it "discharges" down. See Example 4.1.8 for a similar, but not quite identical, example.

The output signals $y(n)$ of the three cases (a-c) are shown below:

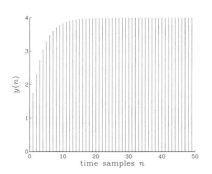
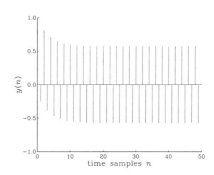

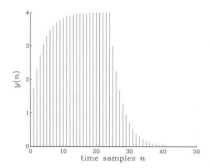

Notice the steady-state behavior of the first two cases, and the input-off transients of the third.   □

Example 4.1.6: We saw in Example 3.4.5 that a filter of the form $h_n = (0.75)^n u(n)$ satisfies the difference equation:

$$y(n) = 0.75y(n-1) + x(n)$$

Verify that the expressions for $y(n)$ obtained in the three cases (a-c) in Example 4.1.5 are solutions of this difference equation, with causal initial conditions.

Solution: In case (a), we have $x(n) = u(n)$ and the difference equation becomes, for $n \geq 0$:

$$y(n) = 0.75y(n-1) + 1$$

For $n = 0$, it gives $y(0) = 0.75y(-1) + 1 = 0.75 \cdot 0 + 1 = 1$, which agrees with the expression $y(n) = 4 - 3(0.75)^n$ evaluated at $n = 0$. For $n \geq 1$, we have, starting with the right-hand side:

$$0.75y(n-1) + 1 = 0.75\left[4 - 3(0.75)^{n-1}\right] + 1 = 4 - 3(0.75)^n = y(n)$$

In case (b), we have $x(n) = (-1)^n u(n)$ and the difference equation becomes for $n \geq 1$:

$$0.75y(n-1) + x(n) = 0.75\left[\frac{4}{7}(-1)^{n-1} + \frac{3}{7}(0.75)^{n-1}\right] + (-1)^n$$

$$= -0.75\frac{4}{7}(-1)^n + \frac{3}{7}(0.75)^n + (-1)^n = \frac{4}{7}(-1)^n + \frac{3}{7}(0.75)^n = y(n)$$

In case (c), we have the difference equations

$$y(n) = 0.75y(n-1) + 1, \qquad \text{for } 0 \leq n \leq 24$$

and

$$y(n) = 0.75y(n-1), \qquad \text{for } n \geq 25$$

The first value at $n = 25$ will be $y(25) = 0.75y(24)$, and therefore, it requires knowledge of the "initial" value $y(24)$. If that is known, then the solution of the homogeneous equation will be

$$y(n) = (0.75)^{n-24} y(24), \quad \text{for} \quad n \geq 25$$

But, $y(24)$ is

$$y(24) = \frac{1 - (0.75)^{25}}{1 - 0.75} = 4 - 3(0.75)^{24}$$

as obtained from the solution of the first difference equation evaluated at the endpoint $n = 24$.   □

The most *general case* that covers any type of filter and input signal— finite or infinite, causal or non-causal—can be defined as follows. Assume the filter's impulse response $h(n)$ is defined over the interval:

$$-M_1 \leq n \leq M_2$$

and the input signal $x(n)$ over the interval:

$$-L_1 \leq n \leq L_2 - 1$$

Any desired case can be obtained by taking appropriate limits in the quantities M_1, M_2, L_1, L_2. We wish to determine the range of the output index n and the limits of summation in the convolutional equation

$$y(n) = \sum_m h(m) x(n - m)$$

The index m of $h(m)$ and $n - m$ of $x(n - m)$ must lie within the given index ranges, that is,

$$-M_1 \leq m \leq M_2 \quad \text{and} \quad -L_1 \leq n - m \leq L_2 - 1$$

From these it follows that the output index n must vary over the range:

$$-M_1 - L_1 \leq n \leq L_2 + M_2 - 1$$

Note that the endpoints of the output index are the *sum* of the corresponding endpoints for $h(n)$ and $x(n)$. For each n in this output range, the summation index must be within the limits:

$$\max(-M_1, n - L_2 + 1) \leq m \leq \min(n + L_1, M_2)$$

Therefore, the I/O equation is in this case:

$$y(n) = \sum_{m=\max(-M_1,n-L_2+1)}^{\min(n+L_1,M_2)} h(m)x(n-m)$$

The results of Eq. (4.1.16) can be recovered as the special case corresponding to $M_1 = 0$, $M_2 = M$, $L_1 = 0$, $L_2 = L$.

4.1.9 Programming Considerations

The following C routine `conv.c` implements the direct form of convolution of Eq. (4.1.16):

```
/* conv.c - convolution of x[n] with h[n], resulting in y[n] */

#include <stdlib.h>                          defines max( ) and min( )

void conv(M, h, L, x, y)
double *h, *x, *y;                           h, x, y = filter, input, output arrays
int M, L;                                    M = filter order, L = input length
{
        int n, m;

        for (n = 0; n < L+M; n++)
                for (y[n] = 0, m = max(0, n-L+1); m <= min(n, M); m++)
                        y[n] += h[m] * x[n-m];
}
```

The quantities h, x, y are *arrays* and must be declared or allocated to proper dimension in the *main* program; for example, using `calloc`:

```
double *h, *x, *y;
h = (double *) calloc(M+1, sizeof(double));        (M+1)-dimensional
x = (double *) calloc(L,   sizeof(double));        L-dimensional
y = (double *) calloc(L+M, sizeof(double));        (L+M)-dimensional
```

In some C implementations,[†] the include file `stdlib.h` contains the definitions of the two macros `max` and `min`; otherwise they must be added in the above routine:

```
#define max(a,b)   (((a) > (b)) ? (a) : (b))
#define min(a,b)   (((a) < (b)) ? (a) : (b))
```

Next, we present a few simulation examples illustrating some of the ideas we discussed, such as input-on and input-off transients, steady state, linearity, and time invariance. A quantity of interest in these examples will be the *DC gain* of a (stable) filter, that is, the steady-state value of its output when the input remains constant for a long period of time. For a unity input, it is given by the sum of the impulse response coefficients:

$$y_{dc} = \sum_m h(m) \tag{4.1.24}$$

It will be derived later on.

[†] For example, Microsoft and Borland C.

Example 4.1.7: Consider an integrator-like FIR filter of order $M = 14$ defined by the following I/O convolutional equation:

$$y(n) = G[x(n) + x(n-1) + x(n-2) + \cdots + x(n-14)]$$

where G is a convenient scale factor, taken to be $G = 0.1$. Such a filter accumulates (integrates) the present and past 14 samples of the input signal. Comparing with Eq. (4.1.23), we identify the impulse response of this filter:

$$h_n = \begin{cases} G, & \text{for } 0 \le n \le 14 \\ 0, & \text{otherwise} \end{cases}$$

The DC gain will be:

$$y_{\text{dc}} = \sum_{m=0}^{14} h(m) = \sum_{m=0}^{14} G = 15G = 1.5$$

To observe the steady-state response of this filter, as well as the input-on and input-off transients, we consider a *square-wave* input signal x_n of length $L = 200$ and period of $K = 50$ samples. Such a signal may be generated by the simple for-loop:

```
for (n=0; n<L; n++)
    if (n%K < K/2)                    n % K is the MOD operation
        x[n] = 1;
    else
        x[n] = 0;
```

The output signal y_n will have length $L_y = L + M = 200 + 14 = 214$ samples. It can be obtained by a single call to the routine `conv`:

```
conv(M, h, L, x, y);
```

The figure below shows the output signal y_n plotted together with the periodic input.

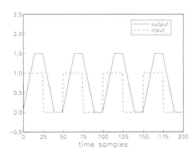

As the square wave periodically goes on and off, we can observe the input-on transient, steady-state, and input-off transient behavior of the filter.

During each on-period of the square wave lasting for 25 samples, the filter exhibits an input-on transient behavior which lasts for 14 samples; then it reaches steady state (equal to its DC gain) lasting only $25 - 14 = 11$ samples, and then the square wave goes off causing the filter to undergo its input-off transient behavior which lasts another 14 samples. The filter's output settles down to zero only after $25 + 14 = 39$ samples and remains zero until the onset of the next on-period of the square wave, and the whole process repeats.   □

Example 4.1.8: Consider the following two FIR filters, one defined in terms of its impulse response and the other in terms of its transfer function:

(a) $h_n = \begin{cases} ba^n, & \text{for } 0 \le n \le M \\ 0, & \text{otherwise} \end{cases}$

Take $M = 14$, $a = 0.75$, and $b = 1 - a = 0.25$. Its DC gain is almost unity:

$$y_{dc} = \sum_{m=0}^{M} h(m) = b \sum_{m=0}^{M} a^m = (1 - a) \cdot \frac{1 - a^{M+1}}{1 - a} = 1 - a^{M+1}$$

or, $y_{dc} = 1 - (0.75)^{15} = 0.987$.

(b) $H(z) = \frac{1}{5}(1 - z^{-1})^5 = 0.2 - z^{-1} + 2z^{-2} - 2z^{-3} + z^{-4} - 0.2z^{-5}$

This filter has $M = 5$ and acts as a 5-fold differentiator. Its impulse response can be extracted from $H(z)$:

$$\mathbf{h} = [0.2, -1, 2, -2, 1, -0.2] = \frac{1}{5}[1, -5, 10, -10, 5, -1]$$

The factor $1/5$ serves only as a convenient scale. Its DC gain is zero.

The square wave input of the previous example is fed into the two filters. The resulting output signals, computed by two calls to conv, are shown in the figure below:

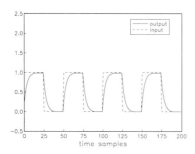 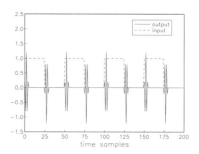

Filter (a) acts more like an RC-type integrator than an accumulator. The exponentially decaying nature of the impulse response causes the *charging/discharging* type of output as the input goes on and off.

Filter (b) acts as a differentiator, differentiating the constant portions (i.e., the on portions) of the input to zero. The input-on and off transients have duration of $M = 5$, but the rest of the output is zero.   □

Example 4.1.9: To demonstrate the concepts of impulse response, linearity, and time invariance, consider an FIR filter with impulse response

$$h(n) = (0.95)^n, \qquad \text{for } 0 \le n \le 24$$

and an input signal

$$x(n) = \delta(n) + 2\delta(n-40) + 2\delta(n-70) + \delta(n-80), \qquad n = 0, 1, \ldots, 120$$

consisting of four impulses of the indicated strengths occurring at the indicated time instants. Note that the first two impulses are separated by more than the duration of the filter, whereas the last two are separated by less.

Using the LTI form of convolution, we obtain the filter output by replacing each delayed impulse by the delayed impulse response, that is,

$$y(n) = h(n) + 2h(n-40) + 2h(n-70) + h(n-80) \qquad (4.1.25)$$

The input signal can be generated with the aid of the following C routine that implements a delta function $\delta(n)$:

```
/* delta.c - delta function */

double delta(n)
int n;
{
        if (n == 0)
                return 1;
        else
                return 0;
}
```

The input signal can be generated by a for-loop of the form

```
for (n=0; n<=120; n++)
    x[n] = delta(n) + 2*delta(n-40) + 2*delta(n-70) + delta(n-80);
```

The corresponding output signal will have length $L_y = L + M = 121 + 24 = 145$, and can be generated by single call to conv:

```
conv(24, h, 121, x, y);
```

The output is shown below, together with the impulsive input.

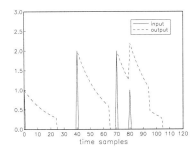

Each impulse of the input generates a copy of the impulse response at the output. The outputs due to the first and second terms of Eq. (4.1.25) do not overlap, but the outputs of the last two terms do.   □

4.1.10 Overlap-Add Block Convolution Method

In the above examples, the entire input signal was passed to conv as a *single* block of samples. This is not feasible in those applications where the input is infinite or extremely long. A practical approach is to divide the long input into *contiguous* non-overlapping blocks of manageable length, say L samples, then filter each block and piece the output blocks together to obtain the overall output, as shown in Fig. 4.1.6. Thus, processing is carried out block by block.

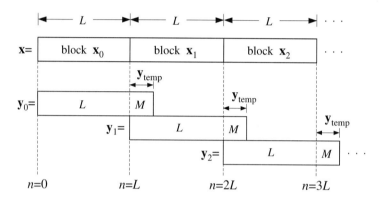

Fig. 4.1.6 Overlap-add block convolution method.

This is known as the *overlap-add* method of block convolution. Each of the input sub-blocks x_0, x_1, x_2, ..., is convolved with the order-M filter h producing the outputs blocks:

$$y_0 = h * x_0$$

$$y_1 = h * x_1 \tag{4.1.26}$$

$$y_2 = h * x_2$$

and so on. The resulting blocks are pieced together according to their absolute timing. Block y_0 starts at absolute time $n = 0$; block y_1 starts at $n = L$ because the corresponding input block x_1 starts then; block y_2 starts at $n = 2L$, and so forth.

Because each output block is longer than the corresponding input block by M samples, the *last* M samples of each output block will *overlap* with the first M outputs of the *next* block. Note that only the next sub-block will be involved if we assume that $2L > L + M$, or, $L > M$. To get the correct output points, the overlapped portions must be added together (hence the name, overlap-add).

Example 4.1.10: Compute the output of Example 4.1.1 using the overlap-add method of block convolution. Use input blocks of length $L = 3$. Perform the required individual convolutions of Eq. (4.1.26) using the convolution table.

Solution: The input is divided into the following three contiguous blocks

$$\mathbf{x} = [\underbrace{1,1,2}_{\mathbf{x}_0}, \underbrace{1,2,2}_{\mathbf{x}_1}, \underbrace{1,1,0}_{\mathbf{x}_2}]$$

where we padded an extra zero at the end of $\mathbf{x}_2$ to get a length-3 block. Convolving each block separately with $\mathbf{h} = [1, 2, -1, 1]$ gives:

$$\mathbf{y}_0 = \mathbf{h} * \mathbf{x}_0 = [1, 3, 3, 4, -1, 2]$$

$$\mathbf{y}_1 = \mathbf{h} * \mathbf{x}_1 = [1, 4, 5, 3, 0, 2]$$

$$\mathbf{y}_2 = \mathbf{h} * \mathbf{x}_2 = [1, 3, 1, 0, 1, 0]$$

These convolutions can be done by separately folding the three convolution subtables:

h\x	block 0			block 1			block 2		
	1	1	2	1	2	2	1	1	0
1	1	1	2	1	2	2	1	1	0
2	2	2	4	2	4	4	2	2	0
-1	-1	-1	-2	-1	-2	-2	-1	-1	0
1	1	1	2	1	2	2	1	1	0

The three sub-blocks begin at the absolute times $n = 0, 3, 6$, respectively. It follows from time invariance that the corresponding output blocks will also begin at the same absolute times. Thus, aligning the output blocks according to their absolute timings and adding them up gives the final result:

n	0	1	2	3	4	5	6	7	8	9	10
$\mathbf{y}_0$	1	3	3	4	-1	2					
$\mathbf{y}_1$				1	4	5	3	0	2		
$\mathbf{y}_2$							1	3	1	0	1
$\mathbf{y}$	1	3	3	5	3	7	4	3	3	0	1

which agrees with Example 4.1.1. □

The method can be implemented by the following algorithm, which reads the input data in blocks $\mathbf{x}$ of length L and outputs the result also in blocks of length L:

> *for each length-L input block* $\mathbf{x}$ *do:*
>   1. *compute length-$(L+M)$ output:* $\mathbf{y} = \mathbf{h} * \mathbf{x}$
>   2. *for $i = 0, 1, \ldots, M-1$:*
> $y(i) = y(i) + y_{\text{temp}}(i)$ (overlap)
> $y_{\text{temp}}(i) = y(i + L)$ (save tail)
>   3. *for $i = 0, 1, \ldots, L-1$:*
>   *output* $y(i)$

It uses a temporary M-dimensional vector $\mathbf{y}_{\text{temp}}$ to store the last M samples of each *previous* block. Before processing the first block, $\mathbf{y}_{\text{temp}}$ must be initialized to zero.

After computing the length-$(L+M)$ filter output $\mathbf{y} = \mathbf{h} * \mathbf{x}$, the first M samples of $\mathbf{y}$ are added to the last M samples of the previous block held in $\mathbf{y}_{temp}$. Then, the last M samples of the currently computed block $\mathbf{y}$ are saved in $\mathbf{y}_{temp}$ for use in the next iteration. Only the first L corrected output samples of $\mathbf{y}$ are sent to the output.

In practice this method is implemented efficiently by computing the individual block convolutions using the FFT instead of time-domain convolution. For an FIR filter of order M and an FFT of length N (which is a power of two), the length of each input block $\mathbf{x}$ is chosen to be $L = N - M$. We will see later that the computational gain of this *fast convolution* method versus the conventional time-domain "slow" method is approximately

$$\frac{\text{fast}}{\text{slow}} = \frac{\log_2 N}{M}$$

For example, for the values $N = 1024 = 2^{10}$ and $M = 100$, we have $\log_2 N/M = 10/100 = 1/10$, gaining a factor of 10 in computational speed. There is also an alternative fast convolution method called the *overlap-save* method that has comparable performance to the overlap-add method. We will also discuss it later.

The following routine `blockcon.c` is an implementation of the above algorithm. It calls the routine `conv` to perform the convolution of each input block. In using this routine, some care must be exercised in handling the very last input block, which in general will have length less than L.

```
/* blockcon.c - block convolution by overlap-add method */

void conv();

void blockcon(M, h, L, x, y, ytemp)
double *h, *x, *y, *ytemp;              ytemp is tail of previous block
int M, L;                               M = filter order, L = block size
{
      int i;

      conv(M, h, L, x, y);              compute output block y

      for (i=0; i<M; i++) {
         y[i] += ytemp[i];              add tail of previous block
         ytemp[i] = y[i+L];             update tail for next call
         }
}
```

The quantities `h, x, y, ytemp` are arrays and must be allocated in the main program, for example, using `calloc`:

```
double *h, *x, *y, *ytemp;
h = (double *) calloc(M+1, sizeof(double));     (M+1)-dimensional
x = (double *) calloc(L,   sizeof(double));     L-dimensional
y = (double *) calloc(L+M, sizeof(double));     (L+M)-dimensional
ytemp = (double *) calloc(M, sizeof(double));   M-dimensional
```

To illustrate the usage of such a routine, suppose the input samples to be filtered have been stored sequentially[†] in a data file `x.dat`. Suppose also that the computed samples will be stored in the output file `y.dat`.

[†] Separated by white space, such as blanks, tabs, or newlines.

The following program segment keeps reading samples from the file `x.dat` in *blocks* of length L. For each such input block, it calls the routine `blockcon` to compute and save the output block in the file `y.dat`. When the end-of-file of `x.dat` is encountered, it determines the length of the last input block and calls `blockcon` one more time to process the last block.

```
for (;;) {                                        keep reading input blocks
        for (N=0; N<L; N++)
             if (fscanf(fpx, "%lf", x+N) == EOF) goto last;

        blockcon(M, h, L, x, y, ytemp);           process input block

        for (i=0; i<L; i++)                       write output block
             fprintf(fpy, "%lf\n", y[i]);
        }
last:
        blockcon(M, h, N, x, y, ytemp);           last block has N ≤ L

        for (i=0; i<N+M; i++)                     last output block
             fprintf(fpy, "%lf\n", y[i]);
```

Note that `x+N` stands for the address of `x[N]`, that is, `&x[N]`. The function `fscanf` returns EOF upon encountering the end of file `x.dat`. The last processed block has length $N \leq L$. The entire last output block of length $(N+M)$ is written into the output file. The last M output samples represent the input-off transients. The file pointers, `fpx`, `fpy`, must be declared and defined in the main program by:

```
FILE *fpx, *fpy;                file pointers
fpx = fopen("x.dat", "r");      open for read
fpy = fopen("y.dat", "w");      open for write
```

4.2 Sample Processing Methods

Convolution methods process the input signal on a *block-by-block* basis. Here, we discuss alternative formulations of FIR filters that operate on a *sample-by-sample* basis. As we mentioned earlier, such methods are convenient for *real-time applications* that require the continuous processing of the incoming input.

Sample processing algorithms are closely related to *block diagram* realizations of the I/O filtering equations. A block diagram is a *mechanization* of the I/O equation in terms of the three basic *building blocks*: adders, multipliers, and delays, shown in Fig. 4.2.1.

In general, a filter may have several *equivalent* block diagram realizations depending on how its I/O equation is organized. Each realization gives rise to its own sample processing algorithm. Some standard filter realizations are the *direct, canonical*, and *cascade* forms and the corresponding *transposed* versions. We will discuss them systematically in a later chapter. In this chapter, we consider only the *direct form* for FIR filters; its transpose is discussed in the Problems.

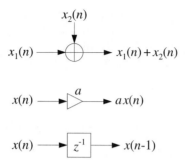

Fig. 4.2.1 Basic building blocks of DSP systems.

4.2.1 Pure Delays

As an introduction to the concept of a sample processing algorithm, consider the case of a single delay, shown in Fig. 4.2.2. It is an LTI system with I/O relationship:

$$y(n) = x(n - 1)$$

It can be thought of as a *register* holding the *previous* input sample $x(n-1)$. At each time instant n, two steps must be carried out: (a) the current content $x(n-1)$ is clocked out to the output and (b) the current input $x(n)$ gets stored in the register, where it will be held for one sampling instant and become the output at the *next* time $n+1$.

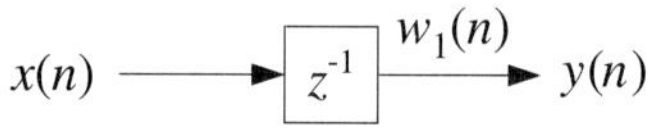

$$x(n) \longrightarrow \boxed{z^{-1}} \overset{w_1(n)}{\longrightarrow} y(n)$$

Fig. 4.2.2 Single delay.

We can think of the content of the delay register at time n as the *internal state* of the filter. Let us denote it by

$$w_1(n) = x(n - 1) \qquad \text{(internal state at time } n)$$

Thus, the output is $y(n) = w_1(n)$. Replacing n by $n+1$, we obtain the content of the register at the next time instant,

$$w_1(n + 1) = x(n) \qquad \text{(internal state at time } n+1)$$

The two processing steps (a) and (b) can be expressed then as follows:

$$\boxed{\begin{aligned} y(n) &= w_1(n) \\ w_1(n + 1) &= x(n) \end{aligned}} \tag{4.2.1}$$

In words, at time n the content of the register $w_1(n)$ becomes the output and the input $x(n)$ is saved and becomes the new content. At time $n+1$, the two steps are repeated:

$$y(n + 1) = w_1(n + 1)$$

$$w_1(n + 2) = x(n + 1)$$

The internal state $w_1(n + 1)$ is available from the *previous* time step when it was saved; the current input $x(n + 1)$ is also available and gets saved for the next time step. Before processing the first input sample, the delay register is typically *initialized* to zero, that is, at time $n = 0$ it contains

$$w_1(0) = 0$$

The following table shows the values of the input $x(n)$, the internal state $w_1(n)$, and the output $y(n)$ at different time instants:

n	$x(n)$	$w_1(n)$	$y(n)$
0	x_0	0	0
1	x_1	x_0	x_0
2	x_2	x_1	x_1
3	x_3	x_2	x_2
4	x_4	x_3	x_3
$\vdots$	$\vdots$	$\vdots$	$\vdots$

Thus, the input sequence gets delayed as a whole by one time unit:

$$[x_0, x_1, x_2, x_3, \dots] \xrightarrow{H} [0, x_0, x_1, x_2, x_3, \dots]$$

The two steps of Eq. (4.2.1), representing the output and the state updating, can be expressed in the following algorithmic form, which applies repetitively to every input sample:

> *for each input sample x do:*
> $\quad y := w_1$
> $\quad w_1 := x$

This is the sample-by-sample processing algorithm implementing a single delay. Consider next a double delay, depicted in Fig. 4.2.3. Its I/O equation is

$$y(n) = x(n - 2)$$

Now, there are two registers holding the *previous two* input samples. Denoting the contents of the two registers by $w_1(n)$ and $w_2(n)$, we note that $w_1(n)$ is the delayed version of the input $x(n)$, and $w_2(n)$ the delayed version of $w_1(n)$:

$$w_2(n) = w_1(n - 1)$$

$$w_1(n) = x(n - 1)$$

(4.2.2)

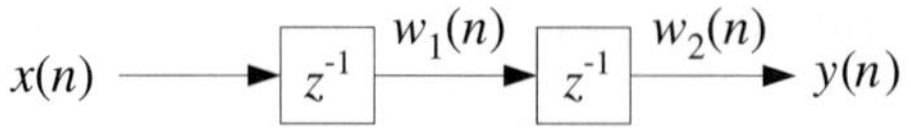

Fig. 4.2.3 Double delay.

Therefore, $w_2(n)$ is the doubly delayed version of $x(n)$:

$$w_2(n) = w_1(n-1) = x((n-1)-1) = x(n-2)$$

At time n, $w_2(n)$ becomes the output, $y(n) = w_2(n)$, and the contents of the two registers are updated in preparation for the next time step, that is,

$$w_2(n+1) = w_1(n)$$

$$w_1(n+1) = x(n)$$

In words, the *next* contents of the two registers are obtained by shifting w_1 into w_2, and x into w_1. In summary, the I/O equations describing the double delay are:

$$\boxed{\begin{aligned} y(n) &= w_2(n) \\ w_2(n+1) &= w_1(n) \\ w_1(n+1) &= x(n) \end{aligned}}$$

The repetitive sample processing algorithm describing these equations is:

$$\boxed{\begin{aligned} &\textit{for each input sample x do:} \\ &\quad y := w_2 \\ &\quad w_2 := w_1 \\ &\quad w_1 := x \end{aligned}}$$

Note, that the *order of updating* the internal states is important: the *last* delay must always be updated first. Once the current value of w_1 has been shifted into w_2, the value of w_1 may be overwritten by x.

The following table shows the values of $x(n)$, the contents of the two registers $w_1(n)$, $w_2(n)$, and the output $y(n)$ at different times (with zero initial values $w_1(0) = w_2(0) = 0$):

n	$x(n)$	$w_1(n)$	$w_2(n)$	$y(n)$
0	x_0	0	0	0
1	x_1	x_0	0	0
2	x_2	x_1	x_0	x_0
3	x_3	x_2	x_1	x_1
4	x_4	x_3	x_2	x_2
5	x_5	x_4	x_3	x_3
$\vdots$	$\vdots$	$\vdots$	$\vdots$	$\vdots$

A triple delay, shown in Fig. 4.2.4, can be described in a similar fashion. Let $w_1(n)$, $w_2(n)$, and $w_3(n)$ denote the contents of the three registers at time n. They are successive delays of each other, that is,

$$w_3(n) = w_2(n-1)$$

$$w_2(n) = w_1(n-1)$$

$$w_1(n) = x(n-1)$$

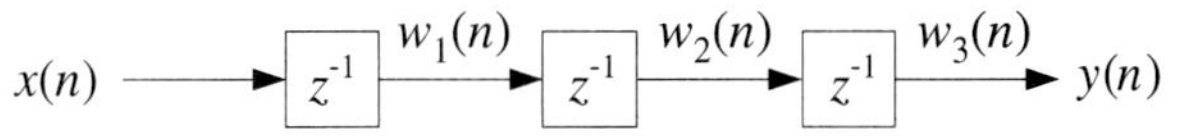

Fig. 4.2.4 Triple delay.

Thus, $w_3(n) = w_2(n-1) = w_1(n-2) = x(n-3)$. Their updates to time $n+1$ are:

$$w_3(n+1) = w_2(n)$$

$$w_2(n+1) = w_1(n) \qquad (4.2.3)$$

$$w_1(n+1) = x(n)$$

Therefore, the I/O equations for a triple delay will be:

$$\boxed{\begin{aligned} y(n) &= w_3(n) \\ w_3(n+1) &= w_2(n) \\ w_2(n+1) &= w_1(n) \\ w_1(n+1) &= x(n) \end{aligned}}$$

And the corresponding sample processing algorithm:

$$\boxed{\begin{aligned} &\textit{for each input sample x do:} \\ &\quad y := w_3 \\ &\quad w_3 := w_2 \\ &\quad w_2 := w_1 \\ &\quad w_1 := x \end{aligned}}$$

In general, for a delay by D units of time, shown in Fig. 4.2.5, the contents of the D registers are denoted by $w_i(n)$, $i = 1, 2, \ldots, D$. For convenience, the input is denoted by $w_0(n)$. The output of each register is the delayed version of its input:

$$\boxed{w_i(n) = w_{i-1}(n-1), \qquad \text{for} \quad i = 1, 2, \ldots, D} \qquad (4.2.4)$$

At time n, the content of the Dth register is output, $y(n) = w_D(n)$. Then, in preparation for the next time step, the content of the w_{D-1} register is shifted into w_D, the content of w_{D-2} is shifted into w_{D-1}, and so on, and finally the current

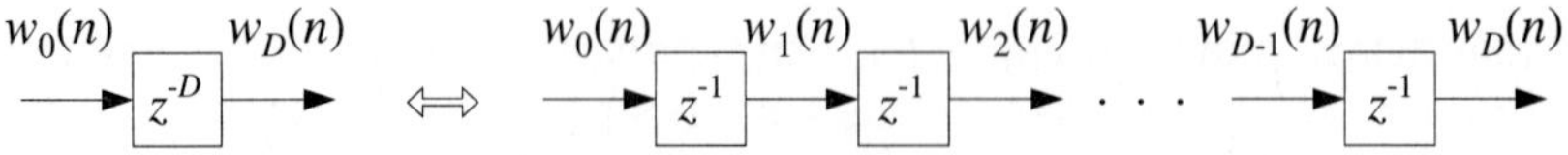

Fig. 4.2.5 *D*-unit delay.

input w_0 is shifted into w_1. These updates may be expressed by replacing n by $n+1$ in Eq. (4.2.4) and reversing the order of the equations. The complete set of I/O equations describing a D-delay becomes:

$$
\begin{aligned}
&y(n) = w_D(n) \\
&w_0(n) = x(n) \\
&w_i(n+1) = w_{i-1}(n), \qquad i = D, D-1, \ldots, 2, 1
\end{aligned}
$$

The corresponding sample processing algorithm will be:

> *for each input sample x do:*
> $y := w_D$
> $w_0 := x$
>   *for* $i = D, D-1, \ldots, 1$ *do:*
> $w_i := w_{i-1}$

or, more simply:

> *for each input sample w_0 do:*
>   *for* $i = D, D-1, \ldots, 1$ *do:*
> $w_i := w_{i-1}$

The following C routine `delay.c` is an implementation of this algorithm:

```
/* delay.c - delay by D time samples */

void delay(D, w)                              w[0] = input, w[D] = output
int D;
double *w;
{
        int i;

        for (i=D; i>=1; i--)                  reverse-order updating
                w[i] = w[i-1];

}
```

The array w has dimension $D+1$ and must be allocated in the main program by a statement of the form:

```
double *w;
w = (double *) calloc(D+1, sizeof(double));       (D+1)-dimensional
```

The array w serves both as input and output of the routine. Upon exit, w is the shifted version of itself. Prior to the first call of this routine, the array w must be initialized to zero. This is indirectly accomplished by `calloc`. The usage of the routine is illustrated by the following program segment, which implements the I/O equation $y(n) = x(n - D)$, for $n = 0, 1, \ldots, N_{\text{tot}} - 1$:

```
for (n = 0; n < Ntot; n++) {
        y[n] = w[D];           write output
        w[0] = x[n];           read input
        delay(D, w);           update delay line
        }
```

We will use this routine to implement FIR and IIR filters and also, in cascaded and feedback arrangements, to implement several digital audio effects, such as digital reverb, and to implement periodic waveform generators.

When used in feedback configurations, we note that its output w[D] is available even *before* the input w[0]. This is illustrated to some degree by the above program segment, where the output is returned before the input is read into w[0]. However, the current input w[0] must be known before the delay line can be updated by the call to `delay`.

4.2.2 FIR Filtering in Direct Form

We saw in Eq. (4.1.23) that the direct form I/O convolutional equation for an FIR filter of order M is given by

$$y(n) = h_0 x(n) + h_1 x(n - 1) + \cdots + h_M x(n - M) \qquad (4.2.5)$$

with impulse response $\mathbf{h} = [h_0, h_1, \ldots, h_M]$. For example, a third-order filter

$$\mathbf{h} = [h_0, h_1, h_2, h_3]$$

will have I/O equation:

$$y(n) = h_0 x(n) + h_1 x(n - 1) + h_2 x(n - 2) + h_3 x(n - 3) \qquad (4.2.6)$$

In order to mechanize this equation, we need to use an *adder* to accumulate the sum of products in the right-hand side; we need *multipliers* to implement the multiplications by the filter weights; and, we need *delays* to implement the delayed terms $x(n - 1), x(n - 2), x(n - 3)$.

Fig. 4.2.6 shows a mechanization of Eq. (4.2.6). It is called a *direct form* realization because it directly realizes all the terms in the right-hand side. The four inputs to the adder are the four terms of the right-hand side of Eq. (4.2.6), and the output of the adder is the left-hand side.

The three delays are equivalent to the triple delay of Fig. 4.2.4, and therefore, we can introduce the same set of three *internal states* $w_1(n)$, $w_2(n)$, $w_3(n)$ to describe the contents of the three registers. Thus, we define:

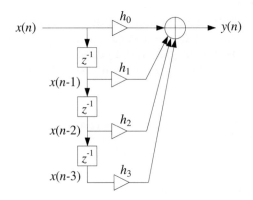

Fig. 4.2.6 Direct form realization of third-order filter.

$$w_0(n) = x(n)$$

$$w_1(n) = x(n-1) = w_0(n-1)$$

$$w_2(n) = x(n-2) = w_1(n-1)$$ (4.2.7)

$$w_3(n) = x(n-3) = w_2(n-1)$$

so that each is a delayed version of the previous one. With these definitions, we can rewrite Eq. (4.2.6) in the form:

$$y(n) = h_0 w_0(n) + h_1 w_1(n) + h_2 w_2(n) + h_3 w_3(n)$$ (4.2.8)

Fig. 4.2.7 shows the realization in this case. The advantage of this equation is that all the terms in the right-hand side refer to the *same* time instant n. All are available for processing at time n; that is, $w_0(n)$ is the current input $x(n)$, and $w_i(n)$, $i = 1, 2, 3$ are the current contents of the delay registers.

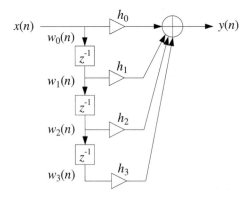

Fig. 4.2.7 Direct form with internal states.

The set of delays is sometimes called a *tapped delay line* because the individual outputs of each delay are tapped out and diverted into the filter multipliers.

Once the current output is computed, the delay registers may be updated to hold the values that will be needed at the next time instant $n+1$. The updating is implemented via Eq. (4.2.4), that is, by shifting from the *bottom up*, w_2 into w_3, w_1 into w_2, and w_0 into w_1. Thus, the I/O equation (4.2.6) is equivalent to the system:

$$
\begin{aligned}
&w_0(n) = x(n) \\
&y(n) = h_0 w_0(n) + h_1 w_1(n) + h_2 w_2(n) + h_3 w_3(n) \\
&w_3(n+1) = w_2(n) \\
&w_2(n+1) = w_1(n) \\
&w_1(n+1) = w_0(n)
\end{aligned}
\qquad (4.2.9)
$$

It can be mechanized by the following sample-by-sample processing algorithm:[†]

$$
\begin{aligned}
&\textit{for each input sample } x \textit{ do:} \\
&\quad w_0 = x \\
&\quad y = h_0 w_0 + h_1 w_1 + h_2 w_2 + h_3 w_3 \\
&\quad w_3 = w_2 \\
&\quad w_2 = w_1 \\
&\quad w_1 = w_0
\end{aligned}
\qquad (4.2.10)
$$

It is shown in Fig. 4.2.8. Thus, each input sample x is subjected to this algorithm and transformed to the output sample y. Before processing the first input sample, the internal states w_1, w_2, and w_3 must be initialized to *zero*.

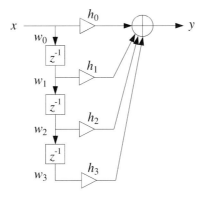

Fig. 4.2.8   Block diagram form of sample processing algorithm.

The input-off and input-on transient behavior of an FIR filter can be understood in terms of the block diagram realization. Initially, the delay registers are cleared

---

[†]For notational simplicity, we used = instead of :=.

to zero. During the input-on transients, the three delays gradually fill up with input samples. It takes $M = 3$ time units for that to happen. Similarly when the input turns off, it takes the last input sample M time units to propagate through the delays, that is, it takes M time units for the delays to empty out their contents and be filled with zeros again.

The following table shows the contents of the delays at different times and the corresponding outputs for the length-5 input given in Eq. (4.1.18):

n	x	w_0	w_1	w_2	w_3	$y = h_0 w_0 + h_1 w_1 + h_2 w_2 + h_3 w_3$
0	x_0	x_0	0	0	0	$h_0 x_0$
1	x_1	x_1	x_0	0	0	$h_0 x_1 + h_1 x_0$
2	x_2	x_2	x_1	x_0	0	$h_0 x_2 + h_1 x_1 + h_2 x_0$
3	x_3	x_3	x_2	x_1	x_0	$h_0 x_3 + h_1 x_2 + h_2 x_1 + h_3 x_0$
4	x_4	x_4	x_3	x_2	x_1	$h_0 x_4 + h_1 x_3 + h_2 x_2 + h_3 x_1$
5	0	0	x_4	x_3	x_2	$h_1 x_4 + h_2 x_3 + h_3 x_2$
6	0	0	0	x_4	x_3	$h_2 x_4 + h_3 x_3$
7	0	0	0	0	x_4	$h_3 x_4$

(4.2.11)

Each column of w's is the delayed (down-shifted) version of the previous one. Each row of w's is the delayed (right-shifted) version of the previous row. The computed outputs agree with Eq. (4.1.18). The three zeros padded at the end of the input correspond to the input-off transients. Note also that the four w columns are essentially the data matrix X of Eq. (4.1.21).

More generally, for an Mth order filter, we may define $w_0(n) = x(n)$ and for $i = 1, 2, \ldots, M$

$$w_i(n) = x(n - i) \tag{4.2.12}$$

They satisfy

$$w_i(n) = w_{i-1}(n - 1), \qquad i = 1, 2, \ldots, M \tag{4.2.13}$$

Indeed, $w_{i-1}(n - 1) = x((n-1)-(i-1)) = x(n-i) = w_i(n)$. Therefore, at the next time instant:

$$w_i(n + 1) = w_{i-1}(n), \qquad i = 1, 2, \ldots, M$$

It follows that Eq. (4.2.5) can be written as

$$y(n) = h_0 w_0(n) + h_1 w_1(n) + \cdots + h_M w_M(n)$$

Thus, the FIR filter Eq. (4.2.5) is described by the following system:

$$\boxed{\begin{aligned} w_0(n) &= x(n) \\ y(n) &= h_0 w_0(n) + h_1 w_1(n) + \cdots + h_M w_M(n) \\ w_i(n+1) &= w_{i-1}(n), \qquad \text{for } i = M, M-1, \ldots, 1 \end{aligned}} \tag{4.2.14}$$

with corresponding sample processing algorithm:

> *for each input sample x do:*
> $w_0 = x$
> $y = h_0 w_0 + h_1 w_1 + \cdots + h_M w_M$
> *for* $i = M, M-1, \ldots, 1$ *do:*
> $w_i = w_{i-1}$

(4.2.15)

Fig. 4.2.9 shows the corresponding direct form realization.

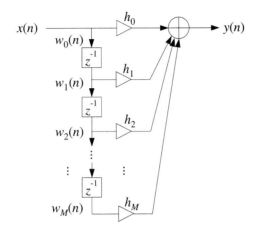

Fig. 4.2.9 Direct form realization of Mth order filter.

Example 4.2.1: Determine the sample processing algorithm of Example 4.1.1, which had filter and input

$$\mathbf{h} = [1, 2, -1, 1], \qquad \mathbf{x} = [1, 1, 2, 1, 2, 2, 1, 1]$$

Then, using the algorithm compute the corresponding output, including the input-off transients.

Solution: The I/O equation of this filter is

$$y(n) = x(n) + 2x(n-1) - x(n-2) + x(n-3)$$

Introducing the internal states $w_i(n) = x(n-i)$, $i = 1, 2, 3$, and setting $w_0(n) = x(n)$, we obtain the following system describing the output equation and the state updating:

$$w_0(n) = x(n)$$
$$y(n) = w_0(n) + 2w_1(n) - w_2(n) + w_3(n)$$
$$w_3(n+1) = w_2(n)$$
$$w_2(n+1) = w_1(n)$$
$$w_1(n+1) = w_0(n)$$

The corresponding block diagram realization and sample processing algorithm are shown below:

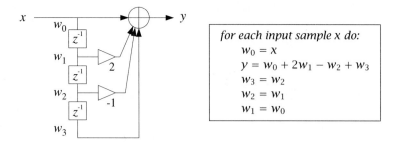

The sample processing algorithm generates the following output samples:

n	x	w_0	w_1	w_2	w_3	$y = w_0 + 2w_1 - w_2 + w_3$
0	1	1	0	0	0	1
1	1	1	1	0	0	3
2	2	2	1	1	0	3
3	1	1	2	1	1	5
4	2	2	1	2	1	3
5	2	2	2	1	2	7
6	1	1	2	2	1	4
7	1	1	1	2	2	3
8	0	0	1	1	2	3
9	0	0	0	1	1	0
10	0	0	0	0	1	1

The first three outputs correspond to the input-on transients (the internal delay registers are still filling up). The period $3 \le n \le 7$ corresponds to steady state (all delays are filled). The last three outputs—in general, the last M outputs for an Mth order FIR filter—are the input-off ($x = 0$) transients (the delays gradually empty out their contents).   □

Example 4.2.2: To illustrate the repetitive nature of the sample processing algorithm, we present a small C program that implements the previous example.

```
/* firexmpl.c - Example of FIR sample processing algorithm */

#include <stdio.h>
#include <stdlib.h>                              declares calloc

double x[8] = {1,1,2,1,2,2,1,1};                 input signal

double filter();

void main()
{
        int n;
        double y, *w;

        w = (double *) calloc(4, sizeof(double)); allocate/initialize w
```

```
            for (n=0; n<8; n++) {                          on-transients & steady state
                    y = filter(x[n], w);                   nth output sample
                    printf("%lf\n", y);
                    }

            for (n=8; n<11; n++) {                         input-off transients
                    y = filter(0.0, w);                    called with x = 0
                    printf("%lf\n", y);
                    }
    }                                                      end of main

double filter(x, w)                                        Usage: y = filter(x, w);
double x, *w;
{
        double y;

        w[0] = x;                                          read input sample

        y = w[0] + 2 * w[1] - w[2] + w[3];                 compute output sample

        w[3] = w[2];                                       update internal states
        w[2] = w[1];
        w[1] = w[0];

        return y;
}
```

The sample processing algorithm is implemented by the routine `filter` whose input is the current input sample and the internal states w. At each call, it returns the computed output sample and the updated state vector w. The routine `filter` is called 8 times (i.e., the length of the input) producing the first 8 outputs. Then, it is called 3 more times ($M = 3$), to generate the input-off transients. The total number of output samples is $L_y = L + M = 11$. □

Example 4.2.3: Draw the direct form realization and write the corresponding sample processing algorithm of the FIR filter defined by the I/O equation:

$$y(n) = x(n) - x(n-4)$$

For the input $\mathbf{x} = [1, 1, 2, 1, 2, 2, 1, 1]$, compute the output using the sample processing algorithm.

Solution: Because the filter has order $M = 4$, we define the following internal states:

$$w_0(n) = x(n)$$
$$w_1(n) = x(n-1) = w_0(n-1)$$
$$w_2(n) = x(n-2) = w_1(n-1)$$
$$w_3(n) = x(n-3) = w_2(n-1)$$
$$w_4(n) = x(n-4) = w_3(n-1)$$

Then, the given I/O equation together with the state-updating equations will read:

$$w_0(n) = x(n)$$
$$y(n) = w_0(n) - w_4(n)$$

and

$$w_4(n+1) = w_3(n)$$
$$w_3(n+1) = w_2(n)$$
$$w_2(n+1) = w_1(n)$$
$$w_1(n+1) = w_0(n)$$

This leads to the following sample processing algorithm and block diagram realization:

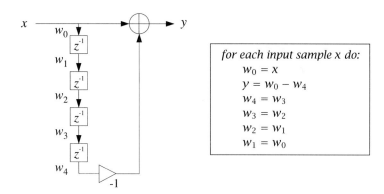

for each input sample x do:
$$w_0 = x$$
$$y = w_0 - w_4$$
$$w_4 = w_3$$
$$w_3 = w_2$$
$$w_2 = w_1$$
$$w_1 = w_0$$

The following table shows the computation of the output.

n	x	w_0	w_1	w_2	w_3	w_4	$y = w_0 - w_4$
0	1	1	0	0	0	0	1
1	1	1	1	0	0	0	1
2	2	2	1	1	0	0	2
3	1	1	2	1	1	0	1
4	2	2	1	2	1	1	1
5	2	2	2	1	2	1	1
6	1	1	2	2	1	2	−1
7	1	1	1	2	2	1	0
8	0	0	1	1	2	2	−2
9	0	0	0	1	1	2	−2
10	0	0	0	0	1	1	−1
11	0	0	0	0	0	1	−1

The 4 zeros padded at the end of the input correspond to the input-off transients, during which the contents of the delays gradually become empty. A similar program as the above `firexmpl.c` can be used to implement this example. The function `filter` will be in this case:

```
double filter(x, w)                          Usage: y = filter(x, w);
double x, *w;
{
        double y;
```

```
        w[0] = x;                              read input sample

        y = w[0] - w[4];                       compute output sample

        w[4] = w[3];                           update internal states
        w[3] = w[2];
        w[2] = w[1];
        w[1] = w[0];

        return y;
}
```

where w must be allocated as a 5-dimensional array in the main program. □

4.2.3 *Programming Considerations*

The following C routine `fir.c` is an implementation of the sample processing algorithm Eq. (4.2.15):

```
/* fir.c - FIR filter in direct form */

double fir(M, h, w, x)                    Usage: y = fir(M, h, w, x);
double *h, *w, x;                         h = filter, w = state, x = input sample
int M;                                    M = filter order
{
        int i;
        double y;                         output sample

        w[0] = x;                         read current input sample x

        for (y=0, i=0; i<=M; i++)
                y += h[i] * w[i];         compute current output sample y

        for (i=M; i>=1; i--)              update states for next call
                w[i] = w[i-1];            done in reverse order

        return y;
}
```

It is the generalization of the example routine `filter` to the Mth order case. It is patterned after the Fortran and C routines in [28,29]. The routine returns the computed output sample into a `double`, so that its typical usage will be of the form:

```
y = fir(M, h, w, x);
```

The filter vector h and internal state w are $(M+1)$-dimensional arrays, which must be defined and allocated in the main program by the statements:

```
double *h, *w;
h = (double *) calloc(M+1, sizeof(double));    (M+1)-dimensional
w = (double *) calloc(M+1, sizeof(double));    (M+1)-dimensional
```

Note that `calloc` initializes w to zero. The following program segment illustrates the usage of the routine. The input samples are read one at a time from an input file `x.dat`, and the output samples are written one at a time into the output file `y.dat`, as they are computed.

```
FILE *fpx, *fpy;
fpx = fopen("x.dat", "r");                    input file
fpy = fopen("y.dat", "w");                    output file

while(fscanf(fpx, "%lf", &x) != EOF) {        read x from x.dat
        y = fir(M, h, w, x);                  process x to get y
        fprintf(fpy, "%lf\n", y);             write y into y.dat
        }

for (i=0; i<M; i++) {                         M input-off transients
        y = fir(M, h, w, 0.0);                with x = 0
        fprintf(fpy, "%lf\n", y);
        }
```

Filtering stops as soon as the end of file of `x.dat` is detected and then the input-off transients are computed by making M additional calls to `fir` with zero input.

The `fir` routine performs three basic operations: (i) reading the current input sample, (ii) computing the current output by the *dot product* of the filter vector with the state vector, and (iii) updating the delay line containing the states. The dot product operation is defined by

$$y = h_0 w_0 + h_1 w_1 + \cdots + h_M w_M = [h_0, h_1, \ldots, h_M] \begin{bmatrix} w_0 \\ w_1 \\ \vdots \\ w_M \end{bmatrix} = \mathbf{h}^T \mathbf{w}$$

and can be implemented by the following routine `dot.c`:

```
/* dot.c - dot product of two length-(M+1) vectors */

double dot(M, h, w)                           Usage: y = dot(M, h, w);
double *h, *w;                                h = filter vector, w = state vector
int M;                                        M = filter order
{
        int i;
        double y;

        for (y=0, i=0; i<=M; i++)             compute dot product
                y += h[i] * w[i];

        return y;
}
```

The updating of the delay line can be implemented by the routine `delay` given earlier. Therefore, a second version of `fir`, which separates these three conceptual parts, is as follows:

```
/* fir2.c - FIR filter in direct form */

double dot();
void delay();

double fir2(M, h, w, x)          Usage: y = fir2(M, h, w, x);
double *h, *w, x;                h = filter, w = state, x = input
int M;                           M = filter order
{
        double y;

        w[0] = x;                read input

        y = dot(M, h, w);        compute output

        delay(M, w);             update states

        return y;
}
```

It has the same usage as `fir`. (See Appendix D for the MATLAB version `fir.m`.) The sample processing algorithm Eq. (4.2.15) reads in this case:

$$
\boxed{
\begin{array}{l}
\textit{for each input sample } x \textit{ do:} \\
\quad w_0 = x \\
\quad y = \mathrm{dot}(M, \mathbf{h}, \mathbf{w}) \\
\quad \mathrm{delay}(M, \mathbf{w})
\end{array}
}
\tag{4.2.16}
$$

4.2.4 Hardware Realizations and Circular Buffers

The FIR filtering algorithms of Eqs. (4.2.15) or (4.2.16) can be realized in hardware using DSP chips or special purpose dedicated hardware.

Modern programmable DSP chips, such as the Texas Instruments TMS320C25, C30, or C50, the Motorola DSP56001, or DSP96002, the AT&T DSP16A or DSP32C, and the Analog Devices ADSP-2101 or ADSP-21020, have architectures that are *optimized* for the specialized repetitive nature of sample-by-sample processing algorithms. They excel at performing the multiplications and accumulations required in computing the dot product $y = \mathbf{h}^T \mathbf{w}$, and at performing the memory moves required in updating the contents of the delay-line registers.

A generic DSP chip is shown in Fig. 4.2.10. The tapped delay-line registers w_i are sequential RAM locations on board the chip and the filter weights h_i reside either in RAM or ROM. In addition, there is program RAM or ROM on board (not shown in the figure) to hold the instructions for the filtering algorithm. A typical DSP chip may have data wordlengths of 16–32 bits, several double-precision accumulators, and on-board RAM and ROM of 512 to 4k words.

The workhorse of the DSP chip is an on-board multiplier accumulator (MAC), which implements the dot product multiplications/accumulations, that is, the operations:

$$y := y + h_i w_i$$

State-of-the-art DSP chips can perform this type of MAC operation in one instruction cycle in about:

$$T_{\text{instr}} = 30\text{--}80 \text{ nanoseconds} \tag{4.2.17}$$

Assuming a MAC operation counts for two floating point operations (one multiplication and one addition), this corresponds to a numerical computation speed of 25–67 million floating point operations per second (MFLOPS).

For an order-M filter having $M+1$ taps, one would require about $(M + 1)\,T_{\text{instr}}$ sec to calculate the required dot product. To this time, one must add the overhead required for shifting the input sample from the input port to the register w_0, the time required to update the delay-line registers, and the time it takes to send y to the output port.

The goal of modern DSP architectures has been to try to minimize this overhead as much as possible. To see the type of computational efficiencies built into DSP chips, let us rewrite the sample processing algorithm for a third-order filter given in Eq. (4.2.10) in the following equivalent form:

for each input sample x do:
$$w_0 := x$$
$$y := h_3 w_3$$
$$w_3 := w_2$$
$$y := y + h_2 w_2$$
$$w_2 := w_1$$
$$y := y + h_1 w_1$$
$$w_1 := w_0$$
$$y := y + h_0 w_0$$

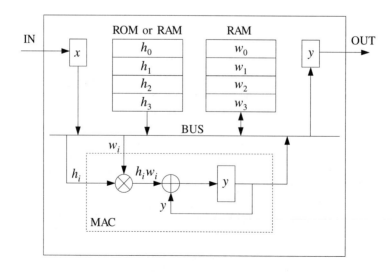

Fig. 4.2.10 Typical DSP chip.

This works because once the multiplication $h_3 w_3$ is performed, the current content of w_3 is no longer needed and can be updated to the next time instant. Similarly, once $h_2 w_2$ has been accumulated into y, w_2 may be updated, and so on. In general, for a filter of order M, we can rewrite Eq. (4.2.15) in the form:

$$
\boxed{
\begin{array}{l}
\textit{for each input sample x do:} \\
\quad w_0 := x \\
\quad y := h_M w_M \\
\quad \textit{for } i = M-1, \ldots, 1, 0 \textit{ do:} \\
\qquad w_{i+1} := w_i \\
\qquad y := y + h_i w_i
\end{array}
}
\tag{4.2.18}
$$

In earlier generations of DSP chips, the two operations:

$$w_{i+1} := w_i$$

$$y := y + h_i w_i$$

were carried out with two instructions, one for the data shifting and the other for the MAC operation. In modern DSP chips, the two operations can be carried out with a *single* instruction, such as MACD of the TMS320C25.

Therefore, the total processing time for each input sample is about T_{instr} per filter tap, or, for an Mth order filter:

$$
\boxed{T_{\text{proc}} = (M + 1)\, T_{\text{instr}}}
\tag{4.2.19}
$$

As discussed in Chapter 1, this imposes a maximum limit on the allowed sampling rate for the application:

$$
T \geq T_{\text{proc}} \quad \Rightarrow \quad f_s \leq \frac{1}{T_{\text{proc}}}
\tag{4.2.20}
$$

Example 4.2.4: What is the longest FIR filter that can be implemented with a 50 nsec per instruction DSP chip for digital audio applications?

Solution: We have from Eq. (4.2.20)

$$
T = (M + 1)\, T_{\text{instr}} \quad \Rightarrow \quad M + 1 = \frac{T}{T_{\text{instr}}} = \frac{1}{f_s T_{\text{instr}}} = \frac{f_{\text{instr}}}{f_s}
$$

where the *instruction rate* is $f_{\text{instr}} = 1/T_{\text{instr}} = 20$ million instructions per second (MIPS). For digital audio at $f_s = 44.1$ kHz, we find

$$
M + 1 = \frac{f_{\text{instr}}}{f_s} = \frac{20 \cdot 10^6}{44.1 \cdot 10^3} = 453 \text{ taps}
$$

This filter length is quite sufficient to implement several digital audio algorithms. □

The following C routine `fir3.c` is yet a third version of the sample-by-sample processing algorithm implementing Eq. (4.2.18). Its usage is the same as `fir`'s:

```
/* fir3.c - FIR filter emulating a DSP chip */

double fir3(M, h, w, x)
double *h, *w, x;
int M;
{
       int i;
       double y;

       w[0] = x;                                              read input

       for (y=h[M]*w[M], i=M-1; i>=0; i--) {
              w[i+1] = w[i];                                  data shift instruction
              y += h[i] * w[i];                               MAC instruction
              }

       return y;
}
```

The sample processing algorithm (4.2.18) and the routine `fir3` assume a *linear* delay-line memory buffer for holding the internal states $\{w_0, w_1, \ldots, w_M\}$. At each time instant, the data in the delay line are shifted one memory location ahead. This arrangement is used in some DSP processors, such as the TMS32020.

An alternative way to update the internal states is to use a *circular delay-line buffer*. This is used, for example, by the Motorola DSP56001/96002, the Texas Instruments TMS320C30–C50, and the Analog Devices ADSP2101–21020 processors. Instead of shifting the data forward while holding the buffer addresses fixed, the data are kept fixed and the addresses are shifted backwards in the circular buffer. The relative movement of data versus addresses remains the same.

To understand this, consider first the conventional linear delay-line buffer case, but wrap it around in a circle, as shown in Fig. 4.2.11 for the case $M = 3$.

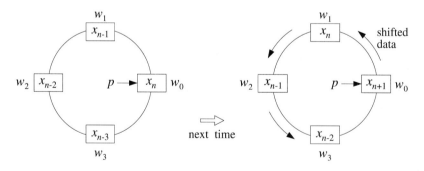

Fig. 4.2.11 Wrapped linear delay-line buffer.

Going from time n to $n+1$ involves shifting the content of each register counterclockwise into the next register. The addresses of the four registers $\{w_0, w_1, w_2, w_3\}$ remain the same, but now they hold the shifted data values, and the first register w_0 receives the next input sample x_{n+1}.

By contrast, in the circular buffer arrangement shown in Fig. 4.2.12, instead of shifting the data counterclockwise, the buffer addresses are decremented, or shifted clockwise once, so that w_3 becomes the new beginning of the circular buffer and will hold the next input x_{n+1}.

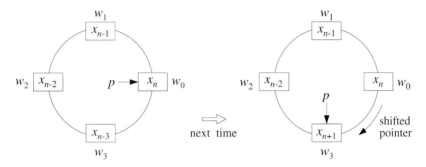

Fig. 4.2.12 Modulo-$(M+1)$ circular delay-line buffer.

The internal state vectors at times n and $n + 1$ are the *same* in both the linear and circular buffer implementations, namely,

$$\mathbf{s}(n) = \begin{bmatrix} x_n \\ x_{n-1} \\ x_{n-2} \\ x_{n-3} \end{bmatrix}, \qquad \mathbf{s}(n + 1) = \begin{bmatrix} x_{n+1} \\ x_n \\ x_{n-1} \\ x_{n-2} \end{bmatrix} \tag{4.2.21}$$

In both the linear and circular implementations, the starting address of the state vector is the current input sample, and from there, the addresses pointing to the rest of the state vector are incremented counterclockwise.

But, whereas in the linear case the starting address is always *fixed* and pointing to w_0, as shown in Fig. 4.2.11, the starting address in the circular case is back-shifted from one time instant to the next. To keep track of this changing address, we introduce a *pointer* variable p which always points to the current input, as shown in Fig. 4.2.12.

At each time instant, the w-register pointed to by p gets loaded with the current input sample, that is, $*p = x$, or $p[0] = x$. After computing the current output, the pointer p is decremented circularly. Figure 4.2.13 shows the position of the pointer p at successive sampling instants.

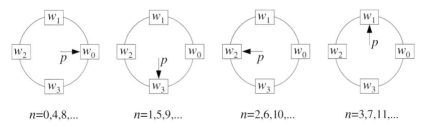

$n=0,4,8,...$ $\qquad$ $n=1,5,9,...$ $\qquad$ $n=2,6,10,...$ $\qquad$ $n=3,7,11,...$

Fig. 4.2.13 Successive positions of address pointer p, repeating modulo-$(M+1)$.

The pointer p is restricted to lie within the pointer range of the linear buffer w, that is, in C notation:

$$w \leq p \leq w + M \qquad (4.2.22)$$

and therefore, it will always point at some w-register, say $w[q]$,

$$p = w + q \quad \Rightarrow \quad *p = p[0] = w[q] \qquad (4.2.23)$$

where q is an integer that gives the offset of p with respect to the fixed beginning of the w-buffer. The restriction of Eq. (4.2.22) translates to the restriction on the range of q:

$$0 \leq q \leq M \qquad (4.2.24)$$

Inspecting the indices of the w's pointed to by the successive p's in Fig. 4.2.13, we may identify the periodically repeating sequence of values of q:

$$q = 0, 3, 2, 1, 0, 3, 2, 1, 0, 3, 2, 1, \ldots$$

and in general, q cycles over the values: $q = 0, M, M-1, \ldots, 1$.

At each time instant, the pointer p, or equivalently the offset q, defines the vector $\mathbf{s}$ of internal states. The sequence of pointers $p, p + 1, \ldots, p + M$, point at the components of the state vector, that is,

$$s_i = p[i] = *(p + i) = *(w + q + i) = w[q + i], \qquad i = 0, 1, \ldots, M$$

This definition is correct as long as the shifted pointer $p + i$ does not exceed the array bounds of w, or equivalently, as long as $q + i \leq M$. If $q + i > M$, it must be *reduced modulo-*$(M+1)$. Therefore, the correct definition of the internal state vector defined by the pointer p is, in C notation:

$$\boxed{s_i = w[(q + i)\%(M + 1)] = w[(p - w + i)\%(M + 1)]} \qquad (4.2.25)$$

for $i = 0, 1, \ldots, M$, where we solved Eq. (4.2.23) for $q = p - w$.[†] In particular, note that the first component of the state vector is:

$$s_0 = w[q] = p[0] = *p = *(w + q) \qquad (4.2.26)$$

and the last component, corresponding to $i = M$:

$$s_M = w[(q + M)\%(M + 1)] = w[(p - w + M)\%(M + 1)] \qquad (4.2.27)$$

Note that $s_M = w[M] = p[M]$ if $q = 0$, and $s_M = w[q - 1] = p[-1]$ otherwise. Therefore, s_M is always in the w-register that circularly *precedes* $w[q]$.

Assuming that the current input sample x has been read into $s_0 = w[q] = *p$, that is, $*p = x$, the corresponding output sample y will be computed by the dot product:

[†]Some C compilers may require the cast: q = (int) (p-w).

$$y = \sum_{i=0}^{M} h_i s_i = [h_0, h_1, \ldots, h_M] \begin{bmatrix} s_0 \\ s_1 \\ \vdots \\ s_M \end{bmatrix} = \mathbf{h}^T \mathbf{s}$$

As an example illustrating Eq. (4.2.25), Fig. 4.2.14 shows p at the time when it points to w_2, so that $p = w + 2$ and $q = 2$. We assume that w_2 has been loaded with the current input sample. In this case, the indices $q + i = 2 + i$ and their mod-4 reductions are, for $i = 0, 1, 2, 3$:

$$q + i = 2 + i = 2, 3, 4, 5 \quad \xrightarrow{\text{mod-4}} \quad 2, 3, 0, 1$$

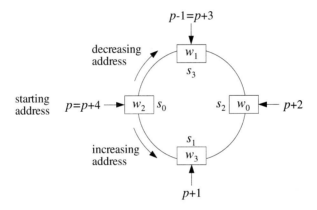

Fig. 4.2.14 Internal states defined by circular pointers $p + i$, $i = 0, 1, \ldots, M$.

Therefore, the state vector defined by this value of p will be

$$\mathbf{s} = \begin{bmatrix} s_0 \\ s_1 \\ s_2 \\ s_3 \end{bmatrix} = \begin{bmatrix} w_2 \\ w_3 \\ w_0 \\ w_1 \end{bmatrix}$$

The following table shows the succession of "rotating" state vectors of Fig. 4.2.13 and their contents, as they fill up with input samples x_n at the successive time instants $n = 0, 1, \ldots, 7$. An equivalent view is shown in Fig. 4.2.15. The w_i columns show the contents of the array $\mathbf{w} = [w_0, w_1, w_2, w_3]$ over which the pointer p circulates. At each time instant, *only one entry* in each row changes as it receives the new input sample, namely, the entry w_q. By contrast, in the linear buffer case given in Eq. (4.2.11), all entries of $\mathbf{w}$ shift. The first four s_i columns show the w-registers selected by p or q. The last four s_i columns show the actual contents of these w-registers:

n	q	w_0	w_1	w_2	w_3	s_0	s_1	s_2	s_3	s_0	s_1	s_2	s_3
0	0	x_0	0	0	0	w_0	w_1	w_2	w_3	x_0	0	0	0
1	3	x_0	0	0	x_1	w_3	w_0	w_1	w_2	x_1	x_0	0	0
2	2	x_0	0	x_2	x_1	w_2	w_3	w_0	w_1	x_2	x_1	x_0	0
3	1	x_0	x_3	x_2	x_1	w_1	w_2	w_3	w_0	x_3	x_2	x_1	x_0
4	0	x_4	x_3	x_2	x_1	w_0	w_1	w_2	w_3	x_4	x_3	x_2	x_1
5	3	x_4	x_3	x_2	x_5	w_3	w_0	w_1	w_2	x_5	x_4	x_3	x_2
6	2	x_4	x_3	x_6	x_5	w_2	w_3	w_0	w_1	x_6	x_5	x_4	x_3
7	1	x_4	x_7	x_6	x_5	w_1	w_2	w_3	w_0	x_7	x_6	x_5	x_4

$$(4.2.28)$$

It is evident that the contents of the columns s_0, s_1, s_2, s_3 are the successively delayed signal samples x_n, x_{n-1}, x_{n-2}, x_{n-3}. Therefore, the state vector $\mathbf{s}(n)$ at each time instant n is generated correctly according to Eq. (4.2.21).

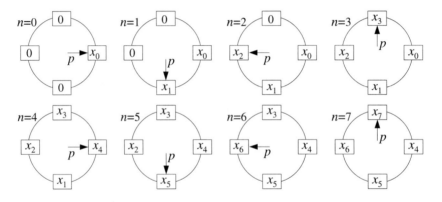

Fig. 4.2.15 Contents of circular buffer at successive time instants.

To make sure that p and its shifts $p+i$ always stay within the address space of w given by Eq. (4.2.22), they must be *wrapped* modulo-$(M+1)$. This is accomplished by the following routine wrap.c, which adjusts the bounds of p, such that if $p = w - 1$, it wraps it around to $p = (w-1)+(M+1) = w+M$, and if $p = w+M+1$, it wraps it to $p = (w+M+1)-(M+1) = w$.

```
/* wrap.c - circular wrap of pointer p, relative to array w */

void wrap(M, w, p)
double *w, **p;
int M;
{
        if (*p > w + M)
                *p -= M + 1;            when *p = w + M + 1, it wraps around to *p = w

        if (*p < w)
                *p += M + 1;            when *p = w - 1, it wraps around to *p = w + M
}
```

Note that p is modified at each call and serves both as an input *and* output of the routine; therefore, it must be passed *by reference*, that is, as a pointer to pointer. If in the main program w and p are declared as ordinary pointers:

```
double *w, *p;
```

then p must be passed into wrap by its address, that is,

```
wrap(M, w, &p);
```

With the help of the routine wrap, an FIR filter can be implemented using a circular delay-line buffer by the following routine cfir.c, which replaces fir:

```
/* cfir.c - FIR filter implemented with circular delay-line buffer */

void wrap();

double cfir(M, h, w, p, x)
double *h, *w, **p, x;                      p = circular pointer to w
int M;                                      M = filter order
{
        int i;
        double y;

        **p = x;                            read input sample x

        for (y=0, i=0; i<=M; i++) {         compute output sample y
                y += (*h++) * (*(*p)++);
                wrap(M, w, p);
                }

        (*p)--;                             update circular delay line
        wrap(M, w, p);

        return y;
}
```

The following three operations are carried out by the routine:

- The current input sample x is read into the w-register pointed to by the current value of the pointer p.

- The for-loop computes the filter's output sample y by accumulating the terms $h_i s_i$ of products of the filter coefficients with the components of the internal state vector defined by the pointer p.

    Each pass through the loop post-increments the h and p pointers and wraps p, if necessary. This loop could as well have been replaced by the following more obscure loop, which uses Eq. (4.2.25) for the states:

```
for (y=0, i=0; i<=M; i++)
        y += h[i] * w[(*p-w+i)%(M+1)];      that is, y = y + hᵢsᵢ
```
that is, $y = y + h_i s_i$

Upon exit from the loop, the pointer p has been circularly incremented $M+1$ times and therefore, it has wrapped around to its original value, that is, pointing again at the current input sample.

The filter pointer h is also incremented $M+1$ times and, after the loop, it points beyond its allowed range, but this does not matter because h will be reset at the next call of cfir. In hardware, h is also stored in a circular buffer, and therefore it wraps back to $h[0]$.

- Finally, the circular delay line is updated by simply decrementing the pointer p and wrapping it modulo $M+1$ if necessary. The pointer p is left pointing at the w-register containing the *last* component s_M of the state vector. This component will be overwritten by the next input sample.

In DSP chips that support circular or modulo addressing, each pass through the above for-loop requires only *one* instruction—the calls to wrap are not necessary because the incrementing pointer wraps around automatically. Therefore, the total number of instructions per call is essentially $M+1$. The total processing time per sample will be $T_{\text{proc}} = (M + 1)T_{\text{instr}}$.

Each call of cfir changes the value of the pointer p, and therefore, p must be passed by reference, as in the routine wrap. The arrays h and w must be declared and allocated in the main program in the usual way, and w must be initialized to zero. The pointer p is initialized to point to $w[0]$, that is, $p = w$. The following program segment illustrates the proper initialization and usage of cfir.

```
double *h, *w, *p;
h = (double *) calloc(M+1, sizeof(double));
w = (double *) calloc(M+1, sizeof(double));        also, initializes w to zero

p = w;                                              initialize p

for (n = 0; n < Ntot; n++)
        y[n] = cfir(M, h, w, &p, x[n]);             p passed by address
```

The routine cfir imitates the hardware implementation of FIR filtering on the Motorola DSP56K chip, as illustrated in Example 4.2.5. A slightly different version of cfir, which essentially emulates the TMS320C30, is given below:

```
/* cfir1.c - FIR filter implemented with circular delay-line buffer */

void wrap();

double cfir1(M, h, w, p, x)
double *h, *w, **p, x;
int M;
{
        int i;
        double y;

        *(*p)-- = x;
        wrap(M, w, p);                              p now points to sM

        for (y=0, h+=M, i=M; i>=0; i--) {           h starts at hM
```

```
y += (*h--) * (*(*p)--);
wrap(M, w, p);
}

return y;
}
```

After loading the current input sample into the w-register pointed to by p, it post-decrements p circularly, so that p becomes the wrapped $p + M$ and points to the last component s_M of the state vector.

The for-loop shifts the h pointer to point to h_M and then runs backwards from $i = M$ down to $i = 0$, accumulating the terms $h_i s_i$ and post-decrementing h and p at each pass.

Upon exit from the loop, p has wrapped around to point back to $p + M$ and is left pointing there upon exit from `cfir1`, but, that is where it should be pointing for processing the next input sample.

Example 4.2.5: It is beyond the scope of this book to discuss architectures and instruction sets for particular DSP chips. However, in order to illustrate the way `fir3`, `cfir`, and `cfir1` emulate assembly code for DSP chips, we present some code examples for the TMS32020, DSP32C, DSP56K, and TMS320C30; see Refs. [79–88] for details.

The following TMS32020 code segment implements the algorithm (4.2.18) for the case $M = 3$:

```
NEWX   IN W0, PA2          read new x into w0
       ZAC                 zero accumulator, y = 0

       LT  W3              load w3
       MPY H3              multiply by h3, y = h3w3

       LTD W2              w3 = w2, load and shift
       MPY H2              y = y + h2w2

       LTD W1              w2 = w1
       MPY H1              y = y + h1w1

       LTD W0              w1 = w0
       MPY H0              y = y + h0w0

       APAC                accumulate final sum
       SACH Y, 1           store accumulator in register Y
       OUT Y, PA2          output Y from output port

       B NEWX              branch back to NEWX to get next input sample
```

where, in the comments: $y = h_3 w_3$; $y = y + h_2 w_2$, $w_3 = w_2$; $y = y + h_1 w_1$, $w_2 = w_1$; $w_1 = w_0$; $y = y + h_0 w_0$.

The shift/MAC pairs of instructions LTD/MPY can be replaced by single MACD instructions.

The same filter would be implemented on the AT&T DSP32C floating point processor by the program segment:

```
a1 = *r4++ * *r2++              y = h3w3
a1 = a1 + (*r3++ = *r4++) * *r2++   y = y + h2w2, w3 = w2
a1 = a1 + (*r3++ = *r4++) * *r2++   y = y + h1w1, w2 = w1
```

where, in the comments: $y = h_3 w_3$; $y = y + h_2 w_2$, $w_3 = w_2$; $y = y + h_1 w_1$, $w_2 = w_1$.

```
a0 = a1 + (*r3 = *r5) * *r2++                y = y + h₀w₀, w₁ = w₀
*r6 = a0 = a0
```

$$a0 = a1 + (*r3 = *r5) * *r2++ \qquad y = y + h_0 w_0, \; w_1 = w_0$$
$$*r6 = a0 = a0$$

The address pointer r4 points to the internal states w_i and r2 to the filter weights h_i. The first line puts the product $h_3 w_3$ into accumulator a1 and increments the pointers to point to w_2, h_2. The second line accumulates the product $h_2 w_2$ into a1 and simultaneously shifts the value of w_2 pointed to by r4 into w_3 pointed to by r3, then post-increments the pointers.

In the fourth line, r5 points to the input data, that is, $w_0 = x$. The sum of the previous value of a1 and product $h_0 w_0$ are put into accumulator a0 and simultaneously w_0 is moved into w_1 pointed to by r3. In the last line, r6 points to the computed output sample y.

The following code implements the same filter on the Motorola DSP56K using a circular buffer. It is essentially equivalent to `cfir`.

```
clr   a         x0,x:(r0)+     y:(r4)+,y0
rep   #M
mac   x0,y0,a   x:(r0)+,x0     y:(r4)+,y0
macr  x0,y0,a   (r0)-
```

Here, the circular buffer resides in the chip's X-memory and is pointed to by the modulo pointer r0. The filter coefficients reside in Y-memory and are pointed to by the modulo pointer r4.

The `clr` instruction clears the accumulator register a, loads the temporary registers x0, y0 with the values w_0 and h_0, and increments the pointers r0, r4.

The `rep` instruction repeats the `mac` instruction M times. During the ith repetition, the registers x0, y0 hold the values of w_{i-1} and h_{i-1}; these values get multiplied and accumulated into the accumulator a, and then x0, y0 are loaded with w_i, h_i, and the modulo pointers r0, r4 are incremented.

Upon exit from this loop, the pointer r0 has been incremented $M+1$ times, and therefore it has wrapped around to point to w_0 again. The last `macr` instruction performs the final accumulation of $w_M h_M$, and then it *decrements* the pointer r0, so that it now points to w_{-1} which is the same as w_M. The value in this register will be overwritten by the next input sample. The total number of instruction cycles for this example is $(M+1)+3$, instead of the nominal $M+1$.

The floating point TMS320C30/C40 processor implements circular addressing in a similar fashion, as illustrated by the following code:

```
NEWX   LDF    IN,  R3                  read new input sample x
       STF    R3,   *AR1++%            put x in w₀ and increment AR1
       LDF    0.0, R0                  initialize accumulators
       LDF    0.0, R2

       RPTS   M                        repeat for i=M down to i=0
       MPYF3  *AR0++%, *AR1++%, R0     hᵢwᵢ → R₀, and in parallel
    || ADDF3  R0, R2, R2               accumulate previous R₀ into R₂

       ADDF   R0, R2                   accumulate last product
       STF    R2, Y                    store R₂ into output register Y
       B      NEWX                     branch to NEWX and repeat
```

Here, the filter coefficients h_i and internal states w_i are stored in reverse order, so that h_M, w_M are in the lowest and h_0, w_0 at the highest address of the modulo-$(M+1)$ circular buffers. Therefore, pointers are incremented, instead of decremented as in cfir1.

Upon entry, the address pointer AR0 points to the beginning of the h-buffer, that is, to h_M, and the pointer AR1 points to the bottom of the w-buffer, that is, to w_0 which receives the current input sample x_n by the first STF instruction. The AR1 pointer is then post-incremented and wraps around to point to the beginning of the w-buffer, that is, to w_M.

The RPTS loop repeats the following instruction $M+1$ times. The multiply instruction MPYF3 and accumulate instruction ADDF3 are done in parallel. The loop accumulates the terms $h_i w_i$, for $i = M, \ldots, 1, 0$. Each repetition post-increments the pointers AR0 and AR1. Therefore, after $M+1$ repetitions, AR0 and AR1 will wrap around and point to the beginning of the circular buffers.

Thus, AR1 is circularly incremented a total of $(M+1)+1$ times and will be left pointing to w_M, which will receive the next input sample x_{n+1}, as shown in Fig. 4.2.12.   □

The final part of cfir, which updates the circular delay line by modulo decrementing p, can be put by itself into a routine that implements the *circular* version of the delay routine delay.c of Section 4.2.1. Denoting M by D in this definition, we have:

```
/* cdelay.c - circular buffer implementation of D-fold delay */

void wrap();

void cdelay(D, w, p)
int D;
double *w, **p;
{
        (*p)--;                          decrement pointer and wrap modulo-(D + 1)
        wrap(D, w, p);                   when *p = w - 1, it wraps around to *p = w + D
}
```

Note that because p is decreasing, only the second half of wrap that tests the lower bound $w \le p$ is effective.

As in the case of the routine delay, the output of the delay line is available even before its input. This output is the last component of the internal state vector and is obtained from Eq. (4.2.27) with $M = D$:

$$s_D = w[(q+D)\%(D+1)] = w[(p-w+D)\%(D+1)]$$

Again, p must be passed by address into cdelay. The usage of the routine is illustrated by the following program segment, which implements the delay equation $y(n) = x(n-D)$:

```
p = w;                          initialize p

for (n = 0; n < Ntot; n++) {
        y[n] = w[(p-w+D)%(D+1)];       write output
        *p = x[n];                     read input; equivalently, p[0] = x[n]
        cdelay(D, w, &p);              update delay line
        }
```

The table in Eq. (4.2.28) illustrates this delay operation, with $D = 3$.

In the linear buffer implementations of `fir` and `delay`, the state vector is w itself, that is, $\mathbf{s} = \mathbf{w}$, and its components are directly accessible as $s_i = w[i]$, for $i = 0, 1, \ldots, D$. In the circular buffer case, the state vector components are given by Eq. (4.2.25). To avoid having to write the complicated expressions of Eq. (4.2.25), we find it convenient to define a routine that returns the ith component s_i, or ith tap, of the circular tapped delay-line state vector:

```
/* tap.c - i-th tap of circular delay-line buffer */

double tap(D, w, p, i)                    usage: si = tap(D, w, p, i);
double *w, *p;                            p passed by value
int D, i;                                 i = 0, 1, ..., D
{
        return w[(p - w + i) % (D + 1)];
}
```

Note that p is not changed by this routine, and therefore, it is passed by value. With the help of this routine, the above example of the D-fold delay would read as follows:

```
p = w;                                    initialize p

for (n = 0; n < Ntot; n++) {
        y[n] = tap(D, w, p, D);           Dth component of state vector
        *p = x[n];                        read input; equivalently, p[0] = x[n]
        cdelay(D, w, &p);                 update delay line
        }
```

The circular buffer implementation of a delay line is very efficient, consisting of just decrementing an address pointer *without* shifting any data (except for the input read into $p[0]$). It is especially useful in implementing digital audio effects, such as reverb, because D can be fairly large in these applications. For example, a 100 msec delay at 44.1 kHz sampling rate corresponds to $D = 100 \times 44.1 = 4410$ samples. It is also used in *wavetable sound synthesis*, where a stored waveform can be generated periodically by cycling over the circular buffer.

Because p is determined uniquely by the offset index q, via $p = w + q$, it is possible to rewrite all of the above routines so that they manipulate the index q instead of the pointer p. These versions can be translated easily into other languages, such as Fortran or MATLAB, that do not support pointer manipulation (see Appendix D).

The following routine `wrap2.c` replaces `wrap`. It simply keeps q within its allowed range, $0 \le q \le M$, by wrapping it modulo-$(M+1)$.

```
/* wrap2.c - circular wrap of pointer offset q, relative to array w */

void wrap2(M, q)
int M, *q;
{
        if (*q > M)
                *q -= M + 1;              when *q = M + 1, it wraps around to *q = 0

        if (*q < 0)
                *q += M + 1;              when *q = -1, it wraps around to *q = M
}
```

Because q is modified by wrap2, it must be passed by reference, that is, as a pointer to integer. The following routine cfir2.c replaces cfir. Note that the current input sample is placed in $p[0] = w[q]$, that is, $w[q] = x$.

```
/* cfir2.c - FIR filter implemented with circular delay-line buffer */

void wrap2();

double cfir2(M, h, w, q, x)
double *h, *w, x;                           q = circular offset index
int M, *q;                                  M = filter order
{
        int i;
        double y;

        w[*q] = x;                          read input sample x

        for (y=0, i=0; i<=M; i++) {         compute output sample y
                y += (*h++) * w[(*q)++];
                wrap2(M, q);
                }

        (*q)--;                             update circular delay line
        wrap2(M, q);

        return y;
}
```

If so desired, the for-loop in this routine can be replaced by the following version, which accesses the ith state via Eq. (4.2.25):

```
for (y=0, i=0; i<=M; i++)
        y += h[i] * w[(*q+i)%(M+1)];        used by cfir2.m of Appendix D
```

The index q must be initialized to $q = 0$, which is equivalent to $p = w$. The usage of cfir2 is illustrated by the following program segment:

```
double *h, *w;
int q;
h = (double *) calloc(M+1, sizeof(double));
w = (double *) calloc(M+1, sizeof(double));    also, initializes w to zero

q = 0;                                      initialize q

for (n = 0; n < Ntot; n++)
        y[n] = cfir2(M, h, w, &q, x[n]);    q passed by address
```

The implementation of the circular delay line is completely trivial. The following routine cdelay2.c replaces cdelay, and consists simply of decrementing q modulo $D+1$, assuming that the current input has been read into $w[q]$, namely, $w[q] = x$.

```
/* cdelay2.c - circular buffer implementation of D-fold delay */

void wrap2();

void cdelay2(D, q)
```

```
int D, *q;
{
        (*q)--;                          decrement offset and wrap modulo-(D + 1)
        wrap2(D, q);                     when *q = −1, it wraps around to *q = D
}
```

Its usage is illustrated by the following program segment. Note, again, that its output, namely, the Dth component of the internal state, is available even before its input:

```
q = 0;                                   initialize q

for (n = 0; n < Ntot; n++) {
        y[n] = w[(q+D)%(D+1)];           alternatively, y[n] = tap2(D, w, q, D);
        w[q] = x[n];                     read input
        cdelay2(D, &q);                  update delay line
        }
```

Finally, the components of the internal state vector given by Eq. (4.2.25) are returned by the following routine `tap2.c` which replaces `tap`:

```
/* tap2.c - i-th tap of circular delay-line buffer */

double tap2(D, w, q, i)                  usage: si = tap2(D, w, q, i);
double *w;
int D, q, i;                             i = 0, 1, ..., D
{
        return w[(q + i) % (D + 1)];
}
```

In summary, the circular buffer implementation of the FIR sample processing algorithm can be stated in the following form (initialized to $p = w$):

> *for each input sample x do:*
> $\quad s_0 = *p = x$
> $\quad$ *for* $i = 1, 2, \ldots, M$ *determine states:*
> $\qquad s_i = \text{tap}(M, \mathbf{w}, p, i)$
> $\quad y = h_0 s_0 + h_1 s_1 + \cdots + h_M s_M$
> $\quad \text{cdelay}(M, \mathbf{w}, \&p)$

where for convenience, we used the routine `tap` to get the current states. In terms of the offset index q (initialized to $q = 0$):

> *for each input sample x do:*
> $\quad s_0 = w[q] = x$
> $\quad$ *for* $i = 1, 2, \ldots, M$ *determine states:*
> $\qquad s_i = \text{tap2}(M, \mathbf{w}, q, i)$
> $\quad y = h_0 s_0 + h_1 s_1 + \cdots + h_M s_M$
> $\quad \text{cdelay2}(M, \&q)$

Example 4.2.6: The circular buffer implementation of Example 4.2.1 is as follows:

> *for each input sample x do:*
> $s_0 = *p = x$
> $s_1 = \text{tap}(3, \mathbf{w}, p, 1)$
> $s_2 = \text{tap}(3, \mathbf{w}, p, 2)$
> $s_3 = \text{tap}(3, \mathbf{w}, p, 3)$
> $y = s_0 + 2s_1 - s_2 + s_3$
> $\text{cdelay}(3, \mathbf{w}, \&p)$

where w is to be declared as a 4-dimensional array and initialized to zero. In terms of the variable q, we have

> *for each input sample x do:*
> $s_0 = w[q] = x$
> $s_1 = \text{tap2}(3, \mathbf{w}, q, 1)$
> $s_2 = \text{tap2}(3, \mathbf{w}, q, 2)$
> $s_3 = \text{tap2}(3, \mathbf{w}, q, 3)$
> $y = s_0 + 2s_1 - s_2 + s_3$
> $\text{cdelay2}(3, \&q)$

For the same input, the output signal samples and internal states computed by either of the above algorithms are exactly those given in the table of Example 4.2.1.

The linear buffer version discussed in Example 4.2.1 can be obtained from the above by freezing the pointer p to always point to w, that is, $p = w$. Then, we have for $i = 0, 1, 2, 3$:

$$s_i = \text{tap}(3, \mathbf{w}, p, i) = w[(p - w + i)\%4] = w[i\%4] = w_i$$

and the algorithm becomes the conventional one:

> *for each input sample x do:*
> $w_0 = x$
> $y = w_0 + 2w_1 - w_2 + w_3$
> $\text{delay}(3, \mathbf{w})$

where `cdelay` was replaced by `delay`. □

4.3 Problems

4.1 Compute the convolution, $\mathbf{y} = \mathbf{h} * \mathbf{x}$, of the filter and input,

$$\mathbf{h} = [1, 1, 2, 1], \qquad \mathbf{x} = [1, 2, 1, 1, 2, 1, 1, 1]$$

using the following three methods: (a) The convolution table. (b) The LTI form of convolution, arranging the computations in a table form. (c) The overlap-add method of block convolution with length-3 input blocks. Repeat using length-5 input blocks.

4.2 Repeat Problem 4.1 for the filter and input:

$$\mathbf{h} = [2, -2, -1, 1], \qquad \mathbf{x} = [2, 2, 0, 1, -1, 0, 1, 2],$$

4.3 The impulse response $h(n)$ of a filter is nonzero over the index range $3 \le n \le 6$. The input signal $x(n)$ to this filter is nonzero over the index range $10 \le n \le 20$. Consider the direct and LTI forms of convolution:

$$y(n) = \sum_m h(m)x(n-m) = \sum_m x(m)h(n-m)$$

a. Determine the overall index range n for the output $y(n)$. For each n, determine the corresponding summation range over m, for both the direct and LTI forms.

b. Assume $h(n) = 1$ and $x(n) = 1$ over their respective index ranges. Calculate and sketch the output $y(n)$. Identify (with an explanation) the input on/off transient and steady state parts of $y(n)$.

4.4 An LTI filter has infinite impulse response $h(n) = a^n u(n)$, where $|a| < 1$. Using the convolution summation formula $y(n) = \sum_m h(m)x(n-m)$, derive closed-form expressions for the output signal $y(n)$ when the input is:

a. A unit step, $x(n) = u(n)$

b. An alternating step, $x(n) = (-1)^n u(n)$.

In each case, determine the steady state and transient response of the filter.

4.5 Consider the IIR filter $h(n) = a^n u(n)$, where $0 < a < 1$. The square pulse $x(n) = u(n) - u(n-L)$ of duration L is applied as input.

Using the time-domain convolution formula, determine a closed-form expression for the output signal $y(n)$ for the two time ranges: $0 \le n \le L - 1$ and $n \ge L$.

4.6 The filter of Problem 4.5 satisfies the difference equation: $y(n) = ay(n-1) + x(n)$. Verify that the solution $y(n)$ that you obtained above satisfies this difference equation for all n.

4.7 *Computer Experiment: Convolution Routines.* Write C or MATLAB routines that implement convolution in: (a) the convolution table form and (b) the LTI form; that is,

$$y_n = \sum_{\substack{i,j \\ i+j=n}} h_i x_j = \sum_m x_m h_{n-m}$$

The routines must have the same input/output variables as conv.c of the text. Write a small main program that tests your routines.

4.8 *Computer Experiment: Filtering by Convolution.* Write small C or MATLAB routines to reproduce all the results and graphs of Examples 4.1.7, 4.1.8, and 4.1.9. The inputs must be treated as single blocks and passed into the routine conv.

4.9 *Computer Experiment: Block-by-Block Processing.* Write a stand-alone C program, say blkfilt.c, that implements the overlap-add block convolution method. The program must have usage:

```
blkfilt h.dat L  < x.dat  > y.dat
```

It must have as command-line inputs a file of impulse response coefficients h.dat (stored one coefficient per line) and the desired input block length L. It must read the input signal samples from stdin or a file x.dat and write the computed output samples to stdout or a file y.dat. It must have the following features built-in: (a) it must allocate storage dynamically for the impulse $h(n)$ read from h.dat (the program should abort with an error message if $L < M$); (b) it must read the input signal in length-L blocks, and call the routine blockcon.c of the text to process each block; (c) it must write the output also in blocks; (d) it must compute correctly both the input-on and input-off transients.

Note that the essence of such a program was already given in Section 4.1.10. Test your program on Example 4.1.10, with blocksizes $L = 3, 4, 5, 6$.

4.10 *Computer Experiment: Sample-by-Sample Processing.* Write a stand-alone C program, say firfilt.c, that implements the FIR sample processing algorithm of Eq. (4.2.15). The program must have usage:

```
firfilt h.dat  < x.dat  > y.dat
```

It must read and dynamically allocate the impulse response vector **h** from an input file, say h.dat, and must allocate the internal state vector **w**. Using the routine fir.c, it must keep processing input samples, reading them one at a time from stdin or from a file x.dat, and writing the computed output samples to stdout or a file y.dat.

It must correctly account for the input-on and input-off transients. Thus, it must produce identical results with convolution or block convolution if applied to a finite input block of samples.

The essence of such a program was given in Section 4.2.3. Test your program on Examples 4.1.7–4.1.10.

Such filters can be cascaded together by piping the output of one into the input to another. For example, the filtering operation by the combined filter $\mathbf{h} = \mathbf{h}_1 * \mathbf{h}_2$ can be implemented by:

```
firfilt h1.dat | firfilt h2.dat  < x.dat  > y.dat
```

Alternatively or additionally, write a MATLAB version, say firfilt.m, with usage:

```
y = firfilt(h, x);
```

It must read the filter and input vectors **h**, **x**, and compute the output vector **y**. The input-off transients must also be part of the output vector. You may use the MATLAB functions delay.m and fir.m of Appendix D.

4.11 *Computer Experiment: Sample Processing with Circular Buffers.* Rewrite the above program, say cfirfilt.c, such that the basic sample-by-sample filtering operation is implemented by the circular buffer routine cfir.c instead of fir.c. The usage of the program will be the same as above:

```
cfirfilt h.dat  < x.dat  > y.dat
```

Test your program on Examples 4.1.7–4.1.10. It must produce identical results as the firfilt program. Rewrite versions of this program that use the alternative circular FIR routines cfir1.c and cfir2.c. You may also write a MATLAB version using cfir2.m.

4.12 *Computer Experiment: Delay Using Circular Buffers.* Write a stand-alone C program, say cdelfilt.c, that implements a plain delay by up to D samples, that is, $y(n) = x(n - i)$, $i = 0, 1, \ldots, D$, and has usage:

```
cdelfilt i D  < x.dat  > y.dat
```

It must read the input samples one at a time from `stdin` and write the delayed samples into `stdout`. It must make use of the circular buffer routines `cdelay.c` and `tap.c`. The delay-line buffer must have length $D + 1$.

Test your program on a length-20 input using the values $D = 5$, and $i = 0, 1, \ldots, 5$. Then, write another version of this program that uses the routines `cdelay2` and `tap2` and test it.

4.13 *Computer Experiment: Alternative Sample Processing Algorithms.* The FIR sample processing algorithm of Eq. (4.2.15) proceeds by (a) reading the current input, (b) processing it, and (c) updating the delay line.

In some texts, the algorithm is structured in a slightly different way, such that the update of the delay line is done *before* the current input is read. Derive the difference equations describing this version of the algorithm. [*Hint:* Use Eq. (4.2.13).]

Translate the difference equations into a sample processing algorithm like Eq. (4.2.15) and then, write a C routine that implements it. Discuss the proper initialization of the internal states in this case. Test your routine to make sure it produces identical results with `fir.c`.

4.14 Consider the filter and input of Problem 4.1. Draw a block diagram realization of the filter, introduce internal states, write the corresponding sample processing algorithm, and convert it into a C routine.

Then, using the sample processing algorithm, compute the full output signal, including the input-off transients (for these, apply $x = 0$ to the algorithm). Display your computations in a table form. Identify on the table which outputs correspond to the input-on transients, to the steady state, and to the input-off transients. [*Note:* The computed output signal should agree exactly with that computed by the convolution method of Problem 4.1.]

4.15 Consider the filter with I/O equation: $y(n) = x(n) - x(n - 3)$.

    a. Determine the impulse response sequence $h(n)$, for all $n \geq 0$.

    b. Draw a *block diagram* realization, introduce appropriate *internal states*, and write the corresponding *sample processing* algorithm. Moreover, implement the algorithm by a C routine. Test your routine on the results of parts (c) and (d).

    c. Send as input the sequence $\mathbf{x} = [1, 1, 2, 2, 4, \ldots]$. Using *convolution*, compute the first five output samples, $y(n)$, $n = 0, 1, \ldots, 4$.

    d. Compute the same outputs using the *sample processing* algorithm. Display your computations in a table that, at each sampling instant n, shows the corresponding input sample $x(n)$, the values of the internal states, and the computed output sample $y(n)$.

4.16 Repeat Problem 4.15 for the filter: $y(n) = 0.8y(n - 1) + x(n)$.

4.17 Repeat Problem 4.15 for the filter: $y(n) = 0.25y(n - 2) + x(n)$.

4.18 Let $\mathbf{x} = [1, 1, 2, 2, 2, 2, 1, 1]$ be an input to the filter described by the I/O equation:

$$y(n) = x(n) - x(n - 2) + 2x(n - 3)$$

    a. Determine the impulse response $h(n)$ of this filter.

    b. Compute the corresponding output signal $y(n)$ using the *LTI form* of convolution. Show your computations in table form.

    c. Compute the same output using the overlap-add method of block convolution by partitioning the input signal into length-4 blocks.

    d. Draw a block diagram realization of this filter. Then, introduce appropriate internal states and write the corresponding sample processing algorithm.

4.19 The length-8 input signal $x(n) = \{8, 7, 6, 5, 4, 3, 2, 1\}$ is applied to the input of a 2-fold delay described by the I/O equation $y(n) = x(n-2)$.

The circular buffer version of the sample processing implementation of the delay operation requires the use of a 3-dimensional *linear buffer array* of internal states $\mathbf{w} = [w_0, w_1, w_2]$ and a pointer p circulating over $\mathbf{w}$.

Make a table of the numerical values of the contents of the array $\mathbf{w}$ for the successive time instants $0 \leq n \leq 10$. In each row, indicate that array element, w_i, which represents the current output $y(n)$ of the delay line. Explain your reasoning in filling this table. Compare with the linear buffer case.

4.20 Figure 4.3.1 shows the transposed realization of the third-order filter with impulse response $\mathbf{h} = [h_0, h_1, h_2, h_3]$. Write the difference equations describing the I/O operation of this realization. Show that the combined effect of these difference equations is equivalent to the standard direct form operation of Eq. (4.2.6).

Write the transposed difference equations as a sample processing algorithm and apply it to the filter and input of Example 4.2.1. Make a table of the values of x, y, and all the variables v_i, for each time instant $0 \leq n \leq 10$.

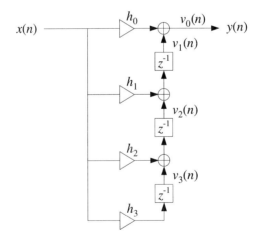

Fig. 4.3.1 Transposed realization of third-order filter.

4.21 *Computer Experiment: Transposed FIR Realization.* Generalize the block diagram of Fig. 4.3.1 to an arbitrary Mth order filter with impulse response $\mathbf{h} = [h_0, h_1, \ldots, h_M]$. Write the corresponding sample processing algorithm and translate it into a C routine `firtr.c` that has usage:

```
y = firtr(M, h, v, x);
```

where $\mathbf{h}$ and $\mathbf{v}$ are $(M+1)$-dimensional vectors, and x, y are the current input and output samples.

To test your routine, write a version of the program `firfilt.c` of Problem 4.10 that uses the routine `firtr.c` instead of `fir.c`. Test your program on Example 4.2.1.

z-Transforms

5.1 Basic Properties

Here, we review briefly z-transforms and their properties. We assume that the reader already has a basic familiarity with the subject. Our usage of z-transforms in this book is basically as a *tool* for the analysis, design, and implementation of digital filters.

Given a discrete-time signal $x(n)$, its z-transform is defined as the following series:

$$\boxed{X(z) = \sum_{n=-\infty}^{\infty} x(n) z^{-n}} \quad \text{(z-transform)} \tag{5.1.1}$$

or, writing explicitly a few of the terms:

$$X(z) = \cdots + x(-2) z^2 + x(-1) z + x(0) + x(1) z^{-1} + x(2) z^{-2} + \cdots$$

There are as many terms as nonzero signal values $x(n)$. The terms z^{-n} can be thought of as place holders for the values $x(n)$. If the signal $x(n)$ is *causal*, only *negative* powers z^{-n}, $n \geq 0$ appear in the expansion. If $x(n)$ is strictly *anticausal*, being nonzero for $n \leq -1$, only *positive* powers will appear in the expansion, that is, $z^{-n} = z^{|n|}$, for $n \leq -1$. And if $x(n)$ is mixed with both causal and anticausal parts, then both negative and positive powers of z will appear.

The definition (5.1.1) can also be applied to the impulse response sequence $h(n)$ of a digital filter. The z-transform of $h(n)$ is called the *transfer function* of the filter and is defined by:

$$\boxed{H(z) = \sum_{n=-\infty}^{\infty} h(n) z^{-n}} \quad \text{(transfer function)} \tag{5.1.2}$$

Example 5.1.1: Determine the transfer function $H(z)$ of the two causal filters of Example 3.4.3, namely,

(a) $\mathbf{h} = [h_0, h_1, h_2, h_3] = [2, 3, 5, 2]$

(b) $\mathbf{h} = [h_0, h_1, h_2, h_3, h_4] = [1, 0, 0, 0, -1]$

Solution: Using the definition (5.1.2), we find:

$$H(z) = h_0 + h_1 z^{-1} + h_2 z^{-2} + h_3 z^{-3} = 2 + 3z^{-1} + 5z^{-2} + 2z^{-3}$$

for case (a), and

$$H(z) = h_0 + h_1 z^{-1} + h_2 z^{-2} + h_3 z^{-3} + h_4 z^{-4} = 1 - z^{-4}$$

for case (b). □

The three most important properties of z-transforms that facilitate the analysis and synthesis of linear systems are:

- linearity property
- delay property
- convolution property

The linearity property simply states that the z-transform of a linear combination of signals is equal to the linear combination of z-transforms, that is, if $X_1(z)$ and $X_2(z)$ are the z transforms of the signals $x_1(n)$ and $x_2(n)$, then the z-transform of the linear combination $a_1 x_1(n) + a_2 x_2(n)$ is

$$\boxed{a_1 x_1(n) + a_2 x_2(n) \xrightarrow{z} a_1 X_1(z) + a_2 X_2(z)} \qquad \text{(linearity)} \qquad (5.1.3)$$

The delay property states that the effect of delaying a signal by D sampling units is equivalent to multiplying its z-transform by a factor z^{-D}, namely,

$$\boxed{x(n) \xrightarrow{z} X(z) \quad \Rightarrow \quad x(n-D) \xrightarrow{z} z^{-D} X(z)} \qquad \text{(delay)} \qquad (5.1.4)$$

Note that D can also be negative, representing a time advance. Finally, the convolution property states that convolution in the time domain becomes multiplication in the z-domain:

$$\boxed{y(n) = h(n) * x(n) \quad \Rightarrow \quad Y(z) = H(z) X(z)} \qquad \text{(convolution)} \qquad (5.1.5)$$

that is, the z-transform of the convolution of two sequences is equal to the product of the z-transforms of the sequences.

Example 5.1.2: The two filters of the above example and of Example 3.4.3 can also be written in the following "closed" forms, valid for all n:

(a) $h(n) = 2\delta(n) + 3\delta(n-1) + 5\delta(n-2) + 2\delta(n-3)$, (b) $h(n) = \delta(n) - \delta(n-4)$

Their transfer functions can be obtained using the linearity and delay properties as follows. First, note that the z-transform of $\delta(n)$ is unity:

$$\delta(n) \xrightarrow{Z} \sum_{n=-\infty}^{\infty} \delta(n)z^{-n} = \delta(0)z^{-0} = 1$$

Then, from the delay property, we have

$$\delta(n-1) \xrightarrow{Z} z^{-1} \cdot 1 = z^{-1}, \qquad \delta(n-2) \xrightarrow{Z} z^{-2}, \qquad \delta(n-3) \xrightarrow{Z} z^{-3}, \quad \text{etc.}$$

Using linearity, we obtain

$$2\delta(n) + 3\delta(n-1) + 5\delta(n-2) + 2\delta(n-3) \xrightarrow{Z} 2 + 3z^{-1} + 5z^{-2} + 2z^{-3}$$

for case (a), and

$$h(n) = \delta(n) - \delta(n-4) \xrightarrow{Z} H(z) = 1 - z^{-4}$$

for case (b).   □

Example 5.1.3: Using the unit-step identity $u(n) - u(n-1) = \delta(n)$, valid for all n, and the z-transform properties, determine the z-transforms of the two signals:

$$\text{(a)} \quad x(n) = u(n), \qquad \text{(b)} \quad x(n) = -u(-n-1)$$

Solution: For case (a), we have the difference equation:

$$x(n) - x(n-1) = u(n) - u(n-1) = \delta(n)$$

Taking z-transforms of both sides and using the linearity and delay properties, we obtain

$$x(n) - x(n-1) = \delta(n) \xrightarrow{Z} X(z) - z^{-1}X(z) = 1 \qquad \Rightarrow \qquad X(z) = \frac{1}{1 - z^{-1}}$$

Similarly, for case (b) we have the difference equation:

$$x(n) - x(n-1) = -u(-n-1) + u(-(n-1)-1) = u(-n) - u(-n-1) = \delta(-n)$$

where in the last equation we used the given identity with n replaced by $-n$. Noting that $\delta(-n) = \delta(n)$, and taking z-transforms of both sides, we find

$$x(n) - x(n-1) = \delta(-n) \xrightarrow{Z} X(z) - z^{-1}X(z) = 1 \qquad \Rightarrow \qquad X(z) = \frac{1}{1 - z^{-1}}$$

Thus, even though the two signals $u(n)$ and $-u(-n-1)$ are completely different in the time domain (one is causal, the other anticausal), their z-transforms are the *same*. We will see in the next section that they can be distinguished in terms of their region of convergence.   □

Example 5.1.4: Compute the output of Example 4.1.1 by carrying out the convolution operation as multiplication in the z-domain.

Solution: The two sequences

$$\mathbf{h} = [1, 2, -1, 1], \qquad \mathbf{x} = [1, 1, 2, 1, 2, 2, 1, 1]$$

have z-transforms:

$$H(z) = 1 + 2z^{-1} - z^{-2} + z^{-3}$$

$$X(z) = 1 + z^{-1} + 2z^{-2} + z^{-3} + 2z^{-4} + 2z^{-5} + z^{-6} + z^{-7}$$

Multiplying these polynomials, we find for the product $Y(z) = H(z)X(z)$:

$$Y(z) = 1 + 3z^{-1} + 3z^{-2} + 5z^{-3} + 3z^{-4} + 7z^{-5} + 4z^{-6} + 3z^{-7} + 3z^{-8} + z^{-10}$$

The coefficients of the powers of z are the convolution output samples:

$$\mathbf{y} = \mathbf{h} * \mathbf{x} = [1, 3, 3, 5, 3, 7, 4, 3, 3, 0, 1]$$

Note that the term z^{-9} is absent, which means that its coefficient is zero.   □

5.2 Region of Convergence

If $x(n)$ has infinite duration, Eq. (5.1.1) becomes an infinite series, and it is possible that certain values of the complex variable z might render it divergent.

The *region of convergence* (ROC) of the z-transform $X(z)$ is defined to be that *subset* of the complex z-plane $\mathbb{C}$ for which the series (5.1.1) *converges*, that is,

$$\text{Region of Convergence} = \left\{ z \in \mathbb{C} \;\middle|\; X(z) = \sum_{n=-\infty}^{\infty} x(n) z^{-n} \neq \infty \right\} \qquad (5.2.1)$$

The ROC is an important concept in many respects: It allows the unique inversion of the z-transform and provides convenient characterizations of the causality and stability properties of a signal or system.

The ROC depends on the signal $x(n)$ being transformed. As an example, consider the following causal signal:

$$x(n) = (0.5)^n u(n) = \{1, 0.5, 0.5^2, 0.5^3, \dots\}$$

Its z-transform will be:

$$X(z) = \sum_{n=-\infty}^{\infty} (0.5)^n u(n) z^{-n} = \sum_{n=0}^{\infty} (0.5)^n z^{-n} = \sum_{n=0}^{\infty} (0.5z^{-1})^n$$

where the summation was restricted over $n \geq 0$ because of the causality of $x(n)$. This infinite sum can be done with the help of the *infinite geometric series* formula:

$$1 + x + x^2 + x^3 + \cdots = \sum_{n=0}^{\infty} x^n = \frac{1}{1-x} \qquad (5.2.2)$$

which is valid only for $|x| < 1$ and diverges otherwise. Setting $x = 0.5z^{-1}$ we find the sum:

$$X(z) = \sum_{n=0}^{\infty} (0.5z^{-1})^n = \sum_{n=0}^{\infty} x^n = \frac{1}{1-x}, \qquad \text{or,}$$

$$X(z) = \frac{1}{1 - 0.5z^{-1}} = \frac{z}{z - 0.5}$$

where the convergence of the geometric series requires

$$|x| = |0.5z^{-1}| < 1 \quad \Rightarrow \quad |z| > 0.5$$

Thus, the ROC is the set of z's in the z-plane that lie strictly *outside* the circle of radius 0.5, as shown below:

$$\text{ROC} = \{z \in \mathbb{C} \mid |z| > 0.5\}$$

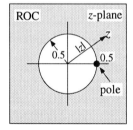

Note, that the z-transform has a *pole* at $z = 0.5$. In summary, we have

$$(0.5)^n u(n) \xrightarrow{z} \frac{1}{1 - 0.5z^{-1}}, \qquad \text{with} \quad |z| > 0.5$$

A z-transform and its ROC are *uniquely* determined by the time signal $x(n)$. However, it is possible for two different time signals $x(n)$ to have the *same z-transform*, as was the case in Example 5.1.3. Such signals can only be distinguished in the z-domain by their region of convergence. Consider, for example, the anti-causal signal

$$x(n) = - (0.5)^n u(-n - 1)$$

The presence of the anti-unit step $u(-n - 1)$ restricts n to be $-n - 1 \geq 0$ or, $n \leq -1$. Its z-transform will be:

$$X(z) = - \sum_{n=-\infty}^{-1} (0.5)^n z^{-n} = - \sum_{n=-\infty}^{-1} ((0.5)^{-1}z)^{-n} = - \sum_{m=1}^{\infty} ((0.5)^{-1}z)^m$$

where we changed summation variables from n to $m = -n$. To sum it, we use the following variant of the infinite geometric series:

$$x + x^2 + x^3 + \cdots = \sum_{m=1}^{\infty} x^m = \frac{x}{1-x}$$

which is valid only for $|x| < 1$ and diverges otherwise. Setting $x = (0.5)^{-1}z$, we have

$$X(z) = -\sum_{m=1}^{\infty} \left((0.5)^{-1}z\right)^m = -\sum_{m=1}^{\infty} x^m = -\frac{x}{1-x} = -\frac{0.5^{-1}z}{1-0.5^{-1}z}, \qquad \text{or,}$$

$$X(z) = \frac{z}{z-0.5} = \frac{1}{1-0.5z^{-1}}$$

which is the same as the causal example above. However, the ROC in this case is different. It is determined by the geometric series convergence condition

$$|x| = |0.5^{-1}z| < 1 \qquad \Rightarrow \qquad |z| < 0.5$$

which is the set of z's that lie strictly *inside* the circle of radius 0.5, as shown below:

$$\text{ROC} = \{z \in \mathbb{C} \mid |z| < 0.5\}$$

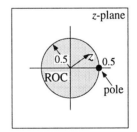

To summarize, we have determined the z-transforms:

$$(0.5)^n u(n) \xrightarrow{z} \frac{1}{1-0.5z^{-1}}, \qquad \text{with } |z| > 0.5$$

$$-(0.5)^n u(-n-1) \xrightarrow{z} \frac{1}{1-0.5z^{-1}}, \qquad \text{with } |z| < 0.5$$

The two signals have the same z-transform but completely disjoint ROCs. More generally, we have the result:

$$a^n u(n) \xrightarrow{z} \frac{1}{1-az^{-1}}, \qquad \text{with } |z| > |a|$$

$$-a^n u(-n-1) \xrightarrow{z} \frac{1}{1-az^{-1}}, \qquad \text{with } |z| < |a|$$

(5.2.3)

where a is any *complex* number. Their ROCs are shown below.

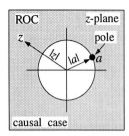
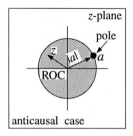

The z-transforms (5.2.3), together with the linearity and delay properties, can be used to construct more complicated transforms.

Example 5.2.1: Setting $a = \pm 1$ in Eq. (5.2.3), we obtain the z-transforms of the causal and anticausal unit-steps and alternating unit-steps:

$$u(n) \xrightarrow{z} \frac{1}{1 - z^{-1}}, \quad \text{with } |z| > 1$$

$$-u(-n-1) \xrightarrow{z} \frac{1}{1 - z^{-1}}, \quad \text{with } |z| < 1$$

$$(-1)^n u(n) \xrightarrow{z} \frac{1}{1 + z^{-1}}, \quad \text{with } |z| > 1$$

$$-(-1)^n u(-n-1) \xrightarrow{z} \frac{1}{1 + z^{-1}}, \quad \text{with } |z| < 1$$

which agree with Example 5.1.3. □

Example 5.2.2: Determine the z-transform and corresponding region of convergence of the following signals:

1. $x(n) = u(n - 10)$

2. $x(n) = (-0.8)^n u(n)$

3. $x(n) = (-0.8)^n [u(n) - u(n - 10)]$

4. $x(n) = \frac{1}{2}[u(n) + (-1)^n u(n)] = \{1, 0, 1, 0, 1, 0, 1, 0, \ldots\}$

5. $x(n) = \frac{1}{2}[(0.8)^n u(n) + (-0.8)^n u(n)]$

6. $x(n) = \cos(\frac{\pi n}{2}) u(n) = \{1, 0, -1, 0, 1, 0, -1, 0, 1, 0, -1, 0, \ldots\}$

7. $x(n) = (0.8)^n \cos(\frac{\pi n}{2}) u(n)$

8. $x(n) = \frac{1}{2}[(0.8j)^n u(n) + (-0.8j)^n u(n)]$

9. $x(n) = \cos(\omega_0 n) u(n)$ and $x(n) = \sin(\omega_0 n) u(n)$

10. $x(n) = \{1, 2, 3, 1, 2, 3, 1, 2, 3, \ldots\}$, periodically repeating $\{1, 2, 3\}$

Solution: Using the delay property, we have in case (1):

$$X(z) = z^{-10}U(z) = \frac{z^{-10}}{1 - z^{-1}}$$

with ROC $|z| > 1$. In case (2), we apply Eq. (5.2.3) with $a = -0.8$ to get

$$X(z) = \frac{1}{1 + 0.8z^{-1}}, \qquad \text{with ROC: } |z| > |-0.8| = 0.8$$

For case (3), we write

$$x(n) = (-0.8)^n u(n) - (-0.8)^{10}(-0.8)^{n-10} u(n - 10)$$

where in the second term we multiplied and divided by the factor $(-0.8)^{10}$ in order to reproduce the delayed (by 10 units) version of the first term. Thus, using the linearity and delay properties and the results of case (2), we get

$$X(z) = \frac{1}{1 + 0.8z^{-1}} - (-0.8)^{10}\frac{z^{-10}}{1 + 0.8z^{-1}} = \frac{1 - (-0.8)^{10}z^{-10}}{1 + 0.8z^{-1}}$$

Here, the ROC is not $|z| > 0.8$ as might appear at first glance. Rather it is the set of all nonzero z's, $z \neq 0$. This follows by recognizing $x(n)$ to be a length-10 finite sequence. Indeed, setting $a = -0.8$, we have

$$x(n) = a^n[u(n) - u(n - 10)] = \{1, a, a^2, a^3, a^4, a^5, a^6, a^7, a^8, a^9, 0, 0, 0, \dots\}$$

and therefore, its z-transform can be computed by the finite sum

$$X(z) = 1 + az^{-1} + a^2z^{-2} + \cdots + a^9 z^{-9}$$

which exists for any $z \neq 0$. Using the finite geometric series

$$1 + x + x^2 + \cdots + x^{N-1} = \frac{1 - x^N}{1 - x}$$

we may sum the above series to

$$X(z) = 1 + az^{-1} + a^2z^{-2} + \cdots + a^9 z^{-9} = \frac{1 - a^{10}z^{-10}}{1 - az^{-1}} = \frac{1 - (-0.8)^{10}z^{-10}}{1 + 0.8z^{-1}}$$

For case (4), we have, using linearity and Eq. (5.2.3) with $a = 1$ and $a = -1$:

$$X(z) = \frac{1}{2}\left[\frac{1}{1 - z^{-1}} + \frac{1}{1 + z^{-1}}\right] = \frac{1}{1 - z^{-2}}$$

with ROC $|z| > 1$. The same result can be obtained using the definition (5.1.1) and summing the series:

$$X(z) = 1 + 0z^{-1} + z^{-2} + 0z^{-3} + z^{-4} + \cdots = 1 + z^{-2} + z^{-4} + z^{-6} + \cdots$$

which is an infinite geometric series of the type of Eq. (5.2.2) with $x = z^{-2}$. Therefore,

$$X(z) = \left. \frac{1}{1-x} \right|_{x=z^{-2}} = \frac{1}{1-z^{-2}}$$

The convergence of the series requires $|x| = |z^{-2}| < 1$ or equivalently, $|z| > 1$. In case (5), we find again using linearity and Eq. (5.2.3):

$$X(z) = \frac{1}{2} \left[\frac{1}{1-0.8z^{-1}} + \frac{1}{1+0.8z^{-1}} \right] = \frac{1}{1-0.64z^{-2}}$$

with ROC $|z| > 0.8$. Case (6) can be handled directly by the definition (5.1.1):

$$X(z) = 1 - z^{-2} + z^{-4} - z^{-6} + z^{-8} + \cdots = 1 + x + x^2 + x^3 + x^4 + \cdots$$

where $x = -z^{-2}$. The series will converge to

$$X(z) = \frac{1}{1-x} = \frac{1}{1+z^{-2}}$$

provided $|x| = |-z^{-2}| < 1$, or equivalently, $|z| > 1$. The same result can be obtained using Euler's formula to split the cosine into exponential signals of the type (5.2.3):

$$x(n) = \cos\left(\frac{\pi n}{2}\right) u(n) = \frac{1}{2} \left[e^{j\pi n/2} u(n) + e^{-j\pi n/2} u(n) \right] = \frac{1}{2} \left[a^n u(n) + a^{*n} u(n) \right]$$

where $a = e^{j\pi/2} = j$ and $a^* = e^{-j\pi/2} = -j$. Thus, we find

$$X(z) = \frac{1}{2} \left[\frac{1}{1-jz^{-1}} + \frac{1}{1+jz^{-1}} \right] = \frac{1}{1+z^{-2}}$$

In case (7), using Euler's formula as above, we find

$$x(n) = (0.8)^n \cos\left(\frac{\pi n}{2}\right) u(n) = \frac{1}{2} \left[(0.8)^n e^{j\pi n/2} u(n) + (0.8)^n e^{-j\pi n/2} u(n) \right]$$

which can be written as the signal of case (8):

$$x(n) = \frac{1}{2} \left[(0.8j)^n u(n) + (-0.8j)^n u(n) \right]$$

Thus, cases (7) and (8) are the same. Their z-transform is obtained using $a = \pm 0.8j$ in Eq. (5.2.3):

$$X(z) = \frac{1}{2} \left[\frac{1}{1-0.8jz^{-1}} + \frac{1}{1+0.8jz^{-1}} \right] = \frac{1}{1+0.64z^{-2}}$$

with ROC $|z| > |0.8j| = 0.8$. The cosinewave in case (9) can be handled in a similar fashion. We write

$$\cos(\omega_0 n)u(n) = \frac{1}{2}\left[e^{j\omega_0 n} + e^{-j\omega_0 n}\right]u(n) \xrightarrow{z} \frac{1}{2}\left[\frac{1}{1 - e^{j\omega_0}z^{-1}} + \frac{1}{1 - e^{-j\omega_0}z^{-1}}\right]$$

which combines to give:

$$X(z) = \frac{1 - \cos(\omega_0)z^{-1}}{1 - 2\cos(\omega_0)z^{-1} + z^{-2}}$$

Setting $\omega_0 = \pi/2$, we recover case (6). Similarly, for a sinewave we have

$$\sin(\omega_0 n)u(n) = \frac{1}{2j}\left[e^{j\omega_0 n} - e^{-j\omega_0 n}\right]u(n) \xrightarrow{z} \frac{1}{2j}\left[\frac{1}{1 - e^{j\omega_0}z^{-1}} - \frac{1}{1 - e^{-j\omega_0}z^{-1}}\right]$$

which combines to give:

$$X(z) = \frac{\sin(\omega_0)z^{-1}}{1 - 2\cos(\omega_0)z^{-1} + z^{-2}}$$

Finally, we consider case (10). Using the definition (5.1.1) and grouping the terms in groups of 3, we obtain:

$$
\begin{aligned}
X(z) &= (1 + 2z^{-1} + 3z^{-2}) + (1 + 2z^{-1} + 3z^{-2})z^{-3} + \\
&\quad + (1 + 2z^{-1} + 3z^{-2})z^{-6} + (1 + 2z^{-1} + 3z^{-2})z^{-9} + \cdots \\
&= (1 + 2z^{-1} + 3z^{-2})(1 + z^{-3} + z^{-6} + z^{-9} + \cdots) \\
&= \frac{1 + 2z^{-1} + 3z^{-2}}{1 - z^{-3}}
\end{aligned}
$$

The infinite geometric series converges for $|z^{-3}| < 1$ or $|z| > 1$. An alternative method is to delay $x(n)$ by one period, that is, 3 time units

$$x(n - 3) = \{0, 0, 0, 1, 2, 3, 1, 2, 3, \ldots\}$$

and subtract it from $x(n)$ to get the difference equation

$$x(n) - x(n - 3) = \{1, 2, 3, 0, 0, 0, 0, 0, 0, \ldots\} = \delta(n) + 2\delta(n - 1) + 3\delta(n - 2)$$

Then, taking z-transforms of both sides, we get

$$X(z) - z^{-3}X(z) = 1 + 2z^{-1} + 3z^{-2} \quad \Rightarrow \quad X(z) = \frac{1 + 2z^{-1} + 3z^{-2}}{1 - z^{-3}}$$

This technique can be generalized to any periodic sequence. It will be used later to implement digital periodic waveform generators. □

Example 5.2.3: Determine the z-transform and corresponding region of convergence of the following signals:

1. $x(n) = (0.8)^n u(n) + (1.25)^n u(n)$

2. $x(n) = (0.8)^n u(n) - (1.25)^n u(-n - 1)$

3. $x(n) = -(0.8)^n u(-n - 1) - (1.25)^n u(-n - 1)$

4. $x(n) = -(0.8)^n u(-n - 1) + (1.25)^n u(n)$

Solution: Using Eq. (5.2.3) with $a = 0.8$ and $a = 1.25$, we note that the first three cases have exactly the *same* z-transform, namely,

$$X(z) = \frac{1}{1 - 0.8z^{-1}} + \frac{1}{1 - 1.25z^{-1}} = \frac{2 - 2.05z^{-1}}{1 - 2.05z^{-1} + z^{-2}}$$

The three cases differ only in their ROCs. In case (1), both terms are causal, and therefore, we must have simultaneously $|z| > 0.8$ and $|z| > 1.25$. Thus, the ROC is $|z| > 1.25$. In case (2), the second term is anticausal and therefore we must have the simultaneous inequalities: $|z| > 0.8$ and $|z| < 1.25$. Thus, the ROC is $0.8 < |z| < 1.25$. In case (3), both terms are anticausal requiring $|z| < 0.8$ and $|z| < 1.25$. Therefore, the ROC is $|z| < 0.8$. The three ROCs are shown below:

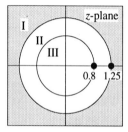 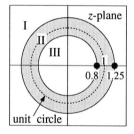 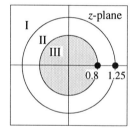

The two poles of $X(z)$ at $z = 0.8$ and $z = 1.25$ divide the z-plane into three non-overlapping regions which are the three possible ROCs.

Note that case (1) is causal but unstable because the term $(1.25)^n$ diverges for large positive n. Case (2) is stable, because the term $(0.8)^n$ converges to zero exponentially for large positive n, and the term $(1.25)^n$ converges to zero exponentially for large negative n. And, case (3) is anticausal and unstable because the term $(0.8)^n$ diverges for large negative n.

The unit circle is contained entirely within the ROC of case (2), in accordance with the general criterion of stability of Section 5.3.

The fourth case, which is unstable both for $n \to \infty$ and $n \to -\infty$, does not have a z-transform because convergence requires $|z| < 0.8$ for the anticausal term and $|z| > 1.25$ for the causal term. Thus, there is no z for which $X(z)$ is converges. The ROC is the empty set.   □

5.3 Causality and Stability

The z-domain characterizations of causality and stability can be obtained with the help of the basic result (5.2.3). A *causal* signal of the form

$$x(n) = A_1 p_1^n u(n) + A_2 p_2^n u(n) + \cdots \tag{5.3.1}$$

will have z-transform

$$X(z) = \frac{A_1}{1 - p_1 z^{-1}} + \frac{A_2}{1 - p_2 z^{-1}} + \cdots \tag{5.3.2}$$

with the restrictions $|z| > |p_1|$, $|z| > |p_2|$, and so forth. Therefore, the common ROC of all the terms will be

$$|z| > \max_i |p_i| \tag{5.3.3}$$

that is, the *outside* of the circle defined by the pole of *maximum magnitude*. Similarly, if the signal is completely *anticausal*

$$x(n) = -A_1 p_1^n u(-n-1) - A_2 p_2^n u(-n-1) - \cdots \tag{5.3.4}$$

its z-transform will be the same as Eq. (5.3.2), but the ROC restrictions on z will be $|z| < |p_1|$, $|z| < |p_2|$, and so forth. Thus, the ROC is in this case:

$$|z| < \min_i |p_i| \tag{5.3.5}$$

that is, the *inside* of the circle defined by the pole of *minimum magnitude*. The ROCs of these two cases are shown in Fig. 5.3.1.

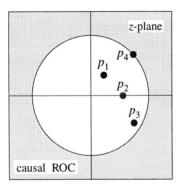 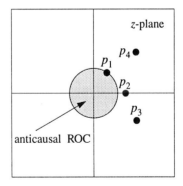

Fig. 5.3.1 Causal and anticausal ROCs.

In summary, causal signals are characterized by ROCs that are outside the maximum pole circle. Anticausal signals have ROCs that are inside the minimum pole circle. Mixed signals have ROCs that are the annular region between two circles—with the poles that lie inside the inner circle contributing causally and the poles that lie outside the outer circle contributing anticausally.

Stability can also be characterized in the z-domain in terms of the choice of the ROC. It can be shown that a *necessary and sufficient* condition for the stability of a signal $x(n)$ is that the ROC of the corresponding z-transform contain the unit circle. For a system $h(n)$, it can be shown that this condition is equivalent to the condition (3.5.4) discussed in Chapter 3.

Stability is not necessarily compatible with causality. For a signal or system to be simultaneously *stable and causal*, it is necessary that *all* its poles lie strictly *inside* the unit circle in the z-plane. This follows from Eq. (5.3.3) which is required for a causal ROC. If this ROC is to also correspond to a stable signal, then it must contain the unit circle. In other words, we may set $|z| = 1$ in Eq. (5.3.3):

$$1 > \max_i |p_i|$$

which implies that all poles must have magnitude less than one. A signal or system can also be simultaneously stable and anticausal, but in this case all its poles must lie strictly *outside* the unit circle. Indeed, the anticausality condition Eq. (5.3.5), together with the stability condition that the ROC contain the points $|z| = 1$, imply

$$1 < \min_i |p_i|$$

which means that all poles must have magnitude greater than one. If some of the poles have magnitude less than one and some greater than one, then it is possible to have a stable signal but it will be of the mixed kind. Those poles that lie inside the unit circle will contribute causally and those that lie outside will contribute anticausally.

Figure 5.3.2 illustrates three such possible stable cases. In all cases, the z-transform has the same form, namely,

$$X(z) = \frac{A_1}{1 - p_1 z^{-1}} + \frac{A_2}{1 - p_2 z^{-1}} + \frac{A_3}{1 - p_3 z^{-1}} + \frac{A_4}{1 - p_4 z^{-1}}$$

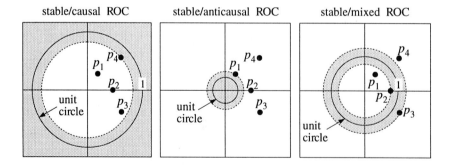

Fig. 5.3.2 Stable ROCs.

In the stable and causal case, all poles must have magnitude less than one, that is, $|p_i| < 1$, $i = 1, 2, 3, 4$ and the signal $x(n)$ will be

$$x(n) = [A_1 p_1^n + A_2 p_2^n + A_3 p_3^n + A_4 p_4^n] u(n)$$

with all terms converging to zero exponentially for large positive n. In the stable/anticausal case, all poles have magnitude greater than one, $|p_i| > 1, i = 1, 2, 3, 4$, and $x(n)$ will be:

$$x(n) = -[A_1 p_1^n + A_2 p_2^n + A_3 p_3^n + A_4 p_4^n] u(-n-1)$$

where because n is negative, each term will tend to zero exponentially for large negative n. This can be seen more clearly by writing a typical term as

$$-A_1 p_1^n u(-n-1) = -A_1 p_1^{-|n|} u(-n-1) = -A_1 \left(\frac{1}{p_1}\right)^{|n|} u(-n-1)$$

where we set $n = -|n|$ for negative n. Because $|p_1| > 1$ it follows that $|1/p_1| < 1$ and its successive powers will tend to zero exponentially. In the mixed case, we have $|p_1| < |p_2| < 1$ and $|p_4| > |p_3| > 1$. Therefore, the stable signal will be

$$x(n) = [A_1 p_1^n + A_2 p_2^n] u(n) - [A_3 p_3^n + A_4 p_4^n] u(-n-1)$$

with p_1, p_2 contributing causally, and p_3, p_4 anticausally. An example of such a stable but mixed signal was given in the second case of Example 5.2.3, namely,

$$x(n) = (0.8)^n u(n) - (1.25)^n u(-n-1)$$

As we emphasized in Chapter 3, stability is more important in DSP than causality in order to avoid numerically divergent computations. Causality can be reconciled exactly if all the poles are inside the unit circle, but only approximately if some of the poles are outside. We will discuss this issue later.

An important class of signals are the so-called *marginally stable* signals, which neither diverge nor converge to zero for large n. Rather, they remain *bounded*. The unit-step, alternating unit-step, and more general sinusoidal signals fall in this class. Such signals have poles that lie *on* the unit circle.

Some examples were cases (1,4,6,9,10) of Example 5.2.2. A simpler example is the case of a complex sinusoid of frequency ω_0

$$\text{(causal)} \qquad x(n) = e^{j\omega_0 n} u(n)$$

$$\text{(anticausal)} \qquad x(n) = -e^{j\omega_0 n} u(-n-1)$$

which is a special case of Eq. (5.2.3) with $a = e^{j\omega_0}$. Note that the plain unit-step $u(n)$ and alternating step $(-1)^n u(n)$ are special cases of this with $\omega_0 = 0$ and $\omega_0 = \pi$. The corresponding z-transform follows from Eq. (5.2.3):

$$X(z) = \frac{1}{1 - e^{j\omega_0} z^{-1}}$$

with ROC being either $|z| > 1$ for the causal case, or $|z| < 1$ for the anticausal one.

5.4 Frequency Spectrum

The *frequency spectrum*, frequency content, or *discrete-time Fourier transform* (DTFT) of a signal $x(n)$ is defined by

$$X(\omega) = \sum_{n=-\infty}^{\infty} x(n) e^{-j\omega n} \qquad \text{(DTFT)} \qquad (5.4.1)$$

It is recognized as the *evaluation* of the z-transform *on the unit circle*, that is, at the z points:

$$z = e^{j\omega} \qquad (5.4.2)$$

Indeed, we have:[†]

$$X(z)\big|_{z=e^{j\omega}} = \sum_{n=-\infty}^{\infty} x(n) z^{-n} \bigg|_{z=e^{j\omega}} = \sum_{n=-\infty}^{\infty} x(n) e^{-j\omega n} = X(\omega)$$

The *frequency response* $H(\omega)$ of a linear system $h(n)$ with transfer function $H(z)$ is defined in the same way, namely,

$$H(\omega) = \sum_{n=-\infty}^{\infty} h(n) e^{-j\omega n} \qquad \text{(frequency response)} \qquad (5.4.3)$$

and it is also the evaluation of $H(z)$ on the unit circle:

$$H(\omega) = H(z)\big|_{z=e^{j\omega}}$$

As discussed in Chapter 1, the *digital frequency* ω is in units of [radians/sample] and is related to the physical frequency f in Hz by

$$\omega = \frac{2\pi f}{f_s} \qquad \text{(digital frequency)} \qquad (5.4.4)$$

The Nyquist interval $[-f_s/2, f_s/2]$ is the following interval in units of ω:

$$-\pi \leq \omega \leq \pi \qquad \text{(Nyquist interval)} \qquad (5.4.5)$$

In Chapter 1, the quantity $X(\omega)$ was denoted by

$$\hat{X}(f) = \sum_{n=-\infty}^{\infty} x(nT) e^{-2\pi j f n / f_s}$$

It was the Fourier spectrum of the sampled signal $x(nT)$ and was given by the periodic replication of the original analog spectrum at multiples of f_s.

[†]Here, we abuse the notation and write $X(\omega)$ instead of $X(e^{j\omega})$.

In units of ω, periodicity in f with period f_s becomes periodicity in ω with period 2π. Therefore, $X(\omega)$ may be considered only over one period, such as the Nyquist interval (5.4.5).

The *inverse DTFT* recovers the time sequence $x(n)$ from its spectrum $X(\omega)$ over the Nyquist interval:

$$\boxed{x(n) = \frac{1}{2\pi} \int_{-\pi}^{\pi} X(\omega)\, e^{j\omega n}\, d\omega} \qquad \text{(inverse DTFT)} \qquad (5.4.6)$$

It expresses $x(n)$ as a linear combination of discrete-time sinusoids $e^{j\omega n}$ of different frequencies. The relative amplitudes and phases of these sinusoidal components are given by the DTFT $X(\omega)$. One quick way to prove Eq. (5.4.6) is to think of Eq. (5.4.1) as the Fourier series expansion of the periodic function $X(\omega)$. Then, Eq. (5.4.6) gives simply the Fourier series expansion coefficients. In terms of the physical frequency f in Hertz, the inverse DTFT reads as

$$\boxed{x(n) = \frac{1}{f_s} \int_{-f_s/2}^{f_s/2} X(f)\, e^{2\pi j f n/f_s}\, df}$$

As an example, consider a (double-sided) complex sinusoid of frequency ω_0:

$$\boxed{x(n) = e^{j\omega_0 n}, \quad -\infty < n < \infty}$$

Then, its DTFT will be given by

$$X(\omega) = 2\pi\delta(\omega - \omega_0) + \text{(Nyquist replicas)}$$

where the term "Nyquist replicas" refers to the periodic replication of the first term at intervals of 2π. This is needed in order to make $X(\omega)$ periodic with period 2π. More precisely, the full expression will be

$$X(\omega) = 2\pi \sum_{m=-\infty}^{\infty} \delta(\omega - \omega_0 - 2\pi m)$$

To verify it, we insert it into the inverse DTFT equation (5.4.6) and recover the given sinusoid. It was also discussed in Chapter 1, Example 1.5.1. Assuming that ω_0 lies in the Nyquist interval $[-\pi, \pi]$, then the restriction of $X(\omega)$ within it will be given only by the $m = 0$ term, that is:

$$\boxed{X(\omega) = 2\pi\delta(\omega - \omega_0), \quad -\pi \le \omega \le \pi}$$

Therefore, Eq. (5.4.6) gives

$$x(n) = \frac{1}{2\pi} \int_{-\pi}^{\pi} X(\omega)\, e^{j\omega n}\, d\omega = \frac{1}{2\pi} \int_{-\pi}^{\pi} 2\pi\delta(\omega - \omega_0)\, e^{j\omega n}\, d\omega = e^{j\omega_0 n}$$

Similarly, for a linear combination of two sinusoids we have:

$$x(n) = A_1 e^{j\omega_1 n} + A_2 e^{j\omega_2 n} \longrightarrow X(\omega) = 2\pi A_1 \delta(\omega - \omega_1) + 2\pi A_2 \delta(\omega - \omega_2)$$

This can be verified in the same way, if we assume that both ω_1 and ω_2 lie in the Nyquist interval. In particular, for real-valued cosine and sine signals, we have:

$$\cos(\omega_0 n) \longrightarrow \pi\delta(\omega - \omega_0) + \pi\delta(\omega + \omega_0)$$
$$\sin(\omega_0 n) \longrightarrow -j\pi\delta(\omega - \omega_0) + j\pi\delta(\omega + \omega_0)$$

Another useful relationship is *Parseval's equation*, which relates the *total energy* of a sequence to its spectrum:

$$\sum_{n=-\infty}^{\infty} |x(n)|^2 = \frac{1}{2\pi} \int_{-\pi}^{\pi} |X(\omega)|^2 \, d\omega \qquad \text{(Parseval)} \qquad (5.4.7)$$

The DTFT can be given a geometric interpretation by recognizing that the points $z = e^{j\omega}$ lie on the unit circle on the z-plane. As ω varies over the Nyquist interval $[-\pi, \pi]$, the complex point $z = e^{j\omega}$ moves around the unit circle, as shown in Fig. 5.4.1. The phase angle of z is ω.

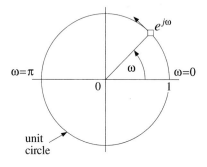

Fig. 5.4.1 Evaluation of z-transform on the unit circle.

In order for the spectrum $X(\omega)$ to exist,[†] the ROC of the z-transform $X(z)$ must contain the unit circle; otherwise the z-transform will diverge at the unit circle points $z = e^{j\omega}$. But if the ROC contains the unit circle, the signal $x(n)$ must be stable. Thus, the Fourier transform $X(\omega)$ exists only for *stable* signals.

Marginally stable signals, such as sinusoids, strictly speaking do not have a spectrum because their poles lie on the unit circle and therefore the evaluation of $X(z)$ on the unit circle will cause $X(z)$ to diverge at certain z's. However, it is intuitively useful to consider their spectra. For example, for the causal complex sinusoid of the previous section we have:

$$x(n) = e^{j\omega_0 n} u(n) \xrightarrow{Z} X(z) = \frac{1}{1 - e^{j\omega_0} z^{-1}}$$

[†] That is, to be finite $X(\omega) \neq \infty$ for all ω.

and therefore the *formal* replacement of z by $e^{j\omega}$ will yield

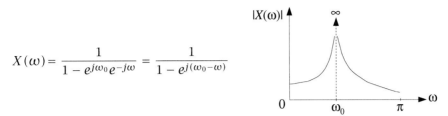

$$X(\omega) = \frac{1}{1 - e^{j\omega_0}e^{-j\omega}} = \frac{1}{1 - e^{j(\omega_0 - \omega)}}$$

which diverges at $\omega = \omega_0$. However, this is to be expected because if the signal were a pure sinusoid $x(n) = e^{j\omega_0 n}$, its spectrum would be a single spectral line concentrated at $\omega = \omega_0$, that is, $X(\omega) = 2\pi\delta(\omega - \omega_0)$ (plus its Nyquist replicas). Here, the signal is not a pure sinusoid; it is a causal, truncated version of a pure sinusoid and therefore additional frequencies are present. However, the dominant frequency is still ω_0.

The shape of the spectrum $X(\omega)$ or $H(\omega)$ is affected by the *pole/zero pattern* of the z-transform $X(z)$ or $H(z)$, that is, by the relative geometric locations of the poles and zeros on the z-plane. To see this, consider a simple z-transform having a single pole at $z = p_1$ and a single zero at $z = z_1$.

$$X(z) = \frac{1 - z_1 z^{-1}}{1 - p_1 z^{-1}} = \frac{z - z_1}{z - p_1}$$

The corresponding spectrum and its magnitude are obtained by replacing z by $e^{j\omega}$:

$$X(\omega) = \frac{e^{j\omega} - z_1}{e^{j\omega} - p_1} \quad \Rightarrow \quad |X(\omega)| = \frac{|e^{j\omega} - z_1|}{|e^{j\omega} - p_1|}$$

Figure 5.4.2 shows the relative locations of the fixed points z_1, p_1 and the moving point $z = e^{j\omega}$. A rough plot of $|X(\omega)|$ based on this pole/zero pattern is also shown. The magnitude spectrum $|X(\omega)|$ is the ratio of the distance of the point $e^{j\omega}$ to the zero z_1, namely, $|e^{j\omega} - z_1|$ divided by the distance of $e^{j\omega}$ to the pole p_1, namely, $|e^{j\omega} - p_1|$.

As $e^{j\omega}$ moves around the unit circle, these distances will vary. As $e^{j\omega}$ passes near the pole, the denominator distance will become small causing the value of $|X(\omega)|$ to increase. If ω_1 is the phase angle of the pole p_1, then the point of closest approach to p_1 will occur at $\omega = \omega_1$ causing a *peak* in $|X(\omega)|$ there. The closer the pole is to the unit circle, the smaller the denominator distance will become at $\omega = \omega_1$, and the sharper the peak of $|X(\omega)|$.

Similarly, as $e^{j\omega}$ passes near the zero z_1, the numerator distance will become small, causing $|X(\omega)|$ to decrease. At the zero's phase angle, say $\omega = \phi_1$, this distance will be smallest, causing a *dip* in $|X(\omega)|$ there. The closer the zero to the unit circle, the sharper the dip. The zero z_1 can also lie *on* the unit circle, in which case $|X(\omega)|$ will vanish exactly at $\omega = \phi_1$.

In summary, we can draw a rough sketch of the spectrum $|X(\omega)|$ by letting $e^{j\omega}$ trace the unit circle and draw peaks as $e^{j\omega}$ passes near poles, and dips as it passes

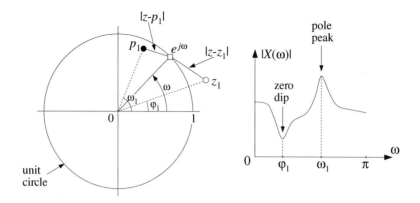

Fig. 5.4.2 Geometric interpretation of frequency spectrum.

near zeros. By proper location of the zeros and poles of $X(z)$ or $H(z)$, one can design any desired shape for $X(\omega)$ or $H(\omega)$.

It is convenient to divide the unit circle into low-, medium-, and high-frequency wedge regions, as shown in Fig. 5.4.3. This subdivision is somewhat arbitrary because what is "low" or "high" frequency depends on the application. It aids, however, in the placement of poles and zeros. For example, to make a lowpass filter that emphasizes low frequencies and attenuates high ones, one would place poles inside the circle somewhere within the low-frequency wedge and/or zeros within the high-frequency wedge.

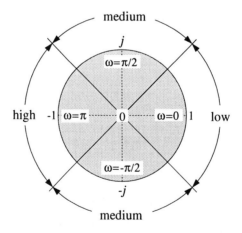

Fig. 5.4.3 Low-, medium-, and high-frequency parts of the unit circle.

Such filter design methods are somewhat crude and are used in practice only for the design of simple and/or specialized filters, such as resonator or notch filters or biquadratic filter sections for digital audio graphic and parametric equalizers. Such design examples will be considered later on.

The DTFT $X(\omega)$ of a signal $x(n)$ is a complex-valued quantity and therefore, it can be characterized also by its real and imaginary parts $\operatorname{Re} X(\omega)$, $\operatorname{Im} X(\omega)$ or, in its polar form, by its magnitude and phase responses $|X(\omega)|$, $\arg X(\omega)$. Thus,

$$X(\omega) = \operatorname{Re} X(\omega) + j \operatorname{Im} X(\omega) = |X(\omega)| \, e^{j \arg X(\omega)}$$

For *real-valued* signals $x(n)$, the quantity $X(\omega)$ satisfies the following so-called *hermitian* property:

$$\boxed{X(\omega)^* = X(-\omega)} \qquad (5.4.8)$$

which translates to the following relationships for the magnitude and phase responses:

$$\boxed{\begin{aligned} |X(\omega)| &= |X(-\omega)| \\ \arg X(\omega) &= -\arg X(-\omega) \end{aligned}} \qquad (5.4.9)$$

that is, the magnitude response is *even* in ω and the phase response *odd*. Similar definitions and results apply to the frequency response $H(\omega)$ of a real-valued system $h(n)$.

We note finally that the multiplicative filtering property $Y(z) = H(z)X(z)$ evaluated on the unit circle takes the following frequency-domain form:

$$\boxed{Y(\omega) = H(\omega)X(\omega)} \qquad \text{(filtering in frequency domain)} \qquad (5.4.10)$$

Its consequences will be explored later on.

5.5 Inverse z-Transforms

The problem of inverting a given z-transform $X(z)$ is to find the time signal $x(n)$ whose z-transform is $X(z)$. As we saw already, the answer for $x(n)$ is not necessarily unique. But it can be made unique by specifying the corresponding ROC.

In inverting a z-transform, it is convenient to break it into its *partial fraction* (PF) expansion form, that is, into a sum of individual pole terms of the type (5.3.2).

Once $X(z)$ is written in the form (5.3.2), one still needs to know how to invert each term, that is, causally or anticausally. This depends on the choice of the ROC.

In general, the circles through the poles at $z = p_1$, $z = p_2$, and so on, divide the z-plane into non-overlapping regions, which are all possible candidates for ROCs. Any one of these ROC regions will result into a different $x(n)$. Among all possible $x(n)$, there will be a *unique* one that is *stable*, because the unit circle lies in exactly one of the possible ROCs.

Example 5.5.1: In Example (5.2.3), the first three signals had a common z-transform:

$$X(z) = \frac{1}{1 - 0.8z^{-1}} + \frac{1}{1 - 1.25z^{-1}}$$

The two circles through the poles at $z = 0.8$ and $z = 1.25$ divide the z-plane into the three regions I, II, III, shown in Example 5.2.3. There are therefore three possible inverse z-transforms, that is, three different signals $x(n)$ corresponding to the three ROC choices. But, only II is stable.   □

The *partial fraction expansion method* can be applied to z-transforms that are ratios of two polynomials in z^{-1} of the form:

$$X(z) = \frac{N(z)}{D(z)}$$

The zeros of the denominator polynomial $D(z)$ are the poles of $X(z)$. Assuming $D(z)$ has degree M, there will be M denominator zeros, say at $p_1, p_2, \ldots, p_M$, and $D(z)$ may be assumed to be in the factored form

$$D(z) = (1 - p_1 z^{-1})(1 - p_2 z^{-1}) \cdots (1 - p_M z^{-1})$$

The partial fraction expansion of $X(z)$ is given by[†]

$$X(z) = \frac{N(z)}{D(z)} = \frac{N(z)}{(1 - p_1 z^{-1})(1 - p_2 z^{-1}) \cdots (1 - p_M z^{-1})}$$

$$= \frac{A_1}{1 - p_1 z^{-1}} + \frac{A_2}{1 - p_2 z^{-1}} + \cdots + \frac{A_M}{1 - p_M z^{-1}}$$

(5.5.1)

For this expansion to be possible as an identity in z^{-1}, the degree of the numerator polynomial $N(z)$ must be *strictly less* than the degree M of the denominator polynomial. The PF expansion coefficients A_i can be computed by the formulas:

$$A_i = \left[(1 - p_i z^{-1}) X(z)\right]_{z = p_i} = \left[\frac{N(z)}{\prod_{j \neq i} (1 - p_j z^{-1})} \right]_{z = p_i}$$

(5.5.2)

for $i = 1, 2, \ldots, M$. In words, the factor $(1 - p_i z^{-1})$ is deleted from the denominator and the remaining expression is evaluated at the pole $z = p_i$.

Example 5.5.2: In Example 5.2.3 the z-transform was written in the form

$$X(z) = \frac{2 - 2.05 z^{-1}}{1 - 2.05 z^{-1} + z^{-2}} = \frac{2 - 2.05 z^{-1}}{(1 - 0.8 z^{-1})(1 - 1.25 z^{-1})}$$

Because the numerator polynomial has degree one in the variable z^{-1}, there is a PF expansion of the form:

$$X(z) = \frac{2 - 2.05 z^{-1}}{(1 - 0.8 z^{-1})(1 - 1.25 z^{-1})} = \frac{A_1}{1 - 0.8 z^{-1}} + \frac{A_2}{1 - 1.25 z^{-1}}$$

The two coefficients are obtained by Eq. (5.5.2) as follows:

[†]We have assumed that all the poles are single poles.

$$A_1 = \left[(1 - 0.8z^{-1})X(z)\right]_{z=0.8} = \left[\frac{2 - 2.05z^{-1}}{1 - 1.25z^{-1}}\right]_{z=0.8} = \frac{2 - 2.05/0.8}{1 - 1.25/0.8} = 1$$

$$A_2 = \left[(1 - 1.25z^{-1})X(z)\right]_{z=1.25} = \left[\frac{2 - 2.05z^{-1}}{1 - 0.8z^{-1}}\right]_{z=1.25} = 1$$

which are as expected. $\square$

If the degree of the numerator polynomial $N(z)$ is exactly *equal* to the degree M of the denominator $D(z)$, then the PF expansion (5.5.1) must be modified by adding an extra term of the form:

$$
\begin{aligned}
X(z) = \frac{N(z)}{D(z)} &= \frac{N(z)}{(1 - p_1 z^{-1})(1 - p_2 z^{-1}) \cdots (1 - p_M z^{-1})} \\
&= A_0 + \frac{A_1}{1 - p_1 z^{-1}} + \frac{A_2}{1 - p_2 z^{-1}} + \cdots + \frac{A_M}{1 - p_M z^{-1}}
\end{aligned}
\tag{5.5.3}
$$

The coefficients A_i, $i = 1, 2, \ldots, M$ are computed in exactly the same way by Eq. (5.5.2). The extra coefficient A_0 is computed by evaluating the z-transform at $z = 0$, that is,

$$A_0 = X(z)\big|_{z=0} \tag{5.5.4}$$

If the degree of $N(z)$ is strictly *greater* than M, one may divide the polynomial $D(z)$ into $N(z)$, finding the *quotient* and *remainder* polynomials, so that

$$N(z) = Q(z)D(z) + R(z)$$

and then writing

$$X(z) = \frac{N(z)}{D(z)} = \frac{Q(z)D(z) + R(z)}{D(z)} = Q(z) + \frac{R(z)}{D(z)}$$

where now the second term will admit an ordinary PF expansion of the form (5.5.1) because the degree of the remainder polynomial $R(z)$ is strictly less than M. Alternatively, one may simply *remove* the numerator polynomial $N(z)$ altogether, then carry out an ordinary PF expansion of the quantity

$$W(z) = \frac{1}{D(z)}$$

and finally *restore* the numerator by writing

$$X(z) = N(z)W(z)$$

We may refer to this method as the "remove/restore" method. Some examples will illustrate these techniques.

Example 5.5.3: We emphasize that a PF expansion may exist in one independent variable, say z^{-1}, but not in another, say z. For example, the z-transform

$$X(z) = \frac{2 - 2.05z^{-1}}{(1 - 0.8z^{-1})(1 - 1.25z^{-1})} = \frac{z(2z - 2.05)}{(z - 0.8)(z - 1.25)}$$

has numerator of degree one with respect to the variable z^{-1}, but degree two with respect to z. Thus, it admits an expansion of the form (5.5.1) with respect to z^{-1}, but not with respect to z.

Many texts prefer to work with z and therefore to make the PF expansion possible, a factor z is divided out to lower the degree of the numerator and then restored at the end, that is,

$$\frac{X(z)}{z} = \frac{(2z - 2.05)}{(z - 0.8)(z - 1.25)} = \frac{A_1}{z - 0.8} + \frac{A_2}{z - 1.25}$$

When z is restored, one gets

$$X(z) = \frac{zA_1}{z - 0.8} + \frac{zA_2}{z - 1.25} = \frac{A_1}{1 - 0.8z^{-1}} + \frac{A_2}{1 - 1.25z^{-1}}$$

It is easily verified that the PF expansion coefficients will be the same in the two approaches. In this book, we prefer to work directly with z^{-1} and avoid the extra algebraic steps required to write everything in terms of z, divide by z, restore z, and rewrite the final answer in terms of z^{-1}.   □

Example 5.5.4: Compute all possible inverse z-transforms of

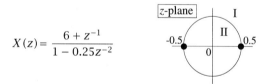

$$X(z) = \frac{6 + z^{-1}}{1 - 0.25z^{-2}}$$

Solution: Because the numerator has degree one in z^{-1}, we have the PF expansion:

$$X(z) = \frac{6 + z^{-1}}{1 - 0.25z^{-2}} = \frac{6 + z^{-1}}{(1 - 0.5z^{-1})(1 + 0.5z^{-1})} = \frac{A_1}{1 - 0.5z^{-1}} + \frac{A_2}{1 + 0.5z^{-1}}$$

where

$$A_1 = \left[\frac{6 + z^{-1}}{1 + 0.5z^{-1}}\right]_{z=0.5} = 4, \qquad A_2 = \left[\frac{6 + z^{-1}}{1 - 0.5z^{-1}}\right]_{z=-0.5} = 2$$

The two poles at ±0.5 have the same magnitude and therefore divide the z-plane into two ROC regions I and II: $|z| > 0.5$ and $|z| < 0.5$. For the first ROC, both terms in the PF expansion are inverted causally giving:

$$x(n) = A_1(0.5)^n u(n) + A_2(-0.5)^n u(n)$$

Because this ROC also contains the unit circle the signal $x(n)$ will be stable. For the second ROC, both PF expansion terms are inverted anticausally giving:

$$x(n) = -A_1 (0.5)^n u(-n-1) - A_2 (-0.5)^n u(-n-1)$$

This answer is unstable, because the ROC does not contain the unit circle. □

Example 5.5.5: Determine all inverse z-transforms of

$$X(z) = \frac{10 + z^{-1} - z^{-2}}{1 - 0.25z^{-2}}$$

Solution: Ordinary partial fraction expansion is not valid in this case because the degree of the numerator is the same as the degree of the denominator. However, we may still have an expansion of the form (5.5.3)

$$X(z) = \frac{10 + z^{-1} - z^{-2}}{1 - 0.25z^{-2}} = \frac{10 + z^{-1} - z^{-2}}{(1 - 0.5z^{-1})(1 + 0.5z^{-1})}$$

$$= A_0 + \frac{A_1}{1 - 0.5z^{-1}} + \frac{A_2}{1 + 0.5z^{-1}}$$

where A_1 and A_2 are determined in the usual manner and A_0 is determined by evaluating $X(z)$ at $z = 0$:

$$A_0 = \left[\frac{10 + z^{-1} - z^{-2}}{1 - 0.25z^{-2}} \right]_{z=0} = \left[\frac{10z^2 + z - 1}{z^2 - 0.25} \right]_{z=0} = \frac{-1}{-0.25} = 4$$

$$A_1 = \left[\frac{10 + z^{-1} - z^{-2}}{1 + 0.5z^{-1}} \right]_{z=0.5} = 4, \qquad A_2 = \left[\frac{10 + z^{-1} - z^{-2}}{1 - 0.5z^{-1}} \right]_{z=-0.5} = 2$$

Again, there are only two ROCs I and II: $|z| > 0.5$ and $|z| < 0.5$. For the first ROC, the A_1 and A_2 terms are inverted causally, and the A_0 term inverts into a simple $\delta(n)$:

$$x(n) = A_0 \delta(n) + A_1 (0.5)^n u(n) + A_2 (-0.5)^n u(n)$$

For the second ROC, we have:

$$x(n) = A_0 \delta(n) - A_1 (0.5)^n u(-n-1) - A_2 (-0.5)^n u(-n-1)$$

Only the first inverse is stable because its ROC contains the unit circle. □

Example 5.5.6: Determine the causal inverse z-transform of

$$X(z) = \frac{6 + z^{-5}}{1 - 0.25z^{-2}}$$

Solution: Here, the degree of the numerator is strictly greater than that of the denominator. The first technique is to divide the denominator into the numerator, giving

$$(6 + z^{-5}) = (1 - 0.25z^{-2})(-16z^{-1} - 4z^{-3}) + (6 + 16z^{-1})$$

where $(6 + 16z^{-1})$ is the remainder polynomial and $(-16z^{-1} - 4z^{-3})$ the quotient. Then,

$$X(z) = \frac{6 + z^{-5}}{1 - 0.25z^{-2}} = -16z^{-1} - 4z^{-3} + \frac{6 + 16z^{-1}}{1 - 0.25z^{-2}}$$

and expanding the last term in PF expansion:

$$X(z) = -16z^{-1} - 4z^{-3} + \frac{19}{1 - 0.5z^{-1}} - \frac{13}{1 + 0.5z^{-1}}$$

The causal inverse, having ROC $|z| > 0.5$, will be:

$$x(n) = -16\delta(n-1) - 4\delta(n-3) + 19(0.5)^n u(n) - 13(-0.5)^n u(n)$$

The second technique is the "remove/restore" method. Ignoring the numerator we have

$$W(z) = \frac{1}{1 - 0.25z^{-2}} = \frac{0.5}{1 - 0.5z^{-1}} + \frac{0.5}{1 + 0.5z^{-1}}$$

which has the causal inverse

$$w(n) = 0.5(0.5)^n u(n) + 0.5(-0.5)^n u(n)$$

Once $w(n)$ is known, one can obtain $x(n)$ by restoring the numerator:

$$X(z) = (6 + z^{-5})W(z) = 6W(z) + z^{-5}W(z)$$

Taking inverse z-transforms of both sides and using the delay property, we find

$$x(n) = 6w(n) + w(n-5) = 3(0.5)^n u(n) + 3(-0.5)^n u(n)$$
$$+ 0.5(0.5)^{n-5} u(n-5) + 0.5(-0.5)^{n-5} u(n-5)$$

The two expressions for $x(n)$ from the two techniques are equivalent. □

Example 5.5.7: Determine all possible inverse z-transforms of

$$X(z) = \frac{7 - 9.5z^{-1} - 3.5z^{-2} + 5.5z^{-3}}{(1 - z^{-2})(1 - 0.5z^{-1})(1 - 1.5z^{-1})}$$

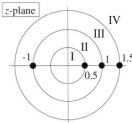

Solution: $X(z)$ admits the PF expansion:

$$X(z) = \frac{1}{1 - z^{-1}} + \frac{1}{1 + z^{-1}} + \frac{3}{1 - 0.5z^{-1}} + \frac{2}{1 - 1.5z^{-1}}$$

where the PF expansion coefficients are easily found. The four poles at $z = 0.5, 1, -1,$ 1.5 divide the z-plane into the four ROC regions I, II, III, IV. Region I corresponds to the completely anticausal inverse and region IV to the completely causal one. For region II, the pole at $z = 0.5$ will be inverted causally and the rest anticausally. For region III, $z = 0.5$ and $z = \pm 1$ will be inverted causally and $z = 1.5$ anticausally. Thus, the four possible inverse z-transforms are:

$$x_1(n) = -\left[1 + (-1)^n + 3(0.5)^n + 2(1.5)^n\right]u(-n-1)$$

$$x_2(n) = 3(0.5)^n u(n) - \left[1 + (-1)^n + 2(1.5)^n\right]u(-n-1)$$

$$x_3(n) = \left[1 + (-1)^n + 3(0.5)^n\right]u(n) - 2(1.5)^n u(-n-1)$$

$$x_4(n) = \left[1 + (-1)^n + 3(0.5)^n + 2(1.5)^n\right]u(n)$$

Strictly speaking there is no stable answer because two of the poles, $z = \pm 1$, lie on the unit circle. However, $x_2(n)$ and $x_3(n)$ are marginally stable, that is, neither diverging nor converging to zero for large n. In both cases, the anticausal term $(1.5)^n$ tends to zero for large negative n. Indeed, because n is negative, we write $n = -|n|$ and

$$(1.5)^n = (1.5)^{-|n|} \to 0 \quad \text{as} \quad n \to -\infty$$

The terms due to the poles $z = \pm 1$ are causal or anticausal in cases III and II, but they remain bounded. The other two signals $x_1(n)$ and $x_4(n)$ are unstable because the unit circle does not lie in their ROCs. $\qquad\square$

The assumption that the numerator and denominator polynomials $N(z)$ and $D(z)$ have *real-valued* coefficients implies that the *complex-valued* poles of $X(z)$ come in *complex-conjugate pairs*. In that case, the PF expansion takes the form

$$X(z) = \frac{A_1}{1 - p_1 z^{-1}} + \frac{A_1^*}{1 - p_1^* z^{-1}} + \frac{A_2}{1 - p_2 z^{-1}} + \cdots$$

where the PF expansion coefficients also come in *conjugate pairs.* Thus, it is necessary to determine only one of them, not both. The corresponding inverse z-transform will be real-valued; indeed, considering the causal case we have

$$x(n) = A_1 p_1^n u(n) + A_1^* p_1^{*n} u(n) + A_2 p_2^n u(n) + \cdots$$

Because the first two terms are complex conjugates of each other, we may use the result that $C + C^* = 2\text{Re}(C)$, for any complex number C, to write the first term as

$$A_1 p_1^n + A_1^* p_1^{*n} = 2\text{Re}\left[A_1 p_1^n\right]$$

Writing A_1 and p_1 in their polar form, say, $A_1 = B_1 e^{j\alpha_1}$ and $p_1 = R_1 e^{j\omega_1}$, with $B_1 > 0$ and $R_1 > 0$, we have

$$\text{Re}\left[A_1 p_1^n\right] = \text{Re}\left[B_1 e^{j\alpha_1} R_1^n e^{j\omega_1 n}\right] = B_1 R_1^n \text{Re}\left[e^{j\omega_1 n + j\alpha_1}\right]$$

and taking the real part of the exponential, we find

$$A_1 p_1^n + A_1^* p_1^{*n} = 2\text{Re}\left[A_1 p_1^n\right] = 2B_1 R_1^n \cos(\omega_1 n + \alpha_1)$$

and for $x(n)$

$$x(n) = 2B_1 R_1^n \cos(\omega_1 n + \alpha_1) u(n) + A_2 p_2^n u(n) + \cdots$$

Thus, complex-valued poles correspond to *exponentially decaying sinusoids* (if $R_1 < 1$). The decay envelope R_1^n and the frequency ω_1 depend on the complex pole by $p_1 = R_1 e^{j\omega_1}$.

The first-order terms in the partial fraction expansion corresponding to complex conjugate poles can be reassembled into second-order terms with *real-valued* coefficients, as follows:

$$\frac{A_1}{1 - p_1 z^{-1}} + \frac{A_1^*}{1 - p_1^* z^{-1}} = \frac{(A_1 + A_1^*) - (A_1 p_1^* + A_1^* p_1) z^{-1}}{(1 - p_1 z^{-1})(1 - p_1^* z^{-1})}$$

Using the identities

$$\boxed{(1 - p_1 z^{-1})(1 - p_1^* z^{-1}) = 1 - 2\text{Re}(p_1) z^{-1} + |p_1|^2 z^{-2}}$$

or,

$$\boxed{(1 - R_1 e^{j\omega_1} z^{-1})(1 - R_1 e^{-j\omega_1} z^{-1}) = 1 - 2R_1 \cos(\omega_1) z^{-1} + R_1^2 z^{-2}}$$

and writing

$$A_1 + A_1^* = 2\text{Re}(A_1) = 2B_1 \cos(\alpha_1)$$

$$A_1 p_1^* + A_1^* p_1 = 2\text{Re}(A_1 p_1^*) = 2B_1 R_1 \cos(\alpha_1 - \omega_1)$$

we find

$$\frac{A_1}{1 - p_1 z^{-1}} + \frac{A_1^*}{1 - p_1^* z^{-1}} = \frac{2B_1 \cos(\alpha_1) - 2B_1 R_1 \cos(\alpha_1 - \omega_1) z^{-1}}{1 - 2R_1 \cos(\omega_1) z^{-1} + R_1^2 z^{-2}}$$

having real-valued coefficients.

Example 5.5.8: Determine all possible inverse z-transforms of

$$X(z) = \frac{4 - 3z^{-1} + z^{-2}}{1 + 0.25z^{-2}}$$

Solution: We write

$$X(z) = \frac{4 - 3z^{-1} + z^{-2}}{1 + 0.25z^{-2}} = \frac{4 - 3z^{-1} + z^{-2}}{(1 - 0.5jz^{-1})(1 + 0.5jz^{-1})}$$

$$= A_0 + \frac{A_1}{1 - 0.5jz^{-1}} + \frac{A_1^*}{1 + 0.5jz^{-1}}$$

with the numerical values:

$$A_0 = \left[\frac{4 - 3z^{-1} + z^{-2}}{1 + 0.25z^{-2}}\right]_{z=0} = 4, \qquad A_1 = \left[\frac{4 - 3z^{-1} + z^{-2}}{1 + 0.5jz^{-1}}\right]_{z=0.5j} = 3j$$

Therefore,

$$X(z) = 4 + \frac{3j}{1 - 0.5jz^{-1}} - \frac{3j}{1 + 0.5jz^{-1}} = 4 - \frac{3z^{-1}}{1 + 0.25z^{-2}}$$

The causal ROC is $|z| > |0.5j| = 0.5$, resulting in

$$x(n) = 4\delta(n) + 3j(0.5j)^n u(n) - 3j(-0.5j)^n u(n)$$

Because the last two terms are complex conjugates of each other, we may write them as

$$x(n) = 4\delta(n) + 2\text{Re}\left[3j(0.5j)^n u(n)\right] = 4\delta(n) + 6(0.5)^n u(n)\text{Re}\left[j^{n+1}\right]$$

Writing $j^{n+1} = e^{j\pi(n+1)/2}$ and taking real parts we find

$$\text{Re}\left[j^{n+1}\right] = \cos\left(\frac{\pi(n+1)}{2}\right) = -\sin\left(\frac{\pi n}{2}\right)$$

and

$$x(n) = 4\delta(n) - 6(0.5)^n \sin\left(\frac{\pi n}{2}\right) u(n)$$

Similarly, we find

$$x(n) = 4\delta(n) + 6(0.5)^n \sin\left(\frac{\pi n}{2}\right) u(-n-1)$$

for the anticausal version with ROC $|z| < 0.5$. Some additional examples with complex conjugate poles were cases (6-9) of Example 5.2.2.   □

5.6 Problems

5.1 Prove the linearity, delay, and convolution properties of z-transforms given by Eqs. (5.1.3)–(5.1.5).

5.2 Compute the z-transform of the following sequences and determine the corresponding region of convergence:

    a. $x(n) = \delta(n-5)$

    b. $x(n) = \delta(n+5)$

    c. $x(n) = u(n-5)$

    d. $x(n) = u(-n-5)$

5.3 Compute the z-transform of the following sequences and determine the corresponding region of convergence:

    a. $x(n) = (-0.5)^n u(n)$

    b. $x(n) = (-0.5)^n [u(n) - u(n-10)]$

    c. $x(n) = (0.5)^n u(n) + (-0.5)^n u(n)$

5.4 Compute the z-transform of the following sequences and determine the corresponding region of convergence:

    a. $x(n) = 2(0.9)^n \cos(\pi n/2) u(n)$

    b. $x(n) = (0.9j)^n u(n) + (-0.9j)^n u(n)$

5.5 Compute the z-transform of the following sequences and determine the corresponding region of convergence:

    a. $x(n) = (0.25)^n u(n) + 4^n u(n)$

    b. $x(n) = (0.25)^n u(n) - 4^n u(-n-1)$

    c. $x(n) = -(0.25)^n u(-n-1) - 4^n u(-n-1)$

    d. Explain why $x(n) = -(0.25)^n u(-n-1) + 4^n u(n)$ does not have a z-transform.

5.6 Using the power series definition of z-transforms, derive the z-transform and its ROC of the signal $x(n) = \cos(\pi n/2) u(n)$.

5.7 Using partial fractions or power series expansions, compute the inverse z-transform of the following z-transforms and determine whether the answer is causal and/or stable:

    a. $X(z) = (1 - 4z^{-2})(1 + 3z^{-1})$

    b. $X(z) = 5 + 3z^3 + 2z^{-2}$

5.8 Using partial fractions or power series expansions, determine all possible inverse z-transforms of the following z-transforms, sketch their ROCs, and discuss their stability and causality properties:

    a. $X(z) = \dfrac{3(1 + 0.3z^{-1})}{1 - 0.81z^{-2}}$

    b. $X(z) = \dfrac{6 - 3z^{-1} - 2z^{-2}}{1 - 0.25z^{-2}}$

    c. $X(z) = \dfrac{6 + z^{-5}}{1 - 0.64z^{-2}}$

d. $X(z) = \dfrac{10 + z^{-2}}{1 + 0.25z^{-2}}$

e. $X(z) = \dfrac{6 - 2z^{-1} - z^{-2}}{(1 - z^{-1})(1 - 0.25z^{-2})}$,   ROC $|z| > 1$

f. $X(z) = -4 + \dfrac{1}{1 + 4z^{-2}}$

g. $X(z) = \dfrac{4 - 0.6z^{-1} + 0.2z^{-2}}{(1 - 0.5z^{-1})(1 + 0.4z^{-1})}$

5.9 Consider the z-transform pair:

$$ x_a(n) = a^n u(n) \quad \Leftrightarrow \quad X_a(z) = \frac{1}{1 - az^{-1}} $$

Applying the derivative operator $\partial/\partial a$ to the pair, derive the z-transform of the sequence $x(n) = na^n u(n)$.

5.10 Consider the differential operator $D = a\dfrac{\partial}{\partial a}$. First, show that its k-fold application gives $D^k a^n = n^k a^n$. Then, use this result to obtain the z-transform of $x(n) = n^k a^n u(n)$. Derive the explicit transforms for the cases $k = 1, 2, 3, 4$.

5.11 Show the z-transform of a triangular signal:

$$ \sum_{n=-L}^{L} \left(1 - \frac{|n|}{L}\right) z^{-n} = \frac{1}{L}\left[\frac{1 - z^{-L}}{1 - z^{-1}}\right]^2 z^{L-1} $$

5.12 Using Euler's formula and the z-transform pair of Problem 5.9, derive the z-transforms of the signals $x(n) = R^n \cos(\omega_0 n) u(n)$ and $x(n) = R^n \sin(\omega_0 n) u(n)$.

5.13 Consider the causal sequence $x(n) = \{a_0, a_1, a_2, a_3, a_0, a_1, a_2, a_3, \cdots\}$, where the dots indicate the periodic repetition of the four samples $\{a_0, a_1, a_2, a_3\}$. Determine the z-transform of $x(n)$ and the corresponding ROC.

5.14 Using partial fraction expansions, determine the inverse z-transform of the z-transform of Problem 5.13. Verify that the sum of the PFE terms generate the periodic sequence $x(n)$ of that problem.

5.15 Consider the z-transform for $|z| > 1$:

$$ X(z) = 1 - z^{-2} + z^{-4} - z^{-6} + z^{-8} - \cdots $$

Derive a rational expression for $X(z)$ in two ways: (a) by summing the above series, and (b) by showing that it satisfies the equation $X(z) = 1 - z^{-2}X(z)$.

Derive also the inverse z-transform $x(n)$ for all n.

5.16 Without using partial fractions, determine the causal inverse z-transforms of:

a. $X(z) = \dfrac{1}{1 + z^{-4}}$

b. $X(z) = \dfrac{1}{1 - z^{-4}}$

c. $X(z) = \dfrac{1}{1 + z^{-8}}$

d. $X(z) = \dfrac{1}{1 - z^{-8}}$

5.17 Using partial fraction expansions, determine the inverse z-transforms of Problem 5.16. Verify that you get the same answers as in that problem.

5.18 Consider a transfer function $H(z) = N(z)/D(z)$, where the numerator and denominator polynomials have real-valued coefficients and degrees L and M in z^{-1}, and assume $L > M$. Show that $H(z)$ can be written in the form:

$$H(z) = Q(z) + \sum_{i=1}^{K} \frac{b_{i0} + z^{-1}b_{i1}}{1 + a_{i1}z^{-1} + a_{i2}z^{-2}}$$

where $Q(z)$ is a polynomial of degree $L - M$ in z^{-1} and the second-order sections have real coefficients. The number of sections K is related to M by $K = M/2$ if M is even and $K = (M-1)/2$ if M is odd. This result forms the basis of the *parallel realization form* of $H(z)$.

5.19 Determine the factorization into first-order zeros of:

$$1 - z^{-D} = \prod_{k=0}^{D-1} (1 - z_k z^{-1})$$

$$1 + z^{-D} = \prod_{k=0}^{D-1} (1 - z_k z^{-1})$$

where D is an integer. What are the zeros z_k in the two cases? For $D = 4$ and $D = 8$, place these zeros on the z-plane with respect to the unit circle.

5.20 Given $a > 0$ and integer D, repeat the previous problem for:

$$1 - az^{-D} = \prod_{k=0}^{D-1} (1 - z_k z^{-1})$$

$$1 + az^{-D} = \prod_{k=0}^{D-1} (1 - z_k z^{-1})$$

5.21 Prove the "modulation" property of z-transforms:

$$x(n) \xrightarrow{z} X(z) \qquad \Rightarrow \qquad a^n x(n) \xrightarrow{z} X(z/a)$$

For $a = e^{j\omega_0}$, show that in the frequency domain this property becomes:

$$x(n) \longrightarrow X(\omega) \qquad \Rightarrow \qquad e^{j\omega_0 n} x(n) \longrightarrow X(\omega - \omega_0)$$

5.22 Given the DTFT equation (5.4.1), prove the inverse DTFT, Eq. (5.4.6).

5.23 Prove the Parseval equation (5.4.7).

5.24 For real-valued signals, prove the hermitian properties (5.4.8) and (5.4.9). What are the hermitian properties satisfied by the real and imaginary parts of the DTFT spectrum?

6

Transfer Functions

6.1 Equivalent Descriptions of Digital Filters

In this chapter, with the aid of z-transforms, we develop several *mathematically equivalent* ways to describe and characterize FIR and IIR filters, namely, in terms of their:

- Transfer function $H(z)$
- Frequency response $H(\omega)$
- Block diagram realization and sample processing algorithm
- I/O difference equation
- Pole/zero pattern
- Impulse response $h(n)$
- I/O convolutional equation

The most important one is the *transfer function* description because from it we can easily obtain all the others. Figure 6.1.1 shows the relationships among these descriptions. The need for such multiple descriptions is that each provides a different insight into the nature of the filter and each serves a different purpose.

In practice, a typical usage of these descriptions is to start by specifying a set of desired *frequency response specifications*, that is, the desired shape of $H(\omega)$ (lower left corner in Fig. 6.1.1). Then, through a *filter design method*, obtain a transfer function $H(z)$ that satisfies the given specifications. From $H(z)$ one can then derive a *block diagram* realization and the corresponding *sample-by-sample* processing algorithm that tells how to operate the designed filter in real time (lower right corner of Fig. 6.1.1). For an FIR filter, one can alternatively obtain the impulse response $h(n)$ and then use one of the convolution-based *block* processing methods to implement the operation of the filter (upper right corner of Fig. 6.1.1).

6.2 Transfer Functions

Here, we illustrate the central role played by the transfer function $H(z)$ of a filter by showing how to pass back and forth from one description to another.

217

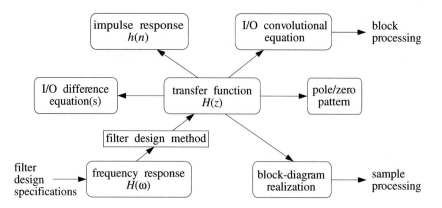

Fig. 6.1.1 Equivalent descriptions of digital filters.

Given a transfer function $H(z)$ one can obtain: (a) the impulse response $h(n)$, (b) the difference equation satisfied by the impulse response, (c) the I/O difference equation relating the output $y(n)$ to the input $x(n)$, (d) the block diagram realization of the filter, (e) the sample-by-sample processing algorithm, (f) the pole/zero pattern, (g) the frequency response $H(\omega)$. Conversely, given any of (a)-(g) as the starting point, one can obtain $H(z)$ and from it the rest of (a)-(g).

As an example, consider the transfer function:

$$H(z) = \frac{5 + 2z^{-1}}{1 - 0.8z^{-1}} \tag{6.2.1}$$

To obtain the *impulse response*, we use partial fraction expansion to write it in the form:

$$H(z) = \frac{5 + 2z^{-1}}{1 - 0.8z^{-1}} = A_0 + \frac{A_1}{1 - 0.8z^{-1}} = -2.5 + \frac{7.5}{1 - 0.8z^{-1}}$$

where A_0 and A_1 are obtained by:

$$A_0 = H(z)\big|_{z=0} = \frac{5 + 2z^{-1}}{1 - 0.8z^{-1}}\bigg|_{z=0} = \frac{5z + 2}{z - 0.8}\bigg|_{z=0} = \frac{2}{-0.8} = -2.5$$

$$A_1 = (1 - 0.8z^{-1})H(z)\big|_{z=0.8} = (5 + 2z^{-1})\big|_{z=0.8} = 5 + 2/0.8 = 7.5$$

Assuming the filter is causal, we find:

$$h(n) = -2.5\delta(n) + 7.5(0.8)^n u(n) \tag{6.2.2}$$

The *difference equation* satisfied by $h(n)$ can be obtained from $H(z)$. The standard approach is to eliminate the denominator polynomial of $H(z)$ and then transfer back to the time domain. Starting with Eq. (6.2.1) and multiplying both sides by the denominator, we find

$$(1 - 0.8z^{-1})H(z) = 5 + 2z^{-1} \quad \Rightarrow \quad H(z) = 0.8z^{-1}H(z) + 5 + 2z^{-1}$$

Taking inverse z-transforms of both sides and using the linearity and delay properties, we obtain the difference equation for $h(n)$:

$$\boxed{h(n) = 0.8h(n-1) + 5\delta(n) + 2\delta(n-1)} \tag{6.2.3}$$

It is easily verified that Eq. (6.2.2) is the causal solution, that is, the solution with the causal initial condition $h(-1) = 0$. Given the impulse response $h(n)$, we can obtain the general I/O convolutional equation for the filter, that is,

$$y_n = h_0 x_n + h_1 x_{n-1} + h_2 x_{n-2} + h_3 x_{n-3} + \cdots$$
$$= 5x_n + 7.5 \big[(0.8)x_{n-1} + (0.8)^2 x_{n-2} + (0.8)^3 x_{n-3} + \cdots \big]$$

It can be rearranged into a difference equation for $y(n)$ using the time-domain techniques of Chapter 3, as in Example 3.4.7. This difference equation can be determined very quickly using z-transforms with the aid of the z-domain equivalent of convolution:

$$Y(z) = H(z)X(z)$$

Again, the standard procedure is to eliminate denominators and go back to the time domain. For this example, we have:

$$Y(z) = H(z)X(z) = \frac{5 + 2z^{-1}}{1 - 0.8z^{-1}}X(z) \quad \Rightarrow \quad (1 - 0.8z^{-1})Y(z) = (5 + 2z^{-1})X(z)$$

which can be written as

$$Y(z) - 0.8z^{-1}Y(z) = 5X(z) + 2z^{-1}X(z)$$

Taking inverse z-transforms of both sides, we have

$$y(n) - 0.8y(n-1) = 5x(n) + 2x(n-1)$$

Therefore, the I/O difference equation is:

$$\boxed{y(n) = 0.8y(n-1) + 5x(n) + 2x(n-1)} \tag{6.2.4}$$

Note that Eq. (6.2.3) is a special case of this, with $x(n) = \delta(n)$ and $y(n) = h(n)$. If the difference equation (6.2.4) was the starting point, we could obtain $H(z)$ by *reversing* all of the above steps, that is, taking z-transforms of both sides

$$Y(z) = 0.8z^{-1}Y(z) + 5X(z) + 2z^{-1}X(z) \quad \Rightarrow$$
$$(1 - 0.8z^{-1})Y(z) = (5 + 2z^{-1})X(z)$$

and solving for the ratio

$$H(z) = \frac{Y(z)}{X(z)} = \frac{5 + 2z^{-1}}{1 - 0.8z^{-1}}$$

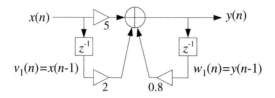

Fig. 6.2.1 Direct form realization of $H(z)$.

Once the I/O difference equation is determined, one can mechanize it by a *block diagram.* For example, Eq. (6.2.4) can be implemented as shown in Fig. 6.2.1. This is referred to as the *direct form realization* because it realizes directly the various terms in the right-hand side of Eq. (6.2.4).

As in the FIR case, the *sample processing algorithm* can be obtained by assigning *internal state variables* to *all* the delays that are present in the block diagram. That is, we may define

$$v_1(n) = x(n-1) \quad \Rightarrow \quad v_1(n+1) = x(n)$$

where $v_1(n)$ is the content of the x-delay at time n. Similarly, we define:

$$w_1(n) = y(n-1) \quad \Rightarrow \quad w_1(n+1) = y(n)$$

so that $w_1(n)$ is the content of the y-delay at time n. In terms of these definitions, we can replace Eq. (6.2.4) by the system of equations:

$$\text{(compute output)} \quad y(n) = 0.8w_1(n) + 5x(n) + 2v_1(n)$$

$$\text{(update states)} \quad v_1(n+1) = x(n)$$

$$w_1(n+1) = y(n)$$

It may be written as the repetitive sample processing algorithm:

$$
\boxed{
\begin{array}{l}
\textit{for each input sample x do:} \\
\quad y = 0.8w_1 + 5x + 2v_1 \\
\quad v_1 = x \\
\quad w_1 = y
\end{array}
}
\qquad \text{(direct form)} \qquad (6.2.5)
$$

The *frequency response* of this particular filter can be obtained by replacing z by $e^{j\omega}$ into $H(z)$. This substitution is valid here because the filter is stable and therefore its ROC, $|z| > 0.8$, contains the unit circle. We find:

$$H(z) = \frac{5(1 + 0.4z^{-1})}{1 - 0.8z^{-1}} \quad \Rightarrow \quad \boxed{H(\omega) = \frac{5(1 + 0.4e^{-j\omega})}{1 - 0.8e^{-j\omega}}}$$

Using the identity

$$|1 - ae^{-j\omega}| = \sqrt{1 - 2a\cos\omega + a^2}$$

which is valid for any real-valued a, we obtain an expression for the *magnitude response*:

$$|H(\omega)| = \frac{5\sqrt{1 + 0.8\cos\omega + 0.16}}{\sqrt{1 - 1.6\cos\omega + 0.64}}$$

This quantity may be plotted with the help of the pole/zero geometric pattern. The filter has a zero at $z = -0.4$ and a pole at $z = 0.8$. Fig. 6.2.2 shows the pole/zero locations relative to the unit circle.

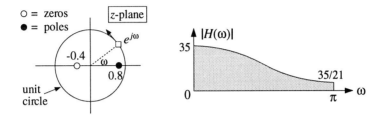

Fig. 6.2.2 Pole/zero pattern and magnitude response.

A quick *sketch* of the magnitude response $|H(\omega)|$ can be obtained by letting the point $e^{j\omega}$ trace the unit circle and drawing peaks when passing near poles and dips when passing near zeros.

The moving point $e^{j\omega}$ is nearest to the pole $z = 0.8$ when $\omega = 0$ and therefore there must be a peak there. Similarly, at $\omega = \pi$ there must be a dip because $e^{j\omega}$ is closest to the zero $z = -0.4$. In particular, setting $z = 1$ or $\omega = 0$, and $z = -1$ or $\omega = \pi$, we can calculate the actual frequency response values at the endpoints of the Nyquist interval:

$$H(\omega)\big|_{\omega=0} = H(z)\big|_{z=1} = \frac{5 + 2}{1 - 0.8} = 35$$

$$H(\omega)\big|_{\omega=\pi} = H(z)\big|_{z=-1} = \frac{5 - 2}{1 + 0.8} = \frac{5}{3} = \frac{35}{21}$$

This filter acts like a *lowpass* filter because it emphasizes low frequencies and attenuates high frequencies. The highest frequency is attenuated by a factor of 21 relative to the lowest one:

$$\frac{|H(\pi)|}{|H(0)|} = \frac{1}{21}$$

or, in decibels:

$$20\log_{10}\left|\frac{H(\pi)}{H(0)}\right| = 20\log_{10}\left(\frac{1}{21}\right) = -26.4 \text{ dB}$$

The block diagram realization of a transfer function is not unique. Different but mathematically equivalent forms of the transfer function may lead to a different set of I/O difference equations which are implemented by a different block diagram and corresponding sample processing algorithm. For our example, the partial fraction expansion form of Eq. (6.2.1)

$$H(z) = \frac{5 + 2z^{-1}}{1 - 0.8z^{-1}} = -2.5 + \frac{7.5}{1 - 0.8z^{-1}}$$

may be thought of as a *parallel* implementation, that is, the sum of two transfer functions

$$H(z) = H_1(z) + H_2(z)$$

where $H_1(z) = -2.5$ and $H_2(z) = 7.5/(1-0.8z^{-1})$. Fig. 6.2.3 shows a block diagram implementation of this form. At first glance, it may not be obvious that the transfer function of this block diagram is the above $H(z)$.

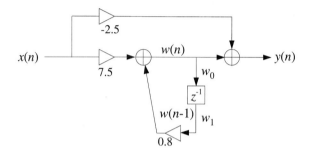

Fig. 6.2.3 Parallel form realization of $H(z)$.

To verify it, we follow the standard procedure of assigning labels, that is, names to all the signal lines that do not already have a label. The output adder has two inputs, one due to the direct connection of the input to the output through the multiplier -2.5, that is, the term $-2.5x(n)$. The other input is assigned a temporary name $w(n)$. Thus, the output adder equation becomes

$$y(n) = w(n) - 2.5x(n) \tag{6.2.6}$$

The quantity $w(n)$ is recognized as the output of the filter $H_2(z)$ with input $x(n)$. The I/O difference equation of $H_2(z)$ is

$$w(n) = 0.8w(n-1) + 7.5x(n) \tag{6.2.7}$$

The two equations (6.2.6) and (6.2.7) together describe the operation of the block diagram in the time domain. Transforming both equations to the z-domain, we obtain

$$Y(z) = W(z) - 2.5X(z)$$

$$W(z) = 0.8z^{-1}W(z) + 7.5X(z) \quad \Rightarrow \quad W(z) = \frac{7.5X(z)}{1 - 0.8z^{-1}}$$

and therefore,

$$Y(z) = W(z) - 2.5X(z) = \frac{7.5X(z)}{1 - 0.8z^{-1}} - 2.5X(z)$$

Solving for the ratio $Y(z)/X(z)$ gives the corresponding transfer function:

$$H(z) = \frac{Y(z)}{X(z)} = \frac{7.5}{1 - 0.8z^{-1}} - 2.5$$

The sample processing algorithm corresponding to this block diagram is obtained by introducing an internal state holding the content of the delay. That is, we define

$$w_0(n) = w(n)$$
$$w_1(n) = w(n-1)$$
$$\Rightarrow \quad w_1(n+1) = w_0(n)$$

Then, Eqs. (6.2.6) and (6.2.7) can be replaced by the system:

$$w_0(n) = 0.8w_1(n) + 7.5x(n)$$

$$y(n) = w_0(n) - 2.5x(n)$$

$$w_1(n+1) = w_0(n)$$

which can be written in the algorithmic form:

> *for each input sample x do:*
> $w_0 = 0.8w_1 + 7.5x$
> $y = w_0 - 2.5x$
> $w_1 = w_0$

(parallel form)   (6.2.8)

Other block diagram realizations can be derived by rearranging the I/O computations differently. A third realization is the so-called *canonical form realization* and is depicted in Fig. 6.2.4. It can be justified as follows. Starting with the z-domain

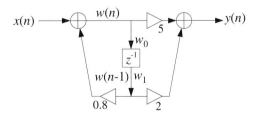

Fig. 6.2.4 Canonical form realization of $H(z)$.

filtering equation

$$Y(z) = H(z)X(z) = \frac{5 + 2z^{-1}}{1 - 0.8z^{-1}}X(z)$$

we separate out the effect of the filter's denominator by defining the temporary quantity

$$W(z) = \frac{1}{1 - 0.8z^{-1}} X(z)$$

then, the output z-transform can be computed by

$$Y(z) = (5 + 2z^{-1}) W(z)$$

Writing these equations in the time domain, we obtain

$$W(z) = \frac{1}{1 - 0.8z^{-1}} X(z) \quad \Rightarrow \quad W(z) = 0.8z^{-1} W(z) + X(z)$$

or,

$$w(n) = 0.8w(n-1) + x(n)$$

Similarly,

$$Y(z) = 5W(z) + 2z^{-1} W(z) \quad \Rightarrow \quad y(n) = 5w(n) + 2w(n-1)$$

Thus, we obtain the system of I/O equations

$$w(n) = 0.8w(n-1) + x(n)$$
$$y(n) = 5w(n) + 2w(n-1)$$

which are mechanized in the block diagram of Fig. 6.2.4. Introducing internal states

$$w_0(n) = w(n)$$
$$w_1(n) = w(n-1)$$
$$\Rightarrow \quad w_1(n+1) = w_0(n)$$

we rewrite the above system as:

$$w_0(n) = 0.8w_1(n) + x(n)$$
$$y(n) = 5w_0(n) + 2w_1(n)$$
$$w_1(n+1) = w_0(n)$$

which can be written in the algorithmic form:

$$
\boxed{
\begin{aligned}
&\textit{for each input sample x do:} \\
&\quad w_0 = 0.8w_1 + x \\
&\quad y = 5w_0 + 2w_1 \\
&\quad w_1 = w_0
\end{aligned}
}
\qquad \text{(canonical form)} \qquad (6.2.9)
$$

A fourth block diagram realization can be obtained by transposing the canonical realization following the transposition rules of replacing adders by nodes, nodes

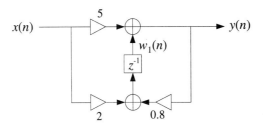

Fig. 6.2.5 Transposed realization of $H(z)$.

by adders, reversing all flows, and exchanging input with output. The resulting *transposed* realization is depicted in Fig. 6.2.5.

Again, we have assigned an internal state variable $w_1(n)$ to hold the contents of the delay register. The input to the delay is the sum $2x(n) + 0.8y(n)$ which gets delayed and becomes $w_1(n)$. Thus,

$$w_1(n) = 2x(n-1) + 0.8y(n-1)$$

The complete I/O description of this realization is then given by the system:

$$y(n) = w_1(n) + 5x(n)$$

$$w_1(n+1) = 2x(n) + 0.8y(n)$$

which translates to the following sample processing algorithm:

$$
\begin{array}{|l|}
\hline
\textit{for each input sample x do:} \\
\quad y = w_1 + 5x \\
\quad w_1 = 2x + 0.8y \\
\hline
\end{array}
\qquad \text{(transposed form)} \qquad (6.2.10)
$$

To verify that this realization describes the same transfer function, we transform the I/O equations to the z-domain:

$$Y(z) = W_1(z) + 5X(z)$$

$$zW_1(z) = 2X(z) + 0.8Y(z)$$

Then, solve the second for $W_1(z)$, insert it in the first, and solve for the ratio $Y(z)/X(z)$. We have:

$$W_1(z) = 0.8z^{-1}Y(z) + 2z^{-1}X(z)$$

and

$$Y(z) = W_1(z) + 5X(z) = 0.8z^{-1}Y(z) + 2z^{-1}X(z) + 5X(z)$$

which gives

$$H(z) = \frac{Y(z)}{X(z)} = \frac{5 + 2z^{-1}}{1 - 0.8z^{-1}}$$

Given a particular block diagram implementation, one can easily translate the corresponding sample processing algorithm into a software or hardware routine. For example, the canonical form of Eq. (6.2.9) can be implemented by the following C routine `filter.c`:

```
/* filter.c - IIR example routine */

double filter(x, w)                        usage: y = filter(x, w);
double x, *w;
{
        double y;

        w[0] = 0.8 * w[1] + x;

        y = 5 * w[0] + 2 * w[1];          compute output

        w[1] = w[0];                      update internal state

        return y;
}
```

The array w must be declared to be a two-dimensional array in the main program. The following program segment illustrates the usage of this routine for processing N input samples:

```
w = (double *) calloc(2, sizeof(double));

for (n=0; n<N; n++)
        y[n] = filter(x[n], w);
```

The internal state array w must be initialized to zero prior to the first call of `filter`. This is indirectly accomplished by `calloc` during the allocation of w.

Our aim in this example was to show not only how to pass from one filter description to another using z-transforms, but also to illustrate how different block diagram realizations correspond to different but equivalent ways of arranging the required I/O filtering equations. A more systematic discussion of filter realizations will be presented in the next chapter.

In general, the transfer function of an IIR filter is given as the ratio of two polynomials of degrees, say L and M:

$$H(z) = \frac{N(z)}{D(z)} = \frac{b_0 + b_1 z^{-1} + b_2 z^{-2} + \cdots + b_L z^{-L}}{1 + a_1 z^{-1} + a_2 z^{-2} + \cdots + a_M z^{-M}} \quad \text{(IIR)} \qquad (6.2.11)$$

Note that by convention, the 0th coefficient of the denominator polynomial has been set to unity $a_0 = 1$. The filter $H(z)$ will have L zeros and M poles. Assuming that the numerator and denominator coefficients are real-valued, then if any of the zeros or poles are complex, they must come in conjugate pairs.

To determine the impulse response $h(n)$ of such a filter, we may use the inverse z-transform techniques of Chapter 5, such as partial fraction expansions. The relative locations of the poles on the z-plane will divide the plane into non-overlapping regions which may be taken as the possible ROCs for $h(n)$.

In particular, to get a stable impulse response, we must pick the ROC that contains the unit circle. Recall that in order for the stable $h(n)$ to also be causal, *all poles* of $H(z)$, that is, the zeros of $D(z)$, must lie strictly *inside* the unit circle. Then, the ROC for inverting $H(z)$ will be the outside of the unit circle.

As the above example showed, there are many different, but mathematically equivalent, I/O difference equations describing such a filter—each leading to a particular block diagram implementation and sample processing algorithm. The simplest one is the direct form obtained by writing

$$Y(z) = H(z)X(z) = \frac{b_0 + b_1 z^{-1} + b_2 z^{-2} + \cdots + b_L z^{-L}}{1 + a_1 z^{-1} + a_2 z^{-2} + \cdots + a_M z^{-M}} X(z)$$

then, multiplying by the denominator:

$$(1 + a_1 z^{-1} + \cdots + a_M z^{-M})Y(z) = (b_0 + b_1 z^{-1} + \cdots + b_L z^{-L})X(z)$$

and finally, transforming back to the time domain:

$$\boxed{y_n + a_1 y_{n-1} + \cdots + a_M y_{n-M} = b_0 x_n + b_1 x_{n-1} + \cdots + b_L x_{n-L}} \qquad (6.2.12)$$

It can also be written as:

$$y_n = -a_1 y_{n-1} - \cdots - a_M y_{n-M} + b_0 x_n + b_1 x_{n-1} + \cdots + b_L x_{n-L}$$

Note also that if the denominator coefficients are zero, that is, $a_i = 0$, $i = 1, 2, \ldots, M$, the denominator polynomial is trivial $D(z) = 1$ and $H(z)$ becomes equal to the numerator polynomial $H(z) = N(z)$, that is, an FIR filter:

$$\boxed{H(z) = N(z) = b_0 + b_1 z^{-1} + b_2 z^{-2} + \cdots + b_L z^{-L}} \qquad \text{(FIR)} \qquad (6.2.13)$$

In this case, the difference equation (6.2.12) becomes the usual I/O convolutional equation for an FIR filter:

$$\boxed{y_n = b_0 x_n + b_1 x_{n-1} + \cdots + b_L x_{n-L}} \qquad \text{(FIR I/O equation)} \qquad (6.2.14)$$

Various implementations of the FIR case were discussed in Chapter 4. The implementations of the IIR case will be discussed in detail in Chapter 7.

Next, we present some further examples. In each case, we determine the transfer function, impulse response, frequency response, pole/zero pattern, block diagram realization and sample processing algorithm.

Example 6.2.1: Determine the transfer function of the following third-order FIR filter with impulse response:

$$\mathbf{h} = [1, 6, 11, 6]$$

Solution: The filter's I/O equation is

$$y(n) = x(n) + 6x(n-1) + 11x(n-2) + 6x(n-3)$$

The z-transform of the finite impulse response sequence is:

$$H(z) = h_0 + h_1 z^{-1} + h_2 z^{-2} + h_3 z^{-3} = 1 + 6z^{-1} + 11z^{-2} + 6z^{-3}$$

Noting that $H(z)$ has a zero at $z = -1$, we may factor it in the form

$$H(z) = (1 + z^{-1})(1 + 5z^{-1} + 6z^{-2}) = (1 + z^{-1})(1 + 2z^{-1})(1 + 3z^{-1})$$

The corresponding frequency response is obtained by the substitution $z = e^{j\omega}$:

$$H(\omega) = 1 + 6e^{-j\omega} + 11e^{-2j\omega} + 6e^{-3j\omega} = (1 + e^{-j\omega})(1 + 2e^{-j\omega})(1 + 3e^{-j\omega})$$

The filter has zeros at $z = -1, -2, -3$. The pole/zero pattern is shown below together with a sketch of the magnitude response $|H(\omega)|$. (The multiple pole at the origin $z = 0$ is not shown.)

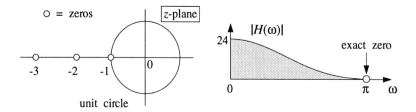

The filter tends to attenuate high frequencies, that is, it will act as a lowpass filter. The filter vanishes exactly at $z = -1$ or $\omega = \pi$. At $\omega = 0$ or $z = 1$, it is equal to $1 + 6 + 11 + 6 = 24$. The block diagram realization and the sample-by-sample processing algorithm are:

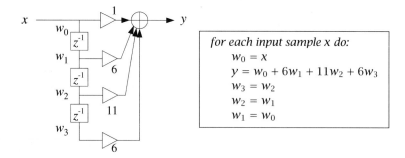

The block diagram and sample processing algorithm correspond to the FIR direct form discussed in Chapter 4.   □

Example 6.2.2: An FIR filter is described by the I/O equation:

$$y(n) = x(n) - x(n-4)$$

Determine its transfer function $H(z)$ and impulse response $h(n)$.

Solution: The I/O equation becomes in the z-domain:

$$Y(z) = X(z) - z^{-4}X(z) \quad \Rightarrow \quad H(z) = \frac{Y(z)}{X(z)} = 1 - z^{-4}$$

It follows that $\mathbf{h} = [1,0,0,0,-1]$. The frequency response is obtained by setting $z = e^{j\omega}$:

$$H(\omega) = 1 - e^{-4j\omega} = (e^{2j\omega} - e^{-2j\omega})e^{-2j\omega} = 2j\sin(2\omega)e^{-2j\omega}$$

Thus, its magnitude response will be $|H(\omega)| = 2|\sin(2\omega)|$. The zeros of $H(z)$ are the fourth roots of unity, obtained by solving $1 - z^{-4} = 0$ or,

$$z^4 = 1 \quad \Rightarrow \quad z = e^{2\pi jk/4}, \ k = 0,1,2,3 \quad \Rightarrow \quad z = 1, j, -1, -j$$

Thus, the magnitude response $|H(\omega)|$ will vanish at $\omega = 2\pi k/4 = 0, \pi/2, \pi, 3\pi/2$, for $k = 0,1,2,3$, as shown below:

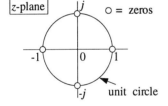 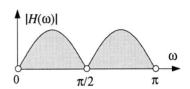

The magnitude response $|H(\omega)|$ is plotted only over the right half of the Nyquist interval, $0 \le \omega \le \pi$, and therefore the zero at $\omega = 3\pi/2$ is not shown—it gets aliased to the negative side: $3\pi/2 - 2\pi = -\pi/2$.

The block diagram realization of the direct form and the corresponding sample processing algorithm are as shown:

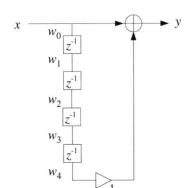

This is a special case of a comb filter with notches at the four frequencies $\omega = 2\pi k/4$, $k = 0, 1, 2, 3$. Comb filters and their applications will be discussed in Chapter 8. □

Example 6.2.3: Determine the transfer function and causal impulse response of the two filters described by the difference equations:

(a) $y(n) = 0.25y(n-2) + x(n)$

(b) $y(n) = -0.25y(n-2) + x(n)$

Solution: For case (a), we take z-transforms of both sides of the difference equation to get:

$$Y(z) = 0.25z^{-2}Y(z) + X(z)$$

Solving for $Y(z)/X(z)$ we find the transfer function:

$$H(z) = \frac{1}{1 - 0.25z^{-2}} = \frac{A_1}{1 - 0.5z^{-1}} + \frac{A_2}{1 + 0.5z^{-1}}$$

with $A_1 = A_2 = 0.5$. Thus, the causal impulse response will be:

$$h(n) = A_1(0.5)^n u(n) + A_2(-0.5)^n u(n)$$

The pole at $z = 0.5$ is in the low-frequency part of the unit circle, and the pole $z = -0.5$ is in the high-frequency part. Thus, the filter will tend to enhance both the low and high frequencies, that is, it will behave as a 2-band bandpass filter, or as a bandstop filter—attenuating the intermediate frequencies between low and high.

Indeed, the value of $H(z)$ at $\omega = 0, \pi$ or $z = \pm1$ is

$$H(0) = H(\pi) = H(z)\big|_{z=\pm1} = \frac{1}{1 - 0.25} = \frac{4}{3}$$

which is larger than the value of $H(z)$ at the intermediate frequency $\omega = \pi/2$ or $z = j$ or $z^2 = -1$, that is, the value

$$H(\pi/2) = H(z)\big|_{z=j} = \frac{1}{1 - 0.25(-1)} = \frac{4}{5}$$

The pole/zero pattern and magnitude spectra are shown below. The peaks at the high-/low-frequency ends are not too high because the poles are not too close to the unit circle.

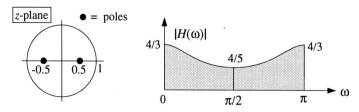

The block diagram implementation of the given difference equation and corresponding sample processing algorithm are:

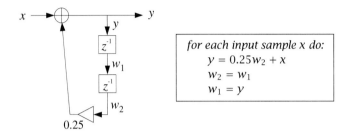

For case (b), the difference equation becomes in the z-domain:

$$Y(z) = -0.25z^{-2}Y(z) + X(z)$$

which can be solved for $Y(z)/X(z)$ to give:

$$H(z) = \frac{1}{1 + 0.25z^{-2}} = \frac{A_1}{1 - 0.5jz^{-1}} + \frac{A_1^*}{1 + 0.5jz^{-1}}$$

with $A_1 = 0.5$. Notice the poles are conjugate pairs as are the PF expansion coefficients. The causal impulse response will be:

$$h(n) = A_1(0.5j)^n u(n) + A_1^*(-0.5j)^n u(n)$$

which can be written in the exponentially decaying form:

$$h(n) = 2\mathrm{Re}[A_1(0.5)^n j^n]u(n) = 2\mathrm{Re}[0.5(0.5)^n e^{j\pi n/2}]u(n)$$

$$= (0.5)^n \cos(\pi n/2)u(n)$$

The two conjugate poles are in the "midfrequency" range, $z = \pm 0.5j = 0.5e^{\pm j\pi/2}$. Thus, the filter will emphasize the middle frequencies, that is, it will act as a bandpass filter.

Again, the value of the magnitude response at $\omega = \pi/2$ or $z = j$ or $z^2 = -1$ is $1/(1 + 0.25(-1)) = 4/3$, whereas the value at $\omega = 0, \pi$ or $z = \pm 1$ or $z^2 = 1$ is $1/(1 + 0.25) = 4/5$.

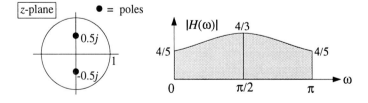

The block diagram and corresponding sample processing algorithm are:

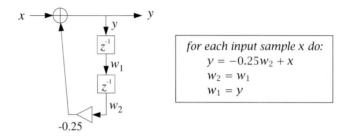

for each input sample x do:
$$y = -0.25w_2 + x$$
$$w_2 = w_1$$
$$w_1 = y$$

The two cases differ by a simple change of sign in the difference equation coefficient 0.25 which leads to drastically different pole locations and frequency responses. □

6.3   Sinusoidal Response

6.3.1 Steady-State Response

The response of a filter to an input sinusoidal signal is referred to as the *sinusoidal response*. Knowing what happens to sinusoids under filtering is important because they are the elementary building blocks of more complicated signals.

Consider an infinitely long, double-sided, complex sinusoid of frequency ω_0 which is applied to the input of a stable filter $h(n)$:

$$x(n) = e^{j\omega_0 n}, \quad -\infty < n < \infty$$

The resulting output can be determined in two ways: (1) using convolution in the time domain, or (2) using multiplication in the frequency domain. Using the first method, we have

$$y(n) = \sum_m h(m)x(n-m) = \sum_m h(m)e^{j(n-m)\omega_0} = e^{j\omega_0 n}\sum_m h(m)e^{-j\omega_0 m}, \quad \text{or,}$$

$$y(n) = H(\omega_0)e^{j\omega_0 n} \tag{6.3.1}$$

where $H(\omega_0)$ is the frequency response of the filter evaluated at $\omega = \omega_0$:

$$H(\omega_0) = \sum_m h(m)e^{-j\omega_0 m}$$

Using the frequency-domain method, we start with the spectrum of the input signal, namely,

$$X(\omega) = 2\pi\delta(\omega - \omega_0) + (\text{replicas})$$

Then, using the frequency-domain multiplication formula (5.4.10), we obtain (the first replica of) the spectrum of the output:

$$Y(\omega) = H(\omega)X(\omega) = H(\omega)2\pi\delta(\omega - \omega_0) = 2\pi H(\omega_0)\delta(\omega - \omega_0)$$

where ω was replaced by ω_0 in the argument of $H(\omega)$, because the delta function $\delta(\omega - \omega_0)$ forces $\omega = \omega_0$. Putting $Y(\omega)$ into the inverse DTFT formula (5.4.6), we find:

$$y(n) = \frac{1}{2\pi} \int_{-\pi}^{\pi} Y(\omega) e^{j\omega n} \, d\omega = \frac{1}{2\pi} \int_{-\pi}^{\pi} 2\pi H(\omega_0) \delta(\omega - \omega_0) e^{j\omega n} \, d\omega$$

The presence of $\delta(\omega - \omega_0)$ causes the integrand to be evaluated at ω_0 resulting in Eq. (6.3.1). To summarize, an infinite double-sided input sinusoid of frequency ω_0 reappears at the output *unchanged* in frequency but modified by the frequency response factor $H(\omega_0)$:

$$\boxed{e^{j\omega_0 n} \xrightarrow{H} H(\omega_0) e^{j\omega_0 n}} \tag{6.3.2}$$

Because $H(\omega)$ is a complex-valued quantity, we can write it in terms of its magnitude and phase as:

$$H(\omega) = |H(\omega)| e^{j \arg H(\omega)}$$

Therefore, Eq. (6.3.2) can be written in the form:

$$\boxed{e^{j\omega_0 n} \xrightarrow{H} |H(\omega_0)| e^{j\omega_0 n + j \arg H(\omega_0)}} \tag{6.3.3}$$

which shows that the filter introduces both a *magnitude* modification by an amount $|H(\omega_0)|$, as well as a *relative phase shift* by an amount $\arg H(\omega_0)$. Taking real or imaginary parts of both sides of this result, we obtain the cosine and sine versions:

$$\boxed{\begin{aligned} \cos(\omega_0 n) &\xrightarrow{H} |H(\omega_0)| \cos(\omega_0 n + \arg H(\omega_0)) \\ \sin(\omega_0 n) &\xrightarrow{H} |H(\omega_0)| \sin(\omega_0 n + \arg H(\omega_0)) \end{aligned}} \tag{6.3.4}$$

Figure 6.3.1 illustrates this result. Note that the phase shift corresponds to the translation of the sinewave as a whole by an amount $\arg H(\omega_0)$ relative to the input sinewave. Typically, $\arg H(\omega_0)$ is negative and therefore it represents a time delay, that is, translation to the right.

The filtering result (6.3.2) is one of the most fundamental results in signal processing. It essentially justifies the use of LTI filters and explains their widespread application. By proper design of the frequency response shape $H(\omega)$, it allows complete control over the frequency content of the input signal.

Using the linearity property of the filter, we can apply Eq. (6.3.2) to a linear combination of two input sinusoids of frequencies ω_1 and ω_2, resulting in the same linear combination of the corresponding outputs, that is,

$$A_1 e^{j\omega_1 n} + A_2 e^{j\omega_2 n} \xrightarrow{H} A_1 H(\omega_1) e^{j\omega_1 n} + A_2 H(\omega_2) e^{j\omega_2 n}$$

which shows that the effect of filtering is to change the *relative amplitudes* and phases of the two sinusoids from the values $\{A_1, A_2\}$ to the values $\{A_1 H(\omega_1),$

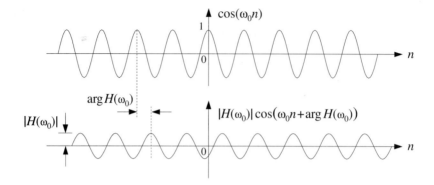

Fig. 6.3.1   Magnitude and phase-shift modification introduced by filtering.

$A_2 H(\omega_2)\}$. In the frequency domain, we have for the spectra of the input and output signals:

$$A_1 \delta(\omega - \omega_1) + A_2 \delta(\omega - \omega_2) \xrightarrow{H} A_1 H(\omega_1) \delta(\omega - \omega_1) + A_2 H(\omega_2) \delta(\omega - \omega_2)$$

where for simplicity, we dropped a common factor of 2π. Figure 6.3.2 shows the input and output spectra and illustrates how the filter alters the relative balance by multiplication by the appropriate frequency response factors.

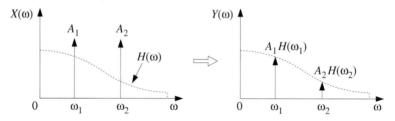

Fig. 6.3.2   Relative amplitudes before and after filtering.

If one of the sinusoids, say ω_1, were a *desired signal* and the other an unwanted *interference*, one could design a filter to remove the interference. For example, the choice:

$$H(\omega_1) = 1, \quad H(\omega_2) = 0$$

would leave the desired signal unaffected and remove the interference. The resulting output signal would be in this case:

$$y(n) = A_1 H(\omega_1) e^{j\omega_1 n} + A_2 H(\omega_2) e^{j\omega_2 n} = A_1 e^{j\omega_1 n}$$

A more general input $x(n)$ with a more complicated spectrum $X(\omega)$ can be resolved into its sinusoidal components by the inverse DTFT formula:

$$x(n) = \frac{1}{2\pi} \int_{-\pi}^{\pi} X(\omega) e^{j\omega n} \, d\omega$$

The filter $H(\omega)$ reshapes the input spectrum $X(\omega)$ into the output spectrum by $Y(\omega) = H(\omega)X(\omega)$. It changes, in a controlled manner, the relative amplitudes and phases of the various frequency components of the input signal. The resulting output signal can be reconstructed from the inverse DTFT formula:

$$y(n) = \frac{1}{2\pi} \int_{-\pi}^{\pi} Y(\omega) e^{j\omega n} \, d\omega = \frac{1}{2\pi} \int_{-\pi}^{\pi} H(\omega) X(\omega) e^{j\omega n} \, d\omega \qquad (6.3.5)$$

Another useful filtering concept is that of the *phase delay* defined in terms of the phase response $\arg H(\omega)$ as follows:

$$\boxed{d(\omega) = -\frac{\arg H(\omega)}{\omega}} \qquad \Rightarrow \qquad \arg H(\omega) = -\omega d(\omega) \qquad (6.3.6)$$

Similarly, the *group delay* of a filter is defined as:

$$\boxed{d_g(\omega) = -\frac{d}{d\omega} \arg H(\omega)} \qquad (6.3.7)$$

The sinusoidal response of Eqs. (6.3.2) or (6.3.3) can be expressed in terms of the phase delay as follows:

$$e^{j\omega n} \xrightarrow{H} |H(\omega)| e^{j\omega(n-d(\omega))} \qquad (6.3.8)$$

which shows that different frequency components get delayed by different amounts, depending on the filter's phase delay.

Linear phase filters have the property that their phase delay $d(\omega)$ is independent of frequency, say $d(\omega) = D$, so that the phase response is linear in ω, $\arg H(\omega) = -\omega D$. Such filters cause every frequency component to be delayed by the *same* amount D, thus corresponding to an overall delay in the output:

$$e^{j\omega n} \xrightarrow{H} |H(\omega)| e^{j\omega(n-D)} \qquad (6.3.9)$$

This overall delay can also be seen by the inverse DTFT formulas:

$$x(n) = \int_{-\pi}^{\pi} X(\omega) e^{j\omega n} \frac{d\omega}{2\pi} \xrightarrow{H} y(n) = \int_{-\pi}^{\pi} |H(\omega)| X(\omega) e^{j\omega(n-D)} \frac{d\omega}{2\pi}$$

The design of FIR linear phase filters will be discussed in Chapter 10. IIR filters that have linear phase over the entire Nyquist interval cannot be designed. However, they can be designed to have approximately linear phase over their *passband* (for example, Bessel filters).

6.3.2 Transient Response

In obtaining the result (6.3.2), we assumed that the input sinusoid had been on for a very long time (since $n = -\infty$), and therefore, Eq. (6.3.2) represents the *steady-state output* resulting after all the filter transients have died out.

In practice, we typically begin processing an input signal at some instant of time, say $n = 0$, and therefore, we must deal with the input-on transients, as well as the input-off transients taking place after the input is turned off. Figure 6.3.3 shows the difference between a double-sided sinewave and a causal one that is turned on at $n = 0$.

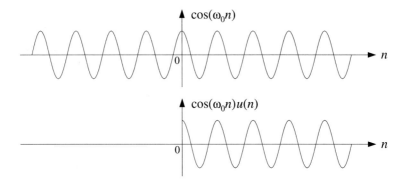

Fig. 6.3.3   Double-sided and one-sided sinewaves.

If we start generating and filtering an input sinewave at $n = 0$, the filter will not "know" immediately that its input is sinusoidal. It takes the filter a certain period of time to settle into its sinusoidal behavior given by Eq. (6.3.2). The analysis of the filter's response in this case can be carried out using z-transforms. Consider the causal sinusoidal input and its z-transform:

$$x(n) = e^{j\omega_0 n} u(n) \xrightarrow{\;z\;} X(z) = \frac{1}{1 - e^{j\omega_0} z^{-1}}$$

having ROC $|z| > |e^{j\omega_0}| = 1$. Assume a filter of the form:

$$H(z) = \frac{N(z)}{D(z)} = \frac{N(z)}{(1 - p_1 z^{-1})(1 - p_2 z^{-1}) \cdots (1 - p_M z^{-1})}$$

with M poles that lie strictly within the unit circle, so that the filter is stable and causal. The output z-transform will be:

$$Y(z) = H(z) X(z) = \frac{N(z)}{(1 - e^{j\omega_0} z^{-1})(1 - p_1 z^{-1})(1 - p_2 z^{-1}) \cdots (1 - p_M z^{-1})}$$

Assuming that the degree of the numerator polynomial $N(z)$ is strictly less than the degree $M+1$ of the denominator,[†] we can write the PF expansion:

$$Y(z) = \frac{C}{1 - e^{j\omega_0} z^{-1}} + \frac{B_1}{1 - p_1 z^{-1}} + \frac{B_2}{1 - p_2 z^{-1}} + \cdots + \frac{B_M}{1 - p_M z^{-1}}$$

[†]This assumption is not critical. The same conclusions can be drawn otherwise.

The PF expansion coefficients are obtained in the usual fashion, via Eq. (5.5.2). In particular, the coefficient of the first term will be:

$$C = (1 - e^{j\omega_0}z^{-1})Y(z)\big|_{z=e^{j\omega_0}} = \left[(1 - e^{j\omega_0}z^{-1})\frac{H(z)}{1 - e^{j\omega_0}z^{-1}}\right]_{z=e^{j\omega_0}}$$

Canceling the $(1 - e^{j\omega_0}z^{-1})$ factors, we find that C is none other than the frequency response $H(\omega)$ evaluated at $\omega = \omega_0$, that is,

$$\boxed{C = H(z)\big|_{z=e^{j\omega_0}} = H(\omega_0)} \tag{6.3.10}$$

Therefore, the PF expansion will read:

$$\boxed{Y(z) = \frac{H(\omega_0)}{1 - e^{j\omega_0}z^{-1}} + \frac{B_1}{1 - p_1 z^{-1}} + \frac{B_2}{1 - p_2 z^{-1}} + \cdots + \frac{B_M}{1 - p_M z^{-1}}}$$

Taking the causal inverse z-transform (with ROC $|z| > 1$), we find for $n \geq 0$:

$$\boxed{y(n) = H(\omega_0)e^{j\omega_0 n} + B_1 p_1^n + B_2 p_2^n + \cdots + B_M p_M^n} \tag{6.3.11}$$

Because the filter was assumed to have all its poles inside the unit circle, namely, $|p_i| < 1$, it follows that in the limit of large n, the p_i^n terms will drop to zero exponentially giving the steady-state output:

$$y(n) \to H(\omega_0)e^{j\omega_0 n} \quad \text{as} \quad n \to \infty$$

For smaller values of n, Eq. (6.3.11) gives the transient response of the filter.

Example 6.3.1: Determine the full transient response of the filter

$$H(z) = \frac{5 + 2z^{-1}}{1 - 0.8z^{-1}}$$

for a causal complex sinusoidal input of frequency ω_0.

Solution: We have the partial fraction expansion for the output z-transform $Y(z) = H(z)X(z)$:

$$Y(z) = \frac{5 + 2z^{-1}}{(1 - e^{j\omega_0}z^{-1})(1 - 0.8z^{-1})} = \frac{H(\omega_0)}{1 - e^{j\omega_0}z^{-1}} + \frac{B_1}{1 - 0.8z^{-1}}$$

where the coefficient B_1 is found by

$$B_1 = (1 - 0.8z^{-1})Y(z)\big|_{z=0.8} = \left[\frac{5 + 2z^{-1}}{1 - e^{j\omega_0}z^{-1}}\right]_{z=0.8} = \frac{7.5}{1 - 1.25e^{j\omega_0}}$$

The causal inverse z-transform will be:

$$y(n) = H(\omega_0)e^{j\omega_0 n} + B_1(0.8)^n, \qquad n \geq 0$$

For large n, the term $(0.8)^n$ drops to zero and the output settles into its steady-state sinusoidal response

$$y(n) \rightarrow H(\omega_0) e^{j\omega_0 n}$$

where $H(\omega_0) = \dfrac{5 + 2e^{-j\omega_0}}{1 - 0.8e^{-j\omega_0}}$.   □

There are four straightforward conclusions that can be drawn from Eq. (6.3.11). First, it shows clearly the requirement of *stability* for the filter. If any of the filter poles, say p_1, were outside the unit circle, such that $|p_1| > 1$, the term p_1^n would be unstable, diverging as $n \rightarrow \infty$. This term would dominate completely the rest of terms of Eq. (6.3.11) and there would be no steady-state response. (Of course, we know that in this case the series definition, Eq. (5.4.3), of the frequency response $H(\omega)$ does not converge because the unit circle does not lie in the causal region of convergence $|z| > |p_1| > 1$.)

Second, assuming the filter is strictly stable, all the transient terms p_i^n will drop to zero exponentially. But some of them will drop to zero faster than others. The effective *time constant* to reach the sinusoidal steady state is dictated by the *slowest* converging pole term, that is, the term with the *largest* magnitude, namely, max $|p_i|$. Equivalently, this is the pole that lies *closest* to the unit circle (from the inside). Denoting the maximum pole magnitude by

$$\rho = \max_i |p_i|$$

we may define the effective time constant to be the time n_{eff} at which the quantity ρ^n has dropped below a certain small value, for example, when it drops below 1% its initial value. We can make this definition more quantitative by defining n_{eff} such that:

$$\rho^{n_{\mathrm{eff}}} = \epsilon$$

where ϵ is the desired level of smallness, for example, $\epsilon = 1\% = 0.01$. It follows that:

$$\boxed{n_{\mathrm{eff}} = \frac{\ln \epsilon}{\ln \rho} = \frac{\ln(1/\epsilon)}{\ln(1/\rho)}} \qquad \text{(time constant)} \qquad\qquad (6.3.12)$$

Because both ϵ and ρ are less than one, their logs are negative, but the ratio is positive. In the last expression, we have the ratio of two positive numbers. The effective time constant n_{eff} becomes larger if the slowest pole is pushed closer to the unit circle, that is, increasing ρ toward one, and also if we require a smaller threshold ϵ.

The value $\epsilon = 1\%$ corresponds to the amplitude of the filter's output falling by a factor of 10^{-2} or 40 dB. The time constant in seconds, $\tau = n_{\mathrm{eff}} T$, is referred to as the 40-dB time constant. In the study of reverberation properties of concert halls, the 60-dB time constants are used, which correspond to $\epsilon = 0.1\% = 10^{-3}$.

In conclusion, the *speed of response* of a stable and causal IIR filter is controlled by the poles nearest to the unit circle. The filter is *slow* reaching steady state if its poles are *near* the unit circle, and fast if they are further away (toward the center).

Example 6.3.2: A sinusoid of frequency $\omega_0 = 0.1\pi$ and duration of 300 samples, that is, $x(n) = \sin(\omega_0 n)$, $0 \le n < 300$, is input to a (causal) filter with transfer function

$$H(z) = \frac{b}{1 - az^{-1}}$$

where $a = 0.97$. Determine the 1% time constant of this filter. Adjust the scale factor b such that the filter's gain at ω_0 is unity. Determine and plot the output of the filter $y(n)$ over the interval $0 \le n < 450$, by iterating the difference equation of the filter.

Solution: The 1% time constant of this filter is computed from Eq. (6.3.12),

$$n_{\text{eff}} = \frac{\ln\epsilon}{\ln a} = \frac{\ln(0.01)}{\ln(0.97)} = 151.2 \text{ samples}$$

The frequency and magnitude responses are

$$H(\omega) = \frac{b}{1 - ae^{-j\omega}} \quad \Rightarrow \quad |H(\omega)| = \frac{b}{\sqrt{1 - 2a\cos\omega + a^2}}$$

The requirement that $|H(\omega_0)| = 1$ leads to the condition on b:

$$|H(\omega_0)| = \frac{b}{\sqrt{1 - 2a\cos\omega_0 + a^2}} = 1 \quad \Rightarrow \quad b = \sqrt{1 - 2a\cos\omega_0 + a^2} = 0.3096$$

The value of the frequency response at ω_0 becomes then,

$$H(\omega_0) = \frac{b}{1 - ae^{-j\omega_0}} = 0.2502 - 0.9682j = 1 \cdot e^{-j1.3179}$$

so that its phase response will be $\arg H(\omega_0) = -1.3179$ radians. The resulting output $y(n)$, shown in Fig. 6.3.4, was computed by the following program segment, which implements the difference equation $y(n) = ay(n-1) + bx(n)$ and sample processing algorithm of this filter:

```
for (y1=0, n=0; n<450; n++) {
        if (n < 300)
                x = sin(w0 * n);
        else
                x = 0;
        y[n] = a * y1 + b * x;          /* y1 = y[n-1] */
        y1 = y[n];
        }
```

Notice the input-on and input-off transients, each lasting approximately $n_{\text{eff}} = 151$ time samples. The time interval $150 \le n \le 300$ corresponds to the steady-state period, during which the output settles to its sinusoidal behavior according to Eq. (6.3.4). The amplitude of the steady-state output is unity because $|H(\omega_0)| = 1$. There is a slight phase delay relative to the input, due to the negative value of the phase response $\arg H(\omega_0) = -1.3179$ radians. □

Example 6.3.3: Derive closed-form expressions for the output $y(n)$ of the previous example in two ways: (a) working with convolution in the time domain, and (b) working with z-transforms.

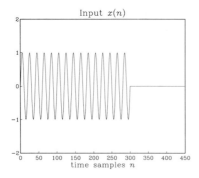 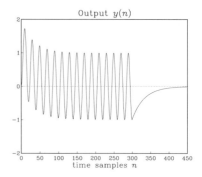

Fig. 6.3.4 Input and output of Example 6.3.2.

Solution: It proves convenient to work with the complex-valued version of the sinusoid, $x(n) = e^{j\omega_0 n}$, $0 \le n \le 299$, and take imaginary parts of the answer at the end. Using convolution, we obtain

$$y(n) = \sum_{m=\max(0,n-299)}^{n} h(m)x(n-m) = \sum_{m=\max(0,n-299)}^{n} ba^m e^{j\omega_0(n-m)}$$

where we used the impulse response $h(m) = ba^m u(m)$. The summation limits were obtained by the requirements that the indices of $h(m)$ and $x(n-m)$ do not exceed the array bounds, that is,

$$0 \le m < \infty, \qquad 0 \le n - m \le 299$$

which are equivalent to $\max(0, n - 299) \le m \le n$. It follows that if $0 \le n \le 299$, then

$$y(n) = e^{j\omega_0 n} \sum_{m=0}^{n} ba^m e^{-j\omega_0 m} = e^{j\omega_0 n} b \frac{1 - a^{n+1} e^{-j\omega_0(n+1)}}{1 - ae^{-j\omega_0}}$$

Setting $H_0 = H(\omega_0) = b/(1 - ae^{-j\omega_0})$ and noting that $(1 - ae^{-j\omega_0})H_0 = b$ which gives $b - H_0 = -H_0 ae^{-j\omega_0}$, we obtain

$$y(n) = H_0 e^{j\omega_0 n} + (b - H_0)a^n, \qquad \text{for } 0 \le n \le 299$$

On the other hand, if $300 \le n \le 449$, then the limits of summation become:

$$y(n) = e^{j\omega_0 n} \sum_{m=n-299}^{n} ba^m e^{-j\omega_0 m}$$

and we must use the finite geometric series formula

$$\sum_{m=m_1}^{m_2} x^m = \frac{x^{m_1} - x^{m_2+1}}{1 - x} \qquad\qquad (6.3.13)$$

with $x = ae^{-j\omega_0}$, to obtain:

$$y(n) = H_0 e^{j\omega_0 n} \left[a^{n-299} e^{-j\omega_0(n-299)} - a^{n+1} e^{-j\omega_0(n+1)} \right]$$

Noting that the factor $e^{j\omega_0 n}$ cancels, we obtain

$$y(n) = H_0 \left[a^{n-299} e^{j299\omega_0} - a^{n+1} e^{-j\omega_0} \right] = H_0 a e^{-j\omega_0} (e^{j300\omega_0} - a^{300}) a^{n-300}$$

At $n = 300$, we have $y(300) = H_0 a e^{-j\omega_0} (e^{j300\omega_0} - a^{300})$, Therefore, we can write

$$y(n) = y(300) a^{n-300}, \qquad \text{for } 300 \le n < \infty$$

Thus, the input-off transients are exponentially decaying as seen in Fig. 6.3.4. Note that the two expressions of $y(n)$ for $n \le 299$ and $n \ge 300$ join smoothly, in the following sense. The difference equation for $y(n)$ gives at $n = 300$, $y(300) = ay(299) + bx(300)$. But $x(300) = 0$ and thus $y(300) = ay(299)$. This condition can be verified for the above expressions. We have, using $b - H_0 = -aH_0 e^{-j\omega_0}$:

$$\begin{aligned} ay(299) &= a \left[H_0 e^{j299\omega_0} + (b - H_0) a^{299} \right] \\ &= a H_0 e^{j300\omega_0} e^{-j\omega_0} - a^2 H_0 e^{-j\omega_0} a^{299} \\ &= a H_0 e^{-j\omega_0} (e^{j300\omega_0} - a^{300}) = y(300) \end{aligned}$$

The real-valued versions can be obtained by taking imaginary parts of these answers. Writing $H_0 = |H_0| e^{j\phi_0}$, with $|H_0| = 1$ and $\phi_0 = \arg H_0 = -1.3179$ radians, we have for $n \le 299$

$$y(n) = H_0 e^{j\omega_0 n} + (b - H_0) a^n = e^{j(\omega_0 n + \phi_0)} + (b - e^{j\phi_0}) a^n$$

Taking imaginary parts, we find

$$y(n) = \sin(\omega_0 n + \phi_0) - a^n \sin\phi_0, \qquad \text{for } 0 \le n \le 299$$

and

$$y(n) = y(300) a^{n-300}, \qquad \text{for } n \ge 300$$

where we calculate $y(300) = ay(299)$. The numerical values of $y(n)$ agree, of course, with the iterated values of Example 6.3.4.

All of the above expressions can be obtained much faster using z-transforms. First, we write the length-300 complex sinusoid in a form which is valid for all n:

$$x(n) = e^{j\omega_0 n} (u(n) - u(n - 300)) = e^{j\omega_0 n} u(n) - e^{j300\omega_0} e^{j\omega_0(n-300)} u(n - 300)$$

where, in the second term we have the delayed version of the first. Taking z-transforms and using the delay property, we find:

$$X(z) = \frac{1 - e^{j300\omega_0} z^{-300}}{1 - e^{j\omega_0} z^{-1}}$$

The output z-transform will be then

$$Y(z) = H(z) X(z) = \frac{b(1 - e^{j300\omega_0} z^{-300})}{(1 - az^{-1})(1 - e^{j\omega_0} z^{-1})}$$

The (causal) inverse z-transform can be found using the "remove/restore" method. Ignoring the numerator temporarily, we have

$$W(z) = \frac{b}{(1 - az^{-1})(1 - e^{j\omega_0} z^{-1})} = \frac{C}{1 - e^{j\omega_0} z^{-1}} + \frac{B}{1 - az^{-1}}$$

where, as we have seen $C = H_0$. Similarly, one can verify that $B = b - H_0$. Therefore, the causal inverse z-transform of $W(z)$ will be:

$$w(n) = H_0 e^{j\omega_0 n} u(n) + (b - H_0) a^n u(n)$$

Restoring the numerator of $Y(z)$, that is, $Y(z) = (1 - e^{j300\omega_0} z^{-300}) W(z)$, we find

$$y(n) = w(n) - e^{j300\omega_0} w(n - 300)$$

which gives rise to the following expression for $y(n)$:

$$y(n) = H_0 e^{j\omega_0 n} u(n) + (b - H_0) a^n u(n) -$$
$$- e^{j300\omega_0} [H_0 e^{j\omega_0 (n-300)} u(n - 300) + (b - H_0) a^{n-300} u(n - 300)]$$

We leave it up to the reader to verify that this expression agrees with the separate expressions given above for $n \le 299$ and $n \ge 300$.   □

A third consequence of Eq. (6.3.11) is its application to two important special cases, namely, the unit-step and the alternating unit-step responses. The *unit-step response* is the output $y(n)$ due to a unit-step input:

$$x(n) = u(n)$$

It is a special case of a sinusoid $e^{j\omega_0 n} u(n)$ with $\omega_0 = 0$. The value $\omega_0 = 0$ corresponds to the complex point $z = e^{j\omega_0} = 1$ on the z-plane. Equation (6.3.11) becomes in this case:

$$y(n) = H(0) + B_1 p_1^n + B_2 p_2^n + \cdots + B_M p_M^n, \qquad n \ge 0$$

Thus, in the limit of large n, the output will settle into a constant value given by

$$y(n) \rightarrow H(0) = H(z)\big|_{z=1} \qquad \text{as} \qquad n \rightarrow \infty$$

We may also refer to it as the DC response of the filter, that is, its response to a constant input. Setting $\omega = 0$ into the definition of $H(\omega)$, Eq. (5.4.3), or $z = 1$ into the definition of $H(z)$, Eq. (5.1.2), we can express $H(0)$ in the alternative form:

$$H(0) = H(z)\big|_{z=1} = \sum_{n=0}^{\infty} h(n) \qquad \text{(DC gain)} \qquad (6.3.14)$$

In Chapter 4, we referred to it as the DC gain of the filter and used it in Eq. (4.1.24) and in the simulation examples. In a similar fashion, we can discuss the alternating step response, namely, the output due to the alternating input:

$$x(n) = (-1)^n u(n)$$

Writing $-1 = e^{j\pi}$, we have $(-1)^n u(n) = e^{j\pi n} u(n)$, and therefore, we recognize this as a special case of Eq. (6.3.11) with $\omega_0 = \pi$, which corresponds to the z-point $z = e^{j\omega_0} = -1$. Eq. (6.3.11) becomes:

$$y(n) = H(\pi)e^{j\pi n} + B_1 p_1^n + B_2 p_2^n + \cdots + B_M p_M^n, \qquad n \geq 0$$

And, in the limit of large n, the output tends to

$$y(n) \to H(\pi)(-1)^n \qquad \text{as} \qquad n \to \infty$$

The quantity $H(\pi)$ may be called the *AC gain* and can be expressed in terms of $h(n)$ by setting $z = -1$ into the definition for $H(z)$:

$$H(\pi) = H(z)\big|_{z=-1} = \sum_{n=0}^{\infty} (-1)^n h(n) \qquad \text{(AC gain)}$$

Example 6.3.4: Determine the DC and alternating-step responses of the filter of Example 6.3.1. Determine also the effective time constant n_{eff} to reach steady state to within one percent.

Solution: Setting $n = 0$ in the expression for $y(n)$ in Example 6.3.1 gives the relationship $y(0) = H(\omega_0) + B_1$. Inspecting the expression for $Y(z)$, the value of $y(0)$ is found to be $y(0) = 5$. More systematically, $y(0)$ can be found by evaluating $Y(z)$ at $z = \infty$. We have therefore, $H(\omega_0) + B_1 = 5$, which can be solved for $B_1 = 5 - H(\omega_0)$. Thus, for a general ω_0 we may write:

$$y(n) = H(\omega_0)e^{j\omega_0 n} + (5 - H(\omega_0))(0.8)^n, \qquad n \geq 0$$

For the DC unit-step response, we have setting $\omega_0 = 0$:

$$H(0) = \left[\frac{5 + 2e^{-j\omega_0}}{1 - 0.8e^{-j\omega_0}}\right]_{\omega_0=0} = \frac{5+2}{1-0.8} = 35$$

Therefore, $B_1 = 5 - H(0) = 5 - 35 = -30$, and the response to a unit-step is:

$$y(n) = 35 - 30(0.8)^n, \quad n \geq 0$$

Similarly, we find $H(\pi) = 5/3$, and $B_1 = 5 - 5/3 = 10/3$. Thus, the alternating unit-step response will be:

$$y(n) = \frac{5}{3}(-1)^n + \frac{10}{3}(0.8)^n, \quad n \geq 0$$

The output signals of the two cases are shown below:

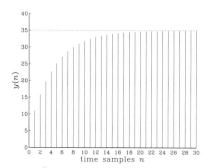 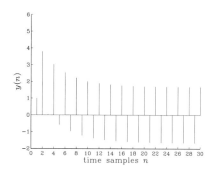

This filter has only one pole; therefore the effective time constant is determined by the quantity $a = 0.8$. At the 1% level, we have $\epsilon = 1\% = 0.01$, and we find

$$n_{\text{eff}} = \frac{\ln \epsilon}{\ln a} = \frac{\ln(0.01)}{\ln(0.8)} = 20.6$$

Thus, the effective duration of the transient behavior is about 20 time steps, after which the transient term $(0.8)^n$ drops by more than 1% its initial value. □

The fourth consequence of Eq. (6.3.11) is its application to *marginally stable* filters. Many filters of practical interest are not strictly stable, but only marginally so, with poles *on* the unit circle. This includes for example, accumulators (integrators), periodic function generators, and others. It is therefore useful to know what happens to the sinusoidal response in such cases.

Suppose the filter $H(z)$ has a pole p_1 on the unit circle at some phase angle θ_1, that is, $p_1 = e^{j\theta_1}$. Of course, the conjugate pole $p_1^* = e^{-j\theta_1}$ is also present. For example, the filters:

$$H(z) = \frac{1}{1 - z^{-1}}, \qquad H(z) = \frac{1}{1 + z^{-1}}, \qquad H(z) = \frac{1}{1 + z^{-2}}$$

have poles at $p_1 = 1 = e^{j0}$, $p_1 = -1 = e^{j\pi}$, and $\pm j = e^{\pm j\pi/2}$, respectively.

Suppose also that all other poles lie inside the unit circle. Then the transient response Eq. (6.3.11) will be:

$$y(n) = H(\omega_0)e^{j\omega_0 n} + B_1 p_1^n + B_1^* p_1^{*n} + B_2 p_2^n + \cdots, \qquad \text{or,}$$

$$y(n) = H(\omega_0) e^{j\omega_0 n} + B_1 e^{j\theta_1 n} + B_1^* e^{-j\theta_1 n} + B_2 p_2^n + \cdots$$

In the limit of large n, the terms p_2^n, p_3^n, etc., will drop to zero exponentially, but the p_1^n being sinusoidal will not. Thus, the filter output tends to

$$y(n) \rightarrow H(\omega_0) e^{j\omega_0 n} + B_1 e^{j\theta_1 n} + B_1^* e^{-j\theta_1 n}$$

for large n. There is no true sinusoidal response even though the output is a sum of sinusoids. Once the sinusoidal pole terms $e^{j\theta_1 n}$ are excited, they will remain in the output forever.

This analysis of Eq. (6.3.11) applies only to the case when $\omega_0 \neq \pm\theta_1$. If $\omega_0 = \pm\theta_1$, one hits a "resonance" of the system and the output $y(n)$ becomes unstable diverging to infinity. In this case, the output z-transform $Y(z)$ has a *double pole* and the discussion must be modified. For example, if $\omega_0 = \theta_1$, then $e^{j\omega_0} = e^{j\theta_1} = p_1$ and $Y(z)$ becomes:

$$Y(z) = H(z)X(z) = \frac{N(z)}{(1 - e^{j\omega_0} z^{-1})(1 - p_1 z^{-1}) \cdots (1 - p_M z^{-1})}$$

$$= \frac{N(z)}{(1 - p_1 z^{-1})^2 (1 - p_2 z^{-1}) \cdots (1 - p_M z^{-1})}$$

The partial fraction expansion takes the form in this case:

$$Y(z) = \frac{B_1}{1 - p_1 z^{-1}} + \frac{B_1'}{(1 - p_1 z^{-1})^2} + \frac{B_2}{1 - p_2 z^{-1}} + \cdots + \frac{B_M}{1 - p_M z^{-1}}$$

Using the causal inverse z-transform (with ROC $|z| > |a|$):

$$\frac{1}{(1 - a z^{-1})^2} \xrightarrow{z^{-1}} (n + 1) a^n u(n)$$

we find for the output signal:

$$y(n) = B_1 p_1^n + B_1' (n + 1) p_1^n + B_2 p_2^n + \cdots + B_M p_M^n, \qquad \text{or,}$$

$$y(n) = B_1 e^{j\theta_1 n} + B_1' (n + 1) e^{j\theta_1 n} + B_2 p_2^n + \cdots + B_M p_M^n$$

which diverges linearly in n.

Until now, the entire discussion of transient response was geared to IIR filters that have nontrivial poles. FIR filters do not have any poles (except at $z = 0$), and therefore, the analysis of their steady versus transient sinusoidal response must be carried out differently.

Consider an FIR filter of order M with impulse response $\mathbf{h} = [h_0, h_1, \cdots, h_M]$. For a causal sinusoidal input $x(n) = e^{j\omega_0 n} u(n)$, the output will be, as discussed in Chapter 4:

$$y(n) = \sum_{m=0}^{\min(n,M)} h(m) x(n-m) = \sum_{m=0}^{\min(n,M)} h(m) e^{j\omega_0(n-m)}$$

or, for any $n \geq 0$:

$$\boxed{y(n) = e^{j\omega_0 n} \sum_{m=0}^{\min(n,M)} h(m) e^{-j\omega_0 m}}$$

When $n \geq M$, the upper summation limit becomes M, giving

$$y(n) = e^{j\omega_0 n} \sum_{m=0}^{M} h(m) e^{-j\omega_0 m} = H(\omega_0) e^{j\omega_0 n}, \qquad n \geq M$$

Therefore, as we have already seen in Chapter 4, the input-on transients last only for the time period $0 \leq n \leq M$. After that period, steady state sets in.

6.4 Pole/Zero Designs

6.4.1 First-Order Filters

Pole/zero placement can be used to design simple filters, such as first-order smoothers, notch filters, and resonators. To illustrate the technique, we design the transfer function

$$H(z) = \frac{5 + 2z^{-1}}{1 - 0.8z^{-1}} = \frac{5(1 + 0.4z^{-1})}{1 - 0.8z^{-1}}$$

discussed in Section 6.2. We begin with the more general transfer function

$$H(z) = \frac{G(1 + bz^{-1})}{1 - az^{-1}} \tag{6.4.1}$$

where both a and b are positive and less than one. The gain factor G is arbitrary. The pole/zero pattern is shown in Fig. 6.4.1.

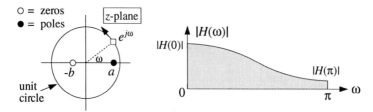

Fig. 6.4.1 Pole/zero pattern and frequency response.

The filter zero at $z = -b$ lies in the left half (the high-frequency part) of the unit circle, and the filter pole at $z = a$ lies in the right half (the low-frequency part). Therefore, the pole emphasizes low frequencies and the zero attenuates high frequencies; in other words, the filter acts as a lowpass filter.

The frequency response values at the lowest and highest frequencies $\omega = 0, \pi$ are found by setting $z = \pm 1$ in Eq. (6.4.1):

$$H(0) = \frac{G(1+b)}{1-a}, \qquad H(\pi) = \frac{G(1-b)}{1+a}$$

Therefore, the attenuation of the highest frequency relative to the lowest one is:

$$\frac{H(\pi)}{H(0)} = \frac{(1-b)(1-a)}{(1+b)(1+a)} \tag{6.4.2}$$

To determine the two unknown parameters a and b in Eq. (6.4.1), we need two *design equations*. One such equation can be Eq. (6.4.2). If a is known, then for a desired level of attenuation $H(\pi)/H(0)$, we can solve for b.

To determine a, we may impose a constraint on the *speed of response* of the filter, that is, we may specify the effective time constant n_{eff}, which is controlled by the value of a. For example, requiring that $n_{\text{eff}} = 20$ time samples and taking $\epsilon = 0.01$, we can solve Eq. (6.3.12) for a:

$$a = \epsilon^{1/n_{\text{eff}}} = (0.01)^{1/20} \simeq 0.8$$

With this value of a, requiring that $H(\pi)/H(0) = 1/21$, Eq. (6.4.2) would give:

$$\frac{(1-b)(1-0.8)}{(1+b)(1+0.8)} = \frac{1}{21} \quad \Rightarrow \quad b = 0.4$$

which gives the desired designed filter, up to the gain G:

$$H(z) = \frac{G(1 + 0.4z^{-1})}{1 - 0.8z^{-1}}$$

Because the parameter b is restricted to the interval $0 \le b \le 1$, we may look at the two extreme designs, namely, for $b = 0$ and $b = 1$. Setting $b = 0$ in Eqs. (6.4.1) and (6.4.2), gives:

$$H(z) = \frac{G}{1 - 0.8z^{-1}}, \qquad \frac{H(\pi)}{H(0)} = \frac{1}{9}$$

and setting $b = 1$,

$$H(z) = \frac{G(1 + z^{-1})}{1 - 0.8z^{-1}}, \qquad \frac{H(\pi)}{H(0)} = 0$$

corresponding to $H(\pi) = 0$. The two design criteria that we used are not the only possible ones. In Section 8.3.1, we will replace the design equation (6.4.2) by an alternative criterion, which is better suited for the design of noise reduction filters.

6.4.2 Parametric Resonators and Equalizers

As another example, consider the design of a simple second-order "resonator" filter whose frequency response is dominated by a single narrow pole peak at some frequency ω_0. Such frequency response is shown in Fig. 6.4.2.

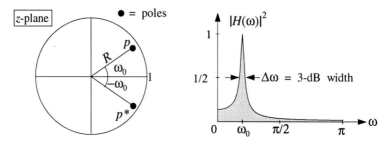

Fig. 6.4.2   Pole/zero pattern and frequency response of resonator filter.

To make a peak at $\omega = \omega_0$, we place a pole inside the unit circle along the ray with phase angle ω_0, that is, at the complex location:

$$p = R e^{j\omega_0}$$

where the pole magnitude is $0 < R < 1$. Together with the conjugate pole $p^* = R e^{-j\omega_0}$, we obtain the transfer function:

$$H(z) = \frac{G}{(1 - R e^{j\omega_0} z^{-1})(1 - R e^{-j\omega_0} z^{-1})} = \frac{G}{1 + a_1 z^{-1} + a_2 z^{-2}} \qquad (6.4.3)$$

where a_1 and a_2 are related to R and ω_0 by

$$a_1 = -2R \cos \omega_0, \quad a_2 = R^2$$

The gain G may be fixed so as to normalize the filter to unity at ω_0, that is, $|H(\omega_0)| = 1$. The frequency response of the filter is obtained by the substitution $z = e^{j\omega}$:

$$H(\omega) = \frac{G}{(1 - R e^{j\omega_0} e^{-j\omega})(1 - R e^{-j\omega_0} e^{-j\omega})} = \frac{G}{1 + a_1 e^{-j\omega} + a_2 e^{-2j\omega}}$$

The normalization requirement $|H(\omega_0)| = 1$ gives the condition:

$$|H(\omega_0)| = \frac{G}{|(1 - R e^{j\omega_0} e^{-j\omega_0})(1 - R e^{-j\omega_0} e^{-j\omega_0})|} = 1$$

which can be solved for G:

$$G = (1 - R)\sqrt{1 - 2R \cos(2\omega_0) + R^2}$$

The magnitude response squared can also be expressed in the form:

$$|H(\omega)|^2 = \frac{G^2}{(1 - 2R\cos(\omega - \omega_0) + R^2)(1 - 2R\cos(\omega + \omega_0) + R^2)}$$

The 3-dB width $\Delta\omega$ of the peak is defined as the *full width at half maximum* of the magnitude squared response. It can be found by solving the equation

$$|H(\omega)|^2 = \frac{1}{2}|H(\omega_0)|^2 = \frac{1}{2}$$

In dB, this condition reads

$$20\log_{10}\left|\frac{H(\omega)}{H(\omega_0)}\right| = 10\log_{10}\left(\frac{1}{2}\right) = -3 \text{ dB}$$

This equation has two solutions, say ω_1 and ω_2, the first to the left of ω_0 and the second to the right. The full width is defined as $\Delta\omega = \omega_2 - \omega_1$. These two frequencies are called the 3-dB frequencies. It can be shown that when p is near the unit circle, that is, $R \lesssim 1$, the full width is given approximately by

$$\boxed{\Delta\omega \simeq 2(1 - R)} \qquad (6.4.4)$$

Thus, the closer R is to one, the sharper the peak, but also the slower the filter will be in reaching its steady-state response, as we discussed in the previous section.

Equation (6.4.4) can be shown geometrically, as follows [15]. In Fig. 6.4.3, the pole p is indicated by the point P whose distance from the origin is $|OP| = R$. Therefore, the distance $|PQ|$ to the unit circle will be $|PQ| = 1 - R$.

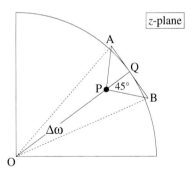

Fig. 6.4.3 Geometric interpretation of 3-dB width.

Assuming the pole P is very near the unit circle, the small 3-dB angle $\Delta\omega$ subtended about the direction OQ will intersect the circle at two points which may be taken to be approximately the points A and B that lie along the tangent to the circle at Q. Denoting by z_A and z_Q the complex numbers represented by the points A and Q, we have the values for the transfer function:

$$|H(z_A)| = \frac{G}{|z_A - p||z_A - p^*|}, \quad |H(z_Q)| = \frac{G}{|z_Q - p||z_Q - p^*|}$$

Assuming P is very near the circle, all four points P, Q, A, and B will be very closely clustered to each other. Therefore, their distances to the conjugate pole p^* will be approximately equal, that is, $|z_A - p^*| \simeq |z_Q - p^*|$. Thus, we have for the ratio:

$$\frac{|H(z_A)|}{|H(z_Q)|} = \frac{|z_Q - p|}{|z_A - p|} = \frac{|PQ|}{|PA|}$$

Then, the 3-dB condition that $|H(z_A)|/|H(z_Q)| = 1/\sqrt{2}$, implies $|PQ|/|PA| = 1/\sqrt{2}$, or, $|PA| = \sqrt{2}|PQ|$, which means that the orthogonal triangle PQA will be equilateral, with a 45° angle $\angle QPA$. A similar argument shows that the triangle PQB is also a 45° orthogonal triangle. It follows that $|AB| = 2|QA| = 2|PQ| = 2(1 - R)$. But the arc subtended by the angle $\Delta\omega$ is equal to the radius of the circle (i.e., 1) times the angle $\Delta\omega$. This arc is approximately equal to $|AB|$ and therefore, $\Delta\omega = |AB| = 2(1 - R)$.

Eq. (6.4.4) can be used as the *design criterion* that determines the value of R for a given bandwidth $\Delta\omega$. The filter's causal impulse response can be obtained from Eq. (6.4.3) using partial fractions. We find, for $n \geq 0$:

$$\boxed{h(n) = \frac{G}{\sin\omega_0} R^n \sin(\omega_0 n + \omega_0)}$$

The difference equation for the filter follows from Eq. (6.4.3). We have:

$$Y(z) = H(z)X(z) = \frac{G}{1 + a_1 z^{-1} + a_2 z^{-2}} X(z)$$

which gives

$$(1 + a_1 z^{-1} + a_2 z^{-2}) Y(z) = GX(z)$$

and in the time domain:

$$y(n) + a_1 y(n - 1) + a_2 y(n - 2) = Gx(n)$$

or,

$$y(n) = -a_1 y(n - 1) - a_2 y(n - 2) + Gx(n) \tag{6.4.5}$$

A block diagram realization is shown in Fig. 6.4.4. The corresponding sample processing algorithm is obtained by introducing the internal states

$$w_1(n) = y(n - 1)$$

$$w_2(n) = y(n - 2) = w_1(n - 1)$$

The difference equation may be replaced by the system:

$$y(n) = -a_1 w_1(n) - a_2 w_2(n) + Gx(n)$$

$$w_2(n + 1) = w_1(n)$$

$$w_1(n + 1) = y(n)$$

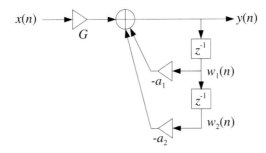

Fig. 6.4.4 Direct form realization of resonator filter.

which gives rise to the following sample processing algorithm:

> *for each input sample x do:*
> $$y = -a_1 w_1 - a_2 w_2 + Gx$$
> $$w_2 = w_1$$
> $$w_1 = y$$

Example 6.4.1: Design a 2-pole resonator filter with peak at $f_0 = 500$ Hz and width $\Delta f = 32$ Hz, operating at the sampling rate of $f_s = 10$ kHz.

Solution: The normalized resonator frequency will be

$$\omega_0 = \frac{2\pi f_0}{f_s} = 0.1\pi \quad \text{[radians/sample]}$$

and the corresponding width:

$$\Delta\omega = \frac{2\pi\Delta f}{f_s} = 0.02$$

Eq. (6.4.4) gives then

$$2(1 - R) = 0.02 \quad \Rightarrow \quad R = 0.99$$

With this value of R, we find the filter parameters:

$$G = 0.0062, \quad a_1 = -1.8831, \quad a_2 = 0.9801$$

and the filter transfer function

$$H(z) = \frac{0.0062}{1 - 1.8831 z^{-1} + 0.9801 z^{-2}}$$

The magnitude response and impulse response $h(n)$ are shown below:

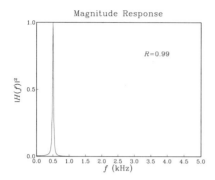
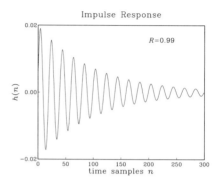

The effective time constant of this filter is about $n_{\text{eff}} = \ln \epsilon / \ln R = 458$ time samples. The graph only plots until $n = 300$.   □

A slight generalization of the resonator filter is to place a pair of zeros near the poles along the same directions as the poles, that is, at locations:

$$z_1 = re^{j\omega_0}, \; z_1^* = re^{-j\omega_0}$$

where r is restricted to the range $0 \leq r \leq 1$. The transfer function becomes:

$$H(z) = \frac{(1 - re^{j\omega_0}z^{-1})(1 - re^{-j\omega_0}z^{-1})}{(1 - Re^{j\omega_0}z^{-1})(1 - Re^{-j\omega_0}z^{-1})} = \frac{1 + b_1 z^{-1} + b_2 z^{-2}}{1 + a_1 z^{-1} + a_2 z^{-2}} \qquad (6.4.6)$$

where the filter coefficients are given in terms of the parameters r, R, and ω_0:

$$\begin{aligned} b_1 &= -2r \cos \omega_0 & b_2 &= r^2 \\ a_1 &= -2R \cos \omega_0 & a_2 &= R^2 \end{aligned} \qquad (6.4.7)$$

The corresponding magnitude squared response is:

$$|H(\omega)|^2 = \frac{(1 - 2r \cos(\omega - \omega_0) + r^2)(1 - 2r \cos(\omega + \omega_0) + r^2)}{(1 - 2R \cos(\omega - \omega_0) + R^2)(1 - 2R \cos(\omega + \omega_0) + R^2)}$$

Figure 6.4.5 shows the pole/zero pattern. When $r < R$, the pole "wins" over the zero, in the sense that it is closer to the unit circle than the zero, giving rise to a peak in the frequency response at $\omega = \omega_0$. The resonator case may be thought of as a special case with $r = 0$. When $r > R$, the zero wins over the pole, giving rise to a dip in the frequency response. In particular, if $r = 1$, one gets an exact zero, a notch, at $\omega = \omega_0$.

When the pole and zero are very near each other, that is, $r \lesssim R$ or $r \gtrsim R$, the frequency response remains essentially flat for frequencies far from $\omega = \pm\omega_0$, because the distances of the moving point $e^{j\omega}$ to the pole/zero pairs are almost equal, giving $|H(\omega)| \simeq 1$. Only near the vicinity of $\pm\omega_0$ does $|H(\omega)|$ vary dramatically, developing a peak or a dip.

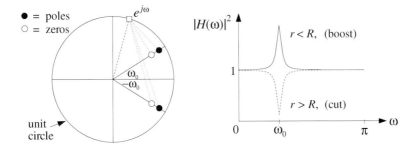

Fig. 6.4.5 Parametric equalizer filter.

Such a filter can be thought of as a simple *parametric equalizer* filter, providing a "boost" if $r < R$, or a "cut" if $r > R$. The height of the boost or cut relative to 1 is controlled by the closeness of r to R. The width of the peaks or dips is controlled by the closeness of R to the unit circle.

Later on, we will reconsider such pole/zero designs of equalization filters for digital audio systems based on analog designs and the bilinear transformation method, and will derive more precise design criteria that are based on the desired values of the bandwidth and gain of the peak or dip. Such second-order filters can be cascaded together to provide boosting or cutting at multiple frequencies.

Example 6.4.2: Using the numerical values $R = 0.98$, $\omega_0 = 0.4\pi$ for the pole, determine the parametric equalizer transfer functions of Eq. (6.4.6) for a boost corresponding to $r = 0.965$, a cut with $r = 0.995$, and an exact notch with $r = 1$.

Solution: The transfer function coefficients are computed by Eq. (6.4.7). The resulting transfer functions are in the three cases of $r = 0.965$, $r = 0.995$, $r = 1$:

$$H(z) = \frac{1 - 0.5964z^{-1} + 0.9312z^{-2}}{1 - 0.6057z^{-1} + 0.9604z^{-2}}$$

$$H(z) = \frac{1 - 0.6149z^{-1} + 0.9900z^{-2}}{1 - 0.6057z^{-1} + 0.9604z^{-2}}$$

$$H(z) = \frac{1 - 0.6180z^{-1} + z^{-2}}{1 - 0.6057z^{-1} + 0.9604z^{-2}}$$

The corresponding magnitude responses $|H(\omega)|$ are shown in Fig. 6.4.6. The case $r = 1$ provides an exact notch at $\omega = \omega_0$. $\qquad\square$

6.4.3 Notch and Comb Filters

The case $r = 1$ for a notch filter deserves some further discussion. In this case, the filter coefficients given by Eq. (6.4.7) can be written as

$$a_1 = Rb_1 = -2R\cos\omega_0, \quad a_2 = R^2b_2 = R^2$$

And, the transfer function takes the form:

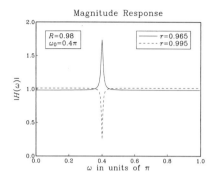
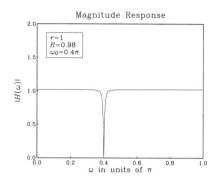

Fig. 6.4.6 Parametric equalizers of Example 6.4.2.

$$H(z) = \frac{1 + b_1 z^{-1} + b_2 z^{-2}}{1 + R b_1 z^{-1} + R^2 b_2 z^{-2}} = \frac{N(z)}{N(R^{-1}z)}$$

where $N(z)$ is the numerator polynomial having zeros at the two notch locations $z = e^{\pm j\omega_0}$:

$$N(z) = 1 + b_1 z^{-1} + b_2 z^{-2} = 1 - 2z^{-1}\cos\omega_0 + z^{-2} = (1 - e^{j\omega_0}z^{-1})(1 - e^{-j\omega_0}z^{-1})$$

This method can be generalized to construct a notch filter with notches at an arbitrary (finite) set of frequencies. The numerator polynomial $N(z)$ is defined as the polynomial whose zeros are *on the unit circle* at the *desired* notch locations. For example, if there are M desired notch frequencies ω_i, $i = 1, 2, \ldots, M$, then $N(z)$ is defined as the Mth degree polynomial with zeros at $z_i = e^{j\omega_i}$, $i = 1, 2, \ldots, M$:

$$\boxed{N(z) = \prod_{i=1}^{M}(1 - e^{j\omega_i}z^{-1})} \qquad \text{(notch polynomial)} \qquad (6.4.8)$$

The denominator polynomial is chosen as $D(z) = N(\rho^{-1}z)$, for some parameter $0 < \rho < 1$, that is,

$$D(z) = N(\rho^{-1}z) = \prod_{i=1}^{M}(1 - e^{j\omega_i}\rho z^{-1})$$

The zeros of $D(z)$ lie at the same directions as the notch zeros, but they are all pushed inside the unit circle at radius ρ. Therefore, for each desired zero $z_i = e^{j\omega_i}$, there is a corresponding pole $p_i = \rho e^{j\omega_i}$. Writing Eq. (6.4.8) in expanded form:

$$N(z) = 1 + b_1 z^{-1} + b_2 z^{-2} + \cdots + b_M z^{-M}$$

we obtain the following transfer function for the notch filter:

$$H(z) = \frac{N(z)}{N(\rho^{-1}z)} = \frac{1 + b_1 z^{-1} + b_2 z^{-2} + \cdots + b_M z^{-M}}{1 + \rho b_1 z^{-1} + \rho^2 b_2 z^{-2} + \cdots + \rho^M b_M z^{-M}} \qquad (6.4.9)$$

that is, the denominator coefficients are chosen as the scaled versions of the numerator coefficients,

$$a_i = \rho^i b_i, \quad i = 1, 2, \cdots, M$$

If ρ is near one, $\rho \lesssim 1$, the distances of the movable point $e^{j\omega}$ to the pole/zero pairs $\{z_i, p_i\} = \{z_i, \rho z_i\}$ are almost equal to each other except in the near vicinity of the pair, that is, except near $\omega = \omega_i$. Thus, $H(\omega)$ remains essentially flat except in the vicinity of the desired notch frequencies.

Example 6.4.3: A DSP system operating at a sampling rate of 600 Hz, is plagued by 60 Hz power frequency interference noise and its harmonics. Design a notch filter that removes all of these harmonics, but remains flat at other frequencies.

Solution: The fundamental harmonic is

$$\omega_1 = \frac{2\pi f_1}{f_s} = \frac{2\pi \cdot 60}{600} = 0.2\pi \quad \text{[radians/sample]}$$

The other harmonics at $f_i = i f_1$ correspond to $\omega_i = i\omega_1$. There are 10 harmonics that lie within the Nyquist interval $[0, f_s]$; namely, f_i, for $i = 0, 1, \cdots, 9$. Because $f_s = 10 f_1$, all the harmonics that lie outside the Nyquist interval (if they have not been filtered out by the antialiasing prefilter) will be aliased onto harmonics inside that interval. For example, the harmonic $f_{11} = 11 f_s$ gets aliased with $f_{11} - f_s = 11 f_1 - 10 f_1 = f_1$, and so on. Therefore, our digital notch filter must be designed to have notches at the 10 frequencies within the Nyquist interval:

$$\omega_i = i\omega_1 = \frac{2\pi i}{10}, \quad i = 0, 1, \ldots, 9$$

These 10 frequencies are none other than the tenth roots of unity, that is, the 10 roots of the polynomial:

$$N(z) = 1 - z^{-10} = \prod_{i=0}^{9} (1 - e^{j\omega_i} z^{-1})$$

Our notch filter is then obtained by

$$H(z) = \frac{N(z)}{N(\rho^{-1}z)} = \frac{1 - z^{-10}}{1 - \rho^{10} z^{-10}} = \frac{1 - z^{-10}}{1 - R z^{-10}}$$

where we set $R = \rho^{10}$. The following figure shows the resulting pole/zero pattern for this transfer function, and the corresponding magnitude response (computed only between $0 \le \omega \le \pi$):

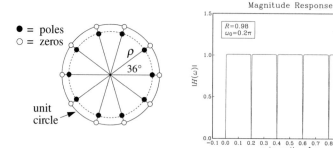

Magnitude Response

where we chose $R = 0.98$, or $\rho = R^{1/10} = (0.98)^{1/10} = 0.9980$. The radius ρ of the poles is very close to the unit circle resulting in very sharp notches at the desired harmonics. At other frequencies the magnitude response is essentially flat.   □

Example 6.4.4: Repeat the previous example when the sampling rate is $f_s = 1200$ Hz. Then, design another notch filter that excludes the DC and AC harmonics at $f = 0$ and $f = f_s/2$.

Solution: Now the fundamental harmonic is $\omega_1 = 2\pi \cdot 60/1200 = 0.1\pi$, and the 20 harmonics in the Nyquist interval will be

$$\omega_i = i\omega_1 = \frac{2\pi i}{20}, \quad i = 0, 1, \ldots, 19$$

They correspond to the 20th roots of unity, that is, the roots of:

$$N(z) = 1 - z^{-20} = \prod_{i=0}^{19}(1 - e^{j\omega_i}z^{-1})$$

The notch filter will be:

$$H(z) = \frac{N(z)}{N(\rho^{-1}z)} = \frac{1 - z^{-20}}{1 - \rho^{20}z^{-20}} = \frac{1 - z^{-20}}{1 - Rz^{-20}}$$

where we set $R = \rho^{20}$. The following figure shows the resulting pole/zero pattern and magnitude response, with the values $R = 0.98$ or $\rho = R^{1/20} = (0.98)^{1/20} = 0.9990$:

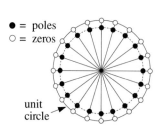
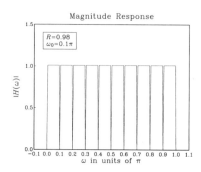

If we want to exclude the $\omega = \omega_0 = 0$ and $\omega = \omega_{10} = \pi$ harmonics from the notch filter, we must divide them out of $N(z)$. They contribute a factor

$$(1 - z^{-1})(1 + z^{-1}) = 1 - z^{-2}$$

Therefore, the new $N(z)$ that has notches at all the harmonics but at $z = \pm 1$ will be:

$$N(z) = \frac{1 - z^{-20}}{1 - z^{-2}} = \frac{1 - z^{-10}}{1 - z^{-2}}(1 + z^{-10}) = (1 + z^{-2} + z^{-4} + z^{-6} + z^{-8})(1 + z^{-10})$$

Therefore, we find for the notch transfer function:

$$H(z) = \frac{N(z)}{N(\rho^{-1}z)} = \frac{(1 + z^{-2} + z^{-4} + z^{-6} + z^{-8})(1 + z^{-10})}{(1 + \rho^2 z^{-2} + \rho^4 z^{-4} + \rho^6 z^{-6} + \rho^8 z^{-8})(1 + \rho^{10} z^{-10})}$$

The resulting pole/zero pattern and magnitude response are shown below:

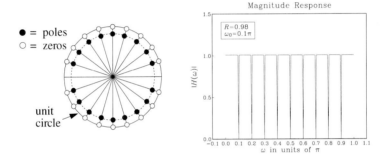

The value of ρ was the same as above, such that $\rho^{20} = R = 0.98$. □

A variant of the above method of constructing notch filters is the generalization of the parametric equalizer filter. It corresponds to moving the notch zeros *into* the unit circle and *behind* the poles, that is, replacing each notch zero by:

$$z_i = e^{j\omega_i} \quad \longrightarrow \quad z_i = r e^{j\omega_i}$$

where $r \lesssim \rho$. This makes the poles win over the zeros, changing all the notch dips into sharp peaks at frequencies $\omega = \omega_i$. The corresponding transfer function is obtained from Eq. (6.4.9) by scaling z in the numerator:

$$H(z) = \frac{N(r^{-1}z)}{N(\rho^{-1}z)} = \frac{1 + rb_1 z^{-1} + r^2 b_2 z^{-2} + \cdots + r^M b_M z^{-M}}{1 + \rho b_1 z^{-1} + \rho^2 b_2 z^{-2} + \cdots + \rho^M b_M z^{-M}} \quad (6.4.10)$$

Example 6.4.5: Starting with the notch polynomial $N(z) = 1 - z^{-20}$ of Example 6.4.4, we obtain the following filter, which will exhibit sharp peaks instead of dips if $r \lesssim \rho$:

$$H(z) = \frac{N(r^{-1}z)}{N(\rho^{-1}z)} = \frac{1 - r^{20} z^{-20}}{1 - \rho^{20} z^{-20}}$$

The pole/zero pattern and magnitude response are shown below:

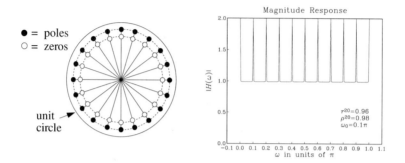

The values of the parameters were $r^{20} = 0.96$ and $\rho^{20} = 0.98$, which correspond to $r = 0.9980$ and $\rho = 0.9990$.   □

The notching and peaking filters of Eqs. (6.4.9) and (6.4.10) are referred to generically as *comb filters*. Notching comb filters are typically used to *cancel periodic interference*, such as the power frequency pickup and its harmonics. Peaking comb filters are used to *enhance periodic signals* in noise. The noise/interference reduction and signal enhancement capabilities of comb filters and their design will be discussed further in Chapters 8 and 11.

Comb filters arise also in the construction of digital reverb processors, where they represent the effects of multiple reflections of a sound wave off the walls of a listening space. They also arise in the design of digital periodic waveform generators. These topics will be discussed in Chapter 8.

6.5 Deconvolution, Inverse Filters, and Stability

In many applications, it is necessary to undo a filtering operation and recover the input signal from the available output signal. The output signal $y(n)$ is related to the input by the convolutional equation:

$$\boxed{y(n) = h(n) * x(n)} \tag{6.5.1}$$

The objective of such "deconvolution" methods is to recover $x(n)$ from the knowledge of $y(n)$ and the filter $h(n)$. In theory, this can be accomplished by *inverse filtering*, that is, filtering $y(n)$ through the inverse filter

$$\boxed{H_{\text{inv}}(z) = \frac{1}{H(z)}} \tag{6.5.2}$$

Indeed, working in the z-domain we have from Eq. (6.5.1):

$$Y(z) = H(z)X(z) \quad \Rightarrow \quad X(z) = \frac{1}{H(z)}Y(z) = H_{\text{inv}}(z)Y(z)$$

which becomes in the time domain:

$$\boxed{x(n) = h_{\mathrm{inv}}(n) * y(n)}$$

(6.5.3)

where $h_{\mathrm{inv}}(n)$ is the impulse response of the inverse filter $H_{\mathrm{inv}}(z)$. This operation is illustrated in Fig. 6.5.1.

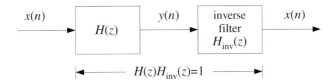

Fig. 6.5.1   Inverse filtering recovers original input.

Two typical applications of inverse filtering are *channel equalization* in digital voice or data transmission and the *equalization* of room or car acoustics in audio systems.

In channel equalization, the effect of a channel can be modeled as a linear filtering operation of the type of Eq. (6.5.1), where the transfer function $H(z)$ incorporates the effects of amplitude and phase distortions introduced by the channel. The signals $x(n)$ and $y(n)$ represent the transmitted and received signals, respectively. The inverse filter—called a *channel equalizer* in this context—is placed at the receiving end and its purpose is to undo the effects of the channel and recover the signal $x(n)$ that was transmitted. The overall processing system is shown in Fig. 6.5.2.

Often the channel itself is not known in advance, as for example in making a phone connection when the channel is established dynamically depending on how the call is routed to its destination. In such cases, the channel's transfer function must be determined (usually using adaptive signal processing techniques) before it can be inverted.

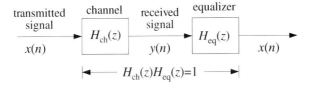

Fig. 6.5.2   Channel equalizer.

The sound generated by an audio system in a listening room is changed by the reverberation and absorption characteristics of the room's wall geometry and objects. The effect of the room can be modeled by a reverberation impulse response $h_{\mathrm{room}}(n)$, so that the actual sound wave impinging on a listener's ears is a distorted version of the original sound wave $x(n)$ produced by the audio system:

$$y_{\mathrm{room}}(n) = h_{\mathrm{room}}(n) * x(n)$$

(6.5.4)

The impulse response $h_{room}(n)$ depends on where one sits in the room, but it can be measured and then deconvolved away by an inverse filtering operation:

$$Y_{room}(z) = H_{room}(z)X(z) \quad \Rightarrow \quad X(z) = \frac{1}{H_{room}(z)} Y_{room}(z)$$

In addition to removing the local reverberation effects of a room, one may want to *add* the reverberation ambience of a concert hall that increases the warmth and richness of the sound. If the same audio signal $x(n)$ were listened to in a concert hall with reverberation impulse response $h_{hall}(n)$, the actual sound wave would be

$$y_{hall}(n) = h_{hall}(n) * x(n) \tag{6.5.5}$$

Available DSP audio effects processors can simulate the reverberation characteristics of typical concert halls and can implement the above filtering operation. An idealized audio effects processor is shown in Fig. 6.5.3.

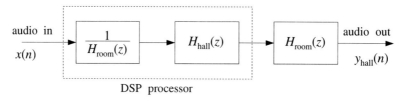

Fig. 6.5.3 An ideal audio effects processor.

First, it *deconvolves* the room acoustics by prefiltering the audio signal $x(n)$ by the inverse filter of the room's transfer function, anticipating the room's effect, and then it *convolves* it with the reverberation response of a desired concert hall. The effective transfer function of the arrangement is:

$$H_{eff}(z) = H_{room}(z) \cdot H_{hall}(z) \cdot \frac{1}{H_{room}(z)} = H_{hall}(z)$$

Thus, with the DSP effects processor, the sound wave produced in a room sounds as it does in a concert hall, Eq. (6.5.5). We will discuss audio effects processors in more detail in Chapter 8.

There are many other applications of inverse filtering in such diverse fields as identification and control systems, communication systems, image enhancement and restoration, digital magnetic recording, oil exploration, geophysics, ocean engineering, electromagnetic theory, scattering theory, radio astronomy, medical tomography, and spectroscopic analysis, as well as in many areas of applied mathematics, such as numerical analysis and statistics.

There are two major issues that arise in the practical application of inverse filtering. One is the requirement of *stability* of the inverse filter $h_{inv}(n)$. Without this requirement, the inverse filtering equation (6.5.3) would be unstable, resulting in numerical nonsense.

The other issue is the presence of *noise* in the data, that is, the available signal $y(n)$ may be (and almost invariably is) contaminated with noise, so that Eq. (6.5.1) is replaced by:

$$\boxed{y(n) = h(n) * x(n) + v(n)} \tag{6.5.6}$$

where $v(n)$ is the noise. Even if there is an exact and stable inverse filter $h_{\text{inv}}(n)$, the deconvolution of the noisy data $y(n)$ will give rise to the signal:

$$\boxed{\hat{x}(n) = h_{\text{inv}}(n) * y(n) = x(n) + \hat{v}(n)} \tag{6.5.7}$$

where $\hat{v}(n)$ is the filtered noise:

$$\hat{v}(n) = h_{\text{inv}}(n) * v(n)$$

Depending on the nature of the inverse filter $h_{\text{inv}}(n)$, even if the measurement noise $v(n)$ is very weak, it is quite possible for the filtered noise $\hat{v}(n)$ to be a much amplified version of $v(n)$, rendering $\hat{x}(n)$ a very poor and almost useless estimate of the desired signal $x(n)$. There exist signal processing techniques that try to address this noise problem to some extent. But they are beyond the scope of this book. See [28] for some discussion and references.

The impulse response $h(n)$ of the system $H(z)$ is assumed to be both stable and causal. This implies that the poles of $H(z)$ must lie strictly *inside* the unit circle. But the zeros of $H(z)$ do not have to lie inside the unit circle—they can be anywhere on the z-plane. Writing $H(z)$ in the ratio of two polynomials,

$$H(z) = \frac{N(z)}{D(z)}$$

we conclude that the zeros of $N(z)$ may be anywhere on the z-plane. Therefore, the inverse filter

$$H_{\text{inv}}(z) = \frac{1}{H(z)} = \frac{D(z)}{N(z)}$$

can have poles outside the unit circle. In this case, the *stable* inverse z-transform $h_{\text{inv}}(n)$ will necessarily be *anticausal*. As an example, consider the case of a filter $H(z)$ to be inverted:

$$H(z) = \frac{1 - 1.25z^{-1}}{1 - 0.5z^{-1}} = 2.5 - \frac{1.5}{1 - 0.5z^{-1}}$$

It has a causal and stable impulse response given by:

$$h(n) = 2.5\delta(n) - 1.5(0.5)^n u(n)$$

The corresponding inverse filter $H_{\text{inv}}(z)$ will be

$$H_{\text{inv}}(z) = \frac{1}{H(z)} = \frac{1 - 0.5z^{-1}}{1 - 1.25z^{-1}} = 0.4 + \frac{0.6}{1 - 1.25z^{-1}}$$

and because it has a pole outside the unit circle, its stable impulse response will be anticausal:

$$h_{\text{inv}}(n) = 0.4\delta(n) - 0.6(1.25)^n u(-n-1) = \begin{cases} 0 & \text{if } n \geq 1 \\ 0.4 & \text{if } n = 0 \\ -0.6(1.25)^n & \text{if } n \leq -1 \end{cases}$$

Such a stable and anticausal impulse response can be handled using the approximation technique discussed in Section 3.5. That is, the infinitely long anticausal tail is clipped off at some large negative time $n = -D$, replacing the exact $h_{\text{inv}}(n)$ by its clipped approximation:

$$\tilde{h}_{\text{inv}}(n) = \begin{cases} h_{\text{inv}}(n) & \text{if } n \geq -D \\ 0 & \text{if } n < -D \end{cases} \tag{6.5.8}$$

The approximate impulse response $\tilde{h}_{\text{inv}}(n)$ has only a finitely anticausal part and can be made causal by delaying it by D time units, as discussed in Section 3.5. The deconvolution error arising from using the approximate response can be determined as follows. Let $\tilde{x}(n)$ be the deconvolved output using $\tilde{h}_{\text{inv}}(n)$, that is,

$$\tilde{x}(n) = \tilde{h}_{\text{inv}}(n) * y(n)$$

Subtracting it from the exact output $x(n)$ of Eq. (6.5.3), we have

$$x(n) - \tilde{x}(n) = h_{\text{inv}}(n) * y(n) - \tilde{h}_{\text{inv}}(n) * y(n) = \left(h_{\text{inv}}(n) - \tilde{h}_{\text{inv}}(n)\right) * y(n)$$

$$= \sum_{m=-\infty}^{\infty} \left(h_{\text{inv}}(m) - \tilde{h}_{\text{inv}}(m)\right) y(n-m)$$

$$= \sum_{m=-\infty}^{-D-1} h_{\text{inv}}(m) y(n-m)$$

where all the terms for $m \geq -D$ were dropped because $\tilde{h}_{\text{inv}}(m)$ agrees with $h_{\text{inv}}(m)$ there, and $\tilde{h}_{\text{inv}}(m) = 0$ for $m < -D$. Assuming the signal $y(n)$ is bounded by some maximum value $|y(n)| \leq A$, we find

$$|x(n) - \tilde{x}(n)| \leq A \sum_{m=-\infty}^{-D-1} |h_{\text{inv}}(m)| \tag{6.5.9}$$

This is a general result for the deconvolution error. The upper bound gets smaller as D is increased. In particular, for the above example, we have:

$$|x(n) - \tilde{x}(n)| \leq A \sum_{m=-\infty}^{-D-1} |0.6(1.25)^m| = 2.4A(1.25)^{-D}$$

which can be made as small as desired by choosing D larger.

For a more general inverse filter $H_{inv}(z)$ having more than one pole *outside* the unit circle, the pole *nearest* the circle controls the decay time constant of the negative-time tail of $h_{inv}(n)$, because it converges to zero the slowest. Therefore, it controls the choice of the delay D. If we denote the minimum magnitude of these poles by

$$a = \min |p_{outside}| > 1$$

then for large D, the upper bound in Eq. (6.5.9) will behave essentially like the term a^{-D}, which gives the approximation error bound, for some constant B:

$$|x(n) - \tilde{x}(n)| \le Ba^{-D}$$

In summary, when the inverse filter transfer function $H_{inv}(z)$ has some poles *outside* the unit circle, one must choose the *anticausal but stable* impulse response $h_{inv}(n)$ and clip it at some large negative time D (and delay it by D to make it causal). The choice of D is dictated by the outside pole closest to the unit circle. The resulting deconvolution error arising from using the clipped filter can be made as small as desired by choosing the clipping delay D larger.

6.6 Problems

6.1 Using z-transforms, determine the transfer function $H(z)$ and from it the *causal* impulse response $h(n)$ of the linear systems described by the following I/O difference equations:

    a. $y(n) = -0.8y(n-1) + x(n)$

    b. $y(n) = 0.8y(n-1) + x(n)$

    c. $y(n) = 0.8y(n-1) + x(n) + x(n-1)$

    d. $y(n) = 0.8y(n-1) + x(n) - 0.5x(n-1)$

    e. $y(n) = 0.8y(n-1) + x(n) + 0.25x(n-2)$

    f. $y(n) = 0.9y(n-1) - 0.2y(n-2) + x(n) + x(n-1) - 6x(n-2)$

In each case, determine also the *frequency response* $H(\omega)$, the *pole/zero* pattern of the transfer function on the z-plane, draw a rough sketch of the *magnitude response* $|H(\omega)|$ over the right half of the Nyquist interval $0 \le \omega \le \pi$, and finally, draw the *direct* and *canonical* block diagram realizations of the difference equation and state the corresponding *sample-by-sample* filtering algorithms.

6.2 A unit-step signal $x(n) = u(n)$ is applied at the input of the linear systems:

    a. $y(n) = x(n) + 6x(n-1) + 11x(n-2) + 6x(n-3)$

    b. $y(n) = x(n) - x(n-4)$

Using z-transforms, determine the corresponding output signals $y(n)$, for all $n \ge 0$. Repeat for the alternating-step input $x(n) = (-1)^n u(n)$.

6.3 Repeat Problem 6.2 for the following systems:

   a. $y(n) = 0.25y(n-2) + x(n)$ b. $y(n) = -0.25y(n-2) + x(n)$

6.4 A unit-step signal $x(n) = u(n)$ is applied at the inputs of the systems of Problem 6.1.

   a. Using z-transforms, derive expressions for the corresponding output signals $y(n)$ for all $n \geq 0$, and determine which part of $y(n)$ is the steady-state part and which the transient part.

   b. Repeat for the input $x(n) = (-1)^n u(n)$.

   c. Repeat for the input $x(n) = (0.5)^n u(n)$ applied only to Problem 6.1(d).

   d. Repeat for the input $x(n) = (0.5)^n \cos(\pi n/2) u(n)$ applied to Problem 6.1(e) only.

   e. Repeat for the *unstable* input $x(n) = 2^n u(n)$ applied only to the system 6.1(f). Why is the output stable in this case?

6.5 Determine the transfer function $H(z)$ and the corresponding I/O difference equation relating $x(n)$ and $y(n)$ of the linear filters having the following impulse responses:

   a. $h(n) = \delta(n-5)$ e. $h(n) = (-0.8)^n[u(n) - u(n-8)]$

   b. $h(n) = u(n-5)$ f. $h(n) = (0.8)^n u(n) + (-0.8)^n u(n)$

   c. $h(n) = (0.8)^n u(n)$ g. $h(n) = 2(0.8)^n \cos(\pi n/2) u(n)$

   d. $h(n) = (-0.8)^n u(n)$ h. $h(n) = (0.8j)^n u(n) + (-0.8j)^n u(n)$

In each case, determine also the *frequency response* $H(\omega)$, the *pole/zero* pattern of the transfer function on the z-plane, draw a rough sketch of the *magnitude response* $|H(\omega)|$ over the right half of the Nyquist interval $0 \leq \omega \leq \pi$, and finally, draw the direct and canonical realizations implementing the I/O difference equation and state the corresponding *sample-by-sample* processing algorithms.

6.6 Find the transfer function $H(z)$ and express it as the *ratio* of two polynomials of the system having impulse response:

$$h(n) = \sum_{m=0}^{\infty} (0.5)^m \delta(n-8m) = \delta(n) + (0.5)\delta(n-8) + (0.5)^2\delta(n-16) + \cdots$$

Then, draw a block diagram realization and write its sample processing algorithm.

6.7 A digital reverberation processor has frequency response:

$$H(\omega) = \frac{-0.5 + e^{-j\omega 8}}{1 - 0.5e^{-j\omega 8}}$$

where ω is the digital frequency in [radians/sample]. Determine the *causal* impulse response $h(n)$, for all $n \geq 0$, and sketch it versus n. [*Hint:* Do not use partial fractions.]

6.8 The first few Fibonacci numbers are:

$$\mathbf{h} = [0, 1, 1, 2, 3, 5, 8, 13, 21, \ldots]$$

where each is obtained by summing the previous two.

   a. Determine the linear system $H(z)$ whose causal impulse response is $\mathbf{h}$, and express it as a rational function in z^{-1}.

b. Using partial fractions, derive an expression for the nth Fibonacci number in terms of the poles of the above filter.

c. Show that the ratio of two successive Fibonacci numbers converges to the *Golden Section*, that is, the positive solution of the quadratic equation $\phi^2 = \phi + 1$, namely, $\phi = (1 + \sqrt{5})/2$.

d. Show that the filter's poles are the two numbers $\{\phi, -\phi^{-1}\}$. Show that the geometric sequence:

$$\mathbf{y} = [0, 1, \phi, \phi^2, \phi^3, \dots]$$

satisfies the same recursion as the Fibonacci sequence (for $n \geq 3$). Show that $\mathbf{y}$ may be considered to be the output of the filter $\mathbf{h}$ for a particular input. What is that input?

See [32,33] for some remarkable applications and properties of Fibonacci numbers.

6.9 Pell's series [32,33] is obtained by summing twice the previous number and the number before (i.e., $h_n = 2h_{n-1} + h_{n-2}$):

$$\mathbf{h} = [0, 1, 2, 5, 12, 29, \dots]$$

Determine the linear system $H(z)$ whose causal impulse response is $\mathbf{h}$, and express it as a rational function in z^{-1}. Using partial fractions, derive an expression for the nth Pell number in terms of the poles of the above filter. Show that the ratio of two successive Pell numbers converges to the positive solution of the quadratic equation $\theta^2 = 2\theta + 1$, that is, $\theta = 1 + \sqrt{2}$. Show that the filter's poles are the two numbers $\{\theta, -\theta^{-1}\}$. Show that the geometric sequence:

$$\mathbf{y} = [0, 1, \theta, \theta^2, \theta^3, \dots]$$

satisfies the same recursion as the Pell sequence (for $n \geq 3$). Show that $\mathbf{y}$ may be considered to be the output of the filter $\mathbf{h}$ for a particular input. What is that input?

6.10 For a particular causal filter, it is observed that the input signal $(0.5)^n u(n)$ produces the output signal $(0.5)^n u(n) + (0.4)^n u(n)$. What *input* signal produces the output signal $(0.4)^n u(n)$?

6.11 For a particular filter, it is observed that the input signal $a^n u(n)$ causes the output signal $a^n u(n) + b^n u(n)$ to be produced. What output signal is produced by the input $c^n u(n)$, where $c = (a + b)/2$?

6.12 The signal $(0.7)^n u(n)$ is applied to the input of an unknown causal LTI filter, and the signal $(0.7)^n u(n) + (0.5)^n u(n)$ is observed at the output. What is the causal input signal that will cause the output $(0.5)^n u(n)$? What is the transfer function $H(z)$ of the system? Determine its causal impulse response $h(n)$, for all $n \geq 0$.

6.13 Design a resonator filter of the form $H(z) = \dfrac{1}{1 + a_1 z^{-1} + a_2 z^{-2}}$, which has a peak at $f_0 = 250$ Hz and a 3-dB width of $\Delta f = 20$ Hz and is operating at a rate of $f_s = 5$ kHz. What are the values of a_1 and a_2? Show that the *time constant* of the resonator is given approximately by

$$n_{\text{eff}} = -\frac{2 \ln \epsilon}{\Delta \omega}$$

which is valid for small $\Delta \omega$. For the designed filter, calculate the 40-dB value of n_{eff}, that is, corresponding to $\epsilon = 10^{-2}$. Compare the approximate and exact values of n_{eff}.

6.14 For *any* stable and causal filter, let τ_{40} and τ_{60} denote its 40-dB and 60-dB time constants, expressed in seconds. Show that they are related by: $\tau_{60} = 1.5\tau_{40}$.

6.15 Show that the 60-dB time constant of a resonator filter is given approximately by:

$$\tau_{60} = \frac{2.2}{\Delta f}$$

where τ_{60} is in seconds and Δf is the 3-dB width in Hz. When is the approximation valid?

6.16 It is desired to generate the following *periodic* waveform:

$$h(n) = [1, 2, 3, 4, 0, 0, 0, 0, 1, 2, 3, 4, 0, 0, 0, 0, \cdots]$$

where the dots indicate the periodic repetition of the 8 samples $[1, 2, 3, 4, 0, 0, 0, 0]$.

 a. Determine the filter $H(z)$ whose impulse response is the above periodic sequence. Express $H(z)$ as a ratio of two polynomials of degree less than 8.

 b. Draw the canonical and direct realization forms of $H(z)$. Write the corresponding sample processing algorithms.

6.17 A digital sawtooth generator filter has a periodic impulse response:

$$\mathbf{h} = [0, 1, 2, 3, 0, 1, 2, 3, 0, 1, 2, 3, \cdots]$$

where the dots indicate the periodic repetition of the length-4 sequence $\{0, 1, 2, 3\}$.

 a. Determine the transfer function $H(z)$.

 b. Draw the *direct* and *canonical* realization forms. Factor $H(z)$ into second-order sections with real coefficients. Draw the corresponding *cascade* realization.

 c. For each of the above three realizations, write the corresponding I/O time-domain difference equations and sample-by-sample processing algorithms.

 d. Using partial fractions, do an inverse z-transform of $H(z)$ and determine a closed form expression for the above impulse response $h(n)$ in the form

$$h(n) = A + B(-1)^n + 2C\cos\left(\frac{\pi n}{2}\right) + 2D\sin\left(\frac{\pi n}{2}\right), \qquad n \geq 0$$

 What are the values of A, B, C, D ?

6.18 Consider the system: $H(z) = \dfrac{1 + z^{-1} + z^{-2} + z^{-3}}{1 - z^{-7}}$.

 a. Without using partial fractions, determine the causal impulse response of the system. Explain your reasoning.

 b. Draw the *canonical* realization form of the system. Write the I/O *difference equations* and the *sample processing algorithm* describing this realization.

 c. The length-3 input signal $\mathbf{x} = [3, 2, 1]$ is applied as input to the system. Using any method, determine the output signal $y(n)$ for all $n \geq 0$. Explain your method.

6.19 A causal filter has transfer function: $H(z) = \dfrac{1 + z^{-1} + z^{-2} + z^{-3}}{1 - z^{-2}}$.

    a. Determine the numerical values of the causal impulse response $h(n)$, for all $n \geq 0$.

    b. Draw the canonical realization form of this filter and write the sample processing algorithm describing it.

6.20 A filter is described by the following sample processing algorithm:

> *for each input x do:*
> $$w_0 = x + w_1$$
> $$y = w_0 + w_2$$
> $$w_2 = w_1$$
> $$w_1 = w_0$$

    a. Determine the transfer function $H(z)$ of this filter.

    b. Show that it is *identically* equal to that of Problem 6.19.

6.21 A biomedical signal, sampled at a rate of 240 Hz, is plagued by 60 Hz power frequency interference and its harmonics.

Design a digital notch filter $H(z)$ that removes all these harmonics, but remains essentially flat at other frequencies.

[*Hint*: You may assume, although it is not necessary, that the signal has been prefiltered by a perfect antialiasing prefilter matched to the above sampling rate. Therefore, only the harmonics that lie in the $0 \leq f < 240$ Hz Nyquist interval are relevant.]

6.22 A digital filter has transfer function, where $0 < a < 1$:

$$H(z) = \frac{1 - z^{-16}}{1 - az^{-16}}$$

    a. What are the poles and zeros of this filter? Show them on the z-plane.

    b. Draw a rough sketch of its magnitude response $|H(\omega)|$ over the frequency interval $0 \leq \omega \leq 2\pi$.

    c. Determine the causal/stable impulse response $h(n)$ for all $n \geq 0$. Sketch it as a function of n. [*Hint*: Do not use PF expansions.]

    d. Draw the *canonical* realization form and write the corresponding sample processing algorithm. (You may make use of the delay routine to simplify the algorithm.)

6.23 Find the *causal* impulse response $h(n)$, for all $n \geq 0$, of $H(z) = \dfrac{0.3 + 0.15z^{-1}}{1 - 0.5z^{-1}}$.

6.24 Let $H(z) = \dfrac{1 - a}{1 - az^{-1}}$ be a first-order lowpass filter (also called a first-order smoother), where $0 < a < 1$. Draw the canonical realization. Draw another realization that uses only one multiplier, (that is, a), one delay, and one adder and one subtractor. For both realizations, write the sample-by-sample processing algorithms. What would you say is the purpose of the chosen gain factor $1 - a$?

6.25 Let $H(z) = \dfrac{3 - 3z^{-1} - z^{-2}}{1 - 1.5z^{-1} - z^{-2}}$. Determine all possible impulse responses $h(n)$, for all n, and the corresponding ROCs.

6.26 A discrete system is described by the difference equation

$$y(n) = 2.5y(n-1) - y(n-2) + 3x(n) + 3x(n-2)$$

Using z-transforms, find *all* possible impulse responses $h(n)$ and indicate their causality and stability properties.

For the causal filter, determine the output $y(n)$ if the input is $x(n) = g(n) - 2g(n-1)$, where $g(n) = \cos(\pi n/2) u(n)$.

6.27 A signal $x(n)$ has frequency bandwidth $0 \le |\omega| \le \omega_c$, where $\omega_c < \pi$. The signal is applied to a lowpass filter $H(\omega)$ resulting in the output $y(n)$. Assuming that the filter has an approximately flat passband over $0 \le |\omega| \le \omega_c$ and is zero outside that band, and assuming that the filter has linear phase with a phase delay $d(\omega) = D$, show that the resulting output will be approximately equal to the delayed input $y(n) = Gx(n-D)$, where G is the filter's passband gain.

6.28 Consider a causal/stable filter $H(z) = \dfrac{N(z)}{(1 - p_1 z^{-1})(1 - p_2 z^{-1}) \cdots (1 - p_M z^{-1})}$, where the M poles are inside the unit circle $|p_i| < 1$, and the numerator $N(z)$ is a polynomial in z^{-1} of degree strictly less than M. Show that the impulse response can be expressed in the form:

$$h(n) = \sum_{i=1}^{M} A_i p_i^n u(n), \qquad \text{where} \quad A_i = \frac{N(p_i)}{\prod_{j \neq i}(1 - p_j p_i^{-1})}$$

6.29 The input-on behavior of the above filter may be studied by applying to it a one-sided sinusoid that starts at $n = 0$ and continues till $n = \infty$. The input-off behavior may be studied by applying a sinusoid that has been on since $n = -\infty$ and turns off at $n = 0$. Using z-transforms, show that the corresponding outputs are in the two cases:

$$e^{j\omega_0 n} u(n) \xrightarrow{H} y(n) = H(\omega_0) e^{j\omega_0 n} u(n) + \sum_{i=1}^{M} B_i p_i^n u(n)$$

$$e^{j\omega_0 n} u(-n-1) \xrightarrow{H} y(n) = H(\omega_0) e^{j\omega_0 n} u(-n-1) - \sum_{i=1}^{M} B_i p_i^n u(n)$$

Thus, the transient behavior for $n \ge 0$ is the same in both cases except for a sign. Show that the coefficients B_i are related to A_i of the previous problem by:

$$B_i = \frac{p_i A_i}{p_i - e^{j\omega_0}}, \qquad i = 1, 2, \ldots, M$$

Using these results and linear superposition, derive the steady-state result of Eq. (6.3.2), which is valid for double-sided sinusoids. See also Problem 8.14.

6.30 Let $H(z) = \dfrac{3 - 5z^{-1} + z^{-2}}{(1 - 0.5z^{-1})(1 - 2z^{-1})}$. Determine the stable but anticausal impulse response $h(n)$ of this system. Let $\tilde{h}(n)$ denote the causal approximation to $h(n)$ obtained by clipping off the anticausal tail of $h(n)$ at some large negative time $n = -D$. What is $\tilde{H}(z)$?

Suppose the signal $x(n) = \delta(n) - 2\delta(n-1)$ is applied at the input of the true and approximate systems resulting in the outputs $y(n)$ and $\tilde{y}(n)$, respectively. Using z-transforms, determine an expression for the output error $e(n) = y(n) - \tilde{y}(n)$.

7

Digital Filter Realizations

7.1 Direct Form

In this section, we discuss in detail the *direct form* realizations of digital filters, otherwise known as *direct form I* realizations. We begin by considering a simple second-order filter with transfer function

$$H(z) = \frac{N(z)}{D(z)} = \frac{b_0 + b_1 z^{-1} + b_2 z^{-2}}{1 + a_1 z^{-1} + a_2 z^{-2}} \qquad (7.1.1)$$

having I/O difference equation:

$$y_n = -a_1 y_{n-1} - a_2 y_{n-2} + b_0 x_n + b_1 x_{n-1} + b_2 x_{n-2} \qquad (7.1.2)$$

The direct form realization is the block diagram representation of this difference equation. It is depicted in Fig. 7.1.1.

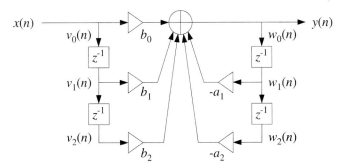

Fig. 7.1.1 Direct form realization of second-order IIR filter.

The main feature of this realization is a single adder that accumulates all the terms in the right-hand side of Eq. (7.1.2) and whose output is $y(n)$.

All the *b*-multiplier terms are *feeding forward*. They correspond to the *numerator* polynomial of $H(z)$ and to the *x*-dependent, *non-recursive* terms of the difference equation (7.1.2).

269

The a-multiplier terms are *feeding back.* They correspond to the *denominator* polynomial of $H(z)$ and to the y-dependent, *recursive* terms of Eq. (7.1.2). Notice the change in sign: The a-multipliers in the block diagram and in the difference equation are the *negatives* of the denominator polynomial coefficients.

The FIR direct form of Chapter 4 is obtained as a special case of this by setting the feedback coefficients to zero $a_1 = a_2 = 0$.

The sample-by-sample processing algorithm can be derived by defining the internal states of the filter to be:

$$v_0(n) = x(n) \qquad\qquad w_0(n) = y(n)$$

$$v_1(n) = x(n-1) = v_0(n-1) \quad \text{and} \quad w_1(n) = y(n-1) = w_0(n-1)$$

$$v_2(n) = x(n-2) = v_1(n-1) \qquad\qquad w_2(n) = y(n-2) = w_1(n-1)$$

The quantities $v_1(n)$, $v_2(n)$, $w_1(n)$, and $w_2(n)$ are the internal states of the filter, representing the *contents* of the delay registers of the block diagram at time n. Replacing n by $n+1$ in the above definitions, we find the time updates:

$$v_1(n+1) = v_0(n) \qquad\qquad w_1(n+1) = w_0(n)$$
$$\text{and}$$
$$v_2(n+1) = v_1(n) \qquad\qquad w_2(n+1) = w_1(n)$$

Therefore, we may replace Eq. (7.1.2) by the system:

$$
\boxed{
\begin{aligned}
&v_0(n) = x(n) \\
&w_0(n) = -a_1 w_1(n) - a_2 w_2(n) + b_0 v_0(n) + b_1 v_1(n) + b_2 v_2(n) \\
&y(n) = w_0(n) \\
&v_2(n+1) = v_1(n), \quad w_2(n+1) = w_1(n) \\
&v_1(n+1) = v_0(n), \quad w_1(n+1) = w_0(n)
\end{aligned}
}
$$

It can be replaced by the following repetitive sample processing algorithm:

$$
\boxed{
\begin{aligned}
&\text{\textit{for each input sample x do:}} \\
&\quad v_0 = x \\
&\quad w_0 = -a_1 w_1 - a_2 w_2 + b_0 v_0 + b_1 v_1 + b_2 v_2 \\
&\quad y = w_0 \\
&\quad v_2 = v_1, \quad w_2 = w_1 \\
&\quad v_1 = v_0, \quad w_1 = w_0
\end{aligned}
}
\qquad (7.1.3)
$$

Note that the state updating must be done in reverse order (from the bottom up in the block diagram).

The direct form realization can be generalized easily to the case of arbitrary numerator and denominator polynomials. One simply extends the structure downward by adding more delays and corresponding multipliers. In the general case, the transfer function is:

$$H(z) = \frac{N(z)}{D(z)} = \frac{b_0 + b_1 z^{-1} + b_2 z^{-2} + \cdots + b_L z^{-L}}{1 + a_1 z^{-1} + a_2 z^{-2} + \cdots + a_M z^{-M}} \qquad (7.1.4)$$

having an Lth degree numerator and Mth degree denominator. The corresponding I/O difference equation is:

$$y_n = -a_1 y_{n-1} - a_2 y_{n-2} - \cdots - a_M y_{n-M} + b_0 x_n + b_1 x_{n-1} + \cdots + b_L x_{n-L} \quad (7.1.5)$$

Figure 7.1.2 shows the case $L = M$. To derive the sample processing algorithm in the general case, we define the internal state signals:

$$v_i(n) = x(n - i), \quad i = 0, 1, \ldots, L$$
$$w_i(n) = y(n - i), \quad i = 0, 1, \ldots, M \qquad (7.1.6)$$

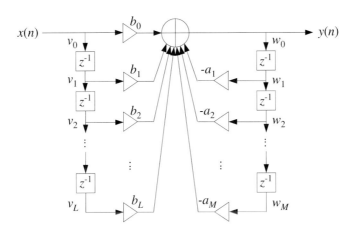

Fig. 7.1.2 Direct form realization of Mth order IIR filter.

They are updated in time by

$$v_i(n + 1) = v_{i-1}(n), \quad i = 1, 2, \ldots, L$$
$$w_i(n + 1) = w_{i-1}(n), \quad i = 1, 2, \ldots, M \qquad (7.1.7)$$

These can be shown easily, for example:

$$w_i(n + 1) = y\big((n + 1) - i\big) = y\big(n - (i - 1)\big) = w_{i-1}(n)$$

Then, the difference equation (7.1.5) is be written as follows:

$$w_0(n) = -a_1 w_1(n) - \cdots - a_M w_M(n) + b_0 v_0(n) + b_1 v_1(n) + \cdots + b_L v_L(n)$$

Together with the time updates (7.1.7), it leads to the following sample processing algorithm for the direct form realization:

$$\boxed{\begin{aligned}
&\textit{for each input sample x do:}\\
&\quad v_0 = x\\
&\quad w_0 = -a_1 w_1 - \cdots - a_M w_M + b_0 v_0 + b_1 v_1 + \cdots + b_L v_L\\
&\quad y = w_0\\
&\quad v_i = v_{i-1}, \quad i = L, L-1, \ldots, 1\\
&\quad w_i = w_{i-1}, \quad i = M, M-1, \ldots, 1
\end{aligned}}
\qquad (7.1.8)$$

Again, the state updating must be done in *reverse* order to avoid overwriting v_i and w_i. Before filtering the first input sample, the internal states must be *initialized to zero*, that is, at time $n = 0$ set:

$$[v_1, v_2, \ldots, v_L] = [0, 0, \ldots, 0], \qquad [w_1, w_2, \ldots, w_M] = [0, 0, \ldots, 0]$$

The following C routine `dir.c` is an implementation of this algorithm:

```
/* dir.c - IIR filtering in direct form */
```

```
double dir(M, a, L, b, w, v, x)          usage: y = dir(M, a, L, b, w, v, x);
double *a, *b, *w, *v, x;                 v, w are internal states
int M, L;                                 denominator and numerator orders
{
        int i;

        v[0] = x;                         current input sample
        w[0] = 0;                         current output to be computed

        for (i=0; i<=L; i++)              numerator part
                w[0] += b[i] * v[i];

        for (i=1; i<=M; i++)              denominator part
                w[0] -= a[i] * w[i];

        for (i=L; i>=1; i--)              reverse-order updating of v
                v[i] = v[i-1];

        for (i=M; i>=1; i--)              reverse-order updating of w
                w[i] = w[i-1];

        return w[0];                      current output sample
}
```

Note that b, a are the numerator and denominator coefficient vectors:

$$\mathbf{b} = [b_0, b_1, b_2, \ldots, b_L], \qquad \mathbf{a} = [1, a_1, a_2, \ldots, a_M] \qquad (7.1.9)$$

They, and the internal state vectors w, v, must be declared and allocated in the main program by

```
double *a, *b, *w, *v;
a = (double *) calloc(M+1, sizeof(double));     (M+1)-dimensional
b = (double *) calloc(L+1, sizeof(double));     (L+1)-dimensional
a[0] = 1;                                       always so
w = (double *) calloc(M+1, sizeof(double));     (M+1)-dimensional
v = (double *) calloc(L+1, sizeof(double));     (L+1)-dimensional
```

Note that `calloc` initializes the internal states w,v to zero. The following program segment illustrates the usage of `dir`:

```
for (n = 0; n < Ntot; n++)
        y[n] = dir(M, a, L, b, w, v, x[n]);
```

Example 7.1.1: Draw the direct form realization of the following filter:

$$H(z) = \frac{2 - 3z^{-1} + 4z^{-3}}{1 + 0.2z^{-1} - 0.3z^{-2} + 0.5z^{-4}}$$

and determine the corresponding difference equation and sample processing algorithm.

Solution: The difference equation is:

$$y_n = -0.2y_{n-1} + 0.3y_{n-2} - 0.5y_{n-4} + 2x_n - 3x_{n-1} + 4x_{n-3}$$

The direct form realization is shown in Fig. 7.1.3. The sample processing algorithm is:

> *for each input sample x do:*
> $v_0 = x$
> $w_0 = -0.2w_1 + 0.3w_2 - 0.5w_4 + 2v_0 - 3v_1 + 4v_3$
> $y = w_0$
> $w_4 = w_3$
> $w_3 = w_2, \qquad v_3 = v_2$
> $w_2 = w_1, \qquad v_2 = v_1$
> $w_1 = w_0, \qquad v_1 = v_0$

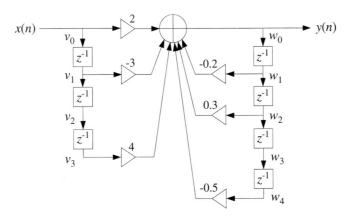

Fig. 7.1.3 Direct form realization of Example 7.1.1.

The filter coefficient and state vectors are in this example:

$$\mathbf{a} = [a_0, a_1, a_2, a_3, a_4] = [1, 0.2, -0.3, 0.0, 0.5]$$

$$\mathbf{b} = [b_0, b_1, b_2, b_3] = [2, -3, 0, 4]$$

$$\mathbf{w} = [w_0, w_1, w_2, w_3, w_4], \quad \mathbf{v} = [v_0, v_1, v_2, v_3]$$

There are four delays for the feedback coefficients and three for the feed forward ones, because the denominator and numerator polynomials have orders 4 and 3. Note also that some coefficients are zero and therefore their connections to the adder are not shown.   □

The direct form algorithm (7.1.8) can be written more simply in terms of the dot product routine dot and delay routine delay in much the same way as was done in Chapter 4 for the FIR filtering routine fir. The output y in Eq. (7.1.8) involves essentially the *dot products* of the filter coefficient vectors with the internal state vectors. These dot products are:

$$\mathrm{dot}(L, \mathbf{b}, \mathbf{v}) = \mathbf{b}^T \mathbf{v} = b_0 v_0 + b_1 v_1 + \cdots + b_L v_L$$

$$\mathrm{dot}(M, \mathbf{a}, \mathbf{w}) = \mathbf{a}^T \mathbf{w} = a_0 w_0 + a_1 w_1 + \cdots + a_M w_M$$

where $a_0 = 1$. If we set $w_0 = 0$, then the second dot product is precisely the contribution of the feedback (denominator) coefficients to the filter output, that is,

$$w_0 = 0 \quad \Rightarrow \quad \mathrm{dot}(M, \mathbf{a}, \mathbf{w}) = a_1 w_1 + \cdots + a_M w_M$$

Therefore, we may replace Eq. (7.1.8) by the more compact version:

$$\boxed{\begin{aligned} &\textit{for each input sample x do:} \\ &\quad v_0 = x \\ &\quad w_0 = 0 \\ &\quad w_0 = \mathrm{dot}(L, \mathbf{b}, \mathbf{v}) - \mathrm{dot}(M, \mathbf{a}, \mathbf{w}) \\ &\quad y = w_0 \\ &\quad \mathrm{delay}(L, \mathbf{v}) \\ &\quad \mathrm{delay}(M, \mathbf{w}) \end{aligned}}$$
(7.1.10)

The following routine dir2.c is an implementation of this algorithm. It is the IIR version of the routine fir2 of Chapter 4.

```
/* dir2.c - IIR filtering in direct form */

double dot();
void delay();

double dir2(M, a, L, b, w, v, x)          usage: y = dir2(M, a, L, b, w, v, x);
double *a, *b, *w, *v, x;
int M, L;
{
        v[0] = x;                          current input sample
        w[0] = 0;                          needed for dot(M,a,w)
```

```
    w[0] = dot(L, b, v) - dot(M, a, w);          current output

    delay(L, v);                                 update input delay line

    delay(M, w);                                 update output delay line

    return w[0];
}
```

7.2 Canonical Form

The *canonical realization form*, otherwise known as *direct form II*, can be obtained from the direct form in the following way. Starting with the second-order filter of Eq. (7.1.1) we may group the five terms in the right-hand side of Eq. (7.1.2) into two subgroups: the recursive terms and the non-recursive ones, that is,

$$y_n = (b_0 x_n + b_1 x_{n-1} + b_2 x_{n-2}) + (-a_1 y_{n-1} - a_2 y_{n-2})$$

This regrouping corresponds to splitting the big adder of the direct form realization of Fig. 7.1.1 into two parts, as shown in Fig. 7.2.1.

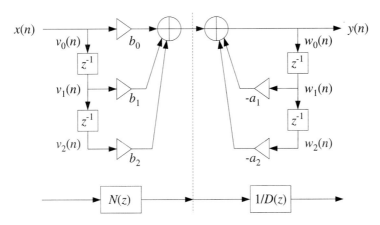

Fig. 7.2.1 Regrouping of direct form terms.

We can think of the resulting realization as the cascade of two filters: one consisting only of the feed-forward terms and the other of the feedback terms. It is easy to verify that these two filters are the numerator $N(z)$ and the inverse of the denominator $1/D(z)$, so that their cascade will be

$$H(z) = N(z) \cdot \frac{1}{D(z)}$$

which is the original transfer function given in Eq. (7.1.1). Mathematically, the order of the cascade factors can be changed so that

$$H(z) = \frac{1}{D(z)} \cdot N(z)$$

which corresponds to changing the order of the block diagrams representing the factors $N(z)$ and $1/D(z)$, as shown in Fig. 7.2.2.

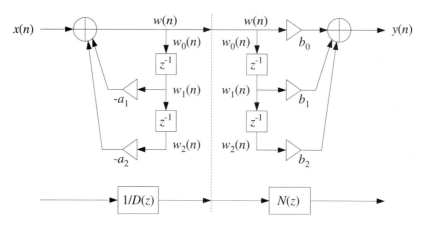

Fig. 7.2.2 Interchanging $N(z)$ and $1/D(z)$.

The output signal of the first filter $1/D(z)$ becomes the input to the second filter $N(z)$. If we denote that signal by $w(n)$, we observe that it gets delayed in the same way by the two sets of delays of the two filters, that is, the two sets of delays have the *same contents*, namely, the numbers $w(n-1)$, $w(n-2)$.

Therefore, there is no need to use two separate sets of delays. The two sets can be merged into one, shared by both the first and second filters $1/D(z)$ and $N(z)$. This leads to the *canonical realization form* depicted in Fig. 7.2.3.

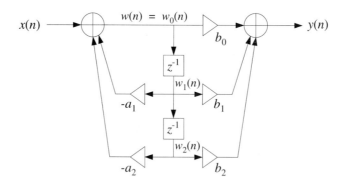

Fig. 7.2.3 Canonical realization form of second-order IIR filter.

The I/O difference equations describing the time-domain operation of this realization can be obtained by writing the conservation equations at each adder, with the input adder written first:

$$\boxed{\begin{aligned} w(n) &= x(n) - a_1 w(n-1) - a_2 w(n-2) \\ y(n) &= b_0 w(n) + b_1 w(n-1) + b_2 w(n-2) \end{aligned}}$$

(7.2.1)

The computed value of $w(n)$ from the first equation is passed into the second to compute the final output $y(n)$. It is instructive also to look at this system in the z-domain. Taking z-transforms of both sides, we find

$$W(z) = X(z) - a_1 z^{-1} W(z) - a_2 z^{-2} W(z)$$

$$Y(z) = b_0 W(z) + b_1 z^{-1} W(z) + b_2 z^{-2} W(z)$$

which can be solved for $W(z)$ and $Y(z)$:

$$W(z) = \frac{1}{1 + a_1 z^{-1} + a_2 z^{-2}} X(z) = \frac{1}{D(z)} X(z)$$

$$Y(z) = (b_0 + b_1 z^{-1} + b_2 z^{-2}) W(z) = N(z) W(z)$$

Eliminating $W(z)$, we find that the transfer function from $X(z)$ to $Y(z)$ is the original one, namely, $N(z)/D(z)$:

$$Y(z) = N(z) W(z) = N(z) \frac{1}{D(z)} X(z) = \frac{N(z)}{D(z)} X(z)$$

At each time n, the quantities $w(n-1)$ and $w(n-2)$ in Eq. (7.2.1) are the contents of the two shared delay registers. Therefore, they are the internal states of the filter. To determine the corresponding sample processing algorithm, we redefine these internal states by:

$$\begin{aligned} w_0(n) &= w(n) \\ w_1(n) &= w(n-1) = w_0(n-1) \quad \Rightarrow \\ w_2(n) &= w(n-2) = w_1(n-1) \end{aligned} \qquad \begin{aligned} w_1(n+1) &= w_0(n) \\ w_2(n+1) &= w_1(n) \end{aligned}$$

Therefore, the system (7.2.1) can be rewritten as:

$$\boxed{\begin{aligned} w_0(n) &= x(n) - a_1 w_1(n) - a_2 w_2(n) \\ y(n) &= b_0 w_0(n) + b_1 w_1(n) + b_2 w_2(n) \\ w_2(n+1) &= w_1(n) \\ w_1(n+2) &= w_0(n) \end{aligned}}$$

which translates to the following sample processing algorithm:

$$\boxed{\begin{aligned} &\textit{for each input sample } x \textit{ do:} \\ &\quad w_0 = x - a_1 w_1 - a_2 w_2 \\ &\quad y = b_0 w_0 + b_1 w_1 + b_2 w_2 \\ &\quad w_2 = w_1 \\ &\quad w_1 = w_0 \end{aligned}}$$

(7.2.2)

where, again, the states w_2 and w_1 must be updated in reverse order. The canonical form for the more general case of Eq. (7.1.4) is obtained following similar steps. That is, we define

$$Y(z) = N(z)W(z) \quad \text{and} \quad W(z) = \frac{1}{D(z)}X(z)$$

and rewrite them in the form:

$$(1 + a_1 z^{-1} + a_2 z^{-2} + \cdots + a_M z^{-M})W(z) = X(z)$$

$$Y(z) = (b_0 + b_1 z^{-1} + \cdots + b_L z^{-L})W(z)$$

or, equivalently

$$W(z) = X(z) - (a_1 z^{-1} + a_2 z^{-2} + \cdots + a_M z^{-M})W(z)$$

$$Y(z) = (b_0 + b_1 z^{-1} + \cdots + b_L z^{-L})W(z)$$

which become in the time domain:

$$\boxed{\begin{aligned} w(n) &= x(n) - a_1 w(n-1) - \cdots - a_M w(n-M) \\ y(n) &= b_0 w(n) + b_1 w(n-1) + \cdots + b_L w(n-L) \end{aligned}}$$

(7.2.3)

The block diagram realization of this system is shown in Fig. 7.2.4 for the case $M = L$. If $M \neq L$ one must draw the *maximum* number of common delays, that is, $K = \max(M, L)$. Defining the internal states by

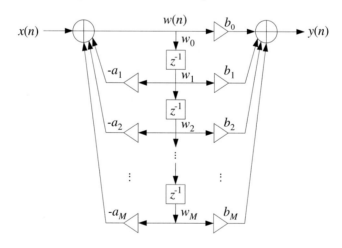

Fig. 7.2.4 Canonical realization form of Mth order IIR filter.

$$w_i(n) = w(n-i) = w_{i-1}(n-1), \quad i = 0, 1, \ldots, K$$

we may rewrite the system (7.2.3) in the form:

$$\boxed{\begin{aligned}
w_0(n) &= x(n) - a_1 w_1(n) - \cdots - a_M w_M(n) \\
y(n) &= b_0 w_0(n) + b_1 w_1(n) + \cdots + b_L w_L(n) \\
w_i(n+1) &= w_{i-1}(n), \quad i = K, K-1, \ldots, 1
\end{aligned}}$$

(7.2.4)

This leads to the following sample-by-sample filtering algorithm:

$$\boxed{\begin{aligned}
&\textit{for each input sample } x \textit{ do:} \\
&\quad w_0 = x - a_1 w_1 - a_2 w_2 - \cdots - a_M w_M \\
&\quad y = b_0 w_0 + b_1 w_1 + \cdots + b_L w_L \\
&\quad w_i = w_{i-1}, \quad i = K, K-1, \ldots, 1
\end{aligned}}$$

(7.2.5)

Again, the state updating must be done in reverse order. Before the first input sample, the internal states must be initialized to zero, that is, $[w_1, w_2, \ldots, w_K] = [0, 0, \ldots, 0]$. The following C routine `can.c` is an implementation of this algorithm:

```
/* can.c - IIR filtering in canonical form */

double can(M, a, L, b, w, x)            usage: y = can(M, a, L, b, w, x);
double *a, *b, *w, x;                   w = internal state vector
int M, L;                               denominator and numerator orders
{
        int K, i;
        double y = 0;

        K = (L <= M) ? M : L;           K = max(M, L)

        w[0] = x;                       current input sample

        for (i=1; i<=M; i++)            input adder
                w[0] -= a[i] * w[i];

        for (i=0; i<=L; i++)            output adder
                y += b[i] * w[i];

        for (i=K; i>=1; i--)            reverse updating of w
                w[i] = w[i-1];

        return y;                       current output sample
}
```

The vectors `a,b` must be allocated just as for the direct form routine `dir`. The state vector w must be allocated to dimension $K+1$, that is,

```
w = (double *) calloc(K+1, sizeof(double));     w = [w_0, w_1, ..., w_K]
```

The same program segment illustrating the usage of `dir` also illustrates the usage of `can`. The only difference is that now there is only one state vector w instead of w,v:

```
for (n = 0; n < Ntot; n++)
        y[n] = can(M, a, L, b, w, x[n]);
```

Comparing Figs. 7.1.2 and 7.2.4, we note that: (a) The direct form requires *twice* as many delays; (b) both have exactly the *same* multiplier coefficients; (c) the direct form has only one adder whose output is the system output; and (d) the canonical form has *two* adders, one at the input and one at the output. Between the two, the canonical form is usually preferred in practice.

Note also that for FIR filters the denominator polynomial is trivial $D(z) = 1$ and thus, the direct and canonical forms are identical to the direct form of Chapter 4.

Example 7.2.1: Draw the canonical realization form of Example 7.1.1 and write the corresponding difference equations and sample processing algorithm.

Solution: Interchanging the feedback and feed-forward parts of Fig. 7.1.3 and merging the common delays, we obtain the canonical realization shown in Fig. 7.2.5. The difference equations describing this block diagram in the time domain are obtained from the two adder equations:

$$w(n) = x(n) - 0.2w(n-1) + 0.3w(n-2) - 0.5w(n-4)$$

$$y(n) = 2w(n) - 3w(n-1) + 4w(n-3)$$

The corresponding sample processing algorithm is:

$$
\boxed{
\begin{aligned}
&\textit{for each input sample x do:}\\
&\quad w_0 = x - 0.2w_1 + 0.3w_2 - 0.5w_4\\
&\quad y = 2w_0 - 3w_1 + 4w_3\\
&\quad w_4 = w_3\\
&\quad w_3 = w_2\\
&\quad w_2 = w_1\\
&\quad w_1 = w_0
\end{aligned}
}
$$

Here, the maximum number of delays is $K = \max(M, L) = \max(4, 3) = 4$. $\square$

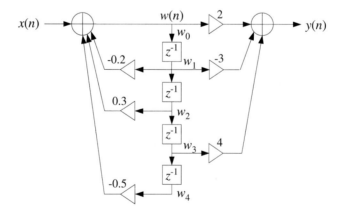

Fig. 7.2.5 Canonical realization form of Example 7.1.1.

Following the discussion of Eq. (7.1.10), we can derive an alternative version of the canonical form algorithm (7.2.5) that uses the dot product and delay routines, as follows:

$$
\boxed{
\begin{aligned}
&\textit{for each input sample } x \textit{ do:}\\
&\quad w_0 = 0\\
&\quad w_0 = x - \mathrm{dot}(M, \mathbf{a}, \mathbf{w})\\
&\quad y = \mathrm{dot}(L, \mathbf{b}, \mathbf{w})\\
&\quad \mathrm{delay}(K, \mathbf{w})
\end{aligned}
}
\qquad (7.2.6)
$$

The C routine `can2.c` is an implementation.

```
/* can2.c - IIR filtering in canonical form */

double dot();
void delay();

double can2(M, a, L, b, w, x)          usage: y = can2(M, a, L, b, w, x);
double *a, *b, *w, x;
int M, L;
{
        int K;
        double y;

        K = (L <= M) ? M : L;          K = max(M,L)

        w[0] = 0;                      needed for dot(M,a,w)

        w[0] = x - dot(M, a, w);       input adder

        y = dot(L, b, w);              output adder

        delay(K, w);                   update delay line

        return y;                      current output sample
}
```

7.3 Cascade Form

A *second-order section* (SOS) is a second-order transfer function of the form (7.1.1). Its canonical realization is depicted in Fig. 7.2.3. In the time domain it operates according to the I/O system of difference equations given by Eq. (7.2.1) and the corresponding sample processing algorithm of Eq. (7.2.2).

It can be implemented by the routine `can` with $M = L = 2$ and three-dimensional coefficient and state arrays `a,b,w`. However, it proves convenient to write a special version of `can` as it applies to this specific case. The following C routine `sos.c` implements a second-order section:

```
/* sos.c - IIR filtering by single second order section */

double sos(a, b, w, x)                 a, b, w are 3-dimensional
double *a, *b, *w, x;                  a[0] = 1 always
```

```
{
        double y;

        w[0] = x - a[1] * w[1] - a[2] * w[2];
        y = b[0] * w[0] + b[1] * w[1] + b[2] * w[2];

        w[2] = w[1];
        w[1] = w[0];

        return y;
}
```

where a, b, w must be declared to be three-dimensional arrays in the main program, for example by

```
a = (double *) calloc(3, sizeof(double));          a = [1, a₁, a₂]
b = (double *) calloc(3, sizeof(double));          b = [b₀, b₁, b₂]
w = (double *) calloc(3, sizeof(double));          w = [w₀, w₁, w₂]
```

The *cascade realization form* of a general transfer function assumes that the transfer function is the product of such second-order sections:

$$H(z) = \prod_{i=0}^{K-1} H_i(z) = \prod_{i=0}^{K-1} \frac{b_{i0} + b_{i1}z^{-1} + b_{i2}z^{-2}}{1 + a_{i1}z^{-1} + a_{i2}z^{-2}} \tag{7.3.1}$$

Any transfer function of the form (7.1.4) can be factored into second-order factors with *real-valued* coefficients, provided Eq. (7.1.4) has real coefficients.

The maximum order of the numerator and denominator polynomials in Eq. (7.3.1) is $2K$, that is, twice the number of second-order sections. By "second order" we really mean "up to second order", and therefore, if some of the z^{-2} coefficients b_{i2} or a_{i2} are zero, the actual numerator and denominator orders will be $L \leq 2K$ and $M \leq 2K$.

A block diagram realization of Eq. (7.3.1) can be obtained by cascading together the block diagram realizations of the SOS filters $H_i(z)$. Each SOS may be realized in its canonical, direct, or transposed realizations. However, the convention is to realize all of them in their *canonical* form, as shown in Fig. 7.3.1.

Let us denote by $x_i(n)$, $y_i(n)$ the input and output signals of the ith section $H_i(z)$. Then, the overall input is the input to $H_0(z)$, namely, $x(n) = x_0(n)$, and the overall output is the output from the last SOS $H_{K-1}(z)$, namely, $y(n) = y_{K-1}(n)$. For the intermediate stages, the output $y_i(n)$ of the ith section becomes the input to the $(i+1)$th section $H_{i+1}(z)$, that is,

$$x_{i+1}(n) = y_i(n), \quad i = 0, 1, \ldots, K - 1$$

Each section has its own internal state vector $\mathbf{w}_i(n) = [w_{i0}(n), w_{i1}(n), w_{i2}(n)]$, $i = 0, 1, \ldots, K - 1$, where the numbers $w_{i1}(n)$, $w_{i2}(n)$ are the contents of the section's delay registers at the nth time instant.

The I/O difference equations describing the time-domain operation of the realization are obtained by writing the difference equations (7.2.1) for each SOS and passing the output of each to the input of the next:

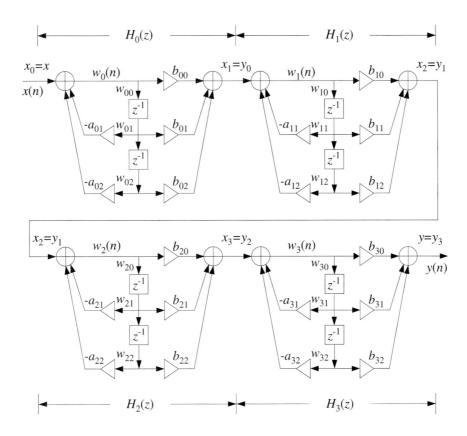

Fig. 7.3.1 Cascade of second-order sections.

$$x_0(n) = x(n)$$
$$for\ i = 0, 1, \ldots, K - 1\ do:$$
$$w_i(n) = x_i(n) - a_{i1} w_i(n - 1) - a_{i2} w_i(n - 2)$$
$$y_i(n) = b_{i0} w_i(n) + b_{i1} w_i(n - 1) + b_{i2} w_i(n - 2)$$
$$x_{i+1}(n) = y_i(n)$$
$$y(n) = y_{K-1}(n)$$

(7.3.2)

It can be translated to the following sample processing algorithm:

> *for each input sample x do:*
> $\quad x_0 = x$
> $\quad$*for $i = 0, 1, \ldots, K - 1$ do:*
> $\qquad w_{i0} = x_i - a_{i1} w_{i1} - a_{i2} w_{i2}$
> $\qquad y_i = b_{i0} w_{i0} + b_{i1} w_{i1} + b_{i2} w_{i2}$
> $\qquad w_{i2} = w_{i1}$
> $\qquad w_{i1} = w_{i0}$
> $\qquad x_{i+1} = y_i$
> $\quad y = y_{K-1}$

$\qquad\qquad$ (7.3.3)

where, the internal state vector $\mathbf{w}_i$ of the ith section is defined at time n by:

$$w_{i0}(n) = w_i(n)$$

$$w_{i1}(n) = w_i(n - 1), \quad \text{for} \quad i = 0, 1, \ldots, K - 1$$

$$w_{i2}(n) = w_i(n - 2)$$

To keep track of the coefficients of the sections and the internal states, we arrange them into $K \times 3$ matrices whose ith rows hold the corresponding parameters of the ith section. For example, if $K = 4$ as in Fig. 7.3.1, we define

$$A = \begin{bmatrix} 1 & a_{01} & a_{02} \\ 1 & a_{11} & a_{12} \\ 1 & a_{21} & a_{22} \\ 1 & a_{31} & a_{32} \end{bmatrix}, \ B = \begin{bmatrix} b_{00} & b_{01} & b_{02} \\ b_{10} & b_{11} & b_{12} \\ b_{20} & b_{21} & b_{22} \\ b_{30} & b_{31} & b_{32} \end{bmatrix}, \ W = \begin{bmatrix} w_{00} & w_{01} & w_{02} \\ w_{10} & w_{11} & w_{12} \\ w_{20} & w_{21} & w_{22} \\ w_{30} & w_{31} & w_{32} \end{bmatrix}$$

The ith rows of these matrices are the three-dimensional coefficient vectors and states of the ith section, that is,

$$\mathbf{a}_i = [1, a_{i1}, a_{i2}]$$

$$\mathbf{b}_i = [b_{i0}, b_{i1}, b_{i2}], \qquad \text{for} \quad i = 0, 1, \ldots, K - 1$$

$$\mathbf{w}_i = [w_{i0}, w_{i1}, w_{i2}]$$

$\qquad\qquad$ (7.3.4)

In this notation, we may rewrite the sample processing algorithm (7.3.3) as K *successive calls* to the basic SOS routine sos:

> *for each input sample x do:*
> $\quad y = x$
> $\quad$*for $i = 0, 1, \ldots, K - 1$ do:*
> $\qquad y = \text{sos}(\mathbf{a}_i, \mathbf{b}_i, \mathbf{w}_i, y)$

$\qquad\qquad$ (7.3.5)

where y denotes both the input and output of each section. The last computed y is the final output. The C implementation of this algorithm is given by the following routine cas.c:

```
/* cas.c - IIR filtering in cascade of second-order sections */

double sos();                                    single second-order section

double cas(K, A, B, W, x)
int K;
double **A, **B, **W, x;                         A, B, W are K×3 matrices
{
        int i;
        double y;

        y = x;                                   initial input to first SOS

        for (i=0; i<K; i++)
                y = sos(A[i], B[i], W[i], y);    output of ith section

        return y;                                final output from last SOS
}
```

The coefficient and state matrices A, B, W must be dimensioned to size $K\times3$ and allocated in the main program, for example, by

```
double **A, **B, **W;

A = (double **) calloc(K, sizeof(double *));         allocate K rows
B = (double **) calloc(K, sizeof(double *));
W = (double **) calloc(K, sizeof(double *));
for (i=0; i<K; i++) {
   A[i] = (double *) calloc(3, sizeof(double));       allocate each row
   B[i] = (double *) calloc(3, sizeof(double));
   W[i] = (double *) calloc(3, sizeof(double));
   }
```

Alternatively, if the value of K is known in advance, we may declare:

```
double A[K][3], B[K][3], W[K][3];
```

In that case, the declarations inside cas must also be modified to read:

```
double A[][3], B[][3], W[][3];
```

The quantities A[i], B[i], W[i] are the ith rows of A, B, W, as given by Eq. (7.3.4). The states W must be initialized to zero before the first call to cas; this is accomplished indirectly by calloc. The usage of cas is the same as can; that is,

```
for (n = 0; n < Ntot; n++)
        y[n] = cas(K, A, B, W, x[n]);
```

Example 7.3.1: Draw the cascade and canonical realizations of the following filter:

$$H(z) = \left[\frac{3 - 4z^{-1} + 2z^{-2}}{1 - 0.4z^{-1} + 0.5z^{-2}}\right]\left[\frac{3 + 4z^{-1} + 2z^{-2}}{1 + 0.4z^{-1} + 0.5z^{-2}}\right] = H_0(z)H_1(z)$$

$$= \frac{9 - 4z^{-2} + 4z^{-4}}{1 + 0.84z^{-2} + 0.25z^{-4}}$$

Write the corresponding I/O difference equations and sample processing algorithms.

Solution: The cascade realization is shown in Fig. 7.3.2 and the canonical one in Fig. 7.3.3. The I/O difference equations describing the cascade realization in the time domain are:

$$w_0(n) = x(n) + 0.4w_0(n-1) - 0.5w_0(n-2)$$

$$x_1(n) = 3w_0(n) - 4w_0(n-1) + 2w_0(n-2)$$

$$w_1(n) = x_1(n) - 0.4w_1(n-1) - 0.5w_1(n-2)$$

$$y(n) = 3w_1(n) + 4w_1(n-1) + 2w_1(n-2)$$

where $x_1(n)$ is the output of $H_0(z)$ and the input to $H_1(z)$. The corresponding sample processing algorithm is:

> for each input sample x do:
> $$w_{00} = x + 0.4w_{01} - 0.5w_{02}$$
> $$x_1 = 3w_{00} - 4w_{01} + 2w_{02}$$
> $$w_{02} = w_{01}$$
> $$w_{01} = w_{00}$$
> $$w_{10} = x_1 - 0.4w_{11} - 0.5w_{12}$$
> $$y = 3w_{10} + 4w_{11} + 2w_{12}$$
> $$w_{12} = w_{11}$$
> $$w_{11} = w_{10}$$

The coefficient and state matrices in the routine cas are in this case:

$$A = \begin{bmatrix} 1 & -0.4 & 0.5 \\ 1 & 0.4 & 0.5 \end{bmatrix}, \quad B = \begin{bmatrix} 3 & -4 & 2 \\ 3 & 4 & 2 \end{bmatrix}, \quad W = \begin{bmatrix} w_{00} & w_{01} & w_{02} \\ w_{10} & w_{11} & w_{12} \end{bmatrix}$$

For the canonical case, we have the coefficient vectors for the numerator and denominator polynomials:

$$\mathbf{b} = [9, 0, -4, 0, 4], \quad \mathbf{a} = [1.00, 0.00, 0.84, 0.00, 0.25]$$

The difference equation at the input and output adders of Fig. 7.3.3 are:

$$w(n) = x(n) - 0.84w(n-2) - 0.25w(n-4)$$

$$y(n) = 9w(n) - 4w(n-2) + 4w(n-4)$$

Defining the internal states as $w_i(n) = w(n-i)$, $i = 0, 1, 2, 3, 4$, we find the sample processing algorithm:

> for each input sample x do:
> $$w_0 = x - 0.84w_2 - 0.25w_4$$
> $$y = 9w_0 - 4w_2 + 4w_4$$
> $$w_4 = w_3$$
> $$w_3 = w_2$$
> $$w_2 = w_1$$
> $$w_1 = w_0$$

The total number of internal states in the cascade and the canonical realizations is the same, namely, four.   □

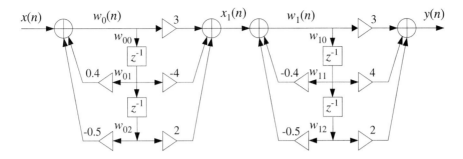

Fig. 7.3.2 Cascade realization of Example 7.3.1.

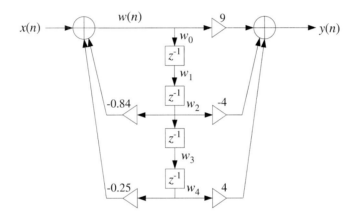

Fig. 7.3.3 Canonical realization of Example 7.3.1.

Example 7.3.2: Consider the filter

$$H(z) = \left[\frac{1 + z^{-1} + z^{-2}}{1 - 0.7z^{-2}}\right]\left[\frac{1 - z^{-2}}{1 - 0.6z^{-1} + 0.4z^{-2}}\right]\left[\frac{1 - z^{-1} + z^{-2}}{1 + 0.5z^{-1} + 0.3z^{-2}}\right]$$

$$= \frac{1 - z^{-6}}{1 - 0.1z^{-1} - 0.3z^{-2} + 0.09z^{-3} - 0.16z^{-4} - 0.014z^{-5} - 0.084z^{-6}}$$

To illustrate the usage of the routines cas and can, we generated and filtered the following "step" input signal of length 100:

$$x(n) = \begin{cases} 2, & \text{if } 0 \leq n \leq 24 \\ 0, & \text{if } 25 \leq n \leq 49 \\ -1, & \text{if } 50 \leq n \leq 74 \\ 1, & \text{if } 75 \leq n \leq 99 \end{cases}$$

The resulting output signal $y(n)$ can be computed either by the routine cas or by can. The cascade realization coefficient matrices are:

$$A = \begin{bmatrix} 1 & 0 & -0.7 \\ 1 & -0.6 & 0.4 \\ 1 & 0.5 & 0.3 \end{bmatrix}, \quad B = \begin{bmatrix} 1 & 1 & 1 \\ 1 & 0 & -1 \\ 1 & -1 & 1 \end{bmatrix}$$

Similarly, the canonical form coefficients of the sixth degree numerator and denominator polynomials of $H(z)$ are:

$$\mathbf{b} = [1, 0, 0, 0, 0, 0, -1]$$

$$\mathbf{a} = [1, -0.1, -0.3, 0.09, -0.16, -0.014, -0.084]$$

These quantities, as well as the cascade state matrix W and canonical internal state vector $\mathbf{w}$, must be declared, allocated, and initialized in the main program as discussed above (with $K = 3$, $L = M = 6$). The output signal can be generated by the for-loop:

```
for (n=0; n<100; n++) {
      ycas[n] = cas(3, A, B, W, x[n]);
      ycan[n] = can(6, a, 6, b, w, x[n]);
      }
```

The two output signals $y_{cas}(n)$ and $y_{can}(n)$ generated by the two routines cas and cas are, of course, the same. This output signal is shown in Fig. 7.3.4.

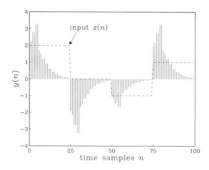

Fig. 7.3.4 Output of Example 7.3.2.

Notice also that for this particular example, the pole closest to the unit circle is that of the first section, that is, $p = \pm\sqrt{0.7} = \pm0.8367$. Therefore, the $\epsilon = 1\% = 0.01$ time constant will be $n_{eff} = \ln\epsilon/\ln(0.8367) \simeq 26$. Because the filter has a zero at $z = 1$, its unit-step response will be $H(0) = H(z)\big|_{z=1} = 0$. As the step input changes level every 25 samples, the output tries to settle to its zero steady-state value with a time constant of about n_{eff}.   □

7.4 Cascade to Canonical

To pass from the direct or canonical realization, Eq. (7.1.4), to the cascade realization, Eq. (7.3.1), requires factoring the numerator and denominator polynomials into their second-order factors.

   This can be done by finding the roots of these polynomials and then pairing them in complex conjugate pairs. The procedure is outlined below. Given the M

zeros p_i, $i = 1, 2, \ldots, M$ of the denominator polynomial of Eq. (7.1.4), we can factor it into its root factors:

$$D(z) = 1 + a_1 z^{-1} + a_2 z^{-2} + \cdots + a_M z^{-M}$$

$$= (1 - p_1 z^{-1})(1 - p_2 z^{-1}) \cdots (1 - p_M z^{-1})$$

The root factors of any real-valued roots can be left as they are or combined in pairs. For example, if both p_1 and p_2 are real, we may combine them into the SOS with *real* coefficients:

$$(1 - p_1 z^{-1})(1 - p_2 z^{-1}) = (1 - (p_1 + p_2) z^{-1} + p_1 p_2 z^{-2})$$

If any roots are complex, they must appear in complex-conjugate pairs, for example, if p_1 is a complex root, then $p_2 = p_1^*$ must also be a root. Combining the root factors of conjugate pairs results into an SOS with *real* coefficients, for example

$$(1 - p_1 z^{-1})(1 - p_1^* z^{-1}) = 1 - (p_1 + p_1^*) z^{-1} + p_1 p_1^* z^{-2}$$

$$= 1 - 2\text{Re}(p_1) z^{-1} + |p_1|^2 z^{-2}$$

This identity was also used in Chapter 5. Using the polar representation of the complex number $p_1 = R_1 e^{j\theta_1}$, we have $\text{Re}(p_1) = R_1 \cos\theta_1$ and $|p_1|^2 = R_1^2$, and we can write the above identity in the alternative form:

$$(1 - p_1 z^{-1})(1 - p_1^* z^{-1}) = 1 - 2\text{Re}(p_1) z^{-1} + |p_1|^2 z^{-2}$$

$$= 1 - 2R_1 \cos(\theta_1) z^{-1} + R_1^2 z^{-2}$$

Once the denominator and numerator polynomials have been factored into their quadratic factors, each quadratic factor from the numerator may be paired with a quadratic factor from the denominator to form a second-order section.

This pairing of numerator and denominator factors and the ordering of the resulting SOSs is *not unique*, but the overall transfer function will be the same. In practice, however, the particular pairing/ordering may make a difference.

In a hardware realization, the internal multiplications in each SOS will generate a certain amount of roundoff error which is then propagated into the next SOS. The net roundoff error at the overall output will depend on the particular pairing/ordering of the quadratic factors. The optimal ordering is the one that generates the *minimum* net roundoff error. Finding this optimal ordering is a difficult problem and is beyond the scope of this book.

Some examples will illustrate the above factoring technique. The most tedious part is finding the actual roots of the numerator and denominator polynomials. For high-order polynomials, one must use a root-finding routine from a numerical software package such as MATLAB or Mathematica. Some special cases of high-order polynomials can be handled by hand, as seen below.

Example 7.4.1: Determine the cascade realization form of the filter:

$$H(z) = \frac{1 - 1.5z^{-1} + 0.48z^{-2} - 0.33z^{-3} + 0.9376z^{-4} - 0.5328z^{-5}}{1 + 2.2z^{-1} + 1.77z^{-2} + 0.52z^{-3}}$$

Solution: Using MATLAB, we find the five roots of the numerator polynomial:

$$z = 0.9, \quad -0.5 \pm 0.7j, \quad 0.8 \pm 0.4j$$

They lead to the following root factors, already paired in conjugate pairs:

$$(1 - 0.9z^{-1})$$

$$(1 - (-0.5 + 0.7j)z^{-1})(1 - (-0.5 - 0.7j)z^{-1}) = (1 + z^{-1} + 0.74z^{-2})$$

$$(1 - (0.8 + 0.4j)z^{-1})(1 - (0.8 - 0.4j)z^{-1}) = (1 - 1.6z^{-1} + 0.8z^{-2})$$

Similarly, we find the roots of the denominator:

$$p = -0.8, \quad -0.7 \pm 0.4j$$

giving the root factors:

$$(1 + 0.8z^{-1})$$

$$(1 - (-0.7 + 0.4j)z^{-1})(1 - (-0.7 - 0.4j)z^{-1}) = (1 + 1.4z^{-1} + 0.65z^{-2})$$

Therefore, a possible pairing/ordering of SOS factors for $H(z)$ will be:

$$H(z) = \frac{1 - 0.9z^{-1}}{1 + 0.8z^{-1}} \cdot \frac{1 + z^{-1} + 0.74z^{-2}}{1 + 1.4z^{-1} + 0.65z^{-2}} \cdot (1 - 1.6z^{-1} + 0.8z^{-2})$$

The coefficient matrices A and B needed for programming this filter by the routine cas will be:

$$A = \begin{bmatrix} 1 & 0.8 & 0 \\ 1 & 1.4 & 0.65 \\ 1 & 0 & 0 \end{bmatrix}, \quad B = \begin{bmatrix} 1 & -0.9 & 0 \\ 1 & 1 & 0.74 \\ 1 & -1.6 & 0.8 \end{bmatrix}$$

The first-order section may be considered as special case of an SOS of the form (7.1.1) with zero z^{-2} coefficients, that is, $b_2 = a_2 = 0$. Similarly, the last quadratic factor is a special case of an FIR SOS, that is, with $a_1 = a_2 = 0$ (but $a_0 = 1$).   □

Example 7.4.2: Determine the cascade form of the filter:

$$H(z) = \frac{1 - 0.48z^{-2} + 0.5476z^{-4}}{1 + 0.96z^{-2} + 0.64z^{-4}}$$

Solution: Even though the polynomials have degree 4, the z^{-1} and z^{-3} terms are missing, and we may think of the polynomials as *quadratic* in the variable z^{-2}. That is, we can find the roots of the denominator by solving the quadratic equation

$$1 + 0.96z^{-2} + 0.64z^{-4} = 0 \quad \Rightarrow \quad (z^2)^2 + 0.96(z^2) + 0.64 = 0$$

which has two solutions:

$$z^2 = \frac{-0.96 \pm \sqrt{0.96^2 - 4 \times 0.64}}{2} = -0.48 \pm 0.64j$$

Taking square roots, we obtain the four roots of the denominator:

$$p = \pm\sqrt{-0.48 \pm 0.64j} = \pm(0.4 \pm 0.8j)$$

Pairing them in conjugate pairs gives the quadratic factors:

$$\left(1 - (0.4 + 0.8j)z^{-1}\right)\left(1 - (0.4 - 0.8j)z^{-1}\right) = 1 - 0.8z^{-1} + 0.8z^{-2}$$
$$\left(1 + (0.4 + 0.8j)z^{-1}\right)\left(1 + (0.4 - 0.8j)z^{-1}\right) = 1 + 0.8z^{-1} + 0.8z^{-2}$$

Similarly, we find for the numerator polynomial:

$$1 - 0.48z^{-2} + 0.5476z^{-4} = 0 \quad \Rightarrow \quad z^2 = 0.24 \pm 0.7j$$

and taking square roots:

$$z = \pm\sqrt{0.24 \pm 0.7j} = \pm(0.7 \pm 0.5j)$$

The quadratic factors are:

$$\left(1 - (0.7 + 0.5j)z^{-1}\right)\left(1 - (0.7 - 0.5j)z^{-1}\right) = 1 - 1.4z^{-1} + 0.74z^{-2}$$
$$\left(1 + (0.7 + 0.5j)z^{-1}\right)\left(1 + (0.7 - 0.5j)z^{-1}\right) = 1 + 1.4z^{-1} + 0.74z^{-2}$$

Thus, we find for $H(z)$:

$$H(z) = \frac{1 - 1.4z^{-1} + 0.74z^{-2}}{1 - 0.8z^{-1} + 0.8z^{-2}} \cdot \frac{1 + 1.4z^{-1} + 0.74z^{-2}}{1 + 0.8z^{-1} + 0.8z^{-2}}$$

which is one possible ordering of the quadratic factors. $\qquad\square$

Example 7.4.3: As another special case, determine the cascade form of the filter:

$$H(z) = \frac{1 + z^{-8}}{1 - 0.0625z^{-8}}$$

Solution: The roots of the numerator are the 8 solutions of:

$$1 + z^{-8} = 0 \quad \Rightarrow \quad z^8 = -1 = e^{j\pi} = e^{j\pi}e^{2\pi jk} = e^{j(2k+1)\pi}$$

where we multiplied by $e^{2\pi jk} = 1$ for integer k. Taking eighth roots of both sides we find:

$$z_k = e^{j(2k+1)\pi/8}, \quad k = 0, 1, \ldots, 7$$

We have the following conjugate pairs, as shown in Fig. 7.4.1: $\{z_0, z_7\}$, $\{z_1, z_6\}$, $\{z_2, z_5\}$, and $\{z_3, z_4\}$, which lead to the quadratic factors:

$$(1 - z_0 z^{-1})(1 - z_7 z^{-1}) = 1 - 2\cos\left(\frac{\pi}{8}\right)z^{-1} + z^{-2} = 1 - 1.8478z^{-1} + z^{-2}$$

$$(1 - z_1 z^{-1})(1 - z_6 z^{-1}) = 1 - 2\cos\left(\frac{3\pi}{8}\right)z^{-1} + z^{-2} = 1 - 0.7654z^{-1} + z^{-2}$$

$$(1 - z_2 z^{-1})(1 - z_5 z^{-1}) = 1 - 2\cos\left(\frac{5\pi}{8}\right)z^{-1} + z^{-2} = 1 + 0.7654z^{-1} + z^{-2}$$

$$(1 - z_3 z^{-1})(1 - z_4 z^{-1}) = 1 - 2\cos\left(\frac{7\pi}{8}\right)z^{-1} + z^{-2} = 1 + 1.8478z^{-1} + z^{-2}$$

Similarly, the filter poles are the roots of the denominator:

$$1 - 0.0625z^{-8} = 0 \quad \Rightarrow \quad z^8 = 0.0625 = 0.0625e^{2\pi jk} = (0.5)^4 e^{2\pi jk}$$

which has the eight solutions:

$$p_k = \sqrt{0.5}\, e^{2\pi jk/8}, \quad k = 0, 1, \ldots, 7$$

Of these, $p_0 = \sqrt{0.5}$ and $p_4 = \sqrt{0.5}\, e^{2\pi j4/8} = -\sqrt{0.5}$ are real and may be paired together into one SOS. The rest are complex and can be paired in conjugate pairs: $\{p_1, p_7\}$, $\{p_2, p_6\}$, $\{p_3, p_5\}$, resulting into the quadratic factors:

$$(1 - p_0 z^{-1})(1 - p_4 z^{-1}) = (1 - \sqrt{0.5}\, z^{-1})(1 + \sqrt{0.5}\, z^{-1}) = 1 - 0.5z^{-2}$$

$$(1 - p_1 z^{-1})(1 - p_7 z^{-1}) = 1 - \sqrt{2}\cos\left(\frac{2\pi}{8}\right)z^{-1} + 0.5z^{-2} = 1 - z^{-1} + 0.5z^{-2}$$

$$(1 - p_2 z^{-1})(1 - p_6 z^{-1}) = 1 - \sqrt{2}\cos\left(\frac{4\pi}{8}\right)z^{-1} + 0.5z^{-2} = 1 + 0.5z^{-2}$$

$$(1 - p_3 z^{-1})(1 - p_5 z^{-1}) = 1 - \sqrt{2}\cos\left(\frac{6\pi}{8}\right)z^{-1} + 0.5z^{-2} = 1 + z^{-1} + 0.5z^{-2}$$

Finally, we obtain the factorization of $H(z)$:

$$H(z) = \left[\frac{1 - 1.8478z^{-1} + z^{-2}}{1 - 0.5z^{-2}}\right] \cdot \left[\frac{1 - 0.7654z^{-1} + z^{-2}}{1 - z^{-1} + 0.5z^{-2}}\right] \cdot$$

$$\cdot \left[\frac{1 + 0.7654z^{-1} + z^{-2}}{1 + 0.5z^{-2}}\right] \cdot \left[\frac{1 + 1.8478z^{-1} + z^{-2}}{1 + z^{-1} + 0.5z^{-2}}\right]$$

The coefficient matrices A and B will be in this case:

$$A = \begin{bmatrix} 1 & 0 & -0.5 \\ 1 & -1 & 0.5 \\ 1 & 0 & 0.5 \\ 1 & 1 & 0.5 \end{bmatrix}, \quad B = \begin{bmatrix} 1 & -1.8478 & 1 \\ 1 & -0.7654 & 1 \\ 1 & 0.7654 & 1 \\ 1 & 1.8478 & 1 \end{bmatrix}$$

This filter acts as a notch/comb filter, where the zero dips are shifted by $\pi/8$ compared to the pole peaks. The pole zero pattern and magnitude response $|H(\omega)|$ are shown in Fig. 7.4.1.

This example was only meant to illustrate the factorization procedure. Its canonical form realization is much more efficient than the cascade one, since it involves only one multiplier and an 8-fold delay. The canonical realization and the corresponding sample processing algorithm are shown in Fig. 7.4.2. Here, $\mathbf{w} = [w_0, w_1, \ldots, w_8]$ is the 9-dimensional internal state vector.   □

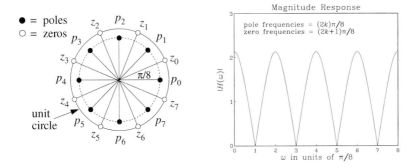

Fig. 7.4.1 Pole/zero pattern and magnitude response of Example 7.4.3.

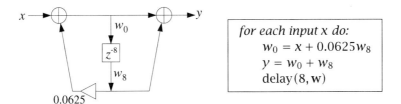

Fig. 7.4.2 Canonical realization of Example 7.4.3.

Example 7.4.4: Sharpen the poles and zeros of the previous filter and determine the cascade form of the resulting filter.

Solution: To sharpen the zeros of the filter, we must place poles "behind" the zeros, that is, replace the numerator polynomial $N(z) = 1 + z^{-8}$ by

$$H_1(z) = \frac{N(z)}{N(\rho^{-1}z)} = \frac{1 + z^{-8}}{1 + \rho^8 z^{-8}}$$

For example, we may choose $\rho^8 = 0.94$, or $\rho = 0.9923$. The factorization of $N(\rho^{-1}z)$ into SOSs is obtained from that of $N(z)$ by replacing z by z/ρ or z^{-1} by ρz^{-1} in each factor. This gives:

$$H_1(z) = \frac{N(z)}{N(\rho^{-1}z)} = \left[\frac{1 - 1.8478z^{-1} + z^{-2}}{1 - 1.8478\rho z^{-1} + \rho^2 z^{-2}}\right] \cdot \left[\frac{1 - 0.7654z^{-1} + z^{-2}}{1 - 0.7654\rho z^{-1} + \rho^2 z^{-2}}\right] \cdot$$

$$\cdot \left[\frac{1 + 0.7654z^{-1} + z^{-2}}{1 + 0.7654\rho z^{-1} + \rho^2 z^{-2}}\right] \cdot \left[\frac{1 + 1.8478z^{-1} + z^{-2}}{1 + 1.8478\rho z^{-1} + \rho^2 z^{-2}}\right]$$

To sharpen the poles, we must do two things: first push the existing poles closer to the unit circle, and second, place zeros "behind" the poles. This can be done by the substitution of the denominator polynomial by

$$\frac{1}{1 - 0.0625z^{-8}} \quad \longrightarrow \quad H_2(z) = \frac{1 - r^8 z^{-8}}{1 - R^8 z^{-8}}$$

where $r \lesssim R$. For example, we may choose $r^8 = 0.96$ or $r = 0.9949$, and $R^8 = 0.98$ or $R = 0.9975$. The SOS factors of the numerator and denominator can be found in the same fashion as for the polynomial $(1 - 0.0625z^{-8})$. The factorization of $H_2(z)$ is then:

$$H_2(z) = \frac{1 - r^8 z^{-8}}{1 - R^8 z^{-8}} = \left[\frac{1 - r^2 z^{-2}}{1 - R^2 z^{-2}}\right] \cdot \left[\frac{1 - \sqrt{2}rz^{-1} + r^2 z^{-2}}{1 - \sqrt{2}Rz^{-1} + R^2 z^{-2}}\right] \cdot$$

$$\cdot \left[\frac{1 + r^2 z^{-2}}{1 + R^2 z^{-2}}\right] \cdot \left[\frac{1 + \sqrt{2}rz^{-1} + r^2 z^{-2}}{1 + \sqrt{2}Rz^{-1} + R^2 z^{-2}}\right]$$

Thus, the new transfer function will be

$$H(z) = H_1(z)H_2(z) = \left[\frac{1 + z^{-8}}{1 + 0.94z^{-8}}\right] \cdot \left[\frac{1 - 0.96z^{-8}}{1 - 0.98z^{-8}}\right]$$

Again, the simplest realization is to realize $H_1(z)$ and $H_2(z)$ in cascade, with each realized in its canonical form. This realization and its sample processing algorithm are shown below; the magnitude response $|H(\omega)|$ is shown in Fig. 7.4.3.   □

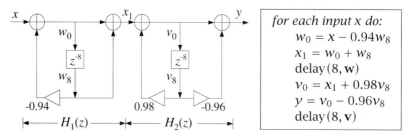

The reverse process of going from the cascade realization, Eq. (7.3.1), to the canonical one, Eq. (7.1.4), is much easier. It amounts to multiplying out the second-order numerator and denominator factors to get the full degree polynomials:

$$N(z) = \prod_{i=0}^{K-1} (b_{i0} + b_{i1}z^{-1} + b_{i2}z^{-2}) = b_0 + b_1 z^{-1} + b_2 z^{-2} + \cdots + b_L z^{-L}$$

$$D(z) = \prod_{i=0}^{K-1} (1 + a_{i1}z^{-1} + a_{i2}z^{-2}) = 1 + a_1 z^{-1} + a_2 z^{-2} + \cdots + a_M z^{-M}$$

where L and M will be at most $2K$, depending on how many sections are full second-order or first-order sections.

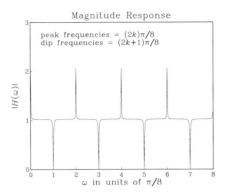

Fig. 7.4.3   Pole/zero sharpening in Example 7.4.4.

The polynomial multiplications may be done in the time domain using convolution. For example, using the definitions (7.1.9) and (7.3.4) for the coefficient vectors, we may write these convolutions in the form:

$$\mathbf{b} = \mathbf{b}_0 * \mathbf{b}_1 * \cdots * \mathbf{b}_{K-1}$$
$$\mathbf{a} = \mathbf{a}_0 * \mathbf{a}_1 * \cdots * \mathbf{a}_{K-1}$$

(7.4.1)

These convolutions may be implemented recursively by defining:

$$\mathbf{d}_i = \mathbf{a}_0 * \mathbf{a}_1 * \cdots * \mathbf{a}_{i-1}$$

and noting that $\mathbf{d}_i$ satisfies the recursion:

$$\mathbf{d}_{i+1} = \mathbf{d}_i * \mathbf{a}_i$$

(7.4.2)

where it must be initialized to $\mathbf{d}_0 = \delta$, that is, a delta function, $d_0(n) = \delta(n)$. We may take $\mathbf{d}_0$ to be the one-dimensional vector: $\mathbf{d}_0 = \delta = [1]$. A few of the recursion steps will be:

$$\mathbf{d}_1 = \mathbf{d}_0 * \mathbf{a}_0 = \delta * \mathbf{a}_0 = \mathbf{a}_0$$
$$\mathbf{d}_2 = \mathbf{d}_1 * \mathbf{a}_1 = \mathbf{a}_0 * \mathbf{a}_1$$
$$\mathbf{d}_3 = \mathbf{d}_2 * \mathbf{a}_2 = (\mathbf{a}_0 * \mathbf{a}_1) * \mathbf{a}_2$$
$$\mathbf{d}_4 = \mathbf{d}_3 * \mathbf{a}_3 = (\mathbf{a}_0 * \mathbf{a}_1 * \mathbf{a}_2) * \mathbf{a}_3, \quad \text{etc.}$$

The recursion ends at $i = K-1$ with the desired answer $\mathbf{a} = \mathbf{d}_K$. Note that the intermediate vector $\mathbf{d}_i$ has order $2i$ and length $2i + 1$; similarly, the resulting vector $\mathbf{d}_{i+1}$ has length $2(i + 1) + 1 = 2i + 3$.

During the recursion (7.4.2), there is no need to save the intermediate vectors $\mathbf{d}_i$. Therefore, the recursion can be stated in the following algorithmic form:

$$\boxed{\begin{array}{l} \mathbf{d} = \delta \\ \textit{for } i = 0, 1, \ldots, K-1 \textit{ do:} \\ \quad \mathbf{d} = \mathbf{a}_i * \mathbf{d} \\ \mathbf{a} = \mathbf{d} \end{array}}$$

A variation, which helps the implementation in terms of the routine conv of Chapter 4, is to keep updating **a** during each step:

$$\boxed{\begin{array}{l} \mathbf{a} = \delta \\ \textit{for } i = 0, 1, \ldots, K-1 \textit{ do:} \\ \quad \mathbf{d} = \mathbf{a}_i * \mathbf{a} \\ \quad \mathbf{a} = \mathbf{d} \end{array}} \qquad (7.4.3)$$

and similarly for the numerator vector **b**. The following C routine cas2can.c is an implementation of this algorithm:

```
/* cas2can.c - cascade to canonical */

#include <stdlib.h>                              declares calloc

void conv();

void cas2can(K, A, a)                            a is (2K + 1)-dimensional
double **A, *a;                                  A is Kx3 matrix
int K;                                           K = no. of sections
{
        int i,j;
        double *d;

        d = (double *) calloc(2*K+1, sizeof(double));

        a[0] = 1;                                initialize

        for(i=0; i<K; i++) {
                conv(2, A[i], 2*i+1, a, d);      d = a[i] * a
                for(j=0; j<2*i+3; j++)           a = d
                        a[j] = d[j];
                }
}
```

Its inputs are the number of sections K and the coefficient matrix A, whose rows hold the coefficients of the successive sections, as in the routine cas. Its output is the $(2K)$-dimensional vector **a**. It must be called separately on the numerator and denominator coefficient matrices.

Example 7.4.5: To illustrate the usage of cas2can, we apply it to the cascade realization of Example 7.4.1:

$$H(z) = \left[\frac{1 - 0.9z^{-1}}{1 + 0.8z^{-1}} \right] \cdot \left[\frac{1 + z^{-1} + 0.74z^{-2}}{1 + 1.4z^{-1} + 0.65z^{-2}} \right] \cdot \left[1 - 1.6z^{-1} + 0.8z^{-2} \right]$$

The routine `cas2can` must be called twice with inputs the coefficient matrices A and B:

$$A = \begin{bmatrix} 1 & 0.8 & 0 \\ 1 & 1.4 & 0.65 \\ 1 & 0 & 0 \end{bmatrix}, \quad B = \begin{bmatrix} 1 & -0.9 & 0 \\ 1 & 1 & 0.74 \\ 1 & -1.6 & 0.8 \end{bmatrix}$$

The quantities A, B, a, b must be dimensioned in the main program as in Section 7.3 or Example 7.3.2. Then, the two calls:

```
cas2can(K, A, a);                    denominator coefficients
cas2can(K, B, b);                    numerator coefficients
```

will return the vectors:

$$\mathbf{a} = [1, 2.2, 1.77, 0.52, 0, 0]$$

$$\mathbf{b} = [1, -1.5, 0.48, -0.33, 0.9376, -0.5328]$$

which define the canonical realization of Example 7.4.1. □

7.5 *Hardware Realizations and Circular Buffers*

Hardware realizations of FIR filters with DSP chips were discussed in Section 4.2.4. IIR filters can be realized in a similar fashion.

Consider, for example, a second-order section (7.1.1) realized in its canonical form shown in Fig. 7.2.3. A hardware realization by a typical DSP chip is shown in Fig. 7.5.1. The filter coefficients $\{b_0, b_1, b_2, a_1, a_2\}$ are stored in RAM or ROM on board the chip; the internal states $\{w_0, w_1, w_2\}$ are stored in RAM.

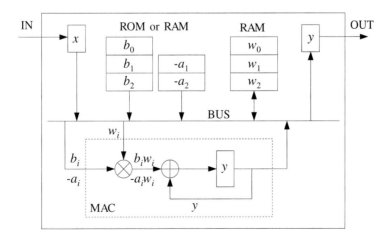

Fig. 7.5.1 Typical DSP chip realization of a second-order section.

As in Section 4.2.4, the sample processing algorithm (7.2.2) can be rewritten in a form that imitates individual instructions of the DSP chip, such as MAC and data shifting instructions:

$$
\begin{aligned}
&\text{for each input sample } x \text{ do:} \\
&\quad w_0 := x - a_1 w_1 \\
&\quad w_0 := w_0 - a_2 w_2 \\
&\quad y := b_2 w_2 \\
&\quad w_2 := w_1, \quad y := y + b_1 w_1 \\
&\quad w_1 := w_0, \quad y := y + b_0 w_0
\end{aligned}
\tag{7.5.1}
$$

In a modern DSP chip, each line in the above algorithm can be implemented with a single MAC-type instruction; therefore, a single SOS can be implemented with *five* instructions per input sample.

Note that the states w_1 and w_2 cannot be updated until after w_0 has been computed. The MAC instructions for computing w_0 proceed forward, that is, from the lowest a_i coefficient to the highest. This is convenient because once w_0 is computed, the combined data shift/MAC instructions for computing y can be started, but proceeding backwards from the highest b_i coefficient to the lowest. In the general case, we can rewrite Eq. (7.2.5), where for simplicity we assumed $L = M$:

$$
\begin{aligned}
&\text{for each input sample } x \text{ do:} \\
&\quad w_0 := x \\
&\quad \text{for } i = 1, 2, \ldots, M \text{ do:} \\
&\qquad w_0 := w_0 - a_i w_i \\
&\quad y := b_M w_M \\
&\quad \text{for } i = M-1, \ldots, 1, 0 \text{ do:} \\
&\qquad w_{i+1} := w_i \\
&\qquad y := y + b_i w_i
\end{aligned}
\tag{7.5.2}
$$

The following C routine `can3.c` is an implementation.

```
/* can3.c - IIR filtering in canonical form, emulating a DSP chip */

double can3(M, a, b, w, x)            usage: y = can3(M, a, b, w, x);
double *a, *b, *w, x;                 w = internal state vector
int M;                                a, b have order M
{
      int i;
      double y;

      w[0] = x;                       read input sample

      for (i=1; i<=M; i++)            forward order
            w[0] -= a[i] * w[i];      MAC instruction

      y = b[M] * w[M];

      for (i=M-1; i>=0; i--) {        backward order
            w[i+1] = w[i];            data shift instruction
            y += b[i] * w[i];         MAC instruction
```

```
            }

     return y;                                    output sample
}
```

Assuming that each MAC operation in Eq. (7.5.2) can be done with one instruction, and in particular that the combined data move/MAC instructions in the second for-loop can be also done with a single instruction, we count the *total number of instructions* for the filtering of each input sample by an Mth order IIR filter to be:

$$N_{\text{instr}} = 2(M + 1) + C \qquad \text{(order-}M\text{ IIR filter)} \qquad (7.5.3)$$

where we have added a constant C to account for any additional overhead (such as loop overhead) in instructions. Its value is typically of the order of 10 or less, depending on the particular DSP chip. In Section 4.2.4, we had arrived at a similar result for an FIR filter, which we rewrite now in the form:

$$N_{\text{instr}} = (M + 1) + C \qquad \text{(order-}M\text{ FIR filter)} \qquad (7.5.4)$$

The total time for processing each input sample will be then

$$\boxed{T_{\text{proc}} = N_{\text{instr}} T_{\text{instr}}} \qquad (7.5.5)$$

where T_{instr} is the time for a basic instruction, such as MAC or MACD. Recall from Section 4.2.4 that T_{instr} is of the order of 30–80 nanoseconds, which corresponds to an instruction rate of $f_{\text{instr}} = 1/T_{\text{instr}} = 12.5$–33.3 MIPS (million instructions per second). The processing time per sample imposes an *upper limit* on the sampling rate f_s at which the filter may be operated, that is,

$$\boxed{f_s = \frac{1}{T_{\text{proc}}} = \frac{1}{N_{\text{instr}} T_{\text{instr}}} = \frac{f_{\text{instr}}}{N_{\text{instr}}}} \qquad (7.5.6)$$

where the quantity $1/T_{\text{proc}}$ is the chip's *computational rate*, that is, the number of samples that can be *processed* per second.

It is impossible to give a processor-independent count of the number of instructions for a particular filter. The precise count, as well as the total processing time T_{proc} per sample, depend on the DSP chip's architecture, instruction set, how memory accessing is used, processor wait states introduced for slow memory, and the way a filter realization is programmed in the chip's assembly language, for example, using in-line code or not.

The above results must be used only as rough guidelines in evaluating the performance of a DSP chip. Our discussion was based on counting the number of MACs in the sample processing algorithm for the particular filter.

The transposed realizations for both IIR and FIR filters can be implemented also by the same number of instructions given by Eqs. (7.5.3) and (7.5.4). The transposed sample processing algorithm uses only plain MAC instructions—not requiring combined data shift/MAC instructions. Therefore, in the early generations of DSP chips, it had a computational advantage in the number of instructions over the canonical realizations.

For a cascade of second-order sections, to find the total processing time we must calculate the time it takes to process a single SOS and then multiply it by the number of sections. We saw in Eq. (7.5.1) that it takes about five instructions per SOS; therefore, the processing time for a single SOS will be approximately (ignoring any other overhead):

$$T_{SOS} \simeq 5 T_{instr} \tag{7.5.7}$$

For K second-order sections that are either cascaded or arranged in parallel, but which are implemented by the *same* DSP, the total number of instructions will be:

$$N_{instr} = 5K + C \qquad (K\text{-section IIR filter}) \tag{7.5.8}$$

where C is any additional overhead for the K-section filter. Therefore, the total processing time will be:

$$T_{proc} = (5K + C) T_{instr} = K T_{SOS} + C T_{instr} \tag{7.5.9}$$

Ignoring the possible small overhead term, we find the maximum sampling rate f_s for implementing K second-order sections:

$$f_s = \frac{1}{T_{proc}} = \frac{1}{K T_{SOS}} = \frac{f_{instr}}{5K} \tag{7.5.10}$$

For *parallel* implementations (see Problem 5.18), we may speed up the throughput rate by using K different DSP chips operating in parallel, each being dedicated to performing a single SOS filtering operation in T_{SOS} seconds. In this case, the total processing time is T_{SOS} because all of the DSPs finish simultaneously, and therefore, the throughput rate is K times *faster* than in the case of a single DSP:

$$T_{proc} = T_{SOS} \quad \Rightarrow \quad f_s = \frac{1}{T_{proc}} = \frac{1}{T_{SOS}} \tag{7.5.11}$$

For a *cascade* implementation, one may also use K DSP chips—one for each SOS—to speed up processing. However, because the output of each section becomes the input of the next section, it is not possible to run all K DSP chips simultaneously. Each DSP must wait T_{SOS} seconds for the DSP before it to finish.

One solution is to *pipeline* the filtering operations of the successive sections, so that all DSPs are working together, but each is processing the input from the previous sampling instant. This can be accomplished by inserting unit delays between the DSPs, as shown in Fig. 7.5.2.

Fig. 7.5.2 Pipelining the operation of multiple DSP processors.

At the nth time instant, while DSP-1 is working on the current input sample $x(n)$, DSP-2 is working on the sample $y_1(n-1)$ which was produced by DSP-1 at the previous time instant and was saved in the delay register until now, and DSP-3 is working on the sample $y_2(n-1)$ which was produced by DSP-2 earlier, and so on. The effect of introducing these delays is only an overall delay in the output. For example, in the case shown in Fig. 7.5.2, the overall transfer function changes from $H(z) = H_1(z)H_2(z)H_3(z)$ to:

$$H(z) = H_1(z)\, z^{-1}\, H_2(z)\, z^{-1}\, H_3(z) = z^{-2} H_1(z) H_2(z) H_3(z)$$

which corresponds to delaying the overall output by two sampling units. For K sections, the overall delay will be $z^{-(K-1)}$.

Example 7.5.1: The AT&T DSP32C floating point DSP chip [87,88] can execute a basic MAC-type instruction in four clock cycles, that is, $T_{instr} = 4T_{clock}$. Therefore, its instruction rate is $f_{instr} = f_{clock}/4$. A typical MAC instruction represents two *floating point operations*: one addition and one multiplication. Therefore, the chip achieves a computational rate of $f_{FLOPS} = 2f_{instr} = f_{clock}/2$ FLOPS.

At a clock rate of $f_{clock} = 50$ MHz, it achieves an instruction rate of $f_{instr} = 50/4 = 12.5$ MIPS, and a computational rate of $f_{FLOPS} = 50/2 = 25$ MFLOPS (megaflops). The time per instruction is $T_{instr} = 1/f_{instr} = 1/12.5 = 80$ nanoseconds.

An order-M FIR filter can be implemented (with in-line code) with

$$N_{instr} = (M+1)+11 = M+12 \qquad \text{(instructions per sample)}$$

Therefore, the processing time per sample will be

$$T_{proc} = (M+12)\,T_{instr}$$

For a 100-tap FIR filter ($M = 99$) with the DSP32C running at 50 MHz, we have $T_{proc} = (99+12)80$ nsec $= 8.9\ \mu$sec, achieving a maximum throughput rate of $f_s = 1/T_{proc} = 112.4$ kHz.

A K-section IIR filter can be implemented (with in-line code) with

$$N_{instr} = 5K+10 \qquad \text{(instructions per sample)}$$

It also requires a number of machine-cycle wait states:

$$N_{wait} = 2K+1 \qquad \text{(wait states per sample)}$$

Therefore, the total processing time for K sections will be:

$$T_{proc} = N_{instr}T_{instr} + N_{wait}T_{clock}$$

Writing $T_{instr} = 4T_{clock} = 4/f_{clock}$, we have

$$T_{proc} = \frac{4N_{instr} + N_{wait}}{f_{clock}} = \frac{4(5K+10)+2K+1}{f_{clock}}$$

For one SOS, $K = 1$, and a 50 MHz clock, we find $T_{proc} = 1.26\ \mu$sec, which translates to maximum sampling rate of $f_s = 1/T_{proc} = 793.6$ kHz. For a 5-section filter $K = 5$, we find $T_{proc} = 3.02\ \mu$sec and $f_s = 331.1$ kHz. And, for a 10-section filter, $K = 10$, we have $T_{proc} = 5.22\ \mu$sec and $f_s = 191.6$ kHz. □

We saw in Section 4.2.4 that *circular addressing* was an efficient way to implement FIR filters and delay lines, and was supported by most of the current DSP chip families. All of the IIR filtering routines—direct, canonical, and cascade—can also be implemented using circular addressing.

The following routine `ccan.c` implements the canonical realization of Fig. 7.2.4 using circular buffers, and replaces `can`. For simplicity, we assume that the numerator and denominator polynomials have the same order M.

```
/* ccan.c - circular buffer implementation of canonical realization */

void wrap();                                    defined in Section 4.2.4

double ccan(M, a, b, w, p, x)                   usage: y = ccan(M, a, b, w, &p, x);
double *a, *b, *w, **p, x;                       p = circular pointer to buffer w
int M;                                           a,b have common order M
{
        int i;
        double y = 0, s0;

        **p = x;                                 read input sample x

        s0  = *(*p)++;                           s0 = x
        wrap(M, w, p);                           p now points to s1

        for (a++, i=1; i<=M; i++) {              start with a incremented to a1
            s0 -= (*a++) * (*(*p)++);
            wrap(M, w, p);
            }

        **p = s0;                                p has wrapped around once

        for (i=0; i<=M; i++) {                   numerator part
            y += (*b++) * (*(*p)++);
            wrap(M, w, p);                        upon exit, p has wrapped
            }                                    around once again

        (*p)--;                                  update circular delay line
        wrap(M, w, p);

        return y;                                output sample
}
```

Like the FIR routine `cfir`, it uses a circular pointer p that always points at the effective starting address of the circular buffer. Here the internal state vector is defined at time n by

$$\mathbf{s}(n) = \begin{bmatrix} s_0(n) \\ s_1(n) \\ \vdots \\ s_M(n) \end{bmatrix} = \begin{bmatrix} w(n) \\ w(n-1) \\ \vdots \\ w(n-M) \end{bmatrix} \quad (7.5.12)$$

Upon entry, the circular pointer p points at the w-register holding $s_0(n) = w(n)$. The value of $s_0(n)$ is not known—only its address. The first for-loop computes the numerical value of $s_0(n)$ and puts it in the correct w-register. This is so because after the loop, p has been post-incremented a total of $M+1$ times and has wrapped completely around.

The second for-loop then computes the contribution of the numerator part. The pointer p is incremented $M+1$ times and cycles around once more. Finally, in preparation for the next time step, p is circularly decremented and points at the w-register that will hold the next value $w(n + 1)$.

As we mentioned in Section 4.2.4, in DSP chips that support circular addressing, the incrementing or decrementing pointer wraps around automatically and there is no need to use the routine `wrap`.

The following program segment illustrates the proper initialization and usage of the routine. Note that p must be passed by address because it is changed after each call:

```
double *a, *b, *w, *p;
a = (double *) calloc(M+1, sizeof(double));
b = (double *) calloc(M+1, sizeof(double));
w = (double *) calloc(M+1, sizeof(double));     initializes w to zero
a[0] = 1;                                        not used in the routine

p = w;                                           initialize p

for (n = 0; n < Ntot; n++)
        y[n] = ccan(M, a, b, w, &p, x[n]);       p is passed by address
```

The operations carried out by the routine `ccan` can be restated in a slightly different form by the following sample processing algorithm:

$$
\begin{aligned}
&\textit{for each input sample x do:}\\
&\quad \textit{for } i = 1, 2, \ldots, M \textit{ determine states:}\\
&\qquad s_i = \text{tap}(M, \mathbf{w}, p, i)\\
&\quad s_0 = x - a_1 s_1 - \cdots - a_M s_M\\
&\quad y = b_0 s_0 + b_1 s_1 + \cdots + b_M s_M\\
&\quad *p = s_0\\
&\quad \text{cdelay}(M, \mathbf{w}, \&p)
\end{aligned}
$$

where for convenience, we used the routine `tap` to compute the current states.

Example 7.5.2: Write the circular-buffer version of the sample processing algorithm of Example 7.2.1 or 7.1.1, whose canonical realization is depicted in the block diagram of Fig. 7.2.5.

Solution: Here, the buffer w is a five-dimensional array, initialized to zero. The circular pointer p is initialized by $p = w$. We have:

> *for each input sample x do:*
> $s_1 = \text{tap}(4, \mathbf{w}, p, 1)$
> $s_2 = \text{tap}(4, \mathbf{w}, p, 2)$
> $s_3 = \text{tap}(4, \mathbf{w}, p, 3)$
> $s_4 = \text{tap}(4, \mathbf{w}, p, 4)$
> $s_0 = x - 0.2s_1 + 0.3s_2 - 0.5s_4$
> $y = 2s_0 - 3s_1 + 4s_3$
> $*p = s_0$
> $\text{cdelay}(4, \mathbf{w}, \&p)$

The statement $*p = s_0$ puts the computed value of the 0th component s_0 into the w-register pointed to by p. Then, `cdelay` decrements p circularly.   □

Example 7.5.3: Determine the circular-buffer version of the sample processing algorithm of Example 7.4.3, whose realization is depicted in Fig. 7.4.2.

Solution: Here, the buffer **w** is nine-dimensional. The algorithm is stated below, where only the output s_8 of the 8-fold delay line is needed:

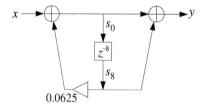

> *for each input sample x do:*
> $s_8 = \text{tap}(8, \mathbf{w}, p, 8)$
> $s_0 = x + 0.0625s_8$
> $y = s_0 + s_8$
> $*p = s_0$
> $\text{cdelay}(8, \mathbf{w}, \&p)$

Note that the output s_8 was available before the input s_0 could be computed.   □

Example 7.5.4: The input signal $\mathbf{x} = [1, 3, 2, 5, 4, 6]$ is applied to the filter:

$$H(z) = \frac{1 + z^{-1} + 2z^{-2}}{1 - z^{-3}}$$

Draw the canonical realization and write its circular buffer sample processing algorithm. Iterate the algorithm nine times for $n = 0, 1, \ldots, 8$ and compute the corresponding output $y(n)$. Make a table of the circular buffer entries **w** and the filter's internal states **s**.

Solution: The block diagram realization and its circular sample processing algorithm are:

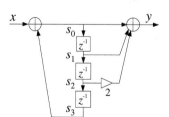

> *for each input sample x do:*
> $s_1 = \text{tap}(3, \mathbf{w}, p, 1)$
> $s_2 = \text{tap}(3, \mathbf{w}, p, 2)$
> $s_3 = \text{tap}(3, \mathbf{w}, p, 3)$
> $s_0 = x + s_3$
> $y = s_0 + s_1 + 2s_2$
> $*p = s_0$
> $\text{cdelay}(3, \mathbf{w}, \&p)$

The entries of the linear buffer $\mathbf{w} = [w_0, w_1, w_2, w_3]$ over which the circular pointer circulates are shown in the following table.

n	x	w_0	w_1	w_2	w_3	s_0	s_1	s_2	s_3	$y = s_0 + s_1 + 2s_2$
0	1	1↑	0	0	0	1	0	0	0	1
1	3	1	0	0	3↑	3	1	0	0	4
2	2	1	0	2↑	3	2	3	1	0	7
3	5	1	6↑	2	3	6	2	3	1	14
4	4	7↑	6	2	3	7	6	2	3	17
5	6	7	6	2	8↑	8	7	6	2	27
6	0	7	6	6↑	8	6	8	7	6	28
7	0	7	7↑	6	8	7	6	8	7	29
8	0	8↑	7	6	8	8	7	6	8	27

In each row, only one w_i changes, that is, the one pointed to by p. These entries, indicated by an up-arrow, can be filled only after the value of s_0 has been calculated from $s_0 = x + s_3$. The internal states s_i are pointed to by $p + i$, $i = 0, 1, 2, 3$ and wrap around if necessary. For example, at time $n = 3$, p is pointing to w_1; therefore, $p + 1$, $p + 2$, point to w_2, w_3, but $p + 3$ wraps around and points to w_0, so that $s_3 = w_0 = 1$.

According to Eq. (7.5.12), the states s_i are the delayed replicas of the signal $w(n)$ running through the intermediate delays. In the z-domain, this signal is

$$W(z) = \frac{1}{D(z)} X(z) = \frac{1 + 3z^{-1} + 2z^{-2} + 5z^{-3} + 4z^{-4} + 6z^{-5}}{1 - z^{-3}}$$

Its inverse z-transform is the period-3 replication of the numerator:

$$
\begin{array}{ccccccccc}
1 & 3 & 2 & 5 & 4 & 6 & & & = x(n) \\
 & & & 1 & 3 & 2 & 5 & 4 & 6 & = x(n-3) \\
 & & & & & & 1 & 3 & 2 & 5 & \cdots & = x(n-6) \\
 & & & & & & & & & 1 & \cdots \\
\hline
1 & 3 & 2 & 6 & 7 & 8 & 6 & 7 & 8 & 6 & \cdots & = w(n)
\end{array}
$$

Thus, the s_0 column holds $w(n)$, s_1 holds $w(n-1)$, and so on. Similarly, the output signal $y(n)$ can be constructed from $w(n)$ by

$$Y(z) = N(z)W(z) = (1 + z^{-1} + 2z^{-2})W(z) \quad \Rightarrow \quad y(n) = w(n) + w(n-1) + 2w(n-2)$$

Adding up the delayed/scaled replicas of $w(n)$, we have:

$$
\begin{array}{ccccccccc}
1 & 3 & 2 & 6 & 7 & 8 & 6 & 7 & 8 & \cdots & = w(n) \\
 & 1 & 3 & 2 & 6 & 7 & 8 & 6 & 7 & \cdots & = w(n-1) \\
 & & 2 & 6 & 4 & 12 & 14 & 16 & 12 & \cdots & = 2w(n-2) \\
\hline
1 & 4 & 7 & 14 & 17 & 27 & 28 & 29 & 27 & \cdots & = y(n)
\end{array}
$$

which agrees with the values computed in the above table. $\square$

A second-order section can be implemented by `ccan` by setting $M = 2$. Alternatively, we can use the following specialized version `csos.c`, which replaces `sos`:

```
/* csos.c - circular buffer implementation of a single SOS */

void wrap();

double csos(a, b, w, p, x)                      a,b,w are 3-dimensional
double *a, *b, *w, **p, x;                       p is circular pointer to w
{
        double y, s0;

        *(*p) = x;                               read input sample x

        s0  = *(*p)++;          wrap(2, w, p);
        s0 -= a[1] * (*(*p)++); wrap(2, w, p);
        s0 -= a[2] * (*(*p)++); wrap(2, w, p);

        *(*p) = s0;                              p has wrapped around once

        y  = b[0] * (*(*p)++);  wrap(2, w, p);
        y += b[1] * (*(*p)++);  wrap(2, w, p);
        y += b[2] * (*(*p));                     p now points to s2

        return y;
}
```

As required, the pointer p points at the w-register containing $w(n)$ and cycles around modulo-3, because the state vector is three-dimensional:

$$\mathbf{s}(n) = \begin{bmatrix} w(n) \\ w(n-1) \\ w(n-2) \end{bmatrix}$$

After the first three post-increments, p cycles around completely. The last two post-increments leave p pointing at the register containing $s_2(n) = w(n-2)$, which is where it should be pointing at the beginning of the next call.

The sample processing algorithm implemented by `csos` can be restated in the following form:

$$
\begin{array}{l}
\textit{for each input sample x do:} \\
\quad s_1 = \mathrm{tap}(2, \mathbf{w}, p, 1) \\
\quad s_2 = \mathrm{tap}(2, \mathbf{w}, p, 2) \\
\quad s_0 = x - a_1 s_1 - a_2 s_2 \\
\quad y = b_0 s_0 + b_1 s_1 + b_2 s_2 \\
\quad *p = s_0 \\
\quad \mathrm{cdelay}(2, \mathbf{w}, \&p)
\end{array}
$$

As in Eq. (7.3.5), the cascade of K second-order sections can be realized by K *successive calls* to the routine `csos`:

$$
\begin{array}{l}
\textit{for each input sample x do:} \\
\quad y = x \\
\quad \textit{for } i = 0, 1, \ldots, K-1 \textit{ do:} \\
\qquad y = \mathrm{csos}(\mathbf{a}_i, \mathbf{b}_i, \mathbf{w}_i, \&p_i, y)
\end{array}
$$

where each of the K sections has its own three-dimensional buffer $\mathbf{w}_i$ and corresponding circular pointer p_i. The following routine `ccas.c` is an implementation, replacing `cas`:

```
/* ccas.c - circular buffer implementation of cascade realization */

double csos();                              circular-buffer version of single SOS

double ccas(K, A, B, W, P, x)
int K;
double **A, **B, **W, **P, x;              P = array of circular pointers
{
        int i;
        double y;

        y = x;

        for (i=0; i<K; i++)
                y = csos(A[i], B[i], W[i], P+i, y);      note, P + i = &P[i]

        return y;
}
```

As in the case of `cas`, we save the individual buffers $\mathbf{w}_i$ as the rows of the matrix W. Similarly, we save the individual pointers p_i in an *array* of pointers P. The declaration and allocation of A, B, and W are the same as in Section 7.3. The declaration of P and initialization and usage of `ccas` is illustrated below:

```
double **P;

P = (double **) calloc(K, sizeof(double *));      array of K pointers
for (i=0; i<K; i++)
        P[i] = W[i];                              P[i] = ith row of W

for (n = 0; n < Ntot; n++)
        y[n] = ccas(K, A, B, W, P, x[n]);
```

Example 7.5.5: Write the circular version of the cascade realization of Example 7.3.1, depicted in Fig. 7.3.2.

Solution: Let p_0 and p_1 denote the circular pointers of the two sections. Initially they point at the first elements of the three-dimensional buffers $\mathbf{w}_0$ and $\mathbf{w}_1$ of the two sections, that is, $p_0 = w_0$ and $p_1 = w_1$. The sample processing algorithm is:

$$
\begin{aligned}
&\textit{for each input sample } x \textit{ do:}\\
&\quad s_1 = \text{tap}\,(2, \mathbf{w}_0, p_0, 1)\\
&\quad s_2 = \text{tap}\,(2, \mathbf{w}_0, p_0, 2)\\
&\quad s_0 = x + 0.4s_1 - 0.5s_2\\
&\quad x_1 = 3s_0 - 4s_1 + 2s_2\\
&\quad *p_0 = s_0\\
&\quad \text{cdelay}\,(2, \mathbf{w}_0, \&p_0)\\
&\quad s_1 = \text{tap}\,(2, \mathbf{w}_1, p_1, 1)\\
&\quad s_2 = \text{tap}\,(2, \mathbf{w}_1, p_1, 2)\\
&\quad s_0 = x_1 - 0.4s_1 - 0.5s_2\\
&\quad y = 3s_0 + 4s_1 + 2s_2\\
&\quad *p_1 = s_0\\
&\quad \text{cdelay}\,(2, \mathbf{w}_1, \&p_1)
\end{aligned}
$$

where the output x_1 of the first section becomes the input to the second.   □

We can also write versions of the routines that manipulate the offset index q instead of the circular pointer p, in the same fashion as the FIR routine `cfir2` of Section 4.2.4. The following routine `ccan2.c` replaces `ccan`:

```
/* ccan2.c - circular buffer implementation of canonical realization */

void wrap2();                                defined in Section 4.2.4

double ccan2(M, a, b, w, q, x)
double *a, *b, *w, x;                        q = circular pointer offset index
int M, *q;                                   a,b have common order M
{
        int i;
        double y = 0;

        w[*q] = x;                                       read input sample x

        for (i=1; i<=M; i++)
             w[*q] -= a[i] * w[(*q+i)%(M+1)];

        for (i=0; i<=M; i++)
             y += b[i] * w[(*q+i)%(M+1)];

        (*q)--;                                   update circular delay line
        wrap2(M, q);

        return y;                                        output sample
}
```

Its usage is illustrated by the following program segment. Note that q must be passed by address:

```
int q;
double *a, *b, *w;
a = (double *) calloc(M+1, sizeof(double));
b = (double *) calloc(M+1, sizeof(double));
```

```
w = (double *) calloc(M+1, sizeof(double));    initializes w to zero
a[0] = 1;                                       not used in the routine

q = 0;                                          initialize q

for (n = 0; n < Ntot; n++)
        y[n] = ccan2(M, a, b, w, &q, x[n]);    p is passed by address
```

Similarly, the following routines `csos2.c` and `ccas2.c` replace `csos` and `ccas`:

```
/* csos2.c - circular buffer implementation of a single SOS */

void wrap2();

double csos2(a, b, w, q, x)
double *a, *b, *w, x;                   a, b, w are 3-dimensional arrays
int *q;                                 q is circular offset relative to w
{
        double y;

        w[*q] = x - a[1] * w[(*q+1)%3] - a[2] * w[(*q+2)%3];

        y = b[0] * w[*q] + b[1] * w[(*q+1)%3] + b[2] * w[(*q+2)%3];

        (*q)--;
        wrap2(2, q);

        return y;
}
```

and

```
/* ccas2.c - circular buffer implementation of cascade realization */

double csos2();                         circular-buffer version of single SOS

double ccas2(K, A, B, W, Q, x)
int K, *Q;                              Q = array of circular pointer offsets
double **A, **B, **W, x;
{
        int i;
        double y;

        y = x;

        for (i=0; i<K; i++)
                y = csos2(A[i], B[i], W[i], Q+i, y);    note, Q + i = &Q[i]

        return y;
}
```

The ith SOS has its own offset index q_i. Therefore, the quantity Q is defined as an array of K integers. The usage and initialization of `ccas2` is illustrated below. The quantities A, B, W are declared and allocated as usual, Q must be declared as:

```
int *Q;
```

```
Q = (double *) calloc(K, sizeof(double));          array of K integers
for (i=0; i<K; i++)
        Q[i] = 0;                                  initialize Q[i]

for (n = 0; n < Ntot; n++)
        y[n] = ccas2(K, A, B, W, Q, x[n]);
```

7.6 Quantization Effects in Digital Filters

There are two types of quantization effects in digital filters besides the quantization of the input and output signals: *roundoff errors* in the internal computations of the filter and *coefficient quantization.*

Coefficient quantization takes place whenever the filter coefficients are *rounded* from their exact values to a finite number of digits (or, bits for hardware implementations). The direct and canonical realizations tend to be *extremely sensitive* to such roundings, whereas the cascade realization remains very *robust.*

For *higher-order* filters whose poles are *closely clustered* in the z-plane, small changes in the denominator coefficients can cause large shifts in the location of the poles. If any of the poles moves outside the unit circle, the filter will become *unstable*, rendering it completely unusable. But even if the poles do not move outside, their large shifts may distort the *frequency response* of the filter so that it no longer satisfies the design specifications.

In practice, one must *always* check that the stability and specifications of the filter are preserved by the rounded coefficients. In using a software package to design a filter, one must always copy the designed coefficients with enough digits to guarantee these requirements. Problems 7.20 and 7.21 explore such quantization effects and some common pitfalls.

We do not mean to imply that the direct and canonical forms are always to be avoided; in fact, we saw in Examples 7.4.3 and 7.4.4 that the canonical forms were much simpler to implement than the cascade ones, and were also very robust under coefficient quantization.

In summary, the cascade form is recommended for the implementation of high-order narrowband lowpass, bandpass, or highpass IIR filters that have closely clustered poles. In this regard, it is convenient that many IIR filter design techniques, such as the bilinear transformation method, give the results of the design already in cascaded form.

There are other realization forms, such as cascaded second-order sections in transposed form, parallel forms, and lattice forms that are also very robust under coefficient quantization [2].

Roundoff errors occur in the internal multiplication and accumulation operations, for example, $y := y + aw$. The product aw requires twice as many bits as the factors to be represented correctly. A roundoff error will occur if this product is rounded to the original wordlength of the two factors. Such roundoff errors can be trapped into the feedback loops of recursive filters and can get amplified, causing too much distortion in the desired output.

Special state-space realizations and the technique of quantization noise shaping (also called in this context *error spectrum shaping*) can be used to minimize the accumulation of roundoff error [14,70–76]. To prevent overflow of intermediate results in filter computations, appropriate scaling factors must be introduced at various points within the filter stages [19,77,78].

Modern DSP chips address the issues of quantization effects in two ways: by using *long* wordlengths, such as 32 bits, for coefficient storage, and by using *double-precision* accumulators that can perform several MAC operations without roundoff error before the final result is rounded out of the accumulator.

7.7 *Problems*

7.1 A system has impulse response:

$$H(z) = \frac{z^{-1} + 2z^{-2} + 3z^{-3} + 4z^{-4}}{1 - z^{-5}}$$

  a. *Without* using partial fractions, determine the *causal* impulse response $h(n)$ of this system, for *all* $n \geq 0$, and sketch it versus n.

  b. Draw the *direct* and *canonical* realization forms. Write the *difference equations* describing these realizations. Then, write the corresponding *sample processing* algorithms.

  c. Factor this transfer function in the form $H(z) = H_1(z)H_2(z)$, where $H_1(z)$ is the ratio of two first-order polynomials, and $H_2(z)$ has numerator of degree 3 and denominator of degree 4. Draw the corresponding cascade realization, with each factor realized in its *canonical* form. Write the *difference equations* describing this realization, and the corresponding *sample processing* algorithm.

7.2 A discrete-time model for a second-order delta-sigma A/D converter is shown below:

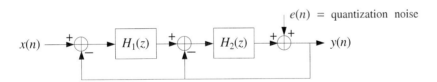

  a. Show that the output z-transform $Y(z)$ is related to $X(z)$ and $E(z)$ by a transfer function relationship of the form:

$$Y(z) = H_x(z)X(z) + H_e(z)E(z)$$

  Express the transfer functions $H_x(z)$ and $H_e(z)$ *in terms of* the loop filters $H_1(z)$ and $H_2(z)$.

  b. Determine $H_1(z)$ and $H_2(z)$ in order for $H_x(z)$ to act as a single delay and $H_e(z)$ as a second-order noise shaper, that is,

$$H_x(z) = z^{-1} \quad \text{and} \quad H_e(z) = (1 - z^{-1})^2$$

7.3 A digital filter has transfer function:

$$H(z) = \frac{z^{-1}(1 + 2z^{-2})(1 + 3z^{-2})}{1 - z^{-6}}$$

a. Draw the *direct* form realization (direct form I). Write the I/O difference equation and corresponding sample processing algorithm for this realization.

b. Draw the *canonical* form realization (direct form II). Write the I/O difference equations and corresponding sample processing algorithm for this realization.

c. Factor $H(z)$ into *second-order sections* with real-valued coefficients, and draw the corresponding *cascade* realization. Write the I/O difference equations and corresponding sample processing algorithm for this realization.

d. *Without* using partial fractions, determine the causal impulse response $h(n)$ of this filter for all n. Explain your reasoning.

7.4 A linear system is described by the system of difference equations:

$$v(n) = x(n) + v(n - 1)$$
$$y(n) = v(n) + v(n - 2) + v(n - 4)$$

Determine the transfer function from $x(n)$ to $y(n)$. Draw the direct, the canonical, and the cascade of SOS realizations (with real coefficients). In each case, state the sample-by-sample processing algorithm.

7.5 Draw the three realizations: (1) direct, (2) canonical, and (3) cascade of second-order sections for the following filter:

$$H(z) = \frac{(2 - 3z^{-1})(1 + z^{-2})}{1 - 0.25z^{-4}}$$

For each realization write the corresponding: (a) I/O difference equations and (b) sample processing algorithm.

7.6 A filter has transfer function:

$$H(z) = \frac{5}{1 + 0.25z^{-2}} - \frac{4}{1 - 0.25z^{-2}} = \frac{1 - 2.25z^{-2}}{(1 + 0.25z^{-2})(1 - 0.25z^{-2})}$$

a. Determine *all possible* impulse responses $h(n)$ and their ROCs.

b. Draw the *direct realization form* of $H(z)$.

c. Draw the *canonical realization form*.

d. Draw the *cascade form*.

In all cases, write all the *I/O difference equations* describing the realization in the time domain, and the *sample processing algorithm* implementing it.

7.7 Draw the *direct* and the *canonical* realizations of the system:

$$H(z) = \frac{1 - 2z^{-2} + z^{-4}}{1 - 0.4096z^{-4}}$$

a. Write the I/O difference equations and state the sample processing algorithms of these two realizations. [*Hint:* $0.4096 = (0.8)^4$.]

b. Factor the above transfer function into second-order sections (with real coefficients). Draw the cascade realization (with each SOS realized in its canonical form). Write *all* the I/O difference equations and state the sample processing algorithm describing this realization.

7.8 An allpass digital reverberator with delay of 10 time units and having input $x(n)$ and overall output $y(n)$, is described by the system of difference equations:

$$w(n) = 0.75w(n-10) + x(n)$$

$$y(n) = -0.75w(n) + w(n-10)$$

a. Draw a block diagram realization of this filter. The realization must use only one 10-fold delay.

b. Write the sample processing algorithm for this filter. Then, convert this algorithm into a C routine that implements it.

c. Show that the magnitude response of the filter is identically equal to one, that is, $|H(\omega)| = 1$ for all ω.

7.9 A filter is described by the following sample processing algorithm relating the input and output samples x and y:

for each input sample x do:
$$w_0 = x + 0.64w_4$$
$$y = w_0 + w_3$$
$$w_4 = w_3$$
$$w_3 = w_2$$
$$w_2 = w_1$$
$$w_1 = w_0$$

Determine the transfer function $H(z)$ of this filter. Factor $H(z)$ into factors of order up to two (with real-valued coefficients) and draw the corresponding cascade realization. State the sample processing algorithm for that realization.

7.10 For the following three filters,

$$H(z) = (1 + z^{-2})^3, \quad H(z) = \frac{1}{1 + 0.81z^{-2}}, \quad H(z) = \frac{1 - z^{-4}}{1 - 0.9z^{-1}}$$

a. Determine *all possible* impulse responses $h(n)$, corresponding ROCs, stability, and causality properties.

b. Draw the *direct, canonical,* and *cascade of SOS* realization forms. Write the I/O difference equations for each realization. State the sample-by-sample processing algorithm for each realization.

c. Determine the corresponding pole/zero plots and then make a rough sketch of the magnitude responses $|H(\omega)|$ versus ω.

7.11 Consider a system with transfer function:

$$H(z) = \frac{(1 - \sqrt{2}z^{-1} + z^{-2})(1 + \sqrt{2}z^{-1} + z^{-2})}{(1 + 0.81z^{-2})(1 - 0.81z^{-2})}$$

a. Determine the poles and zeros of this filter and place them on the z-plane. Then, draw a rough sketch of the *magnitude response* of the filter versus frequency.

b. Draw the *cascade* of second-order sections realization. Write the I/O difference equations for this realization. State the corresponding sample-by-sample processing algorithm.

Repeat for the *canonical* and *direct* form realizations.

7.12 Consider a stable system with transfer function $H(z) = \dfrac{\dfrac{1}{16} + z^{-4}}{1 + \dfrac{1}{16}z^{-4}}$.

a. Determine the poles and zeros of $H(z)$ and place them on the complex z-plane. Pair them in conjugate pairs to write $H(z)$ as a cascade of second-order sections with real coefficients.

b. Draw the direct, canonical, and cascade realization forms. In each case, write the corresponding sample processing algorithm.

c. Determine the impulse response $h(n)$ for all n. And, finally show that $|H(\omega)| = 1$ for all ω, that is, it is an allpass filter.

7.13 *Computer Experiment: IIR Filtering.* A digital filter has transfer function:

$$H(z) = H_0(z)H_1(z)H_2(z)H_3(z)$$

where

$$H_0(z) = \frac{0.313(1+z^{-1})}{1-0.373z^{-1}}, \qquad\qquad H_1(z) = \frac{0.147(1+2z^{-1}+z^{-2})}{1-1.122z^{-1}+0.712z^{-2}}$$

$$H_2(z) = \frac{0.117(1+2z^{-1}+z^{-2})}{1-0.891z^{-1}+0.360z^{-2}}, \quad H_3(z) = \frac{0.103(1+2z^{-1}+z^{-2})}{1-0.780z^{-1}+0.190z^{-2}}$$

a. Draw the cascade realization of $H(z)$ and write all the difference equations required for the time operation of this filter. Write the sample-by-sample processing algorithm implementing the cascade realization.

b. Using the routine cas2can, determine the canonical and direct realizations of $H(z)$ and draw them. Write the corresponding sample processing algorithms and difference equations for the two realizations.

c. Generate a length-100 input signal defined as

$$x(n) = \begin{cases} 1 & \text{if } 0 \le n < 50 \\ 0 & \text{if } 50 \le n < 100 \end{cases}$$

Using the cascade routine cas compute the filter output y_n for $0 \le n \le 99$. Repeat using the routines dir and can. In three parallel columns, print the signal samples y_n computed by the three routines cas, dir, can.

d. On the same graph, plot the two signals x_n and y_n versus n. You will be observing the input-on transients, steady-state response to a constant input, and input-off transients. What is the theoretical value of the steady-state response to a constant input?

e. Send a unit impulse as input. For the cascade realization, using the routine `cas`, compute the corresponding output impulse response for $0 \leq n < 50$, and plot it versus n. Repeat for the canonical realization using the routine `can`. Do you get identical results?

7.14 *Computer Experiment: IIR Filtering in Canonical Form.* Write a stand-alone C program, say `canfilt.c`, that implements the canonical form of the IIR sample processing algorithm, Eq. (7.2.5). The program must have usage:

```
canfilt a.dat b.dat  < x.dat  > y.dat
```

It must read and dynamically allocate the denominator and numerator coefficient vectors **a**, **b** from two input files, say `a.dat` and `b.dat`, and must allocate the internal state vector **w**. Using the routine `can.c`, it must keep processing input samples, reading them one at a time from `stdin` or from a file `x.dat`, and writing the computed output samples to `stdout` or a file `y.dat`. Filtering must stop when the end-of-file of the input file is encountered.

Using this program, calculate the filter outputs required in Problem 7.13.

7.15 *Computer Experiment: IIR Filtering in Cascade Form.* Write a stand-alone C program, say `casfilt.c`, that implements the cascade form of the IIR sample processing algorithm, Eq. (7.3.3). The program must have usage:

```
casfilt A.dat B.dat  < x.dat  > y.dat
```

It must read and dynamically allocate the $K \times 3$ denominator and numerator coefficient matrices A and B from two input files, say `A.dat` and `B.dat` (stored in row-wise fashion). and must allocate the internal $K \times 3$ state matrix W. Using the routine `cas.c`, it must keep processing input samples, reading them one at a time from `stdin` or from a file `x.dat`, and writing the computed output samples to `stdout` or a file `y.dat`. Filtering must stop when the end-of-file of the input file is encountered.

Alternatively or additionally, write a MATLAB version, say `casfilt.m`, that reads the input vector **x** and the matrices A and B and computes the output vector **y**. It may use the MATLAB functions `sos.m` and `cas.m` of Appendix D. Its usage must be:

```
y = casfilt(B, A, x);
```

Using these programs, calculate the filter outputs required in Problem 7.13.

7.16 *Computer Experiment: Comb Filtering.* Consider the two comb filters discussed in Examples 7.4.3 and 7.4.4. To understand the difference in their time-domain operation, consider as input to both filters the "noisy" signal:

$$x(n) = s(n) + v(n), \qquad \text{where} \qquad \begin{aligned} s(n) &= A_0 \cos(w_0 n) + A_2 \cos(w_1 n) \\ v(n) &= A_2 \cos(w_2 n) + A_3 \cos(w_3 n) \end{aligned}$$

where $n = 0, 1, \ldots, 499$, and $A_0 = 1$, $A_1 = A_2 = A_3 = 0.5$. The frequency components of the "desired" signal $s(n)$ are chosen to lie in the flat part, between the zeros, of the frequency response of the sharpened filter shown in Fig. 7.4.3: $w_0 = 0.50\pi/8$, $w_1 = 0.75\pi/8$. The frequency components of the "noise" part $v(n)$ are chosen to be two of the comb's zeros: $w_2 = \pi/8$, $w_3 = 3\pi/8$.

Plot the desired and noisy signals $s(n)$ and $x(n)$. Compute the corresponding output signals $y_1(n)$, $y_2(n)$ for $n = 0, 1, \ldots, 499$ of the two filters, plot them, and compare them with the desired signal $s(n)$.

To implement the filtering operations for the first filter, use the sample processing algorithm of the canonical form given in Example 7.4.3, and for the second filter use the cascade of the two combs given in Example 7.4.4. Moreover, to make a fair comparison *normalize* the two magnitude responses to unity gain at one of the desired frequencies, say at w_0.

7.17 *Computer Experiment: Comb Impulse Response.* First, derive closed-form analytical expressions for the impulse responses of the two comb filters of Examples 7.4.3 and 7.4.4. [*Hint:* You may do partial fractions in the variable z^{-8}.]

Then, using their sample processing algorithms, compute the impulse responses by sending in a unit impulse input. Compare the computed values and the analytical expressions. For each filter, compute the impulse responses $h(n)$ for $0 \leq n < 200$.

7.18 *Computer Experiment: IIR Filtering Using Circular Buffers.* For the canonical and cascade realizations, repeat all the questions of Problem 7.13, using the circular buffer routines ccan.c, ccas.c, and csos.c, instead of the standard linear buffer routines can, cas, and sos.

Repeat this problem using the alternative circular buffer routines ccan2.c, ccas2.c, and csos2.c.

7.19 Consider a filter with transfer function: $H(z) = \dfrac{6 - 2z^{-3}}{1 - 0.5z^{-3}}$.

a. Draw the *canonical realization* form of $H(z)$ and write the corresponding sample processing algorithm both in its *linear* and *circular* buffer versions.

b. Determine the causal impulse response $h(n)$ in two ways: (i) by doing and inverse z-transform on $H(z)$, and (ii) by sending in a unit impulse and iterating the circular buffer sample processing algorithm. Perform seven iterations for $n = 0, 1, \ldots, 6$ filling the entries of the following table:

n	x	w_0	w_1	w_2	w_3	s_0	s_1	s_2	s_3	y
0	1	*	*	*	*	*	*	*	*	*
1	0	*	*	*	*	*	*	*	*	*
⋮	⋮	⋮	⋮	⋮	⋮	⋮	⋮	⋮	⋮	⋮
6	0	*	*	*	*	*	*	*	*	*

where $[w_0, w_1, w_2, w_3]$ is the linear buffer over which the circular pointer circulates and s_i are the internal states (i.e., the tap outputs) of the triple delay z^{-3}.

c. Suppose the length-3 input signal $\mathbf{x} = [1, 2, 3]$ is applied. Compute the corresponding output for $n = 0, 1, \ldots, 6$ by iterating the circular version of the sample processing algorithm and filling the w_i and s_i entries of the above table.

7.20 *Computer Experiment: Coefficient Quantization.* Consider the double resonator filter given in cascade and direct forms:

$$H(z) = \frac{1}{1 - 1.8955z^{-1} + 0.9930z^{-2}} \cdot \frac{1}{1 - 1.6065z^{-1} + 0.9859z^{-2}}$$

$$= \frac{1}{1 - 3.5020z^{-1} + 5.0240z^{-2} - 3.4640z^{-3} + 0.9790z^{-4}}$$

(7.7.1)

using four-digit precision to represent the coefficients.

a. Calculate the zeros of the denominators in polar form and place them on the z-plane with respect to the unit circle. Plot the magnitude response of the filter for $0 \le \omega \le \pi/2$. Then, using the routines cas or can, iterate the sample processing algorithm for $n = 0, 1, \ldots, 599$ to determine the filter's impulse response and plot it versus n.

b. Consider the cascade form and round the coefficients of the individual second-order sections to two-digit accuracy. Then, repeat all questions of part (a), using the routine cas for the impulse response. Discuss the stability of the resulting filter and compare its magnitude response with that of part (a).

c. Consider the fourth-order direct form denominator polynomial and round its co-efficients to two-digit accuracy. Then, compute and plot its impulse response. Discuss its stability.

7.21 *Computer Experiment: Coefficient Quantization.* It is desired to design and implement a digital lowpass Chebyshev type 1 filter having specifications: sampling rate of 10 kHz, passband frequency of 0.5 kHz, stopband frequency of 1 kHz, passband attenuation of 1 dB, and stopband attenuation of 50 dB. Such a filter may be designed by the methods of Section 11.6.6, for example, using Eq. (11.6.65), which are mechanized by the MATLAB routine lhcheb1.m of Appendix D. The design method generates a sixth-order transfer function in cascade form:

$$H(z) = G \cdot \frac{(1 + z^{-1})^2}{1 + a_{01} z^{-1} + a_{02} z^{-2}} \cdot \frac{(1 + z^{-1})^2}{1 + a_{11} z^{-1} + a_{12} z^{-2}} \cdot \frac{(1 + z^{-1})^2}{1 + a_{21} z^{-1} + a_{22} z^{-2}}$$

$$= \frac{G(1 + z^{-1})^6}{1 + a_1 z^{-1} + a_2 z^{-2} + a_3 z^{-3} + a_4 z^{-4} + a_5 z^{-5} + a_6 z^{-6}}$$

where the normalization gain factor is $G = 8.07322364 \times 10^{-7}$. The full precision coefficients are given in terms of the matrix A:

$$A = \begin{bmatrix} 1 & a_{01} & a_{02} \\ 1 & a_{11} & a_{12} \\ 1 & a_{21} & a_{22} \end{bmatrix} = \begin{bmatrix} 1 & -1.86711351 & 0.96228613 \\ 1 & -1.84679822 & 0.89920764 \\ 1 & -1.85182222 & 0.86344488 \end{bmatrix}$$

where "full precision" means eight-digit precision. Let $\mathbf{a}_i = [1, a_{i1}, a_{i2}]$ be the second-order coefficient vectors. Using the routine cas2can, perform the following convolution to obtain the seven-dimensional direct-form denominator vector:

$$\mathbf{a} = \mathbf{a}_0 * \mathbf{a}_1 * \mathbf{a}_2 = [1, a_1, a_2, a_3, a_4, a_5, a_6]$$

a. Replace $\mathbf{a}$ by its quantized version rounded to eight-digit accuracy. Using a root finder routine, such as MATLAB's roots, determine the six zeros of $\mathbf{a}$ and their magnitudes.

Calculate and plot in dB the magnitude response, that is, $20 \log_{10} |H(f)|$, over the interval $0 \le f \le 0.75$ kHz, and verify that it meets the prescribed 1 dB passband specification.

Generate a unit impulse input of length 300, that is, $x(n) = \delta(n), n = 0, 1, \ldots, 299$, and using the canonical routine can, calculate the impulse response $h(n)$, $n = 0, 1, \ldots, 300$ and plot it versus n.

b. Determine the quantized version of **a**, say **â**, rounded to five-digit accuracy and calculate its 6 zeros and their magnitudes. Are all the zeros inside the unit circle? Compare the zeros with the full precision zeros.

Calculate the relative rounding error, that is, the ratio of the vector lengths:

$$\mathcal{E} = \frac{\|\Delta \mathbf{a}\|}{\|\mathbf{a}\|} = \frac{\|\hat{\mathbf{a}} - \mathbf{a}\|}{\|\mathbf{a}\|}$$

Using **â** as the filter denominator vector, calculate and plot in dB the magnitude response and compare it with the full-precision response. Does it meet the specs? Using the routine can, calculate the impulse response $h(n)$, $n = 0, 1, \ldots, 300$, and compare it with the full precision response. Is it stable?

c. Repeat all the questions of part (b), with **â** being the quantized version of **a** rounded to four digits. You should find that the impulse response is unstable and therefore the calculation of the magnitude response is meaningless (why is that?).

d. Round to four digits the *individual* second-order coefficient vectors $\mathbf{a}_i$, $i = 0, 1, 2$, compute the corresponding equivalent direct-form vector by the convolution $\hat{\mathbf{a}}_{cas} = \hat{\mathbf{a}}_0 * \hat{\mathbf{a}}_1 * \hat{\mathbf{a}}_2$, and round it to eight digits. Then, repeat all the questions of part (b) using $\hat{\mathbf{a}}_{cas}$ instead of **â**. You should find that the filter is stable and its impulse and magnitude responses agree very closely with the full precision responses. Also, compare and discuss the relative errors $\mathcal{E}$ in parts (b,c,d).

7.22 Consider a second-order denominator polynomial for an IIR filter with conjugate poles:

$$1 + a_1 z^{-1} + a_2 z^{-2} = (1 - pz^{-1})(1 - p^* z^{-1})$$

Taking differentials of both sides of this identity, show that small changes $\{da_1, da_2\}$ in the polynomial coefficients induce the small change in the pole location:

$$dp = -\frac{p\, da_1 + da_2}{p - p^*}$$

7.23 Consider a fourth-order denominator polynomial for an IIR filter with conjugate poles:

$$1 + a_1 z^{-1} + a_2 z^{-2} + a_3 z^{-3} + a_4 z^{-4} = (1 - p_0 z^{-1})(1 - p_0^* z^{-1})(1 - p_1 z^{-1})(1 - p_1^* z^{-1})$$

Show that small changes $\{da_1, da_2, da_3, da_4\}$ in the coefficients induce the following changes in the pole locations, which may not be too small if p_0 is close to p_1:

$$dp_0 = -\frac{p_0^3 da_1 + p_0^2 da_2 + p_0 da_3 + da_4}{(p_0 - p_0^*)(p_0 - p_1)(p_0 - p_1^*)}$$

$$dp_1 = -\frac{p_1^3 da_1 + p_1^2 da_2 + p_1 da_3 + da_4}{(p_1 - p_0^*)(p_1 - p_0)(p_1 - p_1^*)}$$

Such pole sensitivity formulas can be generalized to polynomials of arbitrary degree; see [2,3,246].

7.24 Consider the transposed realization of a third-order IIR filter shown in Fig. 7.7.1. First, determine the transfer function $H(z)$. Then, using the indicated variables $v_i(n)$, $i = 0, 1, 2, 3$, write the difference equations describing the time-domain operation of this realization. Then, rewrite them as a sample-by-sample processing algorithm that transforms each input x into an output sample y.

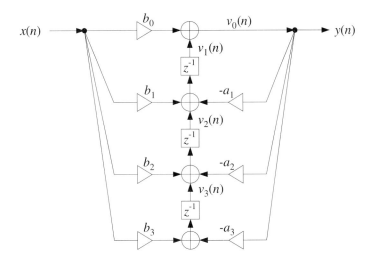

Fig. 7.7.1 Transpose of canonical realization.

7.25 *Computer Experiment: Transposed Realization Form.* Generalize the transposed structure of Fig. 7.7.1 to an arbitrary transfer function with numerator and denominator polynomials of degree M. State the corresponding sample processing algorithm in this case. Then, convert it into a C routine, say `transp.c`, that has usage:

```
y = transp(M, a, b, v, x);
```

where **a**, **b**, and **v** are $(M+1)$-dimensional vectors.

7.26 *Computer Experiment: Filtering in Transposed Form.* Repeat the filtering questions of Problem 7.13. Use the transpose of the canonical form implemented by the routine `transp.c` of Problem 7.25. For the cascade form, realize each second-order section in its transposed form. The output signals must agree exactly with those of Problem 7.13.

7.27 For the filter of Example 7.3.1, draw the transpose realizations of the canonical form. Also draw a cascade realization in which every second-order section is realized in its transposed form. In all cases, write the corresponding I/O difference equations and sample processing algorithms.

7.28 A general feedback system is shown in Fig. 7.7.2, where the output of filter $H_1(z)$ is fed back into filter $H_2(z)$ and then back to the input, and where the delay z^{-1} can be positioned at the four points A, B, C, or D.

For each case, determine the transfer function of the overall closed-loop system, that is, from $x(n)$ to $y(n)$. Assuming that the I/O equations of the filters $H_1(z)$ and $H_2(z)$ are known, state the corresponding sample processing algorithms in the four cases. How does moving the position of the delay change the order of computations? Finally, if the same input $x(n)$ is fed into the four filters, determine the relationships among the four output signals $y_A(n)$, $y_B(n)$, $y_C(n)$, and $y_D(n)$.

7.29 Consider the block diagram of Fig. 7.7.3, where the feedback delay z^{-1} can be positioned at points A or B. For the two cases, introduce appropriate internal state variables and write the difference equations describing the time-domain operations of the overall feedback system. Then, translate these difference equations into sample processing

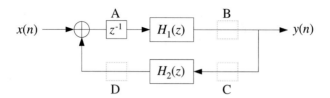

Fig. 7.7.2   General feedback system.

algorithms. What is the effect of moving the delay from A to B on the order of computations? What happens if that delay is removed altogether?

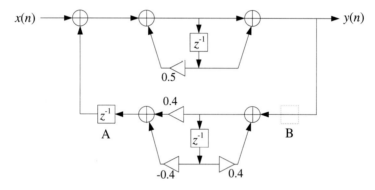

Fig. 7.7.3   Feedback system of Problem 7.29.

7.30 The three block diagrams (a,b,c) shown below are equivalent descriptions of the same
lowpass reverberator filter of the form:

$$H(z) = \frac{1}{1 - G(z)z^{-4}} \qquad \text{where} \quad G(z) = \frac{0.2}{1 - 0.8z^{-1}}$$

a. Draw a realization of the lowpass feedback filter $G(z)$. Replace the block $G(z)$
by its realization.

b. Write the sample processing algorithms describing the time-domain operation of
each of the three block diagrams (a), (b), and (c).

c. In block diagram (a), replace the 4-fold delay z^{-4} by its reverberating version:

$$z^{-4} \;\rightarrow\; \frac{z^{-4}}{1 - 0.8z^{-4}}$$

Draw the block diagram realization of the new overall filter. Your diagram must
be obtained from diagram (a) by replacing the 4-fold delay z^{-4} by the block diagram realization of the given reverberating version. Write the sample processing
algorithm for the new filter.

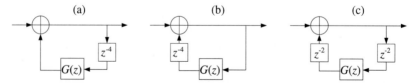

8

Signal Processing Applications

8.1 Digital Waveform Generators

It is often desired to generate various types of waveforms, such as periodic square waves, sawtooth signals, sinusoids, and so on.

A filtering approach to generating such waveforms is to design a filter $H(z)$ whose *impulse response $h(n)$* is the waveform one wishes to generate. Then, sending an impulse $\delta(n)$ as input will generate the desired waveform at the output.

In this approach, generating each sample by running the sample processing algorithm of the filter requires a certain amount of *computational overhead.* A more efficient approach is to *precompute* the samples of the waveform, store them in a table in RAM which is usually implemented as a circular buffer, and access them from the table whenever needed.

The period, or equivalently, the fundamental frequency of the generated waveform is controlled either by varying the speed of cycling around the table or by accessing a subset of the table at a fixed speed. This is the principle of the so-called *wavetable synthesis* which has been used with great success in computer music applications [94–115].

In this section, we discuss both the filtering and wavetable approaches and show how to implement them with circular buffers.

8.1.1 Sinusoidal Generators

The above filtering approach can be used to generate a (causal) sinusoidal signal of frequency f_0 and sampled at a rate f_s. Denoting the digital frequency by $\omega_0 = 2\pi f_0/f_s$, we have the z-transform pair:

$$ h(n) = R^n \sin(\omega_0 n) u(n), \quad H(z) = \frac{R \sin \omega_0 \, z^{-1}}{1 - 2R \cos \omega_0 \, z^{-1} + R^2 z^{-2}} \qquad (8.1.1) $$

For $0 < R < 1$, it corresponds to an exponentially decaying sinusoid of frequency ω_0. A pure sinusoid has $R = 1$. The canonical realization of this transfer function is shown in Fig. 8.1.1. The corresponding sample processing algorithm for the input $x(n) = \delta(n)$ and output $y(n) = h(n)$ is:

> *for* $n = 0, 1, 2, \ldots$ *do:*
> $w_0 = (2R \cos \omega_0) w_1 - R^2 w_2 + \delta(n)$
> $y = (R \sin \omega_0) w_1$
> $w_2 = w_1$
> $w_1 = w_0$

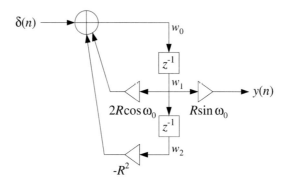

Fig. 8.1.1   Digital sinusoidal generator

In a similar fashion, we can generate an exponentially decaying cosinusoidal signal of frequency ω_0 with the following generator filter:

$$h(n) = R^n \cos(\omega_0 n) u(n), \quad H(z) = \frac{1 - R \cos \omega_0 z^{-1}}{1 - 2R \cos \omega_0 z^{-1} + R^2 z^{-2}} \qquad (8.1.2)$$

The canonical realization is shown in Fig. 8.1.2; its sample processing algorithm is:

> *for* $n = 0, 1, 2, \ldots$ *do:*
> $w_0 = (2R \cos \omega_0) w_1 - R^2 w_2 + \delta(n)$
> $y = w_0 - (R \cos \omega_0) w_1$
> $w_2 = w_1$
> $w_1 = w_0$

Example 8.1.1: A common application of sinusoidal generators is the all-digital touch-tone phone, known as a dual-tone multi-frequency (DTMF) transmitter/receiver [89–93]. Each key-press on the keypad generates the sum of two audible sinusoidal tones, that is, the signal

$$y(n) = \cos(\omega_L n) + \cos(\omega_H n)$$

where the two frequencies $\{\omega_L, \omega_H\}$ uniquely define the key that was pressed. Figure 8.1.3 shows the pairs of frequencies associated with each key.

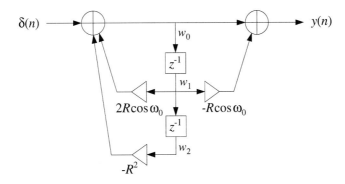

Fig. 8.1.2 Digital cosinusoidal generator

The four frequencies belonging to the low-frequency group select the four rows of the 4×4 keypad,[†] and the four high-group frequencies select the columns. A pair $\{f_L, f_H\}$ with one frequency from the low and one from the high group will select a particular key. With a typical sampling rate of $f_s = 8$ kHz, the corresponding digital frequencies are $\omega_L = 2\pi f_L/f_s$ and $\omega_H = 2\pi f_H/f_s$.

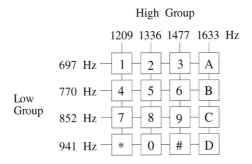

Fig. 8.1.3 DTMF keypad.

The generation of the dual tone can be implemented by using two cosinusoidal generators connected in parallel as in Fig. 8.1.4 and sending an impulse as input.

The particular values of the eight keypad frequencies have been chosen carefully so that they do not interfere with speech. At the receiving end, the dual-tone signal $y(n)$ must be processed to determine which pair of frequencies $\{f_L, f_H\}$ is present. This can be accomplished either by filtering $y(n)$ through a bank of bandpass filters tuned at the eight possible DTMF frequencies, or by computing the DFT of $y(n)$ and determining which pairs of frequency bins contain substantial energy.

Both approaches can be implemented with current DSP chips. We will discuss the DFT detection method further in Chapter 9. □

The poles of the transfer functions (8.1.1) and (8.1.2) are at the complex locations $p = Re^{j\omega_0}$ and $p^* = Re^{-j\omega_0}$. The denominator of these transfer functions factors in the form:

[†]The A,B,C,D keys appear on service keypads.

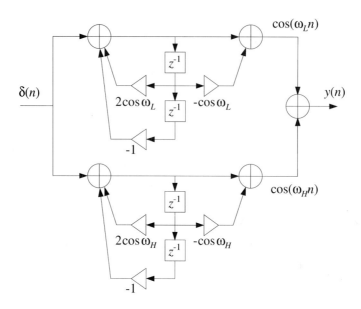

Fig. 8.1.4 DTMF tone generator.

$$1 - 2R \cos \omega_0 \, z^{-1} + R^2 z^{-2} = (1 - p z^{-1})(1 - p^* z^{-1}) \tag{8.1.3}$$

Denoting by a and b the real and imaginary parts of the pole $p = a + jb$, that is, $a = R \cos \omega_0$ and $b = R \sin \omega_0$, we have $R^2 = a^2 + b^2$ and can express the common denominator as

$$1 - 2R \cos \omega_0 \, z^{-1} + R^2 z^{-2} = 1 - 2a z^{-1} + (a^2 + b^2) z^{-2}$$

The cosinusoidal and sinusoidal transfer functions are expressed in terms of a and b as follows:

$$H_1(z) = \frac{1 - a z^{-1}}{1 - 2a z^{-1} + (a^2 + b^2) z^{-2}}$$
$$H_2(z) = \frac{b z^{-1}}{1 - 2a z^{-1} + (a^2 + b^2) z^{-2}} \tag{8.1.4}$$

where $H_1(z)$ corresponds to Eq. (8.1.2) and $H_2(z)$ to Eq. (8.1.1).

Forming the following complex linear combination, and replacing the denominator by its factored form (8.1.3), and noting that the numerators combine to give $(1 - p^* z^{-1})$, with $p^* = a - jb$, we obtain the pole/zero cancellation:

$$H_1(z) + jH_2(z) = \frac{1 - a z^{-1} + jb z^{-1}}{1 - 2a z^{-1} + (a^2 + b^2) z^{-2}} = \frac{1 - p^* z^{-1}}{(1 - p z^{-1})(1 - p^* z^{-1})}$$

$$= \frac{1}{1 - p z^{-1}}$$

Taking causal inverse z-transforms, we find

$$h_1(n) + jh_2(n) = p^n u(n) = R^n e^{j\omega_0 n} u(n)$$

Writing $e^{j\omega_0 n} = \cos(\omega_0 n) + j\sin(\omega_0 n)$ and extracting real and imaginary parts, gives the impulse responses:

$$h_1(n) = R^n \cos(\omega_0 n) u(n), \qquad h_2(n) = R^n \sin(\omega_0 n) u(n)$$

which agree with Eqs. (8.1.2) and (8.1.1), respectively.

The filter coefficients in Figs. 8.1.1 and 8.1.2 involve both the real and imaginary parts a, b, as well as the magnitude squared $R^2 = a^2 + b^2$ of the poles. In a hardware implementation, these coefficients must be quantized to a finite number of bits. One potential drawback is that to be quantized accurately, the coefficient $a^2 + b^2$ will need twice as many bits as the individual coefficients a and b.

An alternative realization [1] that combines both the sinusoidal and cosinusoidal generators is the so-called *coupled form* and is depicted in Fig. 8.1.5. Because only a and b, not their squares, appear as filter coefficients, this form will not suffer from the above quantization drawback.

When $R = 1$, it is impossible in general to find quantized coefficients a, b that satisfy $a^2 + b^2 = 1$ exactly. In that case, one settles for R slightly less than one. There exist filter structures with improved quantization properties when the poles are near the unit circle [73-76].

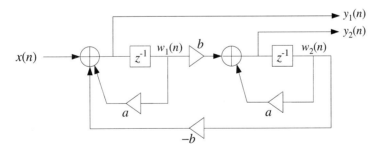

Fig. 8.1.5 Coupled form sine/cosine generator.

Noting that $w_1(n) = y_1(n-1)$ and $w_2(n) = y_2(n-1)$, the difference equations describing this form are in the time and z domains:

$$y_1(n) = ay_1(n-1) - by_2(n-1) + x(n)$$

$$y_2(n) = ay_2(n-1) + by_1(n-1)$$

$$Y_1(z) = az^{-1}Y_1(z) - bz^{-1}Y_2(z) + X(z)$$

$$Y_2(z) = az^{-1}Y_2(z) + bz^{-1}Y_1(z)$$

Solving for the transfer functions $H_1(z) = Y_1(z)/X(z)$ and $H_2(z) = Y_2(z)/X(z)$, we obtain Eq. (8.1.4). The sample processing algorithm that simultaneously generates the two outputs y_1 and y_2 will be:

$$
\begin{array}{l}
\textit{for each input sample x do:} \\
\quad y_1 = aw_1 - bw_2 + x \\
\quad y_2 = aw_2 + bw_1 \\
\quad w_1 = y_1 \\
\quad w_2 = y_2
\end{array}
$$

8.1.2 Periodic Waveform Generators

A periodic analog signal, such as a sinusoid, does not necessarily remain periodic when sampled at a given rate f_s. For example, the samples of $x(t) = \cos(2\pi f t)$, obtained by setting $t = nT$, are:

$$x(n) = \cos(2\pi f nT) = \cos(\omega n)$$

where $\omega = 2\pi f T = 2\pi f / f_s$.

In order for $x(n)$ to be periodic in the time index n with some period, say of D samples, it is necessary that one *whole* period of the sinusoid fit within the D samples, that is, at $n = D$, the sinusoid must cycle by one whole period. This requires that $x(D) = x(0)$, or,

$$\cos(\omega D) = 1$$

which requires that the frequency ω be such that[†]

$$\omega D = 2\pi \quad \Rightarrow \quad \boxed{\omega = \frac{2\pi}{D}} \tag{8.1.5}$$

Writing $\omega = 2\pi f / f_s$ and solving for f, we find the condition that a sampled sinusoid of frequency f is periodic if:

$$\boxed{f = \frac{f_s}{D}} \tag{8.1.6}$$

or, equivalently, if the sampling rate is an *integral multiple* of the frequency:

$$\boxed{f_s = Df} \tag{8.1.7}$$

These results generalize to the case of an arbitrary analog periodic signal, not just a sinusoid. Indeed, if a signal $x(t)$ has period T_D and is sampled at a rate $f_s = 1/T$, then the periodicity condition $x(t + T_D) = x(t)$ implies $x(nT + T_D) = x(nT)$. In order for the sampled signal to be periodic in n, the time $nT + T_D$ must be one of the sampling times, that is,

---

[†]One could also have $\omega = 2\pi c/D$, where c is an integer, but this would correspond to fitting more than one sinusoidal cycles in the D samples, that is, c cycles.

$$nT + T_D = (n + D)T$$

which requires that the period T_D be an integral multiple of the sampling period T:

$$\boxed{T_D = DT} \tag{8.1.8}$$

Because the *fundamental frequency* of such a periodic signal is $f = 1/T_D$, equation (8.1.8) is equivalent to Eq. (8.1.6) or (8.1.7).

In this section, we consider the generation of such discrete-time periodic signals. Because of the periodicity, it is enough to specify the signal over one period only. Denoting the time samples over one period by b_i, $i = 0, 1, \ldots, D - 1$, we have the periodic sequence:

$$\mathbf{h} = [b_0, b_1, \ldots, b_{D-1}, b_0, b_1, \ldots, b_{D-1}, b_0, b_1, \ldots, b_{D-1}, \ldots] \tag{8.1.9}$$

Figure 8.1.6 depicts such a sequence for $D = 4$. The filtering approach to generating such a periodic sequence is to think of it as the impulse response of a filter and then excite the filter with an impulsive input. The following filter has Eq. (8.1.9) as its causal impulse response:

$$H(z) = \frac{b_0 + b_1 z^{-1} + b_2 z^{-2} + \cdots + b_{D-1} z^{-(D-1)}}{1 - z^{-D}} \tag{8.1.10}$$

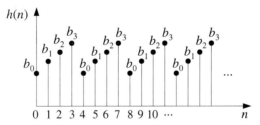

Fig. 8.1.6 Discrete-time periodic signal of period $D = 4$.

As a concrete example, consider the case $D = 4$ with transfer function:

$$H(z) = \frac{b_0 + b_1 z^{-1} + b_2 z^{-2} + b_3 z^{-3}}{1 - z^{-4}} \tag{8.1.11}$$

Its causal impulse response can be obtained by expanding the denominator using the infinite geometric series:

$$H(z) = (b_0 + b_1 z^{-1} + b_2 z^{-2} + b_3 z^{-3})(1 + z^{-4} + z^{-8} + \cdots)$$

$$= (b_0 + b_1 z^{-1} + b_2 z^{-2} + b_3 z^{-3}) \cdot 1$$

$$+ (b_0 + b_1 z^{-1} + b_2 z^{-2} + b_3 z^{-3}) \cdot z^{-4}$$

$$+ (b_0 + b_1 z^{-1} + b_2 z^{-2} + b_3 z^{-3}) \cdot z^{-8} + \cdots$$

Picking out the coefficients of the powers of z^{-1} gives the causal periodic impulse response sequence:

$$\mathbf{h} = [b_0, b_1, b_2, b_3, b_0, b_1, b_2, b_3, b_0, b_1, b_2, b_3, \dots \,] \qquad (8.1.12)$$

Writing

$$(1 - z^{-D})H(z) = b_0 + b_1 z^{-1} + \cdots + b_{D-1} z^{-(D-1)}$$

and transforming it to the time domain gives the difference equation for $h(n)$:

$$h(n) = h(n - D) + b_0 \delta(n) + b_1 \delta(n - 1) + \cdots + b_{D-1} \delta(n - D + 1)$$

For example, with $D = 4$:

$$h(n) = h(n - 4) + b_0 \delta(n) + b_1 \delta(n - 1) + b_2 \delta(n - 2) + b_3 \delta(n - 3) \qquad (8.1.13)$$

which generates Eq. (8.1.12). Indeed, iterating Eq. (8.1.13) with causal initial conditions gives:

$$h(0) = b_0, \quad h(1) = b_1, \quad h(2) = b_2, \quad h(3) = b_3$$

$$h(n) = h(n - 4), \quad \text{for } n \geq 4$$

The transfer function (8.1.11) can be realized in its direct or canonical forms. It is instructive to look at the time-domain operation of these two realizations. The direct form is depicted in Fig. 8.1.7. Note that there are $D = 4$ feedback delays, but only $D - 1 = 3$ feed-forward ones. The corresponding sample processing algorithm will be as follows:

> *for* $n = 0, 1, 2, \dots$ *do:*
> $\quad v_0 = \delta(n)$
> $\quad y = w_0 = w_4 + b_0 v_0 + b_1 v_1 + b_2 v_2 + b_3 v_3$
> $\quad \text{delay}(3, \mathbf{v})$
> $\quad \text{delay}(4, \mathbf{w})$

The following table shows the contents of the delay registers at successive sampling instants. The v and w delays are initialized to zero. Note that the v_0 column is the impulsive input $\delta(n)$. Similarly, the v_1 column represents the delayed version of v_0, that is, $\delta(n - 1)$, and v_2, v_3 represent $\delta(n - 2)$, $\delta(n - 3)$.

n	v_0	v_1	v_2	v_3	w_0	w_1	w_2	w_3	w_4	$y = w_0$
0	1	0	0	0	b_0	0	0	0	0	b_0
1	0	1	0	0	b_1	b_0	0	0	0	b_1
2	0	0	1	0	b_2	b_1	b_0	0	0	b_2
3	0	0	0	1	b_3	b_2	b_1	b_0	0	b_3
4	0	0	0	0	b_0	b_3	b_2	b_1	b_0	b_0
5	0	0	0	0	b_1	b_0	b_3	b_2	b_1	b_1
6	0	0	0	0	b_2	b_1	b_0	b_3	b_2	b_2
7	0	0	0	0	b_3	b_2	b_1	b_0	b_3	b_3
8	0	0	0	0	b_0	b_3	b_2	b_1	b_0	b_0

$$(8.1.14)$$

During the first four sampling instants, $n = 0, 1, 2, 3$, the initial impulse travels through the v-delays and eventually these delays empty out. The only purpose of the first four iterations of the algorithm is to *load* the w-delay registers with the values b_i of the signal. Indeed as can be seen from the table, the contents of the w-registers at time $n = 4$ are the b_i values loaded in *reverse order*:

$$[w_1, w_2, w_3, w_4] = [b_3, b_2, b_1, b_0]$$

and, in general, at time $n = D$ the w-delays will contain the values:

$$w_i = b_{D-i}, \qquad i = 1, 2, \ldots, D \tag{8.1.15}$$

For $n \geq 4$, the input part of the block diagram no longer plays a part because the v-delays are empty, whereas the contents of the w-delays recirculate according to $w_0 = w_4$, or $w_0 = w_D$, in general. Thus, an alternative way to formulate the sample processing algorithm for the generation of a periodic waveform is to break the algorithm into two parts: an initialization part

$$\boxed{\begin{aligned} &\textit{for } n = 0, 1, \ldots, D-1 \ \textit{do:} \\ &\quad w_0 = b_n \\ &\quad \mathrm{delay}(D, \mathbf{w}) \end{aligned}} \tag{8.1.16}$$

and a steady state part

$$\boxed{\begin{aligned} &\textit{repeat forever:} \\ &\quad w_0 = w_D \\ &\quad \mathrm{delay}(D, \mathbf{w}) \end{aligned}} \tag{8.1.17}$$

where $y = w_0$ is the corresponding output, as seen in Fig. 8.1.7.

Equation (8.1.16) effectively loads the w-registers with the b-values in *reverse order*. Then, Eq. (8.1.17) repeatedly recirculates the delay line, producing the periodic output. The canonical realization of the transfer function (8.1.11) is shown in Fig. 8.1.8. Its operation is now by the following sample processing algorithm:

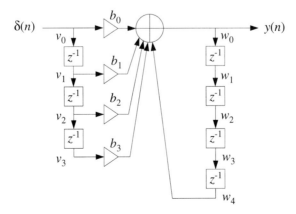

Fig. 8.1.7 Periodic generator in direct form.

$$\boxed{\begin{aligned} &\textit{for } n = 0, 1, 2, \ldots \textit{ do:}\\ &\quad w_0 = w_D + \delta(n)\\ &\quad y = b_0 w_0 + b_1 w_1 + \cdots + b_{D-1} w_{D-1}\\ &\quad \text{delay}(D, \mathbf{w}) \end{aligned}}$$

The contents of the w-register at successive sampling instants are shown below:

n	w_0	w_1	w_2	w_3	w_4	$y = b_0 w_0 + b_1 w_1 + b_2 w_2 + b_3 w_3$
0	1	0	0	0	0	b_0
1	0	1	0	0	0	b_1
2	0	0	1	0	0	b_2
3	0	0	0	1	0	b_3
4	1	0	0	0	1	b_0
5	0	1	0	0	0	b_1
6	0	0	1	0	0	b_2
7	0	0	0	1	0	b_3

The initial impulse gets trapped into the recirculating w-delay line, each time passing through only one of the b_i filter multipliers as it gets shifted from register to register.

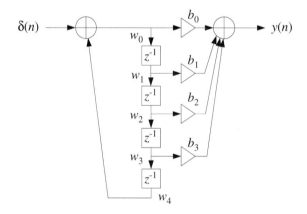

Fig. 8.1.8 Periodic generator in canonical form.

An intuitive way of understanding the operation of the canonical form is to separate out the common set of w-delays and think of the transfer function as the cascade of the two filters:

$$H(z) = \frac{1}{1 - z^{-4}} \cdot N(z) \tag{8.1.18}$$

where $N(z) = b_0 + b_1 z^{-1} + b_2 z^{-2} + b_3 z^{-3}$. Fig. 8.1.9 shows this interpretation. The impulse response of the first factor is a train of pulses separated by the desired period $D = 4$ with $D - 1$ zeros in between, that is, the sequence:

$$[1, 0, 0, 0, 1, 0, 0, 0, 1, 0, 0, 0, \dots]$$

Every time one of these impulses hits the FIR filter $N(z)$, it generates the impulse response $\mathbf{b} = [b_0, b_1, b_2, b_3]$, translated in time to match the time of that input impulse, as required by time invariance. Because the duration of $\mathbf{b}$ is only D samples, there is no overlap of the generated impulse responses, that is, each impulse response ends just before the next one begins.

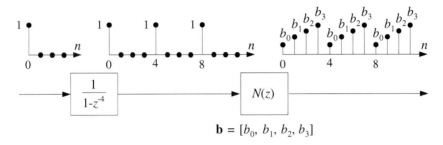

$$\mathbf{b} = [b_0, b_1, b_2, b_3]$$

Fig. 8.1.9 Periodic pulse train causes periodic output.

Both the direct and canonical realizations can be implemented using circular buffers. For example, the direct-form sample processing algorithm described by Eqs.(8.1.16) and (8.1.17) can be written with the help of the circular version of the delay routine, cdelay, as follows:

$$\begin{aligned} &\textit{for } n = 0, 1, \dots, D - 1 \ \textit{do:} \\ &\quad *p = b_n \\ &\quad \text{cdelay}(D, \mathbf{w}, \&p) \end{aligned} \tag{8.1.19}$$

and

$$\begin{aligned} &\textit{repeat forever:} \\ &\quad *p = \text{tap}(D, \mathbf{w}, p, D) \\ &\quad \text{cdelay}(D, \mathbf{w}, \&p) \end{aligned} \tag{8.1.20}$$

As discussed in Chapter 4, the circular pointer must be initialized by $p = w$. Eq. (8.1.19) loads the circular buffer with the D waveform samples, and then Eq. (8.1.20) reproduces them periodically. Alternatively, we may use the routines cdelay2 and tap2 that employ the offset index q such that $p = w + q$. Noting that $*p = p[0] = w[q]$, we have the generation algorithm:

$$\begin{aligned} &\textit{for } n = 0, 1, \dots, D - 1 \ \textit{do:} \\ &\quad w[q] = b_n \\ &\quad \text{cdelay2}(D, \&q) \end{aligned} \tag{8.1.21}$$

and

$$\boxed{\begin{array}{l} \textit{repeat forever:} \\ \quad w[q] = \text{tap2}(D, \mathbf{w}, q, D) \\ \quad \text{cdelay2}(D, \&q) \end{array}} \qquad (8.1.22)$$

where q must be initialized by $q = 0$.

These circular versions provide more efficient implementations than the direct form because at each time instant only the current w-register pointed to by p is updated—being loaded with the value of the *last* state, that is, the Dth state. By contrast, in the linear delay-line implementation, the entire delay line must be shifted at each time instant.

To appreciate how the circular delay line is updated, the table below shows the contents of the vector $\mathbf{w}$ for the case $D = 4$, at successive time instants (grouped every $D + 1 = 5$ samples):

n	q	w_0	w_1	w_2	w_3	w_4	y
0	0	$\uparrow b_0$	0	0	0	0	b_0
1	4	b_0	0	0	0	$\uparrow b_1$	b_1
2	3	b_0	0	0	$\uparrow b_2$	b_1	b_2
3	2	b_0	0	$\uparrow b_3$	b_2	b_1	b_3
4	1	b_0	$\uparrow b_0$	b_3	b_2	b_1	b_0
5	0	$\uparrow b_1$	b_0	b_3	b_2	b_1	b_1
6	4	b_1	b_0	b_3	b_2	$\uparrow b_2$	b_2
7	3	b_1	b_0	b_3	$\uparrow b_3$	b_2	b_3
8	2	b_1	b_0	$\uparrow b_0$	b_3	b_2	b_0
9	1	b_1	$\uparrow b_1$	b_0	b_3	b_2	b_1
10	0	$\uparrow b_2$	b_1	b_0	b_3	b_2	b_2
11	4	b_2	b_1	b_0	b_3	$\uparrow b_3$	b_3
12	3	b_2	b_1	b_0	$\uparrow b_0$	b_3	b_0
13	2	b_2	b_1	$\uparrow b_1$	b_0	b_3	b_1
14	1	b_2	$\uparrow b_2$	b_1	b_0	b_3	b_2
15	0	$\uparrow b_3$	b_2	b_1	b_0	b_3	b_3
16	4	b_3	b_2	b_1	b_0	$\uparrow b_0$	b_0
17	3	b_3	b_2	b_1	$\uparrow b_1$	b_0	b_1
18	2	b_3	b_2	$\uparrow b_2$	b_1	b_0	b_2
19	1	b_3	$\uparrow b_3$	b_2	b_1	b_0	b_3
20	0	$\uparrow b_0$	b_3	b_2	b_1	b_0	b_0
21	4	b_0	b_3	b_2	b_1	$\uparrow b_1$	b_1
22	3	b_0	b_3	b_2	$\uparrow b_2$	b_1	b_2
23	2	b_0	b_3	$\uparrow b_3$	b_2	b_1	b_3
24	1	b_0	$\uparrow b_0$	b_3	b_2	b_1	b_0

The up-arrow symbol ↑ indicates the w-register pointed to by the current value of the output pointer p. The pointer p cycles around every $D + 1 = 5$ samples even though the output is periodic every $D = 4$ samples. Equivalently, the current w-register can be determined by the corresponding value of the offset index q, that is, the register $w[q]$.

In the direct form version of Eq. (8.1.14), the linear delay line recirculates once every $D = 4$ samples, so that at $n = 8$ the state vector w is the same as at time $n = 4$. By contrast, the circular delay line recirculates much more slowly, that is, every $D(D + 1) = 20$ samples, so that the buffer w has the same contents at times $n = 4$ and $n = 24$. In both cases, the first $D = 4$ samples correspond to the initialization phases of Eqs. (8.1.16) and (8.1.19).

The following program segment illustrates the initialization and usage of the circular-buffer generation algorithms. It is assumed that the D-dimensional array of values $b[i]$, $i = 0, 1, \ldots, D - 1$, has already been defined:

```
double *b, *w, *p;

b = (double *) calloc(D, sizeof(double));       definition of b[n] is not shown
w = (double *) calloc(D+1, sizeof(double));      (D+1)-dimensional
p = w;                                            initialize circular pointer

for (n=0; n<D; n++) {                             initialization part
        *p = b[n];                                fill buffer with b[n]'s
        printf("%lf\n", *p);                      current output
        cdelay(D, w, &p);                         update circular delay line
        }

for (n=D; n<Ntot; n++) {                          steady state part
        *p = tap(D, w, p, D);                     first state = last state
        printf("%lf\n", *p);                      current output
        cdelay(D, w, &p);                         update circular delay line
        }
```

For comparison, we also list the linear delay-line version of Eqs. (8.1.16) and (8.1.17):

```
for (n=0; n<D; n++) {                             initialization part
        w[0] = b[n];                              fill buffer with b[n]'s
        printf("%lf\n", w[0]);                    current output
        delay(D, w);                              update linear delay line
        }

for (n=D; n<Ntot; n++) {                          steady state part
        w[0] = w[D];                              first state = last state
        printf("%lf\n", w[0]);                    current output
        delay(D, w);                              update linear delay line
        }
```

The spectra of *double-sided* periodic signals consist of sharp spectral lines at the *harmonics*, which are integral multiples of the fundamental frequency.

One-sided, or causal, periodic sequences such as the sequence (8.1.9), have *comb-like* spectra, as shown in Fig. 8.1.10, with *dominant* peaks at the harmonics.

The spectrum of (8.1.9) can be obtained by setting $z = e^{j\omega} = e^{2\pi j f / f_s}$ in the generating transfer function (8.1.10). Using the trigonometric identity

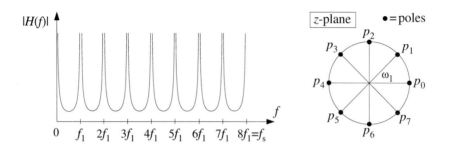

Fig. 8.1.10 Comb-like frequency spectrum, for $D = 8$.

$$1 - z^{-D} = 1 - e^{-j\omega D} = e^{-j\omega D/2}\left(e^{j\omega D/2} - e^{-j\omega D/2}\right) = 2je^{j\omega D/2}\sin\left(\frac{\omega D}{2}\right)$$

we have

$$|H(\omega)| = \frac{|N(\omega)|}{|1 - e^{-j\omega D}|} = \frac{|N(\omega)|}{2\left|\sin\left(\dfrac{\omega D}{2}\right)\right|}$$

or, replacing $\omega = 2\pi f/f_s$ in terms of the physical frequency f:

$$|H(f)| = \frac{|N(f)|}{2\left|\sin\left(\dfrac{\pi f D}{f_s}\right)\right|} \tag{8.1.23}$$

The peaks in the spectrum are due to the zeros of the denominator which vanishes at the harmonics, that is,

$$\sin\left(\frac{\pi f D}{f_s}\right) = 0 \quad \Rightarrow \quad \frac{\pi f D}{f_s} = \pi m$$

with m an integer. Solving for f:

$$f_m = m\frac{f_s}{D} = mf_1 \tag{8.1.24}$$

where we denoted the fundamental frequency by $f_1 = f_s/D$.

Because the spectrum $H(f)$ is periodic in f with period f_s, we may restrict the index m to the D values $m = 0, 1, \ldots, D-1$, which keep f within the Nyquist interval $[0, f_s)$. These harmonics correspond to the *poles* of $H(z)$. Indeed, solving for the zeros of the denominator, we have

$$1 - z^{-D} = 0 \quad \Rightarrow \quad z^D = 1$$

with the D solutions:

$$z = p_m = e^{j\omega_m}, \quad \omega_m = \frac{2\pi f_m}{f_s} = \frac{2\pi m}{D}, \quad m = 0, 1, \ldots, D-1$$

Note that they are the Dth roots of unity on the unit circle.

8.1.3 Wavetable Generators

The linear and circular buffer implementations of the direct form generator both use $(D+1)$-dimensional buffers $\mathbf{w} = [w_0, w_1, \ldots, w_D]$, whereas the periodic waveform has only D samples in one period: $\mathbf{b} = [b_0, b_1, \ldots, b_{D-1}]$. As we saw, this causes the buffer contents to recirculate periodically.

A simpler approach is to use a buffer of length D, that is, $\mathbf{w} = [w_0, w_1, \ldots, w_{D-1}]$ referred to as a *wavetable* and store in it a copy of one period of the desired waveform. The periodic waveform is then generated by repeatedly cycling over the wavetable with the aid of a circular pointer p, which always points at the current *output* sample.

Figure 8.1.11 shows such a wavetable for the case $D = 4$. Also shown are the positions of the circular pointer at successive sampling instants n, and the corresponding values of the offset index q, such that $*p = w[q]$.

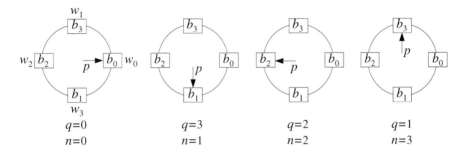

Fig. 8.1.11   Circular pointer cycles over wavetable.

As in the previous section, the waveform samples are stored in *circular reverse order*. The reverse loading of the table with the D waveform samples can be done with the following loop, initialized at $p = w$:

$$\boxed{\begin{aligned} &\textit{for } i = 0, 1, \ldots, D - 1 \ \textit{do:} \\ &\quad *p = b_i \\ &\quad \text{cdelay}(D - 1, \mathbf{w}, \&p) \end{aligned}}$$ (8.1.25)

or, in terms of the offset index q, initialized at $q = 0$:

$$\boxed{\begin{aligned} &\textit{for } i = 0, 1, \ldots, D - 1 \ \textit{do:} \\ &\quad w[q] = b_i \\ &\quad \text{cdelay2}(D - 1, \&q) \end{aligned}}$$ (8.1.26)

Note that the only difference with Eqs. (8.1.19) and (8.1.21) is the dimension of the buffer $\mathbf{w}$, which requires the cdelay routine to have argument $D-1$ instead of D. Upon exit from these initialization loops, the pointer p has wrapped around once and points again at the *beginning* of the buffer, $p = w$ or $q = 0$. Alternatively, the initialization of the wavetable can be done with:

$$\boxed{\begin{aligned} &for\ i = 0, 1, \ldots, D - 1\ \ do{:}\\ &\quad w[i] = b[(D - i)\%D] \end{aligned}} \tag{8.1.27}$$

where the modulo operation is felt only at $i = 0$, giving in this case $w[0] = b[D\%D] = b[0]$. After the wavetable is loaded with the waveform samples, the circular pointer can be made to cycle over the wavetable by successive calls to cdelay:

$$\boxed{\begin{aligned} &repeat\ forever{:}\\ &\quad output\ \ y = *p\\ &\quad \text{cdelay}(D - 1, \text{w}, \&p) \end{aligned}} \tag{8.1.28}$$

Each call to cdelay circularly *decrements* the pointer p to point to the *next* entry in the wavetable. In terms of the offset index q, we have similarly:

$$\boxed{\begin{aligned} &repeat\ forever{:}\\ &\quad output\ \ y = w[q]\\ &\quad \text{cdelay2}(D - 1, \&q) \end{aligned}} \tag{8.1.29}$$

Because the waveform was loaded in reverse order, decrementing the pointer will generate the waveform in forward order, as shown in Fig. 8.1.11.

Traditionally in the computer music literature, the wavetable is loaded in *forward* order, that is, $w[i] = b[i]$, $i = 0, 1, \ldots, D - 1$ and the circular pointer is *incremented* circularly [94,95]. In signal processing language, this corresponds to time advance instead of time delay. We will see how to implement time advances with the help of the generalized circular delay routine gdelay2 discussed below.

The following program segment illustrates the initialization (8.1.25) and the steady-state operation (8.1.28):

```
double *b, *w, *p;

b = (double *) calloc(D, sizeof(double));    definition of b[i] is not shown
w = (double *) calloc(D, sizeof(double));    Note, w is D-dimensional
p = w;                                       initialize circular pointer

for (i=0; i<D; i++) {                        initialization:
      *p = b[i];                             fill buffer with b[i]'s
      cdelay(D-1, w, &p);                    decrement pointer
      }

for (n=0; n<Ntot; n++) {                     steady state operation:
      printf("%lf\n", *p);                   current output
      cdelay(D-1, w, &p);                    decrement pointer
      }
```

Often, it is desired to generate a *delayed* version of the periodic waveform. Instead of loading a delayed period into a new wavetable, we can use the same wavetable, but start cycling over it at a shifted position. For a delay of m time units such that $0 \le m \le D - 1$, the starting pointer p and corresponding offset index q should be:

$$p = w + m, \qquad q = m \qquad\qquad (8.1.30)$$

To understand this, denote by $b(n)$ the original periodic sequence of period D, and let $y(n) = b(n-m)$ be its delayed version by m units. The starting sample will be $y(0) = b(-m)$, but because of the periodicity, we have $y(0) = b(-m) = b(D - m)$, which by the reverse loading of the w-buffer is equal to $w[m]$ according to Eq. (8.1.27). Thus, the starting sample will be $y(0) = w[m]$, corresponding to offset $q = m$. A *time advance* by m units can be implemented by starting at $q = -m$, which wraps to the positive value $q = D - m$.

For example, referring to Fig. 8.1.11, starting at position m and cycling clockwise generates the successively delayed periodic sequences:

$$m = 0: \quad [b_0, b_1, b_2, b_3, b_0, b_1, b_2, b_3, b_0, b_1, b_2, b_3, \ldots]$$

$$m = 1: \quad [b_3, b_0, b_1, b_2, b_3, b_0, b_1, b_2, b_3, b_0, b_1, b_2, \ldots]$$

$$m = 2: \quad [b_2, b_3, b_0, b_1, b_2, b_3, b_0, b_1, b_2, b_3, b_0, b_1, \ldots]$$

$$m = 3: \quad [b_1, b_2, b_3, b_0, b_1, b_2, b_3, b_0, b_1, b_2, b_3, b_0, \ldots]$$

Wavetable synthesis of periodic waveforms lies at the heart of many computer music applications and programs, such as *Music V* and its descendants [94-115]. Generating a single periodic wave is not musically very interesting. However, the generated waveform can be subjected to further operations to create more complex and interesting sounds, such as:

- Varying its amplitude or envelope to imitate the attack and decay of various instruments, or for imposing tremolo-type effects.

- Varying or modulating its frequency to imitate various effects such as vibrato, glissando, or portamento. This also leads to the popular family of FM synthesizers.

- Sending it through linear or nonlinear, time-invariant or time-varying filtering operations, generically known as *waveshaping* operations, which modify it further. They can be used to create models of various instruments, such as plucked strings or drums, or to superimpose various audio effects, such as reverb, stereo imaging, flanging, and chorusing [108-114].

- Adding together the outputs of several wavetables with different amplitudes and frequencies to imitate additive Fourier synthesis of various sounds.

The possibilities are endless. They have been and are actively being explored by the computer music community.

Here, we can only present some simple examples that illustrate the usage of wavetables. We begin by discussing how the *frequency* of the generated waveform may be changed.

Given a wavetable of length D, the period of the generated waveform is given by Eq. (8.1.8), $T_D = DT$, and its fundamental frequency by

$$f = \frac{f_s}{D} \tag{8.1.31}$$

The frequency f can be changed in two ways: by varying the sampling rate f_s or changing the effective length D of the basic period. Changing the sampling rate is not practical, especially when one is dealing with several wavetables of different frequencies, although digital music synthesizers have been built based on this principle. Thus, varying D is of more practical interest and is the preferred approach in most music synthesis programs. Replacing D by a smaller length $d \leq D$ will increase the fundamental frequency to:

$$\boxed{f = \frac{f_s}{d}} \tag{8.1.32}$$

and will decrease the period to $T_d = dT$.

For example, if $d = D/2$, the effective frequency is doubled $f = f_s/(D/2)$ $= 2f_s/D$. Replacing D by $d = D/2$ is equivalent to cycling the pointer p over a *subset* of the circular buffer $\mathbf{w}$ consisting of *every other* point in the buffer. Fig. 8.1.12 shows the positions of the pointer p at successive time instants n for the case of $D = 8$ and $d = D/2 = 4$. Skipping every other point can be accomplished by performing *two* calls to cdelay2 at each time, that is, replacing Eq. (8.1.29) by:

$$\begin{array}{l} \textit{repeat forever:} \\ \quad \textit{output } y = w[q] \\ \quad \text{cdelay2}\,(D-1,\&q) \\ \quad \text{cdelay2}\,(D-1,\&q) \end{array} \tag{8.1.33}$$

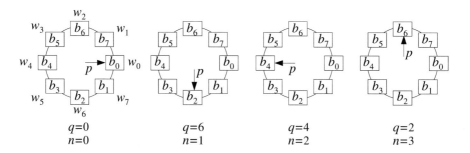

Fig. 8.1.12 Circular pointer is decremented by 2, generating a subsequence of period 4.

The two calls to cdelay2 effectively decrement the offset index q by *two*, that is, $q = q - 2$. The generated q values will be $q = 0$, $q = 0 - 2$, which wraps modulo-8 to $q = 6$, $q = 6 - 2 = 4$, $q = 4 - 2 = 2$, and so on. Thus, the generated subsequence will be the periodic repetition of the contents of the registers $[w_0, w_6, w_4, w_2]$, or, as shown in Fig. 8.1.12:

$$[b_0, b_2, b_4, b_6, b_0, b_2, b_4, b_6, b_0, b_2, b_4, b_6, b_0, b_2, b_4, b_6, \dots\,]$$

which repeats with period $d = 4$. It should be compared to the full wavetable sequence, which repeats every $D = 8$ samples:

$$[b_0, b_1, b_2, b_3, b_4, b_5, b_6, b_7, b_0, b_1, b_2, b_3, b_4, b_5, b_6, b_7, \ldots]$$

Similarly, cycling every four wavetable samples will generate the periodic subsequence of period $d = D/4 = 8/4 = 2$ and frequency $f = 4f_s/D$:

$$[b_0, b_4, b_0, b_4, b_0, b_4, b_0, b_4, b_0, b_4, b_0, b_4, b_0, b_4, b_0, b_4, \ldots]$$

In Fig. 8.1.12, it corresponds to decrementing the pointer p or offset q by 4, that is, $q = q - 4$, and can be implemented by inserting 4 calls to cdelay2 at each iteration:

$$
\boxed{
\begin{array}{l}
\textit{repeat forever:} \\
\quad \textit{output } y = w[q] \\
\quad \text{cdelay2}\,(D - 1, \&q) \\
\quad \text{cdelay2}\,(D - 1, \&q) \\
\quad \text{cdelay2}\,(D - 1, \&q) \\
\quad \text{cdelay2}\,(D - 1, \&q)
\end{array}
}
\tag{8.1.34}
$$

Rather than making multiple calls to cdelay2, we define a *generalized* version of this routine, gdelay2.c, which allows for an arbitrary *real-valued shift* of the pointer index q:

```
/* gdelay2.c - generalized circular delay with real-valued shift */

void gdelay2(D, c, q)
int D;
double c, *q;                    c=shift, q=offset index
{
        *q -= c;                 decrement by c

        if (*q < 0)
                *q += D+1;

        if (*q > D)
                *q -= D+1;
}
```

There are two basic differences with cdelay2. First, the offset index q is allowed to take on real values as opposed to integer values. Second, each call decrements q by the real-valued shift c, that is, $q = q - c$. The reason for allowing real values will become clear shortly. Note that cdelay2 is a special case of gdelay2 with $c = 1.0$. In terms of this routine, Eqs. (8.1.33) or (8.1.34) will read as:

$$
\boxed{
\begin{array}{l}
\textit{repeat forever:} \\
\quad \textit{output } y = w[q] \\
\quad \text{gdelay2}\,(D - 1, c, \&q)
\end{array}
}
\tag{8.1.35}
$$

with $c = 2$ or $c = 4$, respectively. The successive calls to gdelay2 in Eq. (8.1.35) update the offset q according to the iteration:

$$q_{n+1} = (q_n - c)\%D \qquad (8.1.36)$$

where mod-D is used because the dimension of the circular buffer is D. Generally, we have the following relationship between the shift c and the sub-period d:

$$\boxed{c = \frac{D}{d}} \qquad (8.1.37)$$

which gives the *number of times* the sub-period d fits into the full period D, or,

$$\boxed{d = \frac{D}{c}} \qquad (8.1.38)$$

which expresses d as a *fraction* of the full period D, or,

$$\boxed{D = cd} \qquad (8.1.39)$$

Combining Eqs. (8.1.32) and (8.1.38), gives for the frequency of the generated subsequence:

$$\boxed{f = \frac{f_s}{d} = c\frac{f_s}{D}} \qquad (8.1.40)$$

or, in terms of the digital frequency in radians/sample:

$$\omega = \frac{2\pi f}{f_s} = \frac{2\pi}{d} = \frac{2\pi c}{D} \qquad (8.1.41)$$

Equivalently, given a desired frequency f and table length D, we obtain the required value of the shift:

$$\boxed{c = D\frac{f}{f_s} = DF} \qquad (8.1.42)$$

where $F = f/f_s$ is the digital frequency in units of *cycles per sample*. It follows from Eq. (8.1.42) that c is the *number of cycles* of the subsequence that are contained in the D samples of the full wavetable.

So far, our discussion assumed that both the sub-length d and the shift c were integers. Because of the constraint (8.1.39), such restriction would not allow too many choices for c or d, and consequently for f. Therefore, c, d, and q are allowed to take on real values in the definition of gdelay2.

To keep f within the *symmetric* Nyquist interval $|f| \leq f_s/2$, requires that c satisfy the condition: $|cf_s/D| \leq f_s/2$, or,

$$|c| \leq \frac{D}{2} \quad \Rightarrow \quad -\frac{D}{2} \leq c \leq \frac{D}{2} \qquad (8.1.43)$$

Negative values of c correspond to negative frequencies f. This is useful for introducing 180° phase shifts in waveforms. Any value of c in the range $D/2 < c \le D$ is wrapped modulo-D to the value $c - D$, which lies in the negative part of the Nyquist interval (8.1.43). The wrapping $c \to c - D$ is equivalent to the frequency wrapping $f \to f - f_s$.

As we mentioned earlier, in the computer music literature, the circular wavetable is loaded with the waveform in forward order. Cycling over the wavetable at frequency $f = cf_s/D$ is accomplished by *incrementing* the offset index q according to the iteration:

$$q_{n+1} = (q_n + c)\%D \qquad (8.1.44)$$

Such "forward" versions can be implemented easily by the routine `gdelay2` by calling it in Eq. (8.1.35) with c replaced by $-c$. The wrap-around tests in `gdelay2` always force q to lie in the range $0 \le q < D$. If q is not an integer in that range, then it cannot be an array index that defines the output buffer sample $y = w[q]$. However, it can be approximated by an integer, for example, by truncating down, or truncating up, or rounding to the nearest integer:

$$
\begin{aligned}
i &= \lfloor q \rfloor && \text{(truncating down)} \\
j &= \lfloor q + 1 \rfloor \%D && \text{(truncating up)} \\
k &= \lfloor q + 0.5 \rfloor \%D && \text{(rounding)}
\end{aligned}
\qquad (8.1.45)
$$

The modulo-D operation is necessary to keep the index within the circular buffer range $\{0, 1, \ldots, D - 1\}$. The returned output will be in these cases:

$$
\begin{aligned}
y &= w[i] && \text{(truncating down)} \\
y &= w[j] && \text{(truncating up)} \\
y &= w[k] && \text{(rounding)}
\end{aligned}
\qquad (8.1.46)
$$

For example, in the first case, Eq. (8.1.35) will be replaced by:

$$
\boxed{
\begin{aligned}
&\textit{repeat forever:} \\
&\quad i = \lfloor q \rfloor \\
&\quad \textit{output } y = w[i] \\
&\quad \texttt{gdelay2}(D - 1, c, \&q)
\end{aligned}
}
\qquad (8.1.47)
$$

and similarly in the other cases.

Because q lies (circularly) between the integers i and j, a more accurate output can be obtained by *linearly interpolating* between the wavetable values $w[i]$ and $w[j]$, that is, returning the output:

$$y = w[i] + (q - i)(w[j] - w[i]) \qquad (8.1.48)$$

The geometric meaning of Eq. (8.1.48) is depicted in Fig. 8.1.13, where y lies on the straight line connecting $w[i]$ and $w[j]$. Note that $j = i + 1$, except when q falls in the last integer subdivision of the $[0, D)$ interval, that is, when $D - 1 \le q < D$. In that case, $i = D - 1$ and $j = D\%D = 0$, and we must interpolate between the values

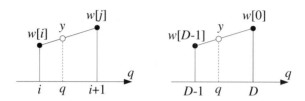

Fig. 8.1.13 Linear interpolation between two successive buffer samples.

$w[D-1]$ and $w[0]$, as shown in Fig. 8.1.13. Equation (8.1.48) correctly accounts for this case with j computed by Eq. (8.1.45), or equivalently by $j = (i+1)\%D$.

The interpolation method produces a more accurate output than the other methods, but at the expense of increased computation. The rounding method is somewhat more accurate than either of the truncation methods. The differences between the methods become unimportant as the length D of the wavetable increases. In computer music applications typical values of D are 512–32768. The nature of the approximation error for the truncation method and the other methods has been studied in [102–104].

The generation algorithm (8.1.47) starts producing the periodic sequence at the beginning of the w buffer, that is, with $q = 0$. If a delayed version of the subsequence is needed, we may shift the initial value of the offset q to a new starting position as in Eq. (8.1.30). However, because q is measured in multiples of the shift c, we must replace Eq. (8.1.30) by

$$q = mc = mDF \tag{8.1.49}$$

Because `gdelay2` decrements q, we can obtain Eq. (8.1.49) by starting with $q = 0$ and calling `gdelay2` once with the *opposite argument*:

$$\text{gdelay2}\,(D-1, -mc, \&q) \tag{8.1.50}$$

This expression implements both time delays ($m > 0$) and time advances ($m < 0$). Because mc must be in the interval $|mc| < D/2$, it follows that the allowed values of m are in the interval $|m| < d/2$. After this call, the generation algorithm Eq. (8.1.47) may be started with the desired value of the shift c.

Example 8.1.2: The eight waveform samples:

$$\mathbf{b} = [b_0, b_1, b_2, b_3, b_4, b_5, b_6, b_7]$$

are stored in (circular) reverse order in the 8-dimensional circular wavetable:

$$\mathbf{w} = [w_0, w_1, w_2, w_3, w_4, w_5, w_6, w_7]$$

It is desired to generate a periodic subsequence of period $d = 3$. Determine this subsequence when the output is obtained by the four methods of: (a) truncating down, (b) truncating up, (c) rounding, and (d) linear interpolation.

Solution: Here, $D = 8$ so that the shift is $c = D/d = 8/3$, which is not an integer. There are $d = 3$ possible values of the offset index q obtained by iterating Eq. (8.1.36):

$$q_0 = 0$$

$$q_1 = q_0 - c = -\frac{8}{3} \equiv 8 - \frac{8}{3} = \frac{16}{3} = 5\frac{1}{3}$$

$$q_2 = q_1 - c = \frac{16}{3} - \frac{8}{3} = \frac{8}{3} = 2\frac{2}{3}$$

The next q will be $q_3 = q_2 - c = (8/3) - (8/3) = 0$, and the above three values will be repeated. Fig. 8.1.14 shows the relative locations of the three q's with respect to the circular buffer indices.

The three q's can be obtained quickly by dividing the circular buffer D into d equal parts and counting clockwise. The direction of the three q-arrows in Fig. 8.1.14 are at relative angles ω as given by Eq. (8.1.41); here, $\omega = 2\pi/3$.

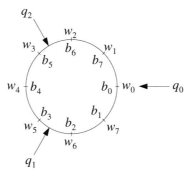

Fig. 8.1.14 Successive positions of q when $d = 3$.

It is seen that q_1 points between the buffer samples $w[5]$ and $w[6]$. If we truncate down, then, we will output the content of $w[5] = b_3$, and if we truncate up, $w[6] = b_2$. Because, q_1 points nearer to $w[5]$ than to $w[6]$, we will output $w[5] = b_3$, if we round to the nearest buffer location. If we interpolate linearly between $w[5]$ and $w[6]$, we will output the value:

$$y = w[5] + (q_1 - i_1)(w[6] - w[5]) = b_3 + \frac{1}{3}(b_2 - b_3) = \frac{1}{3}b_2 + \frac{2}{3}b_3$$

where $i_1 = \lfloor q_1 \rfloor = 5$, and $q_1 - i_1 = 1/3$. Similarly, the next offset index q_2 points between $w[2]$ and $w[3]$. If we truncate down, we will output $w[2] = b_6$, and if we truncate up, $w[3] = b_5$. If we round, we will output $w[3] = b_5$ because q_2 points closer to $w[3]$ than to $w[2]$. And, if we interpolate, we will output the value:

$$y = w[2] + (q_2 - i_2)(w[3] - w[2]) = b_6 + \frac{2}{3}(b_5 - b_6) = \frac{2}{3}b_5 + \frac{1}{3}b_6$$

where $i_2 = \lfloor q_2 \rfloor = 2$, and $q_2 - i_2 = 2/3$.

To summarize, at successive time instants $n = 0, 1, 2, \ldots$, the offset q cycles repeatedly over the three values $\{q_0, q_1, q_2\}$. The output associated with each q depends on the chosen approximation method. For the four methods, we will generate the following period-3 sequences:

$$[b_0, b_3, b_6, b_0, b_3, b_6, \ldots] \qquad \text{(truncate down)}$$
$$[b_0, b_2, b_5, b_0, b_2, b_5, \ldots] \qquad \text{(truncate up)}$$
$$[b_0, b_3, b_5, b_0, b_3, b_5, \ldots] \qquad \text{(round)}$$

and if we interpolate:

$$[b_0, \frac{1}{3}b_2 + \frac{2}{3}b_3, \frac{2}{3}b_5 + \frac{1}{3}b_6, b_0, \frac{1}{3}b_2 + \frac{2}{3}b_3, \frac{2}{3}b_5 + \frac{1}{3}b_6, \ldots]$$

Had we used the computer music convention of forward loading the circular buffer and incrementing q according to Eq. (8.1.44), we would find that the down and up truncated sequences *reverse roles*, that is, the down-truncated sequence would be our up-truncated one.   □

Example 8.1.3: Repeat Example 8.1.2 when the subsequence has period $d = 3$, but with an initial delay of $m = 1/2$ samples.

Solution: The desired delay by m samples (in units of c) can be implemented by an initial call to gdelay2, with an effective negative shift of $-mc$ as in Eq. (8.1.50). This initializes the offset index by shifting it from $q_0 = 0$ to

$$q_0 = mc = \frac{1}{2} \cdot \frac{8}{3} = \frac{4}{3}$$

The other two q's are obtained as in Example 8.1.2:

$$q_1 = q_0 - c = \frac{4}{3} - \frac{8}{3} = -\frac{4}{3} \equiv 8 - \frac{4}{3} = \frac{20}{3} = 6\frac{2}{3}$$

$$q_2 = q_1 - c = \frac{20}{3} - \frac{8}{3} = 4$$

The three q's are depicted in Fig. 8.1.15. The relative angle between the q-arrows is still $\omega = 2\pi c/D$, but the initial arrow for q_0 is displaced by an angle $\omega_0 = 2\pi m/d = 2\pi(mc)/D$ with respect to the horizontal axis. The original q's are shown by the dashed arrows. Note that the delay by $m = 1/2$ sample in units of c, rotates all the q's by half the original angle of $2\pi/d = 2\pi/3$, that is, by $\pi/3$.

Down-truncation gives the following integer values for the q's and corresponding buffer entries:

$$[q_0, q_1, q_2] = [1, 6, 4]$$

$$[w[1], w[6], w[4]] = [b_7, b_2, b_4]$$

Up-truncation gives:

$$[q_0, q_1, q_2] = [2, 7, 4]$$

$$[w[2], w[7], w[4]] = [b_6, b_1, b_4]$$

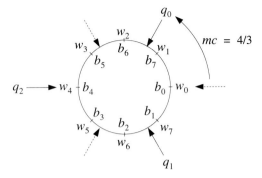

Fig. 8.1.15 Successive positions of q with $d = 3$ and delay $m = 1/2$.

For the rounding case, we have

$$[q_0, q_1, q_2] = [1, 7, 4]$$

$$[w[1], w[7], w[4]] = [b_7, b_1, b_4]$$

For the linear interpolation case, we have the outputs:

$$1 < q_0 < 2 \quad \Rightarrow \quad y_0 = w[1] + (q_0 - 1)(w[2] - w[1]) = \frac{1}{3}b_6 + \frac{2}{3}b_7$$

$$6 < q_1 < 7 \quad \Rightarrow \quad y_1 = w[6] + (q_1 - 6)(w[7] - w[6]) = \frac{2}{3}b_1 + \frac{1}{3}b_2$$

$$q_2 = 4 \quad \Rightarrow \quad y_2 = w[4] = b_4$$

Thus, depending on the output method, the following period-3 delayed subsequences will be generated:

$$[b_7, b_2, b_4, b_7, b_2, b_4, \dots] \qquad \text{(truncate down)}$$
$$[b_6, b_1, b_4, b_6, b_1, b_4, \dots] \qquad \text{(truncate up)}$$
$$[b_7, b_1, b_4, b_7, b_1, b_4, \dots] \qquad \text{(round)}$$
$$[y_0, y_1, y_2, y_0, y_1, y_2, \dots] \qquad \text{(interpolate)}$$

Thus, non-integer delays can be implemented easily. □

The purpose of these examples was to show the mechanisms of producing subsequences of different periods from a fixed wavetable of a given length D.

The generation algorithm of the truncation method given in Eq. (8.1.47), as well as the algorithms of the rounding and interpolation methods, can be programmed easily with the help of the routine `gdelay2`. To this end, we rewrite Eq. (8.1.47) in the following way:

$$
\boxed{
\begin{aligned}
&\textit{repeat forever:} \\
&\quad i = \lfloor q \rfloor \\
&\quad \textit{output } y = Aw[i] \\
&\quad \texttt{gdelay2}\,(D - 1, DF, \&q)
\end{aligned}
}
\qquad (8.1.51)
$$

where we introduced an *amplitude* scale factor A and expressed the shift $c = DF$ in terms of the digital frequency $F = f/f_s$.

With A and F as inputs to the algorithm, we can control the amplitude and frequency of the generated waveform. The following routine wavgen.c is an implementation of Eq. (8.1.51):

```
/* wavgen.c - wavetable generator (truncation method) */

void gdelay2();

double wavgen(D, w, A, F, q)        usage: y = wavgen(D, w, A, F, &q);
int D;                              D = wavetable length
double *w, A, F, *q;                A = amplitude, F = frequency, q = offset index
{
        double y;
        int i;

        i = (int) (*q);                                truncate down

        y = A * w[i];

        gdelay2(D-1, D*F, q);                          shift c = DF

        return y;
}
```

The following routines wavgenr and wavgeni are implementations of the *rounding* and *linear interpolation* generator methods of Eqs. (8.1.46) and (8.1.48), with added amplitude and frequency control:

```
/* wavgenr.c - wavetable generator (rounding method) */

void gdelay2();

double wavgenr(D, w, A, F, q)       usage: y = wavgenr(D, w, A, F, &q);
int D;                              D = wavetable length
double *w, A, F, *q;                A = amplitude, F = frequency, q = offset index
{
        double y;
        int k;

        k = (int) (*q + 0.5);                          round

        y = A * w[k];

        gdelay2(D-1, D*F, q);                          shift c = DF

        return y;
}
```

```
/* wavgeni.c - wavetable generator (interpolation method) */

void gdelay2();

double wavgeni(D, w, A, F, q)       usage: y = wavgeni(D, w, A, F, &q);
```

```
int D;                                  D = wavetable length
double *w, A, F, *q;                    A = amplitude, F = frequency, q = offset index
{
        double y;
        int i, j;

        i = (int) *q;                   interpolate between w[i], w[j]
        j = (i + 1) % D;

        y = A * (w[i] + (*q - i) * (w[j] - w[i]));

        gdelay2(D-1, D*F, q);                   shift c = DF

        return y;
}
```

In computer music, such routines are known as *wavetable oscillators* [94,95]. They are the workhorses of many digital music synthesis algorithms. Our C routines are modeled after [95]. Figure 8.1.16 depicts such an oscillator in computer music notation.

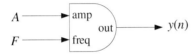

Fig. 8.1.16 Wavetable generator with amplitude and frequency control.

The amplitude and frequency inputs A and F do not have to be constant in time—they can be changing from one sampling instant to the next. In general, the generated signal will be given by:

$$y(n) = \text{wavgen}(D, \mathbf{w}, A(n), F(n), \&q) \qquad (8.1.52)$$

for $n = 0, 1, 2, \ldots$, where $A(n)$ and $F(n)$ can themselves be generated as the outputs of other oscillators to provide amplitude and frequency modulation.

The length-D wavetable $\mathbf{w}$ can be filled with any waveform, such as sinusoids, linear combination of sinusoids of different harmonics, square, triangular, trapezoidal waves, and so on, as long as *one complete period* of the desired waveform fits into the full wavetable. Figure 8.1.17 shows one period of a square, triangular, and trapezoidal wave.

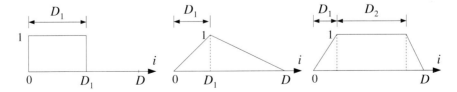

Fig. 8.1.17 Square, triangular, and trapezoidal waveforms.

The following C functions, `sine`, `square`, and `trapez` can be used to fill wavetables with such basic waveforms.

```
/* sine.c - sine wavetable of length D */

#include <math.h>

double sine(D, i)
int D, i;
{
        double pi = 4 * atan(1.0);

        return sin(2 * pi * i / D);
}



/* square.c - square wavetable of length D, with D1 ones */

double square(D1, i)
int D1, i;
{
        if (i < D1)
                return 1;
        else
                return 0;
}



/* trapez.c - trapezoidal wavetable: D1 rising, D2 steady */

double trapez(D, D1, D2, i)
int D, D1, D2, i;
{
        if (i < D1)
                return i/(double) D1;
        else
            if (i < D1+D2)
                    return 1;
            else
                    return (D - i)/(double) (D - D1 - D2);
}
```

To illustrate the usage of the wavetable generator routines, consider a wavetable **w** of length $D = 1000$ and fill it with one period of a sinusoid. The following program segment illustrates the reverse loading of the table using the function `sine`, and the generation of five sinusoids, shown in Fig. 8.1.18. The *same* wavetable is used by all sinusoids, but each is assigned its own offset index q that cycles around the wavetable according to a given frequency.

```
double *w;
w = (double *) calloc(D, sizeof(double));       use: D = 1000

q1 = q2 = q3 = q4 = q5 = 0;                      initialize qs

for (i=0; i<D; i++) {                            load wavetable with a sinusoid
        w[q1]    = sine(D, i);                   may need the cast w[(int)q1]
```

```
            gdelay2(D-1, 1.0, &q1);
            }

    gdelay2(D-1, -m*D*F2, &q4);                     reset q4 = mDF2
    gdelay2(D-1,  m*D*F2, &q5);                      reset q5 = −mDF2

    for (n=0; n<Ntot; n++) {                         use: A = 1, Ntot = 1000
            y1[n] = wavgen(D, w, A, F1, &q1);        use: F1  =  1.0/D
            y2[n] = wavgen(D, w, A, F2, &q2);        use: F2  =  5.0/D
            y3[n] = wavgen(D, w, A, F3, &q3);        use: F3  =  10.5/D
            y4[n] = wavgen(D, w, A, F4, &q4);        use: F4  =  F2
            y5[n] = wavgen(D, w, A, F5, &q5);        use: F5  =  F2
            }
```

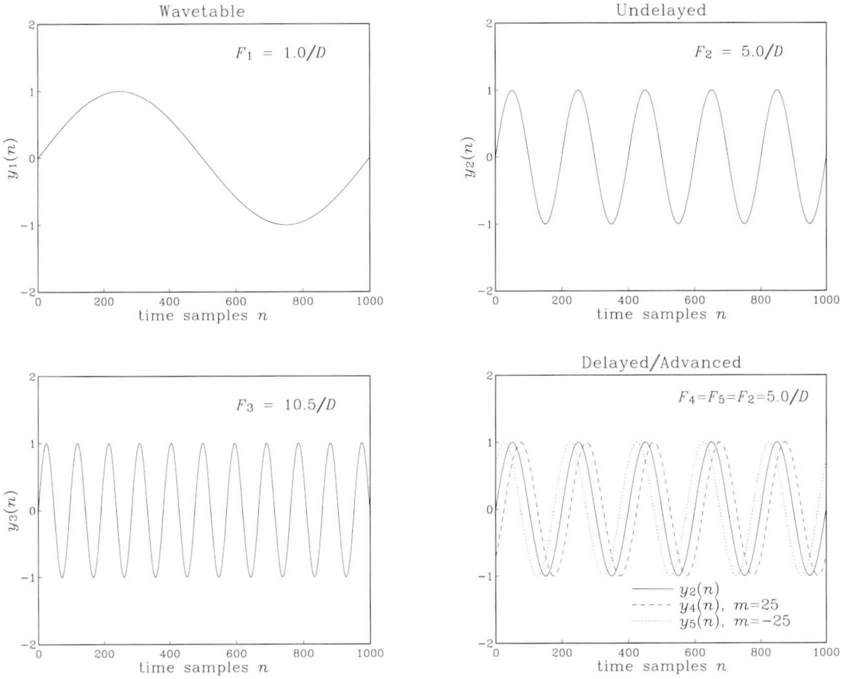

Fig. 8.1.18 Waveforms generated from a common wavetable.

The signal $y_1(n)$ is the sinusoid stored in the wavetable that becomes the source of all the other sinusoids. The first for-loop uses the offset index q_1 to load the wavetable. Upon exit from this loop, q_1 has cycled back to $q_1 = 0$. The frequency of $y_1(n)$ is one cycle in D samples, or,

$$F_1 = \frac{1}{D} = 0.001 \text{ cycles/sample}$$

and the corresponding shift is $c_1 = DF_1 = 1$. The signal $y_2(n)$ is generated from the same wavetable, but with frequency:

$$F_2 = \frac{5}{D} = 0.005 \text{ cycles/sample}$$

which corresponds to $c_2 = DF_2 = 5$ cycles in D samples. The wavetable is cycled over every five of its entries. The signal $y_3(n)$ is also generated from the same wavetable, but has frequency:

$$F_3 = \frac{10.5}{D} = 0.0105 \text{ cycles/sample}$$

which gives $c_3 = DF_3 = 10.5$, a non-integer value. The ten and a half cycles contained in the D samples can be seen in the figure. The wavetable is cycled over every 10.5 of its entries, and the output is obtained by the truncation method.

Finally, the last two signals $y_4(n)$ and $y_5(n)$ are the time-delayed and time-advanced versions of $y_2(n)$ by $m = 25$ samples, that is, $y_4(n) = y_2(n - 25)$ and $y_5(n) = y_2(n + 25)$. They are right- and left-shifted relative to $y_2(n)$ by one-eighth cycle, as can be seen in Fig. 8.1.18, because each F_2-cycle contains $1/F_2 = 200$ samples and therefore $m = 25$ corresponds to a $(1/8)$ of a cycle.

Because they have frequency F_2 and wavetable shift $c_2 = DF_2 = 5$, their effective starting offsets will be $q_4 = mc_2 = 25 \times 5 = 125$, and $q_5 = -mc_2 = -125$ (or, rather $q_5 = 1000 - 125 = 855$). These initial q-values are obtained by the two calls to `gdelay2` preceding the generation loop, with arguments $\mp mc_2$.

More complex waveforms can be generated by using several wavetables in combination. For example, Fig. 8.1.19 connects two wavetables together to implement *amplitude modulation.*

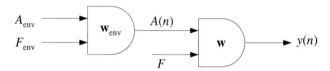

Fig. 8.1.19 Amplitude modulation.

The first generator $\mathbf{w}_{\text{env}}$ produces a time-varying envelope $A(n)$ that becomes the amplitude to the second generator whose wavetable $\mathbf{w}$ stores a copy of the desired signal, such as a note from an instrument. The envelope shape stored in the wavetable $\mathbf{w}_{\text{env}}$ could be triangular or trapezoidal, imitating instrument attack and decay. If we denote the main signal stored in $\mathbf{w}$ by $x(n)$, the configuration of Fig. 8.1.19 generates the modulated signal:

$$y(n) = A(n)x(n)$$

The amplitude input to the envelope generator A_{env} is a constant. Its frequency F_{env} is typically chosen such that the envelope cycles only over *one cycle* during the duration N_{tot} of the signal, that is,

$$\boxed{F_{\text{env}} = \frac{1}{N_{\text{tot}}}} \tag{8.1.53}$$

As an example, consider the generation of a sinusoidal note of frequency $F = 0.01$ cycles/sample:

$$x(n) = \sin(2\pi F n), \qquad n = 0, 1, \dots, N_{tot} - 1$$

with duration of $N_{tot} = 1000$ samples. The signal $x(n)$ is to be modulated by a triangular envelope whose attack portion is one-quarter its duration.

At a 44 kHz sampling rate, the frequency F would correspond to the 440 Hz note, A_{440}. Such a triangular envelope would be characteristic of a piano. Using wavetables of duration $D = 1000$, the following program segment illustrates the loading (in reverse order) of the wavetables with the appropriate waveforms, followed by the generation of the triangular envelope $A(n)$ and the modulated sinusoid $y(n)$. The truncation version, wavgen, of the generator routines was used:

```
double *w, *wenv, q, qenv;
w    = (double *) calloc(D, sizeof(double));      allocate wavetables
wenv = (double *) calloc(D, sizeof(double));      use: D = 1000

q = qenv = 0;                                     initialize offsets

for (i=0; i<D; i++) {                             load wavetables:
    w[q] = sine(D, i);                            may need the cast w[(int)q]
    wenv[qenv] = trapez(D, D/4, 0, i);            triangular envelope
    gdelay2(D-1, 1.0, &q);                        or, cdelay2(D-1, &q);
    gdelay2(D-1, 1.0, &qenv);
    }
                                                  use: Ntot = 1000 or 2000
Fenv = 1.0 / Ntot;                                envelope frequency

for (n=0; n<Ntot; n++) {
    A[n] = wavgen(D, wenv, Aenv, Fenv, &qenv);    use: Aenv = 1.0
    y[n] = wavgen(D, w, A[n], F, &q);             use: F = 0.01
    }
```

Figure 8.1.20 shows the two cases $N_{tot} = 1000, 2000$. Because $F = 0.01$ cycles/sample, there are $F N_{tot}$ cycles of the sinusoid in the duration of N_{tot}, that is, $F N_{tot} = 0.01 \times 1000 = 10$ cycles in the first case, and $F N_{tot} = 0.01 \times 2000 = 20$ cycles in the second. For visual reference, the graphs also plot the triangular envelope $A(n)$ and its negative, $-A(n)$.

The triangular wave was generated from the trapezoidal function by setting $D_1 = D/4$ and $D_2 = 0$. For both values of N_{tot}, the triangular envelope cycles only once, because of the choice (8.1.53) of its frequency. Note that the offset shift c corresponding to the frequency F will be $c = DF = 1000 \times 0.01 = 10$, whereas the shift for the envelope wavetable will be $c_{env} = DF_{env} = D/N_{tot} = 1$ or 0.5 in the two cases.

Figure 8.1.21 shows another example, where the envelope signal was chosen to be varying sinusoidally about a constant value:

$$A(n) = 1 + 0.25 \cos(2\pi F_{env} n)$$

so that the generated waveform will be:

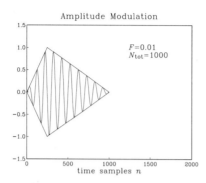
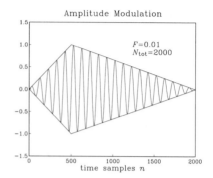

Fig. 8.1.20 Triangularly modulated sinusoid.

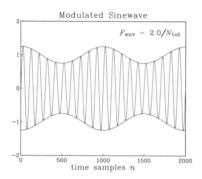
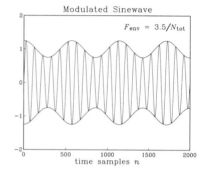

Fig. 8.1.21 Sinusoidally modulated sinusoid.

$$y(n) = A(n)x(n) = \left(1 + 0.25\cos\left(2\pi F_{\mathrm{env}}n\right)\right)\sin\left(2\pi Fn\right)$$

The envelope frequency was chosen to be $F_{\mathrm{env}} = 2/N_{\mathrm{tot}}$ for the first graph and $F_{\mathrm{env}} = 3.5/N_{\mathrm{tot}}$ for the second. These choices correspond to 2 and 3.5 envelope cycles in N_{tot} samples. With these values of F_{env}, the generation part for this example was carried out by exactly the same for-loop as above. The initial loading of the wavetables was carried out by:

```
q = qenv = 0;                          initialize offsets

for (i=0; i<D; i++) {                  load wavetables
        w[q] = sine(D, i);             sinusoidal signal
        wenv[qenv] = 1 + 0.25 * sine(D, i);   sinusoidal envelope
        gdelay2(D-1, 1.0, &q);         or, cdelay2(D-1, &q);
        gdelay2(D-1, 1.0, &qenv);
        }
```

In addition to amplitude modulation, we may introduce *frequency modulation* into the generated waveform. Figure 8.1.22 shows this case, where the first generator produces a periodic output with amplitude A_m and frequency F_m which is added to a carrier frequency F_c and the result becomes the frequency input to the second generator. For example, using a sinusoidal wavetable $\mathbf{w}_m$ will produce the frequency:

$$F(n) = F_c + A_m \sin(2\pi F_m n) \qquad (8.1.54)$$

so that if the signal generator $\mathbf{w}$ is a unit-amplitude sinusoid, then the modulated output will be:

$$y(n) = \sin(2\pi F(n)n) \qquad (8.1.55)$$

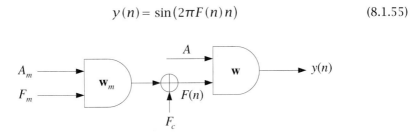

Fig. 8.1.22 Frequency modulation.

The following program segment illustrates the generation of four types of frequency modulated waveforms, shown in Fig. 8.1.23. The four cases can be obtained by uncommenting the applicable statements:

```
double *w, *wm;
w  = (double *) calloc(D, sizeof(double));
wm = (double *) calloc(D, sizeof(double));

q  = qm = 0;

for (i=0; i<D; i++) {                         load wavetables
      w[q]    = sine(D, i);                   signals: y₁(n), y₂(n), y₃(n)
   /* w[q]    = square(D/2, i); */            signal: y₄(n)
      gdelay2(D-1, 1.0, &q);

      wm[qm]  = sine(D, i);                    signal: y₁(n)
   /* wm[qm]  = 2 * square(D/2, i) - 1; */     signal: y₂(n)
   /* wm[qm]  = trapez(D, D, 0, i);    */      signals: y₃(n), y₄(n)
      gdelay2(D-1, 1.0, &qm);
      }

for (n=0; n<Ntot; n++) {                      use: Ntot = 1000
      F[n] = Fc + wavgen(D, wm, Am, Fm, &qm);
      y[n] = wavgen(D, w, A, F[n], &q);        use: A = 1
      }
```

The lengths of the two wavetables $\mathbf{w}$ and $\mathbf{w}_m$ were $D = 1000$ and the signal duration $N_{\text{tot}} = 1000$. The signal $y_1(n)$ was a frequency modulated sinusoid of the form of Eq. (8.1.55) with signal parameters:

$$F_c = 0.02, \quad A_m = 0.5F_c, \quad F_m = 0.003$$

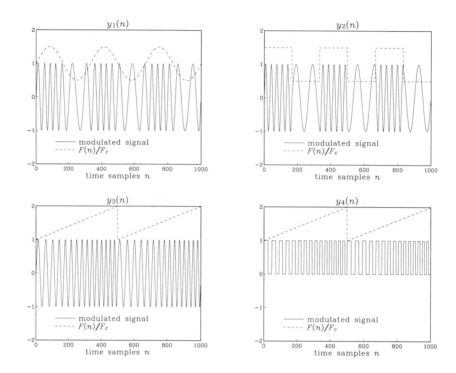

Fig. 8.1.23 Frequency modulated waveforms.

It might be thought of as a vibrato effect. The modulation frequency has $F_m N_{\text{tot}} = 3$ cycles in the N_{tot} samples. The frequency $F(n)$ rises and falls between the limits $F_c - A_m \le F(n) \le F_c + A_m$, or $0.5 F_c \le F(n) \le 1.5 F_c$. The quantity $F(n)/F_c$ is also plotted in order to help visualize the effect of increasing and decreasing frequency.

The signal $y_2(n)$ is a sinusoid whose frequency is modulated by a square wave that switches between the values $F_c + A_m$ and $F_c - A_m$, where again $A_m = 0.5 F_c$. The modulating square wave has frequency of 3 cycles in 1000 samples or $F_m = 0.003$. Note how the modulated signal $y_2(n)$ switches frequency more abruptly than $y_1(n)$.

The signal $y_3(n)$ is a sinusoid whose frequency is linearly swept between the values $F_c \le F(n) \le F_c + A_m$, where here $A_m = F_c$ so that $F(n)$ doubles. It might be thought of as a portamento effect. The sawtooth generator was implemented with the function `trapez`, with arguments $D_1 = D$ and $D_2 = 0$. Its frequency was chosen to be 2 cycles in 1000 samples, or $F_m = 0.002$.

Finally, the signal $y_4(n)$ is a square wave, generated by `square` with $D_1 = D/2$, whose frequency is linearly swept between F_c and $2 F_c$ with a modulation frequency of 2 cycles in 1000 samples, or $F_m = 0.002$.

Complex waveforms with rich sounds can be generated by combining amplitude and frequency modulation, as well as introducing such modulations on more than one level, for example, amplitude and/or frequency modulation of the amplitude generator in which A_{env} and F_{env} are themselves modulated by a third wavetable generator, and so on.

8.2 Digital Audio Effects

Audio effects, such as delay, echo, reverberation, comb filtering, flanging, chorusing, pitch shifting, stereo imaging, distortion, compression, expansion, noise gating, and equalization, are indispensable in music production and performance [115–151]. Some are also available for home and car audio systems.

Most of these effects are implemented using digital signal processors, which may reside in separate modules or may be built into keyboard workstations and tone generators. A typical audio effects signal processor is shown in Fig. 8.2.1.

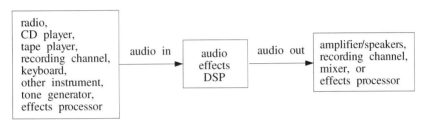

Fig. 8.2.1 Audio effects signal processor.

The processor takes in the "dry" analog input, produced by an instrument such as a keyboard or previously recorded on some medium, and samples it at an appropriate audio rate, such as 44.1 kHz (or less, depending on the effect). The sampled audio signal is then subjected to a DSP effects algorithm and the resulting processed signal is reconstructed into analog form and sent on to the next unit in the audio chain, such as a speaker system, a recording channel, a mixer, or another effects processor.

In all-digital recording systems, the sampling/reconstruction parts can be eliminated and the original audio input can remain in digitized form throughout the successive processing stages that subject it to various DSP effects or mix it with similarly processed inputs from other recording tracks.

In this section, we discuss some basic effects, such as delays, echoes, flanging, chorusing, reverberation, and dynamics processors. The design of equalization filters will be discussed in Chapters 10 and 11.

8.2.1 Delays, Echoes, and Comb Filters

Perhaps the most basic of all effects is that of *time delay* because it is used as the building block of more complicated effects such as reverb.

In a listening space such as a room or concert hall, the sound waves arriving at our ears consist of the *direct* sound from the sound source as well as the waves *reflected* off the walls and objects in the room, arriving with various amounts of time delay and attenuation.

Repeated multiple reflections result in the reverberation characteristics of the listening space that we usually associate with a room, hall, cathedral, and so on.

A *single reflection* or echo of a signal can be implemented by the following filter, which adds to the direct signal an attenuated and delayed copy of itself:

$$\boxed{y(n) = x(n) + ax(n - D)} \quad \text{(echo filter)} \tag{8.2.1}$$

The delay D represents the round-trip travel time from the source to a reflecting wall and the coefficient a is a measure of the reflection and propagation losses, so that $|a| \leq 1$. The transfer function and impulse response of this filter are:

$$H(z) = 1 + az^{-D}, \qquad h(n) = \delta(n) + a\delta(n - D) \tag{8.2.2}$$

Its block diagram realization is shown in Fig. 8.2.2. The frequency response is obtained from Eq. (8.2.2) by setting $z = e^{j\omega}$:

$$H(\omega) = 1 + ae^{-j\omega D}, \qquad |H(\omega)| = \sqrt{1 + 2a\cos(\omega D) + a^2} \tag{8.2.3}$$

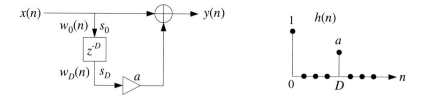

Fig. 8.2.2   Digital echo processor.

Such a filter acts as an FIR *comb filter* whose frequency response exhibits peaks at multiples of the fundamental frequency $f_1 = f_s/D$. The zeros of the transfer function $H(z)$ are the solutions of the equation (assuming $0 < a \leq 1$):

$$1 + az^{-D} = 0 \quad \Rightarrow \quad z_k = \rho e^{\pi j(2k+1)/D}, \quad k = 0, 1, \ldots, D - 1 \tag{8.2.4}$$

where $\rho = a^{1/D}$. The magnitude response and the zero pattern are shown in Fig. 8.2.3, for the case $D = 8$. If $a = 1$, then $\rho = 1$, and the zeros lie on the unit circle corresponding to exact zeros in the frequency response.

At the dip frequencies $\omega_k = (2k + 1)\pi/D$, we have $e^{j\omega_k D} = e^{j\pi} = -1$ giving $H(\omega_k) = 1 - a$. Between the dip frequencies, that is, at $\omega_k = 2\pi k/D$, we have peaks with value $H(\omega_k) = 1 + a$, because $e^{j\omega_k D} = 1$. In units of Hz, these peak frequencies are:

$$f_k = k\frac{f_s}{D} = kf_1, \qquad k = 0, 1, \ldots, D - 1 \tag{8.2.5}$$

The sample processing algorithm for this filter is given below, implemented with both a linear and circular delay line. As we mentioned in Chapter 4, for audio signals the delay D can be very large and therefore the circular delay line is more

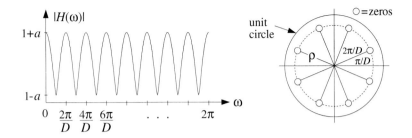

Fig. 8.2.3 FIR comb filter, with peaks at $\omega_k = 2\pi k/D$, $k = 0, 1, \ldots, D - 1$.

efficient. Denoting the $(D+1)$-dimensional delay-line buffer by $\mathbf{w} = [w_0, w_1, \ldots, w_D]$, we have:

for each input sample x do: $\quad y = x + aw_D$ $\quad w_0 = x$ $\quad \text{delay}(D, \mathbf{w})$	*for each input sample x do:* $\quad s_D = \text{tap}(D, \mathbf{w}, p, D)$ $\quad y = x + as_D$ $\quad *p = x$ $\quad \text{cdelay}(D, \mathbf{w}, \&p)$

Note that the quantities w_D in the linear case and $s_D = \text{tap}(D, \mathbf{w}, p, D)$ in the circular one represent the Dth output of the tapped delay line, that is, the signal $x(n - D)$. Comb filters, like the above echo processor, arise whenever the direct signal is mixed with its delayed replicas. For example, instead of adding the echo we can subtract it, obtaining (with $a > 0$):

$$\boxed{y(n) = x(n) - ax(n - D)} \qquad (8.2.6)$$

The transfer function and frequency response are now

$$H(z) = 1 - az^{-D}, \qquad H(\omega) = 1 - ae^{-j\omega D} \qquad (8.2.7)$$

having peaks at $\omega_k = (2k + 1)\pi/D$ and dips at $\omega_k = 2\pi k/D$, $k = 0, 1, \ldots, D - 1$. The magnitude response and zero pattern are shown in Fig. 8.2.4, for $D = 8$. Similarly, if we add three successive echoes, we obtain the filter:

$$\boxed{y(n) = x(n) + ax(n - D) + a^2 x(n - 2D) + a^3 x(n - 3D)} \qquad (8.2.8)$$

Using the finite geometric series, we can express the transfer function as

$$H(z) = 1 + az^{-D} + a^2 z^{-2D} + a^3 z^{-3D} = \frac{1 - a^4 z^{-4D}}{1 - az^{-D}} \qquad (8.2.9)$$

It follows that $H(z)$ vanishes at the zeros of the numerator which are not zeros of the denominator, that is,

$$z^{4D} = a^4, \quad \text{but} \quad z^D \neq a$$

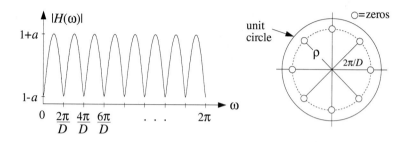

Fig. 8.2.4 Comb filter of Eq. (8.2.6), with dips at $\omega_k = 2\pi k/D$, $k = 0, 1, \ldots, D-1$.

or, equivalently, at

$$z_k = \rho e^{2\pi jk/4D}, \quad k = 0, 1, \ldots, 4D-1, \quad \text{but } k \text{ not a multiple of 4}$$

The filter has peaks at frequencies for which k is a multiple of 4, indeed, if $k = 4m$, $m = 0, 1, \ldots, D-1$, then

$$\omega_k = \frac{2\pi k}{4D} = \frac{2\pi(4m)}{4D} = \frac{2\pi m}{D} \qquad \Rightarrow \qquad e^{j\omega_k D} = 1$$

and the filter's response takes on the maximum value $H(\omega_k) = 1 + a + a^2 + a^3$.

The magnitude response and zero pattern are shown in Fig. 8.2.5, for $D = 8$. The dips occur at the 32nd roots of unity, except at the 8th roots of unity at which there are peaks.

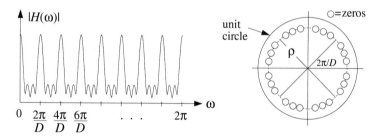

Fig. 8.2.5 Comb filter of Eq. (8.2.8), with peaks at $\omega_k = 2\pi k/D$, $k = 0, 1, \ldots, D-1$.

Adding up an infinite number of successive echoes imitates the reverberating nature of a room and gives rise to an *IIR comb filter*:

$$y(n) = x(n) + ax(n-D) + a^2 x(n-2D) + \cdots \tag{8.2.10}$$

which has impulse response:

$$h(n) = \delta(n) + a\delta(n-D) + a^2 \delta(n-2D) + \cdots \tag{8.2.11}$$

and transfer function:

$$H(z) = 1 + az^{-D} + a^2 z^{-2D} + \cdots$$

which can be summed by the geometric series into the form:

$$H(z) = \frac{1}{1 - az^{-D}} \qquad \text{(plain reverberator)} \qquad (8.2.12)$$

The I/O equation (8.2.10) can then be recast recursively as

$$y(n) = ay(n-D) + x(n) \qquad (8.2.13)$$

A block diagram realization is shown in Fig. 8.2.6. The feedback delay causes a unit impulse input to reverberate at multiples of D, that is, at $n = 0, D, 2D, \ldots$. Such simple recursive comb filters form the elementary building blocks of more complicated reverb processors, and will be discussed further in Section 8.2.3.

The transfer function (8.2.12) has poles at $p_k = \rho e^{j\omega_k}$, $k = 0, 1, \ldots, D-1$, where $\omega_k = 2\pi k/D$ and $\rho = a^{1/D}$. They are spaced equally around the circle of radius ρ, as shown in Fig. 8.2.7, for $D = 8$. At the pole frequencies ω_k, the frequency response develops peaks, just like the FIR comb of Fig. 8.2.3. Here, the sharpness of the peaks depends on how close to the unit circle the radius ρ is.

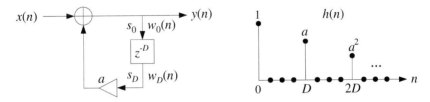

Fig. 8.2.6 Plain reverberator.

The repetition of the echoes every D samples corresponds to the fundamental repetition frequency of $f_1 = f_s/D$ Hz, or $\omega_1 = 2\pi/D$. In music performance, it is sometimes desired to lock the frequency of the decaying echoes to some external frequency, such as a drum beat. If f_1 is known, the proper value of D can be found from $D = f_s/f_1$, or the delay in seconds $T_D = DT = D/f_s = 1/f_1$.

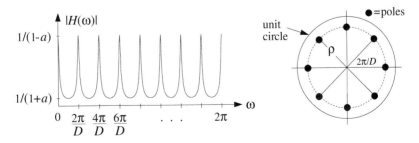

Fig. 8.2.7 IIR comb filter, with peaks at $\omega_k = 2\pi k/D$, $k = 0, 1, \ldots, D-1$.

The sample processing algorithm for the realization of Fig. 8.2.6 can be given in terms of a linear or circular delay-line buffer, as follows:

$$
\boxed{
\begin{aligned}
&\textit{for each input sample x do:} \\
&\quad y = x + aw_D \\
&\quad w_0 = y \\
&\quad \text{delay}(D, \mathbf{w})
\end{aligned}
}
\qquad
\boxed{
\begin{aligned}
&\textit{for each input sample x do:} \\
&\quad s_D = \text{tap}(D, \mathbf{w}, p, D) \\
&\quad y = x + as_D \\
&\quad *p = y \\
&\quad \text{cdelay}(D, \mathbf{w}, \&p)
\end{aligned}
}
\qquad (8.2.14)
$$

Note that, at each time instant, the output of the delay line is available and can be used to compute the filter's output y. The delay line cannot be updated until after y has been computed and fed back into the input of the delay. The quantities w_D and s_D represent the Dth tap output of the delay, that is, the signal $y(n-D)$.

The effective *time constant* for the filter response to decay below a certain level, say ϵ, can be obtained following the discussion of Section 6.3.2. At time $n = mD$ the impulse response has dropped to $\rho^n = \rho^{mD} = a^m$; therefore, the effective time constant $n_{\text{eff}} = m_{\text{eff}}D$ will be such that

$$
\rho^{n_{\text{eff}}} = a^{m_{\text{eff}}} = \epsilon
$$

which can be solved for m_{eff} and n_{eff}:

$$
n_{\text{eff}} = m_{\text{eff}}D = \frac{\ln \epsilon}{\ln a} D = \frac{\ln \epsilon}{\ln \rho} \qquad (8.2.15)
$$

and in seconds:

$$
\boxed{\; \tau_{\text{eff}} = n_{\text{eff}}T = \frac{\ln \epsilon}{\ln a} T_D \;} \qquad (8.2.16)
$$

where T is the sampling interval, such that $f_s = 1/T$, and $T_D = DT$ is the delay D in seconds. The so-called 60 dB reverberation time constant has $\epsilon = 10^{-3}$, which corresponds to a 60 dB attenuation of the impulse response.

8.2.2 Flanging, Chorusing, and Phasing

The value of the delay D in samples, or in seconds $T_D = DT$, can have a drastic effect on the perceived sound [119,120,128]. For example, if the delay is greater than about 100 milliseconds in the echo processor (8.2.1), the delayed signal can be heard as a quick repetition, a "slap". If the delay is less than about 10 msec, the echo blends with the direct sound and because only certain frequencies are emphasized by the comb filter, the resulting sound may have a hollow quality in it.

Delays can also be used to alter the *stereo image* of the sound source and are indispensable tools in stereo mixing. For example, a delay of a few milliseconds applied to one of the speakers can cause shifting and spreading of the stereo image. Similarly, a mono signal applied to two speakers with such a small time delay will be perceived in stereo.

More interesting audio effects, such as flanging and chorusing, can be created by allowing the delay D to *vary* in time [119,120,128]. For example, Eq. (8.2.1) may be replaced by:

$$\boxed{y(n) = x(n) + ax(n - d(n))} \qquad \text{(flanging processor)} \qquad (8.2.17)$$

A *flanging effect* can be created by periodically varying the delay $d(n)$ between 0 and 10 msec with a low frequency such as 1 Hz. For example, a delay varying sinusoidally between the limits $0 \le d(n) \le D$ will be:

$$d(n) = \frac{D}{2}(1 - \cos(2\pi F_d n)) \qquad (8.2.18)$$

where F_d is a low frequency, in units of [cycles/sample].

Its realization is shown in Fig. 8.2.8. The peaks of the frequency response of the resulting time-varying comb filter, occurring at multiples of f_s/d, and its notches at odd multiples of $f_s/2d$, will sweep up and down the frequency axis resulting in the characteristic whooshing type sound called flanging. The parameter a controls the *depth* of the notches. In units of [radians/sample], the notches occur at odd multiples of π/d.

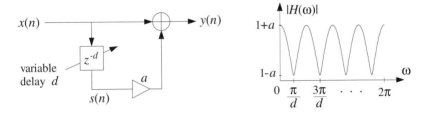

Fig. 8.2.8 Flanging effect, created with a periodically varying delay $d(n)$.

In the early days, the flanging effect was created by playing the music piece simultaneously through two tape players and alternately slowing down each tape by manually pressing the flange of the tape reel.

Because the variable delay d can take *non-integer* values within its range $0 \le d \le D$, the implementation of Eq. (8.2.17) requires the calculation of the output $x(n-d)$ of a delay line at such non-integer values. As we discussed in Section 8.1.3, this can be accomplished easily by truncation, rounding or linear interpolation.

Linear interpolation is the more accurate method, and can be implemented with the help of the following routine `tapi.c`, which is a generalization of the routine `tap` to non-integer values of d.

```
/* tapi.c - interpolated tap output of a delay line */

double tap();

double tapi(D, w, p, d)          usage: sd = tapi(D, w, p, d);
double *w, *p, d;                d = desired non-integer delay
int D;                          p = circular pointer to w
{
        int i, j;
        double si, sj;

        i = (int) d;            interpolate between si and sj
```

```
      j = (i+1) % (D+1);                    if i = D, then j = 0; otherwise, j = i + 1

      si = tap(D, w, p, i);                 note, sᵢ(n) = x(n − i)
      sj = tap(D, w, p, j);                 note, sⱼ(n) = x(n − j)

      return si + (d - i) * (sj - si);
}
```

The input d must always be restricted to the range $0 \le d \le D$. Note that if d is one of the integers $d = 0, 1, \ldots, D$, the routine's output is the same as the output of tap. The mod-$(D+1)$ operation in the definition of j is required to keep j within the array bounds $0 \le j \le D$, and is effective only when $d = D$, in which case the output is the content of the last register of the tapped delay line.

The following routine tapi2.c is a generalization of the routine tap2, which is implemented in terms of the offset index q instead of the circular pointer p, such that $p = w + q$.

```
/* tapi2.c - interpolated tap output of a delay line */

double tap2();

double tapi2(D, w, q, d)              usage: sd = tapi2(D, w, q, d);
double *w, d;                         d = desired non-integer delay
int D, q;                             q = circular offset index
{
      int i, j;
      double si, sj;

      i = (int) d;                        interpolate between sᵢ and sⱼ
      j = (i+1) % (D+1);                  if i = D, then j = 0; otherwise, j = i + 1

      si = tap2(D, w, q, i);              note, sᵢ(n) = x(n − i)
      sj = tap2(D, w, q, j);              note, sⱼ(n) = x(n − j)

      return si + (d - i) * (sj - si);
}
```

Linear interpolation should be adequate for low-frequency inputs, having maximum frequency much less than the Nyquist frequency. For faster varying inputs, more accurate interpolation methods can be used, designed by the methods of Chapter 12.

As an example illustrating the usage of tapi, consider the flanging of a plain sinusoidal signal of frequency $F = 0.05$ cycles/sample with length $N_{tot} = 200$ samples, so that there are $FN_{tot} = 10$ cycles in the 200 samples. The flanged signal is computed by

$$y(n) = \frac{1}{2}[x(n) + x(n - d(n))] \tag{8.2.19}$$

with $d(n)$ given by Eq. (8.2.18), $D = 20$, and $F_d = 0.01$ cycles/sample, so that there are $F_d N_{tot} = 2$ cycles in the 200 samples.

The following program segment implements the calculation of the term $s(n) = x(n - d(n))$ and $y(n)$. A delay-line buffer of maximal dimension $D + 1 = 21$ was used:

```
double *w, *p;
w = (double *) calloc(D+1, sizeof(double));
p = w;

for (n=0; n<Ntot; n++) {
    d = 0.5 * D * (1 - cos(2 * pi * Fd * n));    time-varying delay
    x = cos(2 * pi * F * n);                      input x(n)
    s = tapi(D, w, p, d);                         delay-line output x(n − d)
    y = 0.5 * (x + s);                            filter output
    *p = x;                                       delay-line input
    cdelay(D, w, &p);                             update delay line
}
```

Figure 8.2.9 shows the signals $x(n)$, $s(n) = x(n - d(n))$, $y(n)$, as well as the time-varying delay $d(n)$ normalized by D.

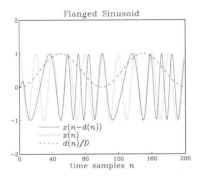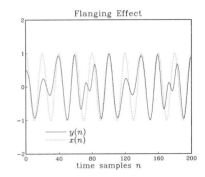

Fig. 8.2.9 Flanged sinusoidal signal.

Recursive versions of flangers can also be used that are based on the all-pole comb filter (8.2.13). The feedback delay D in Fig. 8.2.6 is replaced now by a variable delay d. The resulting flanging effect tends to be somewhat more pronounced than in the FIR case, because the sweeping comb peaks are sharper, as seen in Fig. 8.2.7.

Chorusing imitates the effect of a group of musicians playing the same piece simultaneously. The musicians are more or less synchronized with each other, except for small variations in their strength and timing. These variations produce the chorus effect. A digital implementation of chorusing is shown in Fig. 8.2.10, which imitates a chorus of three musicians.

The small variations in the time delays and amplitudes can be simulated by varying them slowly and randomly [119,120]. A low-frequency random time delay $d(n)$ in the interval $0 \le d(n) \le D$ may be generated by

$$d(n) = D(0.5 + v(n))\qquad(8.2.20)$$

or, if the delay is to be restricted in the interval $D_1 \le d(n) < D_2$

$$d(n) = D_1 + (D_2 - D_1)(0.5 + v(n))\qquad(8.2.21)$$

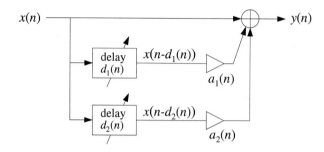

Fig. 8.2.10 Chorus effect, with randomly varying delays and amplitudes.

The signal $v(n)$ is a zero-mean low-frequency random signal varying between $[-0.5, 0.5]$. It can be generated by the linearly interpolated generator routine `ran1` of Appendix B.2. Given a desired rate of variation F_{ran} cycles/sample for $v(n)$, we obtain the period $D_{ran} = 1/F_{ran}$ of the generator `ran1`.

As an example, consider again the signal $y(n)$ defined by Eq. (8.2.19), but with $d(n)$ varying according to Eq. (8.2.20). The input is the same sinusoid of frequency $F = 0.05$ and length $N_{tot} = 200$. The frequency of the random signal $v(n)$ was taken to be $F_{ran} = 0.025$ cycles/sample, corresponding to $N_{tot}F_{ran} = 5$ random variations in the 200 samples. The period of the periodic generator `ran1` was $D_{ran} = 1/F_{ran} = 40$ samples. The same program segment applies here, but with the change:

```
d = D * (0.5 + ran1(Dran, u, &q, &iseed));
```

where the routine parameters `u, q, iseed` are described in Appendix B.2.

Figure 8.2.11 shows the signals $x(n)$, $s(n) = x(n - d(n))$, $y(n)$, as well as the quantity $d(n)/D$.

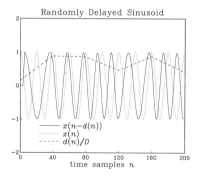

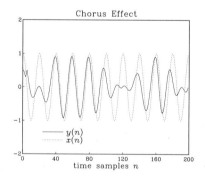

Fig. 8.2.11 Chorusing or doubling of sinusoidal signal.

Phasing or phase shifting is a popular effect among guitarists, keyboardists, and vocalists. It is produced by passing the sound signal through a *narrow notch filter* and combining a proportion of the filter's output with the direct sound.

The frequency of the notch is then varied in a controlled manner, for example, using a low-frequency oscillator, or manually with a foot control. The strong phase shifts that exist around the notch frequency combine with the phases of the direct signal and cause phase cancellations or enhancements that sweep up and down the frequency axis.

A typical overall realization of this effect is shown in Fig. 8.2.12. Multi-notch filters can also be used. The effect is similar to flanging, except that in flanging the sweeping notches are equally spaced along the frequency axis, whereas in phasing the notches can be unequally spaced and independently controlled, in terms of their location and width.

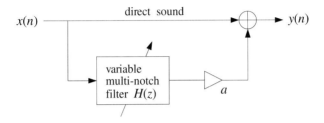

Fig. 8.2.12 Phasing effect with multi-notch filter.

The magnitude and phase responses of a typical single-notch filter are shown in Fig. 8.2.13. Note that the phase response $\arg H(\omega)$ remains essentially zero, except in the vicinity of the notch where it has rapid variations.

In Section 6.4.3, we discussed simple methods of constructing notch filters. The basic idea was to start with the *notch polynomial* $N(z)$, whose zeros are at the desired notch frequencies, and place poles *behind* these zeros inside the unit circle, at some radial distance ρ. The resulting pole/zero notch filter was then $H(z) = N(z)/N(\rho^{-1}z)$.

Such designs are simple and effective, and can be used to construct the multi-notch filter of a phase shifter. Choosing ρ to be near unity gives very narrow notches. However, we cannot have complete and separate control of the widths of the different notches.

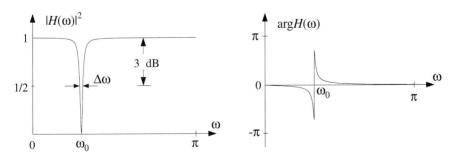

Fig. 8.2.13 Narrow notch filter causes strong phase shifts about the notch frequency.

A design method that gives precise control over the notch frequency and its 3-dB width is the bilinear transformation method, to be discussed in detail in Chapter 11. Using this method, a second-order single-notch filter can be designed as follows:

$$H(z) = b\frac{1 - 2\cos\omega_0 z^{-1} + z^{-2}}{1 - 2b\cos\omega_0 z^{-1} + (2b - 1)z^{-2}} \tag{8.2.22}$$

where the filter parameter b is expressible in terms of the 3-dB width $\Delta\omega$ (in units of radians per sample) as follows:

$$b = \frac{1}{1 + \tan(\Delta\omega/2)} \tag{8.2.23}$$

The Q-factor of a notch filter is another way of expressing the narrowness of the filter. It is related to the 3-dB width and notch frequency by:

$$Q = \frac{\omega_0}{\Delta\omega} \quad \Rightarrow \quad \Delta\omega = \frac{\omega_0}{Q} \tag{8.2.24}$$

Thus, the higher the Q, the narrower the notch. The transfer function (8.2.22) is normalized to unity gain at DC. The basic shape of $H(z)$ is that of Fig. 8.2.13. Because $|H(\omega)|$ is essentially flat except in the vicinity of the notch, several such filters can be *cascaded* together to create a multi-notch filter, with independently controlled notches and widths.

As an example, consider the design of a notch filter with notch frequency $\omega_0 = 0.35\pi$, for the two cases of $Q = 3.5$ and $Q = 35$. The corresponding 3-dB widths are in the two cases:

$$\Delta\omega = \frac{\omega_0}{Q} = \frac{0.35\pi}{3.5} = 0.10\pi \quad \text{and} \quad \Delta\omega = \frac{\omega_0}{Q} = \frac{0.35\pi}{35} = 0.01\pi$$

The filter coefficients are then computed from Eq. (8.2.23), giving the transfer functions in the two cases:

$$H(z) = 0.8633\frac{1 - 0.9080z^{-1} + z^{-2}}{1 - 0.7838z^{-1} + 0.7265z^{-2}}, \quad \text{(for } Q = 3.5\text{)}$$

$$H(z) = 0.9845\frac{1 - 0.9080z^{-1} + z^{-2}}{1 - 0.8939z^{-1} + 0.9691z^{-2}}, \quad \text{(for } Q = 35\text{)}$$

The magnitude squared and phase responses are shown in Fig. 8.2.14.

Given a time-varying notch frequency, say $\omega_0(n)$, and a possibly time-varying width $\Delta\omega(n)$, the filter coefficients in Eq. (8.2.22) will also be time-varying. The time-domain implementation of the filter can be derived using a particular realization, such as the canonical realization. For example, if the notch frequency sweeps sinusoidally between the values $\omega_1 \pm \omega_2$ at a rate ω_{sweep}, that is, $\omega_0(n) = \omega_1 + \omega_2\sin(\omega_{sweep}n)$, then the following sample processing algorithm will determine the filter coefficients on the fly and use them to perform the filtering of the current input sample (here, $\Delta\omega$ and b remain fixed):

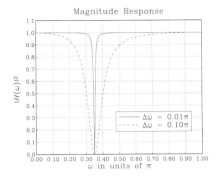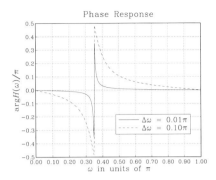

Fig. 8.2.14 Notch filters with $\omega_0 = 0.35\pi$, $Q = 3.5$ and $Q = 35$.

> *for each time instant n and input sample x do:*
> *compute current notch* $\omega_0 = \omega_1 + \omega_2 \sin(\omega_{\text{sweep}}n)$
> $w_0 = bx + 2b\cos\omega_0\, w_1 - (2b - 1)w_2$
> $y = w_0 - 2\cos\omega_0\, w_1 + w_2$
> $w_2 = w_1$
> $w_1 = w_0$

An alternative technique for designing multi-notch phasing filters was proposed by Smith [148]. The method uses a cascade of second-order allpass filters, each having a phase response that looks like that of Fig. 8.2.13 and changes by 180° at the notch. If the output of the allpass filter is added to its input, the 180° phase shifts will introduce notches at the desired frequencies.

The three effects of flanging, chorusing, and phasing are based on simple filter structures that are changed into *time-varying filters* by allowing the filter coefficients or delays to change from one sampling instant to the next.

The subject of *adaptive signal processing* [27] is also based on filters with time-varying coefficients. The time dependence of the coefficients is determined by certain design criteria that force the filter to adjust and optimize itself with respect to its inputs. The implementation of an adaptive algorithm is obtained by augmenting the sample processing algorithm of the filter by adding to it the part that adjusts the filter weights from one time instant to the next [28].

Adaptive signal processing has widespread applications, such as channel equalization, echo cancellation, noise cancellation, adaptive antenna systems, adaptive loudspeaker equalization, adaptive system identification and control, neural networks, and many others.

8.2.3 Digital Reverberation

The reverberation of a listening space is typically characterized by three distinct time periods: the direct sound, the early reflections, and the late reflections [115–

151], as illustrated in Fig. 8.2.15.

The early reflections correspond to the first few reflections off the walls of the room. As the waves continue to bounce off the walls, their density increases and they disperse, arriving at the listener from all directions. This is the late reflection part.

The reverberation time constant is the time it takes for the room's impulse response to decay by 60 dB. Typical concert halls have time constants of about 1.8–2 seconds.

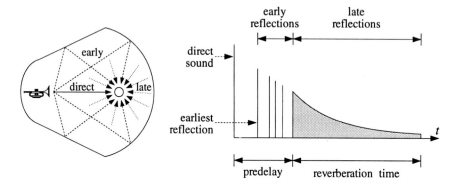

Fig. 8.2.15 Reverberation impulse response of a listening space.

The sound quality of a concert hall depends on the details of its reverberation impulse response, which depends on the *relative locations* of the sound source and the listener. Therefore, simulating digitally the reverb characteristics of any given hall is an almost impossible task. As a compromise, digital reverb processors attempt to simulate a *typical* reverberation impulse response of a hall, and give the user the option of tweaking some of the parameters, such as the duration of the early reflections (the predelay time), or the overall reverberation time.

Other interesting reverb effects can be accomplished digitally that are difficult or impossible to do by analog means. For example, *gated reverb* is obtained by truncating the IIR response to an FIR one, as shown in Fig. 8.2.16, with a user-selectable gate time. This type of reverb is very effective with snare drums [133]. *Time-reversing* a gated response results in a *reverse reverb* that has no parallel in analog signal processing.

The plain reverb filter shown in Fig. 8.2.6 is too simple to produce a realistic reverberation response. However, as suggested by Schroeder [143], it can be used as the building block of more realistic reverb processors that exhibit the discrete early reflections and the diffuse late ones.

In most applications of DSP, we are interested in the steady state response of our filters. Reverberation is an exception. Here, it is the *transient response* of a hall that gives it its particular reverberation characteristics. The steady-state properties, however, do have an effect on the overall perceived sound.

The peaks in the steady-state spectrum of the plain reverb filter of Eq. (8.2.12), shown in Fig. 8.2.7, tend to accentuate those frequencies of the input signal that are near the peak frequencies. To prevent such coloration of the input sound,

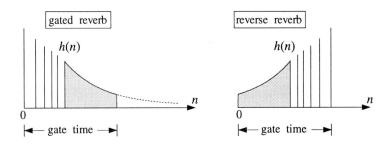

Fig. 8.2.16 Gated and reversed reverberation responses.

Schroeder also proposed [143] an *allpass* version of the plain reverberator that has a *flat* magnitude response for all frequencies:

$$H(z) = \frac{-a + z^{-D}}{1 - az^{-D}} \qquad \text{(allpass reverberator)} \qquad (8.2.25)$$

It has I/O difference equation:

$$y(n) = ay(n - D) - ax(n) + x(n - D) \qquad (8.2.26)$$

Its frequency and magnitude responses are obtained by setting $z = e^{j\omega}$:

$$H(\omega) = \frac{-a + e^{-j\omega D}}{1 - ae^{-j\omega D}} \qquad \Rightarrow \qquad |H(\omega)| = 1, \text{ for all } \omega \qquad (8.2.27)$$

The magnitude response is constant in ω because the numerator and denominator of $H(\omega)$ have the *same* magnitude, as can be seen from the simple identity:

$$|-a + e^{-j\omega D}| = \sqrt{1 - 2a\cos(\omega D) + a^2} = |1 - ae^{-j\omega D}|$$

Although its magnitude response is flat, its transient response exhibits the same exponentially decaying pattern of echoes as the plain reverb. Indeed, the impulse response of Eq. (8.2.25) can be obtained by splitting $H(z)$ into the partial fraction expansion form:

$$H(z) = A + \frac{B}{1 - az^{-D}} \qquad (8.2.28)$$

where $A = -1/a$ and $B = (1 - a^2)/a$. Expanding the B-term into its geometric series, gives

$$H(z) = (A + B) + B(az^{-D} + a^2z^{-2D} + a^3z^{-3D} + \cdots)$$

and taking inverse z-transforms leads to the impulse response:

$$h(n) = (A + B)\delta(n) + Ba\delta(n - D) + Ba^2\delta(n - 2D) + \cdots \qquad (8.2.29)$$

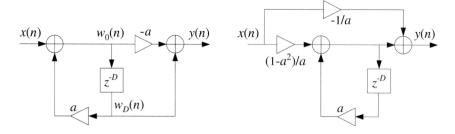

Fig. 8.2.17   Allpass reverberator in canonical and parallel form.

Figure 8.2.17 shows the *canonical* realization of Eq. (8.2.25) realized by a common delay z^{-D}. It also shows the parallel realization of Eq. (8.2.28), which was Schroeder's original realization [143].

The sample processing algorithm of the canonical form, implemented with linear or circular delay lines, is given below:

$$
\boxed{
\begin{array}{l}
\textit{for each input sample x do:} \\
\quad w_0 = x + a w_D \\
\quad y = -a w_0 + w_D \\
\quad \text{delay}(D, \mathbf{w})
\end{array}
}
\qquad
\boxed{
\begin{array}{l}
\textit{for each input sample x do:} \\
\quad s_D = \text{tap}(D, \mathbf{w}, p, D) \\
\quad s_0 = x + a s_D \\
\quad y = -a s_0 + s_D \\
\quad *p = s_0 \\
\quad \text{cdelay}(D, \mathbf{w}, \&p)
\end{array}
}
\qquad (8.2.30)
$$

The circular delay versions of sample processing algorithms of the plain reverberator, Eq. (8.2.14), and the allpass reverberator, Eq. (8.2.30), can be implemented by the following C routines `plain.c` and `allpass.c`:

```
/* plain.c - plain reverberator with circular delay line */

double tap();
void cdelay();

double plain(D, w, p, a, x)            usage: y=plain(D,w,&p,a,x);
double *w, **p, a, x;                   p is passed by address
int D;
{
        double y, sD;

        sD = tap(D, w, *p, D);          Dth tap delay output
        y = x + a * sD;                 filter output
        **p = y;                        delay input
        cdelay(D, w, p);                update delay line

        return y;
}


/* allpass.c - allpass reverberator with circular delay line */
```

```
double tap();
void cdelay();

double allpass(D, w, p, a, x)                usage: y=allpass(D,w,&p,a,x);
double *w, **p, a, x;                         p is passed by address
int D;
{
        double y, s0, sD;

        sD = tap(D, w, *p, D);                Dth tap delay output
        s0 = x + a * sD;
        y  = -a * s0 + sD;                    filter output
        **p = s0;                             delay input
        cdelay(D, w, p);                      update delay line

        return y;
}
```

The linear buffer w is $(D+1)$-dimensional, and the circular pointer p must be initialized to $p = w$, before the first call. The following program segment illustrates their usage:

```
double *w1, *p1;
double *w2, *p2;

w1 = (double *) calloc(D+1, sizeof(double));
w2 = (double *) calloc(D+1, sizeof(double));
p1 = w1; p2 = w2;

for (n=0; n<Ntot; n++) {
        y1[n] = plain(D, w1, &p1, a, x[n]);
        y2[n] = allpass(D, w2, &p2, a, x[n]);
        }
```

The plain and allpass reverberator units can be combined to form more realistic reverb processors. Schroeder's reverberator [143,115,119,137,127,139] consists of several plain units connected in parallel, which are followed by allpass units in cascade, as shown in Fig. 8.2.18. The input signal can also have a direct connection to the output, but this is not shown in the figure.

The implementation of the sample processing reverb algorithm can be carried out with the help of the routines plain and allpass. It is assumed that each unit has its own (D_i+1)-dimensional circular delay-line buffer $\mathbf{w}_i$ and corresponding circular pointer p_i:

$$
\boxed{
\begin{aligned}
&\textit{for each input sample } x \textit{ do:}\\
&\quad x_1 = \text{plain}(D_1, \mathbf{w}_1, \&p_1, a_1, x)\\
&\quad x_2 = \text{plain}(D_2, \mathbf{w}_2, \&p_2, a_2, x)\\
&\quad x_3 = \text{plain}(D_3, \mathbf{w}_3, \&p_3, a_3, x)\\
&\quad x_4 = \text{plain}(D_4, \mathbf{w}_4, \&p_4, a_4, x)\\
&\quad x_5 = b_1 x_1 + b_2 x_2 + b_3 x_3 + b_4 x_4\\
&\quad x_6 = \text{allpass}(D_5, \mathbf{w}_5, \&p_5, a_5, x_5)\\
&\quad y \ = \text{allpass}(D_6, \mathbf{w}_6, \&p_6, a_6, x_6)
\end{aligned}
}
\tag{8.2.31}
$$

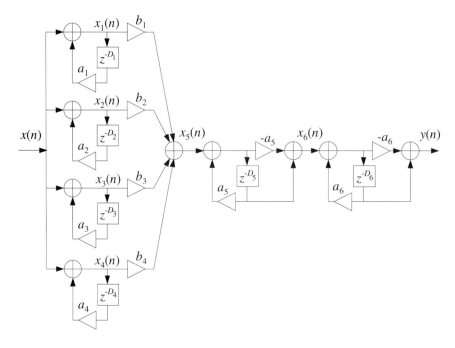

Fig. 8.2.18 Schroeder's reverb processor.

The different delays in the six units cause the density of the reverberating echoes to increase, generating an impulse response that exhibits the typical early and late reflection characteristics. Figure 8.2.19 shows the impulse response of the above filter for the following choices of parameters:

$$D_1 = 29, \ D_2 = 37, \ D_3 = 44, \ D_4 = 50, \ D_5 = 27, \ D_6 = 31$$

$$a_1 = a_2 = a_3 = a_4 = a_5 = a_6 = 0.75$$

$$b_1 = 1, \ b_2 = 0.9, \ b_3 = 0.8, \ b_4 = 0.7$$

Another variation [137,139] of the plain reverb filter of Fig. 8.2.6 is obtained by replacing the simple feedback multiplier a by a nontrivial *lowpass* filter $G(z)$, resulting in the transfer function:

$$H(z) = \frac{1}{1 - z^{-D}G(z)} \qquad \text{(lowpass reverberator)} \qquad (8.2.32)$$

Figure 8.2.20 shows a realization. The presence of the lowpass filter in the feedback loop causes each echo to spread out more and more, resulting in a mellower and more diffuse reverberation response. To see this, expand $H(z)$ using the geometric series formula to get:

$$H(z) = 1 + z^{-D}G(z) + z^{-2D}G^2(z) + z^{-3D}G^3(z) + \cdots$$

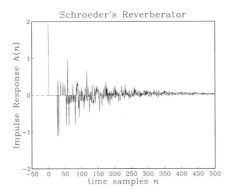

Fig. 8.2.19 Impulse response of Schroeder's reverberator.

giving for the impulse response $h(n)$:

$$h(n) = \delta(n) + g(n-D) + (g * g)(n-2D) + (g * g * g)(n-3D) + \cdots$$

where $g(n)$ is the impulse response of $G(z)$.

It follows that the first echo of the impulse response $h(n)$ at $n = D$ will have the shape of impulse response $g(n)$ the lowpass filter $G(z)$, and will be more spread out than just a single impulse. Similarly, the echo at $n = 2D$ will be the impulse response of $G^2(z)$, which is the convolution $g * g$ of $g(n)$ with itself, and therefore it will be even more spread out than $g(n)$, and so on. The graphs of Fig. 8.2.22 illustrate these remarks.

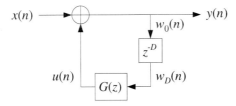

Fig. 8.2.20 Lowpass reverberator.

The feedback filter $G(z)$ can be FIR or IIR. It is described, in general, by the following Mth order transfer function, which also includes the FIR case:

$$G(z) = \frac{N(z)}{D(z)} = \frac{b_0 + b_1 z^{-1} + b_2 z^{-2} + \cdots + b_M z^{-M}}{1 + a_1 z^{-1} + a_2 z^{-2} + \cdots + a_M z^{-M}} \qquad (8.2.33)$$

The filtering operation by $G(z)$ can be implemented by the canonical realization routine can. Assuming a $(D+1)$-dimensional circular buffer **w** for the delay D, and an $(M+1)$-dimensional linear delay-line buffer $\mathbf{v} = [v_0, v_1, \ldots, v_M]$ for $G(z)$, we can write the sample processing algorithm of Eq. (8.2.32), as follows:

$$
\begin{array}{l}
\textit{for each input sample } x \textit{ do:}\\
\quad u = \mathrm{can}\,(M,\mathbf{a},M,\mathbf{b},\mathbf{v},w_D)\\
\quad y = x + u\\
\quad w_0 = y\\
\quad \mathrm{delay}\,(D,\mathbf{w})
\end{array}
\qquad
\begin{array}{l}
\textit{for each input sample } x \textit{ do:}\\
\quad s_D = \mathrm{tap}\,(D,\mathbf{w},p,D)\\
\quad u = \mathrm{can}\,(M,\mathbf{a},M,\mathbf{b},\mathbf{v},s_D)\\
\quad y = x + u\\
\quad *p = y\\
\quad \mathrm{cdelay}\,(D,\mathbf{w},\&p)
\end{array}
\qquad (8.2.34)
$$

where the input to the filter $G(z)$ is the Dth tap output w_D or s_D of the delay line. The following routine `lowpass.c` is an implementation using a circular delay line for D, and a linear delay line and the routine `can` for $G(z)$.

```
/* lowpass.c - lowpass reverberator with feedback filter G(z) */

double tap(), can();
void cdelay();

double lowpass(D, w, p, M, a, b, v, x)
double *w, **p, *a, *b, *v, x;                  v = state vector for G(z)
int D;                                          a, b, v are (M + 1)-dimensional
{
        double y, sD;

        sD = tap(D, w, *p, D);                  delay output is G(z) input
        y = x + can(M, a, M, b, v, sD);         reverb output
        **p = y;                                delay input
        cdelay(D, w, p);                        update delay line

        return y;
}
```

As a simple example, consider the following first-order IIR filter [139] with transfer function:

$$
G(z) = \frac{b_0 + b_1 z^{-1}}{1 + a_1 z^{-1}} = \frac{0.3 + 0.15 z^{-1}}{1 - 0.5 z^{-1}}
\qquad (8.2.35)
$$

and weight vectors $\mathbf{a} = [1, a_1] = [1, -0.5]$ and $\mathbf{b} = [b_0, b_1] = [0.3, 0.15]$.

The corresponding realization of the reverb processor Eq. (8.2.32) is shown in Fig. 8.2.21. The following program segment illustrates the usage `lowpass` for this example:

```
double *w, *p;
double v[2] = {0.0,  0.0};                      G(z) states
double a[2] = {1.0, -0.5};                      G(z) denominator
double b[2] = {0.3,  0.15};                     G(z) numerator

w = (double *) calloc(D+1, sizeof(double));
p = w;

for (n=0; n<Ntot; n++)
        y[n] = lowpass(D, w, &p, M, a, b, v, x[n]);    use M = 1
```

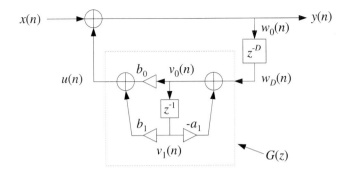

Fig. 8.2.21 Lowpass reverberator, with first-order feedback filter.

Figure 8.2.22 compares the reverberation responses of the plain reverb (8.2.12), allpass reverb (8.2.25), and lowpass reverb (8.2.32) with the loop filter of Eq. (8.2.35), with the parameter values $D = 20$ and $a = 0.75$.

The three inputs were an impulse, a length-5 square pulse, and a length-11 triangular pulse, that is,

$$\mathbf{x} = [1]$$

$$\mathbf{x} = [1, 1, 1, 1, 1]$$

$$\mathbf{x} = [0, 1, 2, 3, 4, 5, 4, 3, 2, 1, 0]$$

The duration of the inputs was chosen to be less than D so that the generated echoes do not overlap, except for the lowpass case in which the echoes become progressively smoother (being successively lowpass filtered) and longer, and eventually will overlap as they decay.

The plain and allpass reverberators have poles that are equidistant from the origin of the unit circle at radius $\rho = a^{1/D}$, and are equally spaced around the circle at the D root-of-unity angles $\omega_k = 2\pi k/D$, $k = 0, 1, \ldots, D - 1$. Therefore, all the poles have the *same* transient response time constants, as given by Eq. (8.2.15).

The reflectivity and absorptivity properties of the walls and air in a real room depend on frequency, with the higher frequencies decaying *faster* than the lower ones.

The lowpass reverberator Eq. (8.2.32) exhibits such frequency-dependent behavior. To understand it, consider the first-order example of Eq. (8.2.35). Its magnitude response $|G(\omega)|$ is shown in Fig. 8.2.23.

The magnitude response and pole locations of the lowpass reverberator (8.2.32) are shown in Fig. 8.2.24. It can be seen that the poles are still approximately equally spaced around the circle, but the high-frequency poles have *shorter* radii and hence *shorter* time constants than the low-frequency ones.

The pole locations of Eq. (8.2.32) are obtained as the roots of the denominator, that is, they are the solutions of

$$z^D = G(z) = \frac{N(z)}{D(z)} \tag{8.2.36}$$

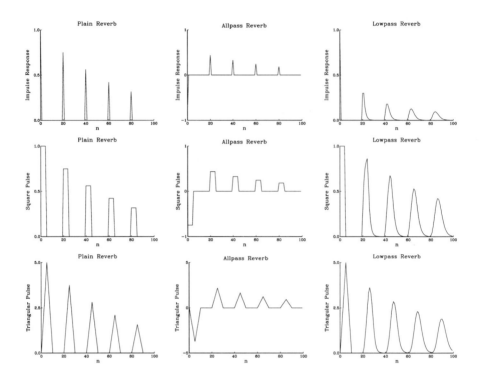

Fig. 8.2.22 Comparison of plain, allpass, and lowpass reverberators.

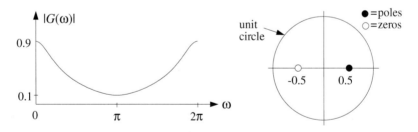

Fig. 8.2.23 Magnitude response of lowpass feedback filter $G(z)$.

For our example, $D(z)$ has order one and therefore Eq. (8.2.36) will have $D+1$ poles, say, p_i, $i = 1, 2, \ldots, D + 1$. Writing p_i in its polar form $p_i = \rho_i e^{j\omega_i}$, we have

$$\rho_i^D e^{j\omega_i D} = G(p_i) = |G(p_i)| e^{j \arg G(p_i)}$$

Defining the *phase delay* of the ith pole by

$$d_i = -\frac{\arg G(p_i)}{\omega_i}$$

we have

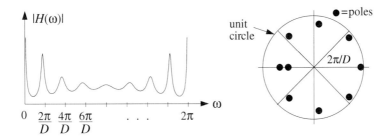

Fig. 8.2.24 Magnitude response of lowpass reverberator, for $D = 8$.

$$\rho_i^D e^{j\omega_i D} = |G(p_i)|e^{-j\omega_i d_i}$$

which can be separated into the two equations:

$$\rho_i^D = |G(p_i)|, \qquad e^{j\omega_i (D+d_i)} = 1$$

and give

$$\rho_i = |G(p_i)|^{1/D}, \qquad \omega_i = \frac{2\pi k_i}{D + d_i} \tag{8.2.37}$$

for some integer k_i.

Although these are coupled equations in the unknowns ρ_i, ω_i, we can see how the angles ω_i will be distributed around the unit circle, near the Dth roots of unity. Similarly, assuming ρ_i is near 1 and replacing $G(p_i) \simeq G(e^{j\omega_i}) = G(\omega_i)$, we have the approximation:

$$\boxed{\rho_i \simeq |G(\omega_i)|^{1/D}} \tag{8.2.38}$$

which by the *lowpass* nature of $G(\omega)$ implies that ρ_i will be *smaller* for higher frequencies ω_i and *larger* for lower ones, in qualitative agreement with the exact pole locations shown in Fig. 8.2.24. Using Eq. (8.2.15), we find for the exact and approximate ϵ-level time constants, in units of the delay time $T_D = TD$:

$$\boxed{\tau_i = \frac{\ln \epsilon}{D \ln \rho_i} T_D \simeq \frac{\ln \epsilon}{\ln |G(\omega_i)|} T_D} \tag{8.2.39}$$

It follows from Eq. (8.2.38) that the stability of the reverb filter $H(z)$, that is, $\rho_i < 1$, will be guaranteed if the feedback filter is normalized such that $|G(\omega)| < 1$, for all ω. Regardless of the above approximation, this condition implies stability by Nyquist's stability criterion or Rouché's theorem [31]. For our example of Eq. (8.2.35), we have $|G(\omega)| \le 0.9$.

Besides the D poles that are approximately equally distributed around the unit circle, there is an extra one that essentially corresponds to the *zero* of the filter $G(z)$. Indeed, for that pole, say p, we have

$$p^D = G(p) = \frac{N(p)}{D(p)}$$

Because p is well inside the unit circle, if D is large, then $p^D \simeq 0$ and therefore, it corresponds to $N(p) \simeq 0$. For our example filter, this extra pole is near the $z = -0.5$ zero of the numerator filter $N(z) = 0.3 + 0.15z^{-1}$.

Table 8.2.1 shows for $D = 8$ the exact poles $p_i = \rho_i e^{j\omega_i}$ of Fig. 8.2.24, their frequencies ω_i and magnitudes ρ_i, as well as the approximate magnitudes given by Eq. (8.2.38), and the exact 60-dB ($\epsilon = 10^{-3}$) time constants τ_i.

The first D pole angles are approximately equal to the Dth root of unity angles. The approximation of Eq. (8.2.38) works well for all but the last pole, which is the one near the zero of $N(z)$.

$p_i = \rho_i e^{j\omega_i}$	ω_i/π	ρ_i	$|G(\omega_i)|^{1/D}$	τ_i/T_D
0.9888	0	0.9888	0.9869	76.594
$0.7282 \pm j0.6026$	± 0.2201	0.9452	0.9412	15.314
$0.1128 \pm j0.8651$	± 0.4587	0.8724	0.8715	6.326
$-0.4866 \pm j0.6303$	± 0.7093	0.7962	0.8047	3.789
-0.6801	1	0.6801	0.7499	2.240
-0.5174	1	0.5174	0.7499	1.310

Table 8.2.1 Reverberator poles and time constants, for $D = 8$.

An alternative way to understand the frequency dependence of the time constants is to look at the input-on and input-off transients and steady-state behavior of the filter $H(z)$ of Eq. (8.2.32). Fig. 8.2.25 compares the plain and lowpass reverberator transient outputs for a sinusoid that is turned on at $n = 0$ and off at $n = 150$. The filter parameters were $D = 30$, $a = 0.75$, and $G(z)$ was given by Eq. (8.2.35). The frequencies of the two sinusoids were $\omega = 0.2\pi$ and $\omega = \pi$ radians/cycle.

At the moment the input is cut off, there are D samples of the sinusoid stored in the delay line. As these samples recirculate around the feedback loop every D samples, they get attenuated effectively by the gain of the loop filter $|G(\omega)|$. For the lowpass reverberator, the loop gain is about 0.9 at low frequencies and 0.1 at high frequencies. Thus, the low-frequency sinusoid dies out slowly, whereas the high-frequency one dies out (and starts up) rapidly, leaving behind the slower but weaker low-frequency mode. For the plain reverberator, both the high- and low-frequency sinusoids die out with the *same* time constants.

Besides its use in reverberation effects, the lowpass reverberator filter (8.2.32) has also been used in computer music to model and synthesize guitar string and drum sounds [108–111]. The Karplus-Strong algorithm [108] for modeling plucked strings uses the following FIR lowpass feedback filter:

$$G(z) = \frac{1}{2}(1 + z^{-1}) \tag{8.2.40}$$

A guitar-string sound is generated by simulating the plucking of the string by

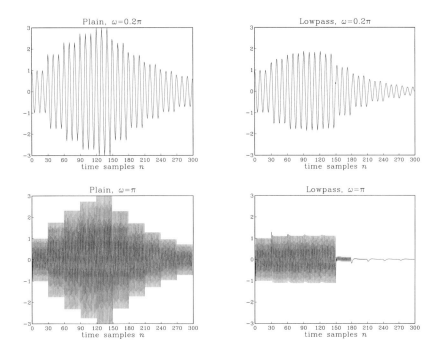

Fig. 8.2.25 Time constants of high- and low-frequency transients, for $D = 30$.

initially filling the delay-line buffer $\mathbf{w} = [w_0, w_1, \ldots, w_D]$ with zero-mean random numbers, and then letting the filter run with zero input. The value D of the delay is chosen to correspond to any desired fundamental frequency f_1, that is, $D = f_s/f_1$.

The recirculating block of random numbers gets lowpass filtered during each pass through the loop filter $G(z)$ and loses its high-frequency content. As a result, the high frequencies in the generated sound decay faster than the low frequencies, as is the case for natural plucked-string sounds.

Physical modeling of instruments is an active research area in computer music. Discrete-time models of the equations describing the physics of an instrument, such as linear or nonlinear wave equations, can be used to generate sounds of the instrument [108–114].

8.2.4 Multitap Delays

Most DSP audio effects processors have built-in a wide class of specialized multiple-delay type effects. They can be obtained from simple low-order FIR or IIR filters by replacing each single unit-delay z^{-1} by the progressively more general substitutions:

$$z^{-1} \longrightarrow z^{-D} \longrightarrow \frac{z^{-D}}{1 - az^{-D}} \longrightarrow \frac{z^{-D}}{1 - z^{-D}G(z)} \tag{8.2.41}$$

which represent a multiple delay, a ringing delay, and a lowpass ringing delay. As a first example, consider the plain ringing delay with transfer function:

$$H(z) = \frac{z^{-D}}{1 - az^{-D}}$$

(8.2.42)

Expanding in powers of z^{-D}, we have

$$H(z) = z^{-D} + az^{-2D} + a^2 z^{-3D} + \cdots$$

The corresponding impulse response will consist of the first delayed impulse $\delta(n - D)$, followed by its successive echoes of exponentially diminishing strength:

$$h(n) = \delta(n - D) + a\delta(n - 2D) + a^2\delta(n - 3D) + \cdots$$

This impulse response and a block diagram realization of Eq. (8.2.42) are shown in Fig. 8.2.26. This is basically the same as the plain reverberator of Fig. 8.2.6, but with the output taken *after* the delay, not before it. Its sample processing algorithm is a variation of Eq. (8.2.14):

for each input sample x do: $\quad y = w_D$ $\quad w_0 = x + aw_D$ $\quad \text{delay}(D, \mathbf{w})$

for each input sample x do: $\quad s_D = \text{tap}(D, \mathbf{w}, p, D)$ $\quad y = s_D$ $\quad *p = x + as_D$ $\quad \text{cdelay}(D, \mathbf{w}, \&p)$

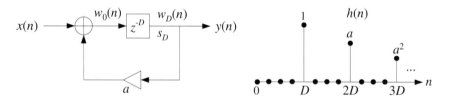

Fig. 8.2.26 Reverberating multi-delay.

As a second example, consider the following second-order FIR filter:

$$H(z) = b_0 + b_1 z^{-1} + b_2 z^{-2}$$

The replacements (8.2.41) lead to the following three multi-delay filters, which are progressively more complicated:

$$H(z) = b_0 + b_1 z^{-D_1} + b_2 z^{-D_1} z^{-D_2}$$

$$H(z) = b_0 + b_1 \left[\frac{z^{-D_1}}{1 - a_1 z^{-D_1}} \right] + b_2 \left[\frac{z^{-D_1}}{1 - a_1 z^{-D_1}} \right] \left[\frac{z^{-D_2}}{1 - a_2 z^{-D_2}} \right]$$

$$H(z) = b_0 + b_1 \left[\frac{z^{-D_1}}{1 - z^{-D_1} G_1(z)} \right] + b_2 \left[\frac{z^{-D_1}}{1 - z^{-D_1} G_1(z)} \right] \left[\frac{z^{-D_2}}{1 - z^{-D_2} G_2(z)} \right]$$

In the last two cases, the reverberating echoes from D_1 are passed into D_2 causing it to reverberate even more densely. Figure 8.2.27 shows the realization of the third case. Its sample processing algorithm can be stated as follows:

$$
\begin{aligned}
&\textit{for each input sample x do:}\\
&\quad y = b_0 x + b_1 w_{1D} + b_2 w_{2D}\\
&\quad u_2 = \mathrm{can}(G_2, w_{2D})\\
&\quad w_{20} = w_{1D} + u_2\\
&\quad \mathrm{delay}(D_2, \mathbf{w}_2)\\
&\quad u_1 = \mathrm{can}(G_1, w_{1D})\\
&\quad w_{10} = x + u_1\\
&\quad \mathrm{delay}(D_1, \mathbf{w}_1)
\end{aligned}
$$

where the statement $u_2 = \mathrm{can}(G_2, w_{2D})$ denotes the generic filtering operation of the filter $G_2(z)$ whose input is w_{2D}, and similarly for $G_1(z)$.

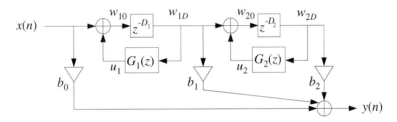

Fig. 8.2.27 Multi-delay effects processor.

Figure 8.2.28 shows the impulse response of such a multi-delay filter, computed by the above sample processing algorithm, with forward taps and delay values:

$$b_0 = 1, \quad b_1 = 0.8, \quad b_2 = 0.6$$

$$D_1 = 30, \quad D_2 = 40$$

and the two cases for the feedback filters:

$$G_1(z) = G_2(z) = 0.75 \qquad \text{(plain)}$$

$$G_1(z) = G_2(z) = \frac{0.3 + 0.15z^{-1}}{1 - 0.5z^{-1}} \qquad \text{(lowpass)}$$

The impulse response exhibits a few early reflections, followed by more dense ones, especially in the lowpass case where successive echoes get spread and overlap more and more with each other. Such multi-delay filters can also be used as preprocessors to reverb units for better modeling of the *early reflection* part of a reverberation response [119,137,127,144].

As a third example, we can start with the simple second-order IIR filter:

$$H(z) = b_0 + \frac{b_1 z^{-1} + b_2 z^{-2}}{1 - a_1 z^{-1} - a_2 z^{-2}}$$

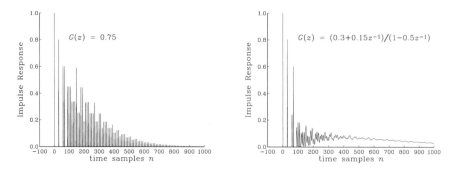

Fig. 8.2.28 Impulse response of plain and lowpass multi-delay.

and replace each single delay z^{-1} by a multiple delay z^{-D}, getting the transfer function:

$$H(z) = b_0 + \frac{b_1 z^{-D_1} + b_2 z^{-D_1 - D_2}}{1 - a_1 z^{-D_1} - a_2 z^{-D_1 - D_2}} \qquad (8.2.43)$$

Its realization is shown in Fig. 8.2.29. It may be thought of as a multitap delay line, tapped at delays D_1 and $D_1 + D_2$. The tap outputs are sent to the overall output and also fed back to the input of the delay line. The b_0 term represents the direct sound.

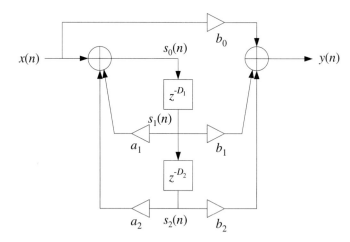

Fig. 8.2.29 Multitap delay effects processor.

Its sample processing algorithm can be implemented with a circular $(D_1 + D_2)$-dimensional delay-line buffer **w** and pointer p, as follows:

> *for each input sample x do:*
> $s_1 = \mathrm{tap}(D_1 + D_2, \mathbf{w}, p, D_1)$
> $s_2 = \mathrm{tap}(D_1 + D_2, \mathbf{w}, p, D_1 + D_2)$
> $y = b_0 x + b_1 s_1 + b_2 s_2$
> $s_0 = x + a_1 s_1 + a_2 s_2$
> $*p = s_0$
> $\mathrm{cdelay}(D_1 + D_2, \mathbf{w}, \&p)$

One potential problem with this arrangement is that the feedback gains can render the filter unstable, if they are taken to be too large. For example, Fig. 8.2.30 shows the impulse response of the filter for the parameter choices

$$b_0 = 1, \quad b_1 = 0.8, \quad b_2 = 0.6$$

$$D_1 = 30, \quad D_2 = 40$$

and for the following two choices of feedback gains, one of which is stable and the other unstable:

$$a_1 = 0.20, \quad a_2 = 0.75 \qquad \text{(stable)}$$
$$a_1 = 0.30, \quad a_2 = 0.75 \qquad \text{(unstable)}$$

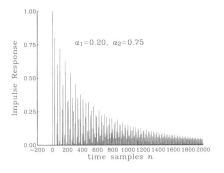 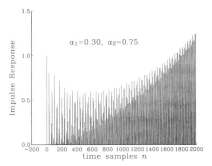

Fig. 8.2.30 Impulse response of multi-tap delay line.

The condition $|a_1| + |a_2| < 1$ guarantees stability by the Nyquist stability criterion or Rouché's theorem [31], because it ensures that $|G(\omega)| < 1$, where $G(z) = a_1 + a_2 z^{-D_2}$.

Typical DSP effects units include both types of delay effects shown in Figures 8.2.27 and 8.2.29, with five or more multiple delay segments and user-selectable feedback and feed-forward multipliers, and delay times D_i adjustable from 0–2000 msec; for example, see [147].

8.2.5 Compressors, Limiters, Expanders, and Gates

Compressors, limiters, expanders, and gates have a wide variety of uses in audio signal processing [124–126,129–132,139,255]. Compressors attenuate strong signals; expanders attenuate weak signals. Because they affect the dynamic range of signals, they are referred to as *dynamics processors.*

Compressors are used mainly to *decrease* the dynamic range of audio signals so that they fit into the dynamic range of the playback or broadcast system; for example, for putting a recording on audio tape. But there are several other applications, such as announcers "ducking" background music, "de-essing" for eliminating excessive microphone sibilance, and other special effects [130].

Expanders are used for *increasing* the dynamic range of signals, for noise reduction, and for various special effects, such as reducing the sustain time of instruments [130].

A typical steady-state input/output relationship for a compressor or expander is as follows, in absolute and decibel units:

$$y = y_0 \left(\frac{x}{x_0} \right)^{\rho} \quad \Rightarrow \quad 20 \log_{10} \left(\frac{y}{y_0} \right) = \rho \, 20 \log_{10} \left(\frac{x}{x_0} \right) \qquad (8.2.44)$$

where x is here a constant input, x_0 a desired threshold, and ρ defines the compression or expansion ratio. A compressor is effective only for $x \geq x_0$ and has $\rho < 1$, whereas an expander is effective for $x \leq x_0$ and has $\rho > 1$. Fig. 8.2.31 shows these relationships in dB, so that a 1 dB change in the input causes ρ dB change in the output, that is, ρ is the slope of the input/output straight lines.

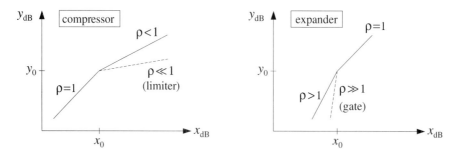

Fig. 8.2.31 Input/output relationship of compressor or expander.

Typical practical values are $\rho = 1/4$–$1/2$ for compression, and $\rho = 2$–4 for expansion. *Limiters* are extreme forms of compressors that prevent signals from exceeding certain maximum thresholds; they have very small slope $\rho \ll 1$, for example, $\rho = 1/10$. *Noise gates* are extreme cases of expanders that infinitely attenuate weak signals, and therefore, can be used to remove weak background noise; they have very large slopes $\rho \gg 1$, for example, $\rho = 10$.

The I/O equation (8.2.44) is appropriate only for constant signals. Writing $y = Gx$, we see that the effective gain of the compressor is a nonlinear function of the input of the form $G = G_0 x^{\rho-1}$. For time-varying signals, the gain must be computed from a *local average* of the signal which is representative of the signal's level.

A model of a compressor/expander is shown in Fig. 8.2.32. The level detector generates a *control signal* c_n that controls the gain G_n of the multiplier through a nonlinear gain processor. Depending on the type of compressor, the control signal may be (1) the instantaneous *peak* value $|x_n|$, (2) the *envelope* of x_n, or (3) the *root-mean-square* value of x_n. A simple model of the envelope detector is as follows:

$$c_n = \lambda c_{n-1} + (1 - \lambda)|x_n| \tag{8.2.45}$$

The difference equation for c_n acts as a *rectifier* followed by a lowpass filter. The time constant of this filter, $n_{\text{eff}} = \ln \epsilon / \ln \lambda$, controls the time to rise or fall to a new input level. The time to rise to a level above the threshold (where the compressor is active) is called the *attack* time constant. This time may be increased further by introducing a delay D in the detector's input, that is, $|x_{n-D}|$. The time to drop to a level below the threshold (where the compressor is inactive) is called the *release* time.

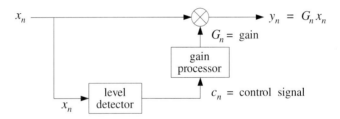

Fig. 8.2.32 Compressor/expander dynamics processor.

For $\lambda = 0$, Eq. (8.2.45) becomes an instantaneous peak detector. This case is useful when the compressor is used as a limiter. If in Eq. (8.2.45) the absolute value $|x_n|$ is replaced by its square, $|x_n|^2$, the control signal will track the *mean-square* value of the input.

The gain processor is a nonlinear function of the control signal imitating the I/O equation (8.2.44). For a compressor, we may define the gain function to be:

$$f(c) = \begin{cases} (c/c_0)^{\rho-1}, & \text{if } c \geq c_0 \\ 1, & \text{if } c \leq c_0 \end{cases} \tag{8.2.46}$$

where c_0 is a desired *threshold* and $\rho < 1$. For an expander, we have $\rho > 1$ and:

$$f(c) = \begin{cases} 1, & \text{if } c \geq c_0 \\ (c/c_0)^{\rho-1}, & \text{if } c \leq c_0 \end{cases} \tag{8.2.47}$$

Thus, the gain G_n and the final output signal y_n are computed as follows:

$$\begin{aligned} G_n &= f(c_n) \\ y_n &= G_n x_n \end{aligned} \tag{8.2.48}$$

Compressors/expanders are examples of *adaptive signal processing* systems, in which the filter coefficients (in this case, the gain G_n) are time-dependent and adapt themselves to the nature of the input signals [27]. The level detector (8.2.45) serves as the "adaptation" equation and its attack and release time constants are the "learning" time constants of the adaptive system; the parameter λ is called the "forgetting factor" of the system [28].

As a simulation example, consider a sinusoid of frequency $\omega_0 = 0.15\pi$ rads per sample whose amplitude changes to the three values $A_1 = 2$, $A_2 = 4$, and $A_3 = 0.5$ every 200 samples, as shown in Fig. 8.2.33, that is, $x_n = A_n \cos(\omega_0 n)$, with:

$$A_n = A_1 (u_n - u_{n-200}) + A_2 (u_{n-200} - u_{n-400}) + A_3 (u_{n-400} - u_{n-600})$$

A compressor is used with parameters $\lambda = 0.9$, $c_0 = 0.5$, and $\rho = 1/2$ (that is, 2:1 compression ratio). The output y_n is shown in Fig. 8.2.33; the control signal c_n and gain G_n in Fig. 8.2.34.

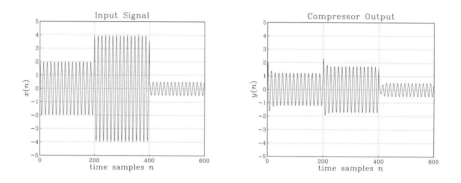

Fig. 8.2.33 Compressor input and output signals ($\rho = 1/2$, $\lambda = 0.9$, $c_0 = 0.5$).

The first two sinusoids A_1 and A_2 lie above the threshold and get compressed. The third one is left unaffected after the release time is elapsed. Although only the stronger signals are attenuated, the overall reduction of the dynamic range will be *perceived* as though the weaker signals also got amplified. This property is the origin of the popular, but somewhat misleading, statement that compressors attenuate strong signals and amplify weak ones.

Jumping between the steady-state levels A_1 and A_2 corresponds to a 6 dB change. Because both levels get compressed, the steady-state output levels will differ by $6\rho = 3$ dB. To eliminate some of the overshoots, an appropriate delay may be introduced in the main signal path, that is, computing the output by $y_n = G_n x_{n-d}$.

Another improvement is to smooth further the nonlinear gain $g_n = f(c_n)$ by a lowpass filter, such as an L-point smoother, so that the final gain is computed by:

$$G_n = \frac{1}{L}[g_n + g_{n-1} + \cdots + g_{n-L+1}] \tag{8.2.49}$$

The overall model for a compressor/expander can be summarized as the following sample processing algorithm, expressed with the help of the routine `fir`:

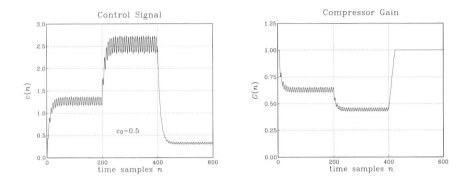

Fig. 8.2.34 Compressor control signal and gain ($\rho = 1/2$, $\lambda = 0.9$, $c_0 = 0.5$).

$$
\boxed{
\begin{aligned}
&\textit{for each input sample x do:} \\
&\quad c = \lambda c_1 + (1 - \lambda)|x| \\
&\quad c_1 = c \\
&\quad G = \mathrm{fir}\,(M, \mathbf{h}, \mathbf{w}, f(c)) \\
&\quad y = Gx
\end{aligned}
}
\tag{8.2.50}
$$

where $\mathbf{h}$ is the impulse response of the L-point smoother, $M = L - 1$ is its order, $\mathbf{w}$ is its L-dimensional delay-line buffer, and c_1 represents c_{n-1}.

Figure 8.2.35 shows the output signal and compressor gain using a seven-point smoother. The initial transients in G_n are caused by the input-on transients of the smoother.

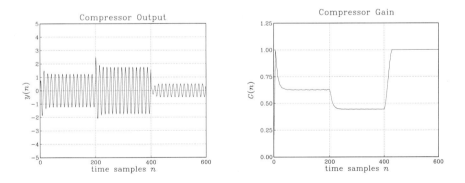

Fig. 8.2.35 Compressor output with smoothed gain ($\rho = 1/2$, $\lambda = 0.9$, $c_0 = 0.5$).

Figure 8.2.36 shows the output signal and compressor gain of a limiter, which has a 10:1 compression ratio, $\rho = 1/10$, and uses also a seven-point smoother. The threshold was increased here to $c_0 = 1.5$, so that only A_2 lies above it and gets compressed.

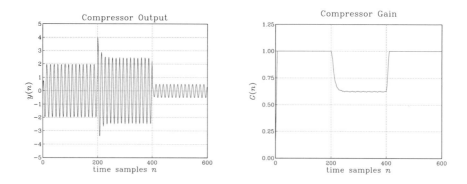

Fig. 8.2.36 Limiter output with smoothed gain ($\rho = 1/10$, $\lambda = 0.9$, $c_0 = 1.5$).

Figure 8.2.37 shows an example of an expander, with parameters $\lambda = 0.9$, $c_0 = 0.5$, $\rho = 2$, and gain function computed by Eq. (8.2.47) and smoothed by a seven-point smoother. Only A_3 lies below the threshold and gets attenuated. This causes the overall dynamic range to increase. Although the expander affects only the weaker signals, the overall increase in the dynamic range is perceived as making the stronger signals louder and the weaker ones quieter.

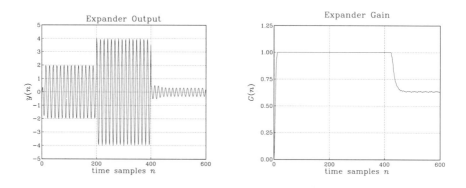

Fig. 8.2.37 Expander output and gain ($\rho = 2$, $\lambda = 0.9$, $c_0 = 0.5$).

Finally, Fig. 8.2.38 shows an example of a noise gate implemented as an expander with a 10:1 expansion ratio, $\rho = 10$, having the same threshold as Fig. 8.2.37. It essentially removes the sinusoid A_3, which might correspond to unwanted noise.

8.3 Noise Reduction and Signal Enhancement

8.3.1 Noise Reduction Filters

One of the most common problems in signal processing is to extract a *desired* signal, say $s(n)$, from a noisy measured signal:

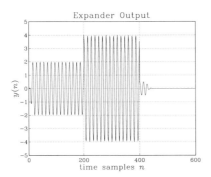
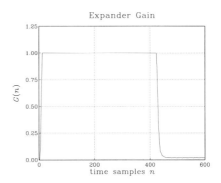

Fig. 8.2.38 Noise gate output and gain ($\rho = 10$, $\lambda = 0.9$, $c_0 = 0.5$).

$$x(n) = s(n) + v(n) \tag{8.3.1}$$

where $v(n)$ is the *undesired* noise component.

The noise signal $v(n)$ depends on the application. For example, it could be (1) a white noise signal, which is typical of the background noise picked up during the measurement process; (2) a periodic interference signal, such as the 60 Hz power-frequency pickup; (3) a low-frequency noise signal, such as radar clutter; (4) any other signal—not necessarily measurement noise—that must be separated from $s(n)$ as, for example, in separating the luminance and chrominance signal components embedded in the composite video signal in a color TV receiver.

The standard method of extracting $s(n)$ from $x(n)$ is to design an appropriate filter $H(z)$ which *removes* the noise component $v(n)$ and at the same time lets the desired signal $s(n)$ go through *unchanged*. Using linearity, we can express the output signal due to the input of Eq. (8.3.1) in the form:

$$y(n) = y_s(n) + y_v(n) \tag{8.3.2}$$

where $y_s(n)$ is the output due to $s(n)$ and $y_v(n)$ the output due to $v(n)$.

The two design conditions for the filter are that $y_v(n)$ be as small as possible and $y_s(n)$ be as similar to $s(n)$ as possible; that is, ideally we require:[†]

$$
\begin{array}{c}
x(n) \\
\xrightarrow{\hspace{2cm}} \boxed{H(z)} \xrightarrow{\hspace{2cm}} \\
s(n)+v(n) \qquad\qquad y_s(n)+y_v(n)
\end{array}
\qquad
\boxed{
\begin{aligned}
y_s(n) &= s(n) \\
y_v(n) &= 0
\end{aligned}
}
\tag{8.3.3}
$$

In general, these conditions cannot be satisfied simultaneously. To determine when they can be satisfied, we express them in the frequency domain in terms of the corresponding frequency spectra as follows: $Y_s(\omega) = S(\omega)$ and $Y_v(\omega) = 0$.

Applying the filtering equation $Y(\omega) = H(\omega)X(\omega)$ separately to the signal and noise components, we have the conditions:

[†]An overall delay in the recovered signal is also acceptable, that is, $y_s(n) = s(n - D)$.

$$Y_s(\omega) = H(\omega)S(\omega) = S(\omega)$$

$$Y_V(\omega) = H(\omega)V(\omega) = 0$$

(8.3.4)

The first requires that $H(\omega) = 1$ at all ω for which the signal spectrum is nonzero, $S(\omega) \neq 0$. The second requires that $H(\omega) = 0$ at all ω for which the noise spectrum is nonzero, $V(\omega) \neq 0$.

These two conditions can be met simultaneously *only if* the signal and noise spectra do *not* overlap, as shown in Fig. 8.3.1. In such cases, the filter $H(\omega)$ must have *passband* that coincides with the signal band, and *stopband* that coincides with the noise band. The filter removes the noise spectrum and leaves the signal spectrum unchanged.

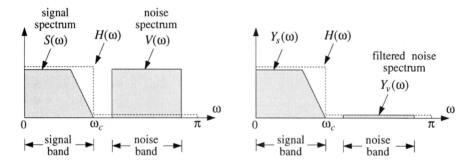

Fig. 8.3.1   Signal and noise spectra before and after filtering.

If the signal and noise spectra overlap, as is the typical case in practice, the above conditions cannot be satisfied simultaneously, because there would be values of ω such that both $S(\omega) \neq 0$ and $V(\omega) \neq 0$ and therefore the conditions (8.3.4) would require $H(\omega) = 1$ and $H(\omega) = 0$ for the same ω.

In such cases, we must compromise between the two design conditions and trade off one for the other. Depending on the application, we may decide to design the filter to remove as much noise as possible, but at the expense of distorting the desired signal. Alternatively, we may decide to leave the desired signal as undistorted as possible, but at the expense of having some noise in the output.

The latter alternative is depicted in Fig. 8.3.2 where a low-frequency signal $s(n)$ exists in the presence of a broadband noise component, such as white noise, having a flat spectrum extending over the entire[†] Nyquist interval, $-\pi \leq \omega \leq \pi$.

The filter $H(\omega)$ is chosen to be an ideal lowpass filter with passband covering the signal bandwidth, say $0 \leq \omega \leq \omega_c$. The noise energy in the filter's stopband $\omega_c \leq \omega \leq \pi$ is removed completely by the filter, thus reducing the strength (i.e., the rms value) of the noise. The spectrum of the desired signal is not affected by the filter, but neither is the portion of the noise spectrum that falls within the signal band. Thus, some noise will survive the filtering process.

The amount of noise reduction achieved by this filter can be calculated using the *noise reduction ratio* (NRR) of Eq. (A.18) of Appendix A.2, which is valid for white

---

[†]For discrete-time signals, the spectra are periodic in ω with period 2π.

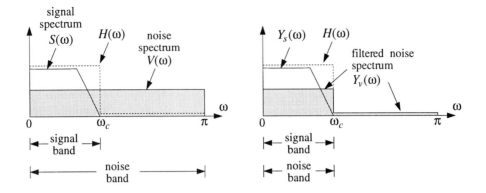

Fig. 8.3.2 Signal enhancement filter with partial noise reduction.

noise input signals. Denoting the input and output *mean-square* noise values by $\sigma_v^2 = E[v(n)^2]$ and $\sigma_{y_v}^2 = E[y_v(n)^2]$, we have

$$
NRR = \frac{\sigma_{y_v}^2}{\sigma_v^2} = \int_{-\pi}^{\pi} |H(\omega)|^2 \frac{d\omega}{2\pi} = \sum_n h_n^2 \tag{8.3.5}
$$

Because $H(\omega)$ is an ideal lowpass filter, the integration range collapses to the filter's passband, that is, $-\omega_c \le \omega \le \omega_c$. Over this range, the value of $H(\omega)$ is unity, giving:

$$
NRR = \frac{\sigma_{y_v}^2}{\sigma_v^2} = \int_{-\omega_c}^{\omega_c} 1 \cdot \frac{d\omega}{2\pi} = \frac{2\omega_c}{2\pi} = \frac{\omega_c}{\pi} \tag{8.3.6}
$$

Thus, the NRR is the *proportion* of the *signal bandwidth* with respect to the Nyquist interval. The same conclusion also holds when the desired signal is a high-frequency or a mid-frequency signal. For example, if the signal spectrum extends only over the mid-frequency band $\omega_a \le |\omega| \le \omega_b$, then $H(\omega)$ can be designed to be unity over this band and zero otherwise. A similar calculation yields in this case:

$$
NRR = \frac{\sigma_{y_v}^2}{\sigma_v^2} = \frac{\omega_b - \omega_a}{\pi} \tag{8.3.7}
$$

The noise reduction/signal enhancement capability of a filter can also be formulated in terms of the signal-to-noise ratio. The SNRs at the input and output of the filter are defined in terms of the mean-square values as:

$$
SNR_{in} = \frac{E[s(n)^2]}{E[v(n)^2]}, \qquad SNR_{out} = \frac{E[y_s(n)^2]}{E[y_v(n)^2]}
$$

Therefore, the relative improvement in the SNR introduced by the filter will be:

$$
\frac{SNR_{out}}{SNR_{in}} = \frac{E[y_s(n)^2]}{E[y_v(n)^2]} \cdot \frac{E[v(n)^2]}{E[s(n)^2]} = \frac{1}{NRR} \cdot \frac{E[y_s(n)^2]}{E[s(n)^2]}
$$

If the desired signal is not changed by the filter, $y_s(n) = s(n)$, then

$$\boxed{\frac{SNR_{\text{out}}}{SNR_{\text{in}}} = \frac{1}{NRR}}$$
(8.3.8)

Thus, *minimizing* the noise reduction ratio is equivalent to *maximizing* the signal-to-noise ratio at the filter's output.

The NRRs computed in Eqs. (8.3.6) or (8.3.7) give the *maximum* noise reductions achievable with *ideal* lowpass or bandpass filters that do not distort the desired signal. Such ideal filters are not realizable because they have double-sided impulse responses with infinite anticausal tails. Thus, in practice, we must use *realizable approximations* to the ideal filters. Chapters 10 and 11 discuss filter design methods that approximate the ideal responses to any desired degree.

The use of realizable noise reduction filters introduces two further design issues that must be dealt with in practice: One is the *transient response* of the filter and the other is the amount of *delay* introduced into the output.

The more closely a filter approximates the sharp transition characteristics of an ideal response, the closer to the unit circle its poles get, and the longer its transient response becomes. Stated differently, maximum noise reduction, approaching the ideal limit (8.3.6), can be achieved only at the expense of introducing long transients in the output.

The issue of the delay introduced into the output has to do with the steady-state response of the filter. We recall from Eq. (6.3.8) of Chapter 6 that after steady state has set in, different frequency components of an input signal suffer different amounts of delay, as determined by the phase delay $d(\omega)$ of the filter.

In particular, if the filter has *linear phase*, then it causes an overall delay in the output. Indeed, assuming that the filter has unity magnitude, $|H(\omega)| = 1$, over its passband (i.e., the signal band) and is zero over the stopband, and assuming a constant phase delay $d(\omega) = D$, we find for the filtered version of the desired signal:

$$y_s(n) = \int_{-\pi}^{\pi} Y_s(\omega) e^{j\omega n} \frac{d\omega}{2\pi} = \int_{-\pi}^{\pi} |H(\omega)| S(\omega) e^{j\omega(n-D)} \frac{d\omega}{2\pi}$$

$$= \int_{-\omega_c}^{\omega_c} S(\omega) e^{j\omega(n-D)} \frac{d\omega}{2\pi} = s(n - D)$$

the last equation following from the inverse DTFT of the desired signal:

$$s(n) = \int_{-\omega_c}^{\omega_c} S(\omega) e^{j\omega n} \frac{d\omega}{2\pi}$$

Essentially all practical FIR noise reduction filters, such as the Savitzky-Golay smoothing filters discussed in Section 8.3.5 and the Kaiser window designs discussed in Section 10.2, have linear phase.

Next, we consider some noise reduction examples based on simple filters, calculate the corresponding noise reduction ratios, discuss the tradeoff between transient response times and noise reduction, and present some simulation examples.

Example 8.3.1: *First-order IIR smoother.* It is desired to extract a constant signal $s(n) = s$ from the noisy measured signal

$$x(n) = s(n) + v(n) = s + v(n)$$

where $v(n)$ is zero-mean white Gaussian noise of variance σ_v^2. To this end, the following IIR lowpass filter is used:

$$H(z) = \frac{b}{1 - az^{-1}}, \quad H(\omega) = \frac{b}{1 - ae^{-j\omega}}, \quad |H(\omega)|^2 = \frac{b^2}{1 - 2a\cos\omega + a^2}$$

where the parameter a is restricted to the range $0 < a < 1$. Because the desired signal $s(n)$ is constant in time, the signal band will only be the DC frequency $\omega = 0$. We require, therefore, that the filter have unity response at $\omega = 0$ or equivalently at $z = 1$. This condition fixes the overall gain b of the filter:

$$H(1) = \frac{b}{1 - a} = 1 \quad \Rightarrow \quad b = 1 - a$$

The NRR of this filter can be calculated from Eq. (8.3.5) by summing the impulse response squared. Here, $h_n = ba^n u(n)$; therefore, using the geometric series, we find

$$\text{NRR} = \frac{\sigma_{y_v}^2}{\sigma_v^2} = \sum_n h_n^2 = b^2 \sum_{n=0}^{\infty} a^{2n} = \frac{b^2}{1 - a^2} = \frac{(1 - a)^2}{1 - a^2} = \frac{1 - a}{1 + a}$$

This ratio is always less than one because a is restricted to $0 < a < 1$. To achieve high noise reduction, a must be chosen near one. But, then the filter's transient time constant, given by Eq. (6.3.12), will become large:

$$n_{\text{eff}} = \frac{\ln \epsilon}{\ln a} \to \infty \quad \text{as} \quad a \to 1$$

The filter's magnitude response, pole-zero pattern, and the corresponding input and output noise spectra are shown in Fig. 8.3.3. The shaded area under the $|H(\omega)|^2$ curve is the same as the NRR computed above.

The filter's 3-dB cutoff frequency ω_c can be calculated by requiring that $|H(\omega_c)|^2$ drops by $1/2$, that is,

$$|H(\omega_c)|^2 = \frac{b^2}{1 - 2a\cos\omega_c + a^2} = \frac{1}{2}$$

which can be solved to give $\cos\omega_c = 1 - (1 - a)^2/2a$. If a is near one, $a \lesssim 1$, we can use the approximation $\cos x \simeq 1 - x^2/2$ and solve for ω_c approximately:

$$\omega_c \simeq 1 - a$$

This shows that as $a \to 1$, the filter becomes a narrower lowpass filter, removing more noise from the input, but at the expense of increasing the time constant.

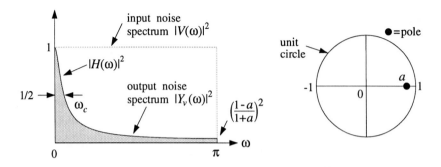

Fig. 8.3.3 Lowpass noise reduction filter of Example 8.3.1.

The tradeoff between noise reduction and speed of response is illustrated in Fig. 8.3.4, where 200 samples of a simulated noisy signal $x(n)$ were filtered using the difference equation of the filter, that is, with $b = 1 - a$

$$y(n) = ay(n-1) + bx(n) \tag{8.3.9}$$

and implemented with the sample processing algorithm, where $w_1(n) = y(n-1)$

> *for each input sample x do:*
> $y = aw_1 + bx$
> $w_1 = y$

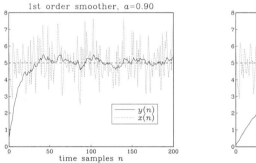
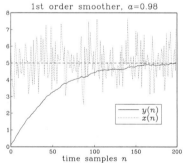

Fig. 8.3.4 Noisy input and smoothed output for Example 8.3.1.

The value of the constant signal was $s = 5$ and the input noise variance $\sigma_v^2 = 1$. The random signal $v(n)$ was generated by successive calls to the Gaussian generator routine gran of Appendix B.1. The figure on the left corresponds to $a = 0.90$, which has 1-percent time constant and NRR:

$$n_{\text{eff}} = \frac{\ln(0.01)}{\ln(0.90)} = 44, \qquad \text{NRR} = \frac{1 - 0.90}{1 + 0.90} = \frac{1}{19}$$

It corresponds to an improvement of the SNR by $10 \log_{10}(1/NRR) = 12.8$ dB. The right figure has $a = 0.98$, with a longer time constant and smaller NRR:

$$n_{\text{eff}} = \frac{\ln(0.01)}{\ln(0.98)} = 228, \qquad NRR = \frac{1 - 0.98}{1 + 0.98} = \frac{1}{99}$$

and an SNR improvement by $10 \log_{10}(1/NRR) = 20$ dB.

To understand how this filter works in the time domain and manages to reduce the noise, we rewrite the difference equation (8.3.9) in its convolutional form:

$$y(n) = b \sum_{m=0}^{n} a^m x(n-m) = b\left(x(n) + ax(n-1) + a^2 x(n-2) + \cdots + a^n x(0)\right)$$

This sum corresponds to the accumulation or averaging of all the past samples up to the present time instant. As a result, the rapid fluctuations of the noise component $v(n)$ are averaged out. The closer a is to 1, the more equal weighting the terms get, resulting in more effective averaging of the noise. The *exponential weighting* deemphasizes the older samples and causes the sum to behave as though it had effectively a finite number of terms, thus, safeguarding the mean-square value of $y(n)$ from diverging. Because of the exponential weights, this filter is also called an *exponential smoother*.

The first-order IIR smoother can be applied to the smoothing of *any* low-frequency signal, not just constants. It is a standard tool in many applications requiring the smoothing of data, such as signal processing, statistics, economics, physics, and chemistry.

In general, one must make sure that the bandwidth of the desired signal $s(n)$ is *narrower* than the filter's lowpass width ω_c, so that the filter will not remove any of the higher frequencies present in $s(n)$. □

Example 8.3.2: *Highpass signal extraction filter.* It is desired to extract a high-frequency signal $s(n) = (-1)^n s$ from the noisy signal

$$x(n) = s(n) + v(n) = (-1)^n s + v(n)$$

where $v(n)$ is zero-mean, white Gaussian noise with variance σ_v^2. Because, the signal band is now at the Nyquist frequency $\omega = \pi$, we may use a first-order *highpass* IIR filter:

$$H(z) = \frac{b}{1 + az^{-1}}, \qquad H(\omega) = \frac{b}{1 + ae^{-j\omega}}, \qquad |H(\omega)|^2 = \frac{b^2}{1 + 2a\cos\omega + a^2}$$

where $0 < a < 1$. The gain b is fixed such that $H(\pi) = 1$, or equivalently $H(z) = 1$ at $z = e^{j\pi} = -1$, which gives the condition:

$$H(-1) = \frac{b}{1-a} = 1 \qquad \Rightarrow \qquad b = 1 - a$$

The impulse response is now $h_n = b(-a)^n u(n)$. The corresponding NRR can be calculated as in the previous example:

$$NRR = \sum_n h_n^2 = b^2 \sum_{n=0}^{\infty} (-a)^{2n} = \frac{b^2}{1-a^2} = \frac{(1-a)^2}{1-a^2} = \frac{1-a}{1+a}$$

The noise reduction frequency characteristics of this highpass filter and its pole/zero pattern are shown in Fig. 8.3.5. Note that the pole is now at $z = -a$. The 3-dB width ω_c is the same as in the previous example.

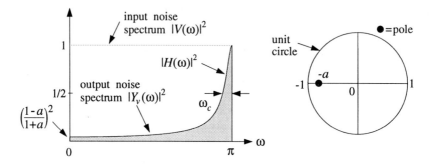

Fig. 8.3.5 Highpass noise reduction filter of Example 8.3.2.

Fig. 8.3.6 shows a simulation of 200 samples $x(n)$ filtered via the difference equation

$$y(n) = -ay(n-1) + (1-a)x(n)$$

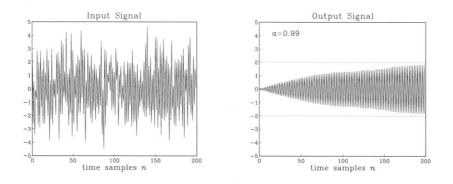

Fig. 8.3.6 Noisy input and high-frequency output for Example 8.3.2.

The following values of the parameters were used: $s = 2$, $a = 0.99$, $\sigma_v^2 = 1$. The corresponding one-percent time constant and NRR are in this case:

$$n_{\text{eff}} = \frac{\ln(0.01)}{\ln(0.99)} = 458, \qquad NRR = \frac{1-0.99}{1+0.99} = \frac{1}{199}$$

which corresponds to an SNR improvement by $10\log_{10}(1/NRR) = 23$ dB.   □

Example 8.3.3: *First-order IIR smoother with prescribed cutoff frequency.* The NRR of Example 8.3.1 can be improved slightly, without affecting the speed of response, by adding a zero in the transfer function at $z = -1$ or equivalently, at $\omega = \pi$. The resulting first-order filter will be:

$$H(z) = \frac{b(1 + z^{-1})}{1 - az^{-1}} \quad \Rightarrow \quad |H(\omega)|^2 = \frac{2b^2(1 + \cos\omega)}{1 - 2a\cos\omega + a^2}$$

where b is fixed by requiring unity gain at DC:

$$H(1) = \frac{2b}{1 - a} = 1 \quad \Rightarrow \quad b = \frac{1 - a}{2}$$

The zero at $\omega = \pi$ suppresses the high-frequency portion of the input noise spectrum even more than the filter of Example 8.3.1, thus, resulting in smaller NRR for the same value of a. The impulse response of this filter can be computed using partial fractions:

$$H(z) = \frac{b(1 + z^{-1})}{1 - az^{-1}} = A_0 + \frac{A_1}{1 - az^{-1}}$$

where

$$A_0 = -\frac{b}{a}, \qquad A_1 = \frac{b(1 + a)}{a}$$

Therefore, the (causal) impulse response will be:

$$h_n = A_0\delta(n) + A_1 a^n u(n)$$

Note, in particular, $h_0 = A_0 + A_1 = b$. It follows that

$$\text{NRR} = \sum_{n=0}^{\infty} h_n^2 = h_0^2 + \sum_{n=1}^{\infty} h_n^2 = b^2 + A_1^2 \frac{a^2}{1 - a^2} = \frac{1 - a}{2}$$

This is slightly smaller than the NRR of Example 8.3.1, because of the inequality:

$$\frac{1 - a}{2} < \frac{1 - a}{1 + a}$$

The 3-dB cutoff frequency can be calculated easily in this example. We have

$$|H(\omega_c)|^2 = \frac{2b^2(1 + \cos\omega_c)}{1 - 2a\cos\omega_c + a^2} = \frac{1}{2}$$

which can be solved for ω_c in terms of a:

$$\cos\omega_c = \frac{2a}{1 + a^2} \qquad\qquad (8.3.10)$$

Conversely, we can solve for a in terms of ω_c:

$$a = \frac{1 - \sin \omega_c}{\cos \omega_c} = \frac{1 - \tan(\omega_c/2)}{1 + \tan(\omega_c/2)} \tag{8.3.11}$$

It is easily checked that the condition $0 < a < 1$ requires that $\omega_c < \pi/2$. We will encounter this example again in Chapter 11 and redesign it using the bilinear transformation. Note also that the replacement $z \to -z$ changes the filter into a highpass one. Such simple first-order lowpass or highpass filters with easily controllable widths are useful in many applications, such as the low- and high-frequency shelving filters of audio equalizers. □

Example 8.3.4: *FIR averaging filters.* The problem of extracting a constant or a low-frequency signal $s(n)$ from the noisy signal $x(n) = s(n) + v(n)$ can also be approached with FIR filters. Consider, for example, the third-order filter

$$H(z) = h_0 + h_1 z^{-1} + h_2 z^{-2} + h_3 z^{-3}$$

The condition that the constant signal $s(n)$ go through the filter unchanged is the condition that the filter have unity gain at DC, which gives the constraint among the filter weights:

$$H(1) = h_0 + h_1 + h_2 + h_3 = 1 \tag{8.3.12}$$

The NRR of this filter will be simply:

$$NRR = \sum_n h_n^2 = h_0^2 + h_1^2 + h_2^2 + h_3^2 \tag{8.3.13}$$

The *best* third-order FIR filter will be the one that *minimizes* this NRR, subject to the lowpass constraint (8.3.12). To solve this minimization problem, we use the constraint to solve for one of the unknowns, say h_3:

$$h_3 = 1 - h_0 - h_1 - h_2$$

Substituting into the NRR, we find

$$NRR = h_0^2 + h_1^2 + h_2^2 + (h_0 + h_1 + h_2 - 1)^2$$

The minimization of this expression can be carried out easily by setting the partial derivatives of *NRR* to zero and solving for the *h*'s:

$$\frac{\partial}{\partial h_0} NRR = 2h_0 + 2(h_0 + h_1 + h_2 - 1) = 2(h_0 - h_3) = 0$$

$$\frac{\partial}{\partial h_1} NRR = 2h_1 + 2(h_0 + h_1 + h_2 - 1) = 2(h_1 - h_3) = 0$$

$$\frac{\partial}{\partial h_2} NRR = 2h_2 + 2(h_0 + h_1 + h_2 - 1) = 2(h_2 - h_3) = 0$$

It follows that all four h's will be equal to each other, $h_0 = h_1 = h_2 = h_3$. But, because they must sum up to 1, we must have the optimum solution:

$$h_0 = h_1 = h_2 = h_3 = \frac{1}{4}$$

and the minimized NRR becomes:

$$NRR_{min} = (\frac{1}{4})^2 + (\frac{1}{4})^2 + (\frac{1}{4})^2 + (\frac{1}{4})^2 = 4 \cdot (\frac{1}{4})^2 = \frac{1}{4}$$

The I/O equation for this optimum smoothing filter becomes:

$$y(n) = \frac{1}{4}(x(n) + x(n-1) + x(n-2) + x(n-3))$$

More generally, the optimum length-N FIR filter with unity DC gain and minimum NRR is the filter with equal weights:

$$\boxed{h_n = \frac{1}{N}, \quad n = 0, 1, \ldots, N-1} \tag{8.3.14}$$

and I/O equation:

$$\boxed{y(n) = \frac{1}{N}(x(n) + x(n-1) + x(n-2) + \cdots + x(n-N+1))} \tag{8.3.15}$$

Its NRR is:

$$\boxed{NRR = h_0^2 + h_1^2 + \cdots + h_{N-1}^2 = N \cdot (\frac{1}{N})^2 = \frac{1}{N}} \tag{8.3.16}$$

Thus, by choosing N large enough, the NRR can be made as small as desired. Again, as the NRR decreases, the filter's time constant increases.

How does the FIR smoother compare with the IIR smoother of Example 8.3.1? First, we note the IIR smoother is very simple computationally, requiring only 2 MACs per output sample, whereas the FIR requires N MACs.

Second, the FIR smoother typically performs better in terms of both the NRR and the transient response, in the sense that for the same NRR value, the FIR smoother has shorter time constant, and for the same time constant, it has smaller NRR.

Given a time constant n_{eff} for an IIR smoother, the "equivalent" FIR smoother should be chosen to have the *same* length, that is,

$$N = n_{eff} = \frac{\ln \epsilon}{\ln a}$$

For example, if $a = 0.90$, then $N = n_{eff} = 44$ as in Example 8.3.1. But then, the NRR of the FIR smoother will be $NRR = 1/N = 1/44$, which is better than that of the IIR filter, $NRR = 1/19$. This case is illustrated in the left graph of Fig. 8.3.7, where the FIR output was computed by Eq. (8.3.15) with $N = 44$, and implemented with the routine

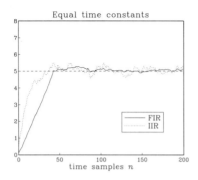 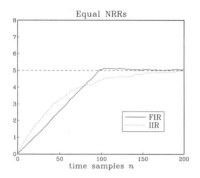

Fig. 8.3.7 Comparison of FIR and IIR smoothing filters.

`cfir` for the same noisy input of Example 8.3.1. The IIR output was already computed in Example 8.3.1.

Similarly, given an IIR smoother that achieves a certain NRR value, the "equivalent" FIR filter with the *same* NRR should have length N such that:

$$NRR = \frac{1-a}{1+a} = \frac{1}{N} \quad \Rightarrow \quad N = \frac{1+a}{1-a}, \quad a = \frac{1-(1/N)}{1+(1/N)}$$

For example, if $a = 0.98$, then we get $N = 99$, which is much shorter than the IIR time constant $n_{\text{eff}} = 228$ computed in Example 8.3.1. The right graph of Fig. 8.3.7 illustrates this case, where the FIR output was computed by Eq. (8.3.15) with $N = 99$.

An approximate relationship between the IIR time constant n_{eff} and N can be derived as follows. Using the small-x approximation $\ln\big((1+x)/(1-x)\big) \simeq 2x$, we have for large N:

$$\ln(1/a) = \ln\left(\frac{1+(1/N)}{1-(1/N)}\right) \simeq \frac{2}{N}$$

It follows that

$$n_{\text{eff}} = \frac{\ln(1/\epsilon)}{\ln(1/a)} \simeq N\,\frac{1}{2}\ln\!\left(\frac{1}{\epsilon}\right)$$

Typically, the factor $(\ln(1/\epsilon)/2)$ is greater than one, resulting in a longer IIR time constant n_{eff} than N. For example, we have:

$$
\begin{aligned}
n_{\text{eff}} &= 1.15\,N, \quad \text{if } \epsilon = 10^{-1} \quad &&(\text{10\% time constant}) \\
n_{\text{eff}} &= 1.50\,N, \quad \text{if } \epsilon = 5 \cdot 10^{-2} \quad &&(\text{5\% time constant}) \\
n_{\text{eff}} &= 2.30\,N, \quad \text{if } \epsilon = 10^{-2} \quad &&(\text{1\% or 40 dB time constant}) \\
n_{\text{eff}} &= 3.45\,N, \quad \text{if } \epsilon = 10^{-3} \quad &&(\text{0.1\% or 60 dB time constant})
\end{aligned}
$$

Finally, we note that a further advantage of the FIR smoother is that it is a *linear phase* filter. Indeed, using the finite geometric series formula, we can write the transfer function of Eq. (8.3.15) in the form:

$$H(z) = \frac{1}{N}\left(1 + z^{-1} + z^{-2} + \cdots + z^{-(N-1)}\right) = \frac{1}{N}\frac{1 - z^{-N}}{1 - z^{-1}} \qquad (8.3.17)$$

Setting $z = e^{j\omega}$, we obtain the frequency response:

$$H(\omega) = \frac{1}{N}\frac{1 - e^{-jN\omega}}{1 - e^{-j\omega}} = \frac{1}{N}\frac{\sin(N\omega/2)}{\sin(\omega/2)}e^{-j\omega(N-1)/2} \qquad (8.3.18)$$

which has a linear phase response. The transfer function (8.3.17) has zeros at the Nth roots of unity, except at $z = 1$, that is,

$$z_k = e^{j\omega_k}, \quad \omega_k = \frac{2\pi k}{N}, \quad k = 1, 2, \ldots, N-1$$

The zeros are distributed equally around the unit circle and tend to suppress the noise spectrum along the Nyquist interval, except at $z = 1$ where there is a pole/zero cancellation and we have $H(z) = 1$.

Fig. 8.3.8 shows the magnitude and phase response of $H(\omega)$ for $N = 16$. Note that the phase response is *piece-wise linear* with slope $(N-1)/2$. It exhibits 180° jumps at $\omega = \omega_k$, where the factor $\sin(N\omega/2)/\sin(\omega/2)$ changes algebraic sign.

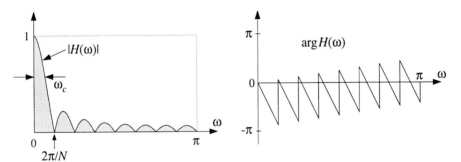

Fig. 8.3.8 Magnitude and phase responses of FIR smoother, for $N = 16$.

The cutoff frequency of the filter can be taken to be approximately *half of the base* of the mainlobe, that is,

$$\omega_c = \frac{\pi}{N}$$

This frequency corresponds to a 3.9 dB drop of the magnitude response. Indeed, setting $\omega = \omega_c = \pi/N$ we have:

$$\left|\frac{1}{N}\frac{\sin(N\pi/2N)}{\sin(\pi/2N)}\right|^2 \simeq \left|\frac{1}{N}\frac{\sin(\pi/2)}{(\pi/2N)}\right|^2 = \left|\frac{2}{\pi}\right|^2$$

where we used the approximation $\sin(\pi/2N) \simeq \pi/2N$, for large N. In decibels, we have $-10\log_{10}\left((2/\pi)^2\right) = 3.9$ dB. (Thus, ω_c is the 3.9-dB frequency.)

Like its IIR counterpart of Example 8.3.1, the FIR averaging filter (8.3.15) can be applied to any low-frequency signal $s(n)$—not just a constant signal. The averaging of the N successive samples in Eq. (8.3.15) tends to smooth out the highly fluctuating noise component $v(n)$, while it leaves the slowly varying component $s(n)$ almost unchanged.

However, if $s(n)$ is not so slowly varying, the filter will also tend to average out these variations, especially when the averaging operation (8.3.15) reaches across many time samples when N is large. In the frequency domain, the same conclusion follows by noting that as N increases, the filter's cutoff frequency ω_c decreases, thus removing more and more of the higher frequencies present in the desired signal.

Thus, there is a limit to the applicability of this type of smoothing filter: Its length must be chosen to be large enough to reduce the noise, but not so large as to start distorting the desired signal by smoothing it too much.

A rough quantitative criterion for the selection of the length N is as follows. If it is known that the desired signal $s(n)$ contains significant frequencies up to a maximum frequency, say ω_{max}, then we may choose N such that $\omega_c \geq \omega_{max}$, which gives $N \leq \pi/\omega_{max}$, and in units of Hz, $N \leq f_s/2f_{max}$.

The FIR smoothing filter (8.3.15) will be considered in further detail in Section 8.3.5 and generalized to include additional linear constraints on the filter weights. Like the IIR smoother, the FIR smoother is widely used in many data analysis applications. □

Example 8.3.5: *Recursive realization of the FIR averager.* The FIR averaging filter can also be implemented in a *recursive form* based on the summed version of the transfer function (8.3.17). For example, the direct form realization of $H(z)$ will be described by the I/O difference equation:

$$y(n) = y(n-1) + \frac{1}{N}(x(n) - x(n-N)) \qquad \text{(direct form)} \qquad (8.3.19)$$

and the *canonical* realization by the system of equations:

$$\begin{aligned} w(n) &= x(n) + w(n-1) \\ y(n) &= \frac{1}{N}(w(n) - w(n-N)) \end{aligned} \qquad \text{(canonical form)} \qquad (8.3.20)$$

These realizations are prone to roundoff accumulation errors and instabilities, and therefore, are not recommended even though they are efficient computationally.

To see the problems that may arise, consider the canonical realization. Assuming that the input $x(n)$ is a white noise signal, the equation $w(n) = w(n-1) + x(n)$ corresponds to the accumulation of $x(n)$ and, as we discuss in Appendix A.2, this causes the mean-square value of $w(n)$ to become unstable. This is unacceptable because $w(n)$ is a required intermediate signal in this realization.

Similarly, considering the direct form realization of Eq. (8.3.19), if $y(n)$ is inadvertently initialized not to zero, but to some other constant, this constant cannot be rid of from the output because it gets canceled from the difference $y(n) - y(n-1)$. Similarly, if the operation $(x(n) - x(n-N))/N$ is done with finite arithmetic precision, which introduces a small roundoff error, this error will get accumulated and eventually grow out of bounds.

The above recursive implementation can be stabilized using the standard procedure of pushing all the marginal poles into the unit circle. The replacement of Eq. (A.21) gives in this case:

$$H(z) = \frac{1}{N} \frac{1 - \rho^N z^{-N}}{1 - \rho z^{-1}}$$

where $0 < \rho \lesssim 1$. If so desired, the filter may be renormalized to unity gain at DC resulting in

$$H(z) = \frac{1 - \rho}{1 - \rho^N} \frac{1 - \rho^N z^{-N}}{1 - \rho z^{-1}}$$

In the limit as $\rho \to 1$, this expression converges to the original filter (8.3.17).

This stabilized version behaves comparably to the first-order smoother of Example 8.3.1. Indeed, if N is taken to be large for a fixed value of ρ, then we can set approximately $\rho^N \simeq 0$. In this limit, the filter reduces to the IIR smoother. □

Example 8.3.6: *Highpass FIR signal extraction.* In general, the substitution $z \to -z$ changes any lowpass filter into a highpass one. It corresponds to a change in the transfer function:

$$H(z) = \sum_n h_n z^{-n} \quad \longrightarrow \quad H(-z) = \sum_n h_n (-z)^{-n} = \sum_n (-1)^n h_n z^{-n}$$

and to the change in the impulse response:

$$h_n \quad \longrightarrow \quad (-1)^n h_n$$

We may think of Example 8.3.2 as being obtained from Example 8.3.1 by this substitution. Similarly, applying the substitution to the lowpass FIR smoother will result into a highpass FIR filter with impulse response:

$$h_n = (-1)^n \frac{1}{N}, \quad n = 0, 1, \dots, N - 1 \tag{8.3.21}$$

and transfer function:

$$H(z) = \frac{1}{N} \sum_{n=0}^{N-1} (-1)^n z^{-n} = \frac{1}{N} \frac{1 - (-1)^N z^{-N}}{1 + z^{-1}}$$

The transfer function has unity gain at $\omega = \pi$, or $z = -1$; indeed,

$$H(-1) = \sum_{n=0}^{N-1} (-1)^n h_n = 1 \tag{8.3.22}$$

The noise reduction ratio remains the same, namely, Eq. (8.3.16). In fact, one can obtain the filter (8.3.21) by minimizing the NRR of (8.3.16) subject to the highpass constraint (8.3.22). □

Example 8.3.7: *Bandpass signal extraction.* A noisy sinusoid of frequency $f_0 = 500$ Hz is sampled at a rate of $f_s = 10$ kHz:

$$x(n) = s(n) + v(n) = \cos(\omega_0 n) + v(n)$$

where $\omega_0 = 2\pi f_0/f_s$, and $v(n)$ is a zero-mean, unit-variance, white Gaussian noise signal. The sinusoid can be extracted by a simple resonator filter of the type discussed in Section 6.4.2. The poles of the filter are placed at $z = Re^{\pm j\omega_0}$, as shown in Fig. 6.4.2.

If R is near 1, the resonator's 3-dB width given by Eq. (6.4.4), $\Delta\omega = 2(1 - R)$, will be small, resulting in a very narrow bandpass filter. The narrower the filter, the more the noise will be reduced. The transfer function and impulse response of the filter were derived in Section 6.4.2:

$$H(z) = \frac{G}{1 + a_1 z^{-1} + a_2 z^{-2}}, \qquad h_n = \frac{G}{\sin\omega_0} R^n \sin(\omega_0 n + \omega_0) u(n)$$

where $a_1 = -2R\cos\omega_0$ and $a_2 = R^2$.

The gain G is adjusted such that the filter's magnitude response is unity at the sinusoid's frequency, that is, $|H(\omega_0)| = 1$. In Section 6.4.2, we found

$$G = (1 - R)\sqrt{1 - 2R\cos(2\omega_0) + R^2}$$

The NRR can be calculated in closed form:

$$NRR = \sum_{n=0}^{\infty} h_n^2 = \frac{(1 - R)(1 + R^2)(1 - 2R\cos(2\omega_0) + R^2)}{(1 + R)(1 - 2R^2\cos(2\omega_0) + R^4)} \qquad (8.3.23)$$

For $R = 0.99$ and $\omega_0 = 0.1\pi$, we have $NRR = 1/99.6$, and filter parameters $a_1 = -1.8831$, $a_2 = 0.9801$, and $G = 6.1502 \times 10^{-3}$. Fig. 8.3.9 shows 300 samples of the noisy sinusoidal input $x(n)$ and the corresponding output signal $y(n)$ plotted together with desired sinusoid $s(n)$. The noise $v(n)$ was generated by the routine gran. The output was computed by the sample processing algorithm of the filter:

$$\begin{array}{l}
\textit{for each input sample x do:}\\
\quad y = -a_1 w_1 - a_2 w_2 + Gx\\
\quad w_2 = w_1\\
\quad w_1 = y
\end{array}$$

The recovered sinusoid is slightly shifted with respect to $s(n)$ by an amount corresponding to the phase delay of the filter at $\omega = \omega_0$, that is, $d(\omega_0) = -\arg H(\omega_0)/\omega_0$. For the given numerical values, we find $d(\omega_0) = 3.95$ samples.   □

8.3.2 Notch and Comb Filters

Two special cases of the signal enhancement/noise reduction problem arise when:

1. The noise signal $v(n)$ in Eq. (8.3.1) is *periodic*. Its spectrum is concentrated at the harmonics of a fundamental frequency. The noise reduction filter is an ideal *notch filter* with notches at these harmonics, as shown in Fig. 8.3.10.

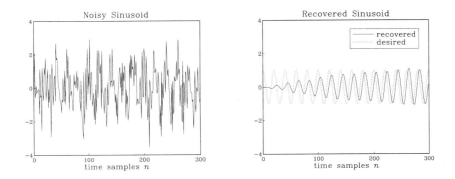

Fig. 8.3.9 Noisy sinusoidal input and extracted sinusoid.

2. The desired signal $s(n)$ is *periodic* and the noise is a wideband signal. Now, the signal enhancement filter is an ideal *comb filter* with peaks at the harmonics of the desired signal, as shown in Fig. 8.3.11.

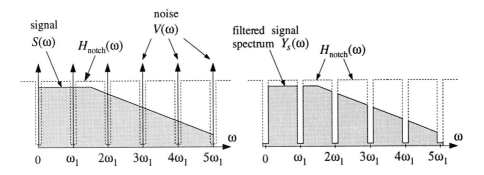

Fig. 8.3.10 Notch filter for reducing periodic interference.

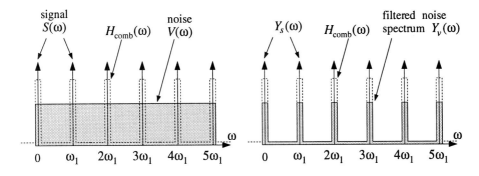

Fig. 8.3.11 Comb filter for enhancing periodic signal.

The ideal notch and comb filters of Figs. 8.3.10 and 8.3.11 are *complementary filters*, in the sense that one is zero where the other is one, so that their frequency responses add up to unity:

$$H_{\text{notch}}(\omega) + H_{\text{comb}}(\omega) = 1 \tag{8.3.24}$$

A typical application of the notch case is the 60 Hz power-frequency interference picked up through insufficiently shielded instrumentation. This problem is especially acute in biomedical applications, such as measuring an electrocardiogram (ECG) by chest electrodes—a procedure which is prone to such interference. The literature on biomedical applications of DSP is extensive, with many filter design methods that are specialized for efficient microprocessor implementations [152–170].

Let f_1 be the fundamental frequency of the periodic noise (e.g., $f_1 = 60$ Hz), or, in radians per sample $\omega_1 = 2\pi f_1/f_s$. If only *one* notch at f_1 must be canceled, then a single-notch filter, such as that given in Eqs. (8.2.22) and (8.2.23) of Section 8.2.2, will be adequate.

Example 8.3.8: *Single-notch filter for ECG processing.* It is desired to design a single-notch filter to cancel the 60 Hz power-frequency pickup in an ECG recording. The ECG is sampled at a rate of 1 kHz, and we assume that the beat rate is 2 beats/sec. Thus, there are 500 samples in each beat.

The digital notch frequency will be:

$$\omega_1 = \frac{2\pi f_1}{f_s} = \frac{2\pi\,60}{1000} = 0.12\pi \text{ radians/sample}$$

Assuming a Q-factor of 60 for the notch filter, we have a 3-dB width:

$$\Delta f = \frac{f_1}{Q} = 1 \text{ Hz} \quad\Rightarrow\quad \Delta\omega = \frac{2\pi\Delta f}{f_s} = 0.002\pi \text{ radians/sample}$$

Using the design equations (8.2.22) and (8.2.23), we find the notch filter:

$$H(z) = 0.99687\,\frac{1 - 1.85955z^{-1} + z^{-2}}{1 - 1.85373z^{-1} + 0.99374z^{-2}}$$

Figure 8.3.12 shows three beats of a simulated ECG with 60 Hz noise generated by

$$x(n) = s(n) + 0.5\cos(\omega_1 n), \quad n = 0, 1, \ldots, 1500$$

The ECG signal $s(n)$ was normalized to maximum value of unity (i.e., unity QRS-peak). Thus, the noise amplitude is 50% the QRS-peak amplitude. Fig. 8.3.12 shows the noisy signal $x(n)$ and the notch filter's magnitude characteristics. The filtered signal $y(n)$ is juxtaposed next to the noise-free ECG for reference.

Except for the initial transients, the filter is very effective in removing the noise. The filter's time constant can be estimated from the filter pole radius, that is, from the last denominator coefficient $a_2 = R^2$:

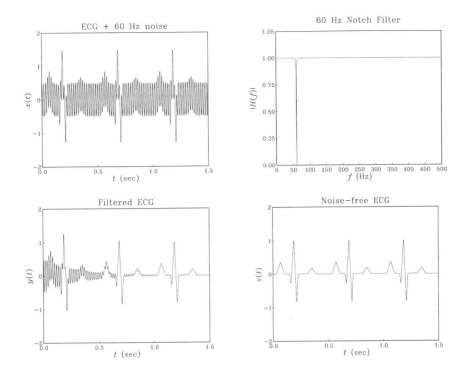

Fig. 8.3.12 Removing 60 Hz noise from ECG signal.

$$R^2 = 0.99374 \quad \Rightarrow \quad R = 0.99686$$

which gives for the 1% time constant $n_{\text{eff}} = \ln(0.01)/\ln(R) = 1464$ samples. In seconds, this is $\tau = n_{\text{eff}}T = 1464/1000 = 1.464$ sec, where $T = 1/f_s = 1$ msec. □

If *all* the harmonics of f_1 must be canceled, then it proves convenient to choose the sampling rate to be a *multiple* of the fundamental frequency, that is, $f_s = Df_1$. Then, the noise harmonics will occur at the Dth roots-of-unity frequencies:

$$f_k = kf_1 = k\frac{f_s}{D}, \quad \text{or,} \quad \omega_k = k\omega_1 = \frac{2\pi k}{D}, \qquad k = 0, 1, \ldots, D-1 \qquad (8.3.25)$$

In this case, we can use the general procedure discussed in Section 6.4.3. The notch polynomial having as roots the Dth roots of unity, $z_k = e^{j\omega_k} = e^{2\pi jk/D}$ is given by $N(z) = 1 - z^{-D}$. The corresponding multi-notch filter is obtained by *sharpening* the zeros, that is, putting poles *behind* the zeros by the replacement $z \to z/\rho$:

$$\boxed{H_{\text{notch}}(z) = \frac{bN(z)}{N(\rho^{-1}z)} = b\,\frac{1 - z^{-D}}{1 - az^{-D}}}, \qquad b = \frac{1+a}{2} \qquad (8.3.26)$$

where $a = \rho^D$.

The choice $b = (1 + a)/2$ ensures that $H(z)$ is normalized to unity half-way between the notches, that is, at $\omega_k = (2k + 1)\pi/D$.

The value of the parameter a depends on the desired 3-dB width $\Delta\omega = 2\pi\Delta f/f_s$ of the notch dips. Using the bilinear transformation method[†] of Chapter 11, we obtain the following design equations, for a given $\Delta\omega$:

$$\boxed{\beta = \tan\left(\frac{D\Delta\omega}{4}\right), \quad a = \frac{1 - \beta}{1 + \beta}, \quad b = \frac{1}{1 + \beta}} \qquad (8.3.27)$$

Because a must be in the interval $0 \le a < 1$, we find the restriction $0 < \beta \le 1$, which translates into the following bound on the desired 3-dB width: $D\Delta\omega/4 \le \pi/4$, or

$$\Delta\omega \le \frac{\pi}{D} \quad \Rightarrow \quad \Delta f \le \frac{f_s}{2D} \qquad (8.3.28)$$

Its maximum value $\Delta\omega = \pi/D$ corresponds to $\beta = 1$ and $a = 0$.

It follows from the bilinear transformation that the magnitude response squared of the filter (8.3.26) can be written in the simple form:

$$\boxed{|H_{\text{notch}}(\omega)|^2 = \frac{\tan^2(\omega D/2)}{\tan^2(\omega D/2) + \beta^2}} \qquad (8.3.29)$$

Example 8.3.9: *Multi-notch filter design.* As an example of the above design method, consider the case $D = 10$. According to Eq. (8.3.28), the 3-dB width is allowed to be in the range $0 < \Delta\omega \le \pi/D = 0.1\pi$.

Using Eqs. (8.3.27), we design the filter for the following three values of $\Delta\omega$:

$$\begin{array}{llll}
\Delta\omega = 0.1\pi & \Rightarrow \quad \beta = 1 & a = 0 & b = 0.5 \\
\Delta\omega = 0.05\pi & \Rightarrow \quad \beta = 0.4142 & a = 0.4142 & b = 0.7071 \\
\Delta\omega = 0.0125\pi & \Rightarrow \quad \beta = 0.0985 & a = 0.8207 & b = 0.9103
\end{array}$$

corresponding to the three transfer functions:

$$H_{\text{notch}}(z) = 0.5(1 - z^{-10})$$

$$H_{\text{notch}}(z) = 0.7071 \frac{1 - z^{-10}}{1 - 0.4142 z^{-10}}$$

$$H_{\text{notch}}(z) = 0.9103 \frac{1 - z^{-10}}{1 - 0.8207 z^{-10}}$$

The magnitude squared responses, $|H_{\text{notch}}(\omega)|^2$, are plotted in Fig. 8.3.13. □

[†]Here, the highpass analog filter $s/(s + \beta)$ is transformed by $s = (1 - z^{-D})/(1 + z^{-D})$.

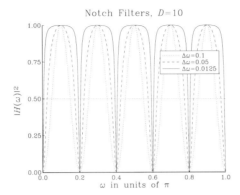

Fig. 8.3.13 Multi-notch filters of different widths ($\Delta\omega$ in units of π).

Example 8.3.10: *Multi-notch filter for ECG processing.* To illustrate the filtering operation with the notch filter of the type (8.3.26), consider the following simulated ECG signal, generated by adding a 60 Hz *square wave* to two beats of a noise-free ECG:

$$x(n) = s(n) + v(n), \qquad n = 0, 1, \dots, 1199$$

where $s(n)$ is a one-beat-per-second ECG signal sampled at a rate of 600 Hz, thus, having 600 samples per beat. The QRS-peak is normalized to unity as in Example 8.3.8. The square wave noise signal $v(n)$ has period

$$D = \frac{f_s}{f_1} = \frac{600}{60} = 10$$

and is defined as follows:

$$v(n) = 1 + 0.5w(n), \qquad w(n) = [1, 1, 1, 1, 1, -1, -1, -1, -1, -1, \dots]$$

where $w(n)$ alternates between $+1$ and -1 every 5 samples. The alternating square wave $w(n)$ has only odd harmonics, namely, f_1, $3f_1$, $5f_1$, and so on.

This particular $v(n)$ is used here only for illustrating the behavior of a multi-notch filter and does not represent any real-life noise. The nonzero mean of $v(n)$ is meant to imitate a typical *baseline shift* that can occur in ECG measurements, in addition to the 60 Hz harmonics.

The filter (8.3.26) was designed by requiring that its Q-factor be 80. This gives the 3-dB width:

$$\Delta f = \frac{f_1}{Q} = \frac{60}{80} = 0.75 \text{ Hz} \quad \Rightarrow \quad \Delta\omega = \frac{2\pi\Delta f}{f_s} = 0.0025\pi \text{ rads/sample}$$

The design equations (8.3.27) give: $a = 0.9615$, $b = 0.9807$, with a resulting transfer function:

$$H_{\text{notch}}(z) = 0.9807 \, \frac{1 - z^{-10}}{1 - 0.9615z^{-10}}$$

Its magnitude response is similar to those of Fig. 8.3.13, but narrower. It can be implemented in its canonical form using the 11-dimensional vector of internal states $\mathbf{w} = [w_0, w_1, \ldots, w_{10}]$ by the sample processing algorithm:

for each input sample x do:
$$w_0 = 0.9615\,w_{10} + 0.9807\,x$$
$$y = w_0 - w_{10}$$
delay(10, **w**)

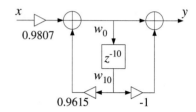

Fig. 8.3.14 shows the input $x(n)$ and the filtered output $y(n)$. To improve the visibility of the graphs, the two beats, for $0 \le t \le 1$ sec and $1 \le t \le 2$ sec, have been split into two side-by-side graphs.

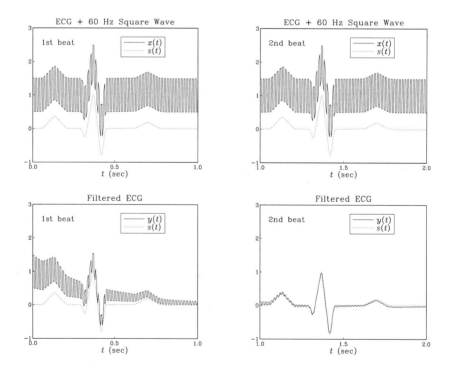

Fig. 8.3.14   Eliminating baseline shifts and 60 Hz harmonics from ECG.

Notice how the filter's zero at DC eliminates the baseline shift, while its notches at the 60 Hz harmonics eliminate the alternating square wave.

A single-notch filter at 60 Hz would not be adequate in this case, because it would not remove the DC and the higher harmonics of the noise. For example, using the method of Example 8.3.8, the single notch filter with the same $Q = 80$ and width Δf as the above multi-notch filter, is found to be:

$$H_1(z) = 0.99609 \, \frac{1 - 1.61803z^{-1} + z^{-2}}{1 - 1.61170z^{-1} + 0.99218z^{-2}}$$

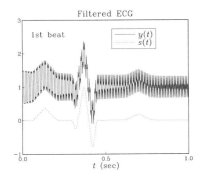 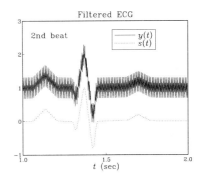

Fig. 8.3.15 Inadequate filtering by the single-notch filter $H_1(z)$.

Fig. 8.3.15 shows the filtered ECG in this case. Only the 60 Hz harmonic of the square wave noise is removed. The DC and the higher harmonics at $3f_1$, $5f_1$, and so on, are still in the output.

Actually, for this example, the highest harmonic is $5f_1$ and coincides with the Nyquist frequency $5f_1 = f_s/2 = 300$ Hz. Because D is even, all the higher odd harmonics will be aliased with one of the three odd Nyquist-interval harmonics: f_1, $3f_1$, or, $5f_1$.

We can attempt to cancel these additional harmonics by designing separate notch filters for each one. For example, using a *common* width $\Delta f = 0.75$ Hz, we design the following notch filters:

$$H_3(z) = 0.99609 \, \frac{1 + 0.61803z^{-1} + z^{-2}}{1 + 0.61562z^{-1} + 0.99218z^{-2}} \qquad \text{(notch at } 3f_1)$$

$$H_5(z) = 0.99609 \, \frac{1 + z^{-1}}{1 + 0.99218z^{-1}} \qquad \text{(notch at } 5f_1)$$

$$H_0(z) = 0.99609 \, \frac{1 - z^{-1}}{1 - 0.99218z^{-1}} \qquad \text{(notch at } 0)$$

Fig. 8.3.16 shows the output of the cascaded filter $H_1(z)H_3(z)H_5(z)$, which removes completely all the harmonics in the square wave, except DC. That can be removed by sending the output through the DC notch filter $H_0(z)$.

Note that the notch filters at DC and Nyquist frequency are first-order filters of the form:

$$H_0(z) = b \, \frac{1 - z^{-1}}{1 - az^{-1}}, \qquad H_5(z) = b \, \frac{1 + z^{-1}}{1 + az^{-1}}$$

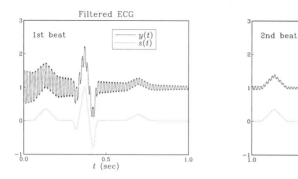

Fig. 8.3.16 Output from the cascade filter $H_1(z)H_3(z)H_5(z)$.

These are limiting cases of the designs of Eq. (8.2.22) for $\omega_0 = 0$ and $\omega_0 = \pi$. In both cases, $b = 1/(1 + \tan(\Delta\omega/2))$ and $a = 2b - 1$.   □

The multi-notch filter (8.3.26) can be obtained from the first-order filter $H(z) = b(1 - z^{-1})/(1 - az^{-1})$ by the substitution

$$\boxed{z \longrightarrow z^D} \tag{8.3.30}$$

that is,

$$H(z) = b\,\frac{1 - z^{-1}}{1 - az^{-1}} \quad \longrightarrow \quad H(z^D) = b\,\frac{1 - z^{-D}}{1 - az^{-D}} \tag{8.3.31}$$

The effect of this substitution is the D-fold replication of the spectrum of the original filter. Indeed, in the frequency domain Eq. (8.3.30) gives:

$$H(\omega) \longrightarrow H(\omega D),$$

This transformation *shrinks* the original spectrum $H(\omega)$ by a factor of D and replicates it D times. Because the spectrum $H(\omega)$ has period $0 \leq \omega \leq 2\pi$, the new spectrum $H(\omega D)$ will have period $0 \leq \omega D \leq 2\pi$, which becomes the scaled period $0 \leq \omega \leq 2\pi/D$ fitting exactly D times into the new Nyquist interval $0 \leq \omega \leq 2\pi$.

The first-order filter in Eq. (8.3.31) has a single notch at $\omega = 0$, which gets replicated D times and becomes a multi-notch filter.

The replicating transformation (8.3.30) can also be applied to any *narrow low-pass* filter, replicating it D times into a *comb filter*. For example, applying it to the filter of Example 8.3.1 we get:

$$H(z) = \frac{1 - a}{1 - az^{-D}} \tag{8.3.32}$$

which has a comb structure similar to that of the plain reverberator shown in Fig. 8.2.7. Similarly, the transformation (8.3.30) applied to the filter of Example 8.3.3, gives the following comb filter with unity-gain peaks at $\omega_k = 2k\pi/D$ and zeros at $\omega_k = (2k+1)\pi/D$:

$$H_{\text{comb}}(z) = b\,\frac{1+z^{-D}}{1-az^{-D}}, \qquad b = \frac{1-a}{2} \tag{8.3.33}$$

This filter can also be designed directly using the bilinear transformation method[†] of Chapter 11. Given a prescribed 3-dB width for the peaks, $\Delta\omega$, the filter parameters can be calculated from the design equations:

$$\beta = \tan\left(\frac{D\Delta\omega}{4}\right), \quad a = \frac{1-\beta}{1+\beta}, \quad b = \frac{\beta}{1+\beta} \tag{8.3.34}$$

where, as in Eq. (8.3.27), the width is constrained to be in the interval: $0 \le \Delta\omega \le \pi/D$. Like Eq. (8.3.29), the magnitude response squared of (8.3.33) can be expressed simply in the form:

$$|H_{\text{comb}}(\omega)|^2 = \frac{\beta^2}{\tan^2(\omega D/2)+\beta^2} \tag{8.3.35}$$

The comb and notch filters of Eqs. (8.3.33) and (8.3.26) are complementary in the sense of Eq. (8.3.24); indeed, we have the identity in z:

$$H_{\text{comb}}(z)+H_{\text{notch}}(z) = \frac{1-a}{2}\frac{1+z^{-D}}{1-az^{-D}} + \frac{1+a}{2}\frac{1-z^{-D}}{1-az^{-D}} = 1$$

It follows by inspecting Eqs. (8.3.29) and (8.3.35) that their magnitude responses squared also add up to one:

$$|H_{\text{comb}}(\omega)|^2 + |H_{\text{notch}}(\omega)|^2 = 1 \tag{8.3.36}$$

This implies that both filters have the same width, as seen in the design equations (8.3.27) and (8.3.34). But, how is it possible to satisfy simultaneously Eq. (8.3.36) and $H_{\text{comb}}(\omega)+H_{\text{notch}}(\omega) = 1$? This happens because their phase responses differ by 90°. Indeed, it is left as an exercise to show that:

$$H_{\text{comb}}(\omega) = j\,H_{\text{notch}}(\omega)\tan(\omega D/2)/\beta$$

Example 8.3.11: *Comb filter design.* As a design example, consider the case $D = 10$. Using Eqs. (8.3.34), we design the filter for the following three values of $\Delta\omega$:

$$\begin{array}{llll}
\Delta\omega = 0.1\pi & \Rightarrow \quad \beta = 1 & a = 0 & b = 0.5 \\
\Delta\omega = 0.05\pi & \Rightarrow \quad \beta = 0.4142 & a = 0.4142 & b = 0.2929 \\
\Delta\omega = 0.0125\pi & \Rightarrow \quad \beta = 0.0985 & a = 0.8207 & b = 0.0897
\end{array}$$

corresponding to the three transfer functions:

[†]Here, the lowpass analog filter $\beta/(s+\beta)$ is transformed by $s = (1-z^{-D})/(1+z^{-D})$.

$$H_{\text{comb}}(z) = 0.5\,(1 + z^{-10})$$

$$H_{\text{comb}}(z) = 0.2929\,\frac{1 + z^{-10}}{1 - 0.4142z^{-10}}$$

$$H_{\text{comb}}(z) = 0.0897\,\frac{1 + z^{-10}}{1 - 0.8207z^{-10}}$$

The magnitude squared responses, $|H(\omega)|^2$, are plotted in Fig. 8.3.17.   □

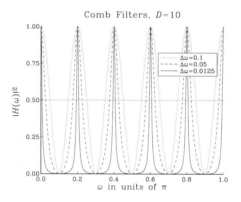

Fig. 8.3.17 Comb filters of different widths ($\Delta\omega$ in units of π).

Either comb filter, (8.3.32) or (8.3.33), can be used to enhance a periodic signal buried in white noise. Their NRRs are the *same* as those of Examples 8.3.1 and 8.3.3, that is, $NRR = (1 - a)/(1 + a)$ and $NRR = (1 - a)/2$, respectively. This follows from the property that the substitution (8.3.30) leaves the NRR *unchanged*. Indeed, in the time domain the transformation is equivalent to inserting $D-1$ zeros between the original impulse response samples:

$$\mathbf{h} = [h_0, h_1, h_2, \dots\,] \longrightarrow \mathbf{h} = [h_0, 0, 0, \dots, 0, h_1, 0, 0, \dots, 0, h_2, \dots\,] \quad (8.3.37)$$

and, therefore, the quantity $\sum h_n^2$ remains invariant.

Example 8.3.12: *Comb filter for periodic signal enhancement.* To illustrate the noise reduction capability of the comb filter (8.3.33), consider the following signal of length 2000:

$$x(n) = s(n) + v(n), \qquad n = 0, 1, \dots, 1999,$$

where $s(n)$ is a periodic *triangular* wave of period $D = 50$, linearly alternating between ± 1 every 25 samples. Thus, there are 40 periods in $x(n)$. The noise signal

$v(n)$ is a zero-mean, white Gaussian noise of rms amplitude equal to 0.5, that is, 50 percent of the triangular wave.

The width of the comb filter is chosen to be $\Delta\omega = 0.0008\pi$ radians/sample, with corresponding Q-factor of:

$$Q = \frac{\omega_1}{\Delta\omega} = \frac{2\pi/D}{\Delta\omega} = 50$$

Using the design equations (8.3.34), we find $a = 0.9391$ and $b = 0.0305$, and the transfer function:

$$H_{\text{comb}}(z) = 0.0305 \frac{1 + z^{-50}}{1 - 0.9391z^{-50}}$$

The filter's magnitude response $|H_{\text{comb}}(\omega)|^2$ and its canonical realization are shown in Fig. 8.3.18. The peaks are at $\omega_k = 2k\pi/50$, and the zeros at $\omega_k = (2k + 1)\pi/50$.

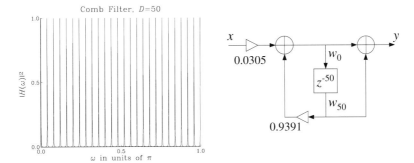

Fig. 8.3.18 Comb filter with $D = 50$ and $Q = 50$.

The canonical form uses a 51-dimensional state vector $\mathbf{w} = [w_0, w_1, \ldots, w_{50}]$ to implement the delay z^{-50}. The corresponding sample processing algorithm can be formulated with a linear or a circular delay line, as follows:

for each input sample x do:
$w_0 = 0.9391\, w_{50} + 0.0305\, x$
$y = w_0 + w_{50}$
delay$(50, \mathbf{w})$

for each input sample x do:
$s_{50} = \text{tap}(50, \mathbf{w}, p, 50)$
$s_0 = 0.9391\, s_{50} + 0.0305\, x$
$y = s_0 + s_{50}$
$*p = s_0$
cdelay$(50, \mathbf{w}, \&p)$

where p is a circular pointer to the linear buffer $\mathbf{w}$, and s_{50}, s_0 denote the 50th and 0th components of the circular state vector pointed to by p.

Fig. 8.3.19 shows the input $x(n)$ and the filtered output $y(n)$. For plotting purposes, the signals have been split into two consecutive segments of length-1000.

The noise reduction ratio of this filter is $NRR = (1 - a)/2 = 0.0305$, which corresponds to a $10\log_{10}(1/NRR) = 15.16$ dB improvement of the SNR, or equivalently, to a suppression of the rms noise value by a factor of $1/\sqrt{NRR} = 5.7$. □

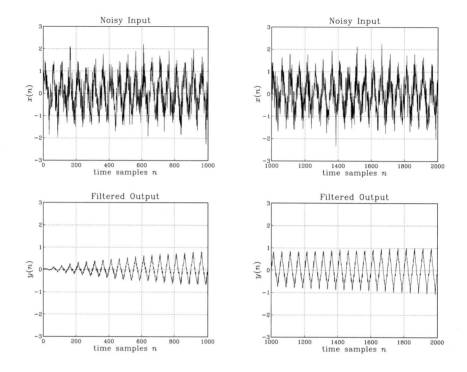

Fig. 8.3.19 Periodic signal enhancement by comb filtering.

The replicating transformation (8.3.30) can also be applied to the FIR averager filter of Example 8.3.4. The resulting periodic comb filter is equivalent to the method of *signal averaging* and is discussed further in Section 8.3.4.

8.3.3 Line and Frame Combs for Digital TV

Another application of notch and comb filters is in the case when *both* signals $s(n)$ and $v(n)$ in Eq. (8.3.1) are *periodic* and must be separated from each other.

To extract $s(n)$, one may use either a comb filter with peaks at the harmonics of $s(n)$, or a notch filter at the harmonics of $v(n)$. Similarly, to extract $v(n)$, one may use a comb at the harmonics of $v(n)$, or a notch at the harmonics of $s(n)$. For the method to work, the harmonics of $s(n)$ may not coincide with the harmonics of $v(n)$.

A major application of this idea is in color TV, digital videodisc systems, and proposed HDTV systems [171–191]. The notch/comb filters are used to separate the luminance (black & white) and chrominance (color) signals from the composite video signal, and also to reduce noise.

Consider a scanned two-dimensional still picture of horizontal and vertical dimensions a and b, as shown in Fig. 8.3.20, and assume there are N horizontal scan lines. If T_H is the time to scan one line, then the time to scan the complete picture (i.e., one frame) is equal to the scanning time for N lines, that is, (ignoring horizontal retrace and blanking times):

$$\boxed{T_F = NT_H} \tag{8.3.38}$$

The quantities T_H and T_F are called the *line and frame delays*, and their inverses are the *line and frame rates*:

$$f_H = \frac{1}{T_H}, \quad f_F = \frac{1}{T_F} \quad \Rightarrow \quad \boxed{f_H = Nf_F} \tag{8.3.39}$$

The frequencies f_H and f_F are related to the horizontal and vertical velocities of the scanning spot by

$$f_H = \frac{v_x}{a}, \quad f_F = \frac{v_y}{b} \tag{8.3.40}$$

The typical spectrum of a video signal, shown in Fig. 8.3.20, has a *macro-structure* consisting of the harmonics of the line rate f_H. About each of these, it has a *micro-structure* consisting of the harmonics of the frame rate f_F. There are N f_F-harmonics between any two f_H-harmonics. The f_H-harmonics represent the *horizontal* variations in the image, and the f_F-harmonics the *vertical* variations.

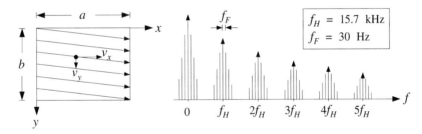

Fig. 8.3.20 Scanned image and corresponding video spectrum.

If there is motion in the image, the sharp spectral lines get smeared somewhat, but the basic macro/micro-structure is preserved. In the rest of this section, we will assume that there is no motion and that we are dealing with a still picture. At the end, we will discuss what happens when motion is introduced.

This type of spectrum can be understood qualitatively as follows: Suppose we have a still picture consisting only of a test pattern of vertical bars. Then, each scan line will be the same and the resulting signal will be periodic in time with period $T_H = 1/f_H$. Its spectrum will consist of the harmonics of f_H. Now, if there is some vertical detail in the picture, the signal will only be periodic with respect to the frame period T_F, and its spectrum will consist of the harmonics of f_F. However, because adjacent lines tend to be similar, the harmonics of f_H will still dominate the macro-structure of the spectrum.

A more mathematical explanation is as follows [178]. Let $g(x, y)$ be the brightness of the picture at position (x, y). Expanding $g(x, y)$ into a double Fourier series, we obtain:

$$g(x, y) = \sum_{k,m} c_{km} e^{2\pi jkx/a} e^{2\pi jmy/b}$$

The indices k and m correspond to the *horizontal and vertical* variations in $g(x, y)$. A scanned picture, with a uniformly moving scanning spot, is obtained by replacing $x = v_x t$ and $y = v_y t$, resulting in the *video signal*:

$$V(t) = g(v_x t, v_y t) = \sum_{k,m} c_{km} e^{2\pi jkv_x t/a} e^{2\pi jmv_y t/b}$$

Using Eq. (8.3.40), we can rewrite:

$$V(t) = \sum_{k,m} c_{km} e^{2\pi jf_{km}t} \tag{8.3.41}$$

where

$$\boxed{f_{km} = kf_H + mf_F = (kN + m)f_F = \left(k + \frac{m}{N}\right)f_H} \tag{8.3.42}$$

Thus, the video signal $V(t)$ will have spectrum with sharp spectral lines at f_{km}. Because of the large difference in value between f_H and f_F, the spectrum will look as in Fig. 8.3.20, that is, exhibiting a coarse structure at the harmonics kf_H and a fine structure at the harmonics mf_F.

In the NTSC[†] TV system used in the U.S., there are $N = 525$ lines in each frame, but they are *interlaced*, with each half (i.e., a field) being presented at double the frame rate, that is, $f_{\text{field}} = 2f_F$. The field rate is approximately 60 Hz in the U.S., and the frame rate approximately 30 Hz. The exact values are [182]:

$$f_H = \frac{4.5 \text{ MHz}}{286} = 15.73426 \text{ kHz}, \qquad f_F = \frac{f_H}{525} = 29.97 \text{ Hz} \tag{8.3.43}$$

where, by convention, the values are derived from the *sound carrier* frequency of 4.5 MHz. The corresponding time delays are $T_H = 63.55 \ \mu\text{sec}$ and $T_F = 33.37$ msec.

In a color TV system, there are three scanning beams for red, green, and blue (RGB), which can be combined to yield other colors. To reduce the transmission bandwidth requirements and maintain compatibility with black and white receivers, appropriate linear combinations of the RGB colors are formed.

The black and white information (brightness and spatial details) is contained in the *luminance* signal defined by:

$$Y = 0.299R + 0.587G + 0.114B$$

Color information (hue and saturation) can be transmitted by the *difference* signals $R - Y$ and $G - Y$. In the NTSC system, the following linear combinations— called the I and Q *chrominance* signals—are transmitted instead:

[†]National Television System Committee.

$$I = 0.736(R - Y) - 0.269(B - Y)$$

$$Q = 0.478(R - Y) + 0.413(B - Y)$$

The three RGB colors can be recovered from the three YIQ signals. The advantage of the IQ linear combinations is that they have *reduced* bandwidth requirements. The luminance bandwidth is 4.2 MHz, whereas the bandwidths of I and Q are 1.3 MHz and 0.5 MHz, respectively.

To transmit the YIQ signals efficiently, the I and Q are placed on a *color sub-carrier* signal by quadrature modulation and added to the luminance component, that is, the following *composite video* signal is transmitted:

$$V(t) = Y(t) + I(t)\cos(2\pi f_{sc}t + \phi) + Q(t)\sin(2\pi f_{sc}t + \phi)$$

where $\phi = 33°$.

To simplify the algebra, we work with the following complex-valued version of the above, with the understanding that we must take real parts:

$$\boxed{V(t) = Y(t) + e^{j2\pi f_{sc}t}C(t) \equiv Y(t) + Z(t)} \tag{8.3.44}$$

where $C(t) \equiv (I(t) - jQ(t))e^{j\phi}$.

The spectra of the separate component signals $\{Y, I, Q\}$ are all similar to the basic video spectrum of Fig. 8.3.20. The subcarrier modulation *shifts* the spectra of the chrominance signals I and Q and *centers* them about the subcarrier frequency f_{sc}. Thus, the frequencies of the modulated chrominance signal $Z(t)$ will be at:

$$f_{sc} + f_{km} = f_{sc} + kf_H + mf_F \tag{8.3.45}$$

By choosing f_{sc} to be a *half-multiple* of the line frequency f_H, the chrominance peaks will fall exactly *half-way* between the luminance peaks, as shown in Fig. 8.3.21. We can take, for example,

$$\boxed{f_{sc} = \left(d_H + \frac{1}{2}\right)f_H = \frac{1}{2}(2d_H + 1)f_H} \tag{8.3.46}$$

Therefore, the chrominance *macro-structure* peaks are centered at half-multiples of f_H:

$$f_{sc} + f_{km} = \left(d_H + k + \frac{1}{2}\right)f_H + mf_F \tag{8.3.47}$$

Moreover, because $f_H = Nf_F$ with N odd, the subcarrier frequency f_{sc} will also be equal to a half-multiple of the *frame* frequency f_F:

$$f_{sc} = \left(d_H + \frac{1}{2}\right)f_H = \left(d_H + \frac{1}{2}\right)Nf_F$$

Setting

$$d_F + \frac{1}{2} = \left(d_H + \frac{1}{2}\right)N \quad \Rightarrow \quad d_F = Nd_H + \frac{N-1}{2}$$

we find,

$$\boxed{f_{sc} = \left(d_F + \frac{1}{2}\right)f_F = \frac{1}{2}(2d_F + 1)f_F} \qquad (8.3.48)$$

It follows that the chrominance *micro-structure* peaks will be centered at half-multiples of f_F (about the kf_H macro-structure peaks), falling half-way between the micro-structure peaks of the luminance signal, as shown in Fig. 8.3.21:

$$f_{sc} + f_{km} = kf_H + \left(d_F + m + \frac{1}{2}\right)f_F \qquad (8.3.49)$$

In the NTSC system, we have the choices:

$$d_H = 227, \qquad d_F = Nd_H + \frac{N-1}{2} = 119437 \qquad (8.3.50)$$

which give the subcarrier frequency:

$$f_{sc} = 227.5f_H = 119437.5f_F = 3.579545 \text{ MHz}$$

In summary, the luminance and modulated chrominance signals have spectra that are *interleaved* both at the *macro-* and *micro-structure* levels. This property makes them ideal candidates for comb filtering.

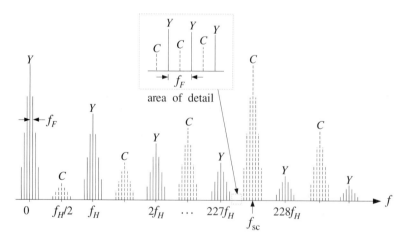

Fig. 8.3.21   Interleaved luminance and chrominance spectra.

In Fig. 8.3.21, the extent of the chrominance spectrum is somewhat exaggerated and shown to reach almost down to zero frequency. In fact, because of the small bandwidth of C, the effective extent of the chrominance spectrum about f_{sc} will be ± 0.5 MHz [182], which translates to about $\pm 32 f_H$ harmonics about f_{sc}; indeed, $227.5 \times 0.5 / 3.58 \simeq 32$.

In a conventional color TV receiver, the luminance part is extracted by lowpass filtering of the composite signal $V(t)$ with a *lowpass* filter having passband from zero to about $f_{sc} - 0.5 = 3.08$ MHz. The chrominance component is extracted by a *bandpass* filter centered at f_{sc}, with passband $f_{sc} \pm 0.5 = 3.58 \pm 0.5$ MHz. The extracted C component is then demodulated and the I and Q parts are linearly combined with Y to form the RGB signals.

These filtering operations can cause various degradations in image quality. For example, the high-frequency part of Y is filtered out, causing some loss of spatial details in the image. Moreover, because of the finite transition widths of the lowpass and bandpass filters, some chrominance will survive the luminance lowpass filter and show up as the so-called cross-luminance or "dot-crawl" effect [183]. Similarly, some high-frequency luminance will survive the chrominance bandpass filter and will appear as the "cross-color" rainbow-type effect around sharp edges [183].

A better method of separation is to take advantage of the interlaced comb-like nature of the composite video spectrum of Fig. 8.3.21 and use digital comb filters to separate the Y and C signals. The development of large-scale digital memories that can be used to store a whole line or a whole frame [174] has made this approach possible.

A common sampling rate for digital video systems is *four* times the color sub-carrier frequency, that is,

$$\boxed{f_s = 4 f_{sc}} \tag{8.3.51}$$

Using Eqs. (8.3.46) and (8.3.48), we may express the sampling rate in terms of the line frequency f_H and the frame frequency f_F:

$$\boxed{f_s = D_H f_H = D_F f_F} \tag{8.3.52}$$

where

$$D_H = 2(2d_H + 1), \qquad D_F = N D_H = 2(2d_F + 1) \tag{8.3.53}$$

For the NTSC system, we have from Eq. (8.3.50)

$$D_H = 910, \qquad D_F = 477750$$

and

$$f_s = 910 f_H = 477750 f_F = 14.31818 \text{ MHz}$$

with a corresponding sampling time interval of $T = 1/f_s = 69.84$ nsec.

Equation (8.3.52) implies that there are D_H samples along each scan line and D_F samples in each frame. In units of *radians per sample*, the subcarrier frequency f_{sc} becomes:

$$\boxed{\omega_{sc} = \frac{2\pi f_{sc}}{f_s} = \frac{\pi}{2}} \tag{8.3.54}$$

Similarly, the frame and line frequencies are in these units:

$$\omega_F = \frac{2\pi f_F}{f_s} = \frac{2\pi}{D_F}, \qquad \omega_H = \frac{2\pi f_H}{f_s} = \frac{2\pi}{D_H} = N\omega_F$$

Using Eq. (8.3.52), the luminance video frequencies f_{km} become:

$$\omega_{km} = \frac{2\pi f_{km}}{f_s} = \frac{2\pi k}{D_H} + \frac{2\pi m}{D_F} = k\omega_H + m\omega_F \qquad (8.3.55)$$

The shifted chrominance frequencies Eq. (8.3.45) can be expressed as half-multiples of either the line or the frame digital frequencies:

$$\omega_{sc} + \omega_{km} = (2d_H + 2k + 1)\frac{\pi}{D_H} + \frac{2\pi m}{D_F}$$

$$= \frac{2\pi k}{D_H} + (2d_F + 2m + 1)\frac{\pi}{D_F} \qquad (8.3.56)$$

$$= (2kN + 2d_F + 2m + 1)\frac{\pi}{D_F}$$

where in the last line we replaced $D_H = D_F/N$.

The comb filters used in video systems are of the type (8.3.33), where D can be either a line delay $D = D_H$ or a frame delay $D = D_F$. Because of the high sampling rates involved, to minimize the computational cost, the filter parameters are chosen to have simple values, such as powers of two. For example, the simplest choice is the following FIR comb filter, obtained by setting $a = 0$ and $b = 1/2$ in Eq. (8.3.33):

$$H_{\text{comb}}(z) = \frac{1}{2}(1 + z^{-D}) \qquad (8.3.57)$$

with a complementary notch filter $H_{\text{notch}}(z) = 1 - H_{\text{comb}}(z)$:

$$H_{\text{notch}}(z) = \frac{1}{2}(1 - z^{-D}) \qquad (8.3.58)$$

Their magnitude responses have been plotted in Figs. 8.3.17 and 8.3.13 for $D = 10$. They have the maximum allowed 3-dB width of all the comb/notch filters of the types (8.3.33) and (8.3.26), that is, $\Delta\omega = \pi/D$.

The comb filter $H_{\text{comb}}(z)$ has (unity-gain) peaks at the multiples $2k\pi/D$ and notches at the half-multiples $(2k + 1)\pi/D$. Conversely, the notch filter $H_{\text{notch}}(z)$ has peaks at the half-multiples $(2k + 1)\pi/D$ and notches at $2k\pi/D$.

If D is a line delay, $D = D_H$, then the peaks of $H_{\text{comb}}(z)$ will coincide with the *macro-structure* line-frequency peaks of the luminance signal Y; and its notches will coincide with the macro-structure peaks of the modulated chrominance signal C. Thus, filtering the composite video signal V through $H_{\text{comb}}(z)$ will tend to remove C and let Y pass through, at least at the macro-structure level which is the dominant part the spectrum.

Conversely, filtering V through the notch filter $H_{\text{notch}}(z)$ will tend to remove Y and let C pass through. Thus, the two filters can be used in parallel to extract the

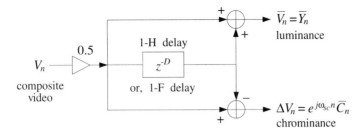

Fig. 8.3.22 Line or frame comb filters.

Y and C components. A block diagram implementation of (8.3.57) and (8.3.58) is shown in Fig. 8.3.22.

The separation of Y and C is not perfect, because the line comb $H_{comb}(z)$ does not remove from the C signal its micro-structure frequencies, that is, the terms mf_F in Eq. (8.3.47). Similarly, $H_{notch}(z)$ does not remove the micro-structure frequencies of Y.

Moreover, because Eq. (8.3.57) is equivalent to the *averaging* of two successive horizontal lines, some vertical detail will be lost or averaged out, resulting in a blurrier Y signal. However, as we see below, the lost vertical detail can be restored by further filtering.

Because the luminance and chrominance spectra are interleaved at their micro-structure frame-frequency level, the delay D can also be chosen to be a frame delay, $D = D_F$. This type of comb/notch filter would do a much better job in separating the Y and C components because $H_{comb}(z)$ now has nulls at *all* the chrominance frequencies, and $H_{notch}(z)$ has nulls at all the luminance frequencies. However, such frame-delay filters can be used only if there is very little motion from frame to frame. The effect of motion is to broaden the f_F-harmonics making the separation of Y and C less than perfect.

Another simple popular type of comb filter uses two D-fold delays and is obtained by squaring Eq. (8.3.57):

$$H_{comb}(z) = \frac{1}{4}(1 + z^{-D})^2 = \frac{1}{4}(1 + 2z^{-D} + z^{-2D}) \qquad (8.3.59)$$

When $D = D_H$, it is referred to as a 2-H comb because it requires the storage of two horizontal lines. It is also known as a 1-2-1 comb because of the particular weights given to the three horizontal lines.

Its peaks and notches are the same as those of the 1-H comb (8.3.57), but here the squaring operation has the effect of making the peaks narrower and the notches flatter. The corresponding complementary notch filter is defined as:

$$H_{notch}(z) = -\frac{1}{4}(1 - z^{-D})^2 = \frac{1}{4}(-1 + 2z^{-D} - z^{-2D}) \qquad (8.3.60)$$

These definitions imply $H_{comb}(z) + H_{notch}(z) = z^{-D}$. This is required because the filters have an inherent delay of D samples. Indeed, if we advance them by

D samples, we get the more symmetric definitions corresponding to truly complementary filters:

$$z^D H_{\text{comb}}(z) = \frac{1}{2} + \frac{1}{4}(z^D + z^{-D})$$

$$z^D H_{\text{notch}}(z) = \frac{1}{2} - \frac{1}{4}(z^D + z^{-D})$$

(8.3.61)

It is instructive to also understand the above comb/notch filtering operations
in the time domain. The sampled version of Eq. (8.3.44) is

$$\boxed{V_n = Y_n + Z_n = Y_n + e^{j\omega_{sc}n}C_n}$$

(8.3.62)

The video time index n can be mapped uniquely onto a particular pixel (i,j) on
the image, as shown in Fig. 8.3.23. The *row* index i corresponds to the quotient of
the division of n by D_H and the *column* index j corresponds to the remainder. That
is, we can write uniquely:

$$(i,j) \longrightarrow n = iD_H + j, \quad j = 0, 1, \ldots, D_H - 1$$

(8.3.63)

The row index i takes on the values $i = 0, 1, \ldots N - 1$, for N lines per frame. The
maximum value of n corresponding to the last pixel of the last line is obtained by
setting $i = N-1$ and $j = D_H - 1$, giving $n = (N-1)D_H + D_H - 1 = ND_H - 1 = D_F - 1$.

Subsequent values of n will map to pixels on the next frame, and so on. Thus,
two values of n separated by D_F samples correspond to the *same* pixel (i,j) on the
image, as shown in Fig. 8.3.23.

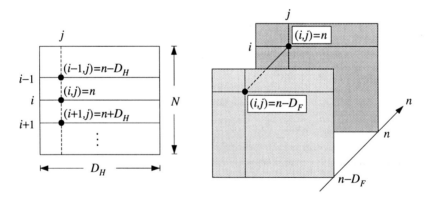

Fig. 8.3.23 Successive lines and successive frames.

Pixels on successive lines on the same column are separated by a time delay of
D_H samples, as shown in Fig. 8.3.23. Indeed, we have from Eq. (8.3.63):

$$n \pm D_H = iD_H + j \pm D_H = (i \pm 1)D_H + j \longrightarrow (i \pm 1, j)$$

With V_n as input, the output signals of the 1-H luminance and chrominance
comb filters, (8.3.57) and (8.3.58), are the sum and difference signals:

$$\boxed{\begin{aligned} \overline{V}_n &= \frac{1}{2}(V_n + V_{n-D}) \\ \Delta V_n &= \frac{1}{2}(V_n - V_{n-D}) \end{aligned}}$$

(8.3.64)

If $D = D_H$, we may think of $\overline{V}_n$ as being the average of the current horizontal line with the previous one. Indeed, using the map (8.3.63), we can rewrite Eq. (8.3.64) in the equivalent form:

$$\overline{V}_{i,j} = \frac{1}{2}(V_{i,j} + V_{i-1,j})$$

$$\Delta V_{i,j} = \frac{1}{2}(V_{i,j} - V_{i-1,j})$$

In a similar fashion, the outputs of the 2-H filters (8.3.61) can be expressed in terms of the video time index n:

$$\overline{V}_n = \frac{1}{2}V_n + \frac{1}{4}(V_{n+D} + V_{n-D})$$

$$\Delta V_n = \frac{1}{2}V_n - \frac{1}{4}(V_{n+D} + V_{n-D})$$

or, in terms of the pixel locations, showing the weighted averaging of the current line with the lines above and below it:

$$\overline{V}_{i,j} = \frac{1}{2}V_{i,j} + \frac{1}{4}(V_{i+1,j} + V_{i-1,j})$$

$$\Delta V_{i,j} = \frac{1}{2}V_{i,j} - \frac{1}{4}(V_{i+1,j} + V_{i-1,j})$$

Using Eq. (8.3.62), we have for the delayed signal V_{n-D}:

$$V_{n-D} = Y_{n-D} + Z_{n-D} = Y_{n-D} + e^{j\omega_{sc}(n-D)}C_{n-D}$$

The property that makes possible the comb filtering separation of the luminance and chrominance is that the subcarrier signal $e^{j\omega_{sc}n}$ changes sign from *line to line* and from *frame to frame*. This follows from the (intentional) choice of D_H and D_F to be even multiples of an odd integer, Eq. (8.3.53). Indeed, assuming that D is of the form $D = 2(2d + 1)$, we find:

$$\omega_{sc}D = \frac{\pi}{2}2(2d + 1) = 2\pi d + \pi$$

which corresponds to a 180° phase shift. Indeed,

$$e^{j\omega_{sc}D} = e^{2\pi jd + j\pi} = e^{j\pi} = -1$$

It follows that:

$$\boxed{V_{n-D} = Y_{n-D} - e^{j\omega_{sc}n}C_{n-D}} \qquad (8.3.65)$$

The outputs of the luminance and chrominance combs can be expressed then in the form:

$$\overline{V}_n = \frac{1}{2}(V_n + V_{n-D}) = \frac{1}{2}(Y_n + Y_{n-D}) + e^{j\omega_{sc}n}\frac{1}{2}(C_n - C_{n-D})$$

$$\Delta V_n = \frac{1}{2}(V_n - V_{n-D}) = \frac{1}{2}(Y_n - Y_{n-D}) + e^{j\omega_{sc}n}\frac{1}{2}(C_n + C_{n-D})$$

which can be written in terms of the corresponding sum and difference signals:

$$\boxed{\begin{aligned} \overline{V}_n &= \overline{Y}_n + e^{j\omega_{sc}n}\Delta C_n \\ \Delta V_n &= \Delta Y_n + e^{j\omega_{sc}n}\overline{C}_n \end{aligned}} \qquad (8.3.66)$$

Consider the line-comb case first, $D = D_H$. Assuming that the chrominance signal C_n does not change much from line to line (i.e., ignoring its micro-structure frequency content), we may set $\Delta C_n \simeq 0$. Similarly, ignoring the micro-structure of Y, we may set $\Delta Y_n \simeq 0$. Then, (8.3.66) simplifies as follows:

$$\boxed{\begin{aligned} \overline{V}_n &= \overline{Y}_n \\ \Delta V_n &= e^{j\omega_{sc}n}\overline{C}_n \end{aligned}} \qquad (8.3.67)$$

Thus, the comb outputs are effectively the desired luminance and chrominance components. The chrominance part $e^{j\omega_{sc}n}\overline{C}_n$ is then sent into a subcarrier demodulator and $\overline{C}_n$ is extracted.

For the frame-comb case, $D = D_F$, the difference signals will be identically zero, $\Delta C_n = \Delta Y_n = 0$, because of the periodicity with period D_F. Thus, the frame combs are capable of separating Y and C exactly. However, they will fail when there is motion in the image which makes the video signal non-periodic. Advanced digital video systems use frame combs when there is no or very little motion, and switch to line combs when substantial motion is detected [174].

In the line-comb case, the approximation $\Delta Y_n \simeq 0$ is more severe than $\Delta C_n \simeq 0$, because the luminance signal carries most of the spatial detail information. Setting $\Delta Y_n = 0$ implies a loss of *vertical detail*, because the output of Eq. (8.3.67) gives a luminance value $\overline{Y}_n$ averaged across two lines, instead of Y_n itself.

It follows from the above that a better approximation may be obtained by setting ΔC_n to zero, but not ΔY_n. The filtering equations (8.3.66) become in this case:

$$\begin{aligned} \overline{V}_n &= \overline{Y}_n \\ \Delta V_n &= \Delta Y_n + e^{j\omega_{sc}n}\overline{C}_n \end{aligned} \qquad (8.3.68)$$

The averaged signal $\overline{Y}_n$ can be expressed in terms of the desired one Y_n and the missing vertical detail signal ΔY_n, as follows:

$$\overline{Y}_n = \frac{1}{2}(Y_n + Y_{n-D}) = Y_n - \frac{1}{2}(Y_n - Y_{n-D}) = Y_n - \Delta Y_n$$

Therefore, we may write Eq. (8.3.68) as:

$$
\begin{aligned}
\overline{V}_n &= Y_n - \Delta Y_n \\
\Delta V_n &= \Delta Y_n + e^{j\omega_{sc}n}\overline{C}_n
\end{aligned}
\tag{8.3.69}
$$

This result suggests a method of restoring the lost vertical detail [171,183,185]. Because the vertical detail signal ΔY_n is common in both outputs, it can be extracted from the second output ΔV_n by *lowpass filtering* and then reinserted into the first to recover Y_n. This works because most of the energy in ΔY_n is between 0–1 MHz, whereas the C-term in ΔV_n is centered around 3.58 MHz. Moreover, the term ΔY_n can be removed from ΔV_n by a *bandpass filter* centered at the subcarrier frequency.

Figure 8.3.24 shows the above method of vertical detail restoration. The delay z^{-M} compensates for the delay introduced by the filter $H_{LP}(z)$. The required lowpass filter $H_{LP}(z)$ must be chosen to remove the C-term from the ΔV_n output and be flat at low frequencies. Therefore, it is really a *bandstop filter* with a zero at the subcarrier frequency $\omega_{sc} = \pi/2$. In the z-domain, the filter must have zeros at $z = e^{\pm j\omega_{sc}} = e^{\pm j\pi/2} = \pm j$.

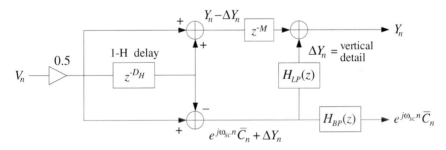

Fig. 8.3.24 Vertical detail reinsertion filters.

Another requirement for such a filter is that it have simple coefficients, expressible as sums of powers of two. Some examples of such filters are given below [171,183,185], normalized to unity gain at DC:

$$H_{LP}(z) = \frac{1}{4}(1 + z^{-2})^2$$

$$H_{LP}(z) = \frac{1}{16}(1 + z^{-2})^2(-1 + 6z^{-2} - z^{-4})$$

$$H_{LP}(z) = \frac{1}{32}(1 + z^{-2})^2(-3 + 14z^{-2} - 3z^{-4})$$

$$H_{LP}(z) = \frac{1}{64}(1 + z^{-2})^4(1 - 4z^{-2} + 5z^{-4} + 5z^{-8} - 4z^{-10} + z^{-12})$$

(8.3.70)

The factors $(1 + z^{-2})$ vanish at $z = \pm j$. The corresponding bandpass filters $H_{BP}(z)$ are obtained by requiring that they be complementary to $H_{LP}(z)$, that is, they satisfy $H_{LP}(z) + H_{BP}(z) = z^{-M}$, where M is the inherent delay introduced by the filters (i.e., half the filter order). Thus, we define:

$$H_{BP}(z) = z^{-M} - H_{LP}(z) \tag{8.3.71}$$

For the above four examples, we have $M = 2, 4, 4, 10$. Using Eqs. (8.3.70) and (8.3.71), we find:

$$H_{BP}(z) = -\frac{1}{4}(1 - z^{-2})^2$$

$$H_{BP}(z) = \frac{1}{16}(1 - z^{-2})^4$$

$$H_{BP}(z) = \frac{1}{32}(1 - z^{-2})^2(3 - 2z^{-2} + 3z^{-4}) \tag{8.3.72}$$

$$H_{BP}(z) = -\frac{1}{64}(1 - z^{-2})^4(1 + 4z^{-2} + 5z^{-4} + 5z^{-8} + 4z^{-10} + z^{-12})$$

all having a common factor $(1 - z^{-2})$, vanishing at DC and the Nyquist frequency.

Figure 8.3.25 shows the magnitude responses of the fourth filters in (8.3.70) and (8.3.72) [171], plotted over the effective video band $0 \le f \le 4.2$ MHz. Over this band, they behave as lowpass and highpass filters. The passband of the lowpass filter coincides with the significant frequency range of the vertical detail signal ΔV_n, whereas the passband of the highpass/bandpass filter coincides with the chrominance band.

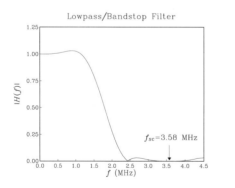 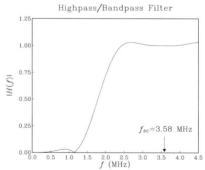

Fig. 8.3.25 Vertical detail restoration filters.

We note finally that in Fig. 8.3.24, the bandpass filter block $H_{BP}(z)$ may be eliminated and replaced by Eq. (8.3.71), which can be used to construct the required bandpass output from the lowpass output and another delay z^{-M}, thus making the implementation more efficient computationally; see, for example, Fig. 10.2.7.

8.3.4 Signal Averaging

As we saw in Section 8.3.2, IIR comb filters of the type (8.3.33) can be used to extract a periodic signal buried noise, provided the teeth of the comb coincide with the harmonics of the periodic signal.

A widely used alternative method is that of *signal averaging*, which is equivalent to comb filtering, but with an FIR comb filter instead of an IIR one.

It can be derived by applying the D-fold replicating transformation (8.3.30) to the length-N FIR averager filter of Example 8.3.4. Replacing z by z^D in the transfer function (8.3.17) gives:

$$H(z) = \frac{1}{N}\left(1 + z^{-D} + z^{-2D} + \cdots + z^{-(N-1)D}\right) = \frac{1}{N}\frac{1 - z^{-ND}}{1 - z^{-D}} \qquad (8.3.73)$$

The frequency response is obtained by setting $z = e^{j\omega}$:

$$H(\omega) = \frac{1}{N}\frac{1 - e^{-jND\omega}}{1 - e^{-j\omega D}} = \frac{1}{N}\frac{\sin(ND\omega/2)}{\sin(D\omega/2)} e^{-j\omega(N-1)D/2} \qquad (8.3.74)$$

The effect of the transformation is to shrink the spectrum of Fig. 8.3.8 by a factor of D and then replicate it D times to fill the Nyquist interval. The first zero at $2\pi/N$ shrinks to become $2\pi/ND$. The cutoff frequency of the main lobe, $\omega_c = \pi/N$, scales down to $\omega_c = \pi/ND$. The resulting widths of the D peaks will be $\Delta\omega = 2\omega_c$, or,

$$\Delta\omega = \frac{2\pi}{ND} \qquad \Rightarrow \qquad \Delta f = \frac{f_s}{ND} = \frac{1}{NDT} = \frac{1}{NT_D} \qquad (8.3.75)$$

that is, the inverse of the total duration of the N periods.

The frequency response $H(\omega)$ vanishes at the (ND)th root-of-unity frequencies, $\omega_k = 2\pi k/ND$, except when they coincide with the Dth roots of unity $\omega_k = 2\pi k/D$ (i.e., the signal harmonics) where it is equal to one.

The NRR of the filter remains the same, that is, $NRR = 1/N$. As the number of periods N increases, the NRR decreases reducing the noise. Equivalently, the widths (8.3.75) of the D peaks become smaller, thus removing more noise.

Figure 8.3.26 shows a plot of the filter's magnitude response squared for the two values $N = 5$ and $N = 10$, and period $D = 10$. The higher the N, the higher the density of the zeros along the Nyquist interval; the narrower the signal harmonic peaks; and the more the filter suppresses the noise.

The time-domain operation of the signal averager comb filter is given by the I/O difference equation:

$$y(n) = \frac{1}{N}\left[x(n) + x(n-D) + x(n-2D) + \cdots + x(n-(N-1)D)\right] \qquad (8.3.76)$$

It is not as efficient as the IIR comb (8.3.33) because it requires $N - 1$ multiple delays z^{-D}, whereas the IIR comb requires only one. However, in practice, the FIR and IIR comb filters are used differently.

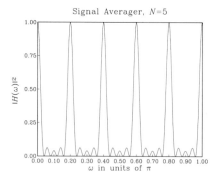 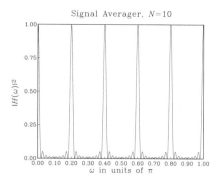

Fig. 8.3.26 Magnitude response of signal averager; $D = 10$, and $N = 5$, $N = 10$.

Because of its efficient implementation, the IIR comb is better suited for *continuous processing* of a noisy periodic signal—the objective being to eliminate the noise from the *on-going* signal. An application of this is in digital color TV systems, where an IIR frame comb [174] can be used to eliminate transmission noise.

On the other hand, the FIR comb is more appropriately used as a signal averager, processing only a *finite number* of periods, that is, N periods of the noisy signal—the objective being to get a better estimate of *one period* of the desired signal $s(n)$.

This approach is used widely in applications where the desired noisy signal can be measured *repeatedly*. For example, evoked brain waves generated by repeated application of visual stimuli can be averaged to reduce the on-going background brain activity, which is typically much stronger than the evoked signal.

The I/O filtering equation (8.3.76) can be cast in a signal averaging form as follows. We assume that the noisy input $x(n)$ has finite length, consisting of N noisy periods, that is, it has length ND samples. The ith period can be defined as follows:

$$x_i(n) = x(iD + n), \qquad n = 0, 1, \ldots, D - 1 \tag{8.3.77}$$

where $i = 0, 1, \ldots, N - 1$ for a total of N periods. Figure 8.3.27 depicts the case $N = 4$, for which the I/O equation becomes:

$$y(n) = \frac{1}{4} [x(n) + x(n - D) + x(n - 2D) + x(n - 3D)]$$

As seen in Fig. 8.3.27, the terms $x(n-iD)$ are successively delayed by D samples at a time. Because of the finite length of the input, the time periods for the input-on, steady-state, and input-off transients will be:

$$
\begin{aligned}
0 &\leq n \leq 3D - 1 & \text{(input-on)} \\
3D &\leq n \leq 4D - 1 & \text{(steady state)} \\
4D &\leq n \leq 7D - 1 & \text{(input-off)}
\end{aligned}
$$

In particular, steady state lasts for only *one period* during which *all* terms in Eq. (8.3.76) are contributing to the output. The contributing periods of each term are shaded in Fig. 8.3.27.

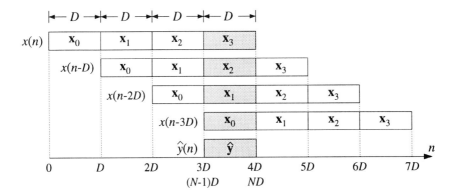

Fig. 8.3.27 Signal averaging as FIR comb filtering.

The time interval $[3D, 4D)$ of the steady state can be parametrized by the shifted time index $(3D + n)$, which runs over this interval provided n takes on the values: $n = 0, 1, \ldots, D - 1$. Thus, replacing n by $3D + n$ in the I/O equation, gives the steady-state output samples of the fourth period:

$$y(3D + n) = \frac{1}{4}[x(3D + n) + x(2D + n) + x(D + n) + x(n)]$$

for $n = 0, 1, \ldots, D - 1$.

Using the definition (8.3.77), we can rewrite the I/O equation in a form that shows the averaging of the four periods. Denoting the steady-state output samples by $\hat{y}(n) = y(3D + n)$, we have:

$$\hat{y}(n) = \frac{1}{4}[x_3(n) + x_2(n) + x_1(n) + x_0(n)], \qquad n = 0, 1, \ldots, D - 1$$

More generally, the outputs $\hat{y}(n) = y((N - 1)D + n), n = 0, 1, \ldots, D - 1$ are the D outputs over the steady-state period covering the time interval $[(N - 1)D, ND)$. They are computed by averaging all N periods in the input:

$$\hat{y}(n) = \frac{1}{N} \sum_{i=0}^{N-1} x_i(n) = \frac{1}{N}[x_0(n) + x_1(n) + \cdots + x_{N-1}(n)] \qquad (8.3.78)$$

The output $\hat{y}(n)$ provides an *estimate* of one period of the desired signal $s(n)$. To see this, we use the periodicity of $s(n)$ to write:

$$x_i(n) = x(iD + n) = s(iD + n) + v(iD + n) = s(n) + v_i(n)$$

where $v_i(n) = v(iD + n)$ are the noise samples during the ith period. Inserting these into Eq. (8.3.78), we find:

$$\hat{y}(n) = \frac{1}{N} \sum_{i=0}^{N-1} (s(n) + v_i(n)) = s(n) + \frac{1}{N} \sum_{i=0}^{N-1} v_i(n)$$

Defining the averaged noise over N periods by

$$\hat{v}(n) = \frac{1}{N} \sum_{i=0}^{N-1} v_i(n)$$

we have

$$\boxed{\hat{y}(n) = s(n) + \hat{v}(n)} \tag{8.3.79}$$

Assuming that the noise samples have zero mean and are mutually independent (at least from period to period), we can verify that the averaged noise $\hat{v}(n)$ will have a mean-square value reduced by the NRR, that is, by a factor of N. Indeed, denoting the variance of the input noise by σ_v^2, we have for the variance of $\hat{v}(n)$:

$$\sigma_{\hat{v}}^2 = \frac{1}{N^2}(\sigma_v^2 + \sigma_v^2 + \cdots + \sigma_v^2) = \frac{1}{N^2}(N\sigma_v^2) = \frac{1}{N}\sigma_v^2$$

The computation of Eq. (8.3.78) can be implemented using the filtering equation (8.3.76), applied to N periods of the input and returning the output only for the last period. Because of the finite length of the input, it proves convenient to work with the recursive version of Eq. (8.3.76) based on the transfer function (8.3.73), that is,

$$y(n) = y(n-D) + \frac{1}{N}(x(n) - x(n-ND))$$

The roundoff noise accumulation problems discussed in Example 8.3.5 do not arise in this case because the input remains finite. In fact, because of the finite length of $x(n)$ over the N periods, $0 \le n \le ND - 1$, the delayed term $x(n-ND)$ will not contribute to the output, giving the recursion:

$$\boxed{y(n) = y(n-D) + \frac{1}{N}x(n), \qquad 0 \le n \le ND - 1} \tag{8.3.80}$$

Figure 8.3.28 shows a block diagram implementation that uses only one D-fold delay z^{-D}. The corresponding sample processing algorithm can be split into two parts: the accumulation of N periods, followed by the output of the averaged values $\hat{y}(n)$. The algorithm, formulated with a linear and a circular delay line, is then:

for each of the ND inputs x do:	for each of the ND inputs x do:
$w_0 = w_D + x/N$	$*p = \text{tap}(D, \mathbf{w}, p, D) + x/N$
$\text{delay}(D, \mathbf{w})$	$\text{cdelay}(D, \mathbf{w}, \&p)$

where $\mathbf{w}$ is the $(D+1)$-dimensional state vector $w_i(n) = y(n-i)$, $i = 0, 1, \ldots, D$.

After the N periods have been accumulated, the last D outputs are returned into the signal $\hat{y}(n)$. Because of the last call to delay, these outputs are now shifted one position ahead into the delay line. They can be returned by reading the delay-line contents from the bottom up:

for $n = 0, 1, \ldots, D-1$,	for $n = 0, 1, \ldots, D-1$,
$\hat{y}(n) = w_{D-n}$	$\hat{y}(n) = \text{tap}(D, \mathbf{w}, p, D-n)$

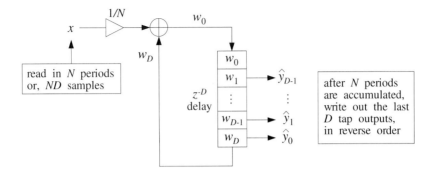

Fig. 8.3.28 Signal averaging implementation with a D-fold delay.

The circular delay line version is the more efficient implementation, especially in hardware, when the period D is large, for example, larger than 1000.

To appreciate the manner in which the difference equation (8.3.80) accumulates the periods of the input and builds up the steady-state output $\hat{y}(n)$, consider a small example where $D = 3$ and $N = 4$. The input signal is assumed to be the noise-free repetition of four periods of the samples $[s_0, s_1, s_2]$. The input and output sequences of Eq. (8.3.80) will be in this case:

$$\mathbf{x} = [s_0 \ s_1 \ s_2 \mid s_0 \ s_1 \ s_2 \mid s_0 \ s_1 \ s_2 \mid s_0 \ s_1 \ s_2]$$

$$\mathbf{y} = \frac{1}{4}[s_0 \ s_1 \ s_2 \mid 2s_0 \ 2s_1 \ 2s_2 \mid 3s_0 \ 3s_1 \ 3s_2 \mid 4s_0 \ 4s_1 \ 4s_2]$$

The last three outputs are the averaged signal $\hat{y}(n)$, which is exactly equal to one period of $s(n)$. Table 8.3.1 shows the successive contents of the delay-line buffer $\mathbf{w} = [w_0, w_1, w_2, w_3]$ (in the linear buffer case).

Note that in each line, the contents of $[w_1, w_2, w_3]$ are obtained by right-shifting the contents of $[w_0, w_1, w_2]$ of the previous line, and w_0 is computed by the recursion $w_0 = w_3 + x$. For convenience, we have moved the multiplier $1/N$ to the output side, that is, the output samples are computed by $y = w_0/4$.

The last call of delay at time $n = ND - 1 = 11$ causes the delay-line contents to shift once more, in preparation for the filtering at time $n = ND = 12$. Thus, as seen in the table, the buffer entries $[w_1, w_2, w_3]$ are left containing the desired period in reverse order.

The following program segment illustrates the usage of the circular version of the sample processing algorithm. The input samples are read sequentially from a data file containing at least N periods and pointed to by the file pointer fpx:

```
double *w, *p;
w = (double *) calloc(D+1, sizeof(double));    D+1 dimensional
p = w;                                          initialize p

for (i=0; i<N; i++)                             N periods
        for (n=0; n<D; n++) {                   D samples per period
                fscanf(fpx, "%lf", &x);         read input x
```

n	x	w_0	w_1	w_2	w_3	$y = w_0/4$
0	s_0	s_0	0	0	0	$s_0/4$
1	s_1	s_1	s_0	0	0	$s_1/4$
2	s_2	s_2	s_1	s_0	0	$s_2/4$
3	s_0	$2s_0$	s_2	s_1	s_0	$2s_0/4$
4	s_1	$2s_1$	$2s_0$	s_2	s_1	$2s_1/4$
5	s_2	$2s_2$	$2s_1$	$2s_0$	s_2	$2s_2/4$
6	s_0	$3s_0$	$2s_2$	$2s_1$	$2s_0$	$3s_0/4$
7	s_1	$3s_1$	$3s_0$	$2s_2$	$2s_1$	$3s_1/4$
8	s_2	$3s_2$	$3s_1$	$3s_0$	$2s_2$	$3s_2/4$
9	s_0	$4s_0$	$3s_2$	$3s_1$	$3s_0$	s_0
10	s_1	$4s_1$	$4s_0$	$3s_2$	$3s_1$	s_1
11	s_2	$4s_2$	$4s_1$	$4s_0$	$3s_2$	s_2
12	$-$	$-$	$4s_2$	$4s_1$	$4s_0$	$-$

Table 8.3.1 Delay line contents for $D = 3$, $N = 4$.

```
*p = tap(D, w, p, D) + x / N;        accumulate x with Dth tap
cdelay(D, w, &p);                    update delay line
}

for (n=0; n<D; n++)                  output D taps
     yhat[n] = tap(D, w, p, D-n);    in reverse order
```

Figure 8.3.29 shows a simulation of the algorithm. The desired periodic signal $s(n)$ was taken to be the following square wave of period $D = 100$:

$$s(n) = \begin{cases} 1, & \text{for } 0 \le n \le D/2 - 1 \\ -1, & \text{for } D/2 \le n \le D - 1 \end{cases}$$

The noise $v(n)$ was zero-mean, unit-variance, white noise generated by the routine gran. Because the rms noise value is unity, the noise is as strong as the square wave. The averaged signal $\hat{y}(n)$ was computed with $N = 1, 16, 64,$ and 256 periods. Note that as N increases by a factor of four, the rms noise value drops by a factor of two, namely, $\sigma_{\hat{v}} = \sigma_v/\sqrt{N}$.

8.3.5 Savitzky-Golay Smoothing Filters*

We mentioned in Example 8.3.4 that there are limits to the applicability of the plain FIR averager filter. In order to achieve a high degree of noise reduction, its length N may be required to be so large that the filter's passband $\omega_c = \pi/N$ becomes smaller than the signal bandwidth, causing the removal of useful high frequencies from the desired signal.

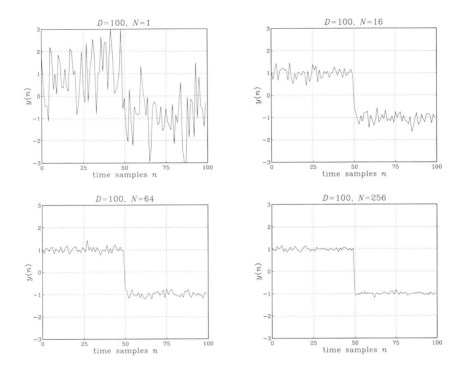

Fig. 8.3.29 Signal averaging with $D = 100$, and $N = 1, 16, 64, 256$ sweeps.

In other words, in its attempt to smooth out the noise $v(n)$, the filter begins to smooth out the desired signal $s(n)$ to an unacceptable degree. For example, if $s(n)$ contains some fairly narrow peaks, corresponding to the higher frequencies present in $s(n)$, and the filter's length N is longer than the duration of the peaks, the filter will tend to smooth the peaks too much, broadening them and reducing their height.

The Savitzky-Golay FIR smoothing filters, also known as *polynomial* smoothing, or *least-squares* smoothing filters [200–216,303], are generalizations of the FIR averager filter that can preserve better the high-frequency content of the desired signal, at the expense of not removing as much noise as the averager.

They can be characterized in three equivalent ways: (1) They are optimal low-pass filters, in the sense that they *minimize* the NRR, but their impulse response coefficients h_n are subject to *additional* constraints than the DC condition (8.3.12); (2) they are the optimal filters whose frequency response $H(\omega)$ satisfies certain *flatness* constraints at DC; (3) they are the filters that optimally fit a set of data points to polynomials of different degrees.

We begin with the third point of view. Figure 8.3.30 shows five noisy signal samples $[x_{-2}, x_{-1}, x_0, x_1, x_2]$ positioned symmetrically about the origin. Later on, we will shift them to an arbitrary position along the time axis.

Polynomial smoothing of the five samples is equivalent to replacing them by the

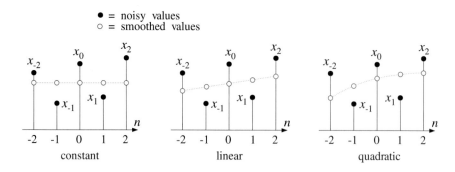

Fig. 8.3.30 Data smoothing with polynomials of degrees $d = 0, 1, 2$.

values that lie on smooth polynomial curves drawn between the noisy samples. In Fig. 8.3.30, we consider fitting the five data to a *constant* signal, a *linear* signal, and a *quadratic* signal. The corresponding smoothed values are given by the 0th, 1st, and 2nd degree polynomials, for $m = -2, -1, 0, 1, 2$:

$$
\begin{array}{lll}
\hat{x}_m = c_0 & \text{(constant)} & \\
\hat{x}_m = c_0 + c_1 m & \text{(linear)} & (8.3.81) \\
\hat{x}_m = c_0 + c_1 m + c_2 m^2 & \text{(quadratic)} &
\end{array}
$$

For each choice of the polynomial order, the coefficients c_i must be determined optimally such that the corresponding polynomial curve best fits the given data. This can be accomplished by a *least-squares fit*, which chooses the c_i that minimize the total mean-square error. For example, in the quadratic case, we have the performance index to be minimized:

$$
\mathcal{J} = \sum_{m=-2}^{2} e_m^2 = \sum_{m=-2}^{2} \left(x_m - (c_0 + c_1 m + c_2 m^2) \right)^2 = \min \qquad (8.3.82)
$$

where the fitting errors are defined as

$$
e_m = x_m - \hat{x}_m = x_m - (c_0 + c_1 m + c_2 m^2), \qquad m = -2, -1, 0, 1, 2
$$

It proves convenient to express Eqs. (8.3.81) and (8.3.82) in a vectorial form, which generalizes to higher polynomial orders and to more than five data points. We define the five-dimensional vectors:

$$
\mathbf{x} = \begin{bmatrix} x_{-2} \\ x_{-1} \\ x_0 \\ x_1 \\ x_2 \end{bmatrix}, \qquad \hat{\mathbf{x}} = \begin{bmatrix} \hat{x}_{-2} \\ \hat{x}_{-1} \\ \hat{x}_0 \\ \hat{x}_1 \\ \hat{x}_2 \end{bmatrix}, \qquad \mathbf{e} = \begin{bmatrix} e_{-2} \\ e_{-1} \\ e_0 \\ e_1 \\ e_2 \end{bmatrix} = \mathbf{x} - \hat{\mathbf{x}}
$$

Similarly, we define the five-dimensional polynomial *basis vectors* $\mathbf{s}_0, \mathbf{s}_1, \mathbf{s}_2$, whose components are:

$$
s_0(m) = 1, \qquad s_1(m) = m, \qquad s_2(m) = m^2, \qquad -2 \le m \le 2
$$

Vectorially, we have:

$$\mathbf{s}_0 = \begin{bmatrix} 1 \\ 1 \\ 1 \\ 1 \\ 1 \end{bmatrix}, \quad \mathbf{s}_1 = \begin{bmatrix} -2 \\ -1 \\ 0 \\ 1 \\ 2 \end{bmatrix}, \quad \mathbf{s}_2 = \begin{bmatrix} 4 \\ 1 \\ 0 \\ 1 \\ 4 \end{bmatrix} \tag{8.3.83}$$

In this notation, we may write the third of Eq. (8.3.81) vectorially:

$$\hat{\mathbf{x}} = c_0 \begin{bmatrix} 1 \\ 1 \\ 1 \\ 1 \\ 1 \end{bmatrix} + c_1 \begin{bmatrix} -2 \\ -1 \\ 0 \\ 1 \\ 2 \end{bmatrix} + c_2 \begin{bmatrix} 4 \\ 1 \\ 0 \\ 1 \\ 4 \end{bmatrix} = c_0 \mathbf{s}_0 + c_1 \mathbf{s}_1 + c_2 \mathbf{s}_2$$

Therefore,

$$\hat{\mathbf{x}} = c_0 \mathbf{s}_0 + c_1 \mathbf{s}_1 + c_2 \mathbf{s}_2 = [\mathbf{s}_0, \mathbf{s}_1, \mathbf{s}_2] \begin{bmatrix} c_0 \\ c_1 \\ c_2 \end{bmatrix} \equiv S\mathbf{c} \tag{8.3.84}$$

The 5×3 basis matrix S has as columns the three basis vectors $\mathbf{s}_0$, $\mathbf{s}_1$, $\mathbf{s}_2$. It is given explicitly as follows:

$$S = [\mathbf{s}_0, \mathbf{s}_1, \mathbf{s}_2] = \begin{bmatrix} 1 & -2 & 4 \\ 1 & -1 & 1 \\ 1 & 0 & 0 \\ 1 & 1 & 1 \\ 1 & 2 & 4 \end{bmatrix} \tag{8.3.85}$$

Writing $\mathbf{e} = \mathbf{x} - \hat{\mathbf{x}} = \mathbf{x} - S\mathbf{c}$, we can express the performance index (8.3.82) as the dot product:

$$\mathcal{J} = \mathbf{e}^T \mathbf{e} = (\mathbf{x} - S\mathbf{c})^T (\mathbf{x} - S\mathbf{c}) = \mathbf{x}^T \mathbf{x} - 2\mathbf{c}^T S^T \mathbf{x} + \mathbf{c}^T S^T S\mathbf{c} \tag{8.3.86}$$

To minimize this expression with respect to $\mathbf{c}$, we must set the gradient $\partial \mathcal{J}/\partial \mathbf{c}$ to zero and solve for $\mathbf{c}$. It is left as an exercise to show that this gradient is:

$$\frac{\partial \mathcal{J}}{\partial \mathbf{c}} = -2S^T \mathbf{e} = -2S^T (\mathbf{x} - S\mathbf{c}) = -2(S^T \mathbf{x} - S^T S\mathbf{c}) \tag{8.3.87}$$

Therefore, the minimization condition gives the so-called *orthogonality equations*:

$$\frac{\partial \mathcal{J}}{\partial \mathbf{c}} = 0 \quad \Rightarrow \quad S^T \mathbf{e} = 0 \tag{8.3.88}$$

which can be written as the *normal equations*:

$$S^T S\mathbf{c} = S^T \mathbf{x} \tag{8.3.89}$$

with optimal solution:

$$\mathbf{c} = (S^T S)^{-1} S^T \mathbf{x} \equiv G^T \mathbf{x} \tag{8.3.90}$$

where we defined the 5×3 matrix G by

$$G = S(S^T S)^{-1} \tag{8.3.91}$$

Inserting the optimal coefficients $\mathbf{c}$ into Eq. (8.3.84), we find the smoothed values:

$$\hat{\mathbf{x}} = S\mathbf{c} = SG^T \mathbf{x} = S(S^T S)^{-1} S^T \mathbf{x} \equiv B\mathbf{x} \tag{8.3.92}$$

where we defined the 5×5 matrix B by

$$B = SG^T = GS^T = S(S^T S)^{-1} S^T \tag{8.3.93}$$

We define the symmetric 3×3 matrix $F = S^T S$, which appears in the expressions for G and B. Its matrix elements are the *dot products* of the basis vectors, that is, the ijth matrix element is $F_{ij} = (S^T S)_{ij} = \mathbf{s}_i^T \mathbf{s}_j$. Indeed, using Eq. (8.3.85), we find

$$F = S^T S = \begin{bmatrix} \mathbf{s}_0^T \\ \mathbf{s}_1^T \\ \mathbf{s}_2^T \end{bmatrix} [\mathbf{s}_0, \mathbf{s}_1, \mathbf{s}_2] = \begin{bmatrix} \mathbf{s}_0^T \mathbf{s}_0 & \mathbf{s}_0^T \mathbf{s}_1 & \mathbf{s}_0^T \mathbf{s}_2 \\ \mathbf{s}_1^T \mathbf{s}_0 & \mathbf{s}_1^T \mathbf{s}_1 & \mathbf{s}_1^T \mathbf{s}_2 \\ \mathbf{s}_2^T \mathbf{s}_0 & \mathbf{s}_2^T \mathbf{s}_1 & \mathbf{s}_2^T \mathbf{s}_2 \end{bmatrix} \tag{8.3.94}$$

Using Eq. (8.3.85), we calculate F and its inverse F^{-1}:

$$F = \begin{bmatrix} 5 & 0 & 10 \\ 0 & 10 & 0 \\ 10 & 0 & 34 \end{bmatrix}, \qquad F^{-1} = \frac{1}{35} \begin{bmatrix} 17 & 0 & -5 \\ 0 & 3.5 & 0 \\ -5 & 0 & 2.5 \end{bmatrix} \tag{8.3.95}$$

Then, we calculate the 5×3 matrix $G = S(S^T S)^{-1} = SF^{-1}$:

$$G = SF^{-1} = \frac{1}{35} \begin{bmatrix} 1 & -2 & 4 \\ 1 & -1 & 1 \\ 1 & 0 & 0 \\ 1 & 1 & 1 \\ 1 & 2 & 4 \end{bmatrix} \begin{bmatrix} 17 & 0 & -5 \\ 0 & 3.5 & 0 \\ -5 & 0 & 2.5 \end{bmatrix} \qquad \text{or,}$$

$$G = \frac{1}{35} \begin{bmatrix} -3 & -7 & 5 \\ 12 & -3.5 & -2.5 \\ 17 & 0 & -5 \\ 12 & 3.5 & -2.5 \\ -3 & 7 & 5 \end{bmatrix} \equiv [\mathbf{g}_0, \mathbf{g}_1, \mathbf{g}_2] \tag{8.3.96}$$

As we see below, the three columns of G have useful interpretations as differentiation filters. Next, using Eq. (8.3.93), we calculate the 5×5 matrix B:

$$B = GS^T = \frac{1}{35} \begin{bmatrix} -3 & -7 & 5 \\ 12 & -3.5 & -2.5 \\ 17 & 0 & -5 \\ 12 & 3.5 & -2.5 \\ -3 & 7 & 5 \end{bmatrix} \begin{bmatrix} 1 & 1 & 1 & 1 & 1 \\ -2 & -1 & 0 & 1 & 2 \\ 4 & 1 & 0 & 1 & 4 \end{bmatrix} \qquad \text{or,}$$

$$B = \frac{1}{35} \begin{bmatrix} 31 & 9 & -3 & -5 & 3 \\ 9 & 13 & 12 & 6 & -5 \\ -3 & 12 & 17 & 12 & -3 \\ -5 & 6 & 12 & 13 & 9 \\ 3 & -5 & -3 & 9 & 31 \end{bmatrix} \equiv [\mathbf{b}_{-2}, \mathbf{b}_{-1}, \mathbf{b}_0, \mathbf{b}_1, \mathbf{b}_2] \tag{8.3.97}$$

Because B is *symmetric*, its rows are the same as its columns. Thus, we can write it either in column-wise or row-wise form:

$$B = [\mathbf{b}_{-2}, \mathbf{b}_{-1}, \mathbf{b}_0, \mathbf{b}_1, \mathbf{b}_2] = \begin{bmatrix} \mathbf{b}_{-2}^T \\ \mathbf{b}_{-1}^T \\ \mathbf{b}_0^T \\ \mathbf{b}_1^T \\ \mathbf{b}_2^T \end{bmatrix} = B^T$$

The five columns or rows of B are the Savitzky-Golay (SG) smoothing filters of length 5 and polynomial order 2. The corresponding smoothed values $\hat{\mathbf{x}}$ can be expressed component-wise in terms of these filters, as follows:

$$\begin{bmatrix} \hat{x}_{-2} \\ \hat{x}_{-1} \\ \hat{x}_0 \\ \hat{x}_1 \\ \hat{x}_2 \end{bmatrix} = \hat{\mathbf{x}} = B\mathbf{x} = B^T\mathbf{x} = \begin{bmatrix} \mathbf{b}_{-2}^T \\ \mathbf{b}_{-1}^T \\ \mathbf{b}_0^T \\ \mathbf{b}_1^T \\ \mathbf{b}_2^T \end{bmatrix} \mathbf{x} = \begin{bmatrix} \mathbf{b}_{-2}^T\mathbf{x} \\ \mathbf{b}_{-1}^T\mathbf{x} \\ \mathbf{b}_0^T\mathbf{x} \\ \mathbf{b}_1^T\mathbf{x} \\ \mathbf{b}_2^T\mathbf{x} \end{bmatrix}$$

or, for $m = -2, -1, 0, 1, 2$:

$$\hat{x}_m = \mathbf{b}_m^T\mathbf{x} \tag{8.3.98}$$

Thus, the mth filter $\mathbf{b}_m$ dotted into the data vector $\mathbf{x}$ generates the mth smoothed data sample. In a similar fashion, we can express the polynomial coefficients c_i as dot products. Using the solution Eq. (8.3.90), we have

$$\begin{bmatrix} c_0 \\ c_1 \\ c_2 \end{bmatrix} = \mathbf{c} = G^T\mathbf{x} = \begin{bmatrix} \mathbf{g}_0^T \\ \mathbf{g}_1^T \\ \mathbf{g}_2^T \end{bmatrix} \mathbf{x} = \begin{bmatrix} \mathbf{g}_0^T\mathbf{x} \\ \mathbf{g}_1^T\mathbf{x} \\ \mathbf{g}_2^T\mathbf{x} \end{bmatrix}$$

Thus, the coefficients c_i can be expressed as the dot products of the columns of G with the data vector $\mathbf{x}$:

$$c_i = \mathbf{g}_i^T\mathbf{x}, \qquad i = 0, 1, 2 \tag{8.3.99}$$

Of the five columns of B, the middle one, $\mathbf{b}_0$, is the most important because it smoothes the value x_0, which is symmetrically placed with respect to the other samples in $\mathbf{x}$, as shown in Fig. 8.3.30.

In smoothing a long block of data, the filter $\mathbf{b}_0$ is used during the *steady-state* period, whereas the other columns of B are used only during the input-on and input-off *transients*. We will refer to $\mathbf{b}_0$ and the other columns of B as the *steady-state* and *transient* Savitzky-Golay filters.

Setting $m = 0$ into Eq. (8.3.81), we note that the middle smoothed value $\hat{x}_0$ is equal to the polynomial coefficient c_0. Denoting it by $y_0 = \hat{x}_0$ and using Eqs. (8.3.98) and (8.3.99), we find

$$y_0 = \hat{x}_0 = c_0 = \mathbf{b}_0^T \mathbf{x} = \mathbf{g}_0^T \mathbf{x} \tag{8.3.100}$$

The dot products are the same because the middle column of B and the first column of G are always the same, $\mathbf{b}_0 = \mathbf{g}_0$. Using the numerical values of $\mathbf{b}_0$, we have:

$$y_0 = \mathbf{b}_0^T \mathbf{x} = \frac{1}{35}[-3,\ 12,\ 17,\ 12,\ -3]\begin{bmatrix} x_{-2} \\ x_{-1} \\ x_0 \\ x_1 \\ x_2 \end{bmatrix} \tag{8.3.101}$$

$$= \frac{1}{35}(-3x_{-2} + 12x_{-1} + 17x_0 + 12x_1 - 3x_2)$$

To express this as a true filtering operation acting on an input sequence x_n, we shift the group of five samples to be centered around the nth time instant, that is, we make the substitution:

$$[x_{-2},\ x_{-1},\ x_0,\ x_1,\ x_2] \longrightarrow [x_{n-2},\ x_{n-1},\ x_n,\ x_{n+1},\ x_{n+2}]$$

The middle sample x_n is then smoothed by forming the same linear combination of the shifted samples as in Eq. (8.3.101):

$$y_n = \frac{1}{35}(-3x_{n-2} + 12x_{n-1} + 17x_n + 12x_{n+1} - 3x_{n+2}) \tag{8.3.102}$$

The filter corresponds to fitting every group of five samples $\{x_{n-2}, x_{n-1}, x_n, x_{n+1}, x_{n+2}\}$ to a quadratic polynomial and replacing the middle sample x_n by its smoothed value y_n. It is a lowpass filter and is normalized to unity gain at DC, because its coefficients add up to one.

Its NRR is the sum of the squared filter coefficients. It can be proved in general that the NRR of any steady-state Savitzky-Golay filter $\mathbf{b}_0$ is equal to the *middle* value of its impulse response, that is, the coefficient $b_0(0)$. Therefore,

$$NRR = \mathbf{b}_0^T \mathbf{b}_0 = \sum_{m=-2}^{2} b_0(m)^2 = b_0(0) = \frac{17}{35} = \frac{17/7}{5} = \frac{2.43}{5} = 0.49$$

By comparison, the length-5 FIR averager operating on the same five samples is:

$$y_n = \frac{1}{5}(x_{n-2} + x_{n-1} + x_n + x_{n+1} + x_{n+2}) \tag{8.3.103}$$

with $NRR = 1/N = 1/5$. Thus, the length-5 SG filter performs 2.43 times worse in reducing noise than the FIR averager. However, the SG filter has other advantages to be discussed later.

We saw in Eq. (8.3.100) that the coefficient c_0 represents the smoothed value of x_0 at $m = 0$. Similarly, the coefficient c_1 represents the slope—*the derivative*—of x_0 at $m = 0$. Indeed, we have from Eq. (8.3.81) by differentiating and setting $m = 0$:

$$\dot{\hat{x}}_0 = \left.\frac{d\hat{x}_m}{dm}\right|_0 = c_1, \qquad \ddot{\hat{x}}_0 = \left.\frac{d^2\hat{x}_m}{dm^2}\right|_0 = 2c_2$$

Thus, c_1 and $2c_2$ represent the polynomial estimates of the first and second derivatives at $m = 0$. Denoting them by $\dot{y}_0 = c_1$ and $\ddot{y}_0 = 2c_2$, and using Eq. (8.3.99) we can express them in terms of the second and third columns of the matrix G:

$$\dot{y}_0 = \dot{\hat{x}}_0 = c_1 = \mathbf{g}_1^T \mathbf{x}$$
$$\ddot{y}_0 = \ddot{\hat{x}}_0 = 2c_2 = 2\mathbf{g}_2^T \mathbf{x}$$

(8.3.104)

Using the numerical values of $\mathbf{g}_1$ and $\mathbf{g}_2$, we have:

$$\dot{y}_0 = \frac{1}{35}(-7x_{-2} - 3.5x_{-1} + 3.5x_1 + 7x_2)$$

$$\ddot{y}_0 = \frac{2}{35}(5x_{-2} - 2.5x_{-1} - 5x_0 - 2.5x_1 + 5x_2)$$

Shifting these to the nth time sample, we find the length-5 Savitzky-Golay filters for estimating the *first and second derivatives* of x_n:

$$\boxed{\begin{aligned} \dot{y}_n &= \frac{1}{35}(-7x_{n-2} - 3.5x_{n-1} + 3.5x_{n+1} + 7x_{n+2}) \\ \ddot{y}_n &= \frac{2}{35}(5x_{n-2} - 2.5x_{n-1} - 5x_n - 2.5x_{n+1} + 5x_{n+2}) \end{aligned}}$$

(8.3.105)

The above designs can be generalized in a straightforward manner to an arbitrary degree d of the polynomial fit and to an arbitrary length N of the data vector $\mathbf{x}$. We require only that $N \geq d + 1$, a restriction to be clarified later.

Assuming that N is odd, say, $N = 2M + 1$, the five-dimensional data vector $\mathbf{x} = [x_{-2}, x_{-1}, x_0, x_1, x_2]^T$ is replaced by an N-dimensional one, having M points on either side of x_0:

$$\mathbf{x} = [x_{-M}, \ldots, x_{-1}, x_0, x_1, \ldots, x_M]^T$$

(8.3.106)

The N data samples in $\mathbf{x}$ are then fitted by a polynomial of degree d, generalizing Eq. (8.3.81):

$$\hat{x}_m = c_0 + c_1 m + \cdots + c_d m^d, \qquad -M \leq m \leq M$$

(8.3.107)

In this case, there are $d+1$ polynomial basis vectors $\mathbf{s}_i$, $i = 0, 1, \ldots, d$, defined to have components:

$$s_i(m) = m^i, \qquad -M \le m \le M \tag{8.3.108}$$

The corresponding $N \times (d+1)$ matrix S is defined to have s_i as columns:

$$S = [s_0, s_1, \ldots, s_d] \tag{8.3.109}$$

The smoothed values (8.3.107) can be written in the vector form:

$$\hat{x} = \sum_{i=0}^{d} c_i s_i = [s_0, s_1, \ldots, s_d] \begin{bmatrix} c_0 \\ c_1 \\ \vdots \\ c_d \end{bmatrix} = Sc \tag{8.3.110}$$

The design steps for the SG filters can be summarized then as follows:

$$\boxed{\begin{aligned} F &= S^T S \quad \Leftrightarrow \quad F_{ij} = s_i^T s_j, \quad i,j = 0,1,\ldots,d \\ G &= SF^{-1} \equiv [g_0, g_1, \ldots, g_d] \\ B &= SG^T = GS^T = SF^{-1}S^T \equiv [b_{-M}, \ldots, b_0, \ldots, b_M] \end{aligned}} \tag{8.3.111}$$

The corresponding coefficient vector c and smoothed data vector will be:

$$\boxed{\begin{aligned} c &= G^T x \quad \Leftrightarrow \quad c_i = g_i^T x, \quad i = 0,1,\ldots,d \\ \hat{x} &= Bx \quad \Leftrightarrow \quad \hat{x}_m = b_m^T x, \quad -M \le m \le M \end{aligned}} \tag{8.3.112}$$

The middle smoothed value $y_0 = \hat{x}_0$ is given in terms of the middle SG filter b_0:

$$y_0 = b_0^T x = \sum_{m=-M}^{M} b_0(m) x_m$$

The N-dimensional vector x can be shifted to the nth time instant by:

$$x \longrightarrow [x_{n-M}, \ldots, x_{n-1}, x_n, x_{n+1}, \ldots, x_{n+M}]^T$$

The resulting length-N, order-d, Savitzky-Golay filter for smoothing a noisy sequence $x(n)$ will be, in its steady-state form:

$$\boxed{y(n) = \sum_{m=-M}^{M} b_0(m) x(n+m) = \sum_{m=-M}^{M} b_0(-m) x(n-m)} \tag{8.3.113}$$

The second equation expresses the output in convolutional form. Because the filter b_0 is symmetric about its middle, we can replace $b_0(-m) = b_0(m)$.

The $d+1$ columns of the $N \times (d+1)$-dimensional matrix G give the SG *differentiation filters*, for derivatives of orders $i = 0,1,\ldots,d$. It follows by differentiating Eq. (8.3.107) i times and setting $m = 0$:

$$y_0^{(i)} = \hat{x}_0^{(i)} = \frac{d^i \hat{x}_m}{dm^i}\bigg|_0 = i!\, c_i = i!\, \mathbf{g}_i^T \mathbf{x}$$

Shifting these to time n, gives the differentiation filtering equations:

$$y^{(i)}(n) = i! \sum_{m=-M}^{M} g_i(-m) x(n-m), \qquad i = 0, 1, \dots, d \qquad (8.3.114)$$

where, as in Eq. (8.3.113), we reversed the order of writing the terms, but here the filters $\mathbf{g}_i$ are not necessarily symmetric. Actually, they are symmetric for even i, and antisymmetric for odd i.

Example 8.3.13: We construct the length-5 SG filters for the cases $d = 0$ and $d = 1$. For $d = 0$, corresponding to the constant $\hat{x}_m = c_0$ in Eq. (8.3.81), there is only one basis vector $\mathbf{s}_0$ defined in Eq. (8.3.83). The basis matrix $S = [\mathbf{s}_0]$ will have just one column, and the matrix F will be the scalar

$$F = S^T S = \mathbf{s}_0^T \mathbf{s}_0 = [1,1,1,1,1] \begin{bmatrix} 1 \\ 1 \\ 1 \\ 1 \\ 1 \end{bmatrix} = 5$$

The matrix G will then be

$$G = SF^{-1} = \frac{1}{5}\mathbf{s}_0 = \frac{1}{5}[1,1,1,1,1]^T$$

resulting in the SG matrix B:

$$B = GS^T = \frac{1}{5}\mathbf{s}_0\mathbf{s}_0^T = \frac{1}{5}\begin{bmatrix} 1 \\ 1 \\ 1 \\ 1 \\ 1 \end{bmatrix}[1,1,1,1,1] = \frac{1}{5}\begin{bmatrix} 1 & 1 & 1 & 1 & 1 \\ 1 & 1 & 1 & 1 & 1 \\ 1 & 1 & 1 & 1 & 1 \\ 1 & 1 & 1 & 1 & 1 \\ 1 & 1 & 1 & 1 & 1 \end{bmatrix}$$

Thus, the steady-state SG filter is the length-5 averager:

$$\mathbf{b}_0 = \frac{1}{5}[1,1,1,1,1]^T$$

For the case $d = 1$, corresponding to the linear fit $\hat{x}_m = c_0 + c_1 m$, we have the two basis vectors $\mathbf{s}_0$ and $\mathbf{s}_1$, given in Eq. (8.3.83). We calculate the matrices S, F, and F^{-1}:

$$S = [\mathbf{s}_0, \mathbf{s}_1] = \begin{bmatrix} 1 & -2 \\ 1 & -1 \\ 1 & 0 \\ 1 & 1 \\ 1 & 2 \end{bmatrix}, \quad F = S^T S = \begin{bmatrix} 5 & 0 \\ 0 & 10 \end{bmatrix}, \quad F^{-1} = \frac{1}{5}\begin{bmatrix} 1 & 0 \\ 0 & 0.5 \end{bmatrix}$$

This gives for G and B:

$$G = SF^{-1} = \frac{1}{5} \begin{bmatrix} 1 & -1 \\ 1 & -0.5 \\ 1 & 0 \\ 1 & 0.5 \\ 1 & 1 \end{bmatrix}, \quad B = GS^T = \frac{1}{5} \begin{bmatrix} 3 & 2 & 1 & 0 & -1 \\ 2 & 1.5 & 1 & 0.5 & 0 \\ 1 & 1 & 1 & 1 & 1 \\ 0 & 0.5 & 1 & 1.5 & 2 \\ -1 & 0 & 1 & 2 & 3 \end{bmatrix}$$

Thus, the steady-state SG filter $\mathbf{b}_0$ is still equal to the length-5 FIR averager. It is a general property of SG filters, that the filter $\mathbf{b}_0$ is the same for successive polynomial orders, that is, for $d = 0, 1$, $d = 2, 3$, $d = 4, 5$, and so on. However, the transient SG filters are different.   □

Example 8.3.14: Here, we construct the Savitzky-Golay filters of length $N = 5$ and order $d = 3$. The smoothed estimates are given by the cubic polynomial:

$$\hat{x}_m = c_0 + c_1 m + c_2 m^2 + c_3 m^3$$

There is an additional basis vector $\mathbf{s}_3$ with components $s_3(m) = m^3$. Therefore, the basis matrix S is:

$$S = [\mathbf{s}_0, \mathbf{s}_1, \mathbf{s}_2, \mathbf{s}_3] = \begin{bmatrix} 1 & -2 & 4 & -8 \\ 1 & -1 & 1 & -1 \\ 1 & 0 & 0 & 0 \\ 1 & 1 & 1 & 1 \\ 1 & 2 & 4 & 8 \end{bmatrix} \Rightarrow F = S^T S = \begin{bmatrix} 5 & 0 & 10 & 0 \\ 0 & 10 & 0 & 34 \\ 10 & 0 & 34 & 0 \\ 0 & 34 & 0 & 130 \end{bmatrix}$$

Because of the checkerboard pattern of this matrix, its inverse can be obtained from the inverses of the two 2×2 interlaced submatrices:

$$\begin{bmatrix} 5 & 10 \\ 10 & 34 \end{bmatrix}^{-1} = \frac{1}{70} \begin{bmatrix} 34 & -10 \\ -10 & 5 \end{bmatrix}, \quad \begin{bmatrix} 10 & 34 \\ 34 & 130 \end{bmatrix}^{-1} = \frac{1}{144} \begin{bmatrix} 130 & -34 \\ -34 & 10 \end{bmatrix}$$

Interlacing these inverses, we obtain:

$$F^{-1} = \begin{bmatrix} 34/70 & 0 & -10/70 & 0 \\ 0 & 130/144 & 0 & -34/144 \\ -10/70 & 0 & 5/70 & 0 \\ 0 & -34/144 & 0 & 10/144 \end{bmatrix}$$

Then, we compute the derivative filter matrix G:

$$G = SF^{-1} = \frac{1}{35} \begin{bmatrix} -3 & 35/12 & 5 & -35/12 \\ 12 & -70/3 & -2.5 & 35/6 \\ 17 & 0 & -5 & 0 \\ 12 & 70/3 & -2.5 & -35/6 \\ -3 & -35/12 & 5 & 35/12 \end{bmatrix}$$

and the SG matrix B:

$$B = SG^T = \frac{1}{35} \begin{bmatrix} 34.5 & 2 & -3 & 2 & -0.5 \\ 2 & 27 & 12 & -8 & 2 \\ -3 & 12 & 17 & 12 & -3 \\ 2 & -8 & 12 & 27 & 2 \\ -0.5 & 2 & -3 & 2 & 34.5 \end{bmatrix}$$

As mentioned above, the steady-state SG filter $\mathbf{b}_0$ is the same as that of case $d = 2$. But, the transient and differentiation filters are different. $\square$

In practice, the most common values of d are $0, 1, 2, 3, 4$. For these ds and arbitrary filter lengths N, the SG matrix B can be constructed in closed form; see references [200-216], as well as the extensive tables in [208]. Denoting the inverse of the $(d+1) \times (d+1)$ matrix $F = S^T S$ by $\Phi = F^{-1}$, we can write

$$B = SF^{-1}S^T = S\Phi S^T = \sum_{i=0}^{d} \sum_{j=0}^{d} \mathbf{s}_i \mathbf{s}_j^T \Phi_{ij}$$

which gives for the mkth matrix element

$$B_{mk} = \sum_{i=0}^{d} \sum_{j=0}^{d} s_i(m) s_j(k) \Phi_{ij} \quad \text{or,}$$

$$\boxed{B_{mk} = \sum_{i=0}^{d} \sum_{j=0}^{d} m^i k^j \Phi_{ij}}, \quad -M \le m, k \le M \tag{8.3.115}$$

Because of symmetry, $B_{mk} = B_{km}$, they are the kth component of the SG filter $\mathbf{b}_m$ or the mth component of the filter $\mathbf{b}_k$, that is,

$$B_{mk} = B_{km} = b_m(k) = b_k(m) = \sum_{i=0}^{d} \sum_{j=0}^{d} m^i k^j \Phi_{ij} \tag{8.3.116}$$

The matrix Φ can be determined easily for the cases $0 \le d \le 4$. The matrix F is what is known as a *Hankel matrix*, that is, having the same entries along each antidiagonal line. Therefore, its matrix elements F_{ij} depend only on the *sum $i + j$* of the indices. To see this, we write F_{ij} as the inner product:

$$F_{ij} = (S^T S)_{ij} = \mathbf{s}_i^T \mathbf{s}_j = \sum_{m=-M}^{M} s_i(m) s_j(m)$$

and because $s_i(m) = m^i$, we have:

$$\boxed{F_{ij} = \sum_{m=-M}^{M} m^{i+j} \equiv F_{i+j}}, \quad 0 \le i, j \le d \tag{8.3.117}$$

Note that because of the symmetric limits of summation, F_{i+j} will be zero whenever $i + j$ is odd. This leads to the checkerboard pattern of alternating zeros in F that we saw in the above examples. Also, because $d \le 4$, the only values of $i + j$ that we need are: $i + j = 0, 2, 4, 6, 8$. For those, the summations over m can be done in closed form, giving:

$$F_0 = \sum_{m=-M}^{M} m^0 = 2M + 1 = N$$

$$F_2 = \sum_{m=-M}^{M} m^2 = \frac{1}{3} M(M+1) F_0$$

$$F_4 = \sum_{m=-M}^{M} m^4 = \frac{1}{5}(3M^2 + 3M - 1)F_2 \tag{8.3.118}$$

$$F_6 = \sum_{m=-M}^{M} m^6 = \frac{1}{7}(3M^4 + 6M^3 - 3M + 1)F_2$$

$$F_8 = \sum_{m=-M}^{M} m^8 = \frac{1}{15}(5M^6 + 15M^5 + 5M^4 - 15M^3 - M^2 + 9M - 3)F_2$$

We can express F in terms of these definitions, for various values of d. For example, for $d = 0, 1, 2, 3$, the F matrices are:

$$[F_0],\quad \begin{bmatrix} F_0 & 0 \\ 0 & F_2 \end{bmatrix},\quad \begin{bmatrix} F_0 & 0 & F_2 \\ 0 & F_2 & 0 \\ F_2 & 0 & F_4 \end{bmatrix},\quad \begin{bmatrix} F_0 & 0 & F_2 & 0 \\ 0 & F_2 & 0 & F_4 \\ F_2 & 0 & F_4 & 0 \\ 0 & F_4 & 0 & F_6 \end{bmatrix}$$

The corresponding inverse matrices $\Phi = F^{-1}$ are obtained by interlacing the inverses of the checkerboard submatrices, as in Example 8.3.14. For $d = 0, 1, 2$, we have for Φ:

$$[1/F_0],\quad \begin{bmatrix} 1/F_0 & 0 \\ 0 & 1/F_2 \end{bmatrix},\quad \begin{bmatrix} F_4/D_4 & 0 & -F_2/D_4 \\ 0 & 1/F_2 & 0 \\ -F_2/D_4 & 0 & F_0/D_4 \end{bmatrix},$$

and for $d = 3$:

$$\Phi = F^{-1} = \begin{bmatrix} F_4/D_4 & 0 & -F_2/D_4 & 0 \\ 0 & F_6/D_8 & 0 & -F_4/D_8 \\ -F_2/D_4 & 0 & F_0/D_4 & 0 \\ 0 & -F_4/D_8 & 0 & F_6/D_8 \end{bmatrix}$$

where the D_4 and D_8 are determinants of the interlaced submatrices:

$$D_4 = F_0 F_4 - F_2^2,\qquad D_8 = F_2 F_6 - F_4^2 \tag{8.3.119}$$

Inserting the above expressions for Φ into Eq. (8.3.116), we determine the corresponding SG filters. For $d = 0$, we find for $-M \le m, k \le M$:

$$b_m(k) = B_{mk} = \frac{1}{F_0} = \frac{1}{N} \tag{8.3.120}$$

For $d = 1$:

$$b_m(k) = B_{mk} = \frac{1}{F_0} + \frac{mk}{F_2} \tag{8.3.121}$$

For $d = 2$:

$$b_m(k) = B_{mk} = \frac{F_4}{D_4} + \frac{1}{F_2} mk - \frac{F_2}{D_4}(m^2 + k^2) + \frac{F_0}{D_4} m^2 k^2 \tag{8.3.122}$$

For $d = 3$:

$$b_m(k) = B_{mk} = \frac{F_4}{D_4} + \frac{F_6}{D_8} mk - \frac{F_2}{D_4}(m^2 + k^2) + \frac{F_0}{D_4} m^2 k^2$$
$$- \frac{F_4}{D_8}(km^3 + mk^3) + \frac{F_2}{D_8} m^3 k^3 \tag{8.3.123}$$

In a similar fashion, we also find for the case $d = 4$:

$$b_m(k) = B_{mk} = \frac{D_{12}}{D} + \frac{F_6}{D_8} mk - \frac{D_{10}}{D}(m^2 + k^2) + \frac{E_8}{D} m^2 k^2$$
$$- \frac{F_4}{D_8}(km^3 + mk^3) + \frac{F_2}{D_8} m^3 k^3 + \frac{D_8}{D}(m^4 + k^4) \tag{8.3.124}$$
$$- \frac{D_6}{D}(m^2 k^4 + k^2 m^4) + \frac{D_4}{D} m^4 k^4$$

where

$$D_6 = F_0 F_6 - F_2 F_4 \qquad\qquad E_8 = F_0 F_8 - F_4^2$$
$$D_{10} = F_2 F_8 - F_4 F_6 \qquad\qquad D_{12} = F_4 F_8 - F_6^2 \tag{8.3.125}$$
$$D = F_0 D_{12} - F_2 D_{10} + F_4 D_8$$

In this case, the matrix F and its two interlaced submatrices are:

$$F = \begin{bmatrix} F_0 & 0 & F_2 & 0 & F_4 \\ 0 & F_2 & 0 & F_4 & 0 \\ F_2 & 0 & F_4 & 0 & F_6 \\ 0 & F_4 & 0 & F_6 & 0 \\ F_4 & 0 & F_6 & 0 & F_8 \end{bmatrix}, \quad \begin{bmatrix} F_0 & F_2 & F_4 \\ F_2 & F_4 & F_6 \\ F_4 & F_6 & F_8 \end{bmatrix}, \quad \begin{bmatrix} F_2 & F_4 \\ F_4 & F_6 \end{bmatrix}$$

Its inverse—obtained by interlacing the inverses of these two submatrices—can be expressed in terms of the determinant quantities of Eq. (8.3.125):

$$\Phi = F^{-1} = \begin{bmatrix} D_{12}/D & 0 & -D_{10}/D & 0 & D_8/D \\ 0 & F_6/D_8 & 0 & -F_4/D_8 & 0 \\ -D_{10}/D & 0 & E_8/D & 0 & -D_6/D \\ 0 & -F_4/D_8 & 0 & F_2/D_8 & 0 \\ D_8/D & 0 & -D_6/D & 0 & D_4/D \end{bmatrix}$$

Setting $m = 0$, we obtain the steady-state SG filters $b_0(k)$. For $d = 0, 1$ and $-M \le k \le M$, we have:

$$\boxed{b_0(k) = \frac{1}{N}} \tag{8.3.126}$$

For $d = 2, 3$:

$$\boxed{b_0(k) = \frac{F_4 - F_2 k^2}{D_4}} \tag{8.3.127}$$

where the ratios F_4/D_4 and F_2/D_4 can be simplified to:

$$\frac{F_4}{D_4} = \frac{3(3M^2 + 3M - 1)}{(2M + 3)(4M^2 - 1)}, \qquad \frac{F_2}{D_4} = \frac{15}{(2M + 3)(4M^2 - 1)} \tag{8.3.128}$$

Finally, for $d = 4, 5$:

$$\boxed{b_0(k) = \frac{D_{12} - D_{10}k^2 + D_8 k^4}{D}} \tag{8.3.129}$$

The coefficient ratios can also be simplified to:

$$\frac{D_{12}}{D} = \frac{15(15M^4 + 30M^3 - 35M^2 - 50M + 12)}{4(2M + 5)(4M^2 - 1)(4M^2 - 9)}$$

$$\frac{D_{10}}{D} = \frac{525(2M^2 + 2M - 3)}{4(2M + 5)(4M^2 - 1)(4M^2 - 9)} \tag{8.3.130}$$

$$\frac{D_8}{D} = \frac{945}{4(2M + 5)(4M^2 - 1)(4M^2 - 9)}$$

Example 8.3.15: Determine the quadratic/cubic SG filters of lengths $N = 5, 7, 9$. Using Eq. (8.3.127) with $M = 2, 3, 4$, we find (for $-M \le k \le M$):

$$b_0(k) = \frac{17 - 5k^2}{35} = \frac{1}{35}[-3, 12, 17, 12, -3]$$

$$b_0(k) = \frac{7 - k^2}{21} = \frac{1}{21}[-2, 3, 6, 7, 6, 3, -2]$$

$$b_0(k) = \frac{59 - 5k^2}{231} = \frac{1}{231}[-21, 14, 39, 54, 59, 54, 39, 14, -21]$$

where the coefficients have been reduced to integers as much as possible.   □

Example 8.3.16: Determine the quartic and quintic SG filters of length $N = 7, 9$. Using Eq. (8.3.129) with $M = 3, 4$, we find:

$$b_0(k) = \frac{131 - 61.25k^2 + 5.25k^4}{231} = \frac{1}{231}[5, -30, 75, 131, 75, -30, 5]$$

$$b_0(k) = \frac{179 - 46.25k^2 + 2.25k^4}{429} = \frac{1}{429}[15, -55, 30, 135, 179, 135, 30, -55, 15]$$

where again the coefficients have been reduced as much as possible. □

The Savitzky-Golay filters admit a nice *geometric* interpretation, which is standard in least-squares problems. Let $\mathbb{X}$ be the vector space of the N-dimensional real-valued vectors $\mathbf{x}$, and let $\mathbb{S}$ be the $(d+1)$-dimensional *subspace* spanned by all linear combinations of the basis vectors $\mathbf{s}_i$, $i = 0, 1, \ldots, d$.

Then, the smoothed vector $\hat{\mathbf{x}}$ of Eq. (8.3.110) will belong to the subspace $\mathbb{S}$. Moreover, because of the orthogonality equations (8.3.88), $\hat{\mathbf{x}}$ is *orthogonal* to the error vector $\mathbf{e}$. Indeed,

$$\hat{\mathbf{x}}^T \mathbf{e} = (S\mathbf{c})^T \mathbf{e} = \mathbf{c}^T S^T \mathbf{e} = 0$$

Solving the equation $\mathbf{e} = \mathbf{x} - \hat{\mathbf{x}}$ for $\mathbf{x}$, we obtain the *orthogonal decomposition*:

$$\mathbf{x} = \hat{\mathbf{x}} + \mathbf{e} \tag{8.3.131}$$

It expresses $\mathbf{x}$ as a sum of a part that belongs to the subspace $\mathbb{S}$ and a part that belongs to the *orthogonal complement* subspace $\mathbb{S}^\perp$. The decomposition is *unique* and represents the *direct sum* decomposition of the full vector space $\mathbb{X}$:

$$\mathbb{X} = \mathbb{S} \oplus \mathbb{S}^\perp$$

This requires that the dimension of the subspace $\mathbb{S}$ not exceed the dimension of the full space $\mathbb{X}$, that is, $d + 1 \leq N$. The component $\hat{\mathbf{x}}$ that lies in $\mathbb{S}$ is the *projection* of $\mathbf{x}$ onto $\mathbb{S}$. The matrix B in Eq. (8.3.92) is the corresponding *projection matrix*. As such, it will be symmetric, $B^T = B$, and *idempotent*:

$$B^2 = B \tag{8.3.132}$$

The proof is straightforward:

$$B^2 = (SF^{-1}S^T)(SF^{-1}S^T) = SF^{-1}(S^T S)F^{-1}S^T = SF^{-1}S^T = B$$

The matrix $(I - B)$, where I is the N-dimensional identity matrix, is also a projection matrix, projecting onto the orthogonal subspace $\mathbb{S}^\perp$. Thus, the error vector $\mathbf{e}$ belonging to $\mathbb{S}^\perp$ can be obtained from $\mathbf{x}$ by the projection:

$$\mathbf{e} = \mathbf{x} - \hat{\mathbf{x}} = (I - B)\mathbf{x}$$

Because $(I - B)$ is also idempotent and symmetric, $(I - B)^2 = (I - B)$, we obtain for the *minimized* value of the performance index $\mathcal{J}$ of Eq. (8.3.86):

$$\mathcal{J}_{\min} = \mathbf{e}^T \mathbf{e} = \mathbf{x}^T (I - B)^2 \mathbf{x} = \mathbf{x}^T (I - B) \mathbf{x} = \mathbf{x}^T \mathbf{x} - \mathbf{x}^T B \mathbf{x} \qquad (8.3.133)$$

The projection properties of B have some other useful consequences. For example, the NRR property mentioned previously follows from Eq. (8.3.132). Using the symmetry of B, we have

$$B^T = B = B^2 = B^T B$$

Taking matrix elements, we have $B_{km} = (B^T)_{mk} = (B^T B)_{mk}$. But, B_{km} is the kth component of the mth column $\mathbf{b}_m$. Using a similar argument as in Eq. (8.3.94), we also have $(B^T B)_{mk} = \mathbf{b}_m^T \mathbf{b}_k$. Therefore,

$$\mathbf{b}_m^T \mathbf{b}_k = b_m(k)$$

For $k = m$, we have the diagonal elements of $B^T B = B$:

$$NRR = \mathbf{b}_m^T \mathbf{b}_m = b_m(m) \qquad (8.3.134)$$

These are recognized as the NRRs of the filters $\mathbf{b}_m$. In particular, for $m = 0$, we have $NRR = \mathbf{b}_0^T \mathbf{b}_0 = b_0(0)$. Setting $k = 0$ in Eqs. (8.3.126)–(8.3.129), we find that the NRRs of the cases $d = 0, 1$, $d = 2, 3$, and $d = 4, 5$ are given by the coefficient ratios $1/F_0$, F_4/D_4, and D_{12}/D. Therefore:

$$(d = 0, 1) \qquad NRR = \frac{1}{N}$$

$$(d = 2, 3) \qquad NRR = \frac{3(3M^2 + 3M - 1)}{(2M + 3)(4M^2 - 1)} \qquad (8.3.135)$$

$$(d = 4, 5) \qquad NRR = \frac{15(15M^4 + 30M^3 - 35M^2 - 50M + 12)}{4(2M + 5)(4M^2 - 1)(4M^2 - 9)}$$

In the limit of large N or M, we have the approximate asymptotic expressions:

$$(d = 0, 1) \qquad NRR = \frac{1}{N}$$

$$(d = 2, 3) \qquad NRR \simeq \frac{9/4}{N} = \frac{2.25}{N} \qquad (8.3.136)$$

$$(d = 4, 5) \qquad NRR \simeq \frac{225/64}{N} = \frac{3.52}{N}$$

Thus, the noise reductions achieved by the quadratic/cubic and quartic/quintic cases are 2.25 and 3.52 times worse than that of the plain FIR averager of the same length N. Another consequence of the projection nature of B is:

$$BS = S, \qquad S^T B = S^T \qquad (8.3.137)$$

Indeed, $BS = S(S^T S)^{-1} S^T S = S$. Column-wise the first equation states:

$$B[\mathbf{s}_0, \mathbf{s}_1, \ldots, \mathbf{s}_d] = [\mathbf{s}_0, \mathbf{s}_1, \ldots, \mathbf{s}_d] \quad \Rightarrow \quad B\mathbf{s}_i = \mathbf{s}_i, \quad i = 0, 1, \ldots, d$$

Thus, the basis vectors $\mathbf{s}_i$ remain *invariant* under projection, but that is to be expected because they already lie in $\mathbb{S}$. In fact, any other linear combination of them, such as Eq. (8.3.110), remains invariant under B, that is, $B\hat{\mathbf{x}} = \hat{\mathbf{x}}$.

This property answers the question: When are the smoothed values equal to the original ones, $\hat{\mathbf{x}} = \mathbf{x}$, or, equivalently, when is the error zero, $\mathbf{e} = 0$? Because $\mathbf{e} = \mathbf{x} - B\mathbf{x}$, the error will be zero if and only if $B\mathbf{x} = \mathbf{x}$, which means that $\mathbf{x}$ already *lies* in $\mathbb{S}$, that is, it is a linear combination of $\mathbf{s}_i$. This implies that the samples x_m are already dth order polynomial functions of m, as in Eq. (8.3.107).

The second equation in (8.3.137) implies certain *constraints* on the filters $\mathbf{b}_m$, which can be used to develop an alternative approach to the SG filter design problem in terms of minimizing the NRR subject to constraints. To see this, write the transposed matrix S^T column-wise:

$$S^T = [\mathbf{u}_{-M}, \ldots, \mathbf{u}_{-1}, \mathbf{u}_0, \mathbf{u}_1, \ldots, \mathbf{u}_M] \tag{8.3.138}$$

For example, in the $N = 5$, $d = 2$ case, we have:

$$S^T = \begin{bmatrix} 1 & 1 & 1 & 1 & 1 \\ -2 & -1 & 0 & 1 & 2 \\ 4 & 1 & 0 & 1 & 4 \end{bmatrix} \equiv [\mathbf{u}_{-2}, \mathbf{u}_{-1}, \mathbf{u}_0, \mathbf{u}_1, \mathbf{u}_2]$$

It is easily verified that the mth column $\mathbf{u}_m$ is simply

$$\mathbf{u}_m = \begin{bmatrix} 1 \\ m \\ m^2 \\ \vdots \\ m^d \end{bmatrix}, \quad -M \le m \le M \tag{8.3.139}$$

Using $B = GS^T$, we can express the SG filters $\mathbf{b}_m$ in terms of $\mathbf{u}_m$, as follows:

$$[\mathbf{b}_{-M}, \ldots, \mathbf{b}_{-1}, \mathbf{b}_0, \mathbf{b}_1, \ldots, \mathbf{b}_M] = B = GS^T = G[\mathbf{u}_{-M}, \ldots, \mathbf{u}_{-1}, \mathbf{u}_0, \mathbf{u}_1, \ldots, \mathbf{u}_M]$$

which implies:

$$\mathbf{b}_m = G\mathbf{u}_m = SF^{-1}\mathbf{u}_m \tag{8.3.140}$$

Multiplying by S^T, we find $S^T\mathbf{b}_m = S^T SF^{-1}\mathbf{u}_m = \mathbf{u}_m$, or,

$$S^T\mathbf{b}_m = \mathbf{u}_m \quad \Rightarrow \quad \begin{bmatrix} \mathbf{s}_0^T\mathbf{b}_m \\ \mathbf{s}_1^T\mathbf{b}_m \\ \vdots \\ \mathbf{s}_d^T\mathbf{b}_m \end{bmatrix} = \begin{bmatrix} 1 \\ m \\ \vdots \\ m^d \end{bmatrix} \tag{8.3.141}$$

These relationships are the column-wise equivalent of $S^T B = S^T$. Thus, each SG filter $\mathbf{b}_m$ satisfies $(d+1)$ linear constraints:

$$\mathbf{s}_i^T \mathbf{b}_m = m^i, \qquad i = 0, 1, \ldots, d \tag{8.3.142}$$

Writing the dot products explicitly, we have equivalently:

$$\sum_{n=-M}^{M} n^i b_m(n) = m^i, \qquad i = 0, 1, \ldots, d \tag{8.3.143}$$

In particular, for the steady-state SG filter $\mathbf{b}_0$, we have $\mathbf{u}_0 = [1, 0, 0, \ldots, 0]^T$, with ith component $\delta(i)$. Therefore, the constraint $S^T \mathbf{b}_0 = \mathbf{u}_0$ reads component-wise:

$$\boxed{\sum_{n=-M}^{M} n^i b_0(n) = \delta(i), \quad i = 0, 1, \ldots, d} \tag{8.3.144}$$

For $i = 0$, this is the usual DC constraint:

$$\sum_{n=-M}^{M} b_0(n) = 1 \tag{8.3.145}$$

and for $i = 1, 2, \ldots, d$:

$$\sum_{n=-M}^{M} n^i b_0(n) = 0 \tag{8.3.146}$$

The quantity in the left-hand side of Eq. (8.3.144) is called the *i*th *moment* of the impulse response $b_0(n)$. Because of the symmetric limits of summation over n and the symmetry of $b_0(n)$ about its middle, the moments (8.3.146) will be zero for odd i, and therefore are not extra constraints. However, for even i, they are nontrivial constraints.

These moments are related to the *derivatives* of the frequency response at $\omega = 0$. Indeed, defining:

$$B_0(\omega) = \sum_{n=-M}^{M} b_0(n) e^{-j\omega n}$$

and differentiating it i times, we have:

$$j^i B_0^{(i)}(\omega) = j^i \frac{d^i}{d\omega^i} B_0(\omega) = \sum_{n=-M}^{M} n^i b_0(n) e^{-j\omega n}$$

Setting $\omega = 0$, we obtain:

$$j^i B_0^{(i)}(0) = j^i \frac{d^i}{d\omega^i} B_0(0) = \sum_{n=-M}^{M} n^i b_0(n) \tag{8.3.147}$$

Thus, the moment constraints (8.3.145) and (8.3.146) are equivalent to the DC constraint and the *flatness* constraints on the frequency response at $\omega = 0$:

$$B_0(0) = 1, \qquad B_0^{(i)}(0) = \frac{d^i}{d\omega^i} B_0(0) = 0, \quad i = 1, 2, \ldots, d \tag{8.3.148}$$

The larger the d, the more derivatives vanish at $\omega = 0$, and the flatter the response $B_0(\omega)$ becomes. This effectively increases the cutoff frequency of the lowpass filter—letting through more noise, but at the same time preserving more of the higher frequencies in the desired signal.

Figure 8.3.31 shows the magnitude response $|B_0(\omega)|$ for the cases $N = 7, 15$ and $d = 0, 2, 4$. The quadratic filters are flatter at DC than the plain FIR averager because of the extra constraint $B_0''(0) = 0$. Similarly, the quartic filters are even flatter because they satisfy two flatness conditions: $B_0''(0) = B_0^{(4)}(0) = 0$. The cutoff frequencies are *approximately* doubled and tripled in the cases $d = 2$ and $d = 4$, as compared to $d = 0$.

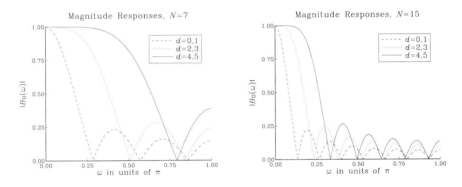

Fig. 8.3.31 Savitzky-Golay filters of lengths $N = 7, 15$, and orders $d = 0, 2, 4$.

A direct consequence of the moment constraints (8.3.144) is that the moments of the input signal $x(n)$ are *preserved* by the filtering operation (8.3.113), that is,

$$\sum_n n^i y(n) = \sum_n n^i x(n), \qquad i = 0, 1, \ldots, d \tag{8.3.149}$$

This can be proved easily working in the frequency domain. Differentiating the filtering equation $Y(\omega) = B_0(\omega) X(\omega)$ i times, and using the product rules of differentiation, we obtain:

$$Y^{(i)}(\omega) = \sum_{j=0}^{i} \binom{i}{j} B_0^{(j)}(\omega) X^{(i-j)}(\omega)$$

Setting $\omega = 0$ and using the moment constraints satisfied by the filter, $B_0^{(j)}(0) = \delta(j)$, we observe that only the $j = 0$ term will contribute to the above sum, giving:

$$Y^{(i)}(0) = B_0(0) X^{(i)}(0) = X^{(i)}(0), \qquad i = 0, 1, \dots, d$$

which implies Eq. (8.3.149), by virtue of Eq. (8.3.147) as applied to $x(n)$ and $y(n)$.

The preservation of moments is a useful property in applications, such as spectroscopic analysis or ECG processing, in which the desired signal has one or more sharp peaks, whose widths must be preserved by the smoothing operation. In particular, the second moment corresponding to $i = 2$ in Eq. (8.3.149) is a measure of the square of the width [200-216,303].

The above moment constraints can be used in a direct way to design the Savitzky-Golay filters. We consider the general problem of designing an optimum length-N filter that *minimizes* the NRR and satisfies $d + 1$ moment constraints. That is, minimize

$$NRR = \mathbf{b}^T \mathbf{b} = \sum_{n=-M}^{M} b(n)^2 = \min \tag{8.3.150}$$

subject to the $d + 1$ constraints, with given $\mathbf{u} = [u_0, u_1, \dots, u_d]^T$:

$$\mathbf{s}_i^T \mathbf{b} = \sum_{n=-M}^{M} n^i b(n) = u_i, \quad i = 0, 1, \dots, d \quad \Rightarrow \quad S^T \mathbf{b} = \mathbf{u} \tag{8.3.151}$$

The minimization of Eq. (8.3.150) subject to (8.3.151) can be carried out with the help of Lagrange multipliers, that is, adding the constraint terms to the performance index:

$$NRR = \mathbf{b}^T \mathbf{b} + 2 \sum_{i=0}^{d} \lambda_i (u_i - \mathbf{s}_i^T \mathbf{b}) \tag{8.3.152}$$

The gradient of *NRR* with respect to the unknown filter $\mathbf{b}$ is:

$$\frac{\partial}{\partial \mathbf{b}} NRR = 2\mathbf{b} - 2 \sum_{i=0}^{d} \lambda_i \mathbf{s}_i$$

Setting the gradient to zero, and solving for $\mathbf{b}$ gives:

$$\mathbf{b} = \sum_{i=0}^{d} \lambda_i \mathbf{s}_i = [\mathbf{s}_0, \mathbf{s}_1, \dots, \mathbf{s}_d] \begin{bmatrix} \lambda_0 \\ \lambda_1 \\ \vdots \\ \lambda_d \end{bmatrix} = S\lambda$$

Component-wise this means that $b(n)$ has the polynomial form:

$$b(n) = \sum_{i=0}^{d} \lambda_i s_i(n) = \sum_{i=0}^{d} \lambda_i n^i, \qquad -M \le n \le M$$

The Lagrange multiplier vector $\boldsymbol{\lambda}$ is determined by imposing the desired constraint, that is,

$$\mathbf{u} = S^T \mathbf{b} = S^T S \boldsymbol{\lambda} = F \boldsymbol{\lambda} \quad \Rightarrow \quad \boldsymbol{\lambda} = F^{-1} \mathbf{u}$$

which gives finally for the optimum $\mathbf{b}$:

$$\mathbf{b} = S \boldsymbol{\lambda} = S F^{-1} \mathbf{u} = G \mathbf{u} \tag{8.3.153}$$

Comparing this solution with Eqs. (8.3.140) and (8.3.141), we conclude that the SG filters $\mathbf{b}_m$ can be thought of as the optimum filters that have minimum NRR with constraint vectors $\mathbf{u} = \mathbf{u}_m$. In particular, the steady-state SG filter $\mathbf{b}_0$ minimizes the NRR with the constraint vector $\mathbf{u} = \mathbf{u}_0 = [1, 0, \ldots, 0]^T$.

Next, we discuss the implementation of the Savitzky-Golay filters and the role played by the steady-state and transient filters $\mathbf{b}_m$. In implementing the smoothing equation Eq. (8.3.113), we would like to use our existing FIR filtering tools, such as the routines fir, delay, and dot, or their circular-buffer versions, as discussed in Chapter 4.

Because of the presence of the terms $x(n + M), \ldots, x(n + 1)$ in (8.3.113), the filtering operation is not completely causal. However, we can do one of two things to make our processing causal: (1) *delay the filter* by M time units to make it causal, which will have the effect of producing the correct output, but with a delay of M samples; (2) *advance the input* by M time units and then filter it causally. In this case, the output comes out undelayed.

We can understand these remarks in the z-domain. We denote the filter by $B_0(z)$ and its delayed causal version by $H_0(z) = z^{-M} B_0(z)$. For example, for $N = 5$ and $d = 2$, we have:

$$B_0(z) = \frac{1}{35}(-3z^2 + 12z + 17 + 12z^{-1} - 3z^{-2})$$

$$H_0(z) = z^{-2} B_0(z) = \frac{1}{35}(-3 + 12z^{-1} + 17z^{-2} + 12z^{-3} - 3z^{-4})$$

The filtering equation (8.3.113) becomes in the z-domain:

$$Y(z) = B_0(z) X(z)$$

Multiplying both sides by z^{-M} gives:

$$z^{-M} Y(z) = z^{-M} B_0(z) X(z) = H_0(z) X(z)$$

which states that if $X(z)$ is processed by the *causal* filter $H_0(z)$, the output will be delayed by M units. Alternatively, we can write:

$$Y(z) = B_0(z) X(z) = z^{-M} B_0(z) \, z^M X(z) = H_0(z) \left(z^M X(z) \right)$$

which states that the *causal* filtering of the *time-advanced* input $z^M X(z)$ will result into the same output as that obtained by the non-causal filter.

For smoothing problems, we prefer to use the second alternative because it produces the smoothed output undelayed. Because the filter length is N, we need an N-dimensional delay-line buffer $\mathbf{w}$ to hold the internal states of the filter. Normally, the internal states are defined as the successively delayed inputs $x(n-i)$. However, because the input is advanced by M units, we must define $w_i(n) = x(n + M - i)$, $i = 0, 1, \ldots, N - 1$, that is,

$$\mathbf{w}(n) = [x(n + M), \ldots, x(n), \ldots, x(n - M)]^T \tag{8.3.154}$$

For example, if $N = 5$ and $M = (N - 1)/2 = 2$,

$$\mathbf{w}(n) = \begin{bmatrix} w_0(n) \\ w_1(n) \\ w_2(n) \\ w_3(n) \\ w_4(n) \end{bmatrix} = \begin{bmatrix} x_{n+2} \\ x_{n+1} \\ x_n \\ x_{n-1} \\ x_{n-2} \end{bmatrix}$$

In particular, at $n = 0$ we recognize $\mathbf{w}$ as being the *reverse* of the vector $\mathbf{x}$, that is,

$$\mathbf{w} = \begin{bmatrix} x_2 \\ x_1 \\ x_0 \\ x_{-1} \\ x_{-2} \end{bmatrix} = \mathbf{x}^R \quad \Rightarrow \quad \mathbf{w}^R = \mathbf{x} = \begin{bmatrix} x_{-2} \\ x_{-1} \\ x_0 \\ x_1 \\ x_2 \end{bmatrix}$$

For implementation purposes, we must rewrite all of the previous dot products in terms of the state vector $\mathbf{w}$. To do this, we use the property that dot products remain unchanged if *both* vectors are reversed; for example,

$$\mathbf{a}^T \mathbf{b} = [a_1, a_2, a_3] \begin{bmatrix} b_1 \\ b_2 \\ b_3 \end{bmatrix} = [a_3, a_2, a_1] \begin{bmatrix} b_3 \\ b_2 \\ b_1 \end{bmatrix} = \mathbf{a}^{RT} \mathbf{b}^R$$

where the superscript RT means the transpose of the reversed vector. To express the smoothed samples (8.3.112) in terms of $\mathbf{w}$, we replace both vectors by their reverses:

$$\hat{x}_m = \mathbf{b}_m^T \mathbf{x} = \mathbf{b}_m^{RT} \mathbf{x}^R = \mathbf{b}_m^{RT} \mathbf{w} \tag{8.3.155}$$

It is a general property of the SG matrix B that its columns are the reverse of each other with respect to the middle column, that is,

$$\mathbf{b}_m^R = \mathbf{b}_{-m} \tag{8.3.156}$$

In particular, $\mathbf{b}_0^R = \mathbf{b}_0$, so that the middle column is a symmetric filter. Using Eqs. (8.3.155) and (8.3.156), we can rewrite Eq. (8.3.112) in the "filtering" form:

$$\hat{x}_m = \mathbf{b}_{-m}^T \mathbf{w} \tag{8.3.157}$$

In practice, we have typically a block of noisy samples x_n, $n = 0, 1, \ldots, L-1$ and wish to replace them by their smoothed versions y_n, $n = 0, 1, \ldots, L-1$. Because the computation of each y_n requires M input samples above x_n and M below, we can use the filtering equation (8.3.113) only *after* the first M and *before* the last M inputs. That is, only for $M \leq n \leq L - 1 - M$.

Thus, the *initial* value of the state vector $\mathbf{w}$ is obtained by setting $n = M$ in Eq. (8.3.154), and the final value by setting $n = L - 1 - M$:

$$
\mathbf{w}(M) = \begin{bmatrix} x_{N-1} \\ x_{N-2} \\ \vdots \\ x_0 \end{bmatrix}, \quad \mathbf{w}(L-1-M) = \begin{bmatrix} x_{L-1} \\ x_{L-2} \\ \vdots \\ x_{L-N} \end{bmatrix} \tag{8.3.158}
$$

In other words, the initial state vector consists of the *first* N input samples, whereas the final one consists of the *last* N samples.

Once the first N input samples have been read into the state vector $\mathbf{w}(M)$, we may use them to compute the first steady output $y_M = \mathbf{b}_0^T \mathbf{w}(M)$, but we can also use them to *estimate* the first M input-on transients using Eq. (8.3.157). For example, the sample x_{M-1} that lies just to the left of x_M will be smoothed by $\hat{x}_m$ with $m = -1$, that is, $y_{M-1} = \mathbf{b}_1^T \mathbf{w}(M)$, and so on. Thus, given the initial state $\mathbf{w}(M)$, we compute the first $M + 1$ outputs:

$$
y_{M-m} = \mathbf{b}_m^T \mathbf{w}(M), \qquad m = 0, 1, \ldots, M
$$

or, re-indexing them from y_0 to y_M:

$$
\boxed{y_i = \mathbf{b}_{M-i}^T \mathbf{w}(M), \qquad i = 0, 1, \ldots, M} \tag{8.3.159}
$$

Similarly, once the last N input samples have been read into the last state vector, we can use them to compute the last steady output y_{L-1-M}, and the last M input-off transients, that is, using Eq. (8.3.157) with $m = 0, 1, \ldots, M$:

$$
\boxed{y_{L-1-M+m} = \mathbf{b}_{-m}^T \mathbf{w}(L-1-M), \qquad m = 0, 1, \ldots, M} \tag{8.3.160}
$$

Note that the computations of the first and last steady-state outputs at $n = M$ and $n = L - 1 - M$ are also included in Eqs. (8.3.159) and (8.3.160). Between these two extremes, we can calculate all of the remaining smoothed samples using the steady-state filter $\mathbf{b}_0$, that is,

$$
\boxed{y_n = \mathbf{b}_0^T \mathbf{w}(n), \qquad M + 1 \leq n \leq L - 2 - M} \tag{8.3.161}
$$

In order that there be at least one steady-state output, the length L of the input signal must be $L \geq N + 1$.

Example 8.3.17: For an input signal of length $L = 10$ and the quadratic, length-5, SG filters given in Eq. (8.3.97), we compute the input-on, steady, and input-off transients as follows:

$$y_0 = \mathbf{b}_2^T \mathbf{w}(2) = \frac{1}{35}[3, -5, -3, 9, 31]\mathbf{w}(2)$$

$$y_1 = \mathbf{b}_1^T \mathbf{w}(2) = \frac{1}{35}[-5, 6, 12, 13, 9]\mathbf{w}(2)$$

$$y_2 = \mathbf{b}_0^T \mathbf{w}(2) = \frac{1}{35}[-3, 12, 17, 12, -3]\mathbf{w}(2)$$

$$y_n = \mathbf{b}_0^T \mathbf{w}(n) = \frac{1}{35}[-3, 12, 17, 12, -3]\mathbf{w}(n), \quad 3 \le n \le 6$$

$$y_7 = \mathbf{b}_0^T \mathbf{w}(7) = \frac{1}{35}[-3, 12, 17, 12, -3]\mathbf{w}(7)$$

$$y_8 = \mathbf{b}_{-1}^T \mathbf{w}(7) = \frac{1}{35}[9, 13, 12, 6, -5]\mathbf{w}(7)$$

$$y_9 = \mathbf{b}_{-2}^T \mathbf{w}(7) = \frac{1}{35}[31, 9, -3, 5, 3]\mathbf{w}(7)$$

where

$$\mathbf{w}(2) = \begin{bmatrix} x_4 \\ x_3 \\ x_2 \\ x_1 \\ x_0 \end{bmatrix}, \quad \mathbf{w}(n) = \begin{bmatrix} x_{n+2} \\ x_{n+1} \\ x_n \\ x_{n-1} \\ x_{n-2} \end{bmatrix}, \quad \mathbf{w}(7) = \begin{bmatrix} x_9 \\ x_8 \\ x_7 \\ x_6 \\ x_5 \end{bmatrix}$$

Figure 8.3.32 illustrates these computations, indicating what filters are used with what outputs.   □

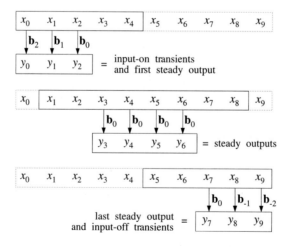

Fig. 8.3.32 Steady-state and transient outputs for $L = 10$, $N = 5$.

Example 8.3.18: Filter the following input signal through the length-5 Savitzky-Golay filters of polynomial orders $d = 0, 1, 2$:

$$\mathbf{x} = [7, 0, 7, 28, 63, 7, 0, 7, 28, 63]$$

This signal was generated by the expression $x_m = 7 - 14m + 7m^2$, $m = 0, 1, 2, 3, 4$ and repeated once. The outputs of the SG filters for $d = 0, 1, 2$ are:

$$\mathbf{y} = [21, 21, 21, 21, 21, 21, 21, 21, 21, 21]$$

$$\mathbf{y} = [-7, 7, 21, 21, 21, 21, 21, 21, 35, 49]$$

$$\mathbf{y} = [7, 0, 7, 37, 42, 22, -3, 7, 28, 63]$$

The $d = 0$ filter gives the same answer because it so happened that the average of any 5 successive samples is 21. The $d = 1$ filter tries to predict the first 2 and last 2 values on the basis of the first 5 and last 5 input samples, but its steady outputs are still 21 because the $\mathbf{b}_0$ filters are the same for $d = 0, 1$. The $d = 2$ filter also predicts the first and last 2 samples, but it predicts them exactly because by design the first 5 and last 5 inputs were taken to be quadratic functions of time. □

The C implementation of the filtering equations (8.3.159)–(8.3.161) requires the $N{\times}N$ SG matrix B and N-dimensional state vector $\mathbf{w}$. They can be declared as follows:

```
double *w, **B;
w = (double *) calloc(N, sizeof(double));

B = (double **) calloc(N, sizeof(double *));      allocate N rows
for (m=0; m<N; m++)
      B[m] = (double *) calloc(N, sizeof(double));    allocate mth row
```

The matrix elements B[m][k] can be defined via the closed form expressions of (8.3.120)–(8.3.124), with appropriate shifting of the indices. For example, the matrix element B[M+m][M+k] represents B_{mk}, $-M \leq m, k \leq M$. Similarly, the mth filter $\mathbf{b}_m$ will be represented by the row B[M+m].

We may assume that the input samples are read sequentially from an input file x.dat and the smoothed data are returned sequentially into the file y.dat. The pointers to these files may be defined by:

```
FILE *fpx, *fpy;
fpx = fopen("x.dat", "r");       note that fpx could also be stdin
fpy = fopen("y.dat", "w");       and fpy could be stdout
```

The filtering steps are as follows: The state vector $\mathbf{w}$ is loaded with the first N input samples, with the following for-loop:

```
for (i=0; i<N; i++) {            read N samples
      fscanf(fpx, "%lf", &x);    read sample x
      delay(N-1, w);             shift delay line
      w[0] = x;
      }
```

We assume that the input file fpx contains *at least* N samples. The reason for shifting the delay *before* $w[0]$ is loaded with x is that after the Nth sample has been read, the state vector **w** will be left as in Eq. (8.3.158).

The state vector is then frozen for the first $M + 1$ output computations, corresponding to Eq. (8.3.159). The outputs are computed by the dot product routine dot and dumped into the output file:

```
for (i=0; i<=M; i++) {
        y = dot(N-1, B[N-1-i], w);                note, B[N – 1 – i] = B[2M – i] = bM–i
        fprintf(fpy, "%lf\n", y);
        }
delay(N-1, w);                                    updates delay after first steady output
```

The last dot product, together with the subsequent call to delay, are effectively equivalent to a call to fir for computing the first steady output y_M.

Then, we keep reading input samples until the end of the file fpx is encountered, and processing them with fir:

```
while(fscanf(fpx, "%lf", &x) != EOF) {
        u = w[N-1];                               needed to unshift the delay line
        y = fir(N-1, B[M], w, x);                 note, B[M] = b0
        fprintf(fpy, "%lf\n", y);
        }
```

We assume that there is at least one such sample to process in fpx; thus, the input signal should have length $L \geq N + 1$. After the last input sample is read and processed, the delay line is updated by fir. Therefore, upon exiting from this loop, we must *unshift* the delay line and restore **w** to its last steady value, given by Eq. (8.3.158). The auxiliary quantity u keeps track of the last tap of the delay line, and enables us to unshift it:

```
for (i=0; i<N-1; i++)                             unshift delay line
        w[i] = w[i+1];                            w[0] = w[1], w[1] = w[2], etc.
w[N-1] = u;                                       restore Nth tap
```

Finally, the unshifted state vector **w** is kept constant and the last M transient outputs are computed using the filters $\mathbf{b}_{-i}$, which are represented in C by the columns B[M-i]:

```
for (i=1; i<=M; i++) {
        y = dot(N-1, B[M-i], w);                  note, B[M – i] = b–i
        fprintf(fpy, "%lf\n", y);
        }
```

The above C implementation is appropriate for sample-by-sample processing. If the length L of the input is known, one could use a block processing implementation based on Eqs. (8.3.159)–(8.3.161).

Using MATLAB, we can implement very efficiently both the design equations for the SG filters and the filtering equations. The M-file sg.m of Appendix D implements the SG design equations (8.3.111) for arbitrary N and d.

The M-file `sgfilt.m` implements the filtering operations (8.3.159)–(8.3.161). It calls `sg.m` to design the filter and, then, filters a given length-L block of data **x**. The returned filtered vector **y** also has length L.

Finally, we present a simulation example illustrating the behavior of the Savitzky-Golay smoothing filters. Figure 8.3.33 shows a noise-free ECG signal of length $L = 500$, whose QRS peak is normalized to unity. The noisy ECG has 30% zero-mean white Gaussian noise added to it; that is, the rms noise value is $\sigma_v = 0.3$.

Figures 8.3.33 and 8.3.34 compare the plain FIR averager, the quadratic and quartic Savitzky-Golay filters corresponding to $d = 0, 2, 4$, for two filter lengths, $N = 11$ and $N = 41$.

As expected, the FIR averager is more effective in removing noise than the quadratic and quartic SG filters, but at the expense of smoothing out the QRS peak too much, especially for $N = 41$ when the filter's length is comparable to the width of the QRS peak. The quadratic and quartic SG filters preserve the peaks much better than the FIR averager.

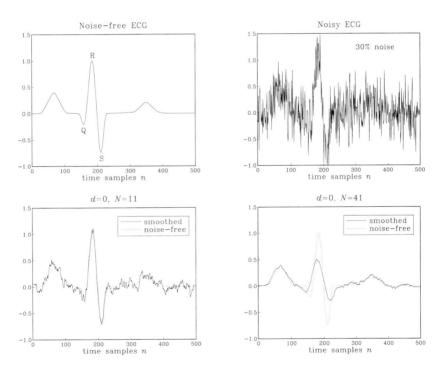

Fig. 8.3.33 Savitzky-Golay filter comparisons, for $N = 11, 41$ and $d = 0$.

Further guidelines on the use of Savitzky-Golay filters may be found in references [200–216,303]. FIR lowpass filters designed by more conventional methods, such as the Kaiser window method, can also be used [217] in data smoothing applications.

Multiple filtering of the data by the same filter has often been used in conjunction with Savitzky-Golay filters, but it can be used with any other lowpass FIR filter.

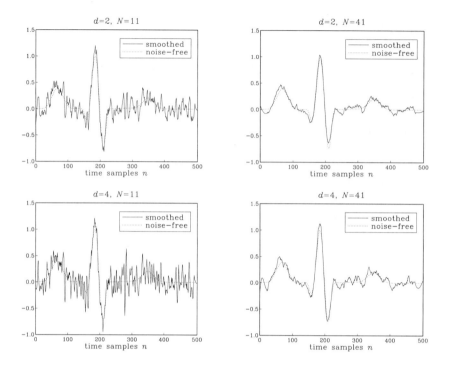

Fig. 8.3.34 Savitzky-Golay filter comparisons, for $N = 11, 41$ and $d = 2, 4$.

A systematic discussion of how to use a given filter repeatedly so that both the passband and stopband of the filter are improved simultaneously has been given by Kaiser and Hamming [218].

8.4 Problems

8.1 It is desired to generate the periodic sequence $\mathbf{h} = [0, 1, 2, 0, 1, 2, 0, 1, 2, \dots]$ of period three. Determine the filter $H(z)$ whose impulse response is $\mathbf{h}$.

    a. Realize the filter in its direct and canonical forms. Write the corresponding sample processing algorithms for generating the periodic sequence. Crank the algorithms for a total of 9 iterations, making a list of the values of the internal states and output of the filter.

    b. For the direct form realization, rewrite the generation algorithm in its circular-buffer form of Eq. (8.1.20) or (8.1.22), and initialized by Eq. (8.1.19).

        Iterate the algorithm 15 times, making a table of the internal states $\mathbf{w}$, the output y, the circular pointer index q, and indicating the buffer entry that holds the current output for each iteration. Why did we choose to iterate 15 times? Do you observe the repetition period of the buffer entries?

8.2 Consider the filter $H(z) = \dfrac{1 + 2z^{-1} + 3z^{-2} - 4z^{-3} - 5z^{-4}}{1 - z^{-5}}$. What is its periodic causal impulse response? Realize the filter in its direct and canonical forms.

    a. For each realization, write the corresponding sample processing algorithm for generating the periodic impulse response. Crank the algorithm for a total of 15 iterations, making a list of the values of the internal states and output of the filter.

    b. For the direct form realization, iterate the generation algorithm in its circular buffer form, making a table as in Problem 8.1(b). How many iterations are needed before we observe the repetition of the buffer entries?

8.3 The eight waveform samples:

$$\mathbf{b} = [b_0, b_1, b_2, b_3, b_4, b_5, b_6, b_7]$$

are stored in reverse order in the eight-dimensional circular wavetable:

$$\mathbf{w} = [w_0, w_1, w_2, w_3, w_4, w_5, w_6, w_7]$$

It is desired to generate a periodic subsequence of period $d = 5$. Determine this subsequence when the output is obtained by the four methods of: (a) truncating down, (b) truncating up, (c) rounding, and (d) linear interpolation.

8.4 Repeat Problem 8.3 when the subsequence has period $d = 6$.

8.5 The waveform samples $\mathbf{b} = [1, 2, 3, 4, 5, 6, 7, 8]$ are stored (in reverse order) into an eight-dimensional circular wavetable $\mathbf{w}$. It is desired to use the wavetable to generate a periodic subsequence of period 3. Determine this subsequence when the output is obtained by the four approximations of: (a) truncating down, (b) truncating up, (c) rounding, and (d) linear interpolation.

8.6 Repeat Problem 8.5, for generating a subsequence of period 5. Repeat for a subsequence of period 6.

8.7 *Computer Experiment: Wavetable Generators.* Using the wavetable generator wavgen, write a C program to reproduce all the graphs of Fig. 8.1.18.

    Then, repeat using the rounding and interpolation versions of the wavetable generator, wavgenr and wavgeni. Compare the outputs of the three generator types.

8.8 *Computer Experiment: Wavetable Amplitude and Frequency Modulation.* Write a program to reproduce all the graphs of Figures 8.1.20–8.1.23.

8.9 Consider the four comb filters:

$$y(n) = x(n) + x(n-8), \qquad y(n) = x(n) + x(n-8) + x(n-16)$$

$$y(n) = x(n) - x(n-8), \qquad y(n) = x(n) - x(n-8) + x(n-16)$$

Determine their transfer functions and their impulse responses. Place their zeros on the z-plane relative to the unit circle. Sketch their magnitude responses. How are they similar or different? Draw their canonical realization forms using 8-fold delays z^{-8}. Write the corresponding sample processing algorithms both in their linear and circular-buffer versions.

8.10 *Computer Experiment: Flanging and Chorusing.* Write a C program to reproduce the graphs of Figures 8.2.9 and 8.2.11.

Repeat the chorusing experiment using the following model for the chorus processor, shown in Fig. 8.2.10:

$$y(n) = \frac{1}{3}\left[x(n) + a_1(n)x(n - d_1(n)) + a_2(n)x(n - d_2(n))\right]$$

where $d_1(n)$ and $d_2(n)$ are generated as in Eq. (8.2.20) by the low-frequency noise routine ran1 of Appendix B.2 using two different seeds. The amplitudes $a_1(n), a_2(n)$ are also low-frequency random numbers with unity mean.

Repeat the flanging experiment using the recursive flanging processor:

$$y(n) = ay(n - d(n)) + x(n)$$

where $a = 0.8$. State the processing algorithm in this case, using a circular buffer for the feedback delay line and the routine tapi to interpolate between buffer entries.

8.11 *Computer Experiment: Reverberation Examples.* Using the circular-buffer reverberator routines plain, allpass, lowpass, write a C program to reproduce all the graphs of Fig. 8.2.22. [*Caution:* Use different circular buffers for the three reverb filters.]

8.12 *Computer Experiment: Schroeder's Reverberator.* Write a C program that implements Schroeder's reverberator shown in Fig. 8.2.18 and uses the sample processing algorithm (8.2.31). Iterate the sample processing algorithm for $0 \le n \le 500$ and reproduce the impulse response shown in Fig. 8.2.19.

8.13 Consider the lowpass reverberator shown in Fig. 8.2.21. Write *explicitly* all the difference equations required for its time-domain implementation. Then, write the corresponding sample processing algorithm, with the D-fold delay implemented circularly.

8.14 Consider the lowpass reverberator $H(z)$ of Eq. (8.2.32) with the first-order feedback filter (8.2.35). Let $p_i, A_i, i = 1, 2, \ldots, D + 1$ be the poles and residues of the $H(z)$, that is,

$$H(z) = \frac{1}{1 - z^{-D}G(z)} = \sum_{i=1}^{D+1} \frac{A_i}{1 - p_i z^{-1}}$$

Assume that all p_i are inside the unit circle. Note that if $b_1 = 0$, then there are only D poles. Suppose a sinusoid of frequency ω and duration L is applied to the input:

$$x(n) = e^{j\omega n}(u(n) - u(n - L))$$

Show that the output signal will be given by:

$$y(n) = H(\omega)e^{j\omega n}(u(n) - u(n - L)) + \sum_{i=1}^{D+1} B_i p_i^n (u(n) - e^{j\omega L}p_i^{-L}u(n - L))$$

where $B_i = p_i A_i / (p_i - e^{j\omega}), i = 1, 2, \ldots, D + 1$. See also Problem 6.29.

8.15 *Computer Experiment: Reverberator Time Constants.* Reproduce all the graphs of Figure 8.2.25 by iterating the sample processing algorithms of the plain and lowpass reverberators. The input is defined as:

$$x(n) = \cos(\omega n)\left(u(n) - u(n-150)\right)$$

with $\omega = 0.2\pi$ and $\omega = \pi$. Generate similar graphs also for the following frequencies: $\omega = 0.4\pi$, 0.6π, 0.8π, and 0.99π.

For the lowpass cases, verify that the output obtained by iterating the sample processing algorithm agrees with (the real part of) the analytical expression given in Problem 8.14. For this part, you will need to use MATLAB to calculate the poles p_i, residues A_i, B_i, and evaluate the expression for $y(n)$, for $0 \le n \le 299$.

8.16 *Computer Experiment: Karplus-Strong String Algorithm.* The Karplus-Strong algorithm for generating plucked-string sounds [108–110] is defined by the lowpass reverberator filter of Eq. (8.2.32) with feedback filter $G(z) = (1 + z^{-1})/2$. It was described in Section 8.2.3.

For the two delay values $D = 25, 50$, initialize the delay-line buffer by filling it with zero-mean random numbers, for example, $w[i] = \text{ran}(\&\text{iseed}) - 0.5$, for $i = 0, 1, \ldots, D$. Then, run the sample processing algorithm (8.2.34) with zero input $x(n) = 0$, for $0 \le n \le 499$. Plot the resulting output signals $y(n)$.

The harshness of the initial plucking of the string is simulated by the initial random numbers stored in the delay line. As these random numbers recirculate the delay line, they get lowpass filtered by $G(z)$, thus losing their high-frequency content and resulting in a decaying signal that is dominated basically by the frequency $f_1 = f_s/D$.

8.17 A prototypical delay effect usually built into commercial audio DSP effects processors is given by the transfer function:

$$H(z) = c + b\frac{z^{-D}}{1 - az^{-D}}$$

where c represents the direct sound path. Draw a block diagram of this filter using only one D-fold delay z^{-D}. Write the difference equations describing it and translate them into a sample processing algorithm implemented with a circular buffer.

8.18 *Computer Experiment: Plain and Lowpass Reverberating Delays.* The basic building blocks of many multi-delay effects are the following plain and lowpass reverberating delays:

$$H(z) = \frac{z^{-D}}{1 - az^{-D}}, \qquad H(z) = \frac{z^{-D}}{1 - z^{-D}G(z)}$$

where $G(z)$ is a lowpass feedback filter. Draw the block diagrams of these filters and write their sample processing algorithms implementing z^{-D} circularly. Then, translate the algorithms into C routines, say `plaindel.c` and `lpdel.c`. How do they differ from the routines `plain` and `lowpass` of Section 8.2.3?

8.19 *Computer Experiment: Multi-Delay Effects.* Commercial audio DSP effects processors have built-in multi-delay effects obtained by cascading several basic reverberating delay of the type of Problem 8.18; for example, see Ref. [147].

A typical example was shown in Fig. 8.2.27. Write a C program that implements this block diagram. The program must make use of the two routines `plaindel` and `lpdel` that you wrote in the previous problem.

Note, that you will need to use two circular buffers $\{w_1, w_2\}$ and their circular pointers $\{p_1, p_2\}$, for the two delays.

Using this program, and the parameter values that were used in Fig. 8.2.28, compute and plot the outputs of the filter, for $0 \le n \le 2000$, for the two inputs:

$$x(n) = \delta(n), \qquad x(n) = u(n) - u(n - 100)$$

8.20 *Computer Experiment: Multi-Tap Delay Effects.* In the electronic music community, a multitap delay is usually defined to have both feed forward and feedback paths, as well as a direct sound path, with user-adjustable gains; for example, see Ref. [147].

Write a C routine that implements the circular-buffer version of the sample processing algorithm of the multitap delay line shown in Fig. 8.2.29. The inputs to the routine should be the current input audio sample x, the values of the forward taps $\{b_0, b_1, b_2\}$, feedback taps $\{a_1, a_2\}$, delay values $\{D_1, D_2\}$, and the $(D_1 + D_2)$-dimensional delay-line buffer w and its associated circular pointer p.

Using this routine, and the parameter values that were used for the stable case of Fig. 8.2.30, compute and plot the outputs of the filter, for $0 \le n \le 1000$, for the two inputs:

$$x(n) = \delta(n), \qquad x(n) = u(n) - u(n - 200)$$

8.21 Show that the condition $|a_1| + |a_2| < 1$ is sufficient to guarantee the stability of the multitap delay line filter of Eq. (8.2.43). [*Hint:* Work with the pole equation $z^{D_1 + D_2} = a_1 z^{D_2} + a_2$.]

8.22 Stereo delay effects can be accomplished by the block diagram of Fig. 8.4.1. Two basic delays of the type of Problem 8.18 are used in the left and right channels and are coupled by introducing cross-feedback coefficients, such that the reverberating output of one is fed into the input of the other; for example, see Ref. [147]. Show that the input/output relationships can be expressed in the z-domain as:

$$Y_L(z) = H_{LL}(z) X_L(z) + H_{LR}(z) X_R(z)$$
$$Y_R(z) = H_{RL}(z) X_L(z) + H_{RR}(z) X_R(z)$$

Determine the direct and cross-transfer functions $H_{LL}(z), H_{LR}(z), H_{RL}(z), H_{RR}(z)$, in terms of the indicated multipliers and feedback filters $G_L(z), G_R(z)$. What conclusions do you draw in the special cases: (1) $d_L = 0$, $d_R \neq 0$; (2) $d_L \neq 0$, $d_R = 0$; (3) $d_L = 0$, $d_R = 0$?

Consider the case of the plain feedback filters: $G_L(z) = a_L$, $G_R(z) = a_R$. Introduce two delay-line buffers w_L and w_R for the indicated delays z^{-L} and z^{-R} and write the *difference equations* describing the time-domain operation of the block diagram. Then, translate the difference equations into a sample processing algorithm that transforms each input stereo pair $\{x_L, x_R\}$ into the corresponding output stereo pair $\{y_L, y_R\}$. Implement the delays circularly; therefore, you will also need to introduce two circular pointers $\{p_L, p_R\}$.

8.23 *Computer Experiment: Stereo Delay Effects.* Write a C routine that implements the stereo sample processing algorithm of the previous problem. Using this routine, compute and plot the left and right output signals $y_L(n)$, $y_R(n)$, for $n = 0, 1, \ldots, 299$, for the case when there is only a left input pulse of duration 5, that is,

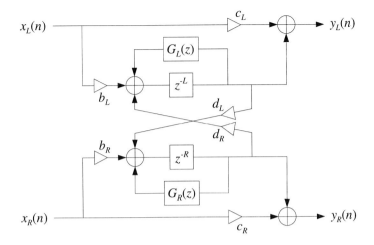

Fig. 8.4.1 Stereo delay effects processor.

$$x_L(n) = u(n) - u(n-5), \qquad x_R(n) = 0$$

Use $L = 30$ and $R = 70$ for the left and right delays, and the multiplier values:

$$a_L = a_R = 0.6, \quad b_L = b_R = 1, \quad c_L = c_R = 0, \quad d_L = d_R = 0.3$$

Identify on your graphs the origin of the various length-5 pulses that appear in the outputs. Next, repeat the experiment using $d_L = 0.3, d_R = 0$, so that only the left output is fed into the right input. Again, identify the origin of the pulses in your outputs.

8.24 *Computer Experiment: Compressors and Limiters.* Consider the compressor and limiter presented in Figures 8.2.33–8.2.36.

   a. Reproduce these graphs. Is it better to apply the smoothing filter to the output of the gain processor $f(c_n)$, rather than to its input c_n?

   b. Given a sinusoid $x(n) = A\cos(\omega_0 n)$, calculate its theoretical mean absolute value $\overline{|x_n|}$ and its rms value $(\overline{|x_n|^2})^{1/2}$, both averaged over one period of the sinusoid.

   Are the steady-state values of the control signal in the above graphs consistent with the theoretical values calculated here? In your program in (a), include the numerical calculation of the mean absolute values of the three output sinusoids, averaged over the three length-200 segments. Are these averages consistent with the given compression ratio?

   c. Redo the graphs in (a), but without using any smoothing filter.

   d. Repeat part (a) using a 3:1 compression ratio, $\rho = 1/3$ and then a 4:1 ratio.

   e. Repeat part (a) using a delay of $d = 40$ samples in the direct signal path, as described in Section 8.2.5. Too much of such a delay can introduce a "backward playing" quality into the output. Can you observe this?

   Repeat using a delay $D = 40$ in the level detector's input (but not in the direct signal path).

f. Repeat part (a) using a seven-point smoother, but with filter parameter $\lambda = 0.99$. Repeat with $\lambda = 0.2$. Do you observe the effect on the attack and release time constants.?

8.25 *Computer Experiment: Expanders and Gates.* Consider the expander and gate of Figures 8.2.37 and 8.2.38.

    a. Redo these graphs using no additional smoothing filter, and then, redo them using a seven-point smoother.

    b. Repeat part (a) using a 3:1 expansion ratio, $\rho = 3$.

    c. Repeat part (a) using a 2:1 expansion ratio, $\rho = 2$, but moving the threshold higher to the value $c_0 = 1.5$. What happens to the three sinusoids in this case? What happens in the case of the noise gate?

8.26 A zero-mean white noise sequence $x(n)$, $n \geq 0$, of variance σ_x^2 is sent through a stable and causal filter. Using convolution, show that the variance of the output sequence $y(n)$ will be given by:

$$\sigma_y^2(n) = E[y(n)^2] = \sigma_x^2 \sum_{m=0}^{n} h(m)^2$$

so that for large n it converges to the theoretical NRR of Eq. (A.18).

8.27 Show that the NRR summation of Eq. (A.18) always converges for a stable and causal filter with rational transfer function. In particular, show that it is bounded by:

$$NRR = \sum_{n=0}^{\infty} h_n^2 \leq \frac{C}{1 - |p_{max}|^2}$$

where p_{max} is the pole of maximum magnitude and C is a constant that depends on the PF expansion coefficients. You may assume the PF expansion: $h_n = \sum_{i=1}^{M} A_i p_i^n u(n)$.

8.28 For an ideal bandpass filter with passband $\omega_a \leq |\omega| \leq \omega_b$, prove the theoretical NRR given by Eq. (8.3.7).

8.29 *Computer Experiment: Exponential Smoother.* Write a C program, say smooth.c, that implements the first-order smoother of Example 8.3.1 with transfer function $H(z) = (1 - a)/(1 - az^{-1})$, where $0 < a < 1$. The program must have usage:

    smooth a   < x.dat > y.dat

where a is a command-line argument. The input signal to be smoothed must be read from stdin or a file x.dat, and the smoothed output must be written to the stdout or a file y.dat.

8.30 *Computer Experiment: Exponential Smoother.* Using the above program smooth.c reproduce the graphs in Fig. 8.3.4. Generate also two similar graphs for the filter parameter values $a = 0.99$ and $a = 0.8$.

In all four cases, compute the experimental NRRs computed from the sample variances based on the $L = 200$ input and output data sequences $x(n), y(n)$ with means m_x, m_y:

$$\hat{\sigma}_x^2 = \frac{1}{L} \sum_{n=0}^{L-1} (x(n) - m_x)^2, \quad \hat{\sigma}_y^2 = \frac{1}{L} \sum_{n=0}^{L-1} (y(n) - m_y)^2, \quad \widehat{NRR} = \frac{\hat{\sigma}_y^2}{\hat{\sigma}_x^2}$$

and compare them with the theoretical values. Explain any discrepancies.

8.31 Normally, you would use a lowpass (or highpass) filter to extract a low- (or high-) frequency signal. Suppose instead you used the lowpass filter $H(z) = b/(1 - az^{-1})$, where $0 < a < 1$, to extract the high-frequency signal $x(n) = s(-1)^n + v(n)$, where $v(n)$ is zero-mean white noise of variance σ_v^2.

How should you choose b so that the part $s(-1)^n$ comes out unchanged? Show that in this case the noise will be amplified. Explain this result by calculating the NRR as well as graphically by sketching the frequency spectra of the signals and filter, as in Fig. 8.3.3.

8.32 Consider the highpass FIR averager filter of Example 8.3.6. Using the minimization techniques outlined in Example 8.3.4, show that the optimum length-N FIR filter that minimizes the NRR subject to the highpass constraint (8.3.22) is given by Eq. (8.3.21).

8.33 Using partial fractions, derive Eq. (8.3.23) for the NRR of the bandpass resonator filter of Example 8.3.7.

8.34 *Computer Experiment: Bandpass Signal Extraction.* An improved version of the bandpass filter of Example 8.3.7, which has prescribed 3-dB width $\Delta\omega$ and center frequency ω_0, can be designed with the methods of Chapter 11. Using the design equations (11.3.21) and (11.3.22), design the following two peaking filters that have specifications:

   a. Center frequency $\omega_0 = 0.1\pi$, 3-dB width $\Delta\omega = 0.05\pi$. Determine the filter's transfer function, write its sample processing algorithm, compute its NRR and its 5% time constant n_{eff}, and plot its magnitude response squared $|H(\omega)|^2$ over 400 equally spaced frequencies over $0 \le \omega < \pi$.

   b. Center frequency $\omega_0 = 0.1\pi$, but with 5% time constant of $n_{\text{eff}} = 300$. Then, repeat all the questions of part (a).

   c. Using the Gaussian generator gran, generate a noisy sinusoidal input of the form:

   $$x(n) = s(n) + v(n) = \cos(\omega_0 n) + v(n), \quad n = 0, 1, \dots, N - 1$$

   where $\omega_0 = 0.1\pi$, $N = 300$, and $v(n) = \text{gran}(0, 1, \&\text{iseed})$ is zero-mean, unit-variance, white Gaussian noise. Send $x(n)$ through the above two filters and compute the output $y(n)$. Plot $x(n)$ versus n. Plot the two outputs $y(n)$ together with the desired signal $s(n)$.

8.35 *Computer Experiment: Single-Notch Filter.* Consider Example 8.3.8, but with a simplified signal instead of the ECG, defined to be a double pulse which is replicated three times at a period of 0.5 sec, with a 60 Hz noise component added to it:

$$f(t) = [u(t - 0.15) - u(t - 0.30)] - 0.75[u(t - 0.30) - u(t - 0.45)]$$

$$s(t) = f(t) + f(t - 0.5) + f(t - 1)$$

$$x(t) = s(t) + 0.5\cos(2\pi f_1 t)$$

where t is in seconds, $u(t)$ is the unit-step function, and $f_1 = 60$ Hz. The signal $x(t)$ is sampled at a rate of 1 kHz for a period of 1.5 seconds. Let $x(n)$ denote the resulting samples. Plot $x(n)$ and the noise-free signal $s(n)$ for $0 \le n \le 1499$.

Using the design method described in Example 8.3.8, design two second-order notch filters with notch frequency at f_1, one having $Q = 6$ and the other $Q = 60$. Determine their filter coefficients and their 1% time constants. Plot their magnitude responses over $0 \le f \le f_s/2$.

Filter the sampled signal $x(n)$ through both filters, and plot the resulting output signals $y(n)$ for $0 \le n \le 1499$. Discuss the capability of the filters in removing the 60 Hz interference. Discuss also the residual ringing that is left in the output after the 60 Hz sinusoid has died out. (To study it, you may use superposition and filter $s(n)$ and the noise part separately; you may also look at the impulse responses.)

8.36 *Computer Experiment: Multi-Notch Filter.* Consider the signal $x(n)$ consisting of three periods of a pulse signal $f(n)$ plus additive noise, defined for $0 \le n < 1800$:

$$f(n) = [u(n - 150) - u(n - 300)] - 0.75[u(n - 300) - u(n - 450)]$$

$$s(n) = f(n) + f(n - 600) + f(n - 1200)$$

$$x(n) = s(n) + v(n)$$

where $v(n)$ is defined as in Example 8.3.10 to be a periodic square wave of period $D = 10$. Therefore, a periodic notch filter with notches at the harmonics of $f_1 = f_s/D$ or $\omega_1 = 2\pi/D$ will remove the noise component. Using the design method of Example 8.3.10, design such a multi-notch filter having $Q = 80$.

Implement the filter using the sample processing algorithm of Example 8.3.10, and process the noisy signal $x(n)$ through it to get the output signal $y(n)$. On separate graphs, plot the signals $s(n), x(n)$, and $y(n)$, for $0 \le n < 1800$. For display purposes, split each graph into three separate graphs that cover the time periods $0 \le n < 600$, $600 \le n < 1200$, and $1200 \le n < 1800$.

The noise is removed fairly well, but you will notice that the filter also distorts the desired signal $s(n)$ rather severely. To understand the origin of this distortion, filter $s(n)$ separately through the filter and plot the corresponding output. Then, design three other periodic notch filters having $Q = 200, 400$, and 800, filter $s(n)$ through them, and plot the outputs. In all cases, compute the 1% time constants of the filters and discuss the tradeoff between speed of response, noise reduction, and non-distortion of the input.

Moreover, for the two cases $Q = 80$ and $Q = 800$, plot the corresponding magnitude responses $|H(f)|$ over one Nyquist interval $0 \le f \le f_s$ assuming $f_s = 600$ Hz, so that $f_1 = f_s/D = 60$ Hz.

8.37 *Computer Experiment: ECG Processing.* Reproduce all the designs, results, and graphs of Example 8.3.8. The simulated ECG data $s(n)$ may be generated by the MATLAB routine `ecg.m` of Appendix D, as follows:

```
s  = ecg(500)';          one beat of length 500
s  = [s; s; s];          three beats
s0 = sgfilt(0, 5, s);    5-point smoother
s  = s0 / max(s0);       normalized to unity maximum
```

8.38 *Computer Experiment: ECG Processing.* Reproduce all the designs, results, and graphs of Example 8.3.10. The simulated ECG data $s(n)$ may be generated by the MATLAB routine `ecg.m` of Appendix D, as follows:

```
s  = ecg(600)';          one beat of length 600
s  = [s; s; s];          three beats
s0 = sgfilt(0, 9, s);    9-point smoother
s  = s0 / max(s0);       normalized to unity maximum
```

8.39 Show that the following periodic comb filter has NRR:

$$H(z) = \frac{1-a}{2}\frac{1+z^{-D}}{1-az^{-D}} \quad \Rightarrow \quad NRR = \frac{1-a}{2}$$

Then show that if we define its Q-factor in terms of its 3-dB width $\Delta\omega$ and its first harmonic $\omega_1 = 2\pi/D$ by $Q = \omega_1/\Delta\omega$, then the parameter a can be calculated as:

$$a = \frac{1 - \tan(\pi/2Q)}{1 + \tan(\pi/2Q)}$$

Finally, determine and sketch its causal impulse response $h(n)$.

8.40 *Computer Experiment: Periodic Signal Enhancement.* Reproduce all the results and graphs of Example 8.3.12. Implement the comb filter using the circular-buffer version of the sample processing algorithm. (This is more appropriate because in practice the signal's period may be large.)

Repeat using Q-factors: $Q = 40$ and $Q = 30$. In all cases, compute the filter's 5% time constant and discuss the tradeoff between speed of response, signal enhancement, and noise reduction.

8.41 *Computer Experiment: TV Vertical Detail Filters.* First, verify that the vertical detail re-insertion filters given in Eqs. (8.3.70) and (8.3.72) satisfy the complementarity property of Eq. (8.3.71).

Then, plot their magnitude response $|H(f)|$ using the same scales as in Fig. 8.3.25.

8.42 Table 8.3.1 illustrating the signal averaging algorithm of Section 8.3.4 uses a linear delay-line buffer $\mathbf{w} = [w_0, w_1, w_2, w_3]$. Prepare a similar table for the circular-buffer version of the algorithm. Your table must also show the successive values of the circular pointer indices pointing to the 0th and Dth taps, q and $q_D = (q + D)\%(D + 1)$, so that $w[q] = w[q_D] + x/N$. Note how only one element of $\mathbf{w}$ changes at each time.

8.43 *Computer Experiment: Signal Averaging.* Write a stand-alone C program, say `sigav.c`, that implements the signal averaging of a signal consisting of N periods each of length D. The program must have usage:

```
sigav D N  < x.dat  > y.dat
```

It must read the required ND input data samples from `stdin` or from a file `x.dat`, and write the computed length-D averaged output into `stdout` or a file `y.dat`.

Note that such a program was essentially given in Section 8.3.4. Test your program on some simple data.

8.44 *Computer Experiment: Signal Averaging.* Using your program `sigav.c` from the previous problem or the MATLAB routine `sigav.m` of Appendix D, reproduce all the graphs in Fig. 8.3.29. In addition, to the values $N = 1, 16, 64, 256$, do also $N = 32, 128$.

8.45 *Computer Experiment: Savitzky-Golay Filters.* Reproduce all the results and graphs of Figures 8.3.33 and 8.3.34. The simulated noise-free ECG can be generated by the MATLAB statements:

```
s0 = sgfilt(0, 15, ecg(500)');        noise-free ECG
s = s0 / max(s0);                      normalize to unity maximum
```

where `ecg.m` and `sgfilt.m` are given in Appendix D. To that, you must add the noise component and filter it with the various cases of `sgfilt.m`.

Often a second pass through the smoothing filter helps. For each of the above cases, filter the output through the same SG filter, plot the results, and compare them with those of the single pass.

DFT/FFT Algorithms

The discrete Fourier transform (DFT) and its fast implementation, the fast Fourier transform (FFT), have three major uses in DSP: (a) the numerical *computation* of the frequency spectrum of a signal; (b) the efficient implementation of *convolution* by the FFT; and (c) the *coding* of waveforms, such as speech or pictures, for efficient transmission and storage [219-244,303]. The *discrete cosine transform*, which is a variant of the DFT, is especially useful for coding applications [238-240].

9.1 Frequency Resolution and Windowing

To compute the spectrum of an analog signal digitally, a finite-duration record of the signal is sampled and the resulting samples are transformed to the frequency domain by a DFT or FFT algorithm. The sampling rate f_s must be fast enough to minimize aliasing effects. If necessary, an analog antialiasing prefilter may precede the sampling operation.

The spectrum of the sampled signal $\hat{X}(f)$ is the replication of the desired analog spectrum $X(f)$ at multiples of the sampling rate f_s, as given by the Poisson summation formula, Eq. (1.5.14) of Chapter 1. We saw there that with the proper choice of sampling rate and prefilter, it can be guaranteed that $\hat{X}(f)$ agree with the desired $X(f)$ over the Nyquist interval, that is, by Eq. (1.5.15):

$$T\hat{X}(f) = X(f), \qquad -\frac{f_s}{2} \le f \le \frac{f_s}{2} \tag{9.1.1}$$

This property is a direct consequence of the sampling theorem, following from the non-overlapping of the spectral replicas in $\hat{X}(f)$. However, if the replicas overlap, they will contribute to the right-hand side of Eq. (9.1.1), making the sampled spectrum different from the desired one:

$$T\hat{X}(f) = X(f) + X(f - f_s) + X(f + f_s) + \cdots, \qquad -\frac{f_s}{2} \le f \le \frac{f_s}{2} \tag{9.1.2}$$

Because digitally we can only compute $\hat{X}(f)$, it is essential that Eq. (9.1.1) be satisfied, or that the extra terms in Eq. (9.1.2) remain small over the Nyquist interval, which happens when $X(f)$ falls off sufficiently fast with f. Example 1.5.2 illustrates the nature of the approximation of Eq. (9.1.2) for a non-bandlimited signal.

Even though $\hat{X}(f)$ is the closest approximation to $X(f)$ that we can achieve by DSP, it is still not computable because generally it requires an infinite number of samples $x(nT)$, $-\infty < n < \infty$. To make it computable, we must make a second approximation to $X(f)$, keeping only a finite number of samples, say, $x(nT)$, $0 \le n \le L - 1$. This *time-windowing* process is illustrated in Fig. 9.1.1.

In terms of the time samples $x(nT)$, the original sampled spectrum $\hat{X}(f)$ and its time-windowed version $\hat{X}_L(f)$ are given by:

$$\hat{X}(f) = \sum_{n=-\infty}^{\infty} x(nT)e^{-2\pi jfnT}$$

$$\hat{X}_L(f) = \sum_{n=0}^{L-1} x(nT)e^{-2\pi jfnT}$$

(9.1.3)

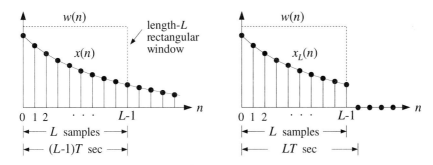

Fig. 9.1.1 Time windowing.

As seen in Fig. 9.1.1, the duration of the windowed data record from the time sample at $n = 0$ to the sample at $n = L - 1$ is $(L - 1)T$ seconds, where T is the sampling time interval $T = 1/f_s$. Because each sample lasts for T seconds, the last sample will last until time LT. Therefore, we may take the duration of the data record to be:

$$\boxed{T_L = LT}$$

(9.1.4)

The windowed signal may be thought of as an infinite signal which is zero outside the range of the window and agrees with the original one within the window. To express this mathematically, we define the *rectangular window* of length L:

$$w(n) = \begin{cases} 1, & \text{if } 0 \le n \le L - 1 \\ 0, & \text{otherwise} \end{cases}$$

(9.1.5)

Then, define the windowed signal as follows:

$$x_L(n) = x(n)w(n) = \begin{cases} x(n), & \text{if } 0 \le n \le L - 1 \\ 0, & \text{otherwise} \end{cases}$$

(9.1.6)

The multiplication by $w(n)$ ensures that $x_L(n)$ vanish outside the window. Equations (9.1.3) can now be expressed more simply in the form:

$$X(\omega) = \sum_{n=-\infty}^{\infty} x(n)e^{-j\omega n}$$

$$X_L(\omega) = \sum_{n=0}^{L-1} x(n)e^{-j\omega n} = \sum_{n=-\infty}^{\infty} x_L(n)e^{-j\omega n} \qquad (9.1.7)$$

where $\omega = 2\pi f/f_s$. Thus, $X_L(\omega)$ is the DTFT of the windowed signal $x_L(n)$ and is computable for any desired value of ω.

As the length L of the data window increases, the windowed signal $x_L(n)$ becomes a better approximation of $x(n)$, and thus, $X_L(\omega)$ a better approximation of $X(\omega)$. Example 1.5.2 illustrates this approximation as L increases.

In general, the windowing process has two major effects: First, it *reduces the frequency resolution* of the computed spectrum, in the sense that the smallest resolvable frequency difference is limited by the length of the data record, that is, $\Delta f = 1/T_L$. This is the well-known "uncertainty principle." Second, it introduces *spurious* high-frequency components into the spectrum, which are caused by the sharp clipping of the signal $x(n)$ at the left and right ends of the rectangular window. This effect is referred to as "frequency leakage."

Both effects can be understood by deriving the precise connection of the windowed spectrum $X_L(\omega)$ to the unwindowed one $X(\omega)$ of Eq. (9.1.7). Using the property that the Fourier transform of the *product* of two time functions is the *convolution* of their Fourier transforms, we obtain the frequency-domain version of $x_L(n) = x(n)w(n)$:

$$X_L(\omega) = \int_{-\pi}^{\pi} X(\omega')W(\omega - \omega')\frac{d\omega'}{2\pi} \qquad (9.1.8)$$

where $W(\omega)$ is the DTFT of the rectangular window $w(n)$, that is,

$$W(\omega) = \sum_{n=0}^{L-1} w(n)e^{-j\omega n}$$

It can be thought of as the evaluation of the z-transform on the unit circle at $z = e^{j\omega}$. Setting $w(n) = 1$ in the sum, we find:

$$W(z) = \sum_{n=0}^{L-1} w(n)z^{-n} = \sum_{n=0}^{L-1} z^{-n} = \frac{1 - z^{-L}}{1 - z^{-1}}$$

Setting $z = e^{j\omega}$, we find for $W(\omega)$:

$$W(\omega) = \frac{1 - e^{-jL\omega}}{1 - e^{-j\omega}} = \frac{\sin(\omega L/2)}{\sin(\omega/2)}e^{-j\omega(L-1)/2} \qquad (9.1.9)$$

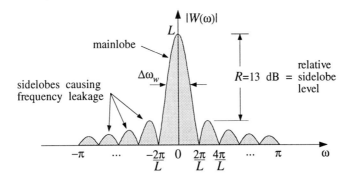

Fig. 9.1.2 Magnitude spectrum of rectangular window.

The magnitude spectrum $|W(\omega)| = |\sin(\omega L/2)/\sin(\omega/2)|$ is depicted in Fig. 9.1.2. It consists of a *mainlobe* of height L and base width $4\pi/L$ centered at $\omega = 0$, and several smaller *sidelobes*.

The sidelobes are between the zeros of $W(\omega)$, which are the zeros of the numerator $\sin(\omega L/2) = 0$, that is, $\omega = 2\pi k/L$, for $k = \pm 1, \pm 2, \ldots$ (with $k = 0$ excluded).

The mainlobe peak at DC dominates the spectrum, because $w(n)$ is essentially a DC signal, except when it cuts off at its endpoints. The higher frequency components that have "leaked" away from DC and lie under the sidelobes represent the sharp transitions of $w(n)$ at the endpoints.

The *width* of the mainlobe can be defined in different ways. For example, we may take it to be the width of the base, $4\pi/L$, or, take it to be the 3-dB width, that is, where $|W(\omega)|^2$ drops by $1/2$. For simplicity, we will define it to be *half* the base width, that is, in units of radians per sample:

$$\boxed{\Delta\omega_w = \frac{2\pi}{L}} \qquad \text{(rectangular window width)} \qquad (9.1.10)$$

In units of Hz, it is defined through $\Delta\omega_w = 2\pi\Delta f_w/f_s$. Using Eq. (9.1.4), we have:

$$\boxed{\Delta f_w = \frac{f_s}{L} = \frac{1}{LT} = \frac{1}{T_L}} \qquad (9.1.11)$$

We will see shortly that the mainlobe width Δf_w determines the *frequency resolution limits* of the windowed spectrum. As L increases, the height of the mainlobe increases and its width becomes narrower, getting more concentrated around DC. However, the height of the sidelobes also increases, but *relative* to the mainlobe height, it remains approximately the same and about 13 dB down.

For example, the peak of the first sidelobe occurs approximately halfway between the two zeros $2\pi/L$ and $4\pi/L$, that is, at $\omega = 3\pi/L$. Using $W(0) = L$, we find that the *relative* heights are essentially independent of L:

$$\left| \frac{W(\omega)}{W(0)} \right|_{\omega=3\pi/L} = \left| \frac{\sin(\omega L/2)}{L\sin(\omega/2)} \right| = \left| \frac{\sin(3\pi/2)}{L\sin(3\pi/2L)} \right| \simeq \frac{1}{L\cdot(3\pi/2L)} = \frac{2}{3\pi}$$

We assumed that L was fairly large (typically, $L \geq 10$), and used the small-x approximation $\sin x \simeq x$ with $x = 3\pi/2L$. In decibels, the *relative sidelobe* level is

$$R = 20\log_{10}\left|\frac{W(\omega)}{W(0)}\right|_{\omega=3\pi/L} \simeq 20\log_{10}\left(\frac{2}{3\pi}\right) = -13.46 \text{ dB}$$

To illustrate the effect of the convolutional equation (9.1.8), we consider the case of a single analog complex sinusoid of frequency f_1 and its sampled version:

$$x(t) = e^{2\pi jf_1t}, \quad -\infty < t < \infty \quad \Rightarrow \quad x(n) = e^{2\pi jf_1nT} = e^{j\omega_1n}, \quad -\infty < n < \infty$$

where $\omega_1 = 2\pi Tf_1 = 2\pi f_1/f_s$. The spectrum of the analog signal $x(t)$ is the Fourier transform:

$$X(f) = \int_{-\infty}^{\infty} x(t)e^{-2\pi jft}\,dt = \int_{-\infty}^{\infty} e^{-2\pi j(f-f_1)t}\,dt = \delta(f-f_1)$$

Therefore, $X(f)$ consists of a single sharp *spectral line* at $f = f_1$. For a real sinusoid $x(t) = \cos(2\pi f_1t)$, we would get *two* half-height lines at $f = \pm f_1$. Indeed, the Fourier transform of the cosine is:

$$\cos(2\pi f_1t) = \frac{1}{2}e^{2\pi jf_1t} + \frac{1}{2}e^{-2\pi jf_1t} \longrightarrow \frac{1}{2}\delta(f-f_1) + \frac{1}{2}\delta(f+f_1)$$

Assuming that f_1 lies within the Nyquist interval, that is, $|f_1| \leq f_s/2$, we may use Eq. (9.1.1) to determine the spectrum of the signal $x(n)$ for $-f_s/2 \leq f \leq f_s/2$:

$$X(\omega) = \hat{X}(f) = \frac{1}{T}X(f) = \frac{1}{T}\delta(f-f_1)$$

Using the delta function property, $|a|\delta(ax) = \delta(x)$, we can express the spectrum in terms of the digital frequency $\omega = 2\pi f/f_s = 2\pi Tf$, as follows:

$$2\pi\delta(\omega-\omega_1) = \frac{1}{T}2\pi T\,\delta(2\pi Tf - 2\pi Tf_1) = \frac{1}{T}\delta(f-f_1)$$

Therefore, the spectrum of the sampled signal will be, over the Nyquist interval:

$$X(\omega) = 2\pi\delta(\omega-\omega_1), \quad -\pi \leq \omega \leq \pi \tag{9.1.12}$$

Outside the Nyquist interval, the spectral line is replicated at multiples of 2π, that is, $2\pi\delta(\omega-\omega_1-2\pi m)$. This was also discussed in Section 5.4. It can be verified that Eq. (9.1.12) generates the sampled sinusoid from the inverse DTFT formula, Eq. (1.5.5):

$$x(n) = \int_{-\pi}^{\pi} X(\omega)e^{j\omega n}\frac{d\omega}{2\pi} = \int_{-\pi}^{\pi} 2\pi\delta(\omega-\omega_1)e^{j\omega n}\frac{d\omega}{2\pi} = e^{j\omega_1n}$$

The windowed sinusoid consists of the L samples:

$$x_L(n) = e^{j\omega_1n}, \quad n = 0, 1, \ldots, L-1$$

Its spectrum is obtained by inserting Eq. (9.1.12) into (9.1.8):

$$X_L(\omega) = \int_{-\pi}^{\pi} X(\omega') W(\omega - \omega') \frac{d\omega'}{2\pi} = \int_{-\pi}^{\pi} 2\pi \, \delta(\omega' - \omega_1) W(\omega - \omega') \frac{d\omega'}{2\pi}$$

Because of the delta function $\delta(\omega' - \omega_1)$ in the integrand, we obtain:

$$X_L(\omega) = W(\omega - \omega_1) \qquad\qquad (9.1.13)$$

This is the translation of $W(\omega)$ centered about ω_1, as shown in Fig. 9.1.3. Thus, the windowing process has the effect of *smearing* the sharp spectral line $\delta(\omega - \omega_1)$ at ω_1 and replacing it by $W(\omega - \omega_1)$.

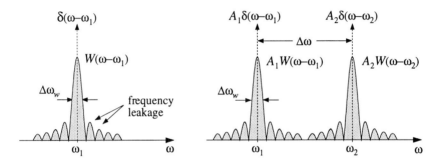

Fig. 9.1.3 Spectra of windowed single and double sinusoids.

A similar analysis can be made in the case when $x(t)$ is a linear combination of *two* complex sinusoids with frequencies f_1 and f_2 and (complex) amplitudes A_1 and A_2. We have for the analog, sampled, and windowed signals and their spectra:

$$x(t) = A_1 e^{2\pi j f_1 t} + A_2 e^{2\pi j f_2 t}, \quad -\infty < t < \infty$$

$$X(f) = A_1 \delta(f - f_1) + A_2 \delta(f - f_2)$$

$$x(n) = A_1 e^{j\omega_1 n} + A_2 e^{j\omega_2 n}, \quad -\infty < n < \infty$$

$$X(\omega) = 2\pi A_1 \delta(\omega - \omega_1) + 2\pi A_2 \delta(\omega - \omega_2), \quad -\pi \le \omega \le \pi$$

$$x_L(n) = A_1 e^{j\omega_1 n} + A_2 e^{j\omega_2 n}, \quad 0 \le n \le L - 1$$

$$X_L(\omega) = A_1 W(\omega - \omega_1) + A_2 W(\omega - \omega_2)$$

Again, the two sharp spectral lines are replaced by their smeared versions, as shown in Fig. 9.1.3. In this figure, we have taken the frequency separation, $\Delta f = |f_2 - f_1|$, or $\Delta\omega = |\omega_2 - \omega_1|$, of the two sinusoids to be large enough so that the mainlobes are distinct and do not overlap. However, if Δf is decreased, the mainlobes will begin merging with each other and will not appear as distinct. This will start to happen when Δf is approximately equal to the mainlobe width Δf_w.

The *resolvability* condition that the two sinusoids appear as two distinct ones is that their frequency separation Δf be *greater* than the mainlobe width:

$$\boxed{\Delta f \geq \Delta f_w = \frac{f_s}{L}} \qquad \text{(frequency resolution)} \qquad (9.1.14)$$

or, in radians per sample:

$$\boxed{\Delta \omega \geq \Delta \omega_w = \frac{2\pi}{L}} \qquad\qquad (9.1.15)$$

These equations can be rewritten to give the *minimum number* of samples required to achieve a desired frequency resolution Δf. The smaller the desired separation, the longer the data record:

$$\boxed{L \geq \frac{f_s}{\Delta f} = \frac{2\pi}{\Delta \omega}} \qquad\qquad (9.1.16)$$

The mainlobe width of $W(\omega)$ determines the amount of achievable frequency resolution. The sidelobes, on the other hand, determine the amount of frequency leakage and are undesirable artifacts of the windowing process. They must be suppressed as much as possible because they may be confused with the mainlobes of *weaker* sinusoids that might be present.

The standard technique for suppressing the sidelobes is to use a *non-rectangular window*—a window that cuts off to zero less sharply and more gradually than the rectangular one. There are literally dozens of possible shapes for such windows, such as trapezoidal, triangular, Gaussian, raised cosine, and many others [219–222].

One of the simplest and most widely used window is the *Hamming window*. It provides a suppression of the sidelobes by at least 40 dB. Another one that allows the user to control the desired amount of sidelobe suppression is the *Kaiser window* [221], which we will discuss later in Section 10.2.2. The Hamming window, depicted in Fig. 9.1.4, is a raised-cosine type of window defined as follows:

$$w(n) = \begin{cases} 0.54 - 0.46\cos\left(\dfrac{2\pi n}{L-1}\right), & \text{if } 0 \leq n \leq L-1 \\ 0, & \text{otherwise} \end{cases} \qquad (9.1.17)$$

At its center, $n = (L-1)/2$, the value of $w(n)$ is $0.54 + 0.46 = 1$, and at its endpoints, $n = 0$ and $n = L-1$, its value is $0.54 - 0.46 = 0.08$. Because of the gradual transition to zero, the high frequencies that are introduced by the windowing process are deemphasized. Fig. 9.1.4 shows the magnitude spectrum $|W(\omega)|$. The sidelobes are still present, but are barely visible because they are suppressed relative to the mainlobe by $R = 40$ dB.

The main tradeoff in using any type of non-rectangular window is that its mainlobe becomes *wider* and shorter, thus, reducing the frequency resolution capability of the windowed spectrum. For any type of window, the effective width of the mainlobe is still *inversely proportional* to the window length:

$$\boxed{\Delta f_w = c\,\frac{f_s}{L} = c\,\frac{1}{T_L}} \qquad\qquad (9.1.18)$$

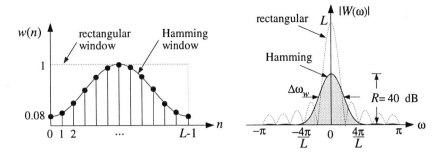

Fig. 9.1.4 Hamming window in the time and frequency domains.

or, in radians per sample:

$$\Delta\omega_w = c\,\frac{2\pi}{L}$$ (9.1.19)

where the constant c depends on the window used and is always $c \geq 1$.

The rectangular window has the *narrowest* width, corresponding to $c = 1$. As seen in Fig. 9.1.4, the Hamming window has approximately $c = 2$, that is, its main-lobe is twice as wide as the rectangular one. The Kaiser window has variable c that depends on the prescribed amount of relative sidelobe level R; see Eq. (10.2.25).

Given a finite data record of L samples, $x(n)$, $n = 0, 1, \ldots, L - 1$, the windowed signal is defined by Eq. (9.1.6); for example, for the Hamming window:

$$x_L(n) = w(n)x(n) = \left[0.54 - 0.46\cos\left(\frac{2\pi n}{L-1}\right) \right] x(n)$$ (9.1.20)

for $n = 0, 1, \ldots, L - 1$.

The corresponding spectrum $X_L(\omega)$ will still be given by Eq. (9.1.8). If $x(n)$ consists of a linear combination of sinusoids, then each sharp spectral line $\delta(\omega - \omega_i)$ of $x(n)$ will be replaced by the Hamming window spectrum $W(\omega - \omega_i)$. The frequency resolution depends now on the width of the Hamming window Δf_w. It follows that the minimum resolvable frequency difference will be:

$$\Delta f \geq \Delta f_w = c\,\frac{f_s}{L} = c\,\frac{1}{T_L}$$ (9.1.21)

This implies that the minimum data record required to achieve a given value of Δf is c-times longer than that of a rectangular window:

$$L \geq c\,\frac{f_s}{\Delta f} = c\,\frac{2\pi}{\Delta\omega}$$ (9.1.22)

In summary, the windowing process introduces artificial high-frequency components, which can be suppressed by using a non-rectangular window, but at the expense of reducing the frequency resolution. The lost frequency resolution can be recovered only by increasing the length L of the data record.

For random signals, such as sinusoids in noise, one must also deal with the *statistical reliability* of the computed spectra. In Appendix A.1, we discuss the *periodogram averaging* method which may be used to reduce the statistical *variability* of the spectrum estimate.

The method consists of dividing the total length-L data record into K segments of length N, such that $L = KN$. In order to reduce frequency leakage, a length-N non-rectangular window, such as a Hamming window, may be applied to each signal segment before its DFT is computed. The resulting reduction in resolution must be compensated for by increasing the length N. For a fixed total length L, this will reduce the number of segments K, thus worsening the spectrum estimate. Therefore, N must be chosen to be large enough to achieve a desired frequency resolution, but not larger.

The relationships $L = KN$ and $N = cf_s/\Delta f$ capture these issues: We want K to be large to get a reliable spectrum estimate, and we want N to be large to give us the desired resolution Δf for the particular window that we chose. Thus, together the two conditions require the total length L to be large. In some applications, this may be impossible to achieve either because we cannot collect more data, or because beyond a certain length L, the signal will no longer remain stationary.

Parametric spectrum estimation methods, such as those based on linear prediction, maximum likelihood, and eigenvector techniques, offer the possibility of obtaining high-resolution spectrum estimates based on short data records [25,26,28].

Example 9.1.1: A signal consisting of four sinusoids of frequencies of 1, 1.5, 2.5, and 2.75 kHz is sampled at a rate of 10 kHz. What is the minimum number of samples that should be collected for the frequency spectrum to exhibit four distinct peaks at these frequencies? How many samples should be collected if they are going to be preprocessed by a Hamming window and then Fourier transformed?

Solution: The *smallest* frequency separation that must be resolved by the DFT is $\Delta f = 2.75 - 2.5 = 0.25$ kHz. Using Eq. (9.1.16) for a rectangular window, we get

$$L \geq \frac{f_s}{\Delta f} = \frac{10}{0.25} = 40 \text{ samples}$$

Because the mainlobe width of the Hamming window is twice as wide as that of the rectangular window, it follows that twice as many samples must be collected, that is, $L = 80$. This value can also be calculated from Eq. (9.1.22) with $c = 2$. ☐

Example 9.1.2: A 10-millisecond portion of a signal is sampled at a rate of 10 kHz. It is known that the signal consists of two sinusoids of frequencies $f_1 = 1$ kHz and $f_2 = 2$ kHz. It is also known that the signal contains a third component of frequency f_3 that lies somewhere between f_1 and f_2. (a) How close to f_1 could f_3 be in order for the spectrum of the collected samples to exhibit three distinct peaks? How close to f_2 could f_3 be? (b) What are the answers if the collected samples are windowed by a Hamming window?

Solution: The total number of samples collected is $L = f_s T_L = 10 \times 10 = 100$. The frequency resolution of the rectangular window is $\Delta f = f_s/L = 10/100 = 0.1$ kHz. Thus, the closest f_3 to f_1 and f_2 will be:

$$f_3 = f_1 + \Delta f = 1.1 \text{ kHz}, \quad \text{and} \quad f_3 = f_2 - \Delta f = 1.9 \text{ kHz}$$

In the Hamming case, the minimum resolvable frequency separation doubles, that is, $\Delta f = c f_s / L = 2 \cdot 10/100 = 0.2$ kHz, which gives $f_3 = 1.2$ kHz or $f_3 = 1.8$ kHz. $\square$

Example 9.1.3: The sinusoid $x(t) = \cos(2\pi f_0 t)$, where $f_0 = 50$ Hz is sampled at a rate of $f_s = 1$ kHz. The sampled signal is $x(n) = \cos(\omega_0 n)$, where $\omega_0 = 2\pi f_0/f_s = 2\pi \cdot 50/1000 = 0.1\pi$ rads/sample. A length-L portion of $x(n)$ is windowed by a rectangular and a Hamming window, that is, for $n = 0, 1, \ldots, L - 1$:

$$x_L(n) = w_{\text{rec}}(n)x(n) = \cos(\omega_0 n)$$

$$x_L(n) = w_{\text{ham}}(n)x(n) = \left[0.54 - 0.46 \cos\left(\frac{2\pi n}{L - 1}\right) \right] \cos(\omega_0 n)$$

Figure 9.1.5 shows the rectangularly windowed signals, for $L = 100$ and $L = 200$. Figure 9.1.6 shows the Hamming windowed signals. Figure 9.1.7 shows the corresponding spectra, $|X_L(\omega)|$, plotted over the Nyquist subinterval, $0 \leq \omega \leq 0.2\pi$. The spectra were computed by successive calls to the routine `dtft` of the next section, for 200 equally spaced values of ω in the interval $0 \leq \omega \leq 0.2\pi$.

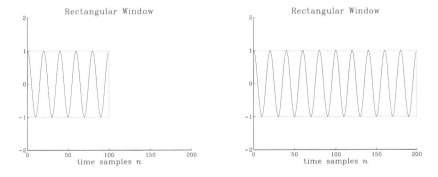

Fig. 9.1.5 Rectangularly windowed sinusoids of lengths $L = 100$ and $L = 200$.

As L doubles, both the rectangular and the Hamming mainlobe widths become narrower, with the Hamming one always lagging behind the rectangular one. Note also that as L doubles, the sidelobes of the rectangular window get more compressed, but also higher so that their *relative* depth compared to the mainlobe remains the same.

The reason why the peak height of the rectangular mainlobe is $L/2$ instead of L is that we are working with a real-valued sinusoid and looking only at its positive-frequency half-height peak. $\square$

Example 9.1.4: The following analog signal consisting of three equal-strength sinusoids of frequencies $f_1 = 2$ kHz, $f_2 = 2.5$ kHz, and $f_3 = 3$ kHz:

$$x(t) = \cos(2\pi f_1 t) + \cos(2\pi f_2 t) + \cos(2\pi f_3 t)$$

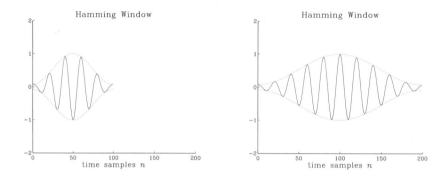

Fig. 9.1.6 Hamming windowed sinusoids of lengths $L = 100$ and $L = 200$.

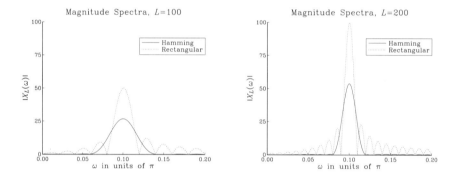

Fig. 9.1.7 Rectangular and Hamming spectra for $L = 100$ and $L = 200$.

where t is in milliseconds, is sampled at a rate of 10 kHz. We consider four data records of lengths $L = 10, 20, 40,$ and 100 samples. They correspond to the time durations of 1, 2, 4, and 10 msec. To facilitate comparison, the same vertical scale has been used in all figures.

Figures 9.1.8 and 9.1.9 show the magnitude spectra of the rectangularly and Hamming windowed signals for the above four values of L. The spectra were computed by calling a 256-point FFT routine and plotted over an entire Nyquist interval, $0 \le f \le f_s$. For each L, the 256-point input to the FFT routine was obtained by padding $(256-L)$ zeros at the end of the L-point signal **x** to make it of length 256. (The padding operation does not affect the DTFT—see Section 9.2.2.)

As we will see in the next section, the three peaks in the right half of the Nyquist interval correspond to the negative-frequency peaks of the sinusoids, but they have been shifted to the right by one f_s, using the periodicity property of the spectra with respect to f_s.

The minimum frequency separation is $\Delta f = 2.5 - 2 = 0.5$ kHz. According to (9.1.16), the minimum length L to resolve all three sinusoids should be $L = f_s / \Delta f = 10/0.5 = 20$ samples for the rectangular window, and $L = 40$ samples for the Hamming case.

In the case $L = 10$, the signal does not have enough length to separate the sinusoids,

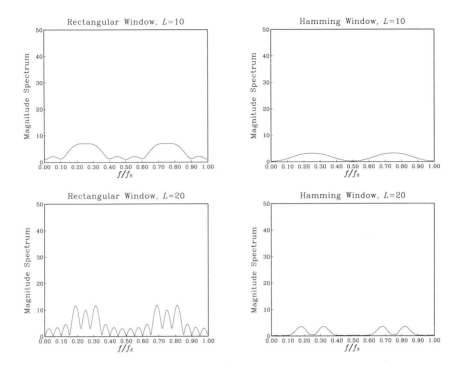

Fig. 9.1.8 Rectangular and Hamming spectra for $L = 10$ and 20.

which appear merged into one wide peak.

For $L = 20$, corresponding to the minimum acceptable length, the sinusoids begin to be separated for the rectangular window, but not yet for the Hamming window. Note also in the Hamming case, the destructive interference taking place exactly at the position of the middle sinusoid, $f_2/f_s = 0.25$.

For $L = 40$, the Hamming windowed spectra are beginning to show the separate peaks. Finally, when $L = 100$, both windows have clearly separated peaks.

The Hamming window spectra lag behind the rectangular ones in resolution, but improve with increasing L, while they provide higher sidelobe suppression.   □

9.2 DTFT Computation

9.2.1 DTFT at a Single Frequency

In this section, we turn our attention to the computational aspects of the DTFT. We consider a length-L signal $x(n)$, $n = 0, 1, \ldots, L - 1$, which may have been prewindowed by a length-L non-rectangular window. Its DTFT, defined by Eq. (9.1.7), can be written in the simplified notation:

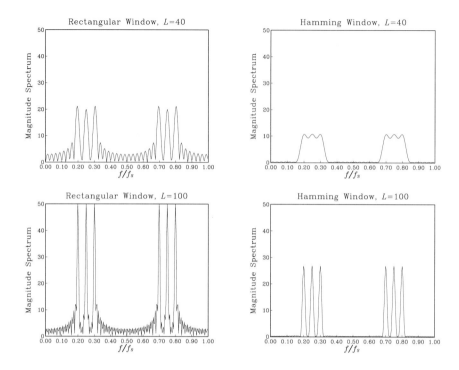

Fig. 9.1.9 Rectangular and Hamming spectra for $L = 40$ and 100.

$$\boxed{X(\omega) = \sum_{n=0}^{L-1} x(n)\,e^{-j\omega n}}$$ (DTFT of length-L signal)   (9.2.1)

This expression may be computed at any desired value of ω in the Nyquist interval $-\pi \le \omega \le \pi$. It is customary in the context of developing computational algorithms to take advantage of the periodicity of $X(\omega)$ and map the conventional symmetric Nyquist interval $-\pi \le \omega \le \pi$ onto the right-sided one $0 \le \omega \le 2\pi$. We will refer to the latter as the *DFT Nyquist interval*. This mapping is shown in Fig. 9.2.1.

The positive-frequency subinterval $0 \le \omega \le \pi$ remains unchanged, but the negative-frequency one, $-\pi \le \omega \le 0$, gets mapped onto the second half of the DFT Nyquist interval, $\pi \le \omega \le 2\pi$.

For example, a cosinusoidal signal $\cos(\omega_1 n)$ with two spectral peaks at $\pm\omega_1$ will be represented by the two shifted peaks:

$$\{\omega_1,\ -\omega_1\} \Leftrightarrow \{\omega_1,\ 2\pi - \omega_1\}$$

or, in Hz

$$\{f_1,\ -f_1\} \Leftrightarrow \{f_1,\ f_s - f_1\}$$

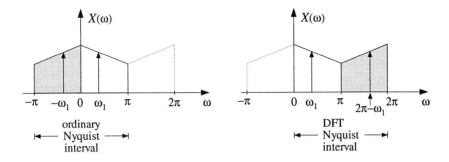

Fig. 9.2.1 Equivalent Nyquist intervals.

As we saw in Section 5.4, the DTFT (9.2.1) can be thought of as the evaluation of the z-transform of the sequence $x(n)$ on the unit circle:

$$X(\omega) = \sum_{n=0}^{L-1} x(n) e^{-j\omega n} = \sum_{n=0}^{L-1} x(n) z^{-n} \bigg|_{z=e^{j\omega}} = X(z) \big|_{z=e^{j\omega}} \qquad (9.2.2)$$

Thus, $X(\omega)$ can be computed by evaluating the polynomial $X(z)$ at $z = e^{j\omega}$. Hörner's rule of synthetic division that was discussed in the problems of Chapter 2 is an efficient polynomial evaluator. It can be adapted in the following form for the evaluation of the z-transform $X(z)$:

$$\boxed{\begin{array}{l} \textit{for each complex } z \textit{ do:} \\ \quad X = 0 \\ \quad \textit{for } n = L-1 \textit{ down to } n = 0 \textit{ do:} \\ \quad\quad X = x_n + z^{-1}X \end{array}} \qquad \text{(Hörner's rule)} \qquad (9.2.3)$$

Upon exit, X is the desired value of $X(z)$. To see how the iterations build up the z-transform, we iterate them for the case $L = 4$. Starting with $X = 0$ at $n = L-1 = 3$, we have:

$$X = x_3 + z^{-1}X = x_3$$
$$X = x_2 + z^{-1}X = x_2 + z^{-1}x_3$$
$$X = x_1 + z^{-1}X = x_1 + z^{-1}x_2 + z^{-2}x_3$$
$$X = x_0 + z^{-1}X = x_0 + z^{-1}x_1 + z^{-2}x_3 + z^{-3}x_3 = X(z)$$

This algorithm can then be applied to any point on the unit circle $z = e^{j\omega}$ to evaluate $X(\omega)$. The following routine `dtft.c` is an implementation:

```
/* dtft.c - DTFT of length-L signal at a single frequency w */

#include <cmplx.h>                          complex arithmetic

complex dtft(L, x, w)                       usage: X=dtft(L, x, w);
double *x, w;                               x is L-dimensional
```

```
int L;
{
        complex z, X;
        int n;

        z = cexp(cmplx(0, -w));                      set z = e^{-jω}

        X = cmplx(0,0);                              initialize X = 0

        for (n=L-1; n>=0; n--)
                X = cadd(cmplx(x[n], 0), cmul(z, X));

        return X;

}
```

The routine uses the suite of complex arithmetic functions defined in the file `complex.c` of Appendix C. The header file `cmplx.h` contains all the necessary declarations. For each value of ω, the routine returns the DTFT value $X(\omega)$ as a complex number. The L-dimensional time data array x is assumed real, although the routine can be modified easily when it is complex.

9.2.2 DTFT over Frequency Range

Often, we must compute the DTFT over a *frequency range*, $\omega_a \leq \omega < \omega_b$. The following routine `dtftr.c` computes the DTFT $X(\omega)$ at N frequencies that are equally spaced over this interval, that is,

$$\omega_k = \omega_a + k\,\frac{\omega_b - \omega_a}{N} = \omega_a + k\Delta\omega_{\mathrm{bin}}, \quad k = 0, 1, \ldots N - 1 \qquad (9.2.4)$$

where $\Delta\omega_{\mathrm{bin}}$ is the *bin width*, that is, the spacing of the frequencies ω_k:

$$\Delta\omega_{\mathrm{bin}} = \frac{\omega_b - \omega_a}{N}, \qquad \text{or, in Hz} \qquad \Delta f_{\mathrm{bin}} = \frac{f_b - f_a}{N} \qquad (9.2.5)$$

The routine returns the N-dimensional complex-valued array $X[k] = X(\omega_k)$ by making N successive calls to `dtft`:

```
/* dtftr.c - N DTFT values over frequency range [wa, wb) */

#include <cmplx.h>                              complex arithmetic

complex dtft();                                 DTFT at one frequency

void dtftr(L, x, N, X, wa, wb)                  usage: dtftr(L, x, N, X, wa, wb);
double *x, wa, wb;                              x is L-dimensional real
complex *X;                                     X is N-dimensional complex
int L, N;
{
        int k;
        double dw = (wb-wa)/N;                  frequency bin width

        for (k=0; k<N; k++)
                X[k] = dtft(L, x, wa + k*dw);    kth DTFT value X(ω_k)
}
```

The usage of these routines is illustrated by the following program segment, which computes the DTFT of Example 9.1.3 for a rectangularly windowed sinusoid of length $L = 100$. The DTFT was computed over $N = 200$ frequencies in the interval $[\omega_a, \omega_b) = [0, 0.2\pi)$.

```
double *x;
complex *X;

x = (double *) calloc(L, sizeof(double));        use L = 100
X = (complex *) calloc(N, sizeof(complex));      use N = 200

for (n=0; n<L; n++)
      x[n] = cos(w0 * n);                        use ω₀ = 0.1π

dtftr(L, x, N, X, wa, wb);                       use ωₐ = 0.0 and ω_b = 0.2π
```

The main program must include the header file `cmplx.h` and must be linked with the arithmetic routines `complex.c` of Appendix C.

The computational cost of the routine `dtft` is L complex MACs for each call. Similarly, the cost of `dtftr` is NL complex MACs, because it calls `dtft` N times. The so-called Goertzel algorithm [2] for computing the DFT is a variant of Hörner's rule that uses mostly real multiplications, and therefore is somewhat more efficient than the above routines.

The routines `dtft` and `dtftr` are useful, and even competitive with the FFT, when we need to know the DTFT at only very few frequencies. A MATLAB version `dtft.m`, which replaces both `dtft.c` and `dtftr.c`, is given in Appendix D.

Example 9.2.1: In Example 8.1.1, we discussed the generation of dual sinusoidal tones for DTMF touch-tone phones. Each keypress generates two frequencies ω_H and ω_L, one from the high and one from the low group of frequencies. A total of $4 \times 4 = 16$ pairs of frequencies can be generated.

Such a signal can be detected by computing its DTFT at the 4 high and 4 low group frequencies and then deciding with the help of a threshold which pair $\{X(\omega_H), X(\omega_L)\}$ of DTFT values has the largest magnitudes. The corresponding pair of frequencies $\{\omega_H, \omega_L\}$ can then be decoded into the appropriate key.

Because the DTFT is needed only at 8 positive frequencies, the use of the routine `dtft` or Goertzel's algorithm is more efficient than using an FFT. Such DTMF detectors can be implemented easily on present-day DSP chips [91–93].

The minimum duration L of the received dual tone may be estimated by requiring that the high and low groups of frequencies remain distinct, so that the DTFT will consist of one peak lying in the high group and one in the low group.

The resolvability condition depends on the minimum frequency difference *between* the groups, that is, from Fig. 8.1.3 we have

$$\Delta f = f_{H,\min} - f_{L,\max} = 1209 - 941 = 268 \text{ Hz}$$

which at sampling rate of $f_s = 8$ kHz and rectangular windowing gives the minimum length $L = f_s / \Delta f = 8000/268 \simeq 30$ samples. □

9.2.3 DFT

The *N-point DFT of a length-L signal* is defined to be the DTFT evaluated at N equally spaced frequencies over the full Nyquist interval, $0 \le \omega \le 2\pi$. These "DFT frequencies" are defined in radians per sample as follows:

$$\boxed{\omega_k = \frac{2\pi k}{N}}, \qquad k = 0, 1, \ldots, N-1 \qquad (9.2.6)$$

or, in Hz

$$\boxed{f_k = \frac{k f_s}{N}}, \qquad k = 0, 1, \ldots, N-1 \qquad (9.2.7)$$

Thus, the N-point DFT will be, for $k = 0, 1, \ldots, N-1$:

$$\boxed{X(\omega_k) = \sum_{n=0}^{L-1} x(n) e^{-j\omega_k n}} \qquad \text{(N-point DFT of length-L signal)} \qquad (9.2.8)$$

The N-dimensional complex DFT array $X[k] = X(\omega_k)$, $k = 0, 1, \ldots, N-1$ can be computed by calling the routine `dtftr` over the frequency range $[\omega_a, \omega_b) = [0, 2\pi)$. The following routine `dft.c` is an implementation:

```
/* dft.c - N-point DFT of length-L real-valued signal */

#include <cmplx.h>                         complex arithmetic

void dtftr();                              DTFT over a frequency range

void dft(L, x, N, X)                       usage: dft(L, x, N, X);
double *x;                                 x is L-dimensional real
complex *X;                                X is N-dimensional complex
int L, N;
{
      double pi = 4 * atan(1.0);

      dtftr(L, x, N, X, 0.0, 2*pi);        N frequencies over [0,2π)
}
```

Note that the value at $k = N$, corresponding to $\omega_N = 2\pi$, is not computed because by periodicity it equals the value at $\omega_0 = 0$, that is, $X(\omega_N) = X(\omega_0)$.

The only difference between `dft` and `dtftr` is that the former has its N frequencies distributed evenly over the full Nyquist interval, $[0, 2\pi)$, as shown in Fig. 9.2.2, whereas the latter has them distributed over any desired subinterval. The bin width (9.2.5) is in the DFT case, in rads/sample or Hz:

$$\Delta\omega_{\text{bin}} = \frac{2\pi}{N} \qquad \text{or,} \qquad \Delta f_{\text{bin}} = \frac{f_s}{N} \qquad (9.2.9)$$

In Fig. 9.2.2, the same number of frequencies N was used for the full Nyquist interval and the subinterval $[\omega_a, \omega_b)$. Therefore, the N frequencies in the `dtftr`

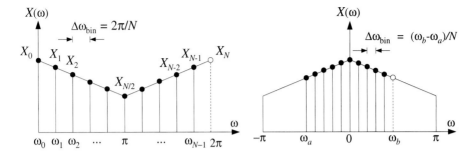

Fig. 9.2.2 N-point DTFTs over $[0, 2\pi)$ and over subinterval $[\omega_a, \omega_b)$, for $N = 10$.

case are more closely spaced. We took the subinterval $[\omega_a, \omega_b)$ to be symmetric with respect to the origin, but it could have been any other subinterval.

The N computed values $X(\omega_k)$ can also be thought of as the evaluation of the z-transform $X(z)$ at the following z-points on the unit circle:

$$X(\omega_k) = X(z_k) = \sum_{n=0}^{L-1} x(n) z_k^{-n} \tag{9.2.10}$$

where

$$\boxed{z_k = e^{j\omega_k} = e^{2\pi jk/N}}, \qquad k = 0, 1, \ldots, N - 1 \tag{9.2.11}$$

These are recognized as the Nth *roots of unity*, that is, the N solutions of the equation $z^N = 1$. They are evenly spaced around the unit circle at relative angle increments of $2\pi/N$, as shown in Fig. 9.2.3.

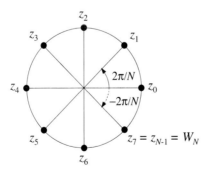

Fig. 9.2.3 Nth roots of unity, for $N = 8$.

Note also that the periodicity of $X(\omega)$ with period 2π is reflected in the periodicity of the DFT $X(k) = X(\omega_k)$ in the index k with period N. This follows from:

$$\omega_{k+N} = \frac{2\pi(k + N)}{N} = \frac{2\pi k}{N} + 2\pi = \omega_k + 2\pi$$

which implies:

$$X(k+N) = X(\omega_{k+N}) = X(\omega_k + 2\pi) = X(\omega_k) = X(k)$$

9.2.4 Zero Padding

In principle, the two lengths L and N can be specified *independently* of each other: L is the number of *time samples* in the data record and can even be infinite; N is the number of *frequencies* at which we choose to evaluate the DTFT.

Most discussions of the DFT assume that $L = N$. The reason for this will be discussed later. If $L < N$, we can pad $N - L$ zeros at the end of the data record to make it of length N. If $L > N$, we may reduce the data record to length N by *wrapping* it modulo-N—a process to be discussed in Section 9.5.

Padding any number of zeros at the *end* of a signal has no effect on its DTFT. For example, padding D zeros will result into a length-$(L+D)$ signal:

$$\mathbf{x} = [x_0, x_1, \ldots, x_{L-1}]$$

$$\mathbf{x}_D = [x_0, x_1, \ldots, x_{L-1}, \underbrace{0, 0, \ldots, 0}_{D \text{ zeros}}]$$

Because $x_D(n) = x(n)$ for $0 \le n \le L - 1$ and $x_D(n) = 0$ for $L \le n \le L + D - 1$, the corresponding DTFTs will remain the same:

$$X_D(\omega) = \sum_{n=0}^{L+D-1} x_D(n) e^{-j\omega n} = \sum_{n=0}^{L-1} x_D(n) e^{-j\omega n} + \sum_{n=L}^{L+D-1} x_D(n) e^{-j\omega n}$$

$$= \sum_{n=0}^{L-1} x(n) e^{-j\omega n} = X(\omega)$$

Therefore, their evaluation at the N DFT frequencies will be the same: $X_D(\omega_k) = X(\omega_k)$. We note also that padding the D zeros to the *front* of the signal will be equivalent to a delay by D samples, which in the z-domain corresponds to multiplication by z^{-D} and in the frequency domain by $e^{-j\omega D}$. Therefore, the signals:

$$\mathbf{x} = [x_0, x_1, \ldots, x_{L-1}]$$

$$\mathbf{x}_D = [\underbrace{0, 0, \ldots, 0}_{D \text{ zeros}}, x_0, x_1, \ldots, x_{L-1}] \tag{9.2.12}$$

will have DTFTs and DFTs:

$$X_D(\omega) = e^{-j\omega D} X(\omega)$$

$$X_D(\omega_k) = e^{-j\omega_k D} X(\omega_k), \quad k = 0, 1, \ldots, N - 1 \tag{9.2.13}$$

9.3 *Physical versus Computational Resolution*

The bin width Δf_{bin} represents the spacing between the DFT frequencies at which the DTFT is computed and must not be confused with the frequency resolution width $\Delta f = f_s/L$ of Eq. (9.1.14), which refers to the minimum resolvable frequency separation between two sinusoidal components. To avoid confusion, we will refer to Eq. (9.1.14) as the *physical* frequency resolution and to Eq. (9.2.9) as the *computational* frequency resolution.

The interplay between physical and computational resolution is illustrated in Fig. 9.3.1 for the triple sinusoidal signal of Example 9.1.4. The $N = 32$ and $N = 64$ point DFTs of the rectangularly windowed signals of lengths $L = 10$ and $L = 20$ are shown together with their full DTFTs (computed here as 256-point DFTs).

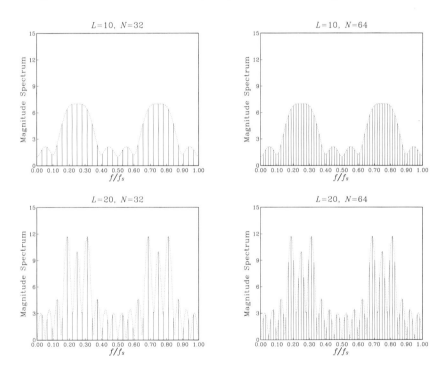

Fig. 9.3.1 Physical versus computational resolution in DTFT computation.

It is evident from these graphs that if the length L of the signal is not large enough to provide sufficient physical resolution, then there is no point increasing the length N of the DFT—that would only put more points on the wrong curve.

Another issue related to physical and computational resolution is the question of how accurately the DFT represents the peaks in the spectrum. For each sinusoid that is present in the signal, say, at frequency f_0, the DTFT will exhibit a mainlobe peak arising from the shifted window $W(f - f_0)$. When we evaluate the N-point DFT, we would like the peak at f_0 to *coincide* with one of the N DFT frequencies (9.2.7). This will happen if there is an integer $0 \le k_0 \le N - 1$, such that

$$f_0 = f_{k_0} = \frac{k_0 f_s}{N} \quad \Rightarrow \quad \boxed{k_0 = N\frac{f_0}{f_s}} \tag{9.3.1}$$

Similarly, the peak at the negative frequency, $-f_0$, or at the equivalent shifted one, $f_s - f_0$, will correspond to the integer, $-k_0$, or to the shifted one $N - k_0$:

$$-f_0 = -k_0\frac{f_s}{N} \quad \Rightarrow \quad f_s - f_0 = f_s - k_0\frac{f_s}{N} = (N - k_0)\frac{f_s}{N}$$

In summary, for each sinusoid with peaks at $\pm f_0$, we would like our DFT to show these peaks at the integers:

$$\{f_0, -f_0\} \quad \Rightarrow \quad \{f_0, f_s - f_0\} \quad \Rightarrow \quad \{k_0, N - k_0\} \tag{9.3.2}$$

In general, this is not possible because k_0 computed from Eq. (9.3.1) is not an integer, and the DFT will miss the exact peaks. However, for large N, we may *round* k_0 to the nearest integer and use the corresponding DFT frequency as an *estimate* of the actual peak frequency.

A pitfall of using the DFT can be seen in the lower two graphs of Fig. 9.3.1, where it appears that the DFT correctly identifies the three peaks in the spectrum, for both $N = 32$ and $N = 64$.

However, this is misleading for two reasons: First, it is a numerical accident in this example that the mainlobe maxima coincide with the DFT frequencies. Second, it can be seen in the figure that these maxima correspond to the *wrong* frequencies and not to the correct ones, which are:

$$\frac{f_1}{f_s} = 0.20, \quad \frac{f_2}{f_s} = 0.25, \quad \frac{f_3}{f_s} = 0.30 \tag{9.3.3}$$

This phenomenon, whereby the maxima of the peaks in the spectrum do not quite correspond to the correct frequencies, is called *biasing* and is caused by the lack of adequate *physical resolution*, especially when the sinusoidal frequencies are too close to each other and the sum of terms $W(f - f_0)$ interact strongly.

Using Eq. (9.3.1), we can calculate the DFT indices k and $N - k$ to which the true frequencies (9.3.3) correspond. For $N = 32$, we have:

$$k_1 = N\frac{f_1}{f_s} = 32 \cdot 0.20 = 6.4, \quad N - k_1 = 25.6$$

$$k_2 = N\frac{f_2}{f_s} = 32 \cdot 0.25 = 8, \quad N - k_2 = 24$$

$$k_3 = N\frac{f_3}{f_s} = 32 \cdot 0.30 = 9.6, \quad N - k_3 = 22.4$$

Similarly, for $N = 64$, we find:

$$k_1 = N\frac{f_1}{f_s} = 64 \cdot 0.20 = 12.8, \quad N - k_1 = 51.2$$

$$k_2 = N\frac{f_2}{f_s} = 64 \cdot 0.25 = 16, \quad\;\; N - k_2 = 48$$

$$k_3 = N\frac{f_3}{f_s} = 64 \cdot 0.30 = 19.2, \quad N - k_3 = 44.8$$

Only the middle one at f_2 corresponds to an integer, and therefore, coincides with a DFT value. The other two are missed by the DFT. We may round k_1 and k_3 to their nearest integers and then compute the corresponding DFT frequencies. We find for $N = 32$:

$$k_1 = 6.4 \quad \Rightarrow \quad k_1 = 6 \quad \Rightarrow \quad \frac{f_1}{f_s} = \frac{k_1}{N} = 0.1875$$

$$k_3 = 9.6 \quad \Rightarrow \quad k_3 = 10 \quad \Rightarrow \quad \frac{f_3}{f_s} = \frac{k_3}{N} = 0.3125$$

and for $N = 64$:

$$k_1 = 12.8 \quad \Rightarrow \quad k_1 = 13 \quad \Rightarrow \quad \frac{f_1}{f_s} = \frac{k_1}{N} = 0.203125$$

$$k_3 = 19.2 \quad \Rightarrow \quad k_3 = 19 \quad \Rightarrow \quad \frac{f_3}{f_s} = \frac{k_3}{N} = 0.296875$$

The *rounding error* in the frequencies remains less than $f_s/2N$. It decreases with increasing DFT length N. The *biasing error*, on the other hand, can only be decreased by increasing the data length L.

Figure 9.3.2 shows the spectrum of the same signal of Example 9.1.4, but with length $L = 100$ samples. Biasing is virtually eliminated with the peak maxima at the correct frequencies. The spectrum is plotted versus the DFT index k, which is proportional to the frequency f via the mapping (9.2.7), or

$$\boxed{k = N\frac{f}{f_s}} \qquad \text{(frequency in units of the DFT index)} \qquad (9.3.4)$$

The Nyquist interval $0 \le f \le f_s$ corresponds to the index interval $0 \le k \le N$. The N-point DFT is at the integer values $k = 0, 1, \ldots, N - 1$. For plotting purposes, the graph of the spectrum over the full interval $0 \le k \le N$ has been split into two side-by-side graphs covering the half-intervals: $0 \le k \le N/2$ and $N/2 \le k \le N$.

In the upper two graphs having $N = 32$, the DFT misses the f_1 and f_3 peaks completely (the peak positions are indicated by the arrows). The actual peaks are so narrow that they fit completely within the computational resolution width Δf_{bin}.

In the lower two graphs having $N = 64$, the DFT still misses these peaks, but less so. Further doubling of N will interpolate half-way between the frequencies of the 64-point case resulting in a better approximation.

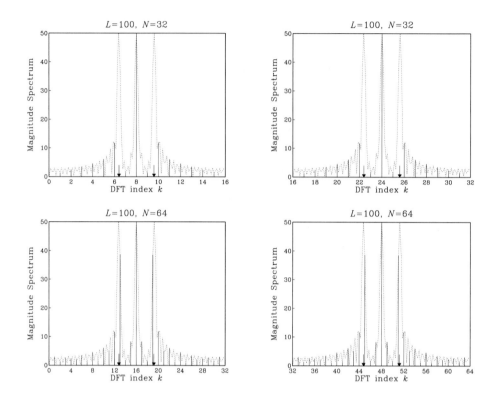

Fig. 9.3.2 DFT can miss peaks in the spectrum.

Example 9.3.1: A 5 kHz sinusoidal signal is sampled at 40 kHz and 128 samples are collected and used to compute the 128-point DFT of the signal. What is the time duration
in seconds of the collected samples? At what DFT indices do we expect to see any
peaks in the spectrum?

Solution: The time duration is $T_N = NT = N/f_s = 128/40 = 3.2$ msec. Using Eq. (9.3.1),
we calculate $k = Nf/f_s = 128 \cdot 5/40 = 16$. The negative frequency -5 kHz is represented by the DFT index $N - k = 128 - 16 = 112$.   □

Example 9.3.2: A 10 msec segment of a signal is sampled at a rate of 10 kHz and the
resulting samples are saved. It is desired to compute the spectrum of that segment
at 128 equally spaced frequencies covering the range $2.5 \le f < 5$ kHz. We would
like to use an off-the-shelf N-point FFT routine to perform this computation. The
routine takes as input an N-dimensional vector **x** of time samples. Its output is an N-
dimensional DFT vector **X**. (a) What value of N should we use? (b) How is the routine's
input vector **x** defined in terms of the time samples that we collected? (c) Exactly what
DFT indices k and DFT values $X[k]$ correspond to the 128 spectral values that we wish
to compute?

Solution: The interval $[2.5, 5]$ kHz is one-quarter the Nyquist interval $[0, 10]$ kHz. Thus,
the DFT size should be $N = 4 \times 128 = 512$. This choice places 128 frequencies over the

[2.5, 5) interval. Another way is to identify the bin width over the [2.5, 5] subinterval with the bin width over the full interval:

$$\Delta f_{\text{bin}} = \frac{5 - 2.5}{128} = \frac{10}{N} \quad \Rightarrow \quad N = 512$$

The number of collected samples is $L = T_L f_s = (10 \text{ msec}) \times (10 \text{ kHz}) = 100$. Thus, the subroutine's 512-dimensional input vector **x** will consist of the 100 input samples with 412 zeros padded at the end.

Because the range [2.5, 5) is the second quarter of the Nyquist interval, it will be represented by the second quarter of DFT indices, that is, $128 \leq k < 256$. □

9.4 Matrix Form of DFT

The N-point DFT (9.2.8) can be thought of as a linear *matrix transformation* of the L-dimensional vector of time data into an N-dimensional vector of frequency data:

$$\mathbf{x} = \begin{bmatrix} x_0 \\ x_1 \\ \vdots \\ x_{L-1} \end{bmatrix} \quad \xrightarrow{\text{DFT}} \quad \mathbf{X} = \begin{bmatrix} X_0 \\ X_1 \\ \vdots \\ X_{N-1} \end{bmatrix}$$

where we denoted the DFT components by $X_k = X(\omega_k)$, $k = 0, 1, \ldots, N - 1$.

The linear transformation is implemented by an $N \times L$ matrix A, to be referred to as the *DFT matrix*, and can be written compactly as follows:

$$\boxed{\mathbf{X} = \text{DFT}(\mathbf{x}) = A\mathbf{x}} \quad \text{(matrix form of DFT)} \qquad (9.4.1)$$

or, component-wise:

$$\boxed{X_k = \sum_{n=0}^{L-1} A_{kn} x_n}, \qquad k = 0, 1, \ldots, N - 1 \qquad (9.4.2)$$

The matrix elements A_{kn} are defined from Eq. (9.2.8):

$$A_{kn} = e^{-j\omega_k n} = e^{-2\pi jkn/N} = W_N^{kn} \qquad (9.4.3)$$

for $0 \leq k \leq N - 1$ and $0 \leq n \leq L - 1$. For convenience, we defined the so-called *twiddle factor* W_N as the complex number:

$$W_N = e^{-2\pi j/N} \qquad (9.4.4)$$

Thus, the DFT matrix for an N-point DFT is built from the powers of W_N. Note that the first row ($k = 0$) and first column ($n = 0$) of A are always unity:

$$A_{0n} = 1, \quad 0 \leq n \leq L - 1 \quad \text{and} \quad A_{k0} = 1, \quad 0 \leq k \leq N - 1$$

The matrix A can be built from its second row ($k = 1$), consisting of the successive powers of W_N:

$$A_{1n} = W_N^n, \qquad n = 0, 1, \ldots, L - 1$$

It follows from the definition that the kth row is obtained by raising the second row to the kth power—element by element:

$$A_{kn} = W_N^{kn} = (W_N^n)^k = A_{1n}^k$$

Some examples of twiddle factors, DFT matrices, and DFTs are as follows: For $L = N$ and $N = 2, 4, 8$, we have:

$$W_2 = e^{-2\pi j/2} = e^{-\pi j} = -1$$

$$W_4 = e^{-2\pi j/4} = e^{-\pi j/2} = \cos(\pi/2) - j\sin(\pi/2) = -j$$

$$W_8 = e^{-2\pi j/8} = e^{-\pi j/4} = \cos(\pi/4) - j\sin(\pi/4) = \frac{1 - j}{\sqrt{2}}$$

(9.4.5)

The corresponding 2-point and 4-point DFT matrices are:

$$A = \begin{bmatrix} 1 & 1 \\ 1 & W_2 \end{bmatrix} = \begin{bmatrix} 1 & 1 \\ 1 & -1 \end{bmatrix}$$

$$A = \begin{bmatrix} 1 & 1 & 1 & 1 \\ 1 & W_4 & W_4^2 & W_4^3 \\ 1 & W_4^2 & W_4^4 & W_4^6 \\ 1 & W_4^3 & W_4^6 & W_4^9 \end{bmatrix} = \begin{bmatrix} 1 & 1 & 1 & 1 \\ 1 & -j & -1 & j \\ 1 & -1 & 1 & -1 \\ 1 & j & -1 & -j \end{bmatrix}$$

(9.4.6)

And, the 2-point and 4-point DFTs of a length-2 and a length-4 signal will be:

$$\begin{bmatrix} X_0 \\ X_1 \end{bmatrix} = \begin{bmatrix} 1 & 1 \\ 1 & -1 \end{bmatrix} \begin{bmatrix} x_0 \\ x_1 \end{bmatrix} = \begin{bmatrix} x_0 + x_1 \\ x_0 - x_1 \end{bmatrix}$$

$$\begin{bmatrix} X_0 \\ X_1 \\ X_2 \\ X_3 \end{bmatrix} = \begin{bmatrix} 1 & 1 & 1 & 1 \\ 1 & -j & -1 & j \\ 1 & -1 & 1 & -1 \\ 1 & j & -1 & -j \end{bmatrix} \begin{bmatrix} x_0 \\ x_1 \\ x_2 \\ x_3 \end{bmatrix}$$

(9.4.7)

Thus, the 2-point DFT is formed by taking the sum and difference of the two time samples. We will see later that the 2-point DFT is a convenient starting point for the merging operation in performing the FFT by hand.

The twiddle factor W_N satisfies $W_N^N = 1$, and therefore it is one of the Nth roots of unity; indeed, in the notation of Eq. (9.2.11), it is the root $W_N = z_{N-1}$ and is shown in Fig. 9.2.3. Actually, all the successive powers W_N^k, $k = 0, 1, \ldots, N - 1$ are Nth roots of unity, but in *reverse order* (i.e., clockwise) than the z_k of Eq. (9.2.11):

$$W_N^k = e^{-2\pi jk/N} = z_{-k} = z_k^{-1}, \quad k = 0, 1, \ldots, N - 1$$

(9.4.8)

Figure 9.4.1 shows W_N and its successive powers for the values $N = 2, 4, 8$. Because $W_N^N = 1$, the exponents in W_N^{kn} can be reduced modulo-N, that is, we may replace them by $W_N^{(nk) \bmod (N)}$.

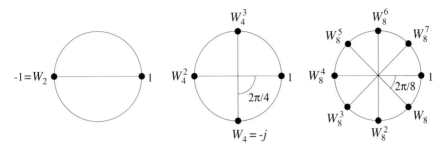

Fig. 9.4.1 Twiddle factor lookup tables for $N = 2, 4, 8$.

For example, using the property $W_4^4 = 1$, we may reduce all the powers of W_4 in the 4-point DFT matrix of Eq. (9.4.6) to one of the four powers W_4^k, $k = 0, 1, 2, 3$ and write it as

$$
A = \begin{bmatrix} 1 & 1 & 1 & 1 \\ 1 & W_4 & W_4^2 & W_4^3 \\ 1 & W_4^2 & W_4^4 & W_4^6 \\ 1 & W_4^3 & W_4^6 & W_4^9 \end{bmatrix} = \begin{bmatrix} 1 & 1 & 1 & 1 \\ 1 & W_4 & W_4^2 & W_4^3 \\ 1 & W_4^2 & 1 & W_4^2 \\ 1 & W_4^3 & W_4^2 & W_4 \end{bmatrix}
$$

The entries in A can be read off from the circular lookup table of powers of W_4 in Fig. 9.4.1, giving

$$
W_4 = -j, \quad W_4^2 = -1, \quad W_4^3 = j
$$

9.5 Modulo-N Reduction

The *modulo-N reduction* or *wrapping* of a signal plays a fundamental part in the theory of the DFT. It is defined by dividing the signal **x** into contiguous non-overlapping blocks of length N, wrapping the blocks around to be time-aligned with the first block, and adding them up. The process is illustrated in Fig. 9.5.1. The resulting wrapped block $\tilde{\mathbf{x}}$ has length N.

The length L of the signal **x** could be finite or infinite. If L is not an integral multiple of N, then the last sub-block will have length less than N; in this case, we may pad enough zeros at the end of the last block to increase its length to N.

The wrapping process can also be thought of as *partitioning* the signal vector **x** into N-dimensional subvectors and adding them up. For example, if $L = 4N$, the signal **x** will consist of four length-N subvectors:

$$
\mathbf{x} = \begin{bmatrix} \mathbf{x}_0 \\ \mathbf{x}_1 \\ \mathbf{x}_2 \\ \mathbf{x}_3 \end{bmatrix} \quad \Rightarrow \quad \tilde{\mathbf{x}} = \mathbf{x}_0 + \mathbf{x}_1 + \mathbf{x}_2 + \mathbf{x}_3 \tag{9.5.1}
$$

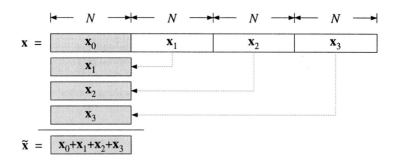

Fig. 9.5.1 Modulo-N reduction of a signal.

Example 9.5.1: Determine the mod-4 and mod-3 reductions of the length-8 signal vector:

$$\mathbf{x} = [1,\ 2,\ -2,\ 3,\ 4,\ -2,\ -1,\ 1]^T$$

For the $N = 4$ case, we may divide $\mathbf{x}$ into two length-4 sub-blocks to get:

$$\tilde{\mathbf{x}} = \begin{bmatrix} 1 \\ 2 \\ -2 \\ 3 \end{bmatrix} + \begin{bmatrix} 4 \\ -2 \\ -1 \\ 1 \end{bmatrix} = \begin{bmatrix} 5 \\ 0 \\ -3 \\ 4 \end{bmatrix}$$

Similarly, for $N = 3$ we divide $\mathbf{x}$ into length-3 blocks:

$$\tilde{\mathbf{x}} = \begin{bmatrix} 1 \\ 2 \\ -2 \end{bmatrix} + \begin{bmatrix} 3 \\ 4 \\ -2 \end{bmatrix} + \begin{bmatrix} -1 \\ 1 \\ 0 \end{bmatrix} = \begin{bmatrix} 3 \\ 7 \\ -4 \end{bmatrix}$$

where we padded a zero at the end of the third sub-block.   □

We may express the sub-block components in terms of the time samples of the signal $x(n)$, $0 \le n \le L - 1$, as follows. For $m = 0, 1, \ldots$

$$x_m(n) = x(mN + n), \quad n = 0, 1, \ldots, N - 1 \tag{9.5.2}$$

Thus, the mth sub-block occupies the time interval $[mN,\ (m+1)N)$. The wrapped vector $\tilde{\mathbf{x}}$ will be in this notation:

$$\begin{aligned} \tilde{x}(n) &= x_0(n) + x_1(n) + x_2(n) + x_3(n) + \cdots \\ &= x(n) + x(N + n) + x(2N + n) + x(3N + n) + \cdots \end{aligned} \tag{9.5.3}$$

for $n = 0, 1, \ldots, N - 1$, or, more compactly

$$\boxed{\ \tilde{x}(n) = \sum_{m=0}^{\infty} x(mN + n)\ }, \quad n = 0, 1, \ldots, N - 1 \tag{9.5.4}$$

This expression can be used to define $\tilde{x}(n)$ for all n, not just $0 \le n \le N - 1$. The resulting double-sided infinite signal is the so-called *periodic extension* of the signal $x(n)$ with period N. More generally, it is defined by

$$\tilde{x}(n) = \sum_{m=-\infty}^{\infty} x(mN + n), \qquad -\infty < n < \infty \tag{9.5.5}$$

The signal $\tilde{x}(n)$ is periodic in n with period N, that is, $\tilde{x}(n + N) = \tilde{x}(n)$. The definition (9.5.4) evaluates only one basic period $0 \le n \le N - 1$ of $\tilde{x}(n)$, which is all that is needed in the DFT.

The periodic extension interpretation of mod-N reduction is shown in Fig. 9.5.2. The terms $x(n + N)$, $x(n + 2N)$, and $x(n + 3N)$ of Eq. (9.5.3) can be thought as the time-advanced or left-shifted versions of $x(n)$ by N, $2N$, and $3N$ time samples. The successive sub-blocks of $x(n)$ get time-aligned one under the other over the basic period $0 \le n \le N - 1$, thus, their sum is the wrapped signal $\tilde{\mathbf{x}}$.

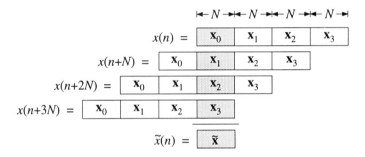

Fig. 9.5.2 Periodic extension interpretation of mod-N reduction of a signal.

The connection of the mod-N reduction to the DFT is the theorem that the length-N wrapped signal $\tilde{\mathbf{x}}$ has the *same* N-point DFT as the original unwrapped signal $\mathbf{x}$, that is,

$$\boxed{\tilde{X}_k = X_k \quad \text{or,} \quad \tilde{X}(\omega_k) = X(\omega_k)}, \qquad k = 0, 1, \dots, N - 1 \tag{9.5.6}$$

where $\tilde{X}_k = \tilde{X}(\omega_k)$ is the N-point DFT of the length-N signal $\tilde{x}(n)$:

$$\tilde{X}_k = \tilde{X}(\omega_k) = \sum_{n=0}^{N-1} \tilde{x}(n) e^{-j\omega_k n}, \qquad k = 0, 1, \dots, N - 1 \tag{9.5.7}$$

In the notation of Eq. (9.4.2), we may write:

$$\tilde{X}_k = \sum_{n=0}^{N-1} W_N^{kn} \tilde{x}(n) = \sum_{n=0}^{N-1} \tilde{A}_{kn} \tilde{x}(n) \tag{9.5.8}$$

where $\tilde{A}$ is the DFT matrix defined as in Eq. (9.4.3):

$$\tilde{A}_{kn} = W_N^{kn}, \qquad 0 \le k \le N - 1, \quad 0 \le n \le N - 1 \tag{9.5.9}$$

The DFT matrices A and $\tilde{A}$ have the same definition, except they differ in their dimensions, which are $N \times L$ and $N \times N$, respectively. We can write the DFT of $\tilde{\mathbf{x}}$ in the compact matrix form:

$$\tilde{\mathbf{X}} = \text{DFT}(\tilde{\mathbf{x}}) = \tilde{A}\tilde{\mathbf{x}} \qquad (9.5.10)$$

Thus, the above theorem can be stated in vector form:

$$\tilde{\mathbf{X}} = \mathbf{X} = A\mathbf{x} = \tilde{A}\tilde{\mathbf{x}} \qquad (9.5.11)$$

Symbolically, we will write $\text{DFT}(\tilde{\mathbf{x}}) = \text{DFT}(\mathbf{x})$ to denote Eqs. (9.5.6) or (9.5.11). The above theorem can be proved in many ways. In matrix form, it follows from the property that the $N \times N$ submatrices of the full $N \times L$ DFT matrix A are all *equal* to the DFT matrix $\tilde{A}$.

These submatrices are formed by grouping the first N columns of A into the first submatrix, the next N columns into the second submatrix, and so on. The matrix elements of the mth submatrix will be:

$$A_{k,mN+n} = W_N^{k(mN+n)} = W_N^{mkN} W_N^{kn}$$

Using the property $W_N^N = 1$, it follows that $W_N^{kmN} = 1$, and therefore:

$$A_{k,mN+n} = W_N^{kn} = A_{kn} = \tilde{A}_{kn}, \qquad 0 \le k, n \le N - 1$$

Thus, in general, A is partitioned in the form:

$$A = [\tilde{A}, \tilde{A}, \tilde{A}, \dots] \qquad (9.5.12)$$

As an example, consider the case $L = 8$, $N = 4$. The 4×8 DFT matrix A can be partitioned into two 4×4 identical submatrices, which are equal to $\tilde{A}$. Using $W_4 = e^{-2\pi j/4} = -j$, we have:

$$A = \begin{bmatrix} 1 & 1 & 1 & 1 & 1 & 1 & 1 & 1 \\ 1 & W_4 & W_4^2 & W_4^3 & W_4^4 & W_4^5 & W_4^6 & W_4^7 \\ 1 & W_4^2 & W_4^4 & W_4^6 & W_4^8 & W_4^{10} & W_4^{12} & W_4^{14} \\ 1 & W_4^3 & W_4^6 & W_4^9 & W_4^{12} & W_4^{15} & W_4^{18} & W_4^{21} \end{bmatrix}$$

$$= \begin{bmatrix} 1 & 1 & 1 & 1 & 1 & 1 & 1 & 1 \\ 1 & W_4 & W_4^2 & W_4^3 & 1 & W_4 & W_4^2 & W_4^3 \\ 1 & W_4^2 & W_4^4 & W_4^6 & 1 & W_4^2 & W_4^4 & W_4^6 \\ 1 & W_4^3 & W_4^6 & W_4^9 & 1 & W_4^3 & W_4^6 & W_4^9 \end{bmatrix}$$

$$= \begin{bmatrix} 1 & 1 & 1 & 1 & 1 & 1 & 1 & 1 \\ 1 & -j & -1 & j & 1 & -j & -1 & j \\ 1 & -1 & 1 & -1 & 1 & -1 & 1 & -1 \\ 1 & j & -1 & -j & 1 & j & -1 & -j \end{bmatrix} = [\tilde{A}, \tilde{A}]$$

where in the second submatrix, we partially reduced the powers of W_4 modulo-4.

The proof of the theorem follows now as a simple consequence of this partitioning property. For example, we have for the N-point DFT of Eq. (9.5.1):

$$\mathbf{X} = A\mathbf{x} = [\tilde{A}, \tilde{A}, \tilde{A}, \tilde{A}] \begin{bmatrix} \mathbf{x}_0 \\ \mathbf{x}_1 \\ \mathbf{x}_2 \\ \mathbf{x}_3 \end{bmatrix} = \tilde{A}\mathbf{x}_0 + \tilde{A}\mathbf{x}_1 + \tilde{A}\mathbf{x}_3 + \tilde{A}\mathbf{x}_3$$

$$= \tilde{A}(\mathbf{x}_0 + \mathbf{x}_1 + \mathbf{x}_2 + \mathbf{x}_3) = \tilde{A}\tilde{\mathbf{x}} = \tilde{\mathbf{X}}$$

Figure 9.5.3 illustrates the relative dimensions of these operations. The DFT (9.5.10) of $\tilde{\mathbf{x}}$ requires N^2 complex multiplications, whereas that of $\mathbf{x}$ requires NL. Thus, if $L > N$, it is more efficient to first wrap the signal mod-N and then take its DFT.

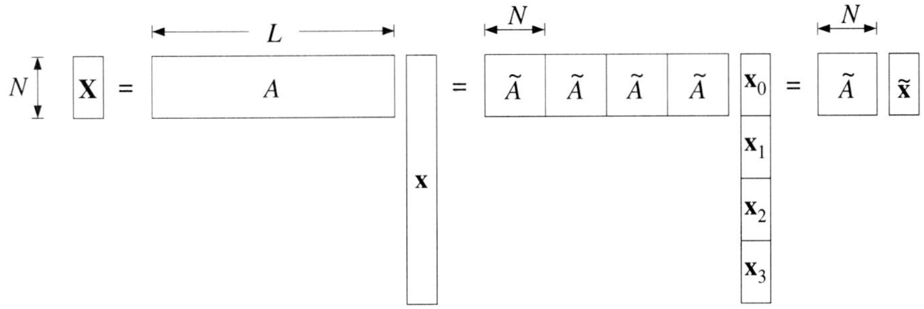

Fig. 9.5.3 *N*-point DFTs of the full and wrapped signals are equal.

Example 9.5.2: Compute the 4-point DFT of the length-8 signal of Example 9.5.1 in two ways: (a) working with the full unwrapped vector $\mathbf{x}$ and (b) computing the DFT of its mod-4 reduction.

Solution: The 4×8 DFT matrix was worked out above. The corresponding DFT is:

$$\mathbf{X} = A\mathbf{x} = \begin{bmatrix} 1 & 1 & 1 & 1 & 1 & 1 & 1 & 1 \\ 1 & -j & -1 & j & 1 & -j & -1 & j \\ 1 & -1 & 1 & -1 & 1 & -1 & 1 & -1 \\ 1 & j & -1 & -j & 1 & j & -1 & -j \end{bmatrix} \begin{bmatrix} 1 \\ 2 \\ -2 \\ 3 \\ 4 \\ -2 \\ -1 \\ 1 \end{bmatrix} = \begin{bmatrix} 6 \\ 8+4j \\ -2 \\ 8-4j \end{bmatrix}$$

The same DFT can be computed by the DFT matrix $\tilde{A}$ acting on the wrapped signal $\tilde{\mathbf{x}}$, determined in Example 9.5.1:

$$\tilde{\mathbf{X}} = \tilde{A}\tilde{\mathbf{x}} = \begin{bmatrix} 1 & 1 & 1 & 1 \\ 1 & -j & -1 & j \\ 1 & -1 & 1 & -1 \\ 1 & j & -1 & -j \end{bmatrix} \begin{bmatrix} 5 \\ 0 \\ -3 \\ 4 \end{bmatrix} = \begin{bmatrix} 6 \\ 8+4j \\ -2 \\ 8-4j \end{bmatrix}$$

The two methods give identical results. □

Example 9.5.3: The length L of the signal **x** can be infinite, as long as the signal is stable, so that the sum (9.5.4) converges. To illustrate the theorem (9.5.6) or (9.5.11), consider the causal signal $x(n) = a^n u(n)$, where $|a| < 1$.

To compute its N-point DFT, we determine its z-transform and evaluate it at the Nth root of unity points $z_k = e^{j\omega_k} = e^{2\pi jk/N}$. This gives:

$$X(z) = \frac{1}{1 - az^{-1}} \quad \Rightarrow \quad X_k = X(z_k) = \frac{1}{1 - az_k^{-1}}, \quad k = 0, 1, \ldots, N - 1$$

Next, we compute its mod-N reduction by the sum (9.5.4):

$$\tilde{x}(n) = \sum_{m=0}^{\infty} x(mN + n) = \sum_{m=0}^{\infty} a^{mN} a^n = \frac{a^n}{1 - a^N}, \quad n = 0, 1, \ldots, N - 1$$

where we used the geometric series sum. Computing its z-transform, we find:

$$\tilde{X}(z) = \sum_{n=0}^{N-1} \tilde{x}(n) z^{-n} = \frac{1}{1 - a^N} \sum_{n=0}^{N-1} a^n z^{-n} = \frac{1 - a^N z^{-N}}{(1 - a^N)(1 - az^{-1})}$$

Evaluating it at $z = z_k$ and using the property that $z_k^N = 1$, we find

$$\tilde{X}_k = \tilde{X}(z_k) = \frac{1 - a^N z_k^{-N}}{(1 - a^N)(1 - az_k^{-1})} = \frac{1 - a^N}{(1 - a^N)(1 - az_k^{-1})} = \frac{1}{1 - az_k^{-1}} = X_k$$

Thus, even though $x(n)$ and $\tilde{x}(n)$ are different and have different z-transforms and DTFTs, their N-point DFTs are the same.   □

The following C routine `modwrap.c` implements the modulo-N reduction operation. If $L < N$, it pads $N - L$ zeros at the end of **x** so that $\tilde{x}$ will have length N. If $L > N$, it determines how many length-N blocks fit into L, and adds them up, also taking into account the few excess points at the end.

```
/* modwrap.c - modulo-N wrapping of length-L signal */

void modwrap(L, x, N, xtilde)           usage: modwrap(L, x, N, xtilde);
int L, N;                               x is L-dimensional
double *x, *xtilde;                     xtilde is N-dimensional
{
    int n, r, m, M;

    r = L % N;                          remainder r = 0, 1, ..., N - 1
    M = (L-r) / N;                      quotient of division L/N

    for (n=0; n<N; n++) {
        if (n < r)                      non-zero part of last block
            xtilde[n] = x[M*N+n];       if L < N, this is the only block
        else
            xtilde[n] = 0;              if L < N, pad N - L zeros at end

        for (m=M-1; m>=0; m--)          remaining blocks
            xtilde[n] += x[m*N+n];      if L < N, this loop is skipped
    }
}
```

Using this routine, we may compute the N-point DFT of a length-L signal by first wrapping it modulo-N and then computing the N-point DFT of the wrapped signal:

```
modwrap(L, x, N, xtilde);          wrap input modulo-N
dft(N, xtilde, N, X);              DFT(x̃) = DFT(x)
```

Assuming L is a multiple of N, $L = MN$, the computational cost of the routine modwrap is $N(M-1) \simeq MN$ MACs, whereas that of the above dft is N^2 MACS. Thus, the total cost of computing the DFT is $N^2 + MN$ MACs. This is to be compared to $LN = MN^2$ MACs for the routine dft acting on the full length-L input. Replacing the above DFT by an FFT routine gives an even more efficient implementation, requiring $N \log_2(N)/2 + MN$ operations.

Example 9.5.4: Compare the cost of computing the 128-point DFT of a length-1024 signal, using a direct DFT, a prewrapped DFT, and a prewrapped FFT.

The number of length-N segments is $M = L/N = 1024/128 = 8$. The cost of wrapping the signal to length 128 is $N(M-1) = 896$. The cost of the three methods will be:

(direct DFT) $LN = 1024 \cdot 128 = 131{,}072$

(wrapped DFT) $N^2 + N(M-1) = 128^2 + 128 \cdot (8-1) = 17{,}280$

(wrapped FFT) $\dfrac{1}{2} N \log_2(N) + N(M-1) = \dfrac{1}{2} \cdot 128 \cdot 7 + 128 \cdot (8-1) = 1{,}344$

where we assumed that all the MAC operations are complex-valued. We may also compare the above with the cost of a direct 1024-point FFT on the 1024-point input:

(1024-point FFT) $\dfrac{1}{2} L \log_2(L) = \dfrac{1}{2} \cdot 1024 \cdot 10 = 5{,}120$

The DFT frequencies of the desired 128-point DFT are a *subset* of the DFT frequencies of the 1024-point DFT; indeed, we have:

$$\omega_k = \frac{2\pi k}{128} = \frac{2\pi (8k)}{1024}, \qquad k = 0, 1, \ldots, 127$$

Thus, the 128-point DFT can be extracted from the 1024-point FFT by taking every eighth entry, that is, $X_{128}(k) = X_{1024}(8k)$.   □

The two signals $\mathbf{x}$ and $\tilde{\mathbf{x}}$ are not the only ones that have a common DFT. Any other signal that has the *same* mod-N reduction as $\mathbf{x}$ will have the same DFT as $\mathbf{x}$. To see this, consider a length-L signal $\mathbf{y}$ such that $\tilde{\mathbf{y}} = \tilde{\mathbf{x}}$; then its N-point DFT can be obtained by applying Eq. (9.5.11):

$$\mathbf{Y} = A\mathbf{y} = \tilde{A}\tilde{\mathbf{y}} = \tilde{A}\tilde{\mathbf{x}} = A\mathbf{x} = \mathbf{X}$$

For example, the following length-8 signals all have the same 4-point DFT,

$$
\begin{bmatrix} x_0 \\ x_1 \\ x_2 \\ x_3 \\ x_4 \\ x_5 \\ x_6 \\ x_7 \end{bmatrix}, \quad
\begin{bmatrix} x_0 + x_4 \\ x_1 \\ x_2 \\ x_3 \\ 0 \\ x_5 \\ x_6 \\ x_7 \end{bmatrix}, \quad
\begin{bmatrix} x_0 + x_4 \\ x_1 + x_5 \\ x_2 \\ x_3 \\ 0 \\ 0 \\ x_6 \\ x_7 \end{bmatrix}, \quad
\begin{bmatrix} x_0 + x_4 \\ x_1 + x_5 \\ x_2 + x_6 \\ x_3 \\ 0 \\ 0 \\ 0 \\ x_7 \end{bmatrix}, \quad
\begin{bmatrix} x_0 + x_4 \\ x_1 + x_5 \\ x_2 + x_6 \\ x_3 + x_7 \\ 0 \\ 0 \\ 0 \\ 0 \end{bmatrix}
$$

because all have the same mod-4 reduction:

$$
\tilde{\mathbf{x}} = \begin{bmatrix} x_0 + x_4 \\ x_1 + x_5 \\ x_2 + x_6 \\ x_3 + x_7 \end{bmatrix}
$$

The above signals have a bottom half that becomes progressively zero, until the last vector which is recognized as the $\tilde{\mathbf{x}}$, viewed as a length-8 vector. In fact, the mod-N wrapped signal $\tilde{\mathbf{x}}$ is *unique* in the above class of signals in the sense that it is *shortest* signal, that is, of length N, that has the same DFT as the signal $\mathbf{x}$.

An equivalent characterization of the class of signals that have a common DFT can be given in the z-domain. Suppose the length-L signals $\mathbf{y}$ and $\mathbf{x}$ have equal mod-N reductions, $\tilde{\mathbf{y}} = \tilde{\mathbf{x}}$ and, therefore, equal DFTs $X_k = Y_k$. We form the difference of their z-transforms:

$$
F(z) = X(z) - Y(z) = \sum_{n=0}^{L-1} x(n)z^{-n} - \sum_{n=0}^{L-1} y(n)z^{-n}
$$

Evaluating $F(z)$ at the Nth roots of unity and using the equality of their N-point DFTs, we find:

$$
F(z_k) = X(z_k) - Y(z_k) = X_k - Y_k = 0, \qquad k = 0, 1, \ldots, N-1
$$

Thus, the N complex numbers z_k are roots of the difference polynomial $F(z)$. Therefore, $F(z)$ will be divisible by the Nth order product polynomial:

$$
1 - z^{-N} = \prod_{k=0}^{N-1} (1 - z_k z^{-1})
$$

which represents the factorization of $1 - z^{-N}$ into its Nth root-of-unity zeros. Therefore, we can write:

$$
X(z) - Y(z) = F(z) = \left(1 - z^{-N}\right) Q(z) \quad \text{or,}
$$

$$
\boxed{X(z) = Y(z) + \left(1 - z^{-N}\right) Q(z)} \tag{9.5.13}
$$

Because $X(z)$ and $Y(z)$ have degree $L - 1$, it follows that $Q(z)$ is an *arbitrary* polynomial of degree $L - 1 - N$. Denoting the coefficients of $Q(z)$ by $q(n)$, $0 \le n \le L - 1 - N$, we may write Eq. (9.5.13) in the time domain:

$$\boxed{x(n) = y(n) + q(n) - q(n - N)}, \qquad n = 0, 1, \dots, L - 1 \qquad (9.5.14)$$

Thus, any two sequences $x(n)$ and $y(n)$ related by Eq. (9.5.14) will have the same N-point DFT. The mod-N reduction $\tilde{\mathbf{x}}$ and its z-transform $\tilde{X}(z)$ are also related by Eq. (9.5.13):

$$\boxed{X(z) = \left(1 - z^{-N}\right) Q(z) + \tilde{X}(z)} \qquad (9.5.15)$$

Because $\tilde{X}(z)$ has degree $N - 1$, Eq. (9.5.15) represents the division of the polynomial $X(z)$ by the DFT polynomial $1 - z^{-N}$, with $\tilde{X}(z)$ being the *remainder* polynomial and $Q(z)$ the *quotient* polynomial. The remainder $\tilde{X}(z)$ is the unique polynomial satisfying Eq. (9.5.15) that has *minimal* degree $N - 1$.

9.6 Inverse DFT

The problem of inverting an N-point DFT is the problem of recovering the original length-L signal $\mathbf{x}$ from its N-point DFT $\mathbf{X}$, that is, inverting the relationship:

$$\mathbf{X} = A\mathbf{x} = \tilde{A}\tilde{\mathbf{x}} \qquad (9.6.1)$$

When $L > N$, the matrix A is not invertible. As we saw, there are in this case several possible solutions $\mathbf{x}$, all satisfying Eq. (9.6.1) and having the same mod-N reduction $\tilde{\mathbf{x}}$.

Among these solutions, the only one that is uniquely obtainable from the knowledge of the DFT vector $\mathbf{X}$ is $\tilde{\mathbf{x}}$. The corresponding DFT matrix $\tilde{A}$ is an $N \times N$ square invertible matrix. Thus, we define the *inverse DFT* by

$$\boxed{\tilde{\mathbf{x}} = \text{IDFT}(\mathbf{X}) = \tilde{A}^{-1}\mathbf{X}} \qquad \text{(inverse DFT)} \qquad (9.6.2)$$

or, component-wise,

$$\tilde{x}_n = \sum_{k=0}^{N-1} (\tilde{A}^{-1})_{nk} X_k, \qquad n = 0, 1, \dots, N - 1 \qquad (9.6.3)$$

The inverse $\tilde{A}^{-1}$ can be obtained *without* having to perform a matrix inversion by using the following *unitarity* property of the DFT matrix $\tilde{A}$:

$$\boxed{\dfrac{1}{N}\tilde{A}\tilde{A}^* = I_N} \qquad (9.6.4)$$

where I_N is the N-dimensional identity matrix and $\tilde{A}^*$ is the complex conjugate of $\tilde{A}$, obtained by conjugating *every* matrix element of $\tilde{A}$. For example, for $N = 4$, we can verify easily:

$$\frac{1}{4}\tilde{A}\tilde{A}^* = \frac{1}{4}\begin{bmatrix} 1 & 1 & 1 & 1 \\ 1 & -j & -1 & j \\ 1 & -1 & 1 & -1 \\ 1 & j & -1 & -j \end{bmatrix}\begin{bmatrix} 1 & 1 & 1 & 1 \\ 1 & j & -1 & -j \\ 1 & -1 & 1 & -1 \\ 1 & -j & -1 & j \end{bmatrix} = \begin{bmatrix} 1 & 0 & 0 & 0 \\ 0 & 1 & 0 & 0 \\ 0 & 0 & 1 & 0 \\ 0 & 0 & 0 & 1 \end{bmatrix}$$

Multiplying both sides of Eq. (9.6.4) by $\tilde{A}^{-1}$, we obtain for the matrix inverse:

$$\boxed{\tilde{A}^{-1} = \frac{1}{N}\tilde{A}^*} \tag{9.6.5}$$

Thus, the IDFT (9.6.2) can be written in the form:

$$\boxed{\tilde{\mathbf{x}} = \text{IDFT}(\mathbf{X}) = \frac{1}{N}\tilde{A}^*\mathbf{X}} \quad \text{(inverse DFT)} \tag{9.6.6}$$

We note also that the IDFT can be thought of as a DFT in the following sense. Introducing a second conjugation instruction, we have:

$$\tilde{A}^*\mathbf{X} = (\tilde{A}\mathbf{X}^*)^* = \left[\text{DFT}(\mathbf{X}^*)\right]^*$$

where the matrix $\tilde{A}$ acting on the conjugated vector $\mathbf{X}^*$ is the DFT of that vector. Dividing by N, we have:

$$\boxed{\text{IDFT}(\mathbf{X}) = \frac{1}{N}\left[\text{DFT}(\mathbf{X}^*)\right]^*} \tag{9.6.7}$$

Replacing DFT by FFT, we get a convenient inverse FFT formula, which uses an FFT to perform the IFFT. It is used in most FFT routines.

$$\boxed{\text{IFFT}(\mathbf{X}) = \frac{1}{N}\left[\text{FFT}(\mathbf{X}^*)\right]^*} \tag{9.6.8}$$

Example 9.6.1: To illustrate Eqs. (9.6.6) and (9.6.7), we calculate the IDFT of the 4-point DFT of Example 9.5.2. We have:

$$\tilde{\mathbf{x}} = \text{IDFT}(\mathbf{X}) = \frac{1}{N}\tilde{A}^*\mathbf{X} = \frac{1}{4}\begin{bmatrix} 1 & 1 & 1 & 1 \\ 1 & j & -1 & -j \\ 1 & -1 & 1 & -1 \\ 1 & -j & -1 & j \end{bmatrix}\begin{bmatrix} 6 \\ 8+4j \\ -2 \\ 8-4j \end{bmatrix} = \begin{bmatrix} 5 \\ 0 \\ -3 \\ 4 \end{bmatrix}$$

and using Eq. (9.6.7), we conjugate $\mathbf{X}$ and transform it:

$$\frac{1}{N}(\tilde{A}\mathbf{X}^*)^* = \frac{1}{4}\begin{bmatrix} 1 & 1 & 1 & 1 \\ 1 & -j & -1 & j \\ 1 & -1 & 1 & -1 \\ 1 & j & -1 & -j \end{bmatrix}\begin{bmatrix} 6 \\ 8-4j \\ -2 \\ 8+4j \end{bmatrix} = \begin{bmatrix} 5 \\ 0 \\ -3 \\ 4 \end{bmatrix}$$

where the final overall conjugation was omitted because $\tilde{\mathbf{x}}$ is real.   □

Using Eq. (9.4.3) the matrix elements of $\tilde{A}^{-1}$ are:

$$(\tilde{A}^{-1})_{nk} = \frac{1}{N}\tilde{A}_{nk}^* = \frac{1}{N}(W_N^{nk})^* = \frac{1}{N}W_N^{-nk}$$

where we used the property $W_N^* = e^{2\pi j/N} = W_N^{-1}$. Then, Eq. (9.6.3) can be written in the form:

$$\text{(IDFT)} \qquad \boxed{\tilde{x}_n = \frac{1}{N}\sum_{k=0}^{N-1}W_N^{-nk}X_k}, \qquad n = 0, 1, \ldots, N-1 \qquad (9.6.9)$$

In terms of the DFT frequencies ω_k, we have $X_k = X(\omega_k)$ and

$$W_N^{-nk} = e^{2\pi jkn/N} = e^{j\omega_k n}$$

Therefore, the inverse DFT can be written in the alternative form:

$$\text{(IDFT)} \qquad \boxed{\tilde{x}(n) = \frac{1}{N}\sum_{k=0}^{N-1}X(\omega_k)e^{j\omega_k n}}, \qquad n = 0, 1, \ldots, N-1 \qquad (9.6.10)$$

It expresses the signal $\tilde{x}(n)$ as a sum of N complex sinusoids of frequencies ω_k, whose *relative amplitudes and phases* are given by the DFT values $X(\omega_k)$.

The forward DFT of Eq. (9.2.8) is sometimes called an *analysis transform*, analyzing a signal $x(n)$ into N Fourier components. The inverse DFT (9.6.10) is called a *synthesis transform*, resynthesizing the signal $\tilde{x}(n)$ from those Fourier components. The forward and inverse N-point DFTs are akin to the more general forward and inverse DTFTs that use all frequencies, not just the N DFT frequencies:

$$X(\omega) = \sum_{n=0}^{L-1}x(n)e^{-j\omega n}, \qquad x(n) = \int_0^{2\pi}X(\omega)e^{j\omega n}\frac{d\omega}{2\pi} \qquad (9.6.11)$$

The difference between this inverse DTFT and (9.6.10) is that (9.6.11) reconstructs the full original signal $x(n)$, whereas (9.6.10) reconstructs only the wrapped signal $\tilde{x}(n)$. Eq. (9.6.10) can be thought of as a numerical approximation of the integral in (9.6.11), obtained by dividing the integration range into N equal bins:

$$\int_0^{2\pi}X(\omega)e^{j\omega n}\frac{d\omega}{2\pi} \simeq \sum_{k=0}^{N-1}X(\omega_k)e^{j\omega_k n}\frac{\Delta\omega_{\text{bin}}}{2\pi}$$

where from the definition (9.2.9), we have $\Delta\omega_{\text{bin}}/2\pi = 1/N$.

In summary, the inverse of an N-point DFT reconstructs only the wrapped version of the original signal that was transformed. This property is shown in Fig. 9.6.1.

In order for the IDFT to generate the original unwrapped signal $\mathbf{x}$, it is necessary to have $\tilde{\mathbf{x}} = \mathbf{x}$. This happens only if the DFT length N is at least L, so that there will be only one length-N sub-block in $\mathbf{x}$ and there will be nothing to wrap around. Thus, we have the condition:

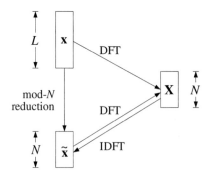

Fig. 9.6.1 Forward and inverse N-point DFTs.

$$\boxed{\tilde{\mathbf{x}} = \mathbf{x} \quad \text{only if} \quad N \geq L}$$

(9.6.12)

If $N = L$, then Eq. (9.6.12) is exact. If $N > L$, then we must pad $N - L$ zeros at the end of $\mathbf{x}$ so that the two sides of Eq. (9.6.12) have compatible lengths. If $N < L$, the wrapped and original signals will be different because there will be several length-N sub-blocks in $\mathbf{x}$ that get wrapped around. Thus, we also have the condition:

$$\boxed{\tilde{\mathbf{x}} \neq \mathbf{x} \quad \text{if} \quad N < L}$$

(9.6.13)

9.7 Sampling of Periodic Signals and the DFT

The inverse DFT (9.6.10) defines the signal $\tilde{x}(n)$ for $n = 0, 1, \ldots, N - 1$. However, the same expression can be used to define it for any value of n. The resulting $\tilde{x}(n)$ will be periodic in n with period N. This follows from the periodicity of the discrete-time sinusoids:

$$e^{j\omega_k (n+N)} = e^{2\pi jk(n+N)/N} = e^{2\pi jkn/N}e^{2\pi jk} = e^{2\pi jkn/N} = e^{j\omega_k n}$$

The periodic signal $\tilde{x}(n)$ is equivalent to the periodic extension of $x(n)$, as discussed in Section 9.5. Therefore, if the original signal $x(n)$ is also *periodic* with period N and we compute its N-point DFT over one period $L = N$, then we will have $\mathbf{x} = \tilde{\mathbf{x}}$, or, $x(n) = \tilde{x}(n)$. It follows that the *periodic* signal $x(n)$ may be represented by the *discrete Fourier series* (DFS):

$$\boxed{x(n) = \frac{1}{N} \sum_{k=0}^{N-1} X(\omega_k) e^{j\omega_k n}} \quad \text{(DFS)}$$

(9.7.1)

with the DFT playing the role of Fourier series coefficients:

$$\boxed{X(\omega_k) = \sum_{n=0}^{N-1} x(n) e^{-j\omega_k n}} \quad \text{(DFS coefficients)}$$

(9.7.2)

These relationships are helpful in the analysis of *analog* periodic signals. We saw in Example 1.4.6 and Section 8.1.2 that for a periodic signal to remain periodic after sampling, it is necessary that the sampling rate be a multiple of the fundamental frequency of the signal:

$$f_s = N f_1$$

The periodic analog signal will have an ordinary Fourier series expansion into a sum of sinusoids at the harmonics of the fundamental, $f_m = m f_1$:

$$x(t) = \sum_{m=-\infty}^{\infty} c_m e^{2\pi j f_m t}$$

In general, an infinite number of harmonics are necessary to represent $x(t)$, and therefore, if the signal is sampled at a rate f_s, all harmonics that lie outside the Nyquist interval will be aliased with the harmonics inside the interval.

Taking the Nyquist interval to be the right-sided one $[0, f_s]$, we note that the harmonics within that interval are none other than the N DFT frequencies:

$$f_k = k f_1 = k \frac{f_s}{N}, \qquad k = 0, 1, \ldots, N-1$$

Given an integer m, we determine its quotient and remainder of the division by N:

$$m = qN + k, \qquad 0 \le k \le N-1$$

and therefore, the corresponding harmonic will be:

$$f_m = m f_1 = qN f_1 + k f_1 = q f_s + f_k$$

which shows that f_m will be aliased with f_k. Therefore, if the signal $x(t)$ is sampled, it will give rise to the samples:

$$x(nT) = \sum_{m=-\infty}^{\infty} c_m e^{2\pi j f_m n / f_s} = \sum_{k=0}^{N-1} \sum_{q=-\infty}^{\infty} c_{qN+k} e^{2\pi j (q f_s + f_k) n / f_s}$$

where we wrote the summation over m as an equivalent double summation over q and k. Noting that,

$$e^{2\pi j q f_s n / f_s} = e^{2\pi j q n} = 1$$

and defining the aliased Fourier series amplitudes,

$$b_k = \sum_{q=-\infty}^{\infty} c_{qN+k}, \qquad k = 0, 1, \ldots, N-1$$

we obtain:

$$x(nT) = \sum_{k=0}^{N-1} b_k e^{2\pi j f_k n/f_s} = \sum_{k=0}^{N-1} b_k e^{j\omega_k n} \tag{9.7.3}$$

Comparing it with Eq. (9.7.1), we may identify the aliased amplitudes:

$$\boxed{b_k = \frac{1}{N} X(\omega_k)}, \qquad k = 0, 1, \ldots, N-1 \tag{9.7.4}$$

Thus, the aliased amplitudes b_k are computable by performing an N-point DFT on the N samples comprising *one period* of the signal $x(nT)$. If the samples $x(nT)$ were to be reconstructed back into analog form using an *ideal* reconstructor, the following aliased analog waveform would be obtained:

$$\boxed{x_{\text{al}}(t) = \sum_{k=0}^{N-1} b_k e^{2\pi j f_k t}} \tag{9.7.5}$$

with the proviso that those harmonics f_k that lie in the right half of the Nyquist interval, $f_s/2 < f_k \le f_s$, will be replaced by their negative selves, $f_k - f_s$.

Example 9.7.1: In Example 1.4.6, we determined the aliased signal $x_{\text{al}}(t)$ resulting by sampling a square wave of frequency $f_1 = 1$ Hz.

For a sampling rate of $f_s = 4$ Hz, we consider one period consisting of $N = 4$ samples and perform its 4-point DFT:

$$\mathbf{x} = \begin{bmatrix} 0 \\ 1 \\ 0 \\ -1 \end{bmatrix} \Rightarrow \mathbf{X} = A\mathbf{x} = \begin{bmatrix} 1 & 1 & 1 & 1 \\ 1 & -j & -1 & j \\ 1 & -1 & 1 & -1 \\ 1 & j & -1 & -j \end{bmatrix} \begin{bmatrix} 0 \\ 1 \\ 0 \\ -1 \end{bmatrix} = \begin{bmatrix} 0 \\ -2j \\ 0 \\ 2j \end{bmatrix}$$

Thus, the Fourier coefficients are:

$$[b_0, b_1, b_2, b_3] = \frac{1}{4}[0, -2j, 0, 2j] = [0, \frac{1}{2j}, 0, -\frac{1}{2j}]$$

corresponding to the harmonics:

$$[f_0, f_1, f_2, f_3] = [0, 1, 2, 3] \equiv [0, 1, 2, -1]$$

where $f_3 = 3$ was replaced by its negative version $f_3 - f_s = 3 - 4 = -1$. It follows that the aliased signal will be:

$$x_{\text{al}}(t) = b_1 e^{2\pi j t} + b_3 e^{-2\pi j t} = \frac{1}{2j} e^{2\pi j t} - \frac{1}{2j} e^{-2\pi j t} = \sin(2\pi t)$$

Similarly, for $N = 8$ corresponding to $f_s = 8$ Hz, we perform the 8-point DFT of one period of the square wave, and divide by 8 to get the aliased amplitudes:

$$\mathbf{x} = \begin{bmatrix} 0 \\ 1 \\ 1 \\ 1 \\ 0 \\ -1 \\ -1 \\ -1 \end{bmatrix} \xrightarrow{\text{DFT}} \mathbf{X} = \begin{bmatrix} 0 \\ -2j(\sqrt{2}+1) \\ 0 \\ -2j(\sqrt{2}-1) \\ 0 \\ 2j(\sqrt{2}-1) \\ 0 \\ 2j(\sqrt{2}+1) \end{bmatrix}, \quad \mathbf{b} = \begin{bmatrix} 0 \\ (\sqrt{2}+1)/4j \\ 0 \\ (\sqrt{2}-1)/4j \\ 0 \\ -(\sqrt{2}-1)/4j \\ 0 \\ -(\sqrt{2}+1)/4j \end{bmatrix}$$

These amplitudes correspond to the frequencies $f_k = kf_1$:

$$[0,1,2,3,4,5,6,7] \equiv [0,1,2,3,4,-3,-2,-1]$$

It follows that the aliased signal will be:

$$x_{\text{al}}(t) = \frac{(\sqrt{2}+1)}{4j} e^{2\pi jt} + \frac{(\sqrt{2}-1)}{4j} e^{2\pi j3t}$$

$$- \frac{(\sqrt{2}-1)}{4j} e^{-2\pi j3t} - \frac{(\sqrt{2}+1)}{4j} e^{-2\pi jt}$$

$$= \frac{\sqrt{2}+1}{2} \sin(2\pi t) + \frac{\sqrt{2}-1}{2} \sin(6\pi t)$$

which agrees with Example 1.4.6. The above 8-point DFT can be done using the 8×8 DFT matrix, or, more quickly using an FFT by hand, as done in Example 9.8.3. □

Example 9.7.2: Without performing any DFT or FFT computations, determine the 16-point DFT of the signal:

$$x(n) = 1 + 2\sin\left(\frac{\pi n}{2}\right) + 2\cos\left(\frac{3\pi n}{4}\right) + \cos(\pi n), \quad n = 0,1,\dots,15$$

Then, determine its 8-point DFT.

Solution: The signal $x(n)$ is already given as a sum of sinusoids at frequencies which are 16-point DFT frequencies. Thus, all we have to do is compare the given expression with the 16-point IDFT formula and identify the DFT coefficients X_k:

$$x(n) = \frac{1}{16} \sum_{k=0}^{15} X_k e^{j\omega_k n}$$

Using Euler's formula, we write the given signal as:

$$x(n) = 1 - je^{j\pi n/2} + je^{-j\pi n/2} + e^{3j\pi n/4} + e^{-3j\pi n/4} + e^{j\pi n}$$

Shifting the negative frequencies by 2π, noting that the 16-point DFT frequencies are $\omega_k = 2\pi k/16 = \pi k/8$, and writing the terms in increasing DFT index k, we have:

$$x(n) = \frac{1}{16} \left[16e^{j\omega_0 n} - 16je^{j\omega_4 n} + 16e^{j\omega_6 n} + 16e^{j\omega_8 n} + 16e^{j\omega_{10} n} + 16je^{j\omega_{12} n} \right]$$

where the frequencies, their negatives, and their relative locations with respect to the 16 DFT roots of unity are as follows:

$$\omega_4 = \frac{2\pi \cdot 4}{16} = \frac{\pi}{2}$$

$$\omega_{12} = \frac{2\pi \cdot 12}{16} = 2\pi - \frac{2\pi \cdot 4}{16} = 2\pi - \omega_4$$

$$\omega_6 = \frac{2\pi \cdot 6}{16} = \frac{3\pi}{4}$$

$$\omega_{10} = \frac{2\pi \cdot 10}{16} = 2\pi - \frac{2\pi \cdot 6}{16} = 2\pi - \omega_6$$

$$\omega_8 = \frac{2\pi \cdot 8}{16} = \pi$$

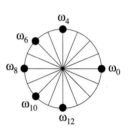

Comparing with the IDFT, we identify the coefficients of the exponentials:

$$X_0 = 16, \ X_4 = -16j, \ X_6 = X_8 = X_{10} = 16, \ X_{12} = 16j$$

Thus, the 16-point DFT vector will be:

$$\mathbf{X} = [16, 0, 0, 0, -16j, 0, 16, 0, 16, 0, 16, 0, 16j, 0, 0, 0]^T$$

The 8-point DFT is obtained by picking every other entry of $\mathbf{X}$, that is,

$$\mathbf{X} = [16, 0, -16j, 16, 16, 16, 16j, 0]^T \tag{9.7.6}$$

This follows because the 8-point DFT frequencies are a subset of the 16-point ones, that is, $\omega_k = 2\pi k / 8 = 2\pi (2k)/16, \ k = 0, 1, \ldots, 7$.   □

Example 9.7.3: The 8-point DFT determined in the previous example was that of the 16-point signal. If the signal $x(n)$ is considered as a length-8 signal over $0 \le n \le 7$, then its 8-point DFT will be different.

To find it, we follow the same method of writing $x(n)$ in its IDFT form, but now we identify the frequencies as 8-point DFT frequencies $\omega_k = 2\pi k / 8$. We have:

$$\omega_2 = \frac{2\pi \cdot 2}{8} = \frac{\pi}{2}, \quad \omega_3 = \frac{2\pi \cdot 3}{8} = \frac{3\pi}{4}, \quad \omega_4 = \frac{2\pi \cdot 4}{8} = \pi$$

and $x(n)$ can be written as:

$$x(n) = \frac{1}{8} \left[8e^{j\omega_0 n} - 8je^{j\omega_2 n} + 8e^{j\omega_3 n} + 8e^{j\omega_4 n} + 8e^{j\omega_5 n} + 8je^{j\omega_6 n} \right]$$

comparing with the 8-point IDFT,

$$x(n) = \frac{1}{8} \sum_{k=0}^{7} X_k e^{j\omega_k n}$$

we obtain:

$$\mathbf{X} = [8,\ 0,\ -8j,\ 8,\ 8,\ 8,\ 8j,\ 0]^T$$

The answer of Eq. (9.7.6) is doubled because the length-16 signal of the previous problem consists of two length-8 periods, which double when wrapped mod-8. □

9.8 FFT

The *fast Fourier transform* is a fast implementation of the DFT. It is based on a divide-and-conquer approach in which the DFT computation is divided into smaller, simpler, problems and the final DFT is rebuilt from the simpler DFTs. For a comprehensive review, history, and recent results, see [237]. For general references, see [223–244,303].

Another application of this divide-and-conquer approach is the computation of *very large FFTs*, in which the time data and their DFT are too large to be stored in main memory. In such cases the FFT is done in parts and the results are pieced together to form the overall FFT, and saved in secondary storage such as on hard disk [241–244,303].

In the simplest Cooley-Tukey version of the FFT, the dimension of the DFT is successively divided in half until it becomes unity. This requires the initial dimension N to be a power of two:

$$\boxed{N = 2^B} \qquad \Rightarrow \qquad B = \log_2(N) \tag{9.8.1}$$

The problem of computing the N-point DFT is replaced by the simpler problems of computing two $(N/2)$-point DFTs. Each of these is replaced by two $(N/4)$-point DFTs, and so on.

We will see shortly that an N-point DFT can be rebuilt from two $(N/2)$-point DFTs by an *additional* cost of $N/2$ complex multiplications. This basic merging step is shown in Fig. 9.8.1.

Thus, if we compute the two $(N/2)$-DFTs directly, at a cost of $(N/2)^2$ multiplications each, the total cost of rebuilding the full N-DFT will be:

$$2 \left(\frac{N}{2} \right)^2 + \frac{N}{2} = \frac{N^2}{2} + \frac{N}{2} \simeq \frac{N^2}{2}$$

where for large N the quadratic term dominates. This amounts to 50 percent savings over computing the N-point DFT directly at a cost of N^2.

Similarly, if the two $(N/2)$-DFTs were computed indirectly by rebuilding each of them from two $(N/4)$-DFTs, the total cost for rebuilding an N-DFT would be:

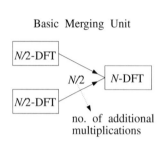

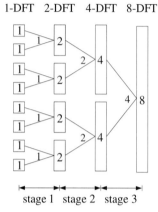

Fig. 9.8.1 Merging two $N/2$-DFTs into an N-DFT and its repeated application.

$$4\left(\frac{N}{4}\right)^2 + 2\frac{N}{4} + \frac{N}{2} = \frac{N^2}{4} + 2\frac{N}{2} \simeq \frac{N^2}{4}$$

Thus, we gain another factor of two, or a factor of four in efficiency over the direct N-point DFT. In the above equation, there are 4 direct $(N/4)$-DFTs at a cost of $(N/4)^2$ each, requiring an additional cost of $N/4$ each to merge them into $(N/2)$-DFTs, which require another $N/2$ for the final merge.

Proceeding in a similar fashion, we can show that if we start with $(N/2^m)$-point DFTs and perform m successive merging steps, the total cost to rebuild the final N-DFT will be:

$$\frac{N^2}{2^m} + \frac{N}{2}m \tag{9.8.2}$$

The first term, $N^2/2^m$, corresponds to performing the initial $(N/2^m)$-point DFTs directly. Because there are 2^m of them, they will require a total cost of $2^m (N/2^m)^2 = N^2/2^m$.

However, if the subdivision process is continued for $m = B$ stages, as shown in Fig. 9.8.1, the final dimension will be $N/2^m = N/2^B = 1$, which requires no computation at all because the 1-point DFT of a 1-point signal is itself.

In this case, the first term in Eq. (9.8.2) will be absent, and the total cost will arise from the second term. Thus, carrying out the subdivision/merging process to its logical extreme of $m = B = \log_2(N)$ stages, allows the computation to be done in:

$$\boxed{\frac{1}{2}NB = \frac{1}{2}N\log_2(N)} \qquad \text{(FFT computational cost)} \tag{9.8.3}$$

It can be seen Fig. 9.8.1 that the total number of multiplications needed to perform all the mergings in each stage is $N/2$, and B is the number of stages. Thus, we may interpret Eq. (9.8.3) as

$$\text{(total multiplications)} = \text{(multiplications per stage)} \times \text{(no. stages)} = \frac{N}{2}B$$

For the $N = 8$ example shown in Fig. 9.8.1, we have $B = \log_2(8) = 3$ stages and $N/2 = 8/2 = 4$ multiplications per stage. Therefore, the total cost is $BN/2 = 3 \cdot 4 = 12$ multiplications.

Next, we discuss the so-called *decimation-in-time radix-2 FFT algorithm.* There is also a decimation-in-frequency version, which is very similar. The term radix-2 refers to the choice of N as a power of 2, in Eq. (9.8.1).

Given a length-N sequence $x(n)$, $n = 0, 1, \dots, N - 1$, its N-point DFT $X(k) = X(\omega_k)$ can be written in the component-form of Eq. (9.4.2):

$$X(k) = \sum_{n=0}^{N-1} W_N^{kn} x(n), \qquad k = 0, 1, \dots, N - 1 \tag{9.8.4}$$

The summation index n ranges over both even and odd values in the range $0 \leq n \leq N - 1$. By grouping the even-indexed and odd-indexed terms, we may rewrite Eq. (9.8.4) as

$$X(k) = \sum_n W_N^{k(2n)} x(2n) + \sum_n W_N^{k(2n+1)} x(2n + 1)$$

To determine the proper range of summations over n, we consider the two terms separately. For the even-indexed terms, the index $2n$ must be within the range $0 \leq 2n \leq N - 1$. But, because N is even (a power of two), the upper limit $N - 1$ will be odd. Therefore, the highest even index will be $N - 2$. This gives the range:

$$0 \leq 2n \leq N - 2 \quad \Rightarrow \quad 0 \leq n \leq \frac{N}{2} - 1$$

Similarly, for the odd-indexed terms, we must have $0 \leq 2n + 1 \leq N - 1$. Now the upper limit can be realized, but the lower one cannot; the smallest odd index is unity. Thus, we have:

$$1 \leq 2n + 1 \leq N - 1 \quad \Rightarrow \quad 0 \leq 2n \leq N - 2 \quad \Rightarrow \quad 0 \leq n \leq \frac{N}{2} - 1$$

Therefore, the summation limits are the same for both terms:

$$X(k) = \sum_{n=0}^{N/2-1} W_N^{k(2n)} x(2n) + \sum_{n=0}^{N/2-1} W_N^{k(2n+1)} x(2n + 1) \tag{9.8.5}$$

This expression leads us to define the two length-$(N/2)$ subsequences:

$$\boxed{\begin{aligned} g(n) &= x(2n) \\ h(n) &= x(2n + 1) \end{aligned}}, \qquad n = 0, 1, \dots, \frac{N}{2} - 1 \tag{9.8.6}$$

and their $(N/2)$-point DFTs:

$$
\begin{aligned}
G(k) &= \sum_{n=0}^{N/2-1} W_{N/2}^{kn} g(n) \\[2mm]
H(k) &= \sum_{n=0}^{N/2-1} W_{N/2}^{kn} h(n)
\end{aligned}
\quad,\qquad k = 0, 1, \ldots, \frac{N}{2} - 1 \qquad (9.8.7)
$$

Then, the two terms of Eq. (9.8.5) can be expressed in terms of $G(k)$ and $H(k)$. We note that the twiddle factors W_N and $W_{N/2}$ of orders N and $N/2$ are related as follows:

$$
W_{N/2} = e^{-2\pi j/(N/2)} = e^{-4\pi j/N} = W_N^2
$$

Therefore, we may write:

$$
W_N^{k(2n)} = (W_N^2)^{kn} = W_{N/2}^{kn}, \quad W_N^{k(2n+1)} = W_N^k W_N^{2kn} = W_N^k W_{N/2}^{kn}
$$

Using the definitions (9.8.6), Eq. (9.8.5) can be written as:

$$
X(k) = \sum_{n=0}^{N/2-1} W_{N/2}^{kn} g(n) + W_N^k \sum_{n=0}^{N/2-1} W_{N/2}^{kn} h(n)
$$

and using Eq. (9.8.7),

$$
\boxed{X(k) = G(k) + W_N^k H(k)}, \qquad k = 0, 1, \ldots, N - 1 \qquad (9.8.8)
$$

This is the basic merging result. It states that $X(k)$ can be rebuilt out of the two $(N/2)$-point DFTs $G(k)$ and $H(k)$. There are N additional multiplications, $W_N^k H(k)$. Using the periodicity of $G(k)$ and $H(k)$, the additional multiplications may be reduced by half to $N/2$. To see this, we split the full index range $0 \le k \le N - 1$ into two half-ranges parametrized by the two indices k and $k + N/2$:

$$
0 \le k \le \frac{N}{2} - 1 \quad \Rightarrow \quad \frac{N}{2} \le k + \frac{N}{2} \le N - 1
$$

Therefore, we may write the N equations (9.8.8) as two groups of $N/2$ equations:

$$
\begin{aligned}
X(k) &= G(k) + W_N^k H(k) \\[2mm]
X(k + N/2) &= G(k + N/2) + W_N^{(k+N/2)} H(k + N/2)
\end{aligned}
\qquad k = 0, 1, \ldots, \frac{N}{2} - 1
$$

Using the periodicity property that any DFT is periodic in k with period its length, we have $G(k + N/2) = G(k)$ and $H(k + N/2) = H(k)$. We also have the twiddle factor property:

$$
W_N^{N/2} = (e^{-2\pi j/N})^{N/2} = e^{-j\pi} = -1
$$

Then, the DFT merging equations become:

$$\boxed{\begin{aligned} X(k) &= G(k) + W_N^k H(k) \\ X(k + N/2) &= G(k) - W_N^k H(k) \end{aligned}} \;, \quad k = 0, 1, \ldots, \frac{N}{2} - 1 \qquad (9.8.9)$$

They are known as the *butterfly* merging equations. The upper group generates the upper half of the N-dimensional DFT vector **X**, and the lower group generates the lower half. The $N/2$ multiplications $W_N^k H(k)$ may be used both in the upper and the lower equations, thus reducing the total extra merging cost to $N/2$. Vectorially, we may write them in the form:

$$\begin{bmatrix} X_0 \\ X_1 \\ \vdots \\ X_{N/2-1} \end{bmatrix} = \begin{bmatrix} G_0 \\ G_1 \\ \vdots \\ G_{N/2-1} \end{bmatrix} + \begin{bmatrix} H_0 \\ H_1 \\ \vdots \\ H_{N/2-1} \end{bmatrix} \times \begin{bmatrix} W_N^0 \\ W_N^1 \\ \vdots \\ W_N^{N/2-1} \end{bmatrix}$$

$$\begin{bmatrix} X_{N/2} \\ X_{N/2+1} \\ \vdots \\ X_{N-1} \end{bmatrix} = \begin{bmatrix} G_0 \\ G_1 \\ \vdots \\ G_{N/2-1} \end{bmatrix} - \begin{bmatrix} H_0 \\ H_1 \\ \vdots \\ H_{N/2-1} \end{bmatrix} \times \begin{bmatrix} W_N^0 \\ W_N^1 \\ \vdots \\ W_N^{N/2-1} \end{bmatrix} \qquad (9.8.10)$$

where the indicated multiplication is meant to be component-wise. Together, the two equations generate the full DFT vector **X**. The operations are shown in Fig. 9.8.2.

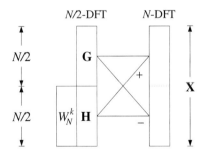

Fig. 9.8.2 Butterfly merging builds upper and lower halves of length-N DFT.

As an example, consider the case $N = 2$. The twiddle factor is now $W_2 = -1$, but only its zeroth power appears $W_2^0 = 1$. Thus, we get two 1-dimensional vectors, making up the final 2-dimensional DFT:

$$\begin{bmatrix} X_0 \end{bmatrix} = \begin{bmatrix} G_0 \end{bmatrix} + \begin{bmatrix} H_0 W_2^0 \end{bmatrix}$$

$$\begin{bmatrix} X_1 \end{bmatrix} = \begin{bmatrix} G_0 \end{bmatrix} - \begin{bmatrix} H_0 W_2^0 \end{bmatrix}$$

For $N = 4$, we have $W_4 = -j$, and only the powers W_4^0, W_4^1 appear:

$$
\begin{bmatrix} X_0 \\ X_1 \end{bmatrix} = \begin{bmatrix} G_0 \\ G_1 \end{bmatrix} + \begin{bmatrix} H_0 W_4^0 \\ H_1 W_4^1 \end{bmatrix}
$$

$$
\begin{bmatrix} X_2 \\ X_3 \end{bmatrix} = \begin{bmatrix} G_0 \\ G_1 \end{bmatrix} - \begin{bmatrix} H_0 W_4^0 \\ H_1 W_4^1 \end{bmatrix}
$$

And, for $N = 8$, we have:

$$
\begin{bmatrix} X_0 \\ X_1 \\ X_2 \\ X_3 \end{bmatrix} = \begin{bmatrix} G_0 \\ G_1 \\ G_2 \\ G_3 \end{bmatrix} + \begin{bmatrix} H_0 W_8^0 \\ H_1 W_8^1 \\ H_2 W_8^2 \\ H_3 W_8^3 \end{bmatrix}
$$

$$
\begin{bmatrix} X_4 \\ X_5 \\ X_6 \\ X_7 \end{bmatrix} = \begin{bmatrix} G_0 \\ G_1 \\ G_2 \\ G_3 \end{bmatrix} - \begin{bmatrix} H_0 W_8^0 \\ H_1 W_8^1 \\ H_2 W_8^2 \\ H_3 W_8^3 \end{bmatrix}
$$

To begin the merging process shown in Fig. 9.8.1, we need to know the starting one-dimensional DFTs. Once these are known, they may be merged into DFTs of dimension 2,4,8, and so on. The starting one-point DFTs are obtained by the so-called *shuffling* or *bit reversal* of the input time sequence. Thus, the typical FFT algorithm consists of three conceptual parts:

1. Shuffling the N-dimensional input into N one-dimensional signals.
2. Performing N one-point DFTs.
3. Merging the N one-point DFTs into one N-point DFT.

Performing the one-dimensional DFTs is only a conceptual part that lets us pass from the time to the frequency domain. Computationally, it is trivial because the one-point DFT $\mathbf{X} = [X_0]$ of a 1-point signal $\mathbf{x} = [x_0]$ is itself, that is, $X_0 = x_0$, as follows by setting $N = 1$ in Eq. (9.8.4).

The shuffling process is shown in Fig. 9.8.3 for $N = 8$. It has $B = \log_2(N)$ stages. During the first stage, the given length-N signal block $\mathbf{x}$ is divided into two length-$(N/2)$ blocks $\mathbf{g}$ and $\mathbf{h}$ by putting every other sample into $\mathbf{g}$ and the remaining samples into $\mathbf{h}$.

During the second stage, the same subdivision is applied to $\mathbf{g}$, resulting into the length-$(N/4)$ blocks $\{\mathbf{a}, \mathbf{b}\}$ and to $\mathbf{h}$ resulting into the blocks $\{\mathbf{c}, \mathbf{d}\}$, and so on. Eventually, the signal $\mathbf{x}$ is time-decimated down to N length-1 subsequences.

These subsequences form the starting point of the DFT merging process, which is depicted in Fig. 9.8.4 for $N = 8$. The butterfly merging operations are applied to each pair of DFTs to generate the next DFT of doubled dimension.

To summarize the operations, the shuffling process generates the smaller and smaller signals:

$$
\mathbf{x} \rightarrow \{\mathbf{g}, \mathbf{h}\} \rightarrow \{\{\mathbf{a}, \mathbf{b}\}, \{\mathbf{c}, \mathbf{d}\}\} \rightarrow \cdots \rightarrow \{1\text{-point signals}\}
$$

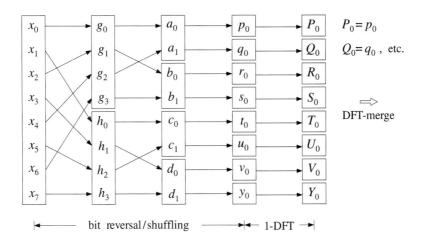

Fig. 9.8.3 Shuffling process generates N 1-dimensional signals.

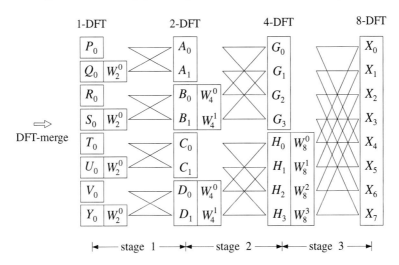

Fig. 9.8.4 DFT merging.

and the merging process rebuilds the corresponding DFTs:

$$\{\text{1-point DFTs}\} \ \rightarrow \ \cdots \ \rightarrow \ \{\{\mathbf{A}, \mathbf{B}\}, \{\mathbf{C}, \mathbf{D}\}\} \ \rightarrow \ \{\mathbf{G}, \mathbf{H}\} \ \rightarrow \ \mathbf{X}$$

The shuffling process may also be understood as a bit-reversal process, shown in Fig. 9.8.5. Given a time index n in the range $0 \le n \le N - 1$, it may be represented in binary by $B = \log_2(N)$ bits. For example, if $N = 8 = 2^3$, we may represent n by three bits $\{b_0, b_1, b_2\}$, which are zero or one:

$$n = (b_2 \, b_1 \, b_0) \equiv b_2 2^2 + b_1 2^1 + b_0 2^0$$

The binary representations of the time index n for x_n are indicated in Fig. 9.8.5, for both the input and the final shuffled output arrays. The bit-reversed version of n is obtained by reversing the order of the bits:

$$r = \text{bitrev}(n) = (b_0\, b_1\, b_2) \equiv b_0 2^2 + b_1 2^1 + b_2 2^0$$

We observe in Fig. 9.8.5 that the overall effect of the successive shuffling stages is to put the nth sample of the input array into the rth slot of the output array, that is, swap the locations of x_n with x_r, where r is the bit-reverse of n. Some slots are reverse-invariant so that $r = n$; those samples remain unmoved. All the others get swapped with the samples at the corresponding bit-reversed positions.

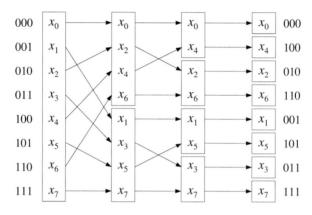

Fig. 9.8.5 Shuffling is equivalent to bit reversal.

The following C routine `fft.c` implements the FFT algorithm, as described above. It consists of two parts: bit-reversal and merging.

```
/* fft.c - decimation-in-time radix-2 FFT */

#include <cmplx.h>

void shuffle(), dftmerge();

void fft(N, X)                              usage: fft(N, X);
complex *X;
int N;
{
        shuffle(N, X);                      bit-reversal
        dftmerge(N, X);                     merging of DFTs
}
```

The bit-reversal operation is implemented by the routine `shuffle.c`, which calls the routines `swap.c` and `bitrev.c` that implement the swapping of the bit-reversed locations:

```
/* shuffle.c - in-place shuffling (bit-reversal) of a complex array */

#include <cmplx.h>
```

```
void swap();
int bitrev();

void shuffle(N, X)
complex *X;
int N;                                          N must be a power of 2
{
        int n, r, B=1;

        while ( (N >> B) > 0 )                  B = number of bits
                B++;

        B--;                                    N = 2^B

        for (n = 0; n < N; n++) {
            r = bitrev(n, B);                   bit-reversed version of n
            if (r < n) continue;                swap only half of the ns
            swap(X+n, X+r);                     swap by addresses
            }
}


/* swap.c - swap two complex numbers (by their addresses) */

#include <cmplx.h>

void swap(a,b)
complex *a, *b;
{
        complex t;

        t = *a;
        *a = *b;
        *b =  t;
}


/* bitrev.c - bit reverse of a B-bit integer n */

#define two(x)          (1 << (x))              2^x by left-shifting

int bitrev(n, B)
int n, B;
{
        int m, r;

        for (r=0, m=B-1; m>=0; m--)
            if ((n >> m) == 1) {                if 2^m term is present, then
                r += two(B-1-m);                add 2^{B-1-m} to r, and
                n -= two(m);                    subtract 2^m from n
                }

        return(r);
}
```

A B-bit number n and its reverse can be expressed in terms of their bits as:

$$n = \sum_{m=0}^{B-1} b_m 2^m$$

$$r = \sum_{m=0}^{B-1} b_m 2^{B-1-m}$$

The routine bitrev builds r by determining if the mth bit b_m is one and adding the corresponding power 2^{B-1-m} to r.

The DFT merging operation is given by the routine dftmerge.c. It is basically a loop that runs over the successive merging stages of dimensions $M = 2, 4, \ldots, N$.

```
/* dftmerge.c - DFT merging for radix 2 decimation-in-time FFT */

#include <cmplx.h>

void dftmerge(N, XF)
complex  *XF;
int N;
{
        double pi = 4. * atan(1.0);
        int k, i, p, q, M;
        complex  A, B, V, W;

        M = 2;
        while (M <= N) {                            two (M/2)-DFTs into one M-DFT
                W = cexp(cmplx(0.0, -2 * pi / M));   order-M twiddle factor
                V = cmplx(1., 0.);                   successive powers of W
                for (k = 0; k < M/2; k++) {          index for an (M/2)-DFT
                        for (i = 0; i < N; i += M) {   ith butterfly; increment by M
                                p = k + i;             absolute indices for
                                q = p + M / 2;         ith butterfly
                                A = XF[p];
                                B = cmul(XF[q], V);    V = W^k
                                XF[p] = cadd(A, B);    butterfly operations
                                XF[q] = csub(A, B);
                        }
                        V = cmul(V, W);               V = VW = W^{k+1}
                }
                M = 2 * M;                            next stage
        }
}
```

The appropriate twiddle factors W_M^k are computed on the fly and updated from stage to stage. For each stage M and value of k, all the butterflies that use the power W_M^k are computed. For example, in Fig. 9.8.4 the butterflies filling the slots $\{G_0, G_2\}$ and $\{H_0, H_2\}$ are done together because they use the same power W_4^0. Then, the butterflies for the slots $\{G_1, G_3\}$ and $\{H_1, H_3\}$ are done, using the power W_4^1.

The routine performs the computations *in place*, that is, the input time data vector X is *overwritten* by its shuffled version, which is repeatedly overwritten by the higher and higher DFTs during the merging process. The final merge produces the desired DFT and stores it in X.

The following routine `ifft.c` implements the inverse FFT via Eq. (9.6.8). The routine conjugates the input DFT, performs its FFT, conjugates the answer, and divides by N.

```
/* ifft.c - inverse FFT */

#include <cmplx.h>

void fft();

void ifft(N, X)
complex *X;
int N;
{
    int k;

    for (k=0; k<N; k++)
        X[k] = conjg(X[k]);                              conjugate input

    fft(N, X);                                           compute FFT of conjugate

    for (k=0; k<N; k++)
        X[k] = rdiv(conjg(X[k]), (double)N);    conjugate and divide by N
}
```

Next, we present some FFT examples. In the merging operations from 2-point to 4-point DFTs and from to 4-DFTs to 8-DFTs, the following twiddle factors are used:

$$\begin{bmatrix} W_4^0 \\ W_4^1 \end{bmatrix} = \begin{bmatrix} 1 \\ -j \end{bmatrix}, \qquad \begin{bmatrix} W_8^0 \\ W_8^1 \\ W_8^2 \\ W_8^3 \end{bmatrix} = \begin{bmatrix} 1 \\ (1-j)/\sqrt{2} \\ -j \\ -(1+j)/\sqrt{2} \end{bmatrix}$$

Example 9.8.1: Using the FFT algorithm, compute the 4-point DFT of the 4-point wrapped signal of Example 9.5.2.

Solution: The sequence of FFT operations are shown in Fig. 9.8.6. The shuffling operation was stopped at dimension 2, and the corresponding 2-point DFTs were computed by taking the sum and difference of the time sequences, as in Eq. (9.4.7).

The DFT merging stage merges the two 2-DFTs into the final 4-DFT. □

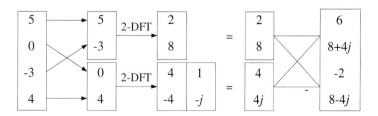

Fig. 9.8.6 4-point FFT of Example 9.8.1.

Example 9.8.2: Using the FFT algorithm, compute the 8-point DFT of the following 8-point signal:

$$\mathbf{x} = [4, -3, 2, 0, -1, -2, 3, 1]^T$$

Then, compute the inverse FFT of the result to recover the original time sequence.

Solution: The required FFT operations are shown in Fig. 9.8.7. Again, the shuffling stages stop with 2-dimensional signals which are transformed into their 2-point DFTs by forming sums and differences.

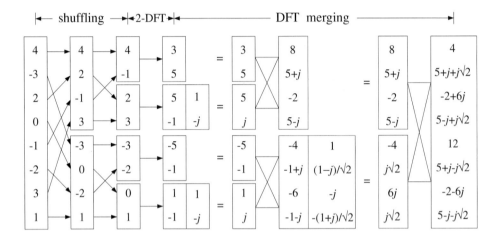

Fig. 9.8.7 8-point FFT of Example 9.8.2.

We find it more convenient to indicate the butterfly operations vectorially, that is, computing the sum and difference of the two 2-dimensional DFT vectors to form the upper and lower parts of the 4-dimensional DFTs, and computing the sum and difference of the two 4-DFT vectors to form the upper and lower parts of the final 8-DFT vector.

The inverse FFT is carried out by the expression (9.6.8). The calculations are shown in Fig. 9.8.8. First, the just computed DFT is complex conjugated. Then, its FFT is computed by carrying out the required shuffling and merging processes. The result must be conjugated (it is real already) and divided by $N = 8$ to recover the original sequence **x**.   □

Example 9.8.3: The 8-point DFT of the square wave of Example 9.7.1 can be calculated easily using the FFT. Figure 9.8.9 shows the details.   □

9.9 Fast Convolution

9.9.1 Circular Convolution

In the frequency domain, convolution of two sequences **h** and **x** is equivalent to multiplication of the respective DTFTs:

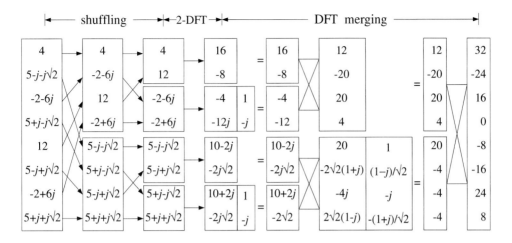

Fig. 9.8.8   8-point inverse FFT of the FFT in Fig. 9.8.7.

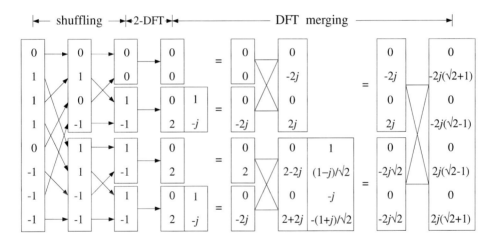

Fig. 9.8.9   8-point FFT of the square wave in Example 9.8.3.

$$\mathbf{y} = \mathbf{h} * \mathbf{x} \qquad \Leftrightarrow \qquad Y(\omega) = H(\omega)X(\omega) \tag{9.9.1}$$

Therefore, $y(n)$ can be recovered by the inverse DTFT of the product of the two DTFTs:

$$y(n) = \int_{-\pi}^{\pi} Y(\omega)e^{j\omega n}\frac{d\omega}{2\pi} = \int_{-\pi}^{\pi} H(\omega)X(\omega)e^{j\omega n}\frac{d\omega}{2\pi} \tag{9.9.2}$$

Symbolically, we write Eq. (9.9.2) as:

$$\mathbf{y} = \text{IDTFT}\big(\text{DTFT}(\mathbf{h}) \cdot \text{DTFT}(\mathbf{x})\big) \tag{9.9.3}$$

Equation (9.9.2) is not a practical method of computing $y(n)$ even in the case of finite-duration signals, because the ω-integration requires knowledge of $Y(\omega)$ at a continuous range of ω's.

A practical approach is to replace all the DTFTs by N-point DFTs. If Eq. (9.9.2) is replaced by an inverse DFT, we saw in Eq. (9.6.10) that it will reconstruct the wrapped signal $\tilde{y}(n)$ instead of the desired one:

$$\tilde{y}(n) = \frac{1}{N}\sum_{k=0}^{N-1} Y(\omega_k)e^{j\omega_k n} = \frac{1}{N}\sum_{k=0}^{N-1} H(\omega_k)X(\omega_k)e^{j\omega_k n} \qquad (9.9.4)$$

for $n = 0, 1, \ldots, N-1$, or, written symbolically:

$$\tilde{\mathbf{y}} = \mathrm{IDFT}(\mathrm{DFT}(\mathbf{h})\cdot\mathrm{DFT}(\mathbf{x})) \qquad (9.9.5)$$

Because the unwrapped $\mathbf{y}$ is the ordinary convolution $\mathbf{y} = \mathbf{h} * \mathbf{x}$, we can write the above as the wrapped convolution:

$$\boxed{\tilde{\mathbf{y}} = \widetilde{\mathbf{h} * \mathbf{x}} = \mathrm{IDFT}(\mathrm{DFT}(\mathbf{h})\cdot\mathrm{DFT}(\mathbf{x}))} \quad \text{(mod-}N\text{ circular convolution)} \qquad (9.9.6)$$

This expression is the definition of the length-N or modulo-N *circular convolution* of the two signals $\mathbf{h}$ and $\mathbf{x}$. A fast version is obtained by replacing DFT by FFT resulting in:

$$\boxed{\tilde{\mathbf{y}} = \widetilde{\mathbf{h} * \mathbf{x}} = \mathrm{IFFT}(\mathrm{FFT}(\mathbf{h})\cdot\mathrm{FFT}(\mathbf{x}))} \qquad (9.9.7)$$

If $\mathbf{h}$ and $\mathbf{x}$ are length-N signals, the computational cost of Eq. (9.9.7) is the cost for three FFTs (i.e., of $\mathbf{x}$, $\mathbf{h}$, and the inverse FFT) plus the cost of the N complex multiplications $Y(\omega_k) = H(\omega_k)X(\omega_k), k = 0, 1, \ldots, N-1$. Thus, the total number of multiplications to implement Eq. (9.9.7) is:

$$3\frac{1}{2}N\log_2(N) + N \qquad (9.9.8)$$

Some alternative ways of expressing $\tilde{\mathbf{y}}$ can be obtained by replacing $\mathbf{h}$ and/or $\mathbf{x}$ by their wrapped versions. This would not change the result because the wrapped signals have the same DFTs as the unwrapped ones, that is, $\mathrm{DFT}(\mathbf{h}) = \mathrm{DFT}(\tilde{\mathbf{h}})$ and $\mathrm{DFT}(\mathbf{x}) = \mathrm{DFT}(\tilde{\mathbf{x}})$. Thus, we can write:

$$\tilde{\mathbf{y}} = \widetilde{\mathbf{h} * \mathbf{x}} = \mathrm{IDFT}(\mathrm{DFT}(\mathbf{h})\cdot\mathrm{DFT}(\mathbf{x}))$$
$$= \widetilde{\tilde{\mathbf{h}} * \tilde{\mathbf{x}}} = \mathrm{IDFT}(\mathrm{DFT}(\tilde{\mathbf{h}})\cdot\mathrm{DFT}(\tilde{\mathbf{x}}))$$
$$= \widetilde{\tilde{\mathbf{h}} * \mathbf{x}} = \mathrm{IDFT}(\mathrm{DFT}(\tilde{\mathbf{h}})\cdot\mathrm{DFT}(\mathbf{x})) \qquad (9.9.9)$$
$$= \widetilde{\mathbf{h} * \tilde{\mathbf{x}}} = \mathrm{IDFT}(\mathrm{DFT}(\mathbf{h})\cdot\mathrm{DFT}(\tilde{\mathbf{x}}))$$

According to Eq. (9.6.12), in order for the circular convolution $\tilde{\mathbf{y}}$ to agree with the ordinary "linear" convolution $\mathbf{y}$, the DFT length N must be chosen to be at least the length L_y of the sequence $\mathbf{y}$. Recall from Eq. (4.1.12) that if a length-L signal $\mathbf{x}$ is convolved with an order-M filter $\mathbf{h}$, the length of the resulting convolution will be $L_y = L + M$. Thus, we obtain the constraint on the choice of N:

$$\boxed{\widetilde{\mathbf{y}} = \mathbf{y} \quad \text{only if} \quad N \geq L_y = L + M} \tag{9.9.10}$$

With this choice of N, Eq. (9.9.7) represents a fast way of computing linear convolution. Because both the filter and input vectors $\mathbf{h}$, $\mathbf{x}$ have lengths less than N (because $L + M = L_y \leq N$), we must increase them to length N by *padding zeros* at their ends, before we actually compute their N-point FFTs.

If $N < L_y$, part of the tail of $\mathbf{y}$ gets wrapped around to ruin the beginning part of $\mathbf{y}$. The following example illustrates the successive improvement of the circular convolution as the length N increases to the value required by (9.9.10).

Example 9.9.1: For the values $N = 3, 5, 7, 9, 11$, compute the mod-N circular convolution of the two signals of Example 4.1.1:

$$\mathbf{h} = [1, 2, -1, 1], \qquad \mathbf{x} = [1, 1, 2, 1, 2, 2, 1, 1]$$

Solution: For this example, we work exclusively in the time domain and perform ordinary convolution and wrap it modulo-N. The convolution table of Example 4.1.1, gives the output signal:

$$\mathbf{y} = \mathbf{x} * \mathbf{h} = [1, 3, 3, 5, 3, 7, 4, 3, 3, 0, 1]$$

The mod-3 circular convolution is obtained by dividing $\mathbf{y}$ into length-3 contiguous blocks, wrapping them around, and summing them to get:

$$\mathbf{y} = [1, 3, 3][5, 3, 7][4, 3, 3][0, 1, 0] \quad \Rightarrow \quad \widetilde{\mathbf{y}} = [10, 10, 13]$$

where we padded a 0 at the end to make the last block of length-3. In a similar fashion, we determine the other cases:

(mod-5):	$\mathbf{y} = [1, 3, 3, 5, 3][7, 4, 3, 3, 0][1]$	$\Rightarrow \quad \widetilde{\mathbf{y}} = [9, 7, 6, 8, 3]$
(mod-7):	$\mathbf{y} = [1, 3, 3, 5, 3, 7, 4][3, 3, 0, 1]$	$\Rightarrow \quad \widetilde{\mathbf{y}} = [4, 6, 3, 6, 3, 7, 4]$
(mod-9):	$\mathbf{y} = [1, 3, 3, 5, 3, 7, 4, 3, 3][0, 1]$	$\Rightarrow \quad \widetilde{\mathbf{y}} = [1, 4, 3, 5, 3, 7, 4, 3, 3]$
(mod-11):	$\mathbf{y} = [1, 3, 3, 5, 3, 7, 4, 3, 3, 0, 1]$	$\Rightarrow \quad \widetilde{\mathbf{y}} = [1, 3, 3, 5, 3, 7, 4, 3, 3, 0, 1]$

As N increases to $L_y = L + M = 8 + 3 = 11$, the lengths of the parts that get wrapped around become less and less, making $\widetilde{\mathbf{y}}$ resemble $\mathbf{y}$ more and more. $\qquad \square$

Example 9.9.2: Recompute the length-3 circular convolution of the previous example by first wrapping mod-3 the signals $\mathbf{h}$ and $\mathbf{x}$, performing their linear convolution, and wrapping it mod-3.

Solution: We find for the mod-3 reductions:

$$\mathbf{h} = [1, 2, -1][1] \quad \Rightarrow \quad \widetilde{\mathbf{h}} = [2, 2, -1]$$

$$\mathbf{x} = [1, 1, 2][1, 2, 2][1, 1] \quad \Rightarrow \quad \widetilde{\mathbf{x}} = [3, 4, 4]$$

The convolution of the wrapped signals is:

$$\tilde{\mathbf{h}} * \tilde{\mathbf{x}} = [2, 2, -1] * [3, 4, 4] = [6, 14, 13, 4, -4]$$

and, its mod-3 reduction:

$$\tilde{\mathbf{h}} * \tilde{\mathbf{x}} = [6, 14, 13][4, -4] \quad \Rightarrow \quad \widetilde{\tilde{\mathbf{h}} * \tilde{\mathbf{x}}} = [10, 10, 13]$$

which agrees with $\tilde{\mathbf{y}}$, in accordance with Eq. (9.9.9).   □

Example 9.9.3: Compute the mod-4 circular convolution of the following signals in two ways: (a) working in the time domain, and (b) using DFTs.

$$\mathbf{h} = [1, 2, 2, 1], \qquad \mathbf{x} = [1, 3, 3, 1]$$

Solution: The linear convolution is:

$$\mathbf{y} = \mathbf{h} * \mathbf{x} = [1, 2, 2, 1] * [1, 3, 3, 1] = [1, 5, 11, 14, 11, 5, 1]$$

wrapping it mod-4, we get:

$$\mathbf{y} = [1, 5, 11, 14][11, 5, 1] \quad \Rightarrow \quad \tilde{\mathbf{y}} = [12, 10, 12, 14]$$

Alternatively, we compute the 4-point DFTs of $\mathbf{h}$ and $\mathbf{x}$:

$$\mathbf{H} = \begin{bmatrix} 1 & 1 & 1 & 1 \\ 1 & -j & -1 & j \\ 1 & -1 & 1 & -1 \\ 1 & j & -1 & -j \end{bmatrix} \begin{bmatrix} 1 \\ 2 \\ 2 \\ 1 \end{bmatrix} = \begin{bmatrix} 6 \\ -1-j \\ 0 \\ -1+j \end{bmatrix}$$

$$\mathbf{X} = \begin{bmatrix} 1 & 1 & 1 & 1 \\ 1 & -j & -1 & j \\ 1 & -1 & 1 & -1 \\ 1 & j & -1 & -j \end{bmatrix} \begin{bmatrix} 1 \\ 3 \\ 3 \\ 1 \end{bmatrix} = \begin{bmatrix} 8 \\ -2-2j \\ 0 \\ -2+2j \end{bmatrix}$$

Multiplying them pointwise, we get:

$$\mathbf{Y} = \begin{bmatrix} Y_0 \\ Y_1 \\ Y_2 \\ Y_3 \end{bmatrix} = \begin{bmatrix} H_0 X_0 \\ H_1 X_1 \\ H_2 X_2 \\ H_3 X_3 \end{bmatrix} = \begin{bmatrix} 48 \\ 4j \\ 0 \\ -4j \end{bmatrix}$$

To take the inverse DFT, we conjugate, take the 4-point DFT, divide by 4, and conjugate the answer:

$$\tilde{\mathbf{y}} = \text{IDFT}(\mathbf{Y}) = \frac{1}{N}[\text{DFT}(\mathbf{Y}^*)]^*$$

$$\tilde{\mathbf{y}} = \frac{1}{4} \begin{bmatrix} 1 & 1 & 1 & 1 \\ 1 & -j & -1 & j \\ 1 & -1 & 1 & -1 \\ 1 & j & -1 & -j \end{bmatrix} \begin{bmatrix} 48 \\ -4j \\ 0 \\ 4j \end{bmatrix} = \begin{bmatrix} 12 \\ 10 \\ 12 \\ 14 \end{bmatrix}$$

The final conjugation is not necessary because $\tilde{\mathbf{y}}$ is real. □

Besides the efficient computation of convolution, the FFT can also be used to determine the impulse response of an unknown system, such as the reverberation impulse response of a room. Given a length-N input and a corresponding length-N measured output, we may compute their N-point DFTs and solve for the DFT of the impulse response of the system:

$$Y(\omega_k) = H(\omega_k) X(\omega_k) \quad \Rightarrow \quad H(\omega_k) = \frac{Y(\omega_k)}{X(\omega_k)}, \quad k = 0, 1, \dots, N-1$$

Then, taking the inverse DFT, we have:

$$\tilde{h}(n) = \frac{1}{N} \sum_{k=0}^{N-1} H(\omega_k) e^{j\omega_k n} = \frac{1}{N} \sum_{k=0}^{N-1} \frac{Y(\omega_k)}{X(\omega_k)} e^{j\omega_k n} \tag{9.9.11}$$

or, symbolically,

$$\tilde{\mathbf{h}} = \text{IDFT}\left[\frac{\text{DFT}(\mathbf{y})}{\text{DFT}(\mathbf{x})}\right] = \text{IFFT}\left[\frac{\text{FFT}(\mathbf{y})}{\text{FFT}(\mathbf{x})}\right] \tag{9.9.12}$$

The result is again the wrapped version $\tilde{h}(n)$ of the desired impulse response. For this type of application, the true impulse response $h(n)$ is typically infinite, and therefore, its wrapped version will be different from $h(n)$. However, if the wrapping length N is sufficiently large, such that the exponentially decaying tails of $h(n)$ can be ignored, then $\tilde{h}(n)$ may be an adequate approximation.

Example 9.9.4: The reverberation impulse response of a room is of the form

$$h(n) = Aa^n + Bb^n, \quad n \geq 0$$

Determine the response $\tilde{h}(n)$ that might be measured by the above procedure of dividing the output DFT by the input DFT and taking the IDFT of the result.

Solution: The mod-N reduction of $h(n)$ can be computed as in Example 9.5.3:

$$\tilde{h}(n) = \sum_{m=0}^{\infty} h(mN + n) = \frac{A}{1-a^N} a^n + \frac{B}{1-b^N} b^n, \quad 0 \leq n \leq N-1$$

If N is large enough such that a^N and b^N are small enough to be ignored, then $\tilde{h}(n) \simeq h(n)$ for $n = 0, 1, \dots, N-1$. □

9.9.2 Overlap-Add and Overlap-Save Methods

When the length L of the input signal $\mathbf{x}$ is infinite or very long, the length $L_y = L + M$ of the output will be infinite and the condition (9.9.10) cannot be satisfied.

In this case, we may apply the overlap-add block convolution method that we discussed in Section 4.1.10. In this method shown in Fig. 4.1.6, the input is partitioned into contiguous non-overlapping blocks of length L, each block is convolved with the filter $\mathbf{h}$, and the resulting output blocks are summed to form the overall output.

A fast version of the method can be obtained by performing the convolutions of the input blocks using circular convolution and the FFT by Eq. (9.9.7). The FFT length N must satisfy Eq. (9.9.10) in order for the output blocks to be correct. Given a desired power of two for the FFT length N, we determine the length of the input segments via:

$$N = L + M \quad \Rightarrow \quad \boxed{L = N - M} \tag{9.9.13}$$

With this choice of N, there would be no wrap-around errors, and the outputs of the successive input blocks $\{\mathbf{x}_0, \mathbf{x}_1, \dots\}$, can be computed by:

$$\mathbf{y}_0 = \tilde{\mathbf{y}}_0 = \text{IFFT}\left(\text{FFT}(\mathbf{h}) \cdot \text{FFT}(\mathbf{x}_0)\right)$$

$$\mathbf{y}_1 = \tilde{\mathbf{y}}_1 = \text{IFFT}\left(\text{FFT}(\mathbf{h}) \cdot \text{FFT}(\mathbf{x}_1)\right) \tag{9.9.14}$$

$$\mathbf{y}_2 = \tilde{\mathbf{y}}_2 = \text{IFFT}\left(\text{FFT}(\mathbf{h}) \cdot \text{FFT}(\mathbf{x}_1)\right)$$

and so on.

In counting the computational cost of this method, the FFT of $\mathbf{h}$ need not be counted. It can be computed once, $\mathbf{H} = \text{FFT}(\mathbf{h})$, and used in all convolutions of Eq. (9.9.14). We must only count the cost of *two* FFTs plus the N pointwise multiplications. Thus, the number of multiplications per input block is:

$$2\frac{1}{2}N \log_2 N + N = N(\log_2 N + 1)$$

This must be compared with the cost of $(M + 1)L = (M + 1)(N - M)$ for performing the ordinary time-domain convolution of each block with the filter. The relative cost of the fast versus the conventional slow method is:

$$\frac{\text{fast}}{\text{slow}} = \frac{N(\log_2 N + 1)}{(M + 1)(N - M)} \simeq \frac{\log_2 N}{M} \tag{9.9.15}$$

where the last equation follows in the limit $N \gg M \gg 1$.

The routine blockconv of Section 4.1.10 can be modified by replacing the internal call to conv by a part that implements the operation:

$$\mathbf{y} = \tilde{\mathbf{y}} = \text{IFFT}\left(\mathbf{H} \cdot \text{FFT}(\mathbf{x})\right)$$

The part of the routine that keeps track of the length-M tails of the output blocks and overlaps them, remains the same.

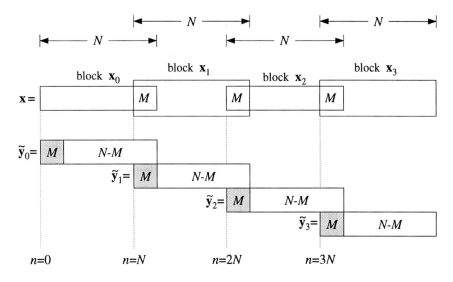

Fig. 9.9.1 Overlap-save method of fast convolution.

The *overlap-save* fast convolution method is an alternative method that also involves partitioning the input into blocks and filtering each block by Eq. (9.9.7). The method is shown in Fig. 9.9.1.

In this method, the input blocks have length equal to the FFT length, $L = N$, but they are made to overlap each other by M points, where M is the filter order. The output blocks will have length $L_y = L + M = N + M$ and therefore, do not satisfy the condition Eq. (9.9.10).

If the output blocks are computed via Eq. (9.9.7), then the last M points of each output block will get wrapped around and be added to the first M output points, ruining them. This is shown in Fig. 9.9.2. Assuming $N > M$, the remaining output points will be correct.

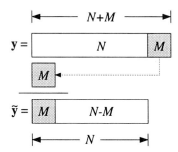

Fig. 9.9.2 Mod-N reduction of output block ruins first M output samples.

As shown in Fig. 9.9.1, because the input blocks overlap by M points, when the wrapped output blocks are aligned according to their absolute timings, the first M points of each block can be ignored because the correct outputs have already been computed from the previous block.

There is only one exception, that is, the very first M points of the output sequence are not computed correctly. This can be corrected by delaying the input by M time units before commencing the filtering operation.

The computational cost of the method is essentially the same as that of the overlap-add method, with the relative performance over conventional convolution given by Eq. (9.9.15).

Example 9.9.5: Using the overlap-save method of fast convolution, implemented in the time domain by mod-8 circular convolutions, compute the linear convolution of the "long" input:

$$\mathbf{x} = [1, 1, 1, 1, 3, 3, 3, 3, 1, 1, 1, 2, 2, 2, 2, 1, 1, 1, 1]$$

with the "short" filter:

$$\mathbf{h} = [1, -1, -1, 1]$$

Solution: For comparison, we compute the linear convolution using the convolution table:

$$\mathbf{y} = [1, 0, -1, 0, 2, 0, -2, 0, -2, 0, 2, 1, 0, -1, 0, -1, 0, 1, 0, -1, 0, 1]$$

For the overlap-save method, we divide the input into length-8 blocks which overlap by $M = 3$ points. These blocks are:

$$\mathbf{x} = [1, 1, 1, 1, 3, (3, 3, 3], 1, 1, [1, 2, 2), 2, 2, (1, 1, 1], 1, 0, 0, 0, 0)$$

Convolving these blocks with $\mathbf{h}$ gives:

$$\mathbf{y}_0 = \mathbf{h} * [1, 1, 1, 1, 3, 3, 3, 3] = [1, 0, -1, 0, 2, 0, -2, 0, -3, 0, 3]$$
$$\mathbf{y}_1 = \mathbf{h} * [3, 3, 3, 1, 1, 1, 2, 2] = [3, 0, -3, -2, 0, 2, 1, 0, -3, 0, 2]$$
$$\mathbf{y}_2 = \mathbf{h} * [1, 2, 2, 2, 2, 1, 1, 1] = [1, 1, -1, -1, 0, -1, 0, 1, -1, 0, 1]$$
$$\mathbf{y}_3 = \mathbf{h} * [1, 1, 1, 1, 0, 0, 0, 0] = [1, 0, -1, 0, -1, 0, 1, 0, 0, 0, 0]$$

Reducing these modulo-8 and ignoring the first $M = 3$ points (indicated by $*$), we obtain:

$$\tilde{\mathbf{y}}_0 = [*, *, *, 0, 2, 0, -2, 0]$$
$$\tilde{\mathbf{y}}_1 = [*, *, *, -2, 0, 2, 1, 0]$$
$$\tilde{\mathbf{y}}_2 = [*, *, *, -1, 0, -1, 0, 1]$$
$$\tilde{\mathbf{y}}_3 = [*, *, *, 0, -1, 0, 1, 0]$$

These would be the outputs computed via the FFT method. Putting them together, we obtain the overall output signal:

$$\mathbf{y} = [*, *, *, 0, 2, 0, -2, 0][-2, 0, 2, 1, 0][-1, 0, -1, 0, 1][0, -1, 0, 1, 0]$$

With the exception of the first 3 points, the answer is correct.   □

9.10 Problems

9.1 A 128-millisecond portion of an analog signal is sampled at a rate of 8 kHz and the resulting L samples are saved for further processing. What is L? The 256-point DFT of these samples is computed. What is the frequency spacing in Hz of the computed DFT values? What is the total number of required multiplications: (a) if the computations are done directly using the definition of the DFT, (b) if the L samples are first wrapped modulo 256 and then the 256-point DFT is computed, and (c) if a 256-point FFT is computed of the wrapped signal?

9.2 A 10 kHz sinusoidal signal is sampled at 80 kHz and 64 samples are collected and used to compute the 64-point DFT of this signal. At what DFT indices $k = 0, 1, \ldots, 63$ would you expect to see any peaks in the DFT?

9.3 A 5 kHz sinusoidal signal is sampled at 40 kHz and 16 periods of the signal are collected. What is the length N of the collected samples? Suppose an N-point DFT is performed. Then, at what DFT indices, $k = 0, 1, \ldots, N - 1$, do you expect to see any peaks in the DFT spectrum?

   In general, how is the number of periods contained in the N samples related to the DFT index at which you get a peak?

9.4 An 18 kHz sinusoid is sampled at a rate of 8 kHz and a 16-point DFT of a finite portion of the signal is computed. At what DFT indices in the range $0 \le k \le 15$ do you expect to see any peaks in the DFT spectrum? Would it matter if first we folded the 18 kHz frequency to lie within the Nyquist interval and then computed the DFT? Explain.

9.5 It is known that the frequency spectrum of a narrowband signal has a peak of width of 20 Hz but it is not known where this peak is located. To find out, an FFT must be computed and plotted versus frequency. If the signal is sampled at a rate of 8 kHz, what would be the minimum number of samples L that must be collected in order for the peak to be resolvable by the length-L data window? What is the duration in seconds of this data segment? What would be the minimum size N of the FFT in order for the N FFT spectral values to represent the L time samples accurately?

9.6 *Computer Experiment: Rectangular and Hamming Windows.* Using the routine `dtftr.c`, reproduce the results and graphs of Example 9.1.3.

9.7 *Computer Experiment: Frequency Resolution and Windowing.* Reproduce the results and graphs of Example 9.1.4. The spectra of the windowed signals must be computed by first windowing them using a length-L window and then padding $256 - L$ zeros at their ends to make them of length-256, and finally calling a 256-point FFT routine.

9.8 *Computer Experiment: Physical versus Computational Resolution.* Reproduce the results of Figs. 9.3.1 and 9.3.2. The theoretical DTFTs may be computed by 256-point FFTs. The 32-point and 64-point DFTs may be extracted from the 256-point FFTs by keeping every 8th point ($256/32 = 8$) and every 4th point ($256/64 = 4$.)

9.9 A dual-tone multi-frequency (DTMF) transmitter (touch-tone phone) encodes each key-press as a sum of two sinusoidal tones, with one frequency taken from group A and one from group B, where:

$$\text{group A} = 697, \ 770, \ 852, \ 941 \text{ Hz}$$

$$\text{group B} = 1209, \ 1336, \ 1477 \text{ Hz}$$

A digital DTMF receiver computes the spectrum of the received dual-tone signal and determines the two frequencies that are present, and thus, the key that was pressed.

What is the *smallest* number of time samples L that we should collect at a sampling rate of 8 kHz, in order for the group-A frequencies to be resolvable from the group-B frequencies? What is L if a Hamming window is used prior to computing the spectrum?

9.10 Suppose we collect 256 samples of the above DTMF signal and compute a 256-point FFT. Explain why each keypress generates substantial signal energy in 2 out of 7 possible DFT frequency bins (and their negatives). What are the indices k for these 7 bins? [*Hint:* Round k to its nearest integer. Do not ignore negative frequencies.]

Note that in practice, it may be more economical to just compute the value of $X(k)$ at those 14 k's instead of computing a full 256-point FFT.

9.11 *Computer Experiment: DTMF Sinusoids.* Consider the following artificial signal consisting of the sum of all seven DTMF sinusoids:

$$x(t) = \sum_{a=1}^{4} \sin(2\pi f_a t) + \sum_{b=1}^{3} \sin(2\pi f_b t)$$

where f_a and f_b are the group A and B frequencies given in Problem 9.9 and t is in seconds. (In practice, of course, only one f_a term and one f_b term will be present.)

The signal $x(t)$ is sampled at a rate of 8 kHz and 256 samples are collected, say, $x(n)$, $n = 0, 1, \ldots, 255$. The spectrum of this signal should consist of seven peaks clustered in two clearly separable groups (and, seven more negative-frequency peaks).

a. Plot the signal $x(n)$ versus n.

b. Compute the 256-point DFT or FFT of the signal $x(n)$ and plot the corresponding magnitude spectrum $|X(f)|$ only over the frequency range $0 \leq f \leq 4$ kHz.

c. Window the signal $x(n)$ by a length-256 Hamming window $w(n)$, that is, $x_{ham}(n) = w(n)x(n)$, and plot it versus n. Then, compute its 256-point DFT and plot the magnitude spectrum $|X_{ham}(f)|$ over $0 \leq f \leq 4$ kHz.

9.12 Let $\mathbf{x} = [1, 2, 2, 1, 2, 1, 1, 2]$. Compute the 4-point DFT of $\mathbf{x}$ using the definition in matrix form. Recompute it by first reducing $\mathbf{x}$ modulo 4 and then computing the 4-DFT of the result. Finally, compute the 4-point IDFT of the result and verify that you recover the mod-4 wrapped version of $\mathbf{x}$.

9.13 Compute the 8-point FFT of the length-8 signal $\mathbf{x} = [5, -1, -3, -1, 5, -1, -3, -1]$. Noting that these samples are the first 8 samples of $x(n) = 4\cos(\pi n/2) + \cos(\pi n)$, discuss whether the 8 computed FFT values accurately represent the expected spectrum of $x(n)$. What FFT indices correspond to the two frequencies of the cosinusoids?

9.14 The 8-point DFT $\mathbf{X}$ of an 8-point sequence $\mathbf{x}$ is given by

$$\mathbf{X} = [0, 4, -4j, 4, 0, 4, 4j, 4]$$

Using the FFT algorithm, compute the inverse DFT: $\mathbf{x} = \text{IFFT}(\mathbf{X})$. Using the given FFT $\mathbf{X}$, express $\mathbf{x}$ as a sum of real-valued (co)sinusoidal signals.

9.15 When a very large FFT of a very large data set is required (e.g., of size 2^{16} or larger), it may be computed in stages by partially decimating the time data down to several data sets of manageable dimension, computing their FFTs, and then rebuilding the desired FFT from the smaller ones. See [241–244,303] for a variety of approaches.

In this context, suppose you want to compute a $(4N)$-point FFT but your FFT hardware can only accommodate N-point FFTs. Explain how you might use this hardware to compute that FFT. Discuss how you must partition the time data, what FFTs must be computed, how they must be combined, and how the partial results must be shipped back and forth from secondary storage to the FFT processor in groups of no more than N samples. What is the total number of complex multiplications with your method? Compare this total to the cost of performing the $(4N)$-point FFT in a single pass? Do you observe anything interesting?

9.16 Compute the length-4 circular convolution of the two signals $\mathbf{h} = [1, 2, 1, 2, 1]$, $\mathbf{x} = [1, 1, 1, 1, 1]$ in two ways: (a) by computing their linear convolution and then reducing the result mod-4, (b) by first reducing $\mathbf{h}$ and $\mathbf{x}$ mod-4, computing the linear convolution of the reduced signals, and reducing the result mod-4.

9.17 Compute the 8-point FFT of $\mathbf{x} = [4, 2, 4, -6, 4, 2, 4, -6]$. Without performing any additional computations, determine the 4-point DFT and the 2-point DFT of the above signal. Explain your reasoning. Using the computed DFT and the inverse DFT formula, express the sequence $x(n)$, $n = 0, 1, \ldots, 7$ as a linear combination of real-valued sinusoidal signals. Does your $x(n)$ agree with the given sequence?

9.18 Let $\mathbf{x} = [1, 2, 3, 4, 5]$. (a) Determine a length-6 signal that has the same 5-point DFT as $\mathbf{x}$. (b) Determine a length-7 signal that has the same 5-point DFT as $\mathbf{x}$. Your answers should be nontrivial, that is, do not increase the length of $\mathbf{x}$ by padding zeros at its end.

9.19 Show the property:

$$\frac{1}{N}\left[1 + W_N^k + W_N^{2k} + W_N^{3k} + \cdots + W_N^{(N-1)k}\right] = \delta(k), \qquad k = 0, 1, \ldots, N-1$$

9.20 Show the following properties:

a. $W_N = W_{2N}^2 = W_{3N}^3 = \cdots = W_{pN}^p$

b. $X_N(k) = X_{pN}(pk), \qquad k = 0, 1, \ldots, N-1$

where W_{pN} is the twiddle factor of order pN, p is any integer, $X_N(k)$ denotes the N-point DFT, and $X_{pN}(k)$ the (pN)-point DFT of a common signal $x(n)$ of length L.

9.21 Consider a 16-point signal x_n, $0 \le n \le 15$, with 16-point DFT X_k, $0 \le k \le 15$, namely,

$$\left[X_0, X_1, X_2, X_3, X_4, X_5, X_6, X_7, X_8, X_9, X_{10}, X_{11}, X_{12}, X_{13}, X_{14}, X_{15}\right]$$

Show that the 8-point DFT of the given 16-point signal is:

$$\left[X_0, X_2, X_4, X_6, X_8, X_{10}, X_{12}, X_{14}\right]$$

9.22 The following analog signal $x(t)$, where t is in msec, is sampled at a rate of 8 kHz:

$$x(t) = \cos(24\pi t) + 2\sin(12\pi t)\cos(8\pi t)$$

a. Determine the signal $x_a(t)$ that is aliased with $x(t)$.

b. Eight consecutive samples of $x(t)$ are collected. *Without* performing any DFT or FFT operations, determine the 8-point DFT of these 8 samples.

9.23 Consider the following 8-point signal, defined for $n = 0, 1, \ldots, 7$:

$$x(n) = 1 + 2\sin\left(\frac{\pi n}{4}\right) - 2\sin\left(\frac{\pi n}{2}\right) + 2\sin\left(\frac{3\pi n}{4}\right) + 3(-1)^n$$

Without performing any DFT or FFT computations, determine the 8-point DFT of this signal.

9.24 Let $x(n) = \cos(\pi n/2) + 2\cos(\pi n/8)$, $n = 0, 1, \ldots, 15$. Without performing any actual DFT/FFT computations, determine the 16-point DFT of this 16-point signal. [*Hint*: Compare $x(n)$ with the 16-point inverse DFT formula.]

9.25 Let $x(n) = \cos(\pi n/2) + 2\cos(\pi n/8)$, $n = 0, 1, \ldots, 31$. Without performing any actual DFT/FFT computations, determine the 32-point DFT of this 32-point signal.

9.26 Consider the following length-16 signal:

$$x(n) = 0.5 + 2\sin(0.5\pi n) + 1.5\cos(\pi n), \qquad n = 0, 1, \ldots, 15$$

a. Determine the DTFT $X(\omega)$ of this finite sequence, and sketch it roughly versus ω in the range $0 \le \omega \le 2\pi$. [*Hint*: Remember that each spectral line gets replaced by the rectangular window's frequency response.]

b. Without performing any DFT or FFT computations, determine the 16-point DFT of this sequence. Then, determine the 8-point DFT of the *same* sequence.

c. Place the 16-point DFT values on the graph of $X(\omega)$ of part (a).

9.27 Let $\mathbf{X} = A\mathbf{x}$ be the N-point DFT of the length-N signal $\mathbf{x}$ expressed in matrix form, where A is the $N \times N$ DFT matrix defined by its matrix elements $A_{kn} = W_N^{kn}$, $k, n = 0, 1, \ldots, N-1$. Show that the inverse of this matrix can be obtained essentially by conjugating A, that is,

$$A^{-1} = \frac{1}{N}A^*$$

Therefore the IDFT can be expressed by $\mathbf{x} = A^{-1}\mathbf{X} = A^*\mathbf{X}/N$. Explain how this result justifies the rule:

$$\text{IFFT}(\mathbf{X}) = \frac{1}{N}\left(\text{FFT}(\mathbf{X}^*)\right)^*$$

9.28 Let $X(k)$ be the N-point DFT of a length-N (complex-valued) signal $x(n)$. Use the results of Problem 9.27 to show the Parseval relation:

$$\sum_{n=0}^{N-1} |x(n)|^2 = \frac{1}{N}\sum_{k=0}^{N-1} |X(k)|^2$$

9.29 Compute the mod-4, mod-5, mod-6, mod-7, and mod-8 circular convolutions of the signals $\mathbf{x} = [2, 1, 1, 2]$ and $\mathbf{h} = [1, -1, -1, 1]$. For what value of N does the mod-N circular convolution agree with the ordinary linear convolution?

9.30 Compute the modulo-8 circular convolution of the two signals

$$\mathbf{h} = [2, 1, 1, 1, 2, 1, 1, 1], \qquad \mathbf{x} = [2, 1, 2, -3, 2, 1, 2, -3]$$

in two ways:

a. Working exclusively in the time domain.

b. Using the formula:

$$\tilde{\mathbf{y}} = \text{IFFT}(\text{FFT}(\mathbf{h}) \cdot \text{FFT}(\mathbf{x}))$$

implemented via 8-point FFTs. All the computational details of the required FFTs must be shown explicitly.

9.31   a. Compute the 8-point FFT of the 8-point signal $\mathbf{x} = [6, 1, 0, 1, 6, 1, 0, 1]$.

       b. Using the inverse DFT formula, express $\mathbf{x}$ as a linear combination of real-valued sinusoids.

       c. Find two other signals, one of length-9 and one of length-10, that have the same 8-point DFT as $\mathbf{x}$. These signals must not begin or end with zeros.

       d. Compute the 4-point FFT of $\mathbf{x}$ by carrying out a *single* 4-point FFT.

9.32 Let $A(k)$ be the N-point DFT of a *real-valued* signal $a(n)$, $n = 0, 1, \ldots, N - 1$. Prove the symmetry property:

$$A(k)^* = A(N - k), \qquad k = 0, 1, \ldots, N - 1$$

If we think of $A(k)$ as an N-dimensional array, then how can we state the above relationship at $k = 0$?

9.33 *Two Real-Valued Signals at a Time.* Let $x(n) = a(n) + jb(n)$ be a length-N complex-valued signal and let $X(k)$ be its N-point DFT. Let $A(k)$ and $B(k)$ denote the N-point DFTs of the real and imaginary parts $a(n)$ and $b(n)$ of $x(n)$. Show that they can be recovered from $X(k)$ by

$$A(k) = \frac{1}{2}[X(k) + X(N - k)^*], \qquad B(k) = \frac{1}{2j}[X(k) - X(N - k)^*]$$

for $k = 0, 1, \ldots, N - 1$. If we think of $X(k)$ as an N-dimensional array, then how can we state the above relationships at $k = 0$?

Thus, the DFTs of *real-valued signals* can be computed *two at a time* by computing the DFT of a *single* complex-valued signal.

9.34 *FFT of Real-Valued Signal.* Using the results of Problem 9.33, show that the N-point FFT $X(k)$ of an N-point real-valued signal $x(n)$, $n = 0, 1, \ldots, N - 1$ can be computed efficiently as follows: First, pack the even and odd parts of $x(n)$ into a complex-valued signal of length $N/2$, that is, define

$$y(n) = x(2n) + jx(2n + 1) \equiv g(n) + jh(n), \qquad n = 0, 1, \ldots, \frac{N}{2} - 1$$

Then, compute the $N/2$-point FFT of $y(n)$, say, $Y(k)$, $k = 0, 1, \ldots, N/2 - 1$, and extract the $N/2$-point FFTs of $g(n)$ and $h(n)$ by

$$G(k) = \frac{1}{2}[Y(k) + Y(\frac{N}{2} - k)^*], \qquad H(k) = \frac{1}{2j}[Y(k) - Y(\frac{N}{2} - k)^*]$$

for $k = 0, 1, \ldots, N/2 - 1$. And finally, construct the desired N-point FFT by

$$X(k) = G(k) + W_N^k H(k), \qquad X(k + \frac{N}{2}) = G(k) - W_N^k H(k)$$

for $k = 0, 1, \ldots, N/2 - 1$. What happens at $k = 0$?

Determine the relative computational savings of this method versus performing the N-point FFT of $x(n)$ directly.

9.35 *Computer Experiment: FFT of Real-Valued Signal.* Write a C routine `fftreal.c` that implements the method of Problem 9.34. The routine must have inputs/output declarations:

```
void fftreal(N, x, X)
int N;                          must be a power of 2
double *x;                      real-valued N-dimensional time data
complex *X;                     complex N-dimensional FFT array
```

The routine must invoke the routine `fft.c` once on the time-decimated, complexified, input. In rebuilding the final DFT $X(k)$, special care must be exercised at $k = 0$.

Write a small main program to test the routine by comparing its output to the output of `fft` called on the full input array as usual.

9.36 Consider the following N-point signal and its reverse:

$$\mathbf{x} = [x_0, x_1, \ldots, x_{N-1}]$$

$$\mathbf{x}_R = [x_{N-1}, \ldots, x_1, x_0]$$

Show that the z-transform and N-point DFT of the reversed signal can be expressed as:

$$X_R(z) = z^{-(N-1)} X(z^{-1})$$

$$X_R(k) = W_N^{-k} X(N - k), \qquad k = 0, 1, \ldots, N - 1$$

Show that in the time domain the reversal process is equivalent to a two-step process of first reflecting the signal around the origin $n = 0$, and then delaying it by $N - 1$ units.

9.37 *Discrete Cosine Transform (DCT).* Consider a length-N real-valued signal $\mathbf{x}$ and its reverse as defined in Problem 9.36. Construct the concatenated signal of length $2N$:

$$\mathbf{y} = [\mathbf{x}, \mathbf{x}_R] = [x_0, x_1, \ldots, x_{N-1}, x_{N-1}, \ldots, x_1, x_0]$$

a. Show that its z-transform can be expressed in terms of the z-transform of $\mathbf{x}$:

$$Y(z) = X(z) + z^{-N} X_R(z) = X(z) + z^{-(2N-1)} X(z^{-1})$$

b. Let Y_k be the $(2N)$-point DFT of $\mathbf{y}$. Show that it can be expressed in the form:

$$Y_k = 2e^{j\omega_k/2} C_k, \qquad k = 0, 1, \ldots, 2N - 1$$

where $\omega_k = 2\pi k / (2N) = \pi k / N$ is the kth frequency for the $(2N)$-point DFT and C_k is one form of the discrete cosine transform of x_n given by:

$$C_k = \sum_{n=0}^{N-1} x_n \cos(\omega_k(n + 1/2)) \qquad (9.10.1)$$

[*Hint:* Evaluate part (a) at the $(2N)$th roots of unity and multiply by $z^{-1/2}$.]

c. Using the results of Problem 9.32, show that C_k satisfies the symmetry property:

$$C_{2N-k} = -C_k, \qquad k = 0, 1, \ldots, 2N - 1$$

In particular, show $C_N = 0$.

d. Applying the inverse DFT equation on Y_k, show the inverse DCT:

$$x_n = \frac{1}{N} \sum_{k=0}^{2N-1} C_k e^{j\omega_k(n+1/2)}, \qquad n = 0, 1, \dots, N-1$$

Using the symmetry property of part (c), show the alternative inverse DCT, which uses only the first N DCT coefficients C_k, $k = 0, 1, \dots, N-1$:

$$x_n = \frac{1}{N} \left[C_0 + 2 \sum_{k=1}^{N-1} C_k \cos(\omega_k(n+1/2)) \right], \qquad n = 0, 1, \dots, N-1 \quad (9.10.2)$$

Together, Eqs. (9.10.1) and (9.10.2) form a forward/inverse DCT pair. The relationship to the doubled signal **y** allows an efficient calculation using $(2N)$-point FFTs [238–240].

9.38   a. Let $X_N(k)$ denote the N-point DFT of a length-L sequence $x(n)$, $n = 0, 1, \dots, L-1$. Show the relationships:

$$X_N(k) = X_{2N}(2k), \qquad k = 0, 1, \dots, N-1$$

b. In particular, we have $X_4(k) = X_8(2k)$, for $k = 0, 1, 2, 3$. That is, the 4-point DFT of a sequence can be obtained by keeping every other entry of the 8-point DFT of that sequence.

9.39 Consider a length-5 sequence and its "circular shifts"

$$\mathbf{x}_0 = [x_0, x_1, x_2, x_3, x_4]$$
$$\mathbf{x}_1 = [x_4, x_0, x_1, x_2, x_3]$$
$$\mathbf{x}_2 = [x_3, x_4, x_0, x_1, x_2]$$
$$\mathbf{x}_3 = [x_2, x_3, x_4, x_0, x_1]$$
$$\mathbf{x}_4 = [x_1, x_2, x_3, x_4, x_0]$$

Show that the 5-point DFT $X_i(k)$ of $\mathbf{x}_i$ is related to the 5-point DFT $X_0(k)$ of $\mathbf{x}_0$ by

$$X_i(k) = W_5^{ik} X_0(k), \qquad \text{for} \quad i = 1, 2, 3, 4$$

Explain this result in terms of ordinary "linear" shifts of the original sequence $\mathbf{x}_0$.

9.40 Show that the following, successively shorter, signals all have the same 4-point DFT:

$$
\begin{bmatrix} x_0 \\ x_1 \\ x_2 \\ x_3 \\ x_4 \\ x_5 \\ x_6 \\ x_7 \end{bmatrix},
\begin{bmatrix} x_0 \\ x_1 \\ x_2 \\ x_3 + x_7 \\ x_4 \\ x_5 \\ x_6 \\ 0 \end{bmatrix},
\begin{bmatrix} x_0 \\ x_1 \\ x_2 + x_6 \\ x_3 + x_7 \\ x_4 \\ x_5 \\ 0 \\ 0 \end{bmatrix},
\begin{bmatrix} x_0 \\ x_1 + x_5 \\ x_2 + x_6 \\ x_3 + x_7 \\ x_4 \\ 0 \\ 0 \\ 0 \end{bmatrix},
\begin{bmatrix} x_0 + x_4 \\ x_1 + x_5 \\ x_2 + x_6 \\ x_3 + x_7 \\ 0 \\ 0 \\ 0 \\ 0 \end{bmatrix}
$$

9.41 Using the *overlap-save* method of fast convolution implemented in the time domain using length-8 circular convolutions, compute the ordinary convolution of the "long" signal

$$\mathbf{x} = [1, 1, 1, 1, 3, 3, 3, 3, 1, 1, 1, 2, 2, 2, 2, 1, 1, 1, 1]$$

with the "short" filter

$$\mathbf{h} = [1, -1, -1, 1]$$

and explain any discrepancies from the correct answer. Repeat using the *overlap-add* method.

9.42 A *periodic* triangular waveform of period $T_0 = 1$ sec is defined over one period $0 \leq t \leq 1$ sec as follows (see also Fig. 1.8.1):

$$x(t) = \begin{cases} t, & \text{if } 0 \leq t \leq 0.25 \\ 0.5 - t, & \text{if } 0.25 \leq t \leq 0.75 \\ t - 1, & \text{if } 0.75 \leq t \leq 1 \end{cases}$$

The signal $x(t)$ is sampled at a rate of 8 Hz and the sampled signal $x(nT)$ is immediately reconstructed into analog form using an *ideal* reconstructor. Because $x(t)$ is not bandlimited, aliasing effects will cause the reconstructed signal to be different from $x(t)$. Show that the aliased reconstructed signal will have the form:

$$x_{\text{al}}(t) = A \sin(2\pi f_1 t) + B \sin(2\pi f_2 t)$$

What are the frequencies f_1 and f_2? Determine the amplitudes A and B by performing an appropriate 8-point FFT by hand. Explain how the negative frequencies in $x_{\text{al}}(t)$ are represented in this FFT.

9.43 A length-L input signal is to be filtered by an order-M FIR filter using the overlap-save method of fast convolution, implemented via N-point FFTs. Assume that $L \gg N$ and $N > M$.

    a. Derive an expression for the total number of multiplications required to compute the output, in terms of L, N, and M.

    b. Repeat part (a) if the overlap-add method is used.

9.44 *Computer Experiment: Overlap-Save Method.* Write a stand-alone C or MATLAB program, say `ovsave.c`, that implements the overlap-save method of fast convolution. The program must have usage:

```
ovsave h.dat N < x.dat > y.dat
```

Like the program `firfilt.c` of Problem 4.10, it must read dynamically the impulse response coefficients from a file `h.dat`. It must keep reading the input samples in blocks of length N (overlapped by M points), processing each block, and writing the output block. The processing of each block must be implemented by N-point FFTs, that is,

$$\tilde{\mathbf{y}} = \text{IFFT}(\mathbf{H} \cdot \text{FFT}(\mathbf{x}))$$

where the FFT of the filter $\mathbf{H} = \text{FFT}(\mathbf{h})$ may be computed once and used in processing all the input blocks.

Care must be exercised in handling the first M inputs, where M zeros must be padded to the beginning of the input. When the end-of-file of the input is detected, the program must calculate correctly the input-off output transients. (The output of this program and that of `firfilt.c` must be identical, up to perhaps some last zeros.)

FIR Digital Filter Design

The *filter design problem* is the problem of constructing the transfer function of a filter that meets *prescribed* frequency response specifications.

The input to any filter design method is the set of desired specifications and the output is the finite impulse response coefficient vector $\mathbf{h} = [h_0, h_1, \ldots, h_{N-1}]$ in the case of FIR filters, or the numerator and denominator coefficient vectors $\mathbf{b} = [b_0, b_1, \ldots, b_M]$, $\mathbf{a} = [1, a_1, \ldots, a_M]$ in the case of IIR filters.

The subject of FIR and IIR digital filter design is very extensive [2-8]. In this and the next chapter, we present only a small cross section of available design methods—our objective being to give the flavor of the subject, while at the same time presenting some practical methods.

The two main advantages of FIR filters are their *linear phase* property and their guaranteed *stability* because of the absence of poles. Their potential disadvantage is that the requirement of sharp filter specifications can lead to long filter lengths N, consequently increasing their computational cost. Recall from Chapter 4 that modern DSP chips require N MACs per output point computed.

The main advantages of IIR filters are their *low computational cost* and their *efficient implementation* in cascade of second-order sections. Their main disadvantage is the potential for instabilities introduced when the quantization of the coefficients pushes the poles outside the unit circle. For IIR filters, linear phase cannot be achieved exactly over the entire Nyquist interval, but it can be achieved *approximately* over the relevant passband of the filter, for example, using Bessel filter designs.

10.1 Window Method

10.1.1 Ideal Filters

The window method is one of the simplest methods of designing FIR digital filters. It is well suited for designing filters with *simple* frequency response shapes, such as ideal lowpass filters. Some typical filter shapes that can be designed are shown in Figs. 10.1.1 and 10.1.2. For arbitrary shapes, a variant of the method, known as

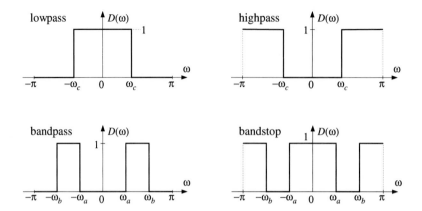

Fig. 10.1.1 Ideal lowpass, highpass, bandpass, and bandstop filters.

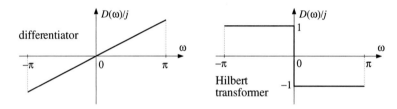

Fig. 10.1.2 Ideal differentiator and Hilbert transformer filters.

frequency sampling method, may be used; it will be discussed in Section 10.3.

A given desired ideal frequency response, say $D(\omega)$, being periodic in ω with period 2π, need only be specified over one complete Nyquist interval $-\pi \leq \omega \leq \pi$. The corresponding impulse response, say $d(k)$, is related to $D(\omega)$ by the DTFT and inverse DTFT relationships:

$$D(\omega) = \sum_{k=-\infty}^{\infty} d(k) e^{-j\omega k} \quad \Leftrightarrow \quad d(k) = \int_{-\pi}^{\pi} D(\omega) e^{j\omega k} \frac{d\omega}{2\pi} \qquad (10.1.1)$$

In general, the impulse response $d(k)$ will be *double-sided and infinite.* For many ideal filter shapes, the ω-integration in Eq. (10.1.1) can be done in closed form. For example, for the *lowpass filter* shown in Fig. 10.1.1, the quantity $D(\omega)$ is defined over the Nyquist interval by

$$D(\omega) = \begin{cases} 1, & \text{if } -\omega_c \leq \omega \leq \omega_c \\ 0, & \text{if } -\pi \leq \omega < -\omega_c, \text{ or } \omega_c < \omega \leq \pi \end{cases}$$

Therefore, Eq. (10.1.1) gives:

$$d(k) = \int_{-\pi}^{\pi} D(\omega) e^{j\omega k} \frac{d\omega}{2\pi} = \int_{-\omega_c}^{\omega_c} 1 \cdot e^{j\omega k} \frac{d\omega}{2\pi}$$

$$= \left[\frac{e^{j\omega k}}{2\pi j k} \right]_{-\omega_c}^{\omega_c} = \frac{e^{j\omega_c k} - e^{-j\omega_c k}}{2\pi j k}$$

which can be rewritten as

(lowpass filter) $\boxed{d(k) = \frac{\sin(\omega_c k)}{\pi k}}, \qquad -\infty < k < \infty$ (10.1.2)

For computational purposes, the case $k = 0$ must be handled separately. Taking the limit $k \to 0$, we find from Eq. (10.1.2):

$$d(0) = \frac{\omega_c}{\pi} \qquad (10.1.3)$$

Similarly, we find for the highpass, bandpass, and bandstop filters of Fig. 10.1.1, defined over $-\infty < k < \infty$

(highpass filter) $\quad d(k) = \delta(k) - \dfrac{\sin(\omega_c k)}{\pi k}$

(bandpass filter) $\quad d(k) = \dfrac{\sin(\omega_b k) - \sin(\omega_a k)}{\pi k}$ (10.1.4)

(bandstop filter) $\quad d(k) = \delta(k) - \dfrac{\sin(\omega_b k) - \sin(\omega_a k)}{\pi k}$

Note that for the same values of the cutoff frequencies ω_c, ω_a, ω_b, the lowpass/highpass and bandpass/bandstop filters are *complementary*, that is, their impulse responses add up to a unit impulse $\delta(k)$ and their frequency responses add up to unity (as can also be seen by inspecting Fig. 10.1.1):

$$d_{LP}(k) + d_{HP}(k) = \delta(k) \quad \Leftrightarrow \quad D_{LP}(\omega) + D_{HP}(\omega) = 1$$
$$d_{BP}(k) + d_{BS}(k) = \delta(k) \quad \Leftrightarrow \quad D_{BP}(\omega) + D_{BS}(\omega) = 1$$
(10.1.5)

As we see below, such complementarity properties can be exploited to simplify the implementation of loudspeaker cross-over networks and graphic equalizers.

The ideal *differentiator* filter of Fig. 10.1.2 has frequency response $D(\omega) = j\omega$, defined over the Nyquist interval. The ideal *Hilbert transformer* response can be expressed compactly as $D(\omega) = -j\,\text{sign}(\omega)$, where $\text{sign}(\omega)$ is the signum function which is equal to ± 1 depending on the algebraic sign of its argument. The ω-integrations in Eq. (10.1.1) give the impulse responses:

(differentiator) $\quad d(k) = \dfrac{\cos(\pi k)}{k} - \dfrac{\sin(\pi k)}{\pi k^2}$

(Hilbert transformer) $\quad d(k) = \dfrac{1 - \cos(\pi k)}{\pi k}$
(10.1.6)

Both filters have $d(0) = 0$, as can be verified by carefully taking the limit $k \to 0$. Both impulse responses $d(k)$ are real-valued and *odd* (antisymmetric) functions of k. By contrast, the filters of Fig. 10.1.1 all have impulse responses that are real and *even* (symmetric) in k. We will refer to the two classes of filters of Figs. 10.1.1 and 10.1.2 as the *symmetric* and *antisymmetric* classes.

In the frequency domain, the symmetric types are characterized by a frequency response $D(\omega)$ which is *real and even* in ω; the antisymmetric ones have $D(\omega)$ which is *imaginary and odd* in ω. One of the main consequences of these frequency properties is the *linear phase* property of the window designs.

10.1.2 Rectangular Window

The window method consists of truncating, or rectangularly windowing, the double-sided $d(k)$ to a finite length. For example, we may keep only the coefficients:

$$d(k) = \int_{-\pi}^{\pi} D(\omega) e^{j\omega k} \frac{d\omega}{2\pi}, \qquad -M \le k \le M \tag{10.1.7}$$

Because the coefficients are taken equally for positive and negative k's, the total number of coefficients will be *odd*, that is, $N = 2M + 1$ (even values of N are also possible, but not discussed in this text). The resulting N-dimensional coefficient vector is the *FIR impulse response* approximating the infinite ideal response:

$$\mathbf{d} = [d_{-M}, \ldots, d_{-2}, d_{-1}, d_0, d_1, d_2, \ldots, d_M] \tag{10.1.8}$$

The time origin $k = 0$ is at the middle d_0 of this vector. To make the filter causal we may shift the time origin to the left of the vector and re-index the entries accordingly:

$$\mathbf{h} = \mathbf{d} = [h_0, \ldots, h_{M-2}, h_{M-1}, h_M, h_{M+1}, h_{M+2}, \ldots, h_{2M}] \tag{10.1.9}$$

where we defined $h_0 = d_{-M}$, $h_1 = d_{-M+1}$, ..., $h_M = d_0$, ..., $h_{2M} = d_M$. Thus, the vectors $\mathbf{d}$ and $\mathbf{h}$ are the same, with the understanding that $\mathbf{d}$'s origin is in its middle and $\mathbf{h}$'s at its left. The definition of $\mathbf{h}$ may be thought of as time-delaying the double-sided sequence $d(k)$, $-M \le k \le M$, by M time units to make it causal:

$$h(n) = d(n - M), \qquad n = 0, 1, \ldots, N - 1 \tag{10.1.10}$$

The operations of windowing and delaying are shown in Fig. 10.1.3. To summarize, the steps of the *rectangular window method* are simply:

1. Pick an odd length $N = 2M + 1$, and let $M = (N - 1)/2$.

2. Calculate the N coefficients $d(k)$ from Eq. (10.1.7), and

3. Make them causal by the delay (10.1.10).

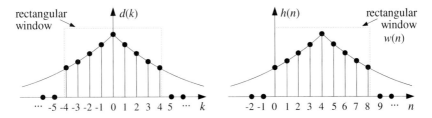

Fig. 10.1.3 Rectangularly windowed impulse response, with $N = 9$, $M = 4$.

For example, the N-dimensional approximation to the ideal lowpass filter of Eq. (10.1.2) will be:

$$h(n) = d(n - M) = \frac{\sin(\omega_c(n - M))}{\pi(n - M)} , \quad n = 0, \ldots, M, \ldots, N - 1 \qquad (10.1.11)$$

where we must calculate separately $h(M) = d(0) = \omega_c/\pi$. For other ideal filter shapes, we can use the functions $d(k)$ of Eqs. (10.1.4) or (10.1.6). Once the impulse response coefficients are calculated, the filter may be implemented by its FIR filtering equation, using the routines `fir` or `cfir` of Chapter 4:

$$y_n = \sum_{m=0}^{N-1} h_m x_{n-m} \qquad (10.1.12)$$

Example 10.1.1: Determine the length-11, rectangularly windowed impulse response that approximates (a) an ideal lowpass filter of cutoff frequency $\omega_c = \pi/4$, (b) the ideal differentiator filter, and (c) the ideal Hilbert transformer filter.

Solution: With $N = 11$, we have $M = (N - 1)/2 = 5$. For the lowpass filter, we evaluate Eq. (10.1.2), that is,

$$d(k) = \frac{\sin(\pi k/4)}{\pi k} , \quad \text{for } -5 \leq k \leq 5$$

We find the numerical values:

$$\mathbf{h} = \mathbf{d} = \left[\frac{\sqrt{2}}{10\pi}, \, 0, \, \frac{\sqrt{2}}{6\pi}, \, \frac{1}{2\pi}, \, \frac{\sqrt{2}}{2\pi}, \, \frac{1}{4}, \, \frac{\sqrt{2}}{2\pi}, \, \frac{1}{2\pi}, \, \frac{\sqrt{2}}{6\pi}, \, 0, \, \frac{\sqrt{2}}{10\pi} \right]$$

For the differentiator filter, the second term, $\sin(\pi k)/\pi k^2$, vanishes for all values $k \neq 0$. Therefore, we find:

$$\mathbf{h} = \mathbf{d} = \left[\frac{1}{5}, \, -\frac{1}{4}, \, \frac{1}{3}, \, -\frac{1}{2}, \, 1, \, 0, \, -1, \, \frac{1}{2}, \, -\frac{1}{3}, \, \frac{1}{4}, \, -\frac{1}{5} \right]$$

And, for the Hilbert transformer:

$$\mathbf{h} = \mathbf{d} = \left[-\frac{2}{5\pi}, \, 0, \, -\frac{2}{3\pi}, \, 0, \, -\frac{2}{\pi}, \, 0, \, \frac{2}{\pi}, \, 0, \, \frac{2}{3\pi}, \, 0, \, \frac{2}{5\pi} \right]$$

Note that the lowpass filter's impulse response is symmetric about its middle, whereas the differentiator's and Hilbert transformer's are antisymmetric. Note also that because of the presence of the factor $1 - \cos(\pi k)$, every other entry of the Hilbert transformer vanishes. This property can be exploited to reduce by half the total number of multiplications required in the convolutional equation Eq. (10.1.12).   □

In the frequency domain, the FIR approximation to $D(\omega)$ is equivalent to truncating the DTFT Fourier series expansion (10.1.1) to the finite sum:

$$\hat{D}(\omega) = \sum_{k=-M}^{M} d(k)e^{-j\omega k} \qquad (10.1.13)$$

Replacing $z = e^{j\omega}$, we may also write it as the double-sided z-transform:

$$\hat{D}(z) = \sum_{k=-M}^{M} d(k)z^{-k} \qquad (10.1.14)$$

The final length-N filter obtained by Eq. (10.1.10) will have transfer function:

$$H(z) = z^{-M}\hat{D}(z) = z^{-M}\sum_{k=-M}^{M} d(k)z^{-k} \qquad (10.1.15)$$

and frequency response:

$$H(\omega) = e^{-j\omega M}\hat{D}(\omega) = e^{-j\omega M}\sum_{k=-M}^{M} d(k)e^{-j\omega k} \qquad (10.1.16)$$

Example 10.1.2: To illustrate the definition of $H(z)$, consider a case with $N = 7$ and $M = (N-1)/2 = 3$. Let the FIR filter weights be $\mathbf{d} = [d_{-3}, d_{-2}, d_{-1}, d_0, d_1, d_2, d_3]$ with truncated z-transform:

$$\hat{D}(z) = d_{-3}z^3 + d_{-2}z^2 + d_{-1}z + d_0 + d_1z^{-1} + d_2z^{-2} + d_3z^{-3}$$

Delaying it by $M = 3$, we get the causal transfer function:

$$H(z) = z^{-3}\hat{D}(z) = z^{-3}(d_{-3}z^3 + d_{-2}z^2 + d_{-1}z + d_0 + d_1z^{-1} + d_2z^{-2} + d_3z^{-3})$$

$$= d_{-3} + d_{-2}z^{-1} + d_{-1}z^{-2} + d_0z^{-3} + d_1z^{-4} + d_2z^{-5} + d_3z^{-6}$$

$$= h_0 + h_1z^{-1} + h_2z^{-2} + h_3z^{-3} + h_4z^{-4} + h_5z^{-5} + h_6z^{-6}$$

where we defined $h(n) = d(n-3)$, $n = 0, 1, 2, 3, 4, 5, 6$.   □

The *linear phase property* of the window design is a direct consequence of Eq. (10.1.16). The truncated $\hat{D}(\omega)$ has the same symmetry/antisymmetry properties as $D(\omega)$. Thus, in the *symmetric* case, $\hat{D}(\omega)$ will be real and even in ω. It follows from Eq. (10.1.16) that the designed FIR filter will have linear phase, arising essentially from the delay factor $e^{-j\omega M}$. More precisely, we may write the real factor $\hat{D}(\omega)$ in terms of its positive magnitude and its sign:

$$\hat{D}(\omega) = \text{sign}(\hat{D}(\omega))\,|\hat{D}(\omega)| = e^{j\pi\beta(\omega)}\,|\hat{D}(\omega)|$$

where $\beta(\omega) = [1 - \text{sign}(\hat{D}(\omega))]/2$, which is zero or one depending on the sign of $\hat{D}(\omega)$. It follows that $H(\omega)$ will be:

$$H(\omega) = e^{-j\omega M}\hat{D}(\omega) = e^{-j\omega M + j\pi\beta(\omega)}\,|\hat{D}(\omega)|$$

Thus, its magnitude and phase responses will be:

$$\boxed{\;|H(\omega)| = |\hat{D}(\omega)|, \quad \arg H(\omega) = -\omega M + \pi\,\beta(\omega)\;}$$

(10.1.17)

making the phase response piece-wise linear in ω with $180°$ jumps at those ω where $\hat{D}(\omega)$ changes sign. For the *antisymmetric* case, $\hat{D}(\omega)$ will be pure imaginary, that is, of the form $\hat{D}(\omega) = jA(\omega)$. The factor j may be made into a phase by writing it as $j = e^{j\pi/2}$. Thus, we have

$$H(\omega) = e^{-j\omega M}\hat{D}(\omega) = e^{-j\omega M}e^{j\pi/2}A(\omega) = e^{-j\omega M}e^{j\pi/2}e^{j\pi\alpha(\omega)}\,|A(\omega)|$$

where $\alpha(\omega) = [1 - \text{sign}(A(\omega))]/2$, which gives for the magnitude and phase responses:

$$\boxed{\;|H(\omega)| = |A(\omega)|, \quad \arg H(\omega) = -\omega M + \frac{\pi}{2} + \pi\,\alpha(\omega)\;}$$

(10.1.18)

How good is the rectangular window design? How well does the truncated $\hat{D}(\omega)$ represent the desired response $D(\omega)$? In other words, how good is the approximation $\hat{D}(\omega) \simeq D(\omega)$?

Intuitively one would expect that $\hat{D}(\omega) \to D(\omega)$ as N increases. This is true for any ω which is a point of *continuity* of $D(\omega)$, but it fails at points of *discontinuity*, such as at the transition edges from passband to stopband. Around these edges one encounters the celebrated *Gibbs phenomenon* of Fourier series, which causes the approximation to be bad regardless of how large N is.

To illustrate the nature of the approximation $\hat{D}(\omega) \simeq D(\omega)$, we consider the design of an ideal lowpass filter of cutoff frequency $\omega_c = 0.3\pi$, approximated by a rectangularly windowed response of length $N = 41$ and then by another one of length $N = 121$. For the case $N = 41$, we have $M = (N-1)/2 = 20$. The designed impulse response is given by Eq. (10.1.10):

$$h(n) = d(n-20) = \frac{\sin(0.3\pi(n-20))}{\pi(n-20)}, \qquad n = 0, 1, \ldots, 40$$

and in particular, $h(20) = d(0) = \omega_c/\pi = 0.3$. The second design has $N = 121$ and $M = 60$. Its impulse response is, with $h(60) = d(0) = 0.3$:

$$h(n) = d(n - 60) = \frac{\sin\left(0.3\pi(n - 60)\right)}{\pi(n - 60)}, \qquad n = 0, 1, \ldots, 120$$

The two impulse responses are plotted in Fig. 10.1.4. Note that the portion of the second response extending ± 20 samples around the central peak at $n = 60$ coincides numerically with the first response. The corresponding magnitude responses are shown in Fig. 10.1.5. An intermediate case having $N = 81$ is shown in Figs. 10.1.6 and 10.1.7.

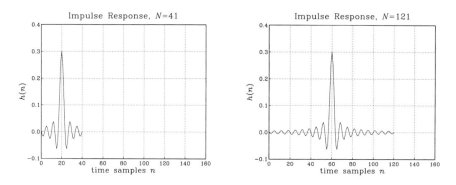

Fig. 10.1.4 Rectangularly windowed impulse responses for $N = 41$ and $N = 121$.

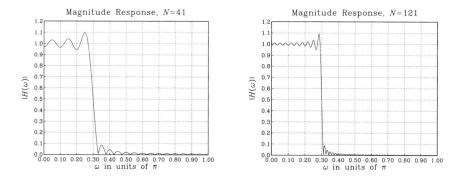

Fig. 10.1.5 Rectangularly windowed magnitude responses for $N = 41$ and $N = 121$.

In Fig. 10.1.5, the magnitude responses were computed by evaluating:

$$H(\omega) = \sum_{n=0}^{N-1} h(n)e^{-j\omega n} \qquad (10.1.19)$$

The length-N impulse response $h(n)$ defined in Eq. (10.1.10) may be thought of formally as the rectangularly windowed double-sided sequence defined by

$$\boxed{h(n) = w(n)d(n - M)}, \qquad -\infty < n < \infty \qquad (10.1.20)$$

where $w(n)$ is the length-N rectangular window. In the frequency domain, this translates to the convolution of the corresponding spectra, as in Eq. (9.1.8):

$$\boxed{H(\omega) = \int_{-\pi}^{\pi} W(\omega - \omega')e^{-j\omega'M}D(\omega')\,\frac{d\omega'}{2\pi}} \qquad (10.1.21)$$

where the $e^{-j\omega'M}$ arises from the delay in $d(n - M)$.

The spectrum $W(\omega)$ of the rectangular window was given in Eq. (9.1.9) (with $L = N$). Thus, the designed filter $H(\omega)$ will be a smeared version of the desired shape $D(\omega)$. In particular, for the ideal lowpass case, because $D(\omega')$ is nonzero and unity only over the subinterval $-\omega_c \le \omega' \le \omega_c$, the frequency convolution integral becomes:

$$H(\omega) = \int_{-\omega_c}^{\omega_c} W(\omega - \omega')e^{-j\omega'M}\,\frac{d\omega'}{2\pi} \qquad (10.1.22)$$

The ripples in the frequency response $H(\omega)$, observed in Fig. 10.1.5, arise from the (integrated) ripples of the rectangular window spectrum $W(\omega)$. As N increases, we observe three effects in Fig. 10.1.5:

1. For ω's that lie well within the passband or stopband (i.e., points of continuity), the ripple size decreases as N increases, resulting in flatter passband and stopband. For such ω, we have $\hat{D}(\omega) \to D(\omega)$ as $N \to \infty$.

2. The transition width decreases with increasing N. Note also that for any N, the windowed response $H(\omega)$ is always equal to 0.5 at the cutoff frequency $\omega = \omega_c$. (This is a standard property of Fourier series.)

3. The largest ripples tend to cluster near the passband-to-stopband discontinuity (from both sides) and do not get smaller with N. Instead, their size remains approximately *constant*, about 8.9 percent, independent of N. Eventually, as $N \to \infty$, these ripples get squeezed onto the discontinuity at $\omega = \omega_c$, occupying a set of measure zero. This behavior is the Gibbs phenomenon.

10.1.3 Hamming Window

To eliminate the 8.9% passband and stopband ripples, we may replace the rectangular window $w(n)$ in Eq. (10.1.20) by a non-rectangular one, which tapers off gradually at its endpoints, thus reducing the ripple effect. There exist dozens of windows [219–222] and among these the Hamming window is a popular choice; it is defined by:

$$\boxed{w(n) = 0.54 - 0.46\cos\left(\frac{2\pi n}{N - 1}\right), \qquad n = 0, 1, \ldots, N - 1} \qquad (10.1.23)$$

In particular, the Hamming windowed impulse response for a length-N lowpass filter will be, where $N = 2M + 1$ and $n = 0, 1, \ldots, N - 1$:

$$h(n) = w(n)d(n - M) = \left[0.54 - 0.46\cos\left(\frac{2\pi n}{N - 1}\right)\right] \cdot \frac{\sin(\omega_c(n - M))}{\pi(n - M)} \quad (10.1.24)$$

As an example, consider the design of a length $N = 81$ lowpass filter with cutoff frequency $\omega_c = 0.3\pi$. Fig. 10.1.6 shows the rectangularly and Hamming windowed impulse responses. Note how the Hamming impulse response tapers off to zero more gradually. It was computed by Eq. (10.1.24) with $N = 81$ and $M = 40$. Fig. 10.1.7 shows the corresponding magnitude responses.

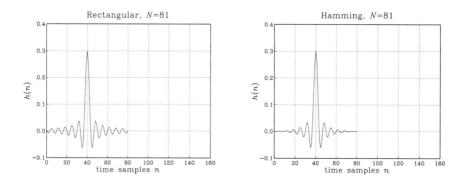

Fig. 10.1.6 Rectangular and Hamming windowed impulse responses for $N = 81$.

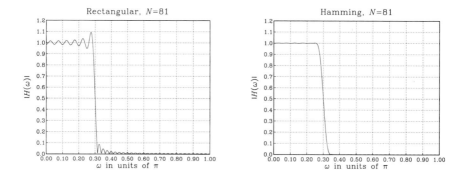

Fig. 10.1.7 Rectangular and Hamming windowed magnitude responses for $N = 81$.

The passband/stopband ripples of the rectangular window design are virtually eliminated from the Hamming window design. Actually, there are small ripples with maximum overshoot of about 0.2%, but they are not visible in the scale of Fig. 10.1.7. The price for eliminating the ripples is loss of resolution, which is reflected into a *wider transition width*.

10.2 Kaiser Window

10.2.1 Kaiser Window for Filter Design

The rectangular and Hamming window designs are very simple, but do not provide good control over the filter design specifications. With these windows, the amount of overshoot is always fixed to 8.9% or 0.2% and cannot be changed to a smaller value if so desired.

A flexible set of specifications is shown in Fig. 10.2.1 in which the designer can arbitrarily specify the amount of passband and stopband overshoot δ_{pass}, δ_{stop}, as well as the transition width Δf.

The passband/stopband frequencies $\{f_{pass}, f_{stop}\}$ are related to the ideal cutoff frequency f_c and transition width Δf by

$$f_c = \frac{1}{2}(f_{pass} + f_{stop}), \quad \Delta f = f_{stop} - f_{pass} \qquad (10.2.1)$$

Thus, f_c is chosen to lie exactly in the middle between f_{pass} and f_{stop}. Eqs. (10.2.1) can be inverted to give:

$$f_{pass} = f_c - \frac{1}{2}\Delta f, \quad f_{stop} = f_c + \frac{1}{2}\Delta f \qquad (10.2.2)$$

The normalized versions of the frequencies are the digital frequencies:

$$\omega_{pass} = \frac{2\pi f_{pass}}{f_s}, \quad \omega_{stop} = \frac{2\pi f_{stop}}{f_s}, \quad \omega_c = \frac{2\pi f_c}{f_s}, \quad \Delta\omega = \frac{2\pi\Delta f}{f_s}$$

In practice, the passband and stopband overshoots are usually expressed in dB:

$$A_{pass} = 20\log_{10}\left(\frac{1 + \delta_{pass}}{1 - \delta_{pass}}\right), \quad A_{stop} = -20\log_{10}\delta_{stop} \qquad (10.2.3)$$

A simplified version of the passband equation can be obtained by expanding it to first order in δ_{pass}, giving:

$$A_{pass} = 17.372\delta_{pass} \qquad (10.2.4)$$

which is valid for small values of δ_{pass}. Eqs. (10.2.3) can be inverted to give:

$$\delta_{pass} = \frac{10^{A_{pass}/20} - 1}{10^{A_{pass}/20} + 1}, \quad \delta_{stop} = 10^{-A_{stop}/20} \qquad (10.2.5)$$

Thus, one can pass back and forth between the specification sets:

$$\{f_{pass}, f_{stop}, A_{pass}, A_{stop}\} \quad \Leftrightarrow \quad \{f_c, \Delta f, \delta_{pass}, \delta_{stop}\}$$

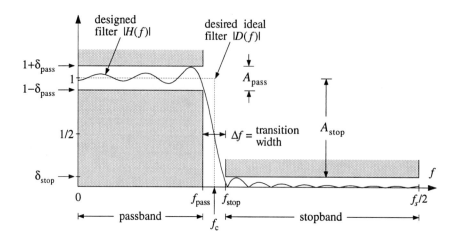

Fig. 10.2.1 Magnitude response specifications for a lowpass filter.

Although δ_{pass} and δ_{stop} can be specified independently of each other, it is a property of *all* window designs that the final designed filter will have *equal* passband and stopband ripples. Therefore, we must design the filter on the basis of the *smaller* of the two ripples, that is,

$$\boxed{\delta = \min\left(\delta_{\text{pass}}, \delta_{\text{stop}}\right)} \qquad (10.2.6)$$

The designed filter will have passband and stopband ripple equal to δ. The value of δ can also be expressed in dB:

$$\boxed{A = -20\log_{10}\delta, \quad \delta = 10^{-A/20}} \qquad (10.2.7)$$

In practice, the design is usually based on the stopband ripple δ_{stop}. This is so because any reasonably good choices for the passband and stopband attenuations (e.g., $A_{\text{pass}} = 0.1$ dB and $A_{\text{stop}} = 60$ dB) will almost always result into $\delta_{\text{stop}} < \delta_{\text{pass}}$, and therefore, $\delta = \delta_{\text{stop}}$, and in dB, $A = A_{\text{stop}}$. Thus, it is useful to think of A as the stopband attenuation.

The main limitation of most windows is that they have a *fixed* value of δ, which depends on the particular window shape. Such windows limit the achievable passband and stopband attenuations $\{A_{\text{pass}}, A_{\text{stop}}\}$ to only certain specific values.

For example, Table 10.2.1 shows the attenuations achievable by the rectangular and Hamming windows, calculated from Eq. (10.2.3) with the values $\delta = \delta_{\text{pass}} = \delta_{\text{stop}} = 0.089$ and $\delta = \delta_{\text{pass}} = \delta_{\text{stop}} = 0.002$, respectively. The table also shows the corresponding value of the transition width parameter D of Eq. (10.2.11).

The only windows that do not suffer from the above limitation are the Kaiser window [245–247], the Dolph-Chebyshev window [248–253], and the Saramäki windows [254]. These windows have an *adjustable shape parameter* that allows the window to achieve any desired value of ripple δ or attenuation A.

The Kaiser window is unique in the above class in that it has near-optimum performance (in the sense of minimizing the sidelobe energy of the window), as

Window	δ	A_{stop}	A_{pass}	D
Rectangular	8.9%	-21 dB	1.55 dB	0.92
Hamming	0.2%	-54 dB	0.03 dB	3.21
Kaiser	variable δ	$-20\log_{10}\delta$	17.372δ	$(A - 7.95)/14.36$

Table 10.2.1 Specifications for rectangular, Hamming, and Kaiser windows.

well as having the simplest implementation. It depends on two parameters: its length N and the shape parameter α. Assuming odd length $N = 2M + 1$, the window is defined, for $n = 0, 1, \ldots, N - 1$, as follows:

$$\text{(Kaiser window)} \qquad w(n) = \frac{I_0\left(\alpha\sqrt{1 - (n - M)^2/M^2}\right)}{I_0(\alpha)} \qquad (10.2.8)$$

where $I_0(x)$ is the *modified Bessel function of the first kind and 0th order*. This function and its evaluation by the routine I0.c are discussed at the end of this section. The numerator in Eq. (10.2.8) can be rewritten in the following form, which is more convenient for numerical evaluation:

$$w(n) = \frac{I_0\left(\alpha\sqrt{n(2M - n)}/M\right)}{I_0(\alpha)}, \quad n = 0, 1, \ldots, N - 1 \qquad (10.2.9)$$

Like all window functions, the Kaiser window is *symmetric* about its middle, $n = M$, and has the value $w(M) = 1$ there. At the endpoints, $n = 0$ and $n = N - 1$, it has the value $1/I_0(\alpha)$ because $I_0(0) = 1$.

Figure 10.2.2 compares a Hamming window of length $N = 51$ to the Kaiser windows of the same length and shape parameters $\alpha = 7$ and $\alpha = 5$. For $\alpha = 5$ the Kaiser and Hamming windows agree closely, except near their endpoints. For $\alpha = 0$ the Kaiser window reduces to the rectangular one.

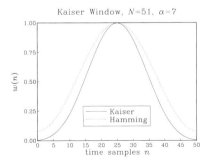 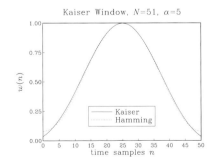

Fig. 10.2.2 Kaiser and Hamming windows for $N = 51$ and $\alpha = 5, 7$.

The window parameters $\{N, \alpha\}$ are computable in terms of the filter specifications, namely, the ripple δ and transition width Δf. The design equations developed by Kaiser [245-247] are as follows. The shape parameter α is calculated from:

$$\alpha = \begin{cases} 0.1102\,(A - 8.7), & \text{if } A \geq 50 \\ 0.5842\,(A - 21)^{0.4} + 0.07886\,(A - 21), & \text{if } 21 < A < 50 \\ 0, & \text{if } A \leq 21 \end{cases} \qquad (10.2.10)$$

where A is the ripple in dB, given by Eq. (10.2.7). The filter length N is inversely related to the transition width:

$$\Delta f = \frac{Df_s}{N - 1} \quad \Leftrightarrow \quad N - 1 = \frac{Df_s}{\Delta f} \qquad (10.2.11)$$

where the factor D is computed also in terms of A by

$$D = \begin{cases} \dfrac{A - 7.95}{14.36}, & \text{if } A > 21 \\ 0.922, & \text{if } A \leq 21 \end{cases} \qquad (10.2.12)$$

The most practical range of these formulas is for $A \geq 50$ dB, for which they simplify to:

$$\alpha = 0.1102\,(A - 8.7), \quad D = \frac{A - 7.95}{14.36} \qquad \text{(for } A \geq 50 \text{ dB)} \qquad (10.2.13)$$

To summarize, the steps for designing a lowpass filter are as follows. Given the specifications $\{f_{\text{pass}}, f_{\text{stop}}, A_{\text{pass}}, A_{\text{stop}}\}$:

1. Calculate f_c and Δf from Eq. (10.2.1). Then, calculate $\omega_c = 2\pi f_c/f_s$.

2. Calculate δ_{pass} and δ_{stop} from Eq. (10.2.5).

3. Calculate $\delta = \min(\delta_{\text{pass}}, \delta_{\text{stop}})$ and $A = -20\log_{10}\delta$ in dB.

4. Calculate α and D from Eqs. (10.2.10) and (10.2.12).

5. Calculate the filter length N from Eq. (10.2.11) and round it up to the next *odd* integer, $N = 2M + 1$, and set $M = (N - 1)/2$.

6. Calculate the window function $w(n)$, $n = 0, 1, \ldots, N - 1$ from Eq. (10.2.8).

7. Calculate the windowed impulse response, for $n = 0, 1, \ldots, N - 1$:

$$h(n) = w(n)d(n - M) = w(n) \cdot \frac{\sin(\omega_c(n - M))}{\pi(n - M)} \qquad (10.2.14)$$

In particular, we have $h(M) = w(M)\omega_c/\pi = \omega_c/\pi$, because $w(M) = 1$.

Note that the window parameters $\{N, \alpha\}$ depend only on the specifications $\{A, \Delta f\}$ and not on f_c. However, $h(n)$ does depend on f_c.

The design steps can be modified easily to design highpass and bandpass filters. For *highpass* filters, the role of f_{pass} and f_{stop} are interchanged; therefore, the only change in the steps is to define $\Delta f = f_{\text{pass}} - f_{\text{stop}}$ and to use the highpass response from Eq. (10.1.4). The highpass impulse response will be:

$$h(n) = w(n)d(n - M) = w(n) \cdot \left[\delta(n - M) - \frac{\sin(\omega_c(n - M))}{\pi(n - M)} \right]$$

The first term can be simplified to $w(n)\delta(n - M) = w(M)\delta(n - M) = \delta(n - M)$ because $w(M) = 1$. Therefore, the designed filter will be:

$$h(n) = \delta(n - M) - w(n) \cdot \frac{\sin(\omega_c(n - M))}{\pi(n - M)} \qquad (10.2.15)$$

For the same value of ω_c, the lowpass and highpass filters are complementary. The sum of Eqs. (10.2.14) and (10.2.15) gives:

$$h_{LP}(n) + h_{HP}(n) = \delta(n - M), \quad n = 0, 1, \ldots, N - 1 \qquad (10.2.16)$$

which becomes in the z-domain:

$$H_{LP}(z) + H_{HP}(z) = z^{-M} \qquad (10.2.17)$$

For *bandpass* filters, the desired specifications may be given as in Fig. 10.2.3. There are now two stopbands and two transition widths. The final design will have equal transition widths, given by Eq. (10.2.11). Therefore, we must design the filter based on the smaller of the two widths, that is,

$$\Delta f = \min(\Delta f_a, \Delta f_b) \qquad (10.2.18)$$

where the left and right transition widths are:

$$\Delta f_a = f_{pa} - f_{sa}, \qquad \Delta f_b = f_{sb} - f_{pb} \qquad (10.2.19)$$

Figure 10.2.3 shows the case where the left transition width is the smaller one and, thus, defines Δf. The ideal cutoff frequencies f_a and f_b can be calculated by taking them to be $\Delta f / 2$ away from the passband or from the stopbands. The *standard definition* is with respect to the passband:

$$f_a = f_{pa} - \frac{1}{2}\Delta f, \qquad f_b = f_{pb} + \frac{1}{2}\Delta f \qquad (10.2.20)$$

This choice makes the passband just right and the stopband somewhat wider than required. The *alternative definition* makes the stopbands right and the passband wider:

$$f_a = f_{sa} + \frac{1}{2}\Delta f, \qquad f_b = f_{sb} - \frac{1}{2}\Delta f \qquad (10.2.21)$$

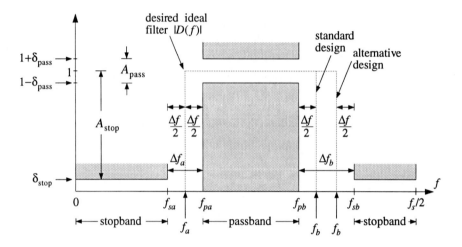

Fig. 10.2.3 Bandpass filter specifications.

Once the cutoff frequencies $\{f_a, f_b\}$ and the window parameters $\{N, \alpha\}$ are calculated, the bandpass impulse response may be defined, for $n = 0, 1, \ldots, N - 1$:

$$h(n) = w(n)d(n - M) = w(n) \cdot \frac{\sin(\omega_b(n - M)) - \sin(\omega_a(n - M))}{\pi(n - M)} \quad (10.2.22)$$

where $h(M) = (\omega_b - \omega_a)/\pi$, and $\omega_a = 2\pi f_a/f_s$, $\omega_b = 2\pi f_b/f_s$. Next, we present a lowpass and a bandpass design example.

Example 10.2.1: *Lowpass Design.* Using the Kaiser window, design a lowpass digital filter with the following specifications:

$$f_s = 20 \text{ kHz}$$
$$f_{\text{pass}} = 4 \text{ kHz}, \quad f_{\text{stop}} = 5 \text{ kHz}$$
$$A_{\text{pass}} = 0.1 \text{ dB}, \quad A_{\text{stop}} = 80 \text{ dB}$$

Solution: First, we calculate δ_{pass} and δ_{stop} from Eq. (10.2.5):

$$\delta_{\text{pass}} = \frac{10^{0.1/20} - 1}{10^{0.1/20} + 1} = 0.0058, \qquad \delta_{\text{stop}} = 10^{-80/20} = 0.0001$$

Therefore, $\delta = \min(\delta_{\text{pass}}, \delta_{\text{stop}}) = \delta_{\text{stop}} = 0.0001$, which in dB is $A = -20\log_{10}\delta = A_{\text{stop}} = 80$. The D and α parameters are computed by:

$$\alpha = 0.1102(A - 8.7) = 0.1102(80 - 8.7) = 7.857, \quad D = \frac{A - 7.95}{14.36} = 5.017$$

The filter width and ideal cutoff frequency are:

$$\Delta f = f_{\text{stop}} - f_{\text{pass}} = 1 \text{ kHz}, \quad f_c = \frac{1}{2}(f_{\text{pass}} + f_{\text{stop}}) = 4.5 \text{ kHz}, \quad \omega_c = \frac{2\pi f_c}{f_s} = 0.45\pi$$

Eq. (10.2.11) gives for the filter length (rounded up to the nearest odd integer):

$$N = 1 + \frac{Df_s}{\Delta f} = 101.35 \quad \Rightarrow \quad N = 103, \quad M = \frac{1}{2}(N-1) = 51$$

The windowed impulse response will be, for $n = 0, 1, \ldots, 102$:

$$h(n) = w(n)d(n-M) = \frac{I_0(7.857\sqrt{n(102-n)}/51)}{I_0(7.857)} \cdot \frac{\sin(0.45\pi(n-51))}{\pi(n-51)}$$

with $h(51) = \omega_c/\pi = 0.45$. Figure 10.2.4 shows the magnitude response in dB of $h(n)$, that is, $20\log_{10}|H(\omega)|$, where $H(\omega)$ was evaluated by Eq. (10.1.19). Note the transition width extending from 4 to 5 kHz and the stopband specification defined by the horizontal grid line at -80 dB. The passband specification is more than satisfied. It is $A_{\text{pass}} \simeq 17.372\delta = 0.0017$ dB.

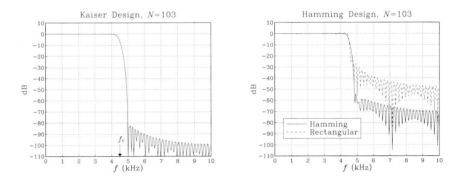

Fig. 10.2.4 Kaiser, Hamming, and rectangular window designs, $N = 103$.

The figure also shows the corresponding Hamming and rectangularly windowed designs for the same length of $N = 103$. They both have a smaller transition width—the rectangular one even more so, but their stopband attenuations are limited to the standard values of 54 dB and 21 dB, respectively. □

Example 10.2.2: *Bandpass Design.* Using the Kaiser window, design a bandpass digital filter with the following specifications:

$$f_s = 20 \text{ kHz}$$
$$f_{sa} = 3 \text{ kHz}, \quad f_{pa} = 4 \text{ kHz}, \quad f_{pb} = 6 \text{ kHz}, \quad f_{sb} = 8 \text{ kHz}$$
$$A_{\text{pass}} = 0.1 \text{ dB}, \quad A_{\text{stop}} = 80 \text{ dB}$$

Solution: The parameters $\{\delta_{\text{pass}}, \delta_{\text{stop}}, \delta, A, \alpha, D\}$ are the same as in the previous example. The two transition widths are:

$$\Delta f_a = f_{pa} - f_{sa} = 4 - 3 = 1 \text{ kHz}, \quad \Delta f_b = f_{sb} - f_{pb} = 8 - 6 = 2 \text{ kHz}$$

Therefore, the minimum width is $\Delta f = \min(\Delta f_a, \Delta f_b) = 1$ kHz, and the filter length:

$$N = 1 + \frac{Df_s}{\Delta f} = 101.35 \quad \Rightarrow \quad N = 103, \quad M = \frac{1}{2}(N-1) = 51$$

Using the *standard definition* of Eq. (10.2.20), we find for the left and right ideal cutoff frequencies:

$$f_a = f_{pa} - \frac{1}{2}\Delta f = 4 - 0.5 = 3.5 \text{ kHz}, \quad f_b = f_{pb} + \frac{1}{2}\Delta f = 6 + 0.5 = 6.5 \text{ kHz}$$

with the normalized values $\omega_a = 2\pi f_a/f_s = 0.35\pi$, $\omega_b = 2\pi f_b/f_s = 0.65\pi$.

For the *alternative definition* of Eq. (10.2.21), we have $f_a = 3 + 0.5 = 3.5$ and $f_b = 8 - 0.5 = 7.5$ kHz, resulting in $\omega_a = 0.35\pi$ and $\omega_b = 0.75\pi$. Figure 10.2.5 shows the magnitude response of the designed filter in dB, both for the standard and the alternative definitions. The standard design has just the right passband extending over [4, 6] kHz and a wider stopband that starts at 7 kHz. The alternative design has a wider passband extending over [4, 7] kHz. □

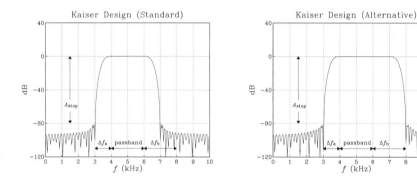

Fig. 10.2.5 Kaiser window design of a bandpass filter.

Next, we discuss three more Kaiser design examples for *digital audio* applications, namely, two-way and three-way loudspeaker crossover filters and a five-band graphic equalizer.

In all three cases, the sampling rate is $f_s = 44.1$ kHz, the stopband attenuation is $A_{\text{stop}} = 65$ dB, and all the transition widths are taken to be equal to $\Delta f = 2$ kHz. This implies that all the filters will have the same length N and Kaiser parameters D and α. With $A = A_{\text{stop}} = 65$ dB, we find

$$D = \frac{A - 7.95}{14.36} = 3.973, \quad \alpha = 0.1102\,(A - 8.7) = 6.204,$$

$$N - 1 = \frac{Df_s}{\Delta f} = \frac{3.973 \times 44.1}{2} = 87.6 \quad \Rightarrow \quad N = 89, \quad M = \frac{1}{2}(N-1) = 44$$

Note that the given value of $A = 65$ dB corresponds to $\delta_{\text{pass}} = \delta_{\text{stop}} = 10^{-65/20} = 0.00056$, which results in the small passband attenuation $A_{\text{pass}} = 0.0097$ dB.

Such filters are to within the capabilities of modern DSP chips. Assuming a typical instruction rate of 20 MIPS, which is 20 instructions per μsec, or, $T_{\text{instr}} = 50$ nsec per instruction, and assuming the sample processing implementation requires N MACs per output sample computed, the total computational time for processing each input sample will be $N T_{\text{instr}} = 89 \times 50 = 4.45$ μsec, which fits well within the time separating each input sample, $T = 1/f_s = 1/44.1 = 22.68$ μsec.

Several such filters can even be implemented simultaneously on the same DSP chip, namely, $22.68/4.45 = 5.1$, or, about five length-89 filters. Conversely, the longest single filter that can be implemented will have length such that $N T_{\text{instr}} = T$, or, $N = T/T_{\text{instr}} = f_{\text{instr}}/f_s = 20000/44.1 \simeq 453$, resulting in the smallest implementable transition width of

$$\Delta f_{\min} \simeq \frac{D f_s}{N} = \frac{D f_s^{\,2}}{f_{\text{instr}}} = 0.386 \text{ kHz}$$

Example 10.2.3: *Two-Way Crossover Filters.* All conventional loudspeakers contain an analog *crossover* network that splits the incoming analog audio signal into its low- and high-frequency components that drive the *woofer* and *tweeter* parts of the loud-speaker. More advanced loudspeakers may contain even a third component for the mid-frequency part of the input [255].

Digital loudspeaker systems operate on the digitized audio input and use (FIR or IIR) digital filters to split it into the appropriate frequency bands, which are then converted to analog format, amplified, and drive the corresponding parts of the loudspeaker [256,257]. Such "digital" loudspeakers have been available for a while in professional digital studios and are becoming commercially available for home use (where typically the digital output of a CD player is connected to the digital input of the loudspeaker).

In this example, we take the cutoff frequency of the lowpass and highpass filters, known as the *crossover frequency*, to be $f_c = 3$ kHz, which leads to the normalized frequency $\omega_c = 2\pi f_c/f_s = 0.136\pi$. (This numerical value of f_c is chosen only for plotting convenience—a more realistic value would be 1 kHz.) The designed low- and high-frequency driver filters are then: For $n = 0, 1, \cdots, N - 1$

$$h_{\text{LP}}(n) = w(n) d_{\text{LP}}(n - M) = w(n) \left[\frac{\sin(\omega_c(n - M))}{\pi(n - M)} \right]$$

$$h_{\text{HP}}(n) = w(n) d_{\text{HP}}(n - M) = w(n) \left[\delta(n - M) - \frac{\sin(\omega_c(n - M))}{\pi(n - M)} \right] =$$

$$= \delta(n - M) - h_{\text{LP}}(n)$$

where $w(n)$ is the Kaiser window given by Eq. (10.2.8). The magnitude responses of the designed filters are shown in Fig. 10.2.6, plotted both in absolute scales, $|H(\omega)|$, and in decibels, $20 \log_{10} |H(\omega)|$.

The complementarity relationship between the impulse responses implies in the z-domain:

$$H_{\text{HP}}(z) = z^{-M} - H_{\text{LP}}(z)$$

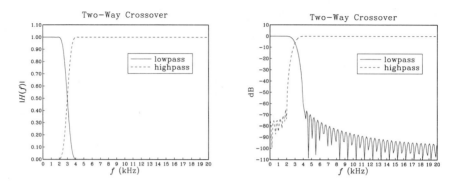

Fig. 10.2.6 Low- and high-frequency magnitude responses.

It leads to the realization of Fig. 10.2.7. Instead of realizing the lowpass and highpass filters separately, it requires only the lowpass filter and one multiple delay.   □

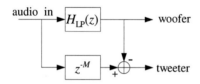

Fig. 10.2.7 Complementary implementation of two-way crossover filters.

Example 10.2.4: *Three-Way Crossover Filters.* In this example, the audio input must be split into its low-, mid-, and high-frequency components. The crossover frequencies are chosen to be $f_a = 3$ kHz and $f_b = 7$ kHz. The midpass filter will be a bandpass filter with these cutoff frequencies. The designed impulse responses will be:

$$h_{\text{LP}}(n) = w(n)d_{\text{LP}}(n-M) = w(n)\left[\frac{\sin(\omega_a(n-M))}{\pi(n-M)}\right]$$

$$h_{\text{MP}}(n) = w(n)d_{\text{MP}}(n-M) = w(n)\left[\frac{\sin(\omega_b(n-M)) - \sin(\omega_a(n-M))}{\pi(n-M)}\right]$$

$$h_{\text{HP}}(n) = w(n)d_{\text{HP}}(n-M) = w(n)\left[\delta(n-M) - \frac{\sin(\omega_b(n-M))}{\pi(n-M)}\right]$$

where, $\omega_a = 2\pi f_a/f_s = 0.136\pi$ and $\omega_b = 2\pi f_b/f_s = 0.317\pi$. Adding the three impulse responses, we find

$$h_{\text{LP}}(n) + h_{\text{MP}}(n) + h_{\text{HP}}(n) = \delta(n-M)$$

and, in the z-domain

$$H_{\mathrm{LP}}(z) + H_{\mathrm{MP}}(z) + H_{\mathrm{HP}}(z) = z^{-M}$$

which allows us to express one of them in terms of the other two, for example

$$H_{\mathrm{HP}}(z) = z^{-M} - H_{\mathrm{LP}}(z) - H_{\mathrm{MP}}(z)$$

The magnitude responses of the designed filters are shown in Fig. 10.2.8. A realization that uses the above complementarity property and requires only two filtering operations instead of three is shown in Fig. 10.2.9. □

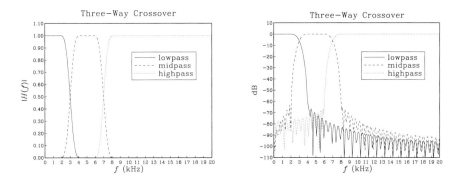

Fig. 10.2.8 Lowpass, midpass, and highpass magnitude responses.

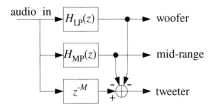

Fig. 10.2.9 Complementary implementation of three-way crossover filters.

Example 10.2.5: *Five-Band Graphic Equalizer.* Present-day graphic equalizers typically employ second-order IIR filters. However, there is no reason not to use FIR filters, if the computational cost is manageable. In this example, we choose the crossover frequencies of the five bands to be $f_a = 3$ kHz, $f_b = 7$ kHz, $f_c = 11$ kHz, $f_d = 15$ kHz, defining the five frequency bands:

$$
\begin{array}{ll}
[0, f_a] & \text{band 1} \\
[f_a, f_b] & \text{band 2} \\
[f_b, f_c] & \text{band 3} \\
[f_c, f_d] & \text{band 4} \\
[f_d, f_s/2] & \text{band 5}
\end{array}
$$

The designed filter impulse responses will be:

$$h_1(n) = w(n)\left[\frac{\sin(\omega_a(n-M))}{\pi(n-M)}\right]$$

$$h_2(n) = w(n)\left[\frac{\sin(\omega_b(n-M)) - \sin(\omega_a(n-M))}{\pi(n-M)}\right]$$

$$h_3(n) = w(n)\left[\frac{\sin(\omega_c(n-M)) - \sin(\omega_b(n-M))}{\pi(n-M)}\right]$$

$$h_4(n) = w(n)\left[\frac{\sin(\omega_d(n-M)) - \sin(\omega_c(n-M))}{\pi(n-M)}\right]$$

$$h_5(n) = w(n)\left[\delta(n-M) - \frac{\sin(\omega_d(n-M))}{\pi(n-M)}\right]$$

where, $\omega_a = 2\pi f_a/f_s = 0.136\pi$, $\omega_b = 2\pi f_b/f_s = 0.317\pi$, $\omega_c = 2\pi f_c/f_s = 0.499\pi$, $\omega_d = 2\pi f_d/f_s = 0.680\pi$. Adding the five filters we find the relationship:

$$h_1(n) + h_2(n) + h_3(n) + h_4(n) + h_5(n) = \delta(n-M)$$

and, in the z-domain

$$H_1(z) + H_2(z) + H_3(z) + H_4(z) + H_5(z) = z^{-M}$$

It can be solved for one of the transfer functions in terms of the other ones:

$$H_5(z) = z^{-M} - H_1(z) - H_2(z) - H_3(z) - H_4(z)$$

The magnitude responses are shown in Fig. 10.2.10. A realization that uses the above complementarity property and requires only four filtering operations instead of five is shown in Fig. 10.2.11.

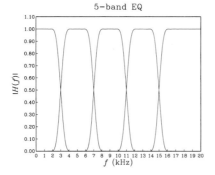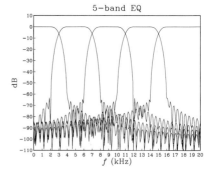

Fig. 10.2.10 Graphic equalizer magnitude responses.

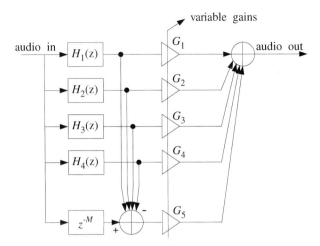

Fig. 10.2.11 Complementary implementation of graphic equalizer.

The outputs of the five filters are weighted by the user-selectable gains G_i and then summed up to form the "equalized" audio signal. The overall transfer function from the input to the overall output is:

$$H(z) = G_1 H_1(z) + G_2 H_2(z) + G_3 H_3(z) + G_4 H_4(z) + G_5 H_5(z)$$

In practice, the crossover frequencies are chosen to follow standard ISO (International Standards Organization) frequencies, dividing the 20 kHz audio range into octaves or fractions of octaves. □

The above three examples are special cases of *parallel filter banks* in which the input is split into several non-overlapping frequency bands covering the Nyquist interval. Applications include multirate filter banks and subband coding of speech, audio, and picture signals in which the outputs of the bank filters are quantized with fewer number of bits and at the same time their sampling rates are dropped, such that the overall bit rate required for the digital transmission or storage of the signal is substantially reduced [275]. For example, in the recent DCC audio cassette system, the allocation of bits in each band is governed by *psychoacoustic perceptual criteria* in which fewer bits are assigned to bands that will be less audible [295–300]. Wavelets and the *discrete wavelet transform* are also examples of filter banks [275].

Finally, we consider the definition and computation of the Bessel function $I_0(x)$. It is defined by its Taylor series expansion:

$$I_0(x) = \sum_{k=0}^{\infty} \left[\frac{(x/2)^k}{k!} \right]^2 \tag{10.2.23}$$

The Kaiser window (10.2.8) requires the evaluation of $I_0(x)$ over the range of argument $0 \le x \le \alpha$. The following C function I0.c evaluates $I_0(x)$ for any x. The routine is essentially a C version of the Fortran routine given by Kaiser in [247].

```
/* I0.c - Modified Bessel Function I0(x)
 *
 * I0(x)  =  Σ∞k=0 [(x/2)^k / k!]^2
 *
 */

#include <math.h>

#define eps    (1.E-9)                          ε = 10^-9

double I0(x)                                    usage: y = I0(x)
double x;
{
        int n = 1;
        double S = 1, D = 1, T;

        while (D > eps * S) {
                T = x / (2 * n++);
                D *= T * T;
                S += D;
                }

        return S;
}
```

The routine is based on the following recursions, which evaluate the power series (10.2.23) by keeping enough terms in the expansion. We define the partial sum of the series (10.2.23):

$$S_n = \sum_{k=0}^{n} \left[\frac{(x/2)^k}{k!} \right]^2$$

It is initialized to $S_0 = 1$ and satisfies the recursion, for $n \geq 1$:

$$S_n = S_{n-1} + D_n, \qquad \text{where } D_n = \left[\frac{(x/2)^n}{n!} \right]^2$$

In turn, D_n itself satisfies a recursion, for $n \geq 1$:

$$D_n = \left[\frac{x}{2n} \right]^2 D_{n-1} = T_n^2 D_{n-1}, \qquad \text{where } T_n = \frac{x}{2n}$$

and it is initialized to $D_0 = 1$. The iteration stops when the successive D_n terms become much smaller than the accumulated terms S_n, that is, when

$$\frac{D_n}{S_n} = \frac{S_n - S_{n-1}}{S_n} < \epsilon$$

where ϵ is a small number, such as, $\epsilon = 10^{-9}$. The Kaiser window itself may be calculated by invoking the routine I0.c; for example:

```
I0a = I0(alpha);

for (n=0; n<N; n++)
        w[n]= I0(alpha * sqrt(n*(2*M-n)) / M) / I0a;
```

10.2.2 Kaiser Window for Spectral Analysis

We saw in Section 9.1 that one of the main issues in spectral analysis was the *tradeoff* between frequency resolution and leakage. The more one tries to suppress the sidelobes, the wider the mainlobe of the window becomes, reducing the amount of achievable resolution.

For example, the Hamming window provides about 40 dB sidelobe suppression at the expense of doubling the mainlobe width of the rectangular window. Recall from Eq. (9.1.18) that the mainlobe width Δf_w of a window depends inversely on the data record length L:

$$\Delta f_w = \frac{cf_s}{L-1} \quad \Leftrightarrow \quad L - 1 = \frac{cf_s}{\Delta f_w} \qquad (10.2.24)$$

where the factor c depends on the window used. For the Kaiser window, we use the more accurate denominator $L - 1$, instead of L of Eq. (9.1.18); the difference is insignificant in spectral analysis where L is large.

The more the sidelobe suppression, the larger the factor c. Thus, to maintain a certain required value for the resolution width Δf_w, one must increase the data length L commensurately with c.

Most windows have fixed values for the amount of sidelobe suppression and width factor c. Table 10.2.2 shows these values for the rectangular and Hamming windows. Adjustable windows, like the Kaiser window, have a variable sidelobe level R that can be chosen as the application requires.

Window	R	c
Rectangular	-13 dB	1
Hamming	-40 dB	2
Kaiser	variable R	$6(R+12)/155$

Table 10.2.2 Relative sidelobe levels.

Kaiser and Schafer [221] have developed simple design equations for the use of the Kaiser window in spectral analysis. Given a desired relative sidelobe level R in dB and a desired amount of resolution Δf_w, the design equations determine the length L and shape parameter α of the window. The length is determined from Eq. (10.2.24), where c is given in terms of R by:

$$c = \frac{6(R+12)}{155} \qquad (10.2.25)$$

Note that our values for c given in Eq. (10.2.25) and Table 10.2.2 are smaller by a factor of 2 than in most texts, because we define Δf_w to be *half the base* of the mainlobe, instead of the base itself. The window shape parameter α can be calculated also in terms of R by:

$$
\alpha = \begin{cases} 0, & R < 13.26 \\ 0.76609\,(R - 13.26)^{0.4} + 0.09834\,(R - 13.26), & 13.26 < R < 60 \\ 0.12438\,(R + 6.3), & 60 < R < 120 \end{cases} \quad (10.2.26)
$$

Once the window parameters $\{L, \alpha\}$ have been determined, the window may be calculated by:

$$
w(n) = \frac{I_0\left(\alpha\sqrt{1 - (n - M)^2/M^2}\right)}{I_0(\alpha)}, \qquad n = 0, 1, \ldots, L - 1 \quad (10.2.27)
$$

where $M = (L - 1)/2$, and then applied to a length-L data record by

$$
x_L(n) = w(n)x(n), \qquad n = 0, 1, \ldots, L - 1 \quad (10.2.28)
$$

The sidelobe level R must be distinguished from the attenuation A of the filter design case. There, the attenuation A and ripple δ arise from the integrated window spectrum $W(\omega)$, as in Eq. (10.1.22), whereas R arises from $W(\omega)$ itself.

Because of the adjustable sidelobe level R, the Kaiser window can be used to pull very weak sinusoids out of the DFT spectrum of a windowed signal, whereas another type of window, such as a Hamming window, might fail. The following example illustrates this remark.

Example 10.2.6: The following analog signal consisting of three sinusoids of frequencies $f_1 = 2$ kHz, $f_2 = 2.5$ kHz, and $f_3 = 3$ kHz is sampled at a rate of $f_s = 10$ kHz:

$$
x(t) = A_1 \cos(2\pi f_1 t) + A_2 \cos(2\pi f_2 t) + A_3 \cos(2\pi f_3 t)
$$

where t is in msec. The relative amplitudes are

$$
A_1 = A_3 = 1, \qquad A_2 = 10^{-50/20} = 0.0032
$$

so that the middle sinusoid is 50 dB below the other two. A finite duration portion of length L is measured and the DFT spectrum is computed for the purpose of detecting the presence of the sinusoids by their peaks.

If we use a length-L Hamming window on the time data, the f_2-component will be lost below the 40 dB sidelobes of the window. Thus, we must use a window whose sidelobe level is well below 50 dB. We choose a Kaiser window with $R = 70$ dB. Moreover, to make the peaks clearly visible, we choose the resolution width to be $\Delta f = (f_2 - f_1)/3 = 0.167$ kHz. The Kaiser window parameters are calculated as follows:

$$
\alpha = 0.12438\,(R + 6.3) = 9.490, \qquad c = \frac{6\,(R + 12)}{155} = 3.174
$$

$$
L = 1 + \frac{cf_s}{\Delta f} = 191.45 \quad \Rightarrow \quad L = 193, \quad M = \frac{1}{2}(L - 1) = 96
$$

The length-L sampled signal is, for $n = 0, 1, \ldots, L - 1$:

$$x(n) = A_1 \cos(2\pi f_1 n/f_s) + A_2 \cos(2\pi f_2 n/f_s) + A_3 \cos(2\pi f_3 n/f_s)$$

The Kaiser and Hamming windowed signals will be, for $n = 0, 1, \ldots, L - 1$:

$$x_K(n) = w(n)x(n) = \frac{I_0\left(\alpha\sqrt{n(2M-n)}/M\right)}{I_0(\alpha)} \cdot x(n)$$

$$x_H(n) = w(n)x(n) = \left[0.54 - 0.46\cos\left(\frac{2\pi n}{L-1}\right)\right] \cdot x(n)$$

The corresponding spectra are:

$$X_K(f) = \sum_{n=0}^{L-1} x_K(n) e^{-2\pi j f n/f_s} \,, \qquad X_H(f) = \sum_{n=0}^{L-1} x_H(n) e^{-2\pi j f n/f_s}$$

Figure 10.2.12 shows these spectra in dB, that is, $20\log_{10}|X_K(f)|$, computed at 256 equally spaced frequencies in the interval $[0, f_s/2]$. (This can be done by the MATLAB function dtft.m of Appendix D.)

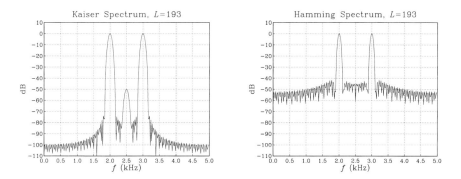

Fig. 10.2.12 Kaiser and Hamming spectra.

Both spectra are normalized to 0 dB at their maximum value. The Kaiser spectrum shows three clearly separated peaks, with the middle one being 50 dB below the other two. The sidelobes are suppressed by at least 70 dB and do not swamp the middle peak, as they do in the Hamming spectrum. That spectrum, on the other hand, has narrower peaks because the length L is somewhat larger than required to resolve the given Δf. The width of the Hamming peaks is $\Delta f = cf_s/(L-1)$ with $c = 2$, or, $\Delta f = 0.104$ kHz. □

10.3 Frequency Sampling Method

The window method is very convenient for designing ideally shaped filters, primarily because the frequency integral in Eq. (10.1.7) can be carried out in closed form.

For arbitrary frequency responses $D(\omega)$, we may use the *frequency sampling* method, in which the integral (10.1.7) is replaced by the approximate sum:

$$\tilde{d}(k) = \frac{1}{N} \sum_{i=-M}^{M} D(\omega_i) e^{j\omega_i k}, \qquad -M \leq k \leq M \tag{10.3.1}$$

where $N = 2M + 1$. The approximation is essentially an inverse N-point DFT, with the DFT frequencies ω_i spanning equally the interval $[-\pi, \pi]$, instead of the standard DFT interval $[0, 2\pi]$:

$$\omega_i = \frac{2\pi i}{N}, \qquad -M \leq i \leq M \tag{10.3.2}$$

The forward DFT applied to Eq. (10.3.1) gives:

$$D(\omega_i) = \sum_{k=-M}^{M} \tilde{d}(k) e^{-j\omega_i k} \tag{10.3.3}$$

The rest of the window method may be applied as before, that is, given an appropriate length-N window $w(n)$, the final designed filter will be the delayed and windowed version of $\tilde{d}(k)$:

$$h(n) = w(n)\tilde{d}(n - M), \qquad n = 0, 1, \ldots, N - 1 \tag{10.3.4}$$

We will discuss some examples of the frequency sampling method in Section 12.4.3, where we will design FIR filters for equalizing the slight passband droop of D/A converters and imperfect analog anti-image postfilters.

10.4 Other FIR Design Methods

The Kaiser window method is simple and flexible and can be applied to a variety of filter design problems. However, it does not always result in the smallest possible filter length N, which may be required in some very stringent applications.

The Parks-McClellan method [2–8] based on the so-called optimum *equiripple Chebyshev approximation* generally results in shorter filters. Kaiser [247] has shown that the filter length can be estimated in such cases by a variant of Eq. (10.2.12) that uses the *geometric mean* of the two ripples, $\delta_g = \sqrt{\delta_{\text{pass}}\delta_{\text{stop}}}$:

$$N - 1 = \frac{Df_s}{\Delta f}, \qquad D = \frac{A_g - 13}{14.6}, \qquad A_g = -20 \log_{10}(\delta_g) \tag{10.4.1}$$

Moreover, it may be desirable at times to design filters that have additional properties, such as convexity constraints, monotonicity constraints in the passband, or a certain degree of flatness at DC. A recent linear-programming-based filter design program called "meteor" by Steiglitz, Parks, and Kaiser [258,259] addresses such type of designs with constraints.

10.5 Problems

10.1 Consider the $d(k)$, $D(\omega)$ pair of Eq. (10.1.1). For the symmetric case, show that the condition that $d(k)$ be *real and even* in k is equivalent to $D(\omega)$ being *real and even* in ω. Similarly for the antisymmetric case, show that the condition that $d(k)$ be *real and odd* in k is equivalent to $D(\omega)$ being *imaginary and odd* in ω.

In both cases, use Euler's formula to write $D(\omega)$ is a form that shows its symmetry properties explicitly.

If you only had the reality condition that $d(k)$ be real-valued (with no other symmetry constraints), what would be the equivalent condition on $D(\omega)$?

10.2 By performing the appropriate integrations in Eq. (10.1.1), verify the expressions for $d(k)$ of the five filters in Eqs. (10.1.4) and (10.1.6).

10.3 Consider the lowpass differentiator and Hilbert transformer filters with ideal frequency responses defined over one Nyquist interval:

$$D(\omega) = \begin{cases} j\omega, & \text{if } |\omega| \le \omega_c \\ 0, & \text{if } \omega_c < |\omega| \le \pi \end{cases} , \quad D(\omega) = \begin{cases} -j\,\text{sign}(\omega), & \text{if } |\omega| \le \omega_c \\ 0, & \text{if } \omega_c < |\omega| \le \pi \end{cases}$$

Show that the corresponding ideal impulse responses are given by:

$$d(k) = \frac{\omega_c \cos(\omega_c k)}{\pi k} - \frac{\sin(\omega_c k)}{\pi k^2} \qquad \text{(differentiator)}$$

$$d(k) = \frac{1 - \cos(\omega_c k)}{\pi k} \qquad \text{(Hilbert transformer)}$$

They reduce to those of Eq. (10.1.6) in the full-band case of $\omega_c = \pi$. Do they have the right value at $k = 0$?

10.4 Determine the ideal impulse response $d(k)$ of the bandpass differentiator defined over one Nyquist interval by:

$$D(\omega) = \begin{cases} j\omega, & \text{if } \omega_a \le |\omega| \le \omega_b \\ 0, & \text{if } 0 \le |\omega| < \omega_a, \text{ or } \omega_b < |\omega| \le \pi \end{cases}$$

10.5 Differentiation is an inherently noisy operation in the sense that it amplifies any noise in the data. To see this, calculate the NRR of the FIR differentiation filter $y(n) = x(n) - x(n-1)$. In what sense is this filter an approximation to the ideal differentiator?

Then, calculate the NRR of the ideal lowpass differentiator of Problem 10.3 and compare it with NRR of the full-band case. How does it vary with the cutoff frequency ω_c?

By choosing ω_c to be the bandwidth of the desired signal, lowpass differentiators strike a compromise between differentiating the data and keeping the noise amplification as low as possible.

Lowpass differentiators designed by the Kaiser window method (see Problem 10.20), perform better than the optimal Savitzky-Golay least-squares differentiators of Section 8.3.5; see Refs. [207,217].

10.6 Show that the mean-square approximation error between the desired and windowed frequency responses of Eqs. (10.1.1) and (10.1.13) can be expressed in the form:

$$\mathcal{E}_M = \int_{-\pi}^{\pi} |D(\omega) - \hat{D}(\omega)|^2 \, \frac{d\omega}{2\pi} = \int_{-\pi}^{\pi} |D(\omega)|^2 \, \frac{d\omega}{2\pi} - \sum_{k=-M}^{M} d(k)^2$$

Then, show the limit $\mathcal{E}_M \to 0$ as $M \to \infty$.

10.7 Using the differentiator and Hilbert transformer filters and the result of Problem 10.6, show the infinite series:

$$\sum_{k=1}^{\infty} \frac{1}{k^2} = \frac{\pi^2}{6}, \qquad \sum_{\substack{k=1 \\ k=\text{odd}}}^{\infty} \frac{1}{k^2} = \frac{\pi^2}{8}, \qquad \sum_{\substack{k=2 \\ k=\text{even}}}^{\infty} \frac{1}{k^2} = \frac{\pi^2}{24}$$

10.8 The ideal Hilbert transformer has frequency response $D(\omega) = -j\,\text{sign}(\omega)$, for $-\pi \le \omega \le \pi$. Show that it acts as $90°$ phase shifter converting a cosinusoidal input into a sinusoidal output and vice versa, that is, show the input/output pairs:

$$\cos(\omega n) \xrightarrow{D} \sin(\omega n), \qquad \sin(\omega n) \xrightarrow{D} -\cos(\omega n)$$

10.9 Consider the length-$(2M + 1)$ FIR filter of Eq. (10.1.14). Show that it satisfies $\hat{D}(z) = \hat{D}(z^{-1})$ in the symmetric case, and $\hat{D}(z) = -\hat{D}(z^{-1})$ in the antisymmetric one.

Then, assuming real coefficients $d(k)$, show that in both the symmetric and antisymmetric cases, if z_0 is a zero of $\hat{D}(z)$ not on the unit circle, then necessarily the complex numbers $\{z_0^{-1}, z_0^*, z_0^{-1*}\}$ are also zeros. Indicate the relative locations of these four zeros on the z-plane with respect to the unit circle.

Moreover, show that in the antisymmetric case, the points $z = \pm 1$ are always zeros of $\hat{D}(z)$. Can you also see this result from the expression of $\hat{D}(\omega)$?

Show that the results of this problem still hold if the impulse response $d(k)$ is windowed with a Hamming, Kaiser, or any other window.

10.10 It is desired to design a linear-phase, odd-length FIR filter having real-valued *symmetric* impulse response. The filter is required to have the *smallest* possible length and to have a zero at the complex location $z = 0.5 + 0.5j$. Determine the impulse response **h** of this filter. [*Hint:* Use the results of Problem 10.9.]

Determine expressions for the *magnitude and phase responses* of this filter. Is the phase response linear in ω?

Repeat the problem if the filter is to have an *antisymmetric* impulse response. Is your answer antisymmetric?

10.11 Determine the (a) symmetric and (b) antisymmetric linear-phase FIR filter that has the *shortest* possible length and has at least one zero at the location $z = 0.5j$.

10.12 Expanding Eq. (10.2.3) to first-order in δ_{pass}, show the approximation of (10.2.4).

10.13 It is desired to design a digital lowpass linear-phase FIR filter using the Kaiser window method. The design specifications are as follows: sampling rate of 10 kHz, passband frequency of 1.5 kHz, stopband frequency of 2 kHz, passband attenuation of 0.1 dB, and stopband attenuation of 80 dB. Determine the number of filter taps N.

10.14 It is desired to design a digital lowpass FIR linear phase filter using the Kaiser window method. The maximum passband attenuation is 0.1 dB and the minimum stopband attenuation is 80 dB. At a sampling rate of 10 kHz, the maximum filter length that can be accommodated by your DSP hardware is 251 taps. What is the narrowest transition width Δf in kHz that you can demand?

10.15 Your DSP chip can accommodate FIR filters of maximum length 129 at audio rates of 44.1 kHz. Suppose such a filter is designed by the Kaiser method.

  a. What would be the *minimum* transition width Δf between passband and stopband that you can demand if the stopband attenuation is to be 80 dB?

b. If the minimum transition width Δf between passband and stopband is taken to be 2 kHz, then what would be the *maximum* stopband attenuation in dB that you can demand? What would be the corresponding passband attenuation in dB of the designed filter in this case?

c. Suppose your DSP chip could handle length-129 FIR filters at four times the audio rate, that is, $4 \times 44.1 = 176.4$ kHz. You wish to use such a filter as a four-times oversampling FIR interpolator filter for a CD player. The filter is required to have passband from 0 kHz to 19.55 kHz and stopband from 24.55 kHz up to the Nyquist frequency $176.4/2 = 88.2$ kHz. Using a Kaiser design, how much stopband attenuation in dB would you have for such a filter?

10.16 A lowpass FIR filter operating at a rate f_s is implemented on a DSP chip that has instruction rate f_{instr}. Suppose the filter is designed using the Kaiser method. Show that the maximum filter length and minimum transition width (in Hz) that can be implemented are given approximately by:

$$N_{max} = \frac{f_{instr}}{f_s}, \qquad \Delta f_{min} = \frac{D f_s^{\;2}}{f_{instr}}$$

where D is the Kaiser design parameter of Eq. (10.2.12). What assumptions were made about the DSP chip?

10.17 A lowpass FIR filter designed by the Kaiser method is required to have transition width Δf Hz and sampling rate f_s. If the filter is to be implemented on a DSP chip that has instruction rate f_{instr}, show that the maximum attainable stopband attenuation for the filter is given by:

$$A_{max} = 14.36 F_{instr} \Delta F + 7.95$$

where we defined the normalized frequencies $F_{instr} = f_{instr}/f_s$, $\Delta F = \Delta f/f_s$.

10.18 *Computer Experiment: Rectangular and Hamming Windows.* For the lengths $N = 11$, 41, 81, and 121, and using a rectangular window, design a lowpass FIR filter of cutoff frequency $\omega_c = 0.3\pi$. Plot the impulse responses $h(n)$ and magnitude responses $|H(\omega)|$ of the designed filters. Repeat using a Hamming window. Compare the two windows.

10.19 *Computer Experiment: Kaiser Window Designs.* Reproduce the designs and graphs of Examples 10.2.1 and 10.2.2. Plot also the phase responses of the designed filters for $0 \le f \le f_s/2$.

You may find useful the MATLAB routines klh.m and kbp.m. Write versions of these routines for the Hamming window and use them in this experiment.

Finally, using a Kaiser window, design a highpass filter with specifications: $f_s = 20$ kHz, $f_{pass} = 5$ kHz, $f_{stop} = 4$ kHz, $A_{pass} = 0.1$ dB, and $A_{stop} = 80$ dB. Plot its magnitude (in dB) and phase response. Compare the Kaiser design with the rectangular and Hamming window designs of the same length.

10.20 *Computer Experiment: Kaiser Window Differentiator Design.* Using the MATLAB routine kdiff, design a lowpass FIR differentiator for the following values of the cutoff frequency, transition width, and stopband attenuation parameters $\{\omega_c, \Delta\omega, A\}$:

$$\omega_c = 0.40\pi, \; 0.80\pi \qquad \text{[rads/sample]}$$

$$\Delta\omega = 0.10\pi, \; 0.05\pi \qquad \text{[rads/sample]}$$

$$A = 30, \; 60 \qquad \qquad \text{[dB]}$$

For each of the eight cases, plot the magnitude response $|H(\omega)|$ of the designed filter over the interval $0 \le \omega \le \pi$.

10.21 *Computer Experiment: Comparison of Kaiser and Savitzky-Golay Differentiators.* For the two cases of Problem 10.20 having $\omega_c = 0.4\pi$, $A = 60$ dB, and $\Delta\omega = \{0.1\pi, 0.05\pi\}$, you will find that the corresponding filter lengths are $N = 75$ and $N = 147$. Using the MATLAB routine sg.m, design the order-2 and order-3 Savitzky-Golay differentiator filters of lengths $N = 75$ and 147. This can be done by the MATLAB statements:

```
[B, S] = sg(d, N);   F = S' * S;   G = S * F^(-1);
```

and extract the second column of G. On the same graph, plot and compare the magnitude responses of the Kaiser design, the order-2, and order-3 SG designs for the two values of N. Use frequency scales $0 \le \omega \le \pi$. Then replot only over the range $0 \le \omega \le 0.1\pi$ and use vertical scales $[0, 0.1]$ to magnify the graphs.

Comment on the bandwidth of the SG designs versus the Kaiser design. See also the comments of Problem 10.5.

10.22 *Computer Experiment: Kaiser Window Hilbert Transformer Design.* Using the MATLAB routine khilb, design a lowpass FIR Hilbert transformer for the following values of the cutoff frequency, transition width, and stopband attenuation parameters $\{\omega_c, \Delta\omega, A\}$:

$$\omega_c = 0.80\pi, \ 1.00\pi \quad \text{[rads/sample]}$$

$$\Delta\omega = 0.10\pi, \ 0.05\pi \quad \text{[rads/sample]}$$

$$A = 30, \ 60 \quad \text{[dB]}$$

For each of the eight cases, plot the magnitude response $|H(\omega)|$ of the designed filter over the interval $0 \le \omega \le \pi$.

10.23 *Computer Experiment: Kaiser Window for Spectral Analysis.* (a) Reproduce all the results and graphs of Example 10.2.6. You may use the MATLAB routine kparm2 to calculate the window parameters and the routine dtft.m to calculate the spectra. (b) Keeping the Kaiser sidelobe level at $R = 70$ dB, repeat part (a) when the middle sinusoid is 35 dB below the other two, and when it is 70 dB below. (c) Repeat parts (a,b) when the transition width is chosen to be $\Delta f = (f_2 - f_1)/6$, and when it is $\Delta f = (f_2 - f_1)/12$.

11

IIR Digital Filter Design

11.1 Bilinear Transformation

One of the simplest and effective methods of designing IIR digital filters with pre-scribed magnitude response specifications is the *bilinear transformation* method.

Instead of designing the digital filter directly, the method maps the digital filter into an *equivalent analog* filter, which can be designed by one of the well-developed analog filter design methods, such as Butterworth, Chebyshev, or elliptic filter designs. The designed analog filter is then mapped back into the desired digital filter. The procedure is illustrated in Fig. 11.1.1.

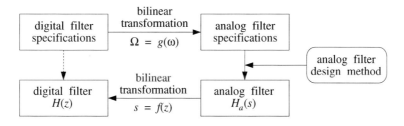

Fig. 11.1.1 Bilinear transformation method.

The z-plane design of the digital filter is replaced by an s-plane design of the equivalent analog filter. The mapping between the s and z planes is carried out by a transformation of the form:

$$\boxed{s = f(z)} \tag{11.1.1}$$

The corresponding mapping between the physical digital frequency $\omega = 2\pi f / f_s$ and the equivalent analog frequency[†] Ω is obtained by replacing $s = j\Omega$ and $z = e^{j\omega}$ into Eq. (11.1.1), giving $j\Omega = f(e^{j\omega})$, which can be written as:

[†]Here, Ω is the frequency of a fictitious equivalent analog filter. It has arbitrary units and should not be confused with the physical frequency $2\pi f$ in radians/sec.

$$\boxed{\Omega = g(\omega)} \qquad (11.1.2)$$

The *bilinear transformation* is a particular case of Eq. (11.1.1) defined by:

$$\boxed{s = f(z) = \frac{1 - z^{-1}}{1 + z^{-1}}} \qquad \text{(bilinear transformation)} \qquad (11.1.3)$$

The corresponding mapping of frequencies is obtained as follows:

$$j\Omega = f(e^{j\omega}) = \frac{1 - e^{-j\omega}}{1 + e^{-j\omega}} = \frac{e^{j\omega/2} - e^{-j\omega/2}}{e^{j\omega/2} + e^{-j\omega/2}} = j\frac{\sin(\omega/2)}{\cos(\omega/2)} = j\tan\left(\frac{\omega}{2}\right)$$

which gives:

$$\boxed{\Omega = g(\omega) = \tan\left(\frac{\omega}{2}\right)} \qquad \text{(bilinear transformation)} \qquad (11.1.4)$$

Because of the nonlinear relationship between the physical frequency ω and the fictitious analog frequency Ω, Eq. (11.1.4) is sometimes referred to as a *frequency prewarping* transformation.

Other versions of the bilinear transformation, which are appropriate for designing highpass, bandpass, or bandstop digital filters by starting from an equivalent lowpass analog filter, are as follows:

$$\text{(highpass)} \quad s = f(z) = \frac{1 + z^{-1}}{1 - z^{-1}}$$

$$\text{(bandpass)} \quad s = f(z) = \frac{1 - 2cz^{-1} + z^{-2}}{1 - z^{-2}} \qquad (11.1.5)$$

$$\text{(bandstop)} \quad s = f(z) = \frac{1 - z^{-2}}{1 - 2cz^{-1} + z^{-2}}$$

with corresponding frequency maps:

$$\text{(highpass)} \quad \Omega = g(\omega) = -\cot\left(\frac{\omega}{2}\right)$$

$$\text{(bandpass)} \quad \Omega = g(\omega) = \frac{c - \cos\omega}{\sin\omega} \qquad (11.1.6)$$

$$\text{(bandstop)} \quad \Omega = g(\omega) = \frac{\sin\omega}{\cos\omega - c}$$

The overall design method can be summarized as follows: Starting with given *magnitude response* specifications for the *digital filter*, the specifications are transformed by the *appropriate* prewarping transformation, Eqs. (11.1.4) or (11.1.6), into the specifications of an equivalent analog filter. Using an analog filter design technique, the equivalent analog filter, say $H_a(s)$, is designed. Using the bilinear transformation, Eqs. (11.1.3) or (11.1.5), the analog filter is mapped back into the desired digital filter $H(z)$, by defining:

$$H(z) = H_a(s)\Big|_{s=f(z)} = H_a\big(f(z)\big) \qquad (11.1.7)$$

The corresponding frequency responses also map in a similar fashion:

$$H(\omega) = H_a(\Omega)\Big|_{\Omega=g(\omega)} = H_a\big(g(\omega)\big) \qquad (11.1.8)$$

A useful property of the bilinear transformation (11.1.3) is that it maps the left-hand s-plane into the inside of the unit circle on the z-plane. Figure 11.1.2 shows this property. Because all analog filter design methods give rise to stable and causal transfer functions $H_a(s)$, this property guarantees that the digital filter $H(z)$ obtained by Eq. (11.1.7) will also be *stable and causal.*

The alternative transformations of Eqs. (11.1.5) also share this property, where in the bandpass and bandstop cases it is required that $|c| \le 1$.

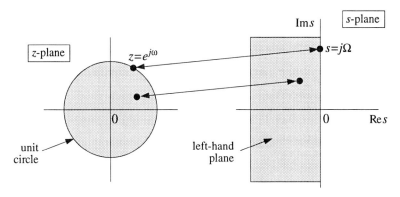

Fig. 11.1.2 Interior of unit z-circle gets mapped onto left-hand s-plane.

A related property of the bilinear transformation is that it maps the s-plane frequency axis, that is, the imaginary axis $s = j\Omega$ onto the z-plane frequency axis, that is, the periphery of the unit circle $z = e^{j\omega}$. The above properties can be proved easily by taking real parts of Eq. (11.1.3):

$$\text{Re}\, s = \frac{1}{2}(s + s^*) = \frac{1}{2}\left[\frac{z-1}{z+1} + \frac{z^*-1}{z^*+1}\right] = \frac{(z-1)(z^*+1) + (z+1)(z^*-1)}{2(z+1)(z^*+1)}$$

or,

$$\text{Re}\, s = \frac{|z|^2 - 1}{|z+1|^2}$$

which shows that

$$\text{Re}\, s < 0 \quad \Leftrightarrow \quad |z| < 1 \quad \text{and} \quad \text{Re}\, s = 0 \quad \Leftrightarrow \quad |z| = 1$$

Next, we apply the bilinear transformation method to the design of simple first- and second-order filters, and then to higher-order filters based on Butterworth and Chebyshev analog designs.

11.2   First-Order Lowpass and Highpass Filters

Perhaps the simplest filter design problem is that of designing a first-order lowpass filter that has a prescribed cutoff frequency, say f_c, and operates at a given sampling rate f_s. Such a filter will have a transfer function of the form:

$$H(z) = \frac{b_0 + b_1 z^{-1}}{1 + a_1 z^{-1}}$$

The design problem is to determine the filter coefficients $\{b_0, b_1, a_1\}$ in terms of the cutoff frequency f_c and rate f_s. The definition of the cutoff frequency is a matter of convention. Roughly speaking, it defines the range of frequencies that pass through, that is, $0 \leq f \leq f_c$, and the range of frequencies that are filtered out, that is, $f_c \leq f \leq f_s/2$. The digital cutoff frequency ω_c in units of radians per sample is defined to be:

$$\boxed{\omega_c = \frac{2\pi f_c}{f_s}}$$

By convention, ω_c is usually taken to be the so-called 3-dB cutoff frequency, that is, the frequency at which the magnitude response squared drops by a factor of two (i.e., 3 dB) compared to its value at DC:

$$\frac{|H(\omega_c)|^2}{|H(0)|^2} = \frac{1}{2} \quad \Rightarrow \quad -10\log_{10}\left[\frac{|H(\omega_c)|^2}{|H(0)|^2}\right] = -10\log_{10}\left[\frac{1}{2}\right] = 3 \text{ dB}$$

Assuming that $H(z)$ is normalized to unity gain at DC, $|H(0)| = 1$, this condition reads equivalently:

$$|H(\omega_c)|^2 = \frac{1}{2} \tag{11.2.1}$$

More generally, we may define ω_c or f_c to be the frequency at which $|H(\omega)|^2$ drops by a factor of $G_c^2 < 1$, or a drop in dB:

$$\boxed{A_c = -10\log_{10}(G_c^2) = -20\log_{10} G_c} \tag{11.2.2}$$

which can be inverted to give:

$$\boxed{G_c = 10^{-A_c/20}} \tag{11.2.3}$$

Thus, in this case, the defining condition for ω_c is:

$$\boxed{|H(\omega_c)|^2 = G_c^2 = 10^{-A_c/10}} \tag{11.2.4}$$

If $A_c = 3$ dB, we have $G_c^2 = 1/2$, and (11.2.4) reduces to (11.2.1). Figure 11.2.1 shows this type of magnitude response specification, both in the general and 3-dB cases. The design problem is then to determine the filter coefficients $\{b_0, b_1, a_1\}$ for given values of the cutoff specifications $\{f_c, A_c\}$.

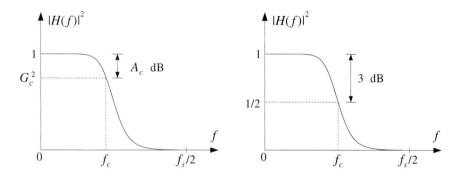

Fig. 11.2.1 Cutoff frequency specifications for lowpass digital filter.

The bilinear transformation method can be applied as follows. First, we prewarp the cutoff frequency to get the cutoff frequency of the equivalent analog filter:

$$\Omega_c = \tan\left(\frac{\omega_c}{2}\right) = \tan\left(\frac{\pi f_c}{f_s}\right)$$

Then, we design a first-order analog filter and adjust its parameters so that its cutoff frequency is Ω_c. Figure 11.2.2 shows the transformation of the specifications. The analog filter's transfer function is taken to be:

$$H_a(s) = \frac{\alpha}{s + \alpha} \qquad (11.2.5)$$

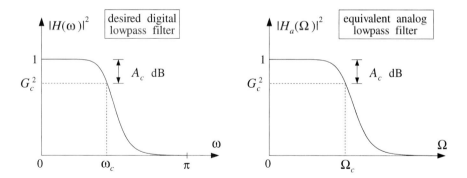

Fig. 11.2.2 Equivalent cutoff specifications of lowpass digital and analog filters.

Note that $H_a(s)$ has been normalized to unity gain at DC, or at $s = 0$. Its frequency and magnitude responses are obtained by setting $s = j\Omega$:

$$H_a(\Omega) = \frac{\alpha}{j\Omega + \alpha} \qquad \Rightarrow \qquad |H_a(\Omega)|^2 = \frac{\alpha^2}{\Omega^2 + \alpha^2} \qquad (11.2.6)$$

Because the design method satisfies Eq. (11.1.8) for the frequency and magnitude responses, we can determine the filter parameter α by requiring the cutoff condition:

$$|H(\omega_c)|^2 = |H_a(\Omega_c)|^2 = \frac{\alpha^2}{\Omega_c^2 + \alpha^2} = G_c^2 \qquad (11.2.7)$$

which can be solved for α:

$$\boxed{\alpha = \frac{G_c}{\sqrt{1 - G_c^2}} \, \Omega_c = \frac{G_c}{\sqrt{1 - G_c^2}} \tan\left(\frac{\omega_c}{2}\right)} \qquad (11.2.8)$$

Once the parameter α of the analog filter is fixed, we may transform the filter to the z-domain by the bilinear transformation (11.1.3):

$$H(z) = H_a(s) = \left.\frac{\alpha}{s + \alpha}\right|_{s=\frac{1-z^{-1}}{1+z^{-1}}} = \frac{\alpha}{\dfrac{1 - z^{-1}}{1 + z^{-1}} + \alpha} = \frac{\alpha(1 + z^{-1})}{1 - z^{-1} + \alpha(1 + z^{-1})}$$

which gives after some algebra:

$$\boxed{H(z) = b\frac{1 + z^{-1}}{1 - az^{-1}}} \qquad (11.2.9)$$

where its coefficients are computed in terms of α:

$$\boxed{a = \frac{1 - \alpha}{1 + \alpha}, \quad b = \frac{\alpha}{1 + \alpha}} \qquad (11.2.10)$$

The overall design is summarized as follows: Given the cutoff frequency ω_c and corresponding gain A_c in dB, compute G_c using Eq. (11.2.3); then compute the analog parameter α and the digital filter coefficients $\{b, a\}$.

Note that because $H_a(s)$ is stable and causal, its pole $s = -\alpha$ lies in the left-hand s-plane. This follows from the fact that $\tan(\omega_c/2) > 0$ for any value of ω_c in the range $0 < \omega_c < \pi$. This pole gets mapped onto the z-plane pole $z = a$, which satisfies $|a| < 1$ for $\alpha > 0$. The zero of the digital filter at $z = -1$ corresponds to the Nyquist frequency $\omega = \pi$ or $f = f_s/2$. Also note that the normalizing gain b can be expressed directly in terms of a, as follows:

$$b = \frac{1 - a}{2} \qquad (11.2.11)$$

If ω_c is taken to be the 3-dB cutoff frequency, then $G_c^2 = 1/2$ and Eq. (11.2.8) simplifies to:

$$\boxed{\alpha = \Omega_c = \tan\left(\frac{\omega_c}{2}\right)} \qquad (11.2.12)$$

The frequency response of the digital filter can be obtained by setting $z = e^{j\omega}$ in Eq. (11.2.9), or more simply in terms of the frequency response of the transformed analog filter, that is, using Eq. (11.1.8):

$$H(\omega) = H_a(\Omega) = \frac{\alpha}{\alpha + j\Omega} = \frac{\alpha}{\alpha + j\tan(\omega/2)}$$

Thus, we have the two equivalent expressions:

$$H(\omega) = b\frac{1 + e^{-j\omega}}{1 - ae^{-j\omega}} = \frac{\alpha}{\alpha + j\tan(\omega/2)}$$

and similarly for the magnitude response:

$$|H(\omega)|^2 = b^2\frac{2(1 + \cos\omega)}{1 - 2a\cos\omega + a^2} = \frac{\alpha^2}{\alpha^2 + \tan^2(\omega/2)} \qquad (11.2.13)$$

The bilinear transformation is not necessary. In fact, using the first of the two expressions in Eq. (11.2.13), the digital filter can be designed directly without the intermediate step of an analog filter, as was done in Example 8.3.3. However, for higher-order filters the bilinear transformation approach is *algebraically simpler* than the direct design.

Example 11.2.1: Design a lowpass digital filter operating at a rate of 10 kHz, whose 3-dB frequency is 1 kHz. Then, redesign it such that at 1 kHz its attenuation is $G_c^2 = 0.9$, corresponding to $A_c = -10\log_{10}(0.9) = 0.46$ dB.

Then, redesign the above two filters such that their cutoff frequency is now 3.5 kHz.

Solution: The digital cutoff frequency is

$$\omega_c = \frac{2\pi f_c}{f_s} = \frac{2\pi \cdot 1\text{ kHz}}{10\text{ kHz}} = 0.2\pi \quad \text{rads/sample}$$

and its prewarped analog version:

$$\Omega_c = \tan\left(\frac{\omega_c}{2}\right) = \tan(0.1\pi) = 0.3249$$

For the first filter, we have $G_c^2 = 0.5$ corresponding to Eq. (11.2.12), which gives the filter parameters:

$$\alpha = \Omega_c = 0.3249, \quad a = \frac{1 - \alpha}{1 + \alpha} = 0.5095, \quad b = \frac{1 - a}{2} = 0.2453$$

and digital filter transfer function:

$$H(z) = 0.2453\frac{1 + z^{-1}}{1 - 0.5095z^{-1}}$$

For the second filter, ω_c corresponds to attenuation $G_c^2 = 0.9$. Using Eq. (11.2.8) we find:

$$\alpha = \Omega_c \frac{G_c}{\sqrt{1 - G_c^2}} = 0.3249 \frac{\sqrt{0.9}}{\sqrt{1 - 0.9}} = 0.9748$$

corresponding to filter coefficients and transfer function:

$$a = \frac{1 - \alpha}{1 + \alpha} = 0.0128, \quad b = \frac{1 - a}{2} = 0.4936, \quad H(z) = 0.4936 \frac{1 + z^{-1}}{1 - 0.0128z^{-1}}$$

The magnitude responses of the two designed filters and their specifications are shown in the left graph of Fig. 11.2.3.

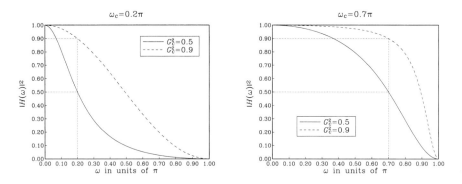

Fig. 11.2.3   First-order lowpass digital filters of Example 11.2.1.

For the next two cases, we have $f_c = 3.5$ kHz, resulting in the digital frequency $\omega_c = 2\pi f_c/f_s = 0.7\pi$, and corresponding prewarped version $\Omega_c = \tan(\omega_c/2) = \tan(0.35\pi) = 1.9626$.

For the 3-dB case, we have $\alpha = \Omega_c = 1.9626$, which gives the filter coefficients and transfer function:

$$a = \frac{1 - \alpha}{1 + \alpha} = -0.3249, \quad b = \frac{1 - a}{2} = 0.6625, \quad H(z) = 0.6625 \frac{1 + z^{-1}}{1 + 0.3249z^{-1}}$$

For the 0.46-dB case having $G_c^2 = 0.9$, we calculate:

$$\alpha = \Omega_c \frac{G_c}{\sqrt{1 - G_c^2}} = 1.9626 \frac{\sqrt{0.9}}{\sqrt{1 - 0.9}} = 5.8878$$

which gives for the digital filter coefficients and transfer function:

$$a = \frac{1 - \alpha}{1 + \alpha} = -0.7096, \quad b = \frac{1 - a}{2} = 0.8548, \quad H(z) = 0.8548 \frac{1 + z^{-1}}{1 + 0.7096z^{-1}}$$

The corresponding magnitude responses are shown in the right graph of Fig. 11.2.3. Note that in the last two cases the cutoff frequency is $\omega_c > \pi/2$ which results in a negative value of the filter pole a.   □

Highpass digital filters can be designed just as easily. Starting with a highpass analog first-order filter of the form:

$$H_a(s) = \frac{s}{s + \alpha} \tag{11.2.14}$$

we can transform it into a highpass digital filter by the bilinear transformation:

$$H(z) = H_a(s) = \left. \frac{s}{s + \alpha} \right|_{s = \frac{1 - z^{-1}}{1 + z^{-1}}} = \frac{\dfrac{1 - z^{-1}}{1 + z^{-1}}}{\dfrac{1 - z^{-1}}{1 + z^{-1}} + \alpha} = \frac{1 - z^{-1}}{1 - z^{-1} + \alpha(1 + z^{-1})}$$

which gives:

$$\boxed{H(z) = b\,\frac{1 - z^{-1}}{1 - az^{-1}}} \tag{11.2.15}$$

where its coefficients are computed in terms of α:

$$\boxed{a = \frac{1 - \alpha}{1 + \alpha}, \quad b = \frac{1}{1 + \alpha} = \frac{1 + a}{2}} \tag{11.2.16}$$

To determine the parameter α, we require that at a given highpass cutoff frequency ω_c the attenuation is A_c dB. That is, we demand

$$|H(\omega_c)|^2 = G_c^2 = 10^{-A_c/10} \tag{11.2.17}$$

Figure (11.2.4) depicts the mapping of the specifications of the digital filter to the equivalent analog filter. Setting $s = j\Omega$ into Eq. (11.2.14), we obtain for the frequency and magnitude responses:

$$H_a(\Omega) = \frac{j\Omega}{j\Omega + \alpha} \quad \Rightarrow \quad |H_a(\Omega)|^2 = \frac{\Omega^2}{\Omega^2 + \alpha^2} \tag{11.2.18}$$

The design condition (11.2.17) gives then:

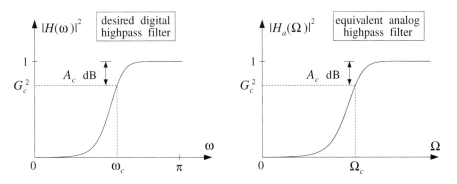

Fig. 11.2.4 Equivalent cutoff specifications of highpass digital and analog filters.

$$|H(\omega_c)|^2 = |H_a(\Omega_c)|^2 = \frac{\Omega_c^2}{\Omega_c^2 + \alpha^2} = G_c^2$$

which can be solved for α:

$$\alpha = \frac{\sqrt{1 - G_c^2}}{G_c}\Omega_c = \frac{\sqrt{1 - G_c^2}}{G_c}\tan\left(\frac{\omega_c}{2}\right) \qquad (11.2.19)$$

In summary, given the specifications $\{\omega_c, A_c\}$, we compute the analog parameter α and then the digital filter parameters $\{b, a\}$ which define the desired transfer function (11.2.15).

As in the lowpass case, we can replace $\Omega = \tan(\omega/2)$ and write the frequency and magnitude responses of the designed filter in the two equivalent forms:

$$H(\omega) = b\frac{1 - e^{-j\omega}}{1 - ae^{-j\omega}} = \frac{j\tan(\omega/2)}{\alpha + j\tan(\omega/2)}$$

$$|H(\omega)|^2 = b^2\frac{2(1 - \cos\omega)}{1 - 2a\cos\omega + a^2} = \frac{\tan^2(\omega/2)}{\alpha^2 + \tan^2(\omega/2)}$$

We note again that if ω_c represents the 3-dB cutoff frequency corresponding to $G_c^2 = 1/2$, then Eq. (11.2.19) simplifies to:

$$\alpha = \Omega_c = \tan\left(\frac{\omega_c}{2}\right) \qquad (11.2.20)$$

This and Eq. (11.2.12) imply that if the lowpass and highpass filters have the *same* 3-dB cutoff frequency ω_c, then they will have the same analog filter parameter α, and hence the same z-plane pole parameter a. Thus, in this case, the analog filters will be:

$$H_{\text{LP}}(s) = \frac{\alpha}{s + \alpha}, \quad H_{\text{HP}}(s) = \frac{s}{s + \alpha} \qquad (11.2.21)$$

and the corresponding digital filters:

$$H_{\text{LP}}(z) = \frac{1 - a}{2}\frac{1 + z^{-1}}{1 - az^{-1}}, \quad H_{\text{HP}}(z) = \frac{1 + a}{2}\frac{1 - z^{-1}}{1 - az^{-1}} \qquad (11.2.22)$$

They are *complementary filters* in the sense that their transfer functions add up to unity:

$$H_{\text{LP}}(s) + H_{\text{HP}}(s) = \frac{\alpha}{s + \alpha} + \frac{s}{s + \alpha} = \frac{s + \alpha}{s + \alpha} = 1$$

and similarly,

$$H_{\text{LP}}(z) + H_{\text{HP}}(z) = 1 \qquad (11.2.23)$$

We have already encountered such complementary filters in the comb and notch filters of Section 8.3.2. The corresponding frequency and magnitude responses squared also add up to unity; for example,

$$|H_{\text{LP}}(\omega)|^2 + |H_{\text{HP}}(\omega)|^2 = \frac{\alpha^2}{\alpha^2 + \tan^2(\omega/2)} + \frac{\tan^2(\omega/2)}{\alpha^2 + \tan^2(\omega/2)} = 1$$

Example 11.2.2: Design a highpass digital filter operating at a rate of 10 kHz, whose 3-dB cutoff frequency is 1 kHz. Then, redesign it such that at 1 kHz its attenuation is $G_c^2 = 0.9$, corresponding to $A_c = -10\log_{10}(0.9) = 0.46$ dB.

Solution: The digital cutoff frequency is as in Example 11.2.1, $\omega_c = 0.2\pi$. Its prewarped analog version is $\Omega_c = \tan(\omega_c/2) = 0.3249$.

For the first filter, we have $G_c^2 = 0.5$ corresponding to Eq. (11.2.20), which gives the filter parameters:

$$\alpha = \Omega_c = 0.3249, \quad a = \frac{1-\alpha}{1+\alpha} = 0.5095, \quad b = \frac{1+a}{2} = 0.7548$$

and transfer function:

$$H(z) = 0.7548 \frac{1 - z^{-1}}{1 - 0.5095z^{-1}}$$

For the second filter, we set $G_c^2 = 0.9$ into Eq. (11.2.19) to get:

$$\alpha = \Omega_c \frac{\sqrt{1 - G_c^2}}{G_c} = 0.1083$$

corresponding to filter coefficients and transfer function:

$$a = \frac{1-\alpha}{1+\alpha} = 0.8046, \quad b = \frac{1-a}{2} = 0.9023, \quad H(z) = 0.9023 \frac{1 - z^{-1}}{1 - 0.8046z^{-1}}$$

The magnitude responses of the two designed filters and their specifications are shown in the left graph of Fig. 11.2.5.

The right graph shows the complementarity property of the highpass and lowpass filters, with 3-dB frequency of $\omega_c = 0.2\pi$. The magnitude responses intersect at precisely the 3-dB point. $\qquad\square$

11.3 Second-Order Peaking and Notching Filters

In Section 6.4, we designed second-order resonator and notch filters using pole/zero placement. For narrow-width filters this technique is adequate. But, it becomes cumbersome for wider peak widths, such as those that might be used in graphic and parametric audio equalizers. The bilinear transformation method offers precise control over the desired specifications of such filters [260–268].

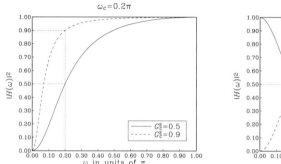
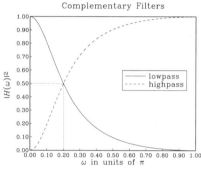

Fig. 11.2.5 Highpass and complementary lowpass digital filters.

Consider first the design of *notch* filters. The desired specifications are the sampling rate f_s, notch frequency f_0, and bandwidth Δf of the notch, or, equivalently, the corresponding digital frequencies:

$$\omega_0 = \frac{2\pi f_0}{f_s}, \quad \Delta\omega = \frac{2\pi \Delta f}{f_s}$$

Alternatively, we may specify ω_0 and the Q-factor, $Q = \omega_0/\Delta\omega = f_0/\Delta f$. The specifications together with their bilinear analog equivalents are shown in Fig. 11.3.1. The bandwidth $\Delta\omega$ is usually defined to be the 3-dB width, that is, the *full width at half maximum* of the magnitude squared response. More generally, it can be defined to be the full width at a level G_B^2, or in decibels:

$$A_B = -10 \log_{10}(G_B^2) \quad \Rightarrow \quad G_B = 10^{-A_B/20} \tag{11.3.1}$$

The bandwidth $\Delta\omega$ is defined as the difference $\Delta\omega = \omega_2 - \omega_1$ of the left and right bandwidth frequencies ω_1 and ω_2 that are solutions of the equation $|H(\omega)|^2 = G_B^2$, as shown in Fig. 11.3.1. For the 3-dB width, we have the condition $|H(\omega)|^2 = 1/2$.

Given the desired specifications $\{\omega_0, \Delta\omega, G_B^2\}$, the design procedure begins with the following expression for the equivalent analog filter, which has a notch at frequency $\Omega = \Omega_0$:

$$H_a(s) = \frac{s^2 + \Omega_0^2}{s^2 + \alpha s + \Omega_0^2} \tag{11.3.2}$$

We will see below that the filter parameters $\{\alpha, \Omega_0\}$ can be calculated from the given specifications by:

$$\Omega_0 = \tan\left(\frac{\omega_0}{2}\right), \quad \alpha = \frac{\sqrt{1 - G_B^2}}{G_B}(1 + \Omega_0^2)\tan\left(\frac{\Delta\omega}{2}\right) \tag{11.3.3}$$

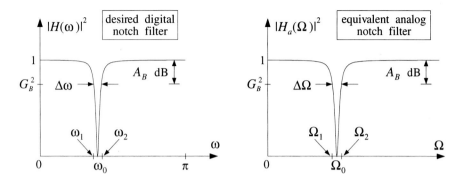

Fig. 11.3.1 Digital notch filter and its analog equivalent.

Then, using the bilinear transformation $s = (1 - z^{-1}) / (1 + z^{-1})$, the filter $H_a(s)$ is transformed into the digital filter $H(z)$ as follows:

$$H(z) = H_a(s) = \frac{s^2 + \Omega_0^2}{s^2 + \alpha s + \Omega_0^2} = \frac{\left(\dfrac{1 - z^{-1}}{1 + z^{-1}}\right)^2 + \Omega_0^2}{\left(\dfrac{1 - z^{-1}}{1 + z^{-1}}\right)^2 + \alpha\left(\dfrac{1 - z^{-1}}{1 + z^{-1}}\right) + \Omega_0^2}$$

$$= \frac{(1 - z^{-1})^2 + \Omega_0^2 (1 + z^{-1})^2}{(1 - z^{-1})^2 + \alpha(1 - z^{-1})(1 + z^{-1}) + \Omega_0^2(1 + z^{-1})^2}$$

$$= \left(\frac{1 + \Omega_0^2}{1 + \Omega_0^2 + \alpha}\right) \frac{1 - 2\left(\dfrac{1 - \Omega_0^2}{1 + \Omega_0^2}\right) z^{-1} + z^{-2}}{1 - 2\left(\dfrac{1 - \Omega_0^2}{1 + \Omega_0^2 + \alpha}\right) z^{-1} + \left(\dfrac{1 + \Omega_0^2 - \alpha}{1 + \Omega_0^2 + \alpha}\right) z^{-2}}$$

The coefficients of the digital filter can be simplified considerably by recognizing that α already has a factor $(1 + \Omega_0^2)$ in its definition (11.3.3). Thus, we may replace it by

$$\alpha = (1 + \Omega_0^2)\beta$$

where

$$\boxed{\beta = \frac{\sqrt{1 - G_B^2}}{G_B} \tan\left(\frac{\Delta\omega}{2}\right)} \tag{11.3.4}$$

Using some trigonometry, we can write also

$$\frac{1 - \Omega_0^2}{1 + \Omega_0^2} = \frac{1 - \tan^2(\omega_0/2)}{1 + \tan^2(\omega_0/2)} = \cos\omega_0$$

Canceling several common factors of $(1+\Omega_0^2)$, we can write the transfer function $H(z)$ in the simplified form:

$$H(z) = \left(\frac{1}{1+\beta} \right) \frac{1 - 2\cos\omega_0\, z^{-1} + z^{-2}}{1 - 2\left(\dfrac{\cos\omega_0}{1+\beta} \right) z^{-1} + \left(\dfrac{1-\beta}{1+\beta} \right) z^{-2}} \tag{11.3.5}$$

Defining the overall normalization gain by

$$\boxed{\; b = \frac{1}{1+\beta} = \frac{1}{1 + \dfrac{\sqrt{1 - G_B^2}}{G_B}\tan\left(\dfrac{\Delta\omega}{2} \right)} \;} \tag{11.3.6}$$

we may write $(1-\beta)/(1+\beta) = 2b - 1$, and therefore, the coefficients of $H(z)$ can be expressed in terms of b as follows:

$$\boxed{\; H(z) = b\,\frac{1 - 2\cos\omega_0\, z^{-1} + z^{-2}}{1 - 2b\cos\omega_0\, z^{-1} + (2b-1)z^{-2}} \;} \qquad \text{(notch filter)} \tag{11.3.7}$$

This is the final design. It expresses the filter coefficients in terms of the design specifications $\{\omega_0, \Delta\omega, G_B^2\}$. Note that the numerator has a notch at the desired frequency ω_o and its conjugate $-\omega_0$, because it factors into:

$$1 - 2\cos\omega_0\, z^{-1} + z^{-2} = (1 - e^{j\omega_0}z^{-1})(1 - e^{-j\omega_0}z^{-1})$$

It remains to justify the design equations (11.3.3). The first one, $\Omega_0 = \tan(\omega_0/2)$, is simply the bilinear transformation of ω_0 and makes the analog filter's notch correspond to the digital filter's notch. The equation for α can be derived as follows. Setting $s = j\Omega$ in Eq. (11.3.2), we obtain the frequency and magnitude responses:

$$H_a(\Omega) = \frac{-\Omega^2 + \Omega_0^2}{-\Omega^2 + j\alpha\Omega + \Omega_0^2} \quad \Rightarrow \quad |H_a(\Omega)|^2 = \frac{(\Omega^2 - \Omega_0^2)^2}{(\Omega^2 - \Omega_0^2)^2 + \alpha^2\Omega^2}$$

It is evident from these expressions that $H_a(\Omega)$ has a notch at $\Omega = \pm\Omega_0$. The analog bandwidth frequencies Ω_1 and Ω_2 are solutions of the equation $|H_a(\Omega)|^2 = G_B^2$, that is,

$$\frac{(\Omega^2 - \Omega_0^2)^2}{(\Omega^2 - \Omega_0^2)^2 + \alpha^2\Omega^2} = G_B^2 \tag{11.3.8}$$

Eliminating the denominator and rearranging terms, we can write it as the quartic equation in Ω:

$$\Omega^4 - \left(2\Omega_0^2 + \frac{G_B^2}{1 - G_B^2}\alpha^2 \right)\Omega^2 + \Omega_0^4 = 0 \tag{11.3.9}$$

It may be thought of as a quadratic equation in the variable $x = \Omega^2$, that is,

$$x^2 - \left(2\Omega_0^2 + \frac{G_B^2}{1 - G_B^2}\alpha^2\right)x + \Omega_0^4 = 0$$

Let $x_1 = \Omega_1^2$ and $x_2 = \Omega_2^2$ be its two solutions. Rather than solving it, we use the properties that the sum and product of the two solutions are related to the first and second coefficients of the quadratic by:

$$\Omega_1^2 + \Omega_2^2 = x_1 + x_2 = 2\Omega_0^2 + \frac{G_B^2}{1 - G_B^2}\alpha^2$$

$$\Omega_1^2\Omega_2^2 = x_1 x_2 = \Omega_0^4$$

(11.3.10)

From the second equation, we obtain:

$$\boxed{\Omega_1\Omega_2 = \Omega_0^2}$$

(11.3.11)

which states that Ω_0 is the *geometric mean* of the left and right bandwidth frequencies. Using this result in the first of (11.3.10), we obtain:

$$\Omega_1^2 + \Omega_2^2 = 2\Omega_1\Omega_2 + \frac{G_B^2}{1 - G_B^2}\alpha^2$$

which allows us to solve for the analog bandwidth:

$$\Delta\Omega^2 = (\Omega_2 - \Omega_1)^2 = \Omega_1^2 + \Omega_2^2 - 2\Omega_1\Omega_2 = \frac{G_B^2}{1 - G_B^2}\alpha^2$$

or,

$$\boxed{\Delta\Omega = \Omega_2 - \Omega_1 = \frac{G_B}{\sqrt{1 - G_B^2}}\alpha} \qquad (A_B\text{-dB width})$$

(11.3.12)

Solving for α, we have:

$$\boxed{\alpha = \frac{\sqrt{1 - G_B^2}}{G_B}\Delta\Omega}$$

(11.3.13)

Note that for the 3-dB case, $G_B^2 = 1/2$, the parameter α is equal to the 3-dB bandwidth:

$$\boxed{\alpha = \Delta\Omega} \qquad (3\text{-dB width})$$

(11.3.14)

Finally, we must relate the analog bandwidth $\Delta\Omega$ to the physical bandwidth $\Delta\omega = \omega_2 - \omega_1$. Using the bilinear transformations $\Omega_1 = \tan(\omega_1/2)$, $\Omega_2 = \tan(\omega_2/2)$, and some trigonometry, we find:

$$\tan\left(\frac{\Delta\omega}{2}\right) = \tan\left(\frac{\omega_2 - \omega_1}{2}\right) = \frac{\tan(\omega_2/2) - \tan(\omega_1/2)}{1 + \tan(\omega_2/2)\tan(\omega_1/2)}$$

$$= \frac{\Omega_2 - \Omega_1}{1 + \Omega_2\Omega_1} = \frac{\Delta\Omega}{1 + \Omega_0^2}$$

where we used $\Omega_1\Omega_2 = \Omega_0^2$. Solving for $\Delta\Omega$, we have:

$$\boxed{\Delta\Omega = (1 + \Omega_0^2)\tan\left(\frac{\Delta\omega}{2}\right)} \tag{11.3.15}$$

Thus, combining Eqs. (11.3.13) and (11.3.15), we obtain Eq. (11.3.3). The design equations (11.3.6) and (11.3.7) and some design examples were discussed also in Sections 8.2.2 and 8.3.2. For example, see Fig. 8.2.14.

In the limit as ω_0 or Ω_0 tend to zero, the notch filter will behave as a high-pass filter. The transfer functions $H(z)$ and $H_a(s)$ become in this case the high-pass transfer functions of the previous section. For example, setting $\Omega_0 = 0$ in Eq. (11.3.2), we have:

$$H_a(s) = \left.\frac{s^2 + \Omega_0^2}{s^2 + \alpha s + \Omega_0^2}\right|_{\Omega_0=0} = \frac{s^2}{s^2 + \alpha s} = \frac{s}{s + \alpha}$$

Peaking or resonator filters can be designed in a similar fashion. The desired specifications are shown in Fig. 11.3.2. The design procedure starts with the second-order analog resonator filter:

$$\boxed{H_a(s) = \frac{\alpha s}{s^2 + \alpha s + \Omega_0^2}} \tag{11.3.16}$$

which has frequency and magnitude responses:

$$H_a(\Omega) = \frac{j\alpha\Omega}{-\Omega^2 + j\alpha\Omega + \Omega_0^2} \quad \Rightarrow \quad |H_a(\Omega)|^2 = \frac{\alpha^2\Omega^2}{(\Omega^2 - \Omega_0^2)^2 + \alpha^2\Omega^2}$$

Note that $H_a(\Omega)$ is normalized to unity gain at the peak frequencies $\Omega = \pm\Omega_0$. The bandwidth frequencies Ω_1 and Ω_2 will satisfy the bandwidth condition:

$$|H_a(\Omega)|^2 = \frac{\alpha^2\Omega^2}{(\Omega^2 - \Omega_0^2)^2 + \alpha^2\Omega^2} = G_B^2$$

It can be written as the quartic:

$$\Omega^4 - \left(2\Omega_0^2 + \frac{1 - G_B^2}{G_B^2}\alpha^2\right)\Omega^2 + \Omega_0^4 = 0$$

which is similar to Eq. (11.3.9). Its two solutions Ω_1^2 and Ω_2^2 satisfy the conditions:

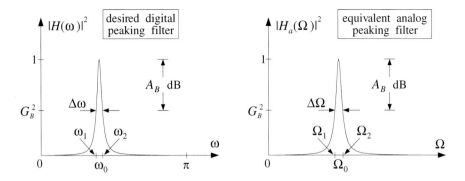

Fig. 11.3.2 Digital peaking filter and its analog equivalent.

$$\Omega_1^2 + \Omega_2^2 = 2\Omega_0^2 + \frac{1 - G_B^2}{G_B^2}\alpha^2$$

$$\Omega_1^2 \Omega_2^2 = \Omega_0^4$$

from which we obtain $\Omega_1\Omega_2 = \Omega_0^2$ and

$$\Delta\Omega = \Omega_2 - \Omega_1 = \frac{\sqrt{1 - G_B^2}}{G_B}\alpha \quad \Rightarrow \quad \boxed{\alpha = \frac{G_B}{\sqrt{1 - G_B^2}}\Delta\Omega}$$

The relationship (11.3.15) between the analog and digital bandwidth remains the same. Therefore, we obtain the analog filter parameters $\{\alpha, \Omega_0\}$ by equations similar to (11.3.3):

$$\boxed{\Omega_0 = \tan\left(\frac{\omega_0}{2}\right), \quad \alpha = \frac{G_B}{\sqrt{1 - G_B^2}}(1 + \Omega_0^2)\tan\left(\frac{\Delta\omega}{2}\right)} \qquad (11.3.17)$$

The digital filter is obtained by the bilinear transformation:

$$H(z) = H_a(s) = \left. \frac{\alpha s}{s^2 + \alpha s + \Omega_0^2} \right|_{s = \frac{1 - z^{-1}}{1 + z^{-1}}}$$

which can be written in the form:

$$H(z) = \left(\frac{\beta}{1 + \beta}\right)\frac{1 - z^{-2}}{1 - 2\left(\dfrac{\cos\omega_0}{1 + \beta}\right)z^{-1} + \left(\dfrac{1 - \beta}{1 + \beta}\right)z^{-2}} \qquad (11.3.18)$$

where β is similar, but not identical, to that in Eq. (11.3.4):

$$\beta = \frac{G_B}{\sqrt{1 - G_B^2}} \tan\left(\frac{\Delta\omega}{2}\right) \tag{11.3.19}$$

Defining the gain b as in Eq. (11.3.6)

$$b = \frac{1}{1 + \beta} = \frac{1}{1 + \dfrac{G_B}{\sqrt{1 - G_B^2}} \tan\left(\dfrac{\Delta\omega}{2}\right)} \tag{11.3.20}$$

we may write $(1 - \beta)/(1 + \beta) = 2b - 1$, and $\beta/(1 + \beta) = 1 - b$, and therefore, the coefficients of $H(z)$ can be expressed in terms of b as follows:

$$H(z) = (1 - b)\frac{1 - z^{-2}}{1 - 2b\cos\omega_0 z^{-1} + (2b - 1)z^{-2}} \quad \text{(peak filter)} \tag{11.3.21}$$

Note that the numerator vanishes at $z = \pm 1$, that is, at DC and the Nyquist frequency. For the 3-dB widths, we have $G_B^2 = 1/2$, and the parameters β or b are the same as those of the notch filter:

$$\beta = \tan\left(\frac{\Delta\omega}{2}\right), \quad b = \frac{1}{1 + \beta} = \frac{1}{1 + \tan(\Delta\omega/2)} \tag{11.3.22}$$

In this case, the notch and peak filters are *complementary* with transfer functions, frequency responses, and magnitude responses squared that add up to unity. For example, adding Eqs. (11.3.7) and (11.3.21), we have:

$$H_{\text{notch}}(z) + H_{\text{peak}}(z) = 1 \tag{11.3.23}$$

Note, finally, that in the limit as ω_0 or Ω_0 tend to zero, the peaking filter will behave as a lowpass filter. The transfer functions $H(z)$ and $H_a(s)$ become in this case the lowpass transfer functions of the previous section. For example, setting $\Omega_0 = 0$ in Eq. (11.3.16), we have:

$$H_a(s) = \left.\frac{\alpha s}{s^2 + \alpha s + \Omega_0^2}\right|_{\Omega_0 = 0} = \frac{\alpha s}{s^2 + \alpha s} = \frac{\alpha}{s + \alpha}$$

Example 11.3.1: Design a peaking digital filter operating at a rate of 10 kHz that has a peak at 1.75 kHz and 3-dB width of 500 Hz. Then, redesign it such that 500 Hz represents its 10-dB width.

For the 3-dB width case, determine also the corresponding complementary notch filter.

Solution: The digital frequencies in radians per sample are:

$$\omega_0 = \frac{2\pi f_0}{f_s} = \frac{2\pi \cdot 1.75}{10} = 0.35\pi, \quad \Delta\omega = \frac{2\pi\Delta f}{f_s} = \frac{2\pi \cdot 0.5}{10} = 0.1\pi$$

For the 3-dB case, we calculate the parameter, $\cos \omega_0 = 0.4540$, and:

$$\beta = \tan\left(\frac{\Delta\omega}{2}\right) = 0.1584, \quad b = \frac{1}{1+\beta} = 0.8633, \quad 1-b = 0.1367$$

For the case $A_B = 10$ dB, we have bandwidth gain $G_B^2 = 10^{-A_B/10} = 0.1$. Then, we calculate:

$$\beta = \frac{G_B}{\sqrt{1-G_B^2}}\tan\left(\frac{\Delta\omega}{2}\right) = 0.0528, \quad b = \frac{1}{1+\beta} = 0.9499, \quad 1-b = 0.0501$$

Inserting the above two sets of parameter values into Eq. (11.3.21), we obtain the transfer functions:

$$H(z) = \frac{0.1367(1-z^{-2})}{1-0.7838z^{-1}+0.7265z^{-2}}, \quad H(z) = \frac{0.0501(1-z^{-2})}{1-0.8624z^{-1}+0.8997z^{-2}}$$

The squared magnitude responses are shown in Fig. 11.3.3. They were calculated using the simpler analog expressions:

$$|H(\omega)|^2 = |H_a(\Omega)|^2 = \frac{\alpha^2\Omega^2}{(\Omega^2-\Omega_0^2)^2+\alpha^2\Omega^2}$$

Replacing $\Omega = \tan(\omega/2) = \tan(\pi f/f_s)$, we have in terms of the physical frequency f in Hz:

$$|H(f)|^2 = \frac{\alpha^2\tan^2(\pi f/f_s)}{\left(\tan^2(\pi f/f_s)-\Omega_0^2\right)^2 + \alpha^2\tan^2(\pi f/f_s)}$$

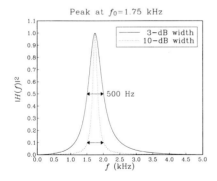
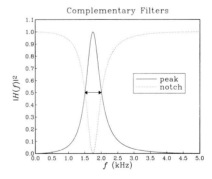

Fig. 11.3.3 Peaking and complementary notch filters.

The right graph of Fig. 11.3.3 shows the complementary peak and notch filters. The parameters β and b of the notch filter were already calculated above. Using Eq. (11.3.7), we find its transfer function:

$$H(z) = 0.8633\frac{1-0.9080z^{-1}+z^{-2}}{1-0.7838z^{-1}+0.7265z^{-2}}$$

The zeros of the denominator, $1 - 0.7838z^{-1} + 0.7265z^{-2} = 0$, determine the poles of the transfer function. They are:

$$p, p^* = 0.3919 \pm j0.7569 = 0.8524e^{\pm j0.3479\pi}$$

The poles are not exactly at the desired frequency $\omega_0 = 0.35\pi$. Naive pole placement would have placed them there. For example, choosing the same radius $R = 0.8524$, we would have in that case:

$$p, p^* = 0.8524e^{\pm j0.35\pi} = 0.3870 \pm j0.7597$$

corresponding to the denominator polynomial $1 - 0.7739z^{-1} + 0.7265z^{-2}$. The bilinear transformation method places the poles at appropriate locations to achieve the desired peak and width.   ☐

11.4 Parametric Equalizer Filters

Frequency equalization (EQ) is a common requirement in audio systems—analog, digital, home, car, public, or studio recording/mixing systems [255].

Graphic equalizers are the more common type, in which the audio band is divided into a *fixed* number of frequency bands, and the amount of equalization in each band is controlled by a bandpass filter whose gain can be varied up and down. The center frequencies of the bands and the filter 3-dB widths are fixed, and the user can vary only the overall gain in each band. Usually, second-order bandpass filters are adequate for audio applications.

A more flexible equalizer type is the *parametric equalizer*, in which all three filter parameters—gain, center frequency, and bandwidth—can be varied. Cascading four or five such filters together can achieve almost any desired equalization effect.

Figure 11.4.1 shows the frequency response of a typical second-order parametric equalizer. The specification parameters are: a *reference* gain G_0 (typically taken to be unity for cascadable filters), the center frequency ω_0 of the boost or cut, the filter gain G at ω_0, and a desired width $\Delta\omega$ at an appropriate bandwidth level G_B that lies between G_0 and G. As shown in Fig. 11.4.1, the relative gains must be chosen as follows, depending on whether we have a boost or a cut:

$$
\begin{array}{ll}
G_0^2 < G_B^2 < G^2 & \text{(boost)} \\[4pt]
G^2 < G_B^2 < G_0^2 & \text{(cut)}
\end{array}
\tag{11.4.1}
$$

The notch and peak filters of the previous section can be thought of as special cases of such a filter. The peaking filter corresponds to $G_0 = 0$, $G = 1$ and the notching filter to $G_0 = 1$, $G = 0$.

The definition of $\Delta\omega$ is arbitrary, and not without ambiguity. For example, we can define it to be the 3-dB width. But, what exactly do we mean by "3 dB"?

For the boosting case, we can take it to mean 3 dB *below the peak*, that is, choose $G_B^2 = G^2/2$; alternatively, we can take it to mean 3 dB *above the reference*, that is,

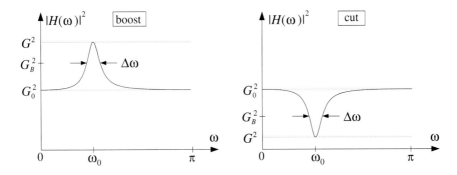

Fig. 11.4.1 Parametric EQ filter with boost or cut.

$G_B^2 = 2G_0^2$. Moreover, because G_B^2 must lie between G_0^2 and G^2, the first alternative implies that $G_0^2 < G_B^2 = G^2/2$, or $2G_0^2 < G^2$, and the second $2G_0^2 = G_B^2 < G^2$. Thus, either alternative requires that $G^2 > 2G_0^2$, that is, the boost gain must be at least 3 dB higher than the reference. So, what do we do when $G_0^2 < G^2 < 2G_0^2$? In that case, any G_B^2 that lies in $G_0^2 < G_B^2 < G^2$ will do. A particularly interesting choice is to take it to be the arithmetic mean of the end values:

$$G_B^2 = \frac{G_0^2 + G^2}{2} \tag{11.4.2}$$

Another good choice is the geometric mean, $G_B^2 = GG_0$, corresponding to the arithmetic mean of the dB values of the endpoints [265,268] (see Problem 11.4.)

Similar ambiguities arise in the cutting case: we can take 3 dB to mean 3 dB *above the dip*, that is, $G_B^2 = 2G^2$, or, alternatively, we can take it to mean 3 dB *below the reference*, $G_B^2 = G_0^2/2$. Either alternative requires that $G^2 < G_0^2/2$, that is, the cut gain must be at least 3 dB below the reference. If $G_0^2/2 < G^2 < G_0^2$, we may again use the average (11.4.2). To summarize, some possible (but not necessary) choices for G_B^2 are as follows:

$$G_B^2 = \begin{cases} G^2/2, & \text{if } G^2 > 2G_0^2 & \text{(boost, alternative 1)} \\ 2G_0^2, & \text{if } G^2 > 2G_0^2 & \text{(boost, alternative 2)} \\ (G_0^2 + G^2)/2, & \text{if } G_0^2 < G^2 < 2G_0^2 & \text{(boost)} \\ \\ 2G^2, & \text{if } G^2 < G_0^2/2 & \text{(cut, alternative 1)} \\ G_0^2/2, & \text{if } G^2 < G_0^2/2 & \text{(cut, alternative 2)} \\ (G_0^2 + G^2)/2, & \text{if } G_0^2/2 < G^2 < G_0^2 & \text{(cut)} \end{cases} \tag{11.4.3}$$

The filter design problem is to determine the filter's transfer function in terms of the specification parameters: $\{G_0, G, G_B, \omega_0, \Delta\omega\}$. In this section, we present a simple design method based on the bilinear transformation, which is a variation of the methods in [260–268].

We define the parametric equalizer filter as the following *linear combination* of the notching and peaking filters of the previous section:

$$H(z) = G_0 H_{\text{notch}}(z) + G H_{\text{peak}}(z) \tag{11.4.4}$$

At ω_0 the gain is G, because the notch filter vanishes and the peak filter has unity gain. Similarly, at DC and the Nyquist frequency, the gain is equal to the reference G_0, because the notch is unity and the peak vanishes. From the complementarity property (11.3.23) it follows that when $G = G_0$ we have $H(z) = G_0$, that is, no equalization. Inserting the expressions (11.3.5) and (11.3.18) into Eq. (11.4.4), we obtain:

$$H(z) = \frac{\left(\dfrac{G_0 + G\beta}{1 + \beta}\right) - 2\left(\dfrac{G_0 \cos \omega_0}{1 + \beta}\right) z^{-1} + \left(\dfrac{G_0 - G\beta}{1 + \beta}\right) z^{-2}}{1 - 2\left(\dfrac{\cos \omega_0}{1 + \beta}\right) z^{-1} + \left(\dfrac{1 - \beta}{1 + \beta}\right) z^{-2}} \tag{11.4.5}$$

The parameter β is a generalization of Eqs. (11.3.4) and (11.3.19) and is given by:

$$\beta = \sqrt{\frac{G_B^2 - G_0^2}{G^2 - G_B^2}} \, \tan\left(\frac{\Delta \omega}{2}\right) \tag{11.4.6}$$

Note that because of the assumed inequalities (11.4.1), the quantity under the square root is always positive. Also, for the special choice of G_B^2 of Eq. (11.4.2), the square root factor is *unity*. This choice (and those of Problem 11.4) allows a smooth transition to the no-equalization limit $G \to G_0$. Indeed, because β does not depend on the G's, setting $G = G_0$ in the numerator of Eq. (11.4.5) gives $H(z) = G_0$.

The design equations (11.4.5) and (11.4.6) can be justified as follows. Starting with the same linear combination of the analog versions of the notching and peaking filters given by Eqs. (11.3.2) and (11.3.16), we obtain the analog version of $H(z)$:

$$H_a(s) = G_0 H_{\text{notch}}(s) + G H_{\text{peak}}(s) = \frac{G_0 (s^2 + \Omega_0^2) + G\alpha s}{s^2 + \alpha s + \Omega_0^2} \tag{11.4.7}$$

Then, the bandwidth condition $|H_a(\Omega)|^2 = G_B^2$ can be stated as:

$$|H_a(\Omega)|^2 = \frac{G_0^2 (\Omega^2 - \Omega_0^2)^2 + G^2 \alpha^2 \Omega^2}{(\Omega^2 - \Omega_0^2)^2 + \alpha^2 \Omega^2} = G_B^2 \tag{11.4.8}$$

It can be cast as the quartic equation:

$$\Omega^4 - \left(2\Omega_0^2 + \frac{G^2 - G_B^2}{G_B^2 - G_0^2} \alpha^2\right) \Omega^2 + \Omega_0^4 = 0$$

Proceeding as in the previous section and using the geometric-mean property $\Omega_1 \Omega_2 = \Omega_0^2$ and Eq. (11.3.15), we find the relationship between the parameter α and the analog bandwidth $\Delta\Omega = \Omega_2 - \Omega_1$:

$$\alpha = \sqrt{\frac{G_B^2 - G_0^2}{G^2 - G_B^2}} \, \Delta\Omega = \sqrt{\frac{G_B^2 - G_0^2}{G^2 - G_B^2}} \, (1 + \Omega_0^2) \tan\left(\frac{\Delta\omega}{2}\right) \equiv (1 + \Omega_0^2)\beta$$

This defines β. Then, the bilinear transformation of Eq. (11.4.7) leads to Eq. (11.4.5).

Example 11.4.1: Design the following six parametric EQ filters operating at 10 kHz rate that satisfy the specifications: $G_0 = 1$ and

(a) center frequency of 1.75 kHz, 9-dB boost gain, and 3-dB width of 500 Hz defined to be 3 dB below the peak (alternative 1).

(b) same as (a), except the width is 3 dB above the reference (alternative 2).

(c) center frequency of 3 kHz, 9-dB cut gain, and 3-dB width of 1 kHz defined to be 3 dB above the dip (alternative 1).

(d) same as (c), except the width is 3 dB below the reference (alternative 2).

(e) center frequency of 1.75 kHz, 2-dB boost, and 500 Hz width defined by Eq. (11.4.2).

(f) center frequency of 3 kHz, 2-dB cut, and 1 kHz width defined by Eq. (11.4.2).

Solution: The boost examples (a), (b), and (e) have digital frequency and width:

$$\omega_0 = \frac{2\pi \cdot 1.75}{10} = 0.35\pi, \quad \Delta\omega = \frac{2\pi \cdot 0.5}{10} = 0.1\pi$$

and the cut examples (c), (d), and (f) have:

$$\omega_0 = \frac{2\pi \cdot 3}{10} = 0.6\pi, \quad \Delta\omega = \frac{2\pi \cdot 1}{10} = 0.2\pi$$

Normally, a "3-dB" change means a change by a factor of 2 in the magnitude square. Here, for plotting purposes, we take "3 dB" to mean literally 3 dB, which corresponds to changes by $10^{3/10} = 1.9953 \simeq 2$. Therefore, in case (a), a boost gain of 9 dB above the reference G_0 corresponds to the value:

$$G = 10^{9/20} G_0 = 2.8184, \quad G^2 = 7.9433 \quad \text{(instead of 8)}$$

The bandwidth level is defined to be 3 dB below the peak, that is, $A_B = 9 - 3 = 6$ dB, and therefore:

$$G_B = 10^{-3/20} G = 10^{6/20} G_0 = 1.9953, \quad G_B^2 = 3.9811$$

With these values of $\{G_0, G, G_B, \omega_0, \Delta\omega\}$, we calculate the value of β from Eq. (11.4.6):

$$\beta = \sqrt{\frac{G_B^2 - G_0^2}{G^2 - G_B^2}} \, \tan\left(\frac{\Delta\omega}{2}\right) = \sqrt{\frac{3.9811 - 1}{7.9433 - 3.9811}} \, \tan\left(\frac{0.1\pi}{2}\right) = 0.1374$$

We calculate also $\cos\omega_0 = \cos(0.35\pi) = 0.4540$. The transfer function of filter (a), obtained from Eq. (11.4.5), is then:

$$H_a(z) = \frac{1.2196 - 0.7983z^{-1} + 0.5388z^{-2}}{1 - 0.7983z^{-1} + 0.7584z^{-2}}$$

For filter (b), the width is defined to be 3 dB above the reference, that is, $A_B = 3$ dB:

$$G_B = 10^{3/20}G_0 = 10^{3/20} = 1.4125, \quad G_B^2 = 1.9953$$

From Eq. (11.4.6), we calculate $\beta = 0.0648$, and from Eq. (11.4.5) the filter:

$$H_b(z) = \frac{1.1106 - 0.8527z^{-1} + 0.7677z^{-2}}{1 - 0.8527z^{-1} + 0.8783z^{-2}}$$

For filter (c), we have a 9-dB cut gain, that is, 9 dB below the reference:

$$G = 10^{-9/20}G_0 = 0.3548, \quad G^2 = 0.1259$$

and the bandwidth level is 3 dB above this dip, that is, $A_B = -9 + 3 = -6$ dB:

$$G_B = 10^{3/20}G = 10^{-6/20}G_0 = 0.5012, \quad G_B^2 = 0.2512$$

Then, we calculate $\cos\omega_0 = \cos(0.6\pi) = -0.3090$ and $\beta = 0.7943$, and the transfer function:

$$H_c(z) = \frac{0.7144 + 0.3444z^{-1} + 0.4002z^{-2}}{1 + 0.3444z^{-1} + 0.1146z^{-2}}$$

For filter (d), the width is 3 dB below the reference, that is, $A_B = 0 - 3 = -3$ dB:

$$G_B = 10^{-3/20}G_0 = 0.7079, \quad G_B^2 = 0.5012$$

We calculate $\beta = 0.3746$ and the transfer function:

$$H_d(z) = \frac{0.8242 + 0.4496z^{-1} + 0.6308z^{-2}}{1 + 0.4496z^{-1} + 0.4550z^{-2}}$$

The four filters (a)-(d) are shown in the left graph of Fig. 11.4.2. The magnitude responses are plotted in dB, that is, $20\log_{10}|H(\omega)|$. The reference level $G_0 = 1$ corresponds to 0 dB. Notice the horizontal grid lines at 6 dB, 3 dB, -3 dB, and -6 dB, whose intersections with the magnitude responses define the corresponding bandwidths $\Delta\omega$.

For filter (e), the boost gain is 2 dB and therefore, the bandwidth level cannot be chosen to be 3 dB below the peak or 3 dB above the reference. We must use an intermediate level between 0 and 2 dB. In particular, we may use Eq. (11.4.2). Thus, we calculate the parameters:

$$G = 10^{2/20}G_0 = 1.2589, \quad G_B^2 = \frac{G_0^2 + G^2}{2} = 1.2924$$

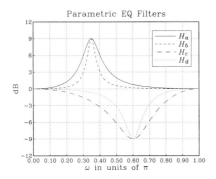 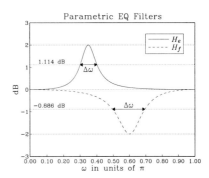

Fig. 11.4.2 Parametric EQ filters of Example 11.4.1.

corresponding to $A_B = 10 \log_{10}(G_B^2) = 1.114$ dB. The square root factor in the definition of β is unity, therefore, we calculate:

$$\beta = \tan\left(\frac{\Delta\omega}{2}\right) = \tan\left(\frac{0.1\pi}{2}\right) = 0.1584$$

and the transfer function:

$$H_e(z) = \frac{1.0354 - 0.7838z^{-1} + 0.6911z^{-2}}{1 - 0.7838z^{-1} + 0.7265z^{-2}}$$

Finally, in case (f), we have a 2-dB cut, giving the values:

$$G = 10^{-2/20}G_0 = 0.7943, \quad G_B^2 = \frac{G_0^2 + G^2}{2} = 0.8155$$

corresponding to $A_B = 10 \log_{10}(G_B^2) = -0.886$ dB. The parameter β is now $\beta = \tan(\Delta\omega/2) = \tan(0.2\pi/2) = 0.3249$, resulting in the transfer function:

$$H_f(z) = \frac{0.9496 + 0.4665z^{-1} + 0.5600z^{-2}}{1 + 0.4665z^{-1} + 0.5095z^{-2}}$$

Filters (e) and (f) are shown in the right graph of Fig. 11.4.2. The vertical scales are expanded compared to those of the left graph. The horizontal lines defining the bandwidth levels $A_B = 1.114$ dB and $A_B = -0.886$ dB are also shown.

In practice, parametric EQ filters for audio have cut and boost gains that vary typically from -18 dB to 18 dB with respect to the reference gain. □

Example 11.4.2: Instead of specifying the parameters $\{\omega_0, \Delta\omega\}$, it is often convenient to specify either one or both of the corner frequencies $\{\omega_1, \omega_2\}$ that define the width $\Delta\omega = \omega_2 - \omega_1$.

Design four parametric EQ filters that have a 2.5-dB cut and bandwidth defined at 1 dB below the reference, and have center or corner frequencies as follows:

(a) Center frequency $\omega_0 = 0.6\pi$ and right corner $\omega_2 = 0.7\pi$. Determine also the left corner ω_1 and the bandwidth $\Delta\omega$.

(b) Center frequency $\omega_0 = 0.6\pi$ and left corner $\omega_1 = 0.5\pi$. Determine also the right corner ω_2 and the bandwidth $\Delta\omega$.

(c) Left and right corner frequencies $\omega_1 = 0.5\pi$ and $\omega_2 = 0.7\pi$. Determine also the center frequency ω_0.

(d) Compare the above to the standard design that has $\omega_0 = 0.6\pi$, and $\Delta\omega = 0.2\pi$. Determine the values of ω_1, ω_2.

Solution: Assuming $G_0 = 1$, the cut and bandwidth gains are:

$$G = 10^{-2.5/20} = 0.7499, \quad G_B = 10^{-1/20} = 0.8913$$

Note that G_B was chosen arbitrarily in this example and not according to Eq. (11.4.2). For case (a), we are given ω_0 and ω_2. Under the bilinear transformation they map to the values:

$$\Omega_0 = \tan(\omega_0/2) = 1.3764, \quad \Omega_2 = \tan(\omega_2/2) = 1.9626$$

Using the geometric-mean property (11.3.11), we may solve for ω_1:

$$\tan\left(\frac{\omega_1}{2}\right) = \Omega_1 = \frac{\Omega_0^2}{\Omega_2} = 0.9653 \quad \Rightarrow \quad \omega_1 = 0.4887\pi$$

Thus, the bandwidth is $\Delta\omega = \omega_2 - \omega_1 = 0.2113\pi$. The design equations (11.4.5) and (11.4.6) give then $\beta = 0.3244$ and the transfer function:

$$H_a(z) = \frac{0.9387 + 0.4666z^{-1} + 0.5713z^{-2}}{1 + 0.4666z^{-1} + 0.5101z^{-2}}$$

For case (b), we are given ω_0 and ω_1 and calculate ω_2:

$$\Omega_1 = \tan\left(\frac{\omega_1}{2}\right) = 1, \quad \tan\left(\frac{\omega_2}{2}\right) = \Omega_2 = \frac{\Omega_0^2}{\Omega_1} = 1.8944 \quad \Rightarrow \quad \omega_2 = 0.6908\pi$$

where Ω_0 was as in case (a). The width is $\Delta\omega = \omega_2 - \omega_1 = 0.1908\pi$. Then, we find $\beta = 0.2910$ and the transfer function:

$$H_b(z) = \frac{0.9436 + 0.4787z^{-1} + 0.6056z^{-2}}{1 + 0.4787z^{-1} + 0.5492z^{-2}}$$

The magnitude responses (in dB) of cases (a) and (b) are shown in the left graph of Fig. 11.4.3. The bandwidths are defined by the intersection of the horizontal grid line at -1 dB and the curves.

For case (c), we are given ω_1 and ω_2. Their bilinear transformations are:

$$\Omega_1 = \tan(\omega_1/2) = 1, \quad \Omega_2 = \tan(\omega_2/2) = 1.9626$$

The center frequency is computed from:

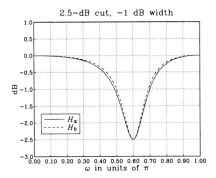 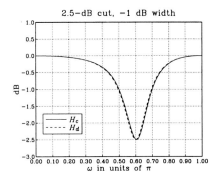

Fig. 11.4.3 Parametric EQ filters.

$$\tan(\omega_0/2) = \Omega_0 = \sqrt{\Omega_1\Omega_2} = 1.4009 \quad \Rightarrow \quad \omega_0 = 0.6053\pi$$

Using the calculated ω_0 and the width $\Delta\omega = \omega_2 - \omega_1 = 0.2\pi$, we find $\cos\omega_0 = -0.3249$, $\beta = 0.3059$, and the transfer function:

$$H_c(z) = \frac{0.9414 + 0.4976z^{-1} + 0.5901z^{-2}}{1 + 0.4976z^{-1} + 0.5315z^{-2}}$$

Finally, in the standard case (d), we start with ω_0 and $\Delta\omega$. We find $\cos\omega_0 = -0.3090$, $\beta = 0.3059$, and the transfer function:

$$H_d(z) = \frac{0.9414 + 0.4732z^{-1} + 0.5901z^{-2}}{1 + 0.4732z^{-1} + 0.5315z^{-2}}$$

With $\Omega_0 = \tan(\omega_0/2) = 1.3764$, the exact values of ω_1 and ω_2 are obtained by solving the system of equations:

$$\Omega_1\Omega_2 = \Omega_0^2 = 1.8944, \quad \Omega_2 - \Omega_1 = \Delta\Omega = (1 + \Omega_0^2)\tan(\Delta\omega/2) = 0.9404$$

which have positive solutions $\Omega_1 = 0.9843$, $\Omega_2 = 1.9247$. It follows that

$$\omega_1 = 2\arctan(\Omega_1) = 0.494951\pi, \quad \omega_2 = 2\arctan(\Omega_2) = 0.694951\pi$$

where as expected $\Delta\omega = \omega_2 - \omega_1 = 0.2\pi$. The magnitude responses are shown in the right graph of Fig. 11.4.3. Note that cases (c) and (d) have the same β because their widths $\Delta\omega$ are the same. But, the values of $\cos\omega_0$ are different, resulting in different values for the coefficients of z^{-1}; the other coefficients are the same. □

In addition to parametric equalizers with variable center frequencies ω_0, in audio applications we also need lowpass and highpass filters, referred to as "shelving" filters, with adjustable gains and cutoff frequencies. Such filters can be obtained from Eq. (11.4.5) by replacing $\omega_0 = 0$ for the lowpass case and $\omega_0 = \pi$ for the highpass one.

In the lowpass limit, $\omega_0 = 0$, we have $\cos\omega_0 = 1$ and the numerator and denominator of Eq. (11.4.5) develop a common factor $(1 - z^{-1})$. Canceling this factor, we obtain the *lowpass shelving filter*:

$$H_{\mathrm{LP}}(z) = \frac{\left(\dfrac{G_0 + G\beta}{1 + \beta}\right) - \left(\dfrac{G_0 - G\beta}{1 + \beta}\right) z^{-1}}{1 - \left(\dfrac{1 - \beta}{1 + \beta}\right) z^{-1}} \tag{11.4.9}$$

where β is still given by Eq. (11.4.6), but with $\Delta\omega$ replaced by the filter's cutoff frequency ω_c and with G_B replaced by the defining level G_c of the cutoff frequency:

$$\beta = \sqrt{\frac{G_c^2 - G_0^2}{G^2 - G_c^2}} \tan\left(\frac{\omega_c}{2}\right) \tag{11.4.10}$$

In the highpass limit, $\omega_0 = \pi$, we have $\cos\omega_0 = -1$ and the numerator and denominator of Eq. (11.4.5) have a common factor $(1 + z^{-1})$. Canceling it, we obtain the *highpass shelving filter*:

$$H_{\mathrm{HP}}(z) = \frac{\left(\dfrac{G_0 + G\beta}{1 + \beta}\right) + \left(\dfrac{G_0 - G\beta}{1 + \beta}\right) z^{-1}}{1 + \left(\dfrac{1 - \beta}{1 + \beta}\right) z^{-1}} \tag{11.4.11}$$

It can also be obtained from Eq. (11.4.9) by the replacement $z \rightarrow -z$. The parameter β is obtained from Eq. (11.4.6) by the replacements $G_B \rightarrow G_c$ and $\Delta\omega \rightarrow \pi - \omega_c$. The latter is necessary because $\Delta\omega$ is measured from the center frequency $\omega_0 = \pi$, whereas ω_c is measured from the origin $\omega = 0$. Noting that $\tan\left((\pi - \omega_c)/2\right) = \cot(\omega_c/2)$, we have:

$$\beta = \sqrt{\frac{G_c^2 - G_0^2}{G^2 - G_c^2}} \cot\left(\frac{\omega_c}{2}\right) \tag{11.4.12}$$

For both the lowpass and highpass cases, the filter specifications are the parameters $\{G_0, G, G_c, \omega_c\}$. They must satisfy Eq. (11.4.1) for boosting or cutting. Figure 11.4.4 depicts these specifications. Some possible choices for G_c^2 are still given by Eq. (11.4.3).

The limiting forms of the corresponding analog filter (11.4.7) can be obtained by taking the appropriate limits in the variable $\Omega_0 = \tan(\omega_0/2)$. For the lowpass case, we have the limit $\Omega_0 \rightarrow 0$ and for the highpass case, the limit $\Omega_0 \rightarrow \infty$. We must also replace $\alpha = (1 + \Omega_0^2)\beta$ before taking these limits.

Taking the limits, we obtain the analog filters whose bilinear transformations are the shelving filters Eq. (11.4.9) and (11.4.11):

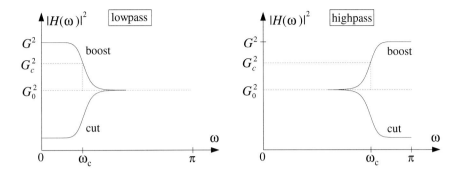

Fig. 11.4.4 Lowpass and highpass shelving filters with boost or cut.

$$H_{\text{LP}}(s) = \frac{G_0 s + G\beta}{s + \beta} \, , \quad H_{\text{HP}}(s) = \frac{G_0 + G\beta s}{1 + \beta s} \tag{11.4.13}$$

The special case $G_0 = 0$, $G = 1$ was considered in Section 11.2. In that section, the lowpass case corresponded to $\alpha = \beta$ and the highpass to $\alpha = 1/\beta$.

Setting $\Omega = \Omega_c = \tan(\omega_c/2)$, the following bandwidth conditions may be solved for β, resulting into the design equations (11.4.10) and (11.4.12):

$$|H_{\text{LP}}(\Omega)|^2 = \frac{G_0^2 \Omega^2 + G^2 \beta^2}{\Omega^2 + \beta^2} = G_c^2 \, , \quad |H_{\text{HP}}(\Omega)|^2 = \frac{G_0^2 + G^2 \beta^2 \Omega^2}{1 + \beta^2 \Omega^2} = G_c^2$$

11.5 Comb Filters

The lowpass and highpass shelving filters of the previous section can be turned into *periodic comb or notch filters* with adjustable gains and peak widths. This can be accomplished by the replicating transformation of Eq. (8.3.30), that is, $z \rightarrow z^D$, which shrinks the frequency response by a factor of D and replicates it D times such that D copies of it fit into the Nyquist interval. Under this transformation, the lowpass filter $H_{\text{LP}}(z)$ of Eq. (11.4.9) becomes the comb filter:

$$\boxed{H(z) = \frac{b - cz^{-D}}{1 - az^{-D}}} \, , \quad a = \frac{1 - \beta}{1 + \beta}, \ b = \frac{G_0 + G\beta}{1 + \beta}, \ c = \frac{G_0 - G\beta}{1 + \beta} \tag{11.5.1}$$

This transfer function can also be obtained from the analog lowpass shelving filter $H_{\text{LP}}(s)$ of Eq. (11.4.13) by the generalized bilinear transformation:

$$s = \frac{1 - z^{-D}}{1 + z^{-D}} \, , \quad \Omega = \tan\left(\frac{\omega D}{2}\right) = \tan\left(\frac{\pi f D}{f_s}\right) \tag{11.5.2}$$

The DC peak of the lowpass filter $H_{\text{LP}}(z)$ has full width $2\omega_c$, counting also its symmetric negative-frequency side. Under D-fold replication, the symmetric DC peak will be shrunk in width by a factor of D and replicated D times, with replicas centered at the Dth root-of-unity frequencies $\omega_k = 2\pi k/D$, $k = 0, 1, \ldots, D - 1$. Thus, the full width $\Delta\omega$ of the individual replicas can be obtained by the substitution $2\omega_c/D \to \Delta\omega$, or $\omega_c \to D\Delta\omega/2$.

Making this substitution in Eq. (11.4.10) and replacing the cutoff frequency gain G_c by the bandwidth gain G_B, the filter parameter β can be expressed as:

$$\beta = \sqrt{\frac{G_B^2 - G_0^2}{G^2 - G_B^2}} \, \tan\left(\frac{D\,\Delta\omega}{4}\right) \tag{11.5.3}$$

The resulting comb filter can be thought of as a "comb equalizer" with variable gain and peak width. The boost or cut choices of the gain G given in Eq. (11.4.1) will correspond to either a peaking or a notching periodic comb filter.

The periodic notch and comb filters of Section 8.3.2 are special cases of the general comb filter (11.5.1). The notch filter of Eq. (8.3.26) is obtained in the limit $G_0 = 1$, $G = 0$, and the comb filter of Eq. (8.3.33) in the limit $G_0 = 0$, $G = 1$. In both cases, the bandwidth gain can be chosen to be $G_B^2 = 1/2$, that is, 3 dB.

The replicating transformation $z \to z^D$ can also be applied to the highpass shelving filter $H_{\text{HP}}(z)$ of Eq. (11.4.11), resulting in the comb filter:

$$H(z) = \frac{b + cz^{-D}}{1 + az^{-D}} \tag{11.5.4}$$

The sign change of the coefficients causes the peaks to shift by π/D, placing them between the peaks of the comb (11.5.1), that is, at the odd-multiple frequencies $\omega_k = (2k + 1)\pi/D$, $k = 0, 1, \ldots, D - 1$. The parameter β is still calculated from Eq. (11.5.3). This type of comb filter can be useful in some applications, as we saw in Section 8.3.3.

Example 11.5.1: Design a peaking comb filter of period 10, reference gain of 0 dB, peak gain of 9 dB, bandwidth gain of 3 dB (above the reference), and bandwidth $\Delta\omega = 0.025\pi$ rads/sample.

Then, design a notching comb filter with dip gain of -12 dB and having the same period, reference gain, and bandwidth as the peaking filer. The bandwidth is defined to be 3 dB below the reference.

Then, redesign both of the above peaking and notching comb filters such that their peaks or dips are shifted to lie exactly between the peaks of the previous filters.

Solution: For the peaking filter, we have:

$$G_0 = 1, \quad G = 10^{9/20}G_0 = 2.8184, \quad G_B = 10^{3/20}G_0 = 1.4125$$

and for the notching filter:

$$G_0 = 1, \quad G = 10^{-12/20}G_0 = 0.2512, \quad G_B = 10^{-3/20}G_0 = 0.7079$$

With $D = 10$ and $\Delta\omega = 0.025\pi$, we find the values: $\beta = 0.0814$ for the peaking filter and $\beta = 0.2123$ for the notching filter. Then, the transfer functions are obtained from Eq. (11.5.1):

$$H_{\text{peak}}(z) = \frac{1.1368 - 0.7127z^{-10}}{1 - 0.8495z^{-10}}, \quad H_{\text{notch}}(z) = \frac{0.8689 - 0.7809z^{-10}}{1 - 0.6498z^{-10}}$$

Their peaks/dips are at the multiples $\omega_k = 2\pi k/D = 0.2\pi k$, $k = 0, 1, \ldots, 9$. The filters with the shifted peaks are obtained by changing the sign of z^{-10}:

$$H_{\text{peak}}(z) = \frac{1.1368 + 0.7127z^{-10}}{1 + 0.8495z^{-10}}, \quad H_{\text{notch}}(z) = \frac{0.8689 + 0.7809z^{-10}}{1 + 0.6498z^{-10}}$$

The magnitude responses of the four filters are shown in Fig. 11.5.1, plotted in dB, that is, $20\log_{10}|H(\omega)|$, over one complete Nyquist interval, $0 \le \omega \le 2\pi$. □

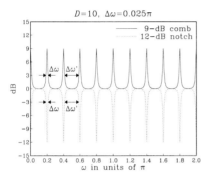
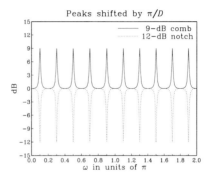

Fig. 11.5.1 Periodic peaking and notching comb filters.

Figure 11.5.1 also shows the peak width $\Delta\omega$ as well as the quantity $\Delta\omega'$, which is a measure of the *separation* of two successive peaks at the bandwidth level. Because of symmetry, the D equal peak widths and the D equal separations must make up the full 2π Nyquist interval. Thus, we have the condition:

$$D\Delta\omega + D\Delta\omega' = 2\pi \quad \Rightarrow \quad \Delta\omega + \Delta\omega' = \frac{2\pi}{D} \tag{11.5.5}$$

Decreasing the peak width increases the separation and vice versa. The maximum possible value of $\Delta\omega$ corresponds to the case when $\Delta\omega' = 0$, that is, zero separation between peaks. This gives $\Delta\omega_{\text{max}} = 2\pi/D$. However, a more useful practical limit is when $\Delta\omega \le \Delta\omega'$, which causes the peaks to be narrower than their separation. This condition requires that $2\Delta\omega \le \Delta\omega + \Delta\omega' = 2\pi/D$, and gives the maximum for $\Delta\omega$:

$$\Delta\omega \le \frac{\pi}{D} \quad \Rightarrow \quad \Delta f \le \frac{f_s}{2D} \tag{11.5.6}$$

This maximum was also discussed in Section 8.3.2.

11.6 Higher-Order Filters

The first- and second-order designs of the above sections are adequate in some applications such as audio equalization, but are too limited when we need filters with very sharp cutoff specifications. Higher-order filters can achieve such sharp cutoffs, but at the price of increasing the filter complexity, that is, the filter order.

Figure 11.6.1 shows the specifications of a typical lowpass filter and its analog equivalent obtained by the bilinear transformation. The specification parameters are the four numbers $\{f_{\text{pass}}, f_{\text{stop}}, A_{\text{pass}}, A_{\text{stop}}\}$, that is, the *passband and stopband frequencies* and the desired *passband and stopband attenuations* in dB.

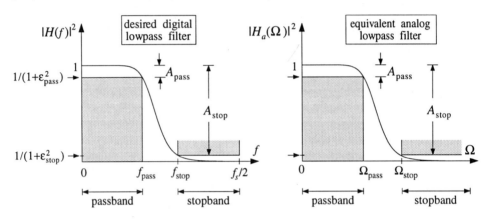

Fig. 11.6.1 Lowpass digital filter and its analog equivalent.

Within the passband range $0 \le f \le f_{\text{pass}}$, the filter's attenuation is required to be *less* than A_{pass} decibels. And, within the stopband $f_{\text{stop}} \le f \le f_s/2$, it is required to be *greater* than A_{stop} decibels. Thus, the quantity A_{pass} is the *maximum* attenuation that can be tolerated in the passband and A_{stop} the *minimum* attenuation that must be achieved in the stopband.

The filter can be made into a *better* lowpass filter in three ways: (1) decreasing A_{pass} so that the passband becomes flatter, (2) increasing A_{stop} so that the stopband becomes deeper, and (3) moving f_{stop} closer to f_{pass} so that the *transition region* between passband and stopband becomes narrower. Thus, by appropriate choice of the specification parameters, the filter can be made as close to an ideal lowpass filter as desired.

Assuming the filter's magnitude response squared $|H(f)|^2$ is normalized to unity at DC, we can express the specification requirements as the following conditions on the filter's attenuation response in dB, defined as $A(f) = -10\log_{10}|H(f)|^2$:

$$\boxed{\begin{aligned} 0 \le A(f) \le A_{\text{pass}}\,, \quad & \text{for} \quad 0 \le f \le f_{\text{pass}} \\ A(f) \ge A_{\text{stop}}\,, \quad & \text{for} \quad f_{\text{stop}} \le f \le f_s/2 \end{aligned}} \tag{11.6.1}$$

Equivalently, in absolute units, the design specifications are:

$$1 \geq |H(f)|^2 \geq \frac{1}{1 + \varepsilon_{\text{pass}}^2}, \quad \text{for} \quad 0 \leq f \leq f_{\text{pass}}$$

$$|H(f)|^2 \leq \frac{1}{1 + \varepsilon_{\text{stop}}^2}, \quad \text{for} \quad f_{\text{stop}} \leq f \leq f_s/2$$

(11.6.2)

where $\{\varepsilon_{\text{pass}}, \varepsilon_{\text{stop}}\}$ are defined in terms of $\{A_{\text{pass}}, A_{\text{stop}}\}$ as follows:

$$|H(f_{\text{pass}})|^2 = \frac{1}{1 + \varepsilon_{\text{pass}}^2} = 10^{-A_{\text{pass}}/10},$$

$$|H(f_{\text{stop}})|^2 = \frac{1}{1 + \varepsilon_{\text{stop}}^2} = 10^{-A_{\text{stop}}/10},$$

(11.6.3)

The quantities $\{\varepsilon_{\text{pass}}, \varepsilon_{\text{stop}}\}$ control the depths of the passband and stopband. They can be written in the equivalent forms:

$$\boxed{\begin{array}{c} \varepsilon_{\text{pass}} = \sqrt{10^{A_{\text{pass}}/10} - 1} \\ \varepsilon_{\text{stop}} = \sqrt{10^{A_{\text{stop}}/10} - 1} \end{array}} \quad \Leftrightarrow \quad \boxed{\begin{array}{c} A_{\text{pass}} = 10\log_{10}(1 + \varepsilon_{\text{pass}}^2) \\ A_{\text{stop}} = 10\log_{10}(1 + \varepsilon_{\text{stop}}^2) \end{array}}$$

(11.6.4)

The specifications of the equivalent analog filter are $\{\Omega_{\text{pass}}, \Omega_{\text{stop}}, A_{\text{pass}}, A_{\text{stop}}\}$, or, $\{\Omega_{\text{pass}}, \Omega_{\text{stop}}, \varepsilon_{\text{pass}}, \varepsilon_{\text{stop}}\}$, where the analog frequencies are obtained by prewarping the digital frequencies:

$$\boxed{\Omega_{\text{pass}} = \tan\left(\frac{\omega_{\text{pass}}}{2}\right), \quad \Omega_{\text{stop}} = \tan\left(\frac{\omega_{\text{stop}}}{2}\right)}$$

(11.6.5)

where

$$\boxed{\omega_{\text{pass}} = \frac{2\pi f_{\text{pass}}}{f_s}, \quad \omega_{\text{stop}} = \frac{2\pi f_{\text{stop}}}{f_s}}$$

(11.6.6)

The parameters $\{\varepsilon_{\text{pass}}, \varepsilon_{\text{stop}}\}$ are useful in the design of both Butterworth and Chebyshev filters. In the next section, we begin with the Butterworth case.

11.6.1 Analog Lowpass Butterworth Filters

Analog lowpass Butterworth filters are characterized by just *two* parameters: the filter order N and the 3-dB normalization frequency Ω_0. Their magnitude response is simply:

$$\boxed{|H(\Omega)|^2 = \frac{1}{1 + \left(\dfrac{\Omega}{\Omega_0}\right)^{2N}}}$$

(11.6.7)

and the corresponding attenuation in decibels:

$$\boxed{A(\Omega) = -10\log_{10}|H(\Omega)|^2 = 10\log_{10}\left[1 + \left(\frac{\Omega}{\Omega_0}\right)^{2N}\right]} \tag{11.6.8}$$

Note that, as N increases for fixed Ω_0, the filter becomes a better lowpass filter. At $\Omega = \Omega_0$, the magnitude response is $|H(\Omega_0)|^2 = 1/2$, or, 3-dB attenuation $A(\Omega_0) = 3$ dB. The two filter parameters $\{N, \Omega_0\}$ can be determined from the given specifications $\{\Omega_{\text{pass}}, \Omega_{\text{stop}}, A_{\text{pass}}, A_{\text{stop}}\}$ by requiring the conditions:

$$A(\Omega_{\text{pass}}) = 10\log_{10}\left[1 + \left(\frac{\Omega_{\text{pass}}}{\Omega_0}\right)^{2N}\right] = A_{\text{pass}} = 10\log_{10}(1 + \varepsilon_{\text{pass}}^2)$$

$$A(\Omega_{\text{stop}}) = 10\log_{10}\left[1 + \left(\frac{\Omega_{\text{stop}}}{\Omega_0}\right)^{2N}\right] = A_{\text{stop}} = 10\log_{10}(1 + \varepsilon_{\text{stop}}^2)$$

Because of the monotonicity of the magnitude response, these conditions are equivalent to the passband/stopband range conditions (11.6.1). To solve them for N and Ω_0, we rewrite them in the form:

$$\left(\frac{\Omega_{\text{pass}}}{\Omega_0}\right)^{2N} = 10^{A_{\text{pass}}/10} - 1 = \varepsilon_{\text{pass}}^2$$

$$\left(\frac{\Omega_{\text{stop}}}{\Omega_0}\right)^{2N} = 10^{A_{\text{stop}}/10} - 1 = \varepsilon_{\text{stop}}^2 \tag{11.6.9}$$

Taking square roots and dividing, we get an equation for N:

$$\left(\frac{\Omega_{\text{stop}}}{\Omega_{\text{pass}}}\right)^N = \frac{\varepsilon_{\text{stop}}}{\varepsilon_{\text{pass}}} = \sqrt{\frac{10^{A_{\text{stop}}/10} - 1}{10^{A_{\text{pass}}/10} - 1}}$$

with exact solution:

$$\boxed{N_{\text{exact}} = \frac{\ln(\varepsilon_{\text{stop}}/\varepsilon_{\text{pass}})}{\ln(\Omega_{\text{stop}}/\Omega_{\text{pass}})} = \frac{\ln(e)}{\ln(w)}} \tag{11.6.10}$$

where we defined the stopband to passband ratios:

$$\boxed{e = \frac{\varepsilon_{\text{stop}}}{\varepsilon_{\text{pass}}} = \sqrt{\frac{10^{A_{\text{stop}}/10} - 1}{10^{A_{\text{pass}}/10} - 1}}, \quad w = \frac{\Omega_{\text{stop}}}{\Omega_{\text{pass}}}} \tag{11.6.11}$$

Since N must be an integer, we choose it to be the *next* integer above N_{exact}, that is,

$$N = \lceil N_{\text{exact}} \rceil \tag{11.6.12}$$

Because N is slightly increased from its exact value, the resulting filter will be slightly better than required. But, because N is different from N_{exact}, we can no longer satisfy simultaneously both of Eqs. (11.6.9). So we choose to satisfy the first one exactly. This determines Ω_0 as follows:

$$\boxed{\Omega_0 = \frac{\Omega_{\text{pass}}}{\left(10^{A_{\text{pass}}/10} - 1\right)^{1/2N}} = \frac{\Omega_{\text{pass}}}{\varepsilon_{\text{pass}}^{1/N}}}$$ (11.6.13)

With these values of N and Ω_0, the stopband specification is more than satisfied, that is, the actual stopband attenuation will be now $A(\Omega_{\text{stop}}) > A_{\text{stop}}$. In summary, given $\{\Omega_{\text{pass}}, \Omega_{\text{stop}}, A_{\text{pass}}, A_{\text{stop}}\}$, we solve Eqs. (11.6.10)–(11.6.13) to get the filter parameters N and Ω_0. We note also that we may rewrite Eq. (11.6.7) in terms of the passband parameters; replacing Ω_0 by Eq. (11.6.13), we have

$$|H(\Omega)|^2 = \frac{1}{1 + \left(\dfrac{\Omega}{\Omega_0}\right)^{2N}} = \frac{1}{1 + \varepsilon_{\text{pass}}^2 \left(\dfrac{\Omega}{\Omega_{\text{pass}}}\right)^{2N}}$$ (11.6.14)

An alternative design can be obtained by matching the stopband specification exactly, resulting in a slightly better passband, that is, $A(\Omega_{\text{stop}}) = A_{\text{stop}}$ and $A(\Omega_{\text{pass}}) < A_{\text{pass}}$. The 3-dB frequency Ω_0 is now computed from the second of Eqs. (11.6.9):

$$\Omega_0 = \frac{\Omega_{\text{stop}}}{\left(10^{A_{\text{stop}}/10} - 1\right)^{1/2N}} = \frac{\Omega_{\text{stop}}}{\varepsilon_{\text{stop}}^{1/N}}$$

In this case, Eq. (11.6.7) can be written in terms of the stopband parameters:

$$|H(\Omega)|^2 = \frac{1}{1 + \varepsilon_{\text{stop}}^2 \left(\dfrac{\Omega}{\Omega_{\text{stop}}}\right)^{2N}} = \frac{\left(\dfrac{\Omega_{\text{stop}}}{\Omega}\right)^{2N}}{\left(\dfrac{\Omega_{\text{stop}}}{\Omega}\right)^{2N} + \varepsilon_{\text{stop}}^2}$$ (11.6.15)

We will see in Section 11.6.6 that the expressions (11.6.14) and (11.6.15) generalize to the Chebyshev type 1 and type 2 filters.

The analog Butterworth transfer function $H(s)$ can be constructed from the knowledge of $\{N, \Omega_0\}$ by the method of *spectral factorization*, as described below. Using $s = j\Omega$ and noting that $(H(\Omega))^* = H^*(-\Omega)$, we may write Eq. (11.6.7) in terms of the variable $s^{\dagger}$

$$H(s)H^*(-s) = \frac{1}{1 + \left(\dfrac{s}{j\Omega_0}\right)^{2N}} = \frac{1}{1 + (-1)^N \left(\dfrac{s}{\Omega_0}\right)^{2N}}$$

Setting $H(s) = \dfrac{1}{D(s)}$, we have

$$\boxed{D(s)D^*(-s) = 1 + (-1)^N \left(\frac{s}{\Omega_0}\right)^{2N}}$$ (11.6.16)

---

†The notation $H^*(-s)$ denotes complex conjugation of the filter coefficients and replacement of s by $-s$, for example, $H^*(-s) = \sum a_n^* (-s)^n$ if $H(s) = \sum a_n s^n$.

Because the right-hand side is a polynomial of degree $2N$ in s, $D(s)$ will be a polynomial of degree N. There exist 2^N *different* polynomials $D(s)$ of degree N satisfying Eq. (11.6.16). But, among them, there is a *unique* one that has *all* its zeros in the *left-hand* s-plane. This is the one we want, because then the transfer function $H(s) = 1/D(s)$ will be *stable and causal*. To find $D(s)$, we first determine all the $2N$ roots of Eq. (11.6.16) and then choose those that lie in the left-hand s-plane. The $2N$ solutions of

$$1 + (-1)^N \left(\frac{s}{\Omega_0} \right)^{2N} = 0 \quad \Rightarrow \quad s^{2N} = (-1)^{N-1} \Omega_0^{2N}$$

are given by

$$\boxed{s_i = \Omega_0 e^{j\theta_i}, \quad \theta_i = \frac{\pi}{2N}(N - 1 + 2i)}, \quad i = 1, 2, \ldots, N, \ldots, 2N \qquad (11.6.17)$$

The index i is chosen such that the first N of the s_i lie in the left-hand s-plane, that is, $\pi/2 < \theta_i < 3\pi/2$ for $i = 1, 2, \ldots, N$. Because $|s_i| = \Omega_0$, all of the zeros lie on a circle of radius Ω_0, called the *Butterworth circle* and shown in Fig. 11.6.2.

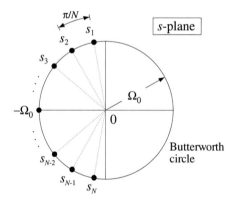

Fig. 11.6.2 Butterworth filter poles lie on Butterworth circle.

It is evident from Eq. (11.6.17) that the s_i can be paired in complex conjugate pairs; that is, $s_N = s_1^*$, $s_{N-1} = s_2^*$, and so on. If N is *even*, say $N = 2K$, then there are exactly K conjugate pairs, namely, $\{s_i, s_i^*\}$, $i = 1, 2, \ldots, K$. In this case, $D(s)$ will factor into second-order sections as follows:

$$D(s) = D_1(s) D_2(s) \cdots D_K(s) \qquad (11.6.18)$$

where

$$D_i(s) = \left(1 - \frac{s}{s_i} \right) \left(1 - \frac{s}{s_i^*} \right), \quad i = 1, 2, \ldots, K \qquad (11.6.19)$$

On the other hand, if N is *odd*, say $N = 2K + 1$, there will be K conjugate pairs and one additional zero that cannot be paired and must necessarily be real-valued.

That zero must lie in the left-hand s-plane and on the Butterworth circle; thus, it must be the point $s = -\Omega_0$. The polynomial $D(s)$ factors now as:

$$D(s) = D_0(s)D_1(s)D_2(s) \cdots D_K(s), \quad \text{where} \quad D_0(s) = \left(1 + \frac{s}{\Omega_0}\right)$$

The remaining factors $D_i(s)$ are the same as in Eq. (11.6.19). They can be rewritten as factors with *real* coefficients as follows. Inserting $s_i = \Omega_0 e^{j\theta_i}$ into Eq. (11.6.19), we find for $i = 1, 2, \ldots, K$:

$$D_i(s) = \left(1 - \frac{s}{s_i}\right)\left(1 - \frac{s}{s_i^*}\right) = 1 - 2\frac{s}{\Omega_0}\cos\theta_i + \frac{s^2}{\Omega_0^2} \tag{11.6.20}$$

Inserting these factors into the Butterworth analog transfer function $H(s) = 1/D(s)$, we can express it as a cascade of second-order sections:

$$\boxed{H(s) = H_0(s)H_1(s)H_2(s) \cdots H_K(s)} \tag{11.6.21}$$

where

$$H_0(s) = \begin{cases} 1, & \text{if } N = 2K \\ \dfrac{1}{1 + \dfrac{s}{\Omega_0}}, & \text{if } N = 2K + 1 \end{cases} \tag{11.6.22}$$

$$H_i(s) = \frac{1}{1 - 2\dfrac{s}{\Omega_0}\cos\theta_i + \dfrac{s^2}{\Omega_0^2}}, \quad i = 1, 2, \ldots, K$$

Example 11.6.1: The Butterworth polynomials $D(s)$ of orders 1–7 and unity 3-dB normalization frequency $\Omega_0 = 1$ are shown in Table 11.6.1. For other values of Ω_0, s must be replaced by s/Ω_0 in each table entry.

The coefficients of s of the second-order sections are the cosine factors, $-2\cos\theta_i$, of Eq. (11.6.20). For example, in the case $N = 7$, we have $K = 3$ and the three θ's are calculated from Eq. (11.6.17):

$$\theta_i = \frac{\pi}{14}(6 + 2i) = \frac{8\pi}{14}, \frac{10\pi}{14}, \frac{12\pi}{14}, \quad \text{for} \quad i = 1, 2, 3$$

$$-2\cos\theta_i = 0.4450, \; 1.2470, \; 1.8019$$

The corresponding Butterworth filters $H(s)$ of orders 1–7 are obtained as the inverses of the table entries. □

Example 11.6.2: Determine the 2^N possible Nth degree polynomials $D(s)$ satisfying Eq. (11.6.16), for the cases $N = 2$ and $N = 3$. Take $\Omega_0 = 1$.

Solution: For $N = 2$, we must find all second-degree polynomials that satisfy Eq. (11.6.16), $D(s)D^*(-s) = 1 + (-1)^2 s^4$. They are:

N	K	$\theta_1, \theta_2, \ldots, \theta_K$	$D(s)$
1	0		$(1 + s)$
2	1	$\dfrac{3\pi}{4}$	$(1 + 1.4142s + s^2)$
3	1	$\dfrac{4\pi}{6}$	$(1 + s)(1 + s + s^2)$
4	2	$\dfrac{5\pi}{8}, \dfrac{7\pi}{8}$	$(1 + 0.7654s + s^2)(1 + 1.8478s + s^2)$
5	2	$\dfrac{6\pi}{10}, \dfrac{8\pi}{10}$	$(1 + s)(1 + 0.6180s + s^2)(1 + 1.6180s + s^2)$
6	3	$\dfrac{7\pi}{12}, \dfrac{9\pi}{12}, \dfrac{11\pi}{12}$	$(1 + 0.5176s + s^2)(1 + 1.4142s + s^2)(1 + 1.9319s + s^2)$
7	3	$\dfrac{8\pi}{14}, \dfrac{10\pi}{14}, \dfrac{12\pi}{14}$	$(1 + s)(1 + 0.4450s + s^2)(1 + 1.2470s + s^2)(1 + 1.8019s + s^2)$

Table 11.6.1 Butterworth polynomials.

$$D(s) = 1 + \sqrt{2}s + s^2 \qquad D^*(-s) = 1 - \sqrt{2}s + s^2$$
$$D(s) = 1 - \sqrt{2}s + s^2 \qquad D^*(-s) = 1 + \sqrt{2}s + s^2$$
$$\Rightarrow$$
$$D(s) = 1 + js^2 \qquad\qquad D^*(-s) = 1 - js^2$$
$$D(s) = 1 - js^2 \qquad\qquad D^*(-s) = 1 + js^2$$

Only the first one has all of its zeros in the left-hand s-plane. Similarly, for $N = 3$ the $2^3 = 8$ different third-degree polynomials $D(s)$ are:

$$D(s) = (1 + s)(1 + s + s^2) \qquad D^*(-s) = (1 - s)(1 - s + s^2)$$
$$D(s) = (1 + s)(1 - s + s^2) \qquad D^*(-s) = (1 - s)(1 + s + s^2)$$
$$D(s) = (1 + s)(1 - s^2 e^{2j\pi/3}) \qquad D^*(-s) = (1 - s)(1 - s^2 e^{-2j\pi/3})$$
$$D(s) = (1 + s)(1 - s^2 e^{-2j\pi/3}) \qquad D^*(-s) = (1 - s)(1 - s^2 e^{2j\pi/3})$$
$$\Rightarrow$$
$$D(s) = (1 - s)(1 - s^2 e^{-2j\pi/3}) \qquad D^*(-s) = (1 + s)(1 - s^2 e^{2j\pi/3})$$
$$D(s) = (1 - s)(1 - s^2 e^{2j\pi/3}) \qquad D^*(-s) = (1 + s)(1 - s^2 e^{-2j\pi/3})$$
$$D(s) = (1 - s)(1 + s + s^2) \qquad D^*(-s) = (1 + s)(1 - s + s^2)$$
$$D(s) = (1 - s)(1 - s + s^2) \qquad D^*(-s) = (1 + s)(1 + s + s^2)$$

They all satisfy $D(s)D^*(-s) = 1 + (-1)^3 s^6$ but, only the first one has its zeros in the left-hand s-plane.

Note also that not all solutions of Eq. (11.6.16) have real coefficients. If we restrict our search to those with real coefficients (pairing the zeros in conjugate pairs), then there are 2^K such polynomials $D(s)$ if $N = 2K$, and 2^{K+1} if $N = 2K + 1$. □

11.6.2 Digital Lowpass Filters

Under the bilinear transformation, the lowpass analog filter will be transformed into a lowpass digital filter. Each analog second-order section will be transformed into a second-order section of the digital filter, as follows:

$$H_i(z) = \left. \frac{1}{1 - 2\dfrac{s}{\Omega_0}\cos\theta_i + \dfrac{s^2}{\Omega_0^2}} \right|_{s = \frac{1-z^{-1}}{1+z^{-1}}} = \frac{G_i(1 + z^{-1})^2}{1 + a_{i1}z^{-1} + a_{i2}z^{-2}} \qquad (11.6.23)$$

where the filter coefficients G_i, a_{i1}, a_{2i} are easily found to be:

$$G_i = \frac{\Omega_0^2}{1 - 2\Omega_0\cos\theta_i + \Omega_0^2}$$

$$a_{i1} = \frac{2(\Omega_0^2 - 1)}{1 - 2\Omega_0\cos\theta_i + \Omega_0^2} \qquad (11.6.24)$$

$$a_{i2} = \frac{1 + 2\Omega_0\cos\theta_i + \Omega_0^2}{1 - 2\Omega_0\cos\theta_i + \Omega_0^2}$$

for $i = 1, 2, \ldots, K$. If N is odd, then there is also a first-order section:

$$H_0(z) = \left. \frac{1}{1 + \dfrac{s}{\Omega_0}} \right|_{s = \frac{1-z^{-1}}{1+z^{-1}}} = \frac{G_0(1 + z^{-1})}{1 + a_{01}z^{-1}} \qquad (11.6.25)$$

where

$$G_0 = \frac{\Omega_0}{\Omega_0 + 1}, \qquad a_{01} = \frac{\Omega_0 - 1}{\Omega_0 + 1} \qquad (11.6.26)$$

If N is even, we may set $H_0(z) = 1$. The overall transfer function of the designed lowpass digital filter is given by:

$$\boxed{H(z) = H_0(z)H_1(z)H_2(z)\cdots H_K(z)} \qquad (11.6.27)$$

with the factors given by Eqs. (11.6.23)–(11.6.26). Note that the 3-dB frequency f_0 in Hz is related to the Butterworth parameter Ω_0 by

$$\Omega_0 = \tan\left(\frac{\omega_0}{2}\right) = \tan\left(\frac{\pi f_0}{f_s}\right) \quad \Rightarrow \quad \boxed{f_0 = \frac{f_s}{\pi}\arctan(\Omega_0)} \qquad (11.6.28)$$

Note that the filter sections have zeros at $z = -1$, that is, the Nyquist frequency $\omega = \pi$. Setting $\Omega = \tan(\omega/2)$, the magnitude response of the designed digital filter can be expressed simply via Eq. (11.6.7), as follows:

$$|H(\omega)|^2 = \frac{1}{1 + (\Omega/\Omega_0)^{2N}} = \frac{1}{1 + (\tan(\omega/2)/\Omega_0)^{2N}} \qquad (11.6.29)$$

Note also that each second-order section has unity gain at zero frequency, $f = 0$, $\omega = 0$, or $z = 1$. Indeed, setting $z = 1$ in Eq. (11.6.23), we obtain the following condition, which can be verified from the definitions (11.6.24):

$$\frac{4G_i}{1 + a_{i1} + a_{i2}} = 1 \quad \text{and} \quad \frac{2G_0}{1 + a_{01}} = 1$$

In summary, the design steps for a lowpass digital filter with given specifications $\{f_{\text{pass}}, f_{\text{stop}}, A_{\text{pass}}, A_{\text{stop}}\}$ are:

1. Calculate the digital frequencies $\{\omega_{\text{pass}}, \omega_{\text{stop}}\}$ and the corresponding pre-warped versions $\{\Omega_{\text{pass}}, \Omega_{\text{stop}}\}$ from Eqs. (11.6.6) and (11.6.5).

2. Calculate the order N and 3-dB frequency Ω_0 of the equivalent lowpass analog Butterworth filter based on the transformed specifications $\{\Omega_{\text{pass}}, \Omega_{\text{stop}}, A_{\text{pass}}, A_{\text{stop}}\}$ by Eqs. (11.6.10)–(11.6.13).

3. The transfer function of the desired lowpass digital filter is then obtained from Eq. (11.6.27), where the SOS coefficients are calculated from Eqs. (11.6.24) and (11.6.26).

Example 11.6.3: Using the bilinear transformation and a lowpass analog Butterworth prototype, design a lowpass digital filter operating at a rate of 20 kHz and having passband extending to 4 kHz with maximum passband attenuation of 0.5 dB, and stopband starting at 5 kHz with a minimum stopband attenuation of 10 dB.

Then, redesign it such that its magnitude response satisfies $1 \geq |H(f)|^2 \geq 0.98$ in the passband, and $|H(f)|^2 \leq 0.02$ in the stopband.

Solution: The digital frequencies in radians per sample are:

$$\omega_{\text{pass}} = \frac{2\pi f_{\text{pass}}}{f_s} = \frac{2\pi \cdot 4}{20} = 0.4\pi, \quad \omega_{\text{stop}} = \frac{2\pi f_{\text{stop}}}{f_s} = \frac{2\pi \cdot 5}{20} = 0.5\pi$$

and their prewarped versions:

$$\Omega_{\text{pass}} = \tan\left(\frac{\omega_{\text{pass}}}{2}\right) = 0.7265, \quad \Omega_{\text{stop}} = \tan\left(\frac{\omega_{\text{stop}}}{2}\right) = 1$$

Eq. (11.6.4) can be used with $A_{\text{pass}} = 0.5$ dB and $A_{\text{stop}} = 10$ dB to calculate the parameters $\{\varepsilon_{\text{pass}}, \varepsilon_{\text{stop}}\}$:

$$\varepsilon_{\text{pass}} = \sqrt{10^{A_{\text{pass}}/10} - 1} = \sqrt{10^{0.5/10} - 1} = 0.3493$$

$$\varepsilon_{\text{stop}} = \sqrt{10^{A_{\text{stop}}/10} - 1} = \sqrt{10^{10/10} - 1} = 3$$

Then, Eq. (11.6.10) gives:

$$N_{\text{exact}} = \frac{\ln(e)}{\ln(w)} = \frac{\ln(\varepsilon_{\text{stop}}/\varepsilon_{\text{pass}})}{\ln(\Omega_{\text{stop}}/\Omega_{\text{pass}})} = \frac{\ln(3/0.3493)}{\ln(1/0.7625)} = 6.73 \quad \Rightarrow \quad N = 7$$

Thus, there is one first-order section $H_0(z)$ and three second-order sections. Eq. (11.6.13) gives for Ω_0 and its value in Hz:

$$\Omega_0 = \frac{\Omega_{\text{pass}}}{\varepsilon_{\text{pass}}^{1/N}} = \frac{0.7265}{(0.3493)^{1/7}} = 0.8443$$

$$f_0 = \frac{f_s}{\pi} \arctan(\Omega_0) = \frac{20}{\pi} \arctan(0.8443) = 4.4640 \text{ kHz}$$

The Butterworth angles $\theta_1, \theta_2, \theta_3$ were calculated in Example 11.6.1. The SOS coefficients are calculated from Eqs. (11.6.26) and (11.6.24):

i	G_i	a_{i1}	a_{i2}
0	0.4578	-0.0844	
1	0.3413	-0.2749	0.6402
2	0.2578	-0.2076	0.2386
3	0.2204	-0.1775	0.0592

resulting in the transfer function:

$$H(z) = H_0(z)H_1(z)H_2(z)H_3(z)$$

$$= \frac{0.4578(1+z^{-1})}{1-0.0844z^{-1}} \cdot \frac{0.3413(1+z^{-1})^2}{1-0.2749z^{-1}+0.6402z^{-2}}$$

$$\cdot \frac{0.2578(1+z^{-1})^2}{1-0.2076z^{-1}+0.2386z^{-2}} \cdot \frac{0.2204(1+z^{-1})^2}{1-0.1775z^{-1}+0.0592z^{-2}}$$

It can be implemented in the time domain by the routines `cas` or `ccas`. The left graph of Fig. 11.6.3 shows the magnitude response squared, $|H(f)|^2$. The brick-wall specifications and the 3-dB line intersecting the response at $f = f_0$ are shown on the graph. The magnitude response was calculated using the simpler formula Eq. (11.6.29), with ω expressed in kHz, $\omega = 2\pi f/f_s$:

$$|H(f)|^2 = \frac{1}{1+(\tan(\pi f/f_s)/\Omega_0)^{2N}} = \frac{1}{1+(\tan(\pi f/20)/0.8443)^{14}}$$

The passband attenuation in absolute units is $10^{-0.5/10} = 0.89125$ and the stopband attenuation $10^{-10/10} = 0.1$. Note that the actual stopband attenuation at $f = f_{\text{stop}} = 5$ kHz is slightly better than required, that is, $A(f_{\text{stop}}) = 10.68$ dB.

The second filter has more stringent specifications. The desired passband attenuation is $A_{\text{pass}} = -10\log_{10}(0.98) = 0.0877$ dB, and the stopband attenuation $A_{\text{stop}} = -10\log_{10}(0.02) = 16.9897$ dB. With these values, we find the design parameters $\{\varepsilon_{\text{pass}}, \varepsilon_{\text{stop}}\} = \{0.1429, 7\}$ and:

$$N_{\text{exact}} = 12.18, \quad N = 13, \quad \Omega_0 = 0.8439, \quad f_0 = 4.4622 \text{ kHz}$$

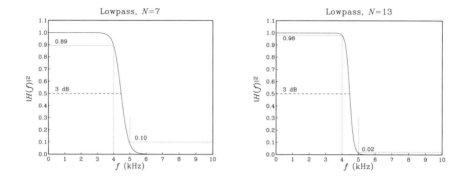

Fig. 11.6.3 Digital lowpass Butterworth filters.

The digital filter will have one first-order and six second-order sections. The SOS coefficients were calculated with the MATLAB function `lhbutt.m` of Appendix D:

i	G_i	a_{i1}	a_{i2}
0	0.4577	−0.0847	
1	0.3717	−0.3006	0.7876
2	0.3082	−0.2492	0.4820
3	0.2666	−0.2156	0.2821
4	0.2393	−0.1935	0.1508
5	0.2221	−0.1796	0.0679
6	0.2125	−0.1718	0.0219

Its magnitude response is shown in the right graph of Fig. 11.6.3. As is always the case, making the specifications more stringent results in higher order N.   □

11.6.3 Digital Highpass Filters

There are two possible approaches one can follow to design a highpass digital filter with the bilinear transformation: One is to use the transformation (11.1.3) to map the given specifications onto the specifications of an equivalent *highpass* analog filter. The other is to use the *highpass* version of the bilinear transformation given in Eq. (11.1.5) to map the given highpass specifications onto equivalent analog *lowpass* specifications.

The first approach was used in the design of the first-order highpass filters of Section 11.2. Here, we will follow the second method, which is more convenient for high-order designs because we can use the lowpass Butterworth design we developed already. The mapping of the highpass specifications to the equivalent analog lowpass ones is depicted in Fig. 11.6.4.

The mapping is accomplished by the highpass version of the bilinear transformation, given in Eqs. (11.1.5) and (11.1.6):

$$s = \frac{1 + z^{-1}}{1 - z^{-1}}, \qquad \Omega = -\cot\left(\frac{\omega}{2}\right), \qquad \omega = \frac{2\pi f}{f_s} \qquad (11.6.30)$$

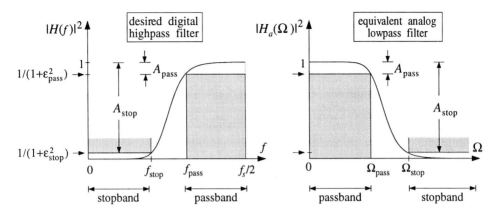

Fig. 11.6.4 Highpass digital filter and its analog lowpass equivalent.

It maps the point $z = -1$ to $s = 0$, or equivalently, the center of the passband of the highpass filter at $\omega = \pi$ to the center of the passband of the lowpass filter at $\Omega = 0$. The prewarped versions of the passband and stopband frequencies are computed as follows:

$$\Omega_{\text{pass}} = \cot\left(\frac{\omega_{\text{pass}}}{2}\right) = \cot\left(\frac{\pi f_{\text{pass}}}{f_s}\right)$$

$$\Omega_{\text{stop}} = \cot\left(\frac{\omega_{\text{stop}}}{2}\right) = \cot\left(\frac{\pi f_{\text{stop}}}{f_s}\right)$$

(11.6.31)

According to Eq. (11.6.30), we should have used $\Omega_{\text{pass}} = -\cot(\omega_{\text{pass}}/2)$. However, as far as the determination of the parameters N and Ω_0 is concerned, it does not matter whether we use positive or negative signs because we are working only with the magnitude response of the analog filter, which is even as a function of Ω.

Using Eqs. (11.6.10)-(11.6.13), we determine the parameters N and Ω_0, and the corresponding analog filter sections given by Eq. (11.6.21). Under the highpass bilinear transformation of Eq. (11.6.30), each SOS of the analog filter will be transformed into an SOS of the digital filter, as follows:

$$H_i(z) = \left.\frac{1}{1 - 2\dfrac{s}{\Omega_0}\cos\theta_i + \dfrac{s^2}{\Omega_0^2}}\right|_{s=\frac{1+z^{-1}}{1-z^{-1}}} = \frac{G_i(1 - z^{-1})^2}{1 + a_{i1}z^{-1} + a_{i2}z^{-2}}$$

(11.6.32)

where the filter coefficients G_i, a_{i1}, a_{i2} are easily found to be

$$G_i = \frac{\Omega_0^2}{1 - 2\Omega_0 \cos\theta_i + \Omega_0{}^2}$$

$$a_{i1} = -\frac{2(\Omega_0^2 - 1)}{1 - 2\Omega_0 \cos\theta_i + \Omega_0{}^2} \tag{11.6.33}$$

$$a_{i2} = \frac{1 + 2\Omega_0 \cos\theta_i + \Omega_0^2}{1 - 2\Omega_0 \cos\theta_i + \Omega_0{}^2}$$

for $i = 1, 2, \ldots, K$. If N is odd, then there is also a first-order section given by

$$H_0(z) = \left. \frac{1}{1 + \dfrac{s}{\Omega_0}} \right|_{s=\frac{1+z^{-1}}{1-z^{-1}}} = \frac{G_0(1 - z^{-1})}{1 + a_{01}z^{-1}} \tag{11.6.34}$$

where

$$G_0 = \frac{\Omega_0}{\Omega_0 + 1}, \qquad a_{01} = -\frac{\Omega_0 - 1}{\Omega_0 + 1} \tag{11.6.35}$$

If N is even, we may set $H_0(z) = 1$. The overall transfer function of the designed highpass digital filter will be given by

$$\boxed{H(z) = H_0(z)H_1(z)H_2(z)\cdots H_K(z)} \tag{11.6.36}$$

with the factors given by Eqs. (11.6.32–11.6.35). The 3-dB frequency f_0 of the designed filter may be calculated from:

$$\Omega_0 = \cot\left(\frac{\omega_0}{2}\right) = \cot\left(\frac{\pi f_0}{f_s}\right) \quad \Rightarrow \quad \boxed{f_0 = \frac{f_s}{\pi} \arctan\left(\frac{1}{\Omega_0}\right)}$$

and the magnitude response from:

$$|H(\omega)|^2 = \frac{1}{1 + \left(\cot(\omega/2)/\Omega_0\right)^{2N}}$$

Note the similarities and differences between the highpass and lowpass cases: The coefficients G_i and a_{i2} are the same, but a_{i1} has reverse sign. Also, the numerator of the SOS is now $(1 - z^{-1})^2$ instead of $(1 + z^{-1})^2$, resulting in a zero at $z = 1$ or $\omega = 0$. These changes are easily understood by noting that the lowpass bilinear transformation (11.1.3) becomes the highpass one given by Eq. (11.6.30) under the substitution $z \to -z$.

Example 11.6.4: Using the bilinear transformation and a lowpass analog Butterworth prototype, design a highpass digital filter operating at a rate of 20 kHz and having passband starting at 5 kHz with maximum passband attenuation of 0.5 dB, and stopband ending at 4 kHz with a minimum stopband attenuation of 10 dB.

Then, redesign it such that its magnitude response satisfies $1 \geq |H(f)|^2 \geq 0.98$ in the passband, and $|H(f)|^2 \leq 0.02$ in the stopband.

Solution: The digital frequencies and their prewarped versions are:

$$\omega_{pass} = \frac{2\pi f_{pass}}{f_s} = \frac{2\pi \cdot 5}{20} = 0.5\pi, \qquad \Omega_{pass} = \cot\left(\frac{\omega_{pass}}{2}\right) = 1$$

$$\Rightarrow$$

$$\omega_{stop} = \frac{2\pi f_{stop}}{f_s} = \frac{2\pi \cdot 4}{20} = 0.4\pi, \qquad \Omega_{stop} = \cot\left(\frac{\omega_{stop}}{2}\right) = 1.3764$$

The dB attenuations $\{A_{pass}, A_{stop}\} = \{0.5, 10\}$ correspond to $\{\varepsilon_{pass}, \varepsilon_{stop}\} = \{0.3493, 3\}$. Then, Eq. (11.6.10) can be solved for the filter order:

$$N_{exact} = \frac{\ln(\varepsilon_{stop}/\varepsilon_{pass})}{\ln(\Omega_{stop}/\Omega_{pass})} = \frac{\ln(3/0.3493)}{\ln(1.3764/1)} = 6.73 \quad \Rightarrow \quad N = 7$$

Thus, there is one first-order section $H_0(z)$ and three second-order sections. Eq. (11.6.13) gives for Ω_0:

$$\Omega_0 = \frac{\Omega_{pass}}{\left(10^{A_{pass}/10} - 1\right)^{1/2N}} = \frac{\Omega_{pass}}{\varepsilon_{pass}^{1/N}} = \frac{1}{(0.3493)^{1/7}} = 1.1621$$

The SOS coefficients are calculated from Eqs. (11.6.26) and (11.6.24):

i	G_i	a_{i1}	a_{i2}
0	0.5375	−0.0750	
1	0.4709	−0.2445	0.6393
2	0.3554	−0.1845	0.2372
3	0.3039	−0.1577	0.0577

resulting in the transfer function:

$$H(z) = H_0(z)H_1(z)H_2(z)H_3(z)$$

$$= \frac{0.5375(1 - z^{-1})}{1 - 0.0750z^{-1}} \cdot \frac{0.4709(1 - z^{-1})^2}{1 - 0.2445z^{-1} + 0.6393z^{-2}}$$

$$\cdot \frac{0.3554(1 - z^{-1})^2}{1 - 0.1845z^{-1} + 0.2372z^{-2}} \cdot \frac{0.3039(1 - z^{-1})^2}{1 - 0.1577z^{-1} + 0.0577z^{-2}}$$

As in Example 11.6.3, the second filter has passband and stopband attenuations: $A_{pass} = -10\log_{10}(0.98) = 0.0877$ dB and $A_{stop} = -10\log_{10}(0.02) = 16.9897$ dB. With these values, we find the design parameters $\{\varepsilon_{pass}, \varepsilon_{stop}\} = \{0.1429, 7\}$ and:

$$N_{exact} = 12.18, \quad N = 13, \quad \Omega_0 = 1.1615, \quad f_0 = 4.5253 \text{ kHz}$$

The coefficients of the first- and second-order sections are:

i	G_i	a_{i1}	a_{i2}
0	0.5374	−0.0747	
1	0.5131	−0.2655	0.7870
2	0.4252	−0.2200	0.4807
3	0.3677	−0.1903	0.2806
4	0.3300	−0.1708	0.1493
5	0.3062	−0.1584	0.0663
6	0.2930	−0.1516	0.0203

The magnitude responses of the two designs are shown in Fig. 11.6.5. □

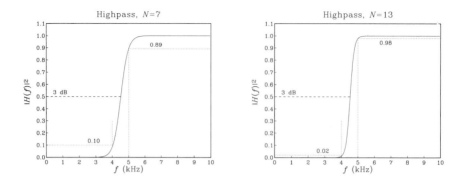

Fig. 11.6.5 Digital highpass Butterworth filters.

11.6.4 Digital Bandpass Filters

As in the highpass case, we can follow two possible approaches to the design of a digital bandpass filter. We can map the digital bandpass filter onto an equivalent analog bandpass filter using the transformation (11.1.3). Alternatively, we can use the bandpass version of the transformation (11.1.5) to map the bandpass digital filter onto an equivalent *lowpass* analog filter.

The first method was used in Sections 11.3 to design bandpass peaking filters. The second method is, however, more convenient because it reduces the bandpass design problem to a standard lowpass analog design problem. Figure 11.6.6 shows the bandpass specifications and their analog equivalents.

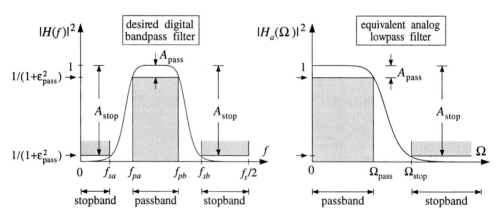

Fig. 11.6.6 Bandpass digital filter and its analog lowpass equivalent.

The specifications are the quantities $\{f_{pa}, f_{pb}, f_{sa}, f_{sb}, A_{\text{pass}}, A_{\text{stop}}\}$, defining the passband range $f_{pa} \leq f \leq f_{pb}$, the left stopband $0 \leq f \leq f_{sa}$, and the right stopband $f_{sb} \leq f \leq f_s/2$. The stopband attenuations were assumed to be equal in the two stopbands; if they are not, we may design the filter based on the maximum of the two.

The bandpass version of the bilinear[†] transformation and the corresponding frequency mapping are in this case:

$$s = \frac{1 - 2cz^{-1} + z^{-2}}{1 - z^{-2}}, \qquad \Omega = \frac{c - \cos \omega}{\sin \omega}, \qquad \omega = \frac{2\pi f}{f_s} \qquad (11.6.37)$$

A new parameter c has been introduced. Note that $c = 1$ recovers the lowpass case, and $c = -1$ the highpass one. The parameter c is required to be $|c| \le 1$ in order to map the left-hand s-plane into the inside of the unit circle in the z-plane.

Therefore, we may set $c = \cos \omega_c$, for some value of ω_c. The center of the analog passband $\Omega = 0$ corresponds to $\cos \omega = c = \cos \omega_c$, or, $\omega = \omega_c$. Therefore, ω_c may be thought of as the "center" frequency of the bandpass filter (although it need not be exactly at the center of the passband).

The given bandpass specifications, must be mapped onto the specifications of the equivalent analog filter, $\{\Omega_{\text{pass}}, \Omega_{\text{stop}}, A_{\text{pass}}, A_{\text{stop}}\}$. This can be done as follows. We require that the passband $[f_{pa}, f_{pb}]$ of the digital filter be mapped onto the entire passband $[-\Omega_{\text{pass}}, \Omega_{\text{pass}}]$ of the analog filter. This requires that:

$$-\Omega_{\text{pass}} = \frac{c - \cos \omega_{pa}}{\sin \omega_{pa}}$$

$$\Omega_{\text{pass}} = \frac{c - \cos \omega_{pb}}{\sin \omega_{pb}}$$

where $\omega_{pa} = 2\pi f_{pa}/f_s$ and $\omega_{pb} = 2\pi f_{pb}/f_s$. By adding them, we solve for c. Then, inserting the computed value of c into one or the other we find Ω_{pass}. The resulting solution is:

$$c = \frac{\sin(\omega_{pa} + \omega_{pb})}{\sin \omega_{pa} + \sin \omega_{pb}}, \qquad \Omega_{\text{pass}} = \left| \frac{c - \cos \omega_{pb}}{\sin \omega_{pb}} \right| \qquad (11.6.38)$$

Note that for ω_{pa}, ω_{pb} in the interval $[0, \pi]$, the above expression for c implies $|c| \le 1$, as required for stability. Next, we compute the two numbers:

$$\Omega_{sa} = \frac{c - \cos \omega_{sa}}{\sin \omega_{sa}}, \qquad \Omega_{sb} = \frac{c - \cos \omega_{sb}}{\sin \omega_{sb}}$$

where $\omega_{sa} = 2\pi f_{sa}/f_s$ and $\omega_{sb} = 2\pi f_{sb}/f_s$.

Ideally, the stopband of the digital filter should map exactly onto the stopband of the analog filter so that we should have $\Omega_{sb} = \Omega_{\text{stop}}$ and $\Omega_{sa} = -\Omega_{\text{stop}}$. But this is impossible because c has already been determined from Eq. (11.6.38).

Because the Butterworth magnitude response is a *monotonically decreasing* function of Ω, it is enough to choose the smallest of the two stopbands defined above. Thus, we define:

$$\Omega_{\text{stop}} = \min(|\Omega_{sa}|, |\Omega_{sb}|) \qquad (11.6.39)$$

With the computed values of Ω_{pass} and Ω_{stop}, we proceed to compute the Butterworth parameters N and Ω_0 and the corresponding SOSs of Eq. (11.6.21). Because s

---

[†]It should really be called "biquadratic" in this case.

is quadratic in z, the substitution of s into these SOSs will give rise to *fourth-order* sections in z:

$$H_i(z) = \left.\frac{1}{1 - 2\dfrac{s}{\Omega_0}\cos\theta_i + \dfrac{s^2}{\Omega_0^2}}\right|_{s = \frac{1 - 2cz^{-1} + z^{-2}}{1 - z^{-2}}} \tag{11.6.40}$$

$$= \frac{G_i(1 - z^{-2})^2}{1 + a_{i1}z^{-1} + a_{i2}z^{-2} + a_{i3}z^{-3} + a_{i4}z^{-4}}$$

where, for $i = 1, 2, \ldots, K$:

$$G_i = \frac{\Omega_0^2}{1 - 2\Omega_0\cos\theta_i + \Omega_0^2}$$

$$a_{i1} = \frac{4c(\Omega_0\cos\theta_i - 1)}{1 - 2\Omega_0\cos\theta_i + \Omega_0^2} \qquad a_{i2} = \frac{2(2c^2 + 1 - \Omega_0^2)}{1 - 2\Omega_0\cos\theta_i + \Omega_0^2}$$

$$a_{i3} = -\frac{4c(\Omega_0\cos\theta_i + 1)}{1 - 2\Omega_0\cos\theta_i + \Omega_0^2} \qquad a_{i4} = \frac{1 + 2\Omega_0\cos\theta_i + \Omega_0^2}{1 - 2\Omega_0\cos\theta_i + \Omega_0^2} \tag{11.6.41}$$

If N is odd, then there is also a first-order section in s which becomes a second-order section in z:

$$H_0(z) = \left.\frac{1}{1 + \dfrac{s}{\Omega_0}}\right|_{s = \frac{1 - 2cz^{-1} + z^{-2}}{1 - z^{-2}}} = \frac{G_0(1 - z^{-2})}{1 + a_{01}z^{-1} + a_{02}z^{-2}} \tag{11.6.42}$$

where

$$G_0 = \frac{\Omega_0}{1 + \Omega_0}, \qquad a_{01} = -\frac{2c}{1 + \Omega_0}, \qquad a_{02} = \frac{1 - \Omega_0}{1 + \Omega_0} \tag{11.6.43}$$

The overall transfer function of the designed bandpass digital filter will be given as the cascade of fourth-order sections with the possibility of one SOS:

$$H(z) = H_0(z)H_1(z)H_2(z)\cdots H_K(z)$$

The order of the digital filter is $2N$, because s is quadratic in z. The filter sections have zeros at $z = \pm 1$, that is, $\omega = 0$ and $\omega = \pi$. The left and right 3-dB frequencies can be calculated from the equations:

$$\frac{c - \cos\omega_0}{\sin\omega_0} = \mp\Omega_0$$

They can be solved by writing $\cos\omega_0$ and $\sin\omega_0$ in terms of $\tan(\omega_0/2)$, solving the resulting quadratic equation, and picking the positive solutions:

$$\tan\left(\frac{\omega_{0a}}{2}\right) = \tan\left(\frac{\pi f_{0a}}{f_s}\right) = \frac{\sqrt{\Omega_0^2 + 1 - c^2} - \Omega_0}{1 + c}$$

$$\tan\left(\frac{\omega_{0b}}{2}\right) = \tan\left(\frac{\pi f_{0b}}{f_s}\right) = \frac{\sqrt{\Omega_0^2 + 1 - c^2} + \Omega_0}{1 + c}$$

(11.6.44)

Example 11.6.5: Using the bilinear transformation and a lowpass analog Butterworth prototype, design a bandpass digital filter operating at a rate of 20 kHz and having left and right passband frequencies of 2 and 4 kHz, and left and right stopband frequencies of 1.5 and 4.5 kHz. The maximum passband attenuation is required to be 0.5 dB, and the minimum stopband attenuation 10 dB.

Then, redesign it such that its magnitude response satisfies $1 \geq |H(f)|^2 \geq 0.98$ in the passband, and $|H(f)|^2 \leq 0.02$ in the stopbands.

Solution: The digital passband frequencies are:

$$\omega_{pa} = \frac{2\pi f_{pa}}{f_s} = \frac{2\pi \cdot 2}{20} = 0.2\pi, \quad \omega_{pb} = \frac{2\pi f_{pb}}{f_s} = \frac{2\pi \cdot 4}{20} = 0.4\pi$$

Then, we calculate c and Ω_{pass}:

$$c = \frac{\sin(\omega_{pa} + \omega_{pb})}{\sin\omega_{pa} + \sin\omega_{pb}} = 0.6180, \quad \Omega_{\text{pass}} = \left|\frac{c - \cos\omega_{pb}}{\sin\omega_{pb}}\right| = 0.3249$$

With the stopband digital frequencies:

$$\omega_{sa} = \frac{2\pi f_{sa}}{f_s} = \frac{2\pi \cdot 1.5}{20} = 0.15\pi, \quad \omega_{sb} = \frac{2\pi f_{sb}}{f_s} = \frac{2\pi \cdot 4.5}{20} = 0.45\pi$$

we calculate:

$$\Omega_{sa} = \frac{c - \cos\omega_{sa}}{\sin\omega_{sa}} = -0.6013, \quad \Omega_{sb} = \frac{c - \cos\omega_{sb}}{\sin\omega_{sb}} = 0.4674$$

and $\Omega_{\text{stop}} = \min(|\Omega_{sa}|, |\Omega_{sb}|) = 0.4674$. The analog filter with the specifications $\{\Omega_{\text{pass}}, \Omega_{\text{stop}}, A_{\text{pass}}, A_{\text{stop}}\} = \{0.3493, 3\}$ and:

$$N_{\text{exact}} = 5.92, \quad N = 6, \quad \Omega_0 = 0.3872$$

The left-right 3-dB frequencies are calculated from Eq. (11.6.44) to be: $f_{0a} = 1.8689$ kHz, $f_{0b} = 4.2206$ kHz. The coefficients of the three fourth-order sections of the digital filter are (computed by the MATLAB function `bpsbutt.m`):

i	G_i	a_{i1}	a_{i2}	a_{i3}	a_{i4}
1	0.1110	-2.0142	2.3906	-1.6473	0.7032
2	0.0883	-1.8551	1.9017	-1.0577	0.3549
3	0.0790	-1.7897	1.7009	-0.8154	0.2118

The magnitude response can be calculated from:

$$|H(\omega)|^2 = \frac{1}{1 + \left(\dfrac{\Omega}{\Omega_0}\right)^{2N}} = \frac{1}{1 + \left(\dfrac{c - \cos\omega}{\Omega_0 \sin\omega}\right)^{2N}}$$

The magnitude response is shown in the left graph of Fig. 11.6.7. The passband specifications are met exactly by design. Because the maximum stopband frequency was on the right, $\Omega_{\text{stop}} = |\Omega_{sb}|$, the right stopband specification is met stringently. The left stopband specification is more than required.

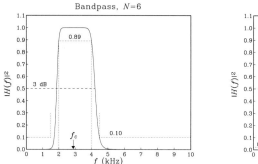
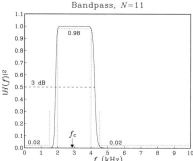

Fig. 11.6.7 Digital bandpass Butterworth filters.

For the second set of specifications, we have $A_{\text{pass}} = -10\log_{10}(0.98) = 0.0877$ dB, and $A_{\text{stop}} = -10\log_{10}(0.02) = 16.9897$ dB and $\{\varepsilon_{\text{pass}}, \varepsilon_{\text{stop}}\} = \{0.1429, 7\}$. The design has the same c, Ω_{pass}, and Ω_{stop}, which lead to the Butterworth parameters:

$$N_{\text{exact}} = 10.71, \quad N = 11, \quad \Omega_0 = 0.3878$$

The left and right 3-dB frequencies are now $f_{0a} = 1.8677$ kHz, $f_{0b} = 4.2228$ kHz. The digital filter coefficients of the second- and fourth-order sections are:

i	G_i	a_{i1}	a_{i2}	a_{i3}	a_{i4}
0	0.2794	−0.8907	0.4411		
1	0.1193	−2.0690	2.5596	−1.8526	0.8249
2	0.1021	−1.9492	2.1915	−1.4083	0.5624
3	0.0907	−1.8694	1.9460	−1.1122	0.3874
4	0.0834	−1.8186	1.7900	−0.9239	0.2762
5	0.0794	−1.7904	1.7033	−0.8193	0.2144

Again, the right stopband specification is more stringently met than the left one. The "center" frequency of the passband is the same for both filters and can be obtained by inverting $\cos\omega_c = c$. In Hz, we have $f_c = f_s \arccos(c)/(2\pi) = 2.8793$ kHz. The magnitude response is normalized to unity at f_c.   □

11.6.5 Digital Bandstop Filters

The specifications of a *bandstop* digital filter are shown in Fig. 11.6.8, together with their analog equivalents. There are now two passbands, that is, $0 \leq f \leq f_{pa}$ and $f_{pb} \leq f \leq f_s/2$, and one stopband $f_{sa} \leq f \leq f_{sb}$.

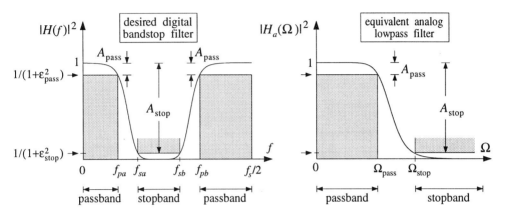

Fig. 11.6.8 Bandstop digital filter and its analog lowpass equivalent.

The bandstop version of the bilinear transformation and the corresponding frequency mapping were given in Eqs. (11.1.5) and (11.1.6):

$$S = \frac{1 - z^{-2}}{1 - 2cz^{-1} + z^{-2}}, \qquad \Omega = \frac{\sin \omega}{\cos \omega - c}, \qquad \omega = \frac{2\pi f}{f_s} \tag{11.6.45}$$

The design steps are summarized as follows. First, we compute the digital frequencies in radians per sample:

$$\omega_{pa} = \frac{2\pi f_{pa}}{f_s}, \qquad \omega_{pb} = \frac{2\pi f_{pb}}{f_s}, \qquad \omega_{sa} = \frac{2\pi f_{sa}}{f_s}, \qquad \omega_{sb} = \frac{2\pi f_{sb}}{f_s}$$

Then, we calculate c and Ω_{pass} by requiring:

$$-\Omega_{\text{pass}} = \frac{\sin \omega_{pa}}{\cos \omega_{pa} - c}, \qquad \Omega_{\text{pass}} = \frac{\sin \omega_{pb}}{\cos \omega_{pb} - c}$$

which may be solved as follows:

$$c = \frac{\sin(\omega_{pa} + \omega_{pb})}{\sin \omega_{pa} + \sin \omega_{pb}}, \qquad \Omega_{\text{pass}} = \left| \frac{\sin \omega_{pb}}{\cos \omega_{pb} - c} \right|$$

Next, we compute the two possible stopbands:

$$\Omega_{sa} = \frac{\sin \omega_{sa}}{\cos \omega_{sa} - c}, \qquad \Omega_{sb} = \frac{\sin \omega_{sb}}{\cos \omega_{sb} - c}$$

and define:

$$\Omega_{\text{stop}} = \min\left(|\Omega_{sa}|, |\Omega_{sb}|\right)$$

Then, use the analog specifications $\{\Omega_{\text{pass}}, \Omega_{\text{stop}}, A_{\text{pass}}, A_{\text{stop}}\}$ to compute the Butterworth parameters $\{N, \Omega_0\}$. And finally, transform the analog filter sections into fourth-order sections by Eq. (11.6.45):

$$H_i(z) = \left. \frac{1}{1 - 2\dfrac{s}{\Omega_0}\cos\theta_i + \dfrac{s^2}{\Omega_0^2}} \right|_{s = \frac{1-z^{-2}}{1-2cz^{-1}+z^{-2}}}$$

$$= \frac{G_i(1 - 2cz^{-1} + z^{-2})^2}{1 + a_{i1}z^{-1} + a_{i2}z^{-2} + a_{i3}z^{-3} + a_{i4}z^{-4}}$$

where the coefficients are given for $i = 1, 2, \ldots, K$:

$$G_i = \frac{\Omega_0^2}{1 - 2\Omega_0\cos\theta_i + \Omega_0{}^2}$$

$$a_{i1} = \frac{4c\Omega_0(\cos\theta_i - \Omega_0)}{1 - 2\Omega_0\cos\theta_i + \Omega_0{}^2} \qquad a_{i2} = \frac{2(2c^2\Omega_0^2 + \Omega_0^2 - 1)}{1 - 2\Omega_0\cos\theta_i + \Omega_0{}^2}$$

$$a_{i3} = -\frac{4c\Omega_0(\cos\theta_i + \Omega_0)}{1 - 2\Omega_0\cos\theta_i + \Omega_0{}^2} \qquad a_{i4} = \frac{1 + 2\Omega_0\cos\theta_i + \Omega_0^2}{1 - 2\Omega_0\cos\theta_i + \Omega_0{}^2}$$

If N is odd, we also have a second-order section in z:

$$H_0(z) = \left. \frac{1}{1 + \dfrac{s}{\Omega_0}} \right|_{s = \frac{1-z^{-2}}{1-2cz^{-1}+z^{-2}}} = \frac{G_0(1 - 2cz^{-1} + z^{-2})}{1 + a_{01}z^{-1} + a_{02}z^{-2}}$$

where

$$G_0 = \frac{\Omega_0}{1 + \Omega_0}, \qquad a_{01} = -\frac{2c\Omega_0}{1 + \Omega_0}, \qquad a_{02} = -\frac{1 - \Omega_0}{1 + \Omega_0}$$

Note that each section has zeros at $1 - 2cz^{-1} + z^{-2} = 0$, which correspond to the angles $\omega = \pm\omega_c$, where $\cos\omega_c = c$. The 3-dB frequencies at the edges of the passbands can be determined by solving for the positive solutions of the equations:

$$\frac{\sin\omega_0}{\cos\omega_0 - c} = \pm\Omega_0$$

which give:

$$\tan\left(\frac{\omega_{0a}}{2}\right) = \tan\left(\frac{\pi f_{0a}}{f_s}\right) = \frac{\sqrt{1 + \Omega_0^2(1 - c^2)} - 1}{\Omega_0(1 + c)}$$

$$\tan\left(\frac{\omega_{0b}}{2}\right) = \tan\left(\frac{\pi f_{0b}}{f_s}\right) = \frac{\sqrt{1 + \Omega_0^2(1 - c^2)} + 1}{\Omega_0(1 + c)}$$

Example 11.6.6: Using the bilinear transformation and a lowpass analog Butterworth prototype, design a bandstop digital filter operating at a rate of 20 kHz and having left and right passband frequencies of 1.5 and 4.5 kHz, and left and right stopband frequencies of 2 and 4 kHz. The maximum passband attenuation is required to be 0.5 dB, and the minimum stopband attenuation 10 dB.

Then, redesign it such that its magnitude response satisfies $1 \geq |H(f)|^2 \geq 0.98$ in the passbands, and $|H(f)|^2 \leq 0.02$ in the stopband.

Solution: The digital passband and stopband frequencies are:

$$\omega_{pa} = \frac{2\pi f_{pa}}{f_s} = \frac{2\pi \cdot 1.5}{20} = 0.15\pi, \quad \omega_{pb} = \frac{2\pi f_{pb}}{f_s} = \frac{2\pi \cdot 4.5}{20} = 0.45\pi$$

$$\omega_{sa} = \frac{2\pi f_{sa}}{f_s} = \frac{2\pi \cdot 2}{20} = 0.2\pi, \quad \omega_{sb} = \frac{2\pi f_{sb}}{f_s} = \frac{2\pi \cdot 4}{20} = 0.4\pi$$

Then, we calculate c and Ω_{pass}:

$$c = \frac{\sin(\omega_{pa} + \omega_{pb})}{\sin \omega_{pa} + \sin \omega_{pb}} = 0.6597, \quad \Omega_{\text{pass}} = \left| \frac{\sin \omega_{pb}}{\cos \omega_{pb} - c} \right| = 1.9626$$

Then, we calculate the stopband frequencies:

$$\Omega_{sa} = \frac{\sin \omega_{sa}}{\cos \omega_{sa} - c} = 3.9361, \quad \Omega_{sb} = \frac{\sin \omega_{sb}}{\cos \omega_{sb} - c} = -2.7121$$

and define $\Omega_{\text{stop}} = \min(|\Omega_{sa}|, |\Omega_{sb}|) = 2.7121$. The analog filter parameters are:

$$N_{\text{exact}} = 6.65, \quad N = 7, \quad \Omega_0 = 2.2808$$

The left-right 3-dB frequencies are calculated to be $f_{0a} = 1.6198$ kHz, $f_{0b} = 4.2503$ kHz. The coefficients of the SOS and the three fourth-order sections of the digital filter are:

i	G_i	a_{i1}	a_{i2}	a_{i3}	a_{i4}
0	0.6952	−0.9172	−0.3904		
1	0.7208	−2.0876	2.4192	−1.7164	0.7187
2	0.5751	−1.9322	1.9301	−1.1026	0.3712
3	0.5045	−1.8570	1.6932	−0.8053	0.2029

The magnitude response of the designed filter is shown in the left graph of Fig. 11.6.9.

For the second set of specifications, we have $A_{\text{pass}} = -10 \log_{10}(0.98) = 0.0877$ dB, and $A_{\text{stop}} = -10 \log_{10}(0.02) = 16.9897$ dB. The design has the same c, Ω_{pass}, and Ω_{stop}, which lead to the Butterworth parameters:

$$N_{\text{exact}} = 12.03, \quad N = 13, \quad \Omega_0 = 2.2795$$

The left-right 3-dB frequencies are now $f_{0a} = 1.6194$ kHz, $f_{0b} = 4.2512$ kHz. The digital filter coefficients of the second- and fourth-order sections are:

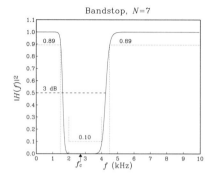 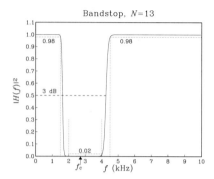

Fig. 11.6.9 Digital bandstop Butterworth filters.

i	G_i	a_{i1}	a_{i2}	a_{i3}	a_{i4}
0	0.6951	−0.9171	−0.3902		
1	0.7703	−2.1401	2.5850	−1.9251	0.8371
2	0.6651	−2.0280	2.2319	−1.4820	0.5862
3	0.5914	−1.9495	1.9847	−1.1717	0.4105
4	0.5408	−1.8956	1.8148	−0.9584	0.2897
5	0.5078	−1.8604	1.7041	−0.8194	0.2110
6	0.4892	−1.8406	1.6415	−0.7410	0.1666

The magnitude response is shown on the right graph of Fig. 11.6.9. The rounding of the exact N of 12.03 to 13 is perhaps overkill in this case. It causes the actual stopband attenuation at the right edge of the stopband to be $A(\Omega_{\text{stop}}) = 19.67$ dB, corresponding to a magnitude square of 0.011 instead of the required 0.02.

For both designs, the "center" notch frequency of the stopband can be obtained by inverting $\cos \omega_c = c$. In Hz, we have $f_c = f_s \arccos(c) / (2\pi) = 2.7069$ kHz. □

11.6.6 Chebyshev Filter Design*

In designing the equivalent analog lowpass filters one can use alternative filter prototypes, such as Chebyshev or elliptic filters [269–271]. For the same set of specifications, they provide steeper transition widths and lead to smaller filter orders than the Butterworth filters.

Chebyshev filters come in two varieties. Type 1 has equiripple passband and monotonic stopband, and type 2, also known as inverse Chebyshev, has equiripple stopband and monotonic passband. A typical Chebyshev magnitude response is shown in Fig. 11.6.10.

It is the equiripple property that is responsible for the narrower transition widths of these filters. For example, for the type 1 case, because the passband response is allowed to go slightly up near the edge of the passband, it can fall off more steeply.

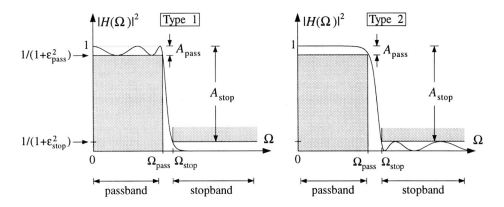

Fig. 11.6.10 Magnitude square responses of type 1 and type 2 Chebyshev filters.

The specifications of the filter are $\{\Omega_{\text{pass}}, \Omega_{\text{stop}}, A_{\text{pass}}, A_{\text{stop}}\}$ and are obtained by prewarping the desired digital filter specifications using the appropriate bilinear transformation (lowpass, highpass, bandpass, or bandstop). Two important design parameters are the quantities $\{\varepsilon_{\text{pass}}, \varepsilon_{\text{stop}}\}$ that were defined in Eq. (11.6.4), and are shown in Figs. 11.6.1 and 11.6.10.

The magnitude response squared of an Nth order Chebyshev filter is expressible in terms of these parameters as follows. For the type 1 case:

$$|H(\Omega)|^2 = \frac{1}{1 + \varepsilon_{\text{pass}}^2 C_N^2\left(\dfrac{\Omega}{\Omega_{\text{pass}}}\right)} \tag{11.6.46}$$

and, for the type 2 case:

$$|H(\Omega)|^2 = \frac{C_N^2\left(\dfrac{\Omega_{\text{stop}}}{\Omega}\right)}{C_N^2\left(\dfrac{\Omega_{\text{stop}}}{\Omega}\right) + \varepsilon_{\text{stop}}^2} \tag{11.6.47}$$

where $C_N(x)$ is the Chebyshev polynomial [272] of degree N, defined by

$$C_N(x) = \begin{cases} \cos(N\cos^{-1}(x)), & \text{if } |x| \le 1 \\ \cosh(N\cosh^{-1}(x)), & \text{if } |x| > 1 \end{cases} \tag{11.6.48}$$

Chebyshev polynomials can be understood by defining the angle $\theta = \cos^{-1} x$, so that $x = \cos\theta$ and $C_N(x) = \cos(N\theta)$. When $|x| > 1$, the equation $x = \cos\theta$ requires θ to be imaginary, say $\theta = j\beta$, so that $x = \cos(j\beta) = \cosh(\beta)$ and

$$C_N(x) = \cos(N\theta) = \cos(Nj\beta) = \cosh(N\beta) = \cosh(N\cosh^{-1}x)$$

Using trigonometric identities, it can be shown that $\cos(N\theta)$ is expressible as an Nth order polynomial in $\cos\theta$, that is,

$$\cos(N\theta) = \sum_{i=0}^{N} c_i (\cos\theta)^i$$

The c_i are the coefficients of the Chebyshev polynomials:

$$C_N(x) = \sum_{i=0}^{N} c_i x^i$$

For example, we have $C_1(x) = \cos\theta = x$, and

$$\cos(2\theta) = 2\cos^2\theta - 1 \qquad\qquad C_2(x) = 2x^2 - 1$$

$$\cos(3\theta) = 4\cos^3\theta - 3\cos\theta \qquad \Rightarrow \qquad C_3(x) = 4x^3 - 3x$$

$$\cos(4\theta) = 8\cos^4\theta - 8\cos^2\theta + 1 \qquad C_4(x) = 8x^4 - 8x^2 + 1$$

The following routine `cheby.c` returns the value of the Nth order Chebyshev polynomial for non-negative values of x and can be used in the numerical evaluation of the magnitude responses:

```
/* cheby.c - Chebyshev polynomial C_N(x) */

#include <math.h>

double cheby(N, x)                          usage: y = cheby(N, x);
int N;                                      N = polynomial order
double x;                                   x must be non-negative
{
        if (x <= 1)
                return cos(N * acos(x));
        else
                return cosh(N * log(x + sqrt(x*x-1)));
}
```

For $x > 1$, the values are computed by the alternative expression:

$$\cosh(N \cosh^{-1} x) = \cosh\left(N \ln\left(x + \sqrt{x^2 - 1}\right)\right)$$

Next, we consider the details of the type 1 case. The argument of $C_N(x)$ in Eq. (11.6.46) is $x = \Omega/\Omega_{\text{pass}}$. Therefore, within the passband range $0 \le \Omega \le \Omega_{\text{pass}}$ we have $0 \le x \le 1$, which makes $C_N(x)$ oscillatory and results in the passband ripples.

Within the passband, the magnitude response remains bounded between the values 1 and $1/(1 + \varepsilon_{\text{pass}}^2)$. At the edge of the passband, corresponding to $x = \Omega/\Omega_{\text{pass}} = 1$, we have $C_N(x) = 1$, giving the value $|H(\Omega_{\text{pass}})|^2 = 1/(1 + \varepsilon_{\text{pass}}^2)$. The value at $\Omega = 0$ depends on N. Because $C_N(0)$ equals zero for odd N and unity for even N, we have:

$$|H(0)|^2 = 1 \quad (\text{odd } N), \qquad |H(0)|^2 = \frac{1}{1 + \varepsilon_{\text{pass}}^2} \quad (\text{even } N) \qquad\qquad (11.6.49)$$

The order N can be determined by imposing the stopband specification, that is, $|H(\Omega)|^2 \le 1/(1 + \varepsilon_{stop}^2)$ for $\Omega \ge \Omega_{stop}$. Because of the monotonicity of the stopband, this condition is equivalent to the stopband edge condition:

$$|H(\Omega_{stop})|^2 = \frac{1}{1 + \varepsilon_{stop}^2}$$

Using Eq. (11.6.46), we obtain:

$$\frac{1}{1 + \varepsilon_{pass}^2 \cosh^2\left(N \cosh^{-1}(\Omega_{stop}/\Omega_{pass})\right)} = \frac{1}{1 + \varepsilon_{stop}^2}$$

which gives:

$$\cosh\left(N \cosh^{-1}(\Omega_{stop}/\Omega_{pass})\right) = \varepsilon_{stop}/\varepsilon_{pass} \quad \Rightarrow \quad \cosh(N \cosh^{-1} w) = e$$

where, as in Eq. (11.6.11), we used the stopband to passband ratios:

$$e = \frac{\varepsilon_{stop}}{\varepsilon_{pass}} = \sqrt{\frac{10^{A_{stop}/10} - 1}{10^{A_{pass}/10} - 1}}, \quad w = \frac{\Omega_{stop}}{\Omega_{pass}} \tag{11.6.50}$$

Thus, solving for N, we find:

$$\boxed{N_{exact} = \frac{\cosh^{-1} e}{\cosh^{-1} w} = \frac{\ln\left(e + \sqrt{e^2 - 1}\right)}{\ln\left(w + \sqrt{w^2 - 1}\right)}} \tag{11.6.51}$$

The final value of N is obtained by rounding N_{exact} up to the next integer, that is, $N = \lceil N_{exact} \rceil$. As in the Butterworth case, increasing N slightly from its exact value results in a slightly better stopband than required, that is, $|H(\Omega_{stop})|^2 < 1/(1 + \varepsilon_{stop}^2)$. The 3-dB frequency can be calculated by requiring $|H(\Omega)|^2 = 1/2$, which can be solved to give:

$$\frac{1}{1 + \varepsilon_{pass}^2 C_N^2(\Omega_{3dB}/\Omega_{pass})} = \frac{1}{2} \quad \Rightarrow \quad \cosh(N \cosh^{-1}(\Omega_{3dB}/\Omega_{pass})) = \frac{1}{\varepsilon_{pass}}$$

or,

$$\tan\left(\frac{\pi f_{3dB}}{f_s}\right) = \Omega_{3dB} = \Omega_{pass} \cosh\left(\frac{1}{N} \cosh^{-1}\left(\frac{1}{\varepsilon_{pass}}\right)\right) \tag{11.6.52}$$

The transfer function $H(s)$ of the Chebyshev filter can be constructed by determining the left-hand-plane poles of Eq. (11.6.46) and pairing them in conjugate pairs to form second-order sections. These conjugate pairs are $\{s_i, s_i^*\}$, where

$$s_i = \Omega_{pass} \sinh a \cos \theta_i + j\Omega_{pass} \cosh a \sin \theta_i, \quad i = 1, 2, \ldots, K \tag{11.6.53}$$

where $N = 2K$ or $N = 2K + 1$. In the odd case, there is also a real pole at

$$s_0 = -\Omega_{pass} \sinh a \tag{11.6.54}$$

where the parameter a is the solution of

$$\sinh(Na) = \frac{1}{\varepsilon_{\text{pass}}} \tag{11.6.55}$$

that is,

$$a = \frac{1}{N}\sinh^{-1}\left(\frac{1}{\varepsilon_{\text{pass}}}\right) = \frac{1}{N}\ln\left(\frac{1}{\varepsilon_{\text{pass}}} + \sqrt{\frac{1}{\varepsilon_{\text{pass}}^2} + 1}\right) \tag{11.6.56}$$

The angles θ_i are the *same* as the Butterworth angles of Eq. (11.6.17):

$$\theta_i = \frac{\pi}{2N}(N - 1 + 2i), \quad i = 1, 2, \ldots, K \tag{11.6.57}$$

The second-quadrant values of these angles place the s_i in the left-hand s-plane. The second-order sections are then:

$$H_i(s) = \frac{1}{\left(1 - \dfrac{s}{s_i}\right)\left(1 - \dfrac{s}{s_i^*}\right)} = \frac{|s_i|^2}{s^2 - (2\text{Re}s_i)s + |s_i|^2}$$

For convenience, we define the parameters:

$$\Omega_0 = \Omega_{\text{pass}}\sinh a, \quad \Omega_i = \Omega_{\text{pass}}\sin\theta_i, \quad i = 1, 2, \ldots, K \tag{11.6.58}$$

Then, we may express the second-order sections in the form:

$$H_i(s) = \frac{\Omega_0^2 + \Omega_i^2}{s^2 - 2\Omega_0\cos\theta_i\, s + \Omega_0^2 + \Omega_i^2}, \quad i = 1, 2, \ldots, K \tag{11.6.59}$$

The first-order factor $H_0(s)$ is defined by

$$H_0(s) = \begin{cases} \sqrt{\dfrac{1}{1 + \varepsilon_{\text{pass}}^2}} & \text{if } N \text{ is even, } N = 2K \\ \dfrac{\Omega_0}{s + \Omega_0} & \text{if } N \text{ is odd, } N = 2K + 1 \end{cases} \tag{11.6.60}$$

If N is odd, all filter sections are normalized to unity gain at DC, as required by Eq. (11.6.49). If N is even, the overall gain is $1/(1 + \varepsilon_{\text{pass}}^2)^{1/2}$. It follows that the overall transfer function will be the cascade:

$$\boxed{H(s) = H_0(s)H_1(s)H_2(s)\cdots H_K(s)} \tag{11.6.61}$$

Next, we verify that the poles are properly given by Eq. (11.6.53). Replacing $s = j\Omega$ or $\Omega = -js$ into Eq. (11.6.46), we see that the zeros of the denominator are the solutions of the equation:

$$1 + \varepsilon_{\text{pass}}^2\cosh^2(N\cosh^{-1}(-js/\Omega_{\text{pass}})) = 0, \quad \text{or}$$

$$\cosh(N\cosh^{-1}(-js/\Omega_{\text{pass}})) = \frac{\pm j}{\varepsilon_{\text{pass}}} \tag{11.6.62}$$

Replacing $\theta_i = \phi_i + \pi/2$, where $\phi_i = (2i-1)\pi/(2N)$, into Eq. (11.6.53), we find

$$-js_i/\Omega_{\text{pass}} = \cosh a \, \sin \theta_i - j \sinh a \, \cos \theta_i = \cosh a \, \cos \phi_i + j \sinh a \, \sin \phi_i$$

Using the trigonometric identity

$$\cosh(x+jy) = \cosh x \, \cos y + j \sinh x \, \sin y$$

we find

$$-js_i/\Omega_{\text{pass}} = \cosh(a+j\phi_i) \quad \Rightarrow \quad \cosh^{-1}(-js_i/\Omega_{\text{pass}}) = a + j\phi_i$$

and therefore,

$$\cosh\left(N \cosh^{-1}(-js_i/\Omega_{\text{pass}})\right) = \cosh(Na + jN\phi_i)$$

$$= \cosh(Na)\cos(N\phi_i) + j \sinh(Na)\sin(N\phi_i) = \frac{\pm j}{\varepsilon_{\text{pass}}}$$

where we used $\cos(N\phi_i) = \cos\left((2i-1)\pi/2\right) = 0$, $\sin(N\phi_i) = \sin\left((2i-1)\pi/2\right) = \pm 1$, and Eq. (11.6.55). Thus, the s_i are solutions of the root equation Eq. (11.6.62).

Once the analog transfer function is constructed, each second-order section may be transformed into a digital second-order section by the appropriate bilinear transformation. For example, applying the lowpass version of the bilinear transformation $s = (1 - z^{-1})/(1 + z^{-1})$, we find the digital transfer function:

$$\boxed{H(z) = H_0(z)H_1(z)H_2(z)\cdots H_K(z)} \tag{11.6.63}$$

where $H_i(z)$ are the transformations of Eq. (11.6.59):

$$H_i(z) = \frac{G_i(1 + z^{-1})^2}{1 + a_{i1}z^{-1} + a_{i2}z^{-2}}, \quad i = 1, 2, \dots, K \tag{11.6.64}$$

where the coefficients are computed in terms of the definitions Eq. (11.6.58):

$$G_i = \frac{\Omega_0^2 + \Omega_i^2}{1 - 2\Omega_0 \cos \theta_i + \Omega_0^2 + \Omega_i^2}$$

$$a_{i1} = \frac{2(\Omega_0^2 + \Omega_i^2 - 1)}{1 - 2\Omega_0 \cos \theta_i + \Omega_0^2 + \Omega_i^2} \tag{11.6.65}$$

$$a_{i2} = \frac{1 + 2\Omega_0 \cos \theta_i + \Omega_0^2 + \Omega_i^2}{1 - 2\Omega_0 \cos \theta_i + \Omega_0^2 + \Omega_i^2}$$

The first-order factor is given by

$$H_0(z) = \begin{cases} \sqrt{\dfrac{1}{1 + \varepsilon_{\text{pass}}^2}} & \text{if } N \text{ is even} \\[2ex] \dfrac{G_0(1 + z^{-1})}{1 + a_{01}z^{-1}} & \text{if } N \text{ is odd} \end{cases} \tag{11.6.66}$$

where

$$G_0 = \frac{\Omega_0}{\Omega_0 + 1}, \quad a_{01} = \frac{\Omega_0 - 1}{\Omega_0 + 1} \tag{11.6.67}$$

Example 11.6.7: Using the bilinear transformation and a lowpass analog Chebyshev type 1 prototype, design a lowpass digital filter operating at a rate of 20 kHz and having passband extending to 4 kHz with maximum passband attenuation of 0.5 dB, and stopband starting at 5 kHz with a minimum stopband attenuation of 10 dB.

Then, redesign it such that its magnitude response satisfies $1 \geq |H(f)|^2 \geq 0.98$ in the passband, and $|H(f)|^2 \leq 0.02$ in the stopband.

Solution: The specifications and the prewarped digital frequencies are the same as in Example 11.6.3, that is, $\Omega_{\text{pass}} = 0.7265$ and $\Omega_{\text{stop}} = 1$.

We calculate $\varepsilon_{\text{pass}} = 0.3493$ and $\varepsilon_{\text{stop}} = 3$ and the quantities in Eq. (11.6.50) $e = \varepsilon_{\text{stop}}/\varepsilon_{\text{pass}} = 8.5883$, $w = \Omega_{\text{stop}}/\Omega_{\text{pass}} = 1.3764$. Then, Eq. (11.6.51) gives $N_{\text{exact}} = 3.37$, which is rounded up to $N = 4$. Thus, there are $K = 2$ second-order sections. The 3-dB frequency can be calculated by inverting the bilinear transformation and Eq. (11.6.52):

$$\tan\left(\frac{\pi f_{3\text{dB}}}{f_s}\right) = \Omega_{3\text{dB}} = \Omega_{\text{pass}} \cosh\left(\frac{1}{N} \cosh^{-1}\left(\frac{1}{\varepsilon_{\text{pass}}}\right)\right)$$

which gives $f_{3\text{dB}} = 4.2729$ kHz. The actual stopband attenuation is larger than A_{stop} because of the increased value of N. We calculate:

$$A(\Omega_{\text{stop}}) = 10\log_{10}\left(1 + \varepsilon_{\text{pass}}^2 C_N^2(\Omega_{\text{stop}}/\Omega_{\text{pass}})\right) = 14.29 \text{ dB}$$

The parameter a is calculated from Eq. (11.6.55) to be $a = 0.4435$. Then, the coefficients of the digital filter are calculated from Eqs. (11.6.65) and (11.6.67), resulting in the transfer function:

$$H(z) = 0.9441 \cdot \frac{0.3091(1 + z^{-1})^2}{1 - 0.4830z^{-1} + 0.7194z^{-2}} \cdot \frac{0.1043(1 + z^{-1})^2}{1 - 0.9004z^{-1} + 0.3177z^{-2}}$$

The magnitude response squared is shown in the left graph of Fig. 11.6.11. It was computed by inserting the bilinear transformation into Eq. (11.6.46) and evaluating it with cheby, that is,

$$|H(f)|^2 = \frac{1}{1 + \varepsilon_{\text{pass}}^2 \left(\text{cheby}(N, \tan(\pi f/f_s)/\Omega_{\text{pass}})\right)^2}$$

For the more stringent specifications, we have $A_{\text{pass}} = 0.08774$ dB and $A_{\text{stop}} = 16.9897$ dB. We calculate the parameters:

$$\varepsilon_{\text{pass}} = 0.1429, \quad \varepsilon_{\text{stop}} = 7, \quad N_{\text{exact}} = 5.44, \quad N = 6, \quad a = 0.4407$$

The 3-dB frequency is found to be $f_{3\text{dB}} = 4.2865$ kHz. The actual stopband attenuation is $A(\Omega_{\text{stop}}) = 21.02$ dB, instead of the nominal one of $A_{\text{stop}} = 16.9897$ dB. The digital filter coefficients are then:

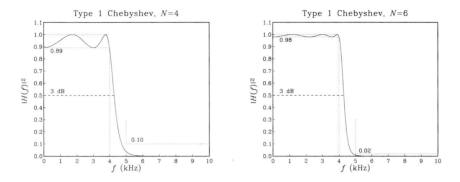

Fig. 11.6.11 Digital lowpass Chebyshev type 1 filters.

i	G_i	a_{i1}	a_{i2}
0	0.9899		
1	0.3394	-0.4492	0.8069
2	0.2028	-0.6809	0.4920
3	0.0811	-0.9592	0.2837

The gain factor G_0 represents here the overall gain $1/\sqrt{1 + \varepsilon_{\text{pass}}^2} = \sqrt{0.98} = 0.9899$. The magnitude response is shown in the right graph of Fig. 11.6.11. By comparison, recall that the two Butterworth filters of Example 11.6.3 had filter orders of 7 and 13, respectively. $\qquad\Box$

Example 11.6.8: Redesign the highpass digital filters of Example 11.6.4 using a Chebyshev type 1 analog lowpass prototype filter.

Solution: The equivalent analog lowpass specifications $\{\Omega_{\text{pass}}, \Omega_{\text{stop}}, A_{\text{pass}}, A_{\text{stop}}\}$ are the same as in Example 11.6.4. We have $\Omega_{\text{pass}} = 1$ and $\Omega_{\text{stop}} = 1.3764$. Based on the first set of specifications, we find the Chebyshev design parameters:

$$\varepsilon_{\text{pass}} = 0.3493, \quad \varepsilon_{\text{stop}} = 3, \quad N_{\text{exact}} = 3.37, \quad N = 4, \quad a = 0.4435$$

and based on the second set:

$$\varepsilon_{\text{pass}} = 0.1429, \quad \varepsilon_{\text{stop}} = 7, \quad N_{\text{exact}} = 5.44, \quad N = 6, \quad a = 0.4407$$

The 3-dB frequencies can be calculated by inverting:

$$\cot\left(\frac{\pi f_{\text{3dB}}}{f_s}\right) = \Omega_{\text{3dB}} = \Omega_{\text{pass}} \cosh\left(\frac{1}{N}\cosh^{-1}\left(\frac{1}{\varepsilon_{\text{pass}}}\right)\right)$$

which gives for the two specification sets: $f_0 = 4.7170$ and 4.7031 kHz. The actual stopband attenuations attained by the designed filter are in the two cases: $A(\Omega_{\text{stop}}) = 14.29$ and 21.02 dB.

The digital transfer functions are obtained by transforming the analog filter sections by the highpass bilinear transformation $s = (1 + z^{-1})/(1 - z^{-1})$.

The digital filter coefficients are obtained from Eqs. (11.6.65) and (11.6.67) by changing the sign of the a_{i1} coefficients. We have for the two specification sets:

i	G_i	a_{i1}	a_{i2}
0	0.9441		
1	0.4405	−0.0526	0.7095
2	0.1618	0.5843	0.2314

i	G_i	a_{i1}	a_{i2}
0	0.9899		
1	0.4799	−0.1180	0.8017
2	0.3008	0.2492	0.4524
3	0.1273	0.6742	0.1834

Thus, for example, the first transfer function will be:

$$H(z) = 0.9441 \cdot \frac{0.4405\,(1 - z^{-1})^2}{1 - 0.0526 z^{-1} + 0.7095 z^{-2}} \cdot \frac{0.1618\,(1 - z^{-1})^2}{1 + 0.5843 z^{-1} + 0.2314 z^{-2}}$$

The designed magnitude responses are shown in Fig. 11.6.12.   □

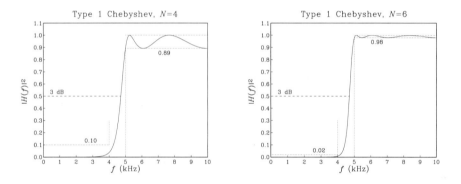

Fig. 11.6.12 Digital highpass Chebyshev type 1 filters.

Next, we discuss type 2 Chebyshev filters. The argument of the Chebyshev polynomials in Eq. (11.6.47) is now $x = \Omega_{\text{stop}}/\Omega$. Therefore, the stopband range $\Omega \geq \Omega_{\text{stop}}$ maps to $0 \leq x \leq 1$ where the Chebyshev polynomial is oscillatory resulting in the stopband ripples.

At the edge of the stopband, $\Omega = \Omega_{\text{stop}}$, we have $x = 1$, $C_N(x) = 1$, and magnitude response equal to $|H(\Omega_{\text{stop}})|^2 = 1/\sqrt{1 + \varepsilon_{\text{stop}}^2}$. At large frequencies, $\Omega \to \infty$, we have $x \to 0$. Because the value of $C_N(0)$ depends on N being zero for odd N and unity for even N, it follows that the magnitude response will either tend to zero for odd N or to $1/\sqrt{1 + \varepsilon_{\text{stop}}^2}$ for even N.

The zero frequency $\Omega = 0$ corresponds to the limit $x \to \infty$ which causes the Chebyshev polynomials to grow like a power x^N. It follows that the numerator and denominator of the magnitude response (11.6.47) will both diverge but in such a way that $|H(0)|^2 = 1$. Thus, type 2 filters are always normalized to unity at DC. The filter order N can be determined by requiring the passband specification:

$$|H(\Omega_{\text{pass}})|^2 = \frac{C_N^2\left(\dfrac{\Omega_{\text{stop}}}{\Omega_{\text{pass}}}\right)}{C_N^2\left(\dfrac{\Omega_{\text{stop}}}{\Omega_{\text{pass}}}\right) + \varepsilon_{\text{stop}}^2} = \frac{1}{1 + \varepsilon_{\text{pass}}^2}$$

It has the *same* solution Eq. (11.6.51) as the type 1 case. Once N is fixed, the 3-dB frequency can be obtained by solving $|H(\Omega_{3\text{dB}})|^2 = 1/2$ which gives the solution:

$$\Omega_{3\text{dB}} = \frac{\Omega_{\text{stop}}}{\cosh(\dfrac{1}{N}\cosh^{-1}(\varepsilon_{\text{stop}}))} \tag{11.6.68}$$

Because of the non-trivial numerator in Eq. (11.6.47), the filter will have both zeros and poles. They are solutions of the following equations obtained by replacing $\Omega = -js$:

$$\cosh^2\left(N\cosh^{-1}(\frac{\Omega_{\text{stop}}}{-js})\right) = 0, \quad \cosh^2\left(N\cosh^{-1}(\frac{j\Omega_{\text{stop}}}{-js})\right) + \varepsilon_{\text{stop}}^2 = 0$$

The zeros are the conjugate pairs $\{z_i, z_i^*\}$:

$$z_i = \frac{j\Omega_{\text{stop}}}{\sin\theta_i}, \quad i = 1, 2, \ldots, K \tag{11.6.69}$$

where $N = 2K$ or $N = 2K + 1$. The poles are essentially the reciprocals of the type 1 poles, that is, the pairs $\{s_i, s_i^*\}$:

$$s_i = \frac{\Omega_{\text{stop}}}{\sinh a\,\cos\theta_i + j\cosh a\,\sin\theta_i}, \quad i = 1, 2, \ldots, K \tag{11.6.70}$$

In the odd-N case, there is also a real pole at

$$s_0 = -\frac{\Omega_{\text{stop}}}{\sinh a} \tag{11.6.71}$$

where the parameter a is the solution of

$$\sinh(Na) = \varepsilon_{\text{stop}} \tag{11.6.72}$$

that is,

$$a = \frac{1}{N}\sinh^{-1}(\varepsilon_{\text{stop}}) = \frac{1}{N}\ln\left(\varepsilon_{\text{stop}} + \sqrt{\varepsilon_{\text{stop}}^2 + 1}\right) \tag{11.6.73}$$

The angles θ_i are the *same* as in the type 1 case and given by Eq. (11.6.57). The second-order sections are formed by pairing the zeros and poles in conjugate pairs:

$$H_i(s) = \frac{\left(1 - \dfrac{s}{z_i}\right)\left(1 - \dfrac{s}{z_i^*}\right)}{\left(1 - \dfrac{s}{s_i}\right)\left(1 - \dfrac{s}{s_i^*}\right)} \tag{11.6.74}$$

For convenience, we define the parameters:

$$\Omega_0 = \frac{\Omega_{\text{stop}}}{\sinh a}, \qquad \Omega_i = \frac{\Omega_{\text{stop}}}{\sin \theta_i}, \quad i = 1, 2, \ldots, K \qquad (11.6.75)$$

Then, we may express the second-order sections in the form:

$$H_i(s) = \frac{1 + \Omega_i^{-2} s^2}{1 - 2\Omega_0^{-1} \cos \theta_i s + (\Omega_0^{-2} + \Omega_i^{-2})s^2}, \quad i = 1, 2, \ldots, K \qquad (11.6.76)$$

The first-order factor $H_0(s)$ is defined by

$$H_0(s) = \begin{cases} 1, & \text{if } N \text{ is even} \\ \dfrac{\Omega_0}{\Omega_0 + s}, & \text{if } N \text{ is odd} \end{cases} \qquad (11.6.77)$$

Again, all filter sections are normalized to unity gain at DC. Under the bilinear transformation $s = (1 - z^{-1})/(1 + z^{-1})$, the analog sections transform to digital versions of the form:

$$H_i(z) = \frac{G_i(1 + b_{i1} z^{-1} + z^{-2})}{1 + a_{i1} z^{-1} + a_{i2} z^{-2}}, \quad i = 1, 2, \ldots, K \qquad (11.6.78)$$

where the coefficients are computed in terms of the definitions in Eq. (11.6.75):

$$G_i = \frac{1 + \Omega_i^{-2}}{1 - 2\Omega_0^{-1} \cos \theta_i + \Omega_0^{-2} + \Omega_i^{-2}}, \quad b_{i1} = 2\frac{1 - \Omega_i^{-2}}{1 + \Omega_i^{-2}}$$

$$a_{i1} = \frac{2(1 - \Omega_0^{-2} - \Omega_i^{-2})}{1 - 2\Omega_0^{-1} \cos \theta_i + \Omega_0^{-2} + \Omega_i^{-2}} \qquad (11.6.79)$$

$$a_{i2} = \frac{1 + 2\Omega_0^{-1} \cos \theta_i + \Omega_0^{-2} + \Omega_i^{-2}}{1 - 2\Omega_0^{-1} \cos \theta_i + \Omega_0^{-2} + \Omega_i^{-2}}$$

The first-order factor, if present, is given by

$$H_0(z) = \frac{G_0(1 + z^{-1})}{1 + a_{01} z^{-1}} \qquad (11.6.80)$$

where

$$G_0 = \frac{\Omega_0}{\Omega_0 + 1}, \quad a_{01} = \frac{\Omega_0 - 1}{\Omega_0 + 1} \qquad (11.6.81)$$

The overall digital transfer function is then:

$$\boxed{H(z) = H_0(z) H_1(z) H_2(z) \cdots H_K(z)} \qquad (11.6.82)$$

Example 11.6.9: Redesign the two filters of Example 11.6.7 using a type 2 Chebyshev design.

Solution: The values of $\varepsilon_{\text{pass}}$, $\varepsilon_{\text{stop}}$, N_{exact}, N remain the same as in Example 11.6.7. The parameter a, calculated from Eq. (11.6.73), is for the two specification sets: $a = 0.4546$ and 0.4407.

The actual passband attenuations are slightly higher than the specified ones. Evaluating Eq. (11.6.47) for the two designs, we find the values $A(\Omega_{\text{pass}}) = 0.18$ and 0.03 dB, instead of the required ones 0.5 and 0.08774. The 3-dB frequencies are obtained by inverting:

$$\tan\left(\frac{\pi f_{3\text{dB}}}{f_s}\right) = \Omega_{3\text{dB}} = \frac{\Omega_{\text{stop}}}{\cosh\left(\frac{1}{N}\cosh^{-1}(\varepsilon_{\text{stop}})\right)}$$

which gives the values $f_{3\text{dB}} = 4.7009$ and 4.7031 kHz. For the first specification set, we have two second-order sections whose coefficients are calculated from Eq. (11.6.79), resulting in the transfer function:

$$H(z) = \frac{0.7612(1 + 0.1580z^{-1} + z^{-2})}{1 - 0.0615z^{-1} + 0.7043z^{-2}} \cdot \frac{0.5125(1 + 1.4890z^{-1} + z^{-2})}{1 + 0.5653z^{-1} + 0.2228z^{-2}}$$

For the second specification set, we have three sections with coefficients:

i	G_i	b_{i1}	a_{i1}	a_{i2}
1	0.8137	0.0693	−0.1180	0.8017
2	0.6381	0.6667	0.2492	0.4524
3	0.4955	1.7489	0.6742	0.1834

The designed magnitude responses are shown in Fig. 11.6.13. □

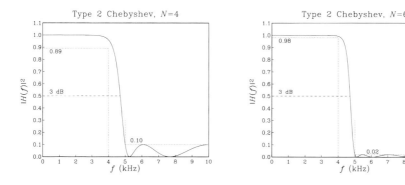

Fig. 11.6.13 Digital lowpass Chebyshev type 2 filters.

The Chebyshev filter can also be transformed by the bandpass or bandstop versions of the bilinear transformation to design bandpass or bandstop digital filters. For example, transforming a type 2 filter by the bandpass transformation in Eq. (11.1.5) gives rise to fourth-order sections of the form:

$$H_i(z) = G_i \frac{1 + b_{i1}z^{-1} + b_{i2}z^{-2} + b_{i1}z^{-3} + z^{-4}}{1 + a_{i1}z^{-1} + a_{i2}z^{-2} + a_{i3}z^{-3} + a_{i4}z^{-4}}, \quad i = 1, 2, \ldots, K \qquad (11.6.83)$$

where by symmetry, the numerator coefficients of z^{-1} and z^{-3} are the same. The coefficients are given by:

$$G_i = \frac{1 + \Omega_i^{-2}}{1 - 2\Omega_0^{-1}\cos\theta_i + \Omega_0^{-2} + \Omega_i^{-2}}$$

$$b_{i1} = -\frac{4c\Omega_i^{-2}}{1 + \Omega_i^{-2}}, \quad b_{i2} = \frac{2c(\Omega_i^{-2}(2c^2+1)-1)}{1 + \Omega_i^{-2}}$$

$$a_{i1} = \frac{4c(\Omega_0^{-1}\cos\theta_i - \Omega_0^{-2} - \Omega_i^{-2})}{1 - 2\Omega_0^{-1}\cos\theta_i + \Omega_0^{-2} + \Omega_i^{-2}}, \quad a_{i2} = \frac{2((\Omega_0^{-2} + \Omega_i^{-2})(2c^2+1)-1)}{1 - 2\Omega_0^{-1}\cos\theta_i + \Omega_0^{-2} + \Omega_i^{-2}}$$

$$a_{i3} = \frac{4c(\Omega_0^{-1}\cos\theta_i + \Omega_0^{-2} + \Omega_i^{-2})}{1 - 2\Omega_0^{-1}\cos\theta_i + \Omega_0^{-2} + \Omega_i^{-2}}, \quad a_{i4} = \frac{1 + 2\Omega_0^{-1}\cos\theta_i + \Omega_0^{-2} + \Omega_i^{-2}}{1 - 2\Omega_0^{-1}\cos\theta_i + \Omega_0^{-2} + \Omega_i^{-2}}$$

If N is odd, the first-order analog section $H_0(s)$ is transformed to the same quadratic section $H_0(z)$ given in Section 11.6.4. Similarly, applying the bandstop transformation of Eq. (11.6.45), the type 2 Chebyshev second-order sections in s are transformed into fourth-order sections in z in the same form of Eq. (11.6.83), but with coefficients given by:

$$G_i = \frac{1 + \Omega_i^{-2}}{1 - 2\Omega_0^{-1}\cos\theta_i + \Omega_0^{-2} + \Omega_i^{-2}}$$

$$b_{i1} = -\frac{4c}{1 + \Omega_i^{-2}}, \quad b_{i2} = \frac{2c(2c^2+1 - \Omega_i^{-2})}{1 + \Omega_i^{-2}}$$

$$a_{i1} = -\frac{4c(1 - \Omega_0^{-1}\cos\theta_i)}{1 - 2\Omega_0^{-1}\cos\theta_i + \Omega_0^{-2} + \Omega_i^{-2}}, \quad a_{i2} = \frac{2(2c^2+1 - \Omega_0^{-2} - \Omega_i^{-2})}{1 - 2\Omega_0^{-1}\cos\theta_i + \Omega_0^{-2} + \Omega_i^{-2}}$$

$$a_{i3} = -\frac{4c(1 + \Omega_0^{-1}\cos\theta_i)}{1 - 2\Omega_0^{-1}\cos\theta_i + \Omega_0^{-2} + \Omega_i^{-2}}, \quad a_{i4} = \frac{1 + 2\Omega_0^{-1}\cos\theta_i + \Omega_0^{-2} + \Omega_i^{-2}}{1 - 2\Omega_0^{-1}\cos\theta_i + \Omega_0^{-2} + \Omega_i^{-2}}$$

The first-order section $H_0(s)$ transforms to the same $H_0(z)$ as that of Section 11.6.5, but with Ω_0 given by Eq. (11.6.75).

Example 11.6.10: Redesign the bandpass digital filters of Example 11.6.5 using a type 2 Chebyshev analog prototype.

Solution: The prewarped frequencies and bilinear transformation parameter c are as in that example: $\Omega_{\text{pass}} = 0.3249$, $\Omega_{\text{stop}} = 0.4674$, $c = 0.6180$.

The Chebyshev design parameters are for the two specification sets:

$$\varepsilon_{\text{pass}} = 0.3493, \quad \varepsilon_{\text{stop}} = 3, \quad N_{\text{exact}} = 3.14, \quad N = 4, \quad a = 0.4546$$

$$\varepsilon_{\text{pass}} = 0.1429, \quad \varepsilon_{\text{stop}} = 7, \quad N_{\text{exact}} = 5.07, \quad N = 6, \quad a = 0.4407$$

For the first set, there are two fourth-order sections with coefficients:

i	G_i	b_{i1}	b_{i2}	a_{i1}	a_{i2}	a_{i3}	a_{i4}
1	0.7334	−1.9684	2.4015	−1.9604	2.2956	−1.6758	0.7697
2	0.3677	−0.9922	0.2187	−1.4221	0.8671	−0.4101	0.1813

For the second set, there are three fourth-order sections:

i	G_i	b_{i1}	b_{i2}	a_{i1}	a_{i2}	a_{i3}	a_{i4}
1	0.7840	−2.0032	2.4793	−2.0118	2.4413	−1.8265	0.8501
2	0.5858	−1.7205	1.8472	−1.7286	1.6780	−1.1223	0.5094
3	0.3159	−0.5802	−0.7026	−1.3122	0.5868	−0.1879	0.0904

The designed magnitude responses are shown in Fig. 11.6.14. ☐

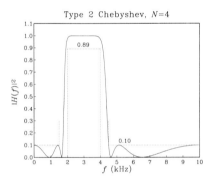 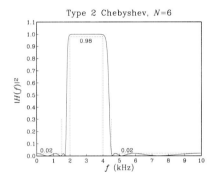

Fig. 11.6.14 Digital bandpass Chebyshev type 2 filters.

Example 11.6.11: Redesign the bandstop digital filters of Example 11.6.6 using a type 2 Chebyshev analog prototype.

Solution: The prewarped frequencies and bilinear transformation parameter c are as in that example: $\Omega_{\text{pass}} = 1.9626$, $\Omega_{\text{stop}} = 2.7121$, $c = 0.6597$.

For the two specification sets, the exact Chebyshev orders are $N_{\text{exact}} = 3.35$ and 5.40, which round up to $N = 4$ and 6, respectively. The other Chebyshev parameters remain the same as in Example 11.6.10. For the first set, the fourth-order sections have coefficients:

i	G_i	b_{i1}	b_{i2}	a_{i1}	a_{i2}	a_{i3}	a_{i4}
1	0.8727	−2.3644	3.1438	−2.2003	2.6965	−1.9264	0.7924
2	0.7442	−2.5872	3.6287	−2.2339	2.6565	−1.6168	0.5323

For the second set, there are three fourth-order sections:

i	G_i	b_{i1}	b_{i2}	a_{i1}	a_{i2}	a_{i3}	a_{i4}
1	0.9074	−2.3417	3.0945	−2.2171	2.7626	−2.0326	0.8601
2	0.8009	−2.4708	3.3754	−2.2137	2.6612	−1.7441	0.6441
3	0.7412	−2.6149	3.6889	−2.2524	2.6929	−1.6241	0.5238

The designed magnitude responses are shown in Fig. 11.6.15. ☐

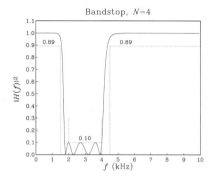 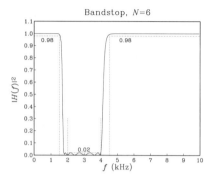

Fig. 11.6.15   Digital bandstop Chebyshev type 2 filters.

11.7 Problems

11.1 Consider the peaking filter of Eq. (11.3.21). Derive an analytical expression for its impulse response $h(n)$ in terms of the parameters ω_0 and $\Delta\omega$. Show that its transient (ϵ-level) time constant is given by:

$$n_{\mathrm{eff}} = \frac{\ln\epsilon}{\ln\rho}, \qquad \rho = \sqrt{\frac{1-\beta}{1+\beta}}$$

11.2 Consider the peaking filter of Eq. (11.3.21). Taking the limit $\Delta\omega \to 0$ and keeping only the lowest-order terms in $\Delta\omega$, show that the pole radius R is related approximately to $\Delta\omega$ by Eq. (6.4.4), that is, $\Delta\omega \simeq 2(1-R)$. Show also that $n_{\mathrm{eff}} \simeq -2\ln\epsilon/\Delta\omega$.

11.3 Verify the complementarity properties for the filters of Eqs. (11.3.7) and (11.3.21):

$$H_{\mathrm{notch}}(z) + H_{\mathrm{peak}}(z) = 1, \quad |H_{\mathrm{notch}}(\omega)|^2 + |H_{\mathrm{peak}}(\omega)|^2 = 1, \qquad (\text{when } G_B^2 = 1/2)$$

11.4 Consider the three choices for the bandwidth reference gain in parametric equalizer filters: $G_B^2 = (G_0^2 + G^2)/2$ (arithmetic mean), $G_B^2 = G_0G$ (geometric mean), and $G_B^2 = 2G_0^2G^2/(G_0^2+G^2)$ (harmonic mean). For each case, discuss how the design parameter β of Eq. (11.4.6) simplifies. For the geometric mean case [268], show that if you design two boost and cut digital filters centered at the same frequency and having equal bandwidths and equal and opposite boost/cut gains (in dB), then their magnitude responses will be related by $|H_{\mathrm{boost}}(\omega)|^2|H_{\mathrm{cut}}(\omega)|^2 = G_0^4$. Show that the more general weighted geometric mean choice $G_B^2 = G_0^{1-c}G^{1+c}$, $0 \le c < 1$ also satisfies this property.

11.5 *Computer Experiment: Peaking and Notching Filters.* Reproduce the results and graphs of Example 11.3.1. Plot also the phase responses of all the filters. For each filter, draw its canonical realization and write the corresponding sample processing algorithm. (You may use the MATLAB function parmeq.m to design them.)

Calculate the 5% time constants of the filters. Send a unit impulse $\delta(n)$ into the input and calculate and plot the impulse responses $h(n)$ of these filters. You must compute $h(n)$ for a period of at least two 5% time constants. (You may use the routines sos.c or sos.m to implement them.)

11.6 *Computer Experiment: Parametric EQ Filter Design.* Reproduce the results and graphs of Examples 11.4.1 and 11.4.2. Plot also the phase responses of all the filters. (You may use the MATLAB function `parmeq.m` to design them.)

For the filters of Example 11.4.1, compute their 5% time constants and compute and plot their impulse responses $h(n)$ versus n.

11.7 *Computer Experiment: Lowpass/Highpass EQ Filters.* Write a MATLAB function (similar to `parmeq.m`) to design the lowpass or highpass shelving equalizer filters defined by Eqs. (11.4.9) and (11.4.11).

Use the function to reproduce the results and graphs of Examples 11.2.1 and 11.2.2.

11.8 *Computer Experiment: Periodic Comb/Notch Parametric EQ Design.* Reproduce the results and graphs of Example 11.5.1. Plot also the phase responses of all the filters. (You may use the MATLAB function `combeq.m` to design them.)

Write a C or MATLAB function, say `combfilt`, that implements the sample processing algorithm of the time domain operation of the comb filter. It must have usage:

```
        y = combfilt(a, b, c, D, w, x);      (C version)
    [y, w] = combfilt(a, b, c, D, w, x);      (MATLAB version)
```

where **w** is the $(D+1)$-dimensional delay-line buffer, and $\{x, y\}$ are the input and output samples. The parameters $\{a, b, c\}$ are generated by `combeq`. For the C case, you may use a circular buffer. Using this function, calculate and plot the impulse response $h(n)$ of all the designed filters.

11.9 The passband and stopband specifications are defined somewhat differently in FIR and IIR designs. Discuss these differences and explain why the parameter sets $\{\delta_{\text{pass}}, \delta_{\text{stop}}\}$ of Eq. (10.2.5) and $\{\varepsilon_{\text{pass}}, \varepsilon_{\text{stop}}\}$ of Eq. (11.6.4) are appropriate for FIR and IIR designs.

11.10 The parameters N and Ω_0 of an analog Butterworth filter are determined by solving the two specification equations $A(\Omega_{\text{pass}}) = A_{\text{pass}}$, $A(\Omega_{\text{stop}}) = A_{\text{stop}}$. The resulting filter order is then rounded up to the next integer value N.

Using this slightly larger N, show that if Ω_0 is found from the passband specification, that is, by solving $A(\Omega_{\text{pass}}) = A_{\text{pass}}$, then the stopband specification is more than satisfied, that is, $A(\Omega_{\text{stop}}) > A_{\text{stop}}$.

Similarly, show that if we find Ω_0 from the stopband specification $A(\Omega_{\text{stop}}) = A_{\text{stop}}$, then the passband specification is more than satisfied, that is, $A(\Omega_{\text{pass}}) < A_{\text{pass}}$.

11.11 Using the bilinear transformation and a lowpass analog Butterworth prototype filter, design a *lowpass* digital filter operating at a rate of 40 kHz and having the following specifications: $f_{\text{pass}} = 10$ kHz, $A_{\text{pass}} = 3$ dB, $f_{\text{stop}} = 15$ kHz, $A_{\text{stop}} = 35$ dB. Carry out all the design steps by hand.

Draw the cascade realization form and write the difference equations and the corresponding sample processing algorithm implementing this realization in the time domain.

11.12 Using the bilinear transformation method and a Butterworth lowpass analog prototype, design (by hand) a digital *highpass* filter operating at a rate of 10 kHz and having passband and stopband frequencies of 3 kHz and 2 kHz, respectively. The maximum passband and minimum stopband attenuations are required to be 0.5 dB and 10 dB respectively.

   a. What are the actual maximum passband and minimum stopband attenuations in dB achieved by the designed filter?

b. Draw the cascade realization and write the corresponding difference equations describing the time-domain operation of the filter.

c. Give a simple closed-form expression of the magnitude response of the designed filter as a function of the variable $\cot(\omega/2)$. Sketch the magnitude response over the frequency interval $0 \le f \le 20$ kHz.

11.13 Show that the generalized bilinear transformation

$$ s = \frac{1 - 2cz^{-1} + z^{-2}}{1 - z^{-2}} \tag{11.7.1} $$

maps the left-hand s-plane onto the inside of the unit circle of the z-plane, provided the real constant c is such that $|c| \le 1$.

11.14 In the design of bandpass/bandstop filters, we found that the constant c of the generalized bilinear transformation of Eq. (11.7.1) was given by an expression of the form:

$$ c = \frac{\sin(\omega_1 + \omega_2)}{\sin(\omega_1) + \sin(\omega_2)} $$

Show that it satisfies the stability condition $|c| \le 1$, regardless of the values of ω_1 and ω_2 in the interval $[0, \pi]$.

11.15 Using a third-order analog lowpass Butterworth prototype filter and the bandpass bilinear transformation of Eq. (11.7.1), design a *digital bandpass* filter operating at 20 kHz and having attenuation of 0.5 dB at the frequencies 2 kHz and 8 kHz.

What is the transfer function of the designed filter? What are the upper and lower 3-dB frequencies of this filter? What is its center frequency in kHz? Sketch roughly the magnitude response over the range $0 \le f \le 30$ kHz.

11.16 Carry out the algebra in Eq. (11.6.23) to show the coefficient equations (11.6.24) for designing a digital lowpass Butterworth filter. Verify also Eq. (11.6.26).

Repeat for the highpass case, Eqs. (11.6.32)—(11.6.35). Repeat for the bandpass case, Eqs. (11.6.40)–(11.6.43).

11.17 Prove Eqs. (11.6.44) for the left and right 3-dB frequencies of a bandpass Butterworth design.

11.18 The bandpass and bandstop Butterworth designs discussed in Sections 11.6.4 and 11.6.5 match the *passband* specifications exactly and use the more conservative of the two stopbands. Instead, if we were to match the stopbands exactly and use the more conservative passband, what changes in the design procedure should we make?

11.19 Carry out the algebra of the bilinear transformation to verify the design equations of Eqs. (11.6.65) and (11.6.79) for designing digital lowpass type 1 and type 2 Chebyshev filters.

11.20 Equations (11.6.53) and (11.6.58) define the Chebyshev poles s_i and the quantities $\{\Omega_0, \Omega_i\}$. Show that they satisfy the following relationship, which is used in the second-order Chebyshev sections (11.6.59):

$$ |s_i|^2 = \Omega_0^2 + \Omega_i^2 $$

11.21 The IIR cascade filtering routines cas.c or cas.m are appropriate for cascading second-order sections and can be used in the lowpass and highpass designs. In bandpass and bandstop designs, however, we have the cascade of fourth-order sections whose coefficients are stored in the $K \times 5$ matrices A and B that are generated by the filter design functions, such as bpcheb2.m, where K is the number of fourth-order sections.

Write C and/or MATLAB versions of the routine cas, say cas4, that works with fourth-order sections. Its inputs must be the matrices A and B, a $K \times 5$ state matrix W whose rows are the state vectors of the cascaded fourth-order sections, and the current input sample x. Its outputs must be the current output sample y and the updated state matrix W. It must have usage:

```
y = cas4(K, B, A, W, x);        (C version)
[y, W] = cas4(K, B, A, W, x);   (MATLAB version)
```

It must call K times a single fourth-order section routine, like the sos routine. Then, write C and/or MATLAB filtering routines, like casfilt of Problem 7.15, that can filter a vector of input samples producing a vector of output samples.

11.22 *Computer Experiment: Butterworth Digital Filter Designs.* Reproduce all the results and graphs of the lowpass, highpass, bandpass, and bandstop Butterworth design Examples 11.6.3–11.6.6. You may use the MATLAB functions lhbutt.m and bpsbutt.m to design the filters.

For each design, also do the following: Plot the phase response of the filter. Compute the filter's 5% time constant (you will need to use MATLAB's root finder roots). Then, using the routines cas.c or cas.m, (or cas4.c, cas4.m of Problem 11.21), compute and plot the impulse response $h(n)$ of the filter over a period lasting two time constants.

11.23 *Computer Experiment: Chebyshev Digital Filter Designs.* Reproduce all the results and graphs of the lowpass, highpass, bandpass, and bandstop Chebyshev design Examples 11.6.7–11.6.11. You may use the MATLAB functions lhcheb1, lhcheb2, bpcheb2, and bscheb2 to design the filters.

For each design, also do the following: Plot the phase response of the filter. Compute the filter's 5% time constant. Then, compute and plot the impulse response $h(n)$ of the filter over a period lasting two time constants.

Since the specifications of the filters are the same as those of Problem 11.22, compare the Butterworth and Chebyshev designs in terms of their order N and their phase response.

In both problems, the frequency responses can be computed with the included MATLAB functions cas2can and dtft. For example, the frequency response of a type 2 bandstop design can be computed as follows:

```
[A, B, P] = bscheb2(fs, fpa, fpb, fsa, fsb, Apass, Astop);
a = cas2can(A);              direct-form denominator
b = cas2can(B);              direct-form numerator
w = (0:NF-1) * pi / NF;      NF frequencies over [0, π]
H = dtft(b, w) ./ dtft(a, w);   compute H(ω) = N(ω)/D(ω)
```

12

Interpolation, Decimation, and Oversampling

12.1 Interpolation and Oversampling

Sampling rate changes are useful in many applications, such as interconnecting digital processing systems operating at different rates [273-276]. Sampling rate increase is accomplished by *interpolation*, that is, the process of inserting additional samples between the original low-rate samples. The inserted, or interpolated, samples are *calculated* by an FIR digital filter.[†] This is illustrated in Fig. 12.1.1 for the case of a 4-fold interpolator which increases the sampling rate by a factor of four, that is, $f_s' = 4f_s$.

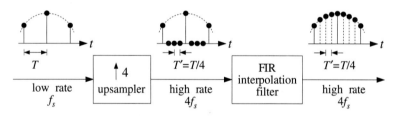

Fig. 12.1.1 Sampling rate increase with digital interpolation.

With respect to the fast time scale, the low-rate samples may be thought of as being separated by three zero samples. The 4-fold *rate expander* or *upsampler* simply inserts three zero samples for every low-rate sample. The job of the FIR filter is to replace the three zeros by the calculated interpolated values.

The interpolating filter is sometimes called an *oversampling digital filter* because it operates at the fast rate $4f_s$. However, because only one out of every four input samples is non-zero, the required filtering operations may be rearranged in such a way as to operate only on the low-rate samples, thus, effectively reducing the computational requirements of the filter—by a factor of four in this case.

[†]IIR filters can also be used, but are less common in practice.

This is accomplished by replacing the high-rate interpolating FIR filter by four shorter FIR subfilters, known as *polyphase* filters, operating at the *low* rate f_s. The length of each subfilter is one-quarter that of the original filter. Because each low-rate input sample generates four high-rate interpolated outputs (itself and three others), each of the four low-rate subfilters is dedicated to computing only one of the four outputs. Such realization is computationally efficient and lends itself naturally to parallel multiprocessor hardware implementations in which a different DSP chip may be used to implement each subfilter.

An interesting application of interpolation is the use of oversampling digital filters in CD or DAT players, where they help to alleviate the need for high-quality analog anti-image postfilters in the playback system. Moreover, each high-rate sample can be *requantized* without loss of quality to fewer number of bits (even as low as 1 bit per sample) using appropriate *noise shaping* quantizers, thus, trading off bits for samples and simplifying the structure of the analog part of the playback system.

To understand the motivation behind this application, consider an analog signal sampled at a rate f_s, such as 44.1 kHz for digital audio. The analog signal is prefiltered by an analog lowpass *antialiasing prefilter* having cutoff frequency $f_c \leq f_s/2$ and then sampled at rate f_s and quantized. This operation is shown in Fig. 12.1.2.

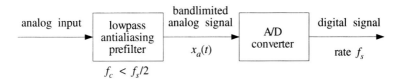

Fig. 12.1.2 Prefiltering and sampling of analog signal.

The prefilter ensures that the spectral images generated by the sampling process at integral multiples of f_s do not overlap, as required by the sampling theorem. This is shown in Fig. 12.1.3 (we ignore here the scaling factor $1/T$).

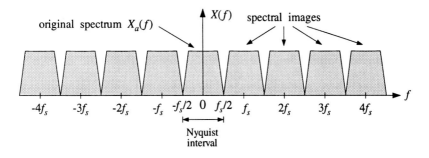

Fig. 12.1.3 Spectrum of signal sampled at low rate f_s.

After digital processing, the sampled signal is reconstructed back to analog form by a D/A *staircase* reconstructor, followed by an analog *anti-image lowpass postfilter* with effective cutoff $f_s/2$, as seen in Fig. 12.1.4.

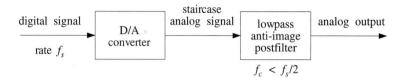

Fig. 12.1.4 Analog reconstruction of sampled signal.

The D/A converter, with its typical $\sin x/x$ response, removes the spectral images partially; the postfilter completes their removal. The combination of the staircase DAC and the postfilter emulates the *ideal* reconstructing analog filter. The ideal reconstructor is a lowpass filter with cutoff the Nyquist frequency $f_s/2$. It has a very sharp transition between its passband, that is, the Nyquist interval, and its stopband, as shown in Fig. 12.1.5.

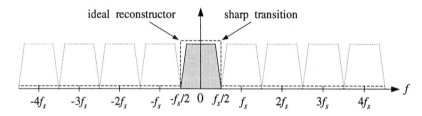

Fig. 12.1.5 Ideal reconstructor removes spectral images due to sampling.

In hi-fi applications such as digital audio, to maintain high quality in the resulting reconstructed analog signal, a very high quality analog postfilter is required, which may be expensive. One way to alleviate the need for a high quality postfilter is to increase the sampling rate. This would cause the spectral images to be more widely separated and, therefore, require a less stringent, simpler, lowpass postfilter. This is depicted in Fig. 12.1.6, for a new sampling rate that is four times higher than required, $f_s' = 4f_s$.

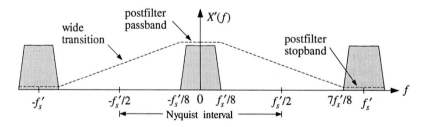

Fig. 12.1.6 Spectrum of signal resampled at high rate $4f_s$, and postfilter requirements.

The *passband* of the postfilter extends up to $f_{\text{pass}} = f_s'/8 = f_s/2$, but its *stopband* need only begin at $f_{\text{stop}} = f_s' - f_s'/8 = 7f_s'/8$. It is this wide transition region between passband and stopband that allows the use of a less stringent postfilter.

For example, in oversampled digital audio applications, simple third-order Butterworth or Bessel analog postfilters are used. See Section 12.4.4.

The same conclusion can also be drawn in the time domain. Figure 12.1.7 shows the staircase output of the D/A converter for the two sampling rates f_s and $f_s' = 4f_s$. It is evident from this figure that the higher the sampling rate, the more closely the staircase output approximates the true signal, and the easier it is for the postfilter to smooth out the staircase levels.

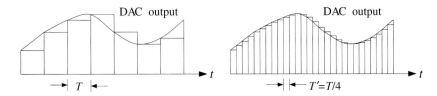

Fig. 12.1.7 Staircase DAC output is smoothed more easily in the oversampled case.

The above approach, however, is impractical because it requires the actual *resampling* of the analog signal at the higher rate f_s'. For example, in a CD player the low rate samples are already stored on the CD at the prescribed rate of 44.1 kHz and the audio signal cannot be resampled.

The philosophy of oversampling is to increase the sampling rate *digitally* using an interpolation filter which operates only on the available *low-rate* input samples. With respect to the new rate f_s' and new Nyquist interval $[-f_s'/2, f_s'/2]$, the spectrum of the low-rate samples depicted in Fig. 12.1.3 will be as shown in Fig. 12.1.8. This is also the spectrum of the high-rate upsampled signal at the output of the rate expander in Fig. 12.1.1.

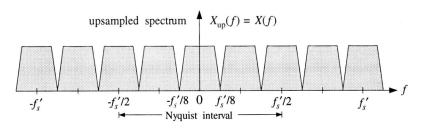

Fig. 12.1.8 Spectrum of low-rate samples with respect to the high rate $4f_s$.

A digital lowpass FIR filter with cutoff frequency $f_s'/8$ and operating at the high rate f_s', would eliminate the three spectral replicas that lie between replicas at multiples of f_s', resulting in a spectrum that is *identical* to that of a signal sampled at the high rate f_s', like that shown in Fig. 12.1.6.

The digital filter, being periodic in f with period f_s', cannot of course remove the spectral replicas that are centered at integral multiples of f_s'. Those are removed later by the D/A reconstructor and the anti-image analog postfilter.

The effect of such a digital filter on the spectrum of the low-rate samples is shown in Fig. 12.1.9, both with respect to the physical frequency f in Hz and the

12. INTERPOLATION, DECIMATION, AND OVERSAMPLING

corresponding digital frequency, $\omega' = 2\pi f / f_s'$, in radians/sample.

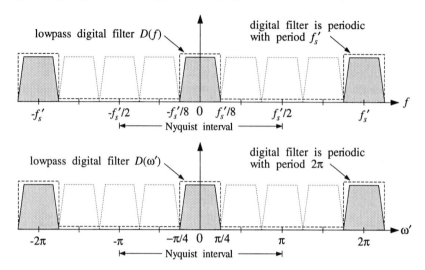

Fig. 12.1.9 High-rate FIR interpolator removes intermediate spectral images.

In summary, a substantial part of the analog reconstruction process is accomplished by DSP methods, that is, using a digital oversampling filter to remove several adjacent spectral replicas and thereby easing the requirements of the analog postfilter. The required sharp transition characteristics of the overall reconstructor are provided by the digital filter. Thus, the high-quality *analog* postfilter is traded off for a high-quality *digital* filter operating at a higher sampling rate. The overall system is depicted in Fig. 12.1.10.

Fig. 12.1.10 4-times oversampling digital filter helps analog reconstruction.

How does an interpolation filter operate in the time domain and calculate the missing signal values between low-rate samples? To illustrate the type of operations it must carry out, consider a 4-fold interpolator and a set of six successive low-rate samples $\{A, B, C, D, E, F\}$ as shown in Fig. 12.1.11.

The filter calculates three intermediate samples, such as $\{X, Y, Z\}$, between any two low-rate samples by forming *linear combinations* of the surrounding low-rate samples. Depending on the type of interpolator and desired quality of the calculated values, several different ways of calculating $\{X, Y, Z\}$ are possible. For example, the simplest one is to keep the value of the previous sample C constant throughout the sampling interval and define:

$$X = Y = Z = C$$

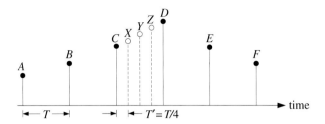

Fig. 12.1.11 Filter calculates missing samples from the surrounding low-rate samples.

This choice corresponds to the so-called *hold* interpolator. Another simple possibility is to interpolate *linearly* between samples $\{C, D\}$ calculating $\{X, Y, Z\}$ as follows:

$$X = 0.75C + 0.25D$$
$$Y = 0.50C + 0.50D \tag{12.1.1}$$
$$Z = 0.25C + 0.75D$$

Indeed, the straight line connecting C and D is parametrized as $C + (D - C)t/T$, for $0 \le t \le T$. Setting $t = T'$, $2T'$, $3T'$ with $T' = T/4$ gives the above expressions for $\{X, Y, Z\}$. For more accurate interpolation, more surrounding samples must be taken into account. For example, using four samples we have:

$$X = -0.18B + 0.90C + 0.30D - 0.13E$$
$$Y = -0.21B + 0.64C + 0.64D - 0.21E \tag{12.1.2}$$
$$Z = -0.13B + 0.30C + 0.90D - 0.18E$$

corresponding to a length-17 FIR approximation to the ideal interpolation filter. Similarly, a length-25 approximation to the ideal interpolator uses six surrounding low-rate samples as follows:

$$X = 0.10A - 0.18B + 0.90C + 0.30D - 0.13E + 0.08F$$
$$Y = 0.13A - 0.21B + 0.64C + 0.64D - 0.21E + 0.13F \tag{12.1.3}$$
$$Z = 0.08A - 0.13B + 0.30C + 0.90D - 0.18E + 0.10F$$

In general, the more the surrounding samples, the more accurate the calculated values. In typical CD players with 4-times oversampling filters, about 20–30 surrounding low-rate samples are used.

The above expressions do not quite look like the convolutional equations of linear filtering. They are special cases of the polyphase realizations of the interpolation filters and are equivalent to convolution. They will be discussed in detail in the next section, where starting with the frequency domain specifications of the filter, its impulse response and corresponding direct and polyphase realization forms are derived. See also Section 12.4.1.

12.2 Interpolation Filter Design*

12.2.1 Direct Form

Consider the general case of an L-fold interpolator, which increases the sampling rate by a factor of L, that is, $f_s' = Lf_s$. The L-fold rate expander inserts $L - 1$ zeros between adjacent low-rate samples and the corresponding $L - 1$ interpolated values are calculated by an FIR digital filter operating at the high rate Lf_s, as shown in Fig. 12.2.1.

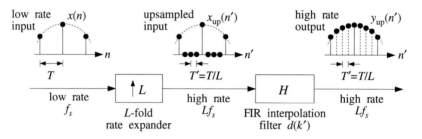

Fig. 12.2.1 L-fold digital interpolator.

Let $x(n)$ denote the low-rate samples that are input to the rate expander and let $x_{\text{up}}(n')$ be its high-rate output, consisting of the low-rate samples separated by $L - 1$ zeros. With respect to the high-rate time index n', the low-rate samples occur every L high-rate ones, that is, at integral multiples of L, $n' = nL$,

$$x_{\text{up}}(nL) = x(n) \tag{12.2.1}$$

The $L - 1$ intermediate samples between $x_{\text{up}}(nL)$ and $x_{\text{up}}(nL + L)$ are zero:

$$x_{\text{up}}(nL + i) = 0, \qquad i = 1, 2, \ldots, L - 1 \tag{12.2.2}$$

This is shown in Fig. 12.2.2. More compactly, the upsampled signal $x_{\text{up}}(n')$ can be defined with respect to the high-rate time index n' by:

$$x_{\text{up}}(n') = \begin{cases} x(n), & \text{if } n' = nL \\ 0, & \text{otherwise} \end{cases} \tag{12.2.3}$$

Given an arbitrary value of the high-rate index n', we can always write it *uniquely* in the form $n' = nL + i$, where i is restricted to the range of values $i = 0, 1, \ldots, L-1$.

Mathematically, the integers n and i are the quotient and remainder of the division of n' by L. Intuitively, this means that n' will either fall exactly on a low-rate sample (when $i = 0$), or will fall strictly between two of them ($i \neq 0$). Using $T = LT'$, we find the absolute time in seconds corresponding to n'

$$t = n'T' = nLT' + iT' = nT + iT'$$

that is, it will be offset from a low-rate sampling time by i high-rate sampling units T'. The interpolated values must be computed at these times.

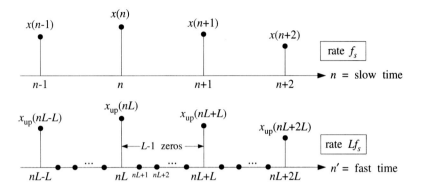

Fig. 12.2.2 Low-rate samples with respect to the slow and fast time scales.

The ideal L-fold interpolation filter is a lowpass filter, operating at the fast rate f_s', with cutoff frequency equal to the *low-rate* Nyquist frequency $f_c = f_s/2$, or in terms of f_s',

$$\boxed{f_c = \frac{f_s}{2} = \frac{f_s'}{2L}} \qquad (12.2.4)$$

and expressed in units of the digital frequency, $\omega' = 2\pi f/f_s'$:

$$\boxed{\omega_c' = \frac{2\pi f_c}{f_s'} = \frac{\pi}{L}} \qquad (12.2.5)$$

The frequency response of this filter is shown in Fig. 12.2.3. Its passband *gain* is taken to be L instead of unity. This is justified below. The ideal impulse response coefficients are obtained from the inverse Fourier transform:

$$d(k') = \int_{-\pi}^{\pi} D(\omega') e^{j\omega'k'} \frac{d\omega'}{2\pi} = \int_{-\pi/L}^{\pi/L} L e^{j\omega'k'} \frac{d\omega'}{2\pi} = \frac{\sin(\pi k'/L)}{\pi k'/L}$$

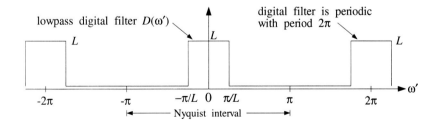

Fig. 12.2.3 Ideal lowpass digital filter operating at high rate Lf_s.

An *FIR approximation* to the ideal interpolator is obtained by truncating $d(k')$ to finite length, say $N = 2LM + 1$:

$$\boxed{d(k') = \frac{\sin(\pi k'/L)}{\pi k'/L}}, \qquad -LM \le k' \le LM \qquad (12.2.6)$$

A *causal* version of the filter may be obtained by delaying it by LM samples:

$$h(n') = d(n' - LM) = \frac{\sin(\pi(n' - LM)/L)}{\pi(n' - LM)/L}, \qquad n' = 0, 1, \dots, N - 1$$

And a *windowed* version is obtained by:

$$\boxed{h(n') = w(n')d(n' - LM)}, \qquad n' = 0, 1, \dots, N - 1 \qquad (12.2.7)$$

where $w(n')$ is an appropriate length-N window, such as a Hamming window:

$$w(n') = 0.54 - 0.46 \cos\left(\frac{2\pi n'}{N - 1}\right), \qquad n' = 0, 1, \dots, N - 1$$

or a Kaiser window. The output of the ideal FIR interpolation filter is obtained by the convolution of the upsampled input $x_{\text{up}}(n')$ with the impulse response $d(k')$:

$$\boxed{y_{\text{up}}(n') = \sum_{k'=-LM}^{LM} d(k')x_{\text{up}}(n' - k')}, \qquad n' = 0, 1, \dots, N - 1 \qquad (12.2.8)$$

12.2.2 Polyphase Form

The interpolated values between the low-rate samples $x_{\text{up}}(nL)$ and $x_{\text{up}}(nL + L)$, that is, the values at the high-rate time instants $n' = nL + i$, are calculated by the filter as follows:

$$y_{\text{up}}(nL + i) = \sum_{k'=-LM}^{LM} d(k')x_{\text{up}}(nL + i - k'), \qquad i = 0, 1, \dots, L - 1 \qquad (12.2.9)$$

Writing uniquely $k' = kL + j$, with $0 \le j \le L - 1$, and replacing the single summation over k' by a double summation over k and j, we find

$$y_{\text{up}}(nL + i) = \sum_{k=-M}^{M-1} \sum_{j=0}^{L-1} d(kL + j)x_{\text{up}}(nL + i - kL - j)$$

To be precise, for the case $i = 0$, the summation over k should be over the range $-M \le k \le M$. But as we will see shortly, the term $k = M$ does not contribute to the sum. Defining the ith *polyphase subfilter* by[†]

$$\boxed{d_i(k) = d(kL + i)}, \qquad -M \le k \le M - 1 \qquad (12.2.10)$$

[†]For $i = 0$, the range of k is $-M \le k \le M$.

for $i = 0, 1, \ldots, L - 1$, we can rewrite the ith interpolated sample value as:

$$y_{\mathrm{up}}(nL + i) = \sum_{k=-M}^{M-1} \sum_{j=0}^{L-1} d_j(k) x_{\mathrm{up}}(nL - kL + i - j)$$

But the upsampled input signal is non-zero only at times that are integral multiples of L. Therefore, using Eqs. (12.2.1) and (12.2.2), we have

$$x_{\mathrm{up}}(nL - kL + i - j) = 0, \qquad \text{if } i \neq j$$

This follows from the fact that $|i - j| \leq L - 1$. Thus, keeping only the $j = i$ term in the above convolution sum, we obtain

$$y_{\mathrm{up}}(nL + i) = \sum_{k=-M}^{M-1} d_i(k) x_{\mathrm{up}}(nL - kL), \qquad i = 0, 1, \ldots, L - 1 \qquad (12.2.11)$$

or, in terms of the low-rate samples:

$$\boxed{y_i(n) = \sum_{k=-M}^{M-1} d_i(k) x(n - k)}, \qquad i = 0, 1, \ldots, L - 1 \qquad (12.2.12)$$

where we set $y_i(n) = y_{\mathrm{up}}(nL + i)$. Thus, the ith interpolated value, $y_{\mathrm{up}}(nL + i)$, is computed by the ith polyphase subfilter, $d_i(k)$, which has length $2M$ and is acting only on the *low-rate* input samples $x(n)$. Each interpolated value is computed as a linear combination of M low-rate samples above and M below the desired interpolation time, as shown in Fig. 12.2.4.

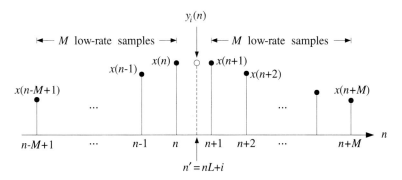

Fig. 12.2.4 Interpolator uses M low-rate samples before and after $y_{\mathrm{up}}(nL + i)$.

Using the L subfilters, interpolation is performed at a *reduced* computational cost as compared with the cost of the full, length-N, interpolation filter $d(k')$ acting on the upsampled signal $x_{\mathrm{up}}(n')$ by Eq. (12.2.9).

The computational cost of Eq. (12.2.9) is essentially $2LM$ multiplications per interpolated value, or, $2L^2M$ multiplications for computing L interpolated values.

By contrast, Eq. (12.2.11) requires $2M$ multiplications per polyphase subfilter, or, $2LM$ multiplications for L interpolated values. Thus, the polyphase subfilter implementation achieves a factor of L in computational savings.

Another way to view the computational rate is in terms of the total number of *multiplications per second* required for the filtering operations, that is,

$$R = N(Lf_s) = NLf_s \qquad \text{(direct form)}$$

$$R = L(2M)f_s = Nf_s \qquad \text{(polyphase form)}$$

where in the *direct* form we have a single filter of length N operating at rate Lf_s and in the *polyphase* form we have L filters operating at f_s, each having computational rate $(2M)f_s$ multiplications per second.[†]

The polyphase implementation is depicted in Fig. 12.2.5, where during each low-rate sampling period T, the commutator reads, in sequence of $T' = T/L$ seconds, the L interpolated values at the outputs of the subfilters.

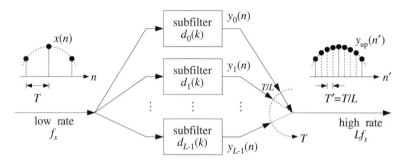

Fig. 12.2.5 Polyphase subfilter implementation of digital interpolator.

This can be seen more formally, as follows. Let ζ^{-1} denote the unit delay with respect to the high rate Lf_s and let z^{-1} denote the low-rate delay. Since L high-rate delays equal one low-rate delay, that is, $LT' = T$, we will have:

$$z = \zeta^L \qquad \Leftrightarrow \qquad \zeta = z^{1/L} \tag{12.2.13}$$

The ζ-transform of the high-rate filter output $y_{\text{up}}(n')$ can be expressed in terms of the z-transforms of the L low-rate output signals $y_i(n)$ as follows. Writing uniquely $n' = nL + i$ with $0 \le i \le L - 1$, we have

$$Y_{\text{up}}(\zeta) = \sum_{n'=-\infty}^{\infty} y_{\text{up}}(n')\zeta^{-n'} = \sum_{i=0}^{L-1}\sum_{n=-\infty}^{\infty} y_{\text{up}}(nL + i)\zeta^{-nL-i}$$

$$= \sum_{i=0}^{L-1}\zeta^{-i}\sum_{n=-\infty}^{\infty} y_i(n)\zeta^{-Ln}, \qquad \text{or,}$$

[†]Actually, there are $L-1$ subfilters of length $(2M)$ and one (the filter d_0) of length $(2M+1)$, giving rise to $R = (L-1)(2M)f_s + (2M+1)f_s = Nf_s$, where $N = 2LM + 1$.

$$Y_{up}(\zeta) = \sum_{i=0}^{L-1} \zeta^{-i} Y_i(\zeta^L) = \sum_{i=0}^{L-1} z^{-i/L} Y_i(z) \tag{12.2.14}$$

which shows how the L low-rate output signals are put together, in sequence of T/L high-rate delays to make up the high-rate interpolated output signal. In a similar fashion, we can derive the relationship between the ζ-transform of the high-rate filter (12.2.6) and the z-transforms of its polyphase subfilters (12.2.10), justifying the realization of Fig. 12.2.5:

$$D(\zeta) = \sum_{i=0}^{L-1} \zeta^{-i} D_i(\zeta^L) = \sum_{i=0}^{L-1} z^{-i/L} D_i(z) \tag{12.2.15}$$

Next, we consider the 0th polyphase subfilter, $d_0(k)$, which plays a special role. It follows from Eqs. (12.2.6) and (12.2.10) that:

$$d_0(k) = d(kL) = \frac{\sin(\pi k)}{\pi k} = \delta(k), \qquad -M \le k \le M$$

and therefore, its output will be trivially equal to its input, that is, the low-rate input sample $x(n) = x_{up}(nL)$. We have from Eq. (12.2.12):

$$y_0(n) = y_{up}(nL) = \sum_{k=-M}^{M-1} d_0(k) x(n-k) = \sum_{k=-M}^{M-1} \delta(k) x(n-k) = x(n)$$

This property is preserved even for the windowed case of Eq. (12.2.7), because all windows $w(n')$ are equal to unity at their middle. This result justifies the requirement for the passband gain of the interpolator filter in Fig. 12.2.3 to be L instead of 1. If the gain were 1, we would have $y_{up}(nL) = x(n)/L$. An alternative, frequency domain justification is given in Section 12.2.3.

The causal filter implementation of Eq. (12.2.12) requires that we either delay the output or *advance the input* by M units. We choose the latter. The polyphase subfilters in Eqs. (12.2.12) can be made causal by a delay of M low-rate samples:

$$h_i(n) = d_i(n-M) = d((n-M)L + i) = d(nL + i - LM) \tag{12.2.16}$$

for $n = 0, 1, \ldots, 2M - 1$. For the windowed case, we have:

$$h_i(n) = d(nL + i - LM) w(nL + i), \qquad n = 0, 1, \ldots, 2M - 1 \tag{12.2.17}$$

In terms of the causal subfilters $h_i(n)$, Eq. (12.2.12) becomes

$$y_i(n) = \sum_{k=-M}^{M-1} d_i(k) x(n-k) = \sum_{k=-M}^{M-1} h_i(k+M) x(n-k)$$

or, setting $m = k + M$ and $k = m - M$,

$$\boxed{y_i(n) = \sum_{m=0}^{P} h_i(m)x(M+n-m)}, \qquad i = 0, 1, \ldots, L-1 \qquad (12.2.18)$$

where $P = 2M - 1$ denotes the *order* of each polyphase subfilter. In other words, the interpolated samples are obtained by ordinary *causal* FIR filtering of the *time-advanced* low-rate input samples. The same result can also be obtained by z-transforms (recall also the implementation of Savitzky-Golay filters in Section 8.3.5.) Definition (12.2.16) reads in the z-domain:

$$H_i(z) = z^{-M}D_i(z)$$

where z^{-1} represents a low-rate delay. Similarly, Eq. (12.2.12) reads

$$Y_i(z) = D_i(z)X(z)$$

Writing $D_i(z) = z^M H_i(z)$, we obtain the z-domain equivalent of Eq. (12.2.18):

$$Y_i(z) = D_i(z)X(z) = H_i(z)\left(z^M X(z)\right)$$

The sample-by-sample processing implementation of Eq. (12.2.18) requires a *common* low-rate tapped delay line which is used in sequence by *all* the subfilters $h_i(n)$ before its contents are updated. Figure 12.4.5 shows a concrete example when $L = 4$ and $M = 2$. The required time-advance by M samples is implemented by initially filling the delay line with the first M low-rate samples. The internal states of the tapped delay line can be defined as

$$w_m(n) = x(M+n-m), \qquad m = 0, 1, \ldots, P$$

Then, Eq. (12.2.18) can be written in the dot-product notation of Chapter 4:

$$y_i(n) = \text{dot}(P, \mathbf{h}_i, \mathbf{w}(n)) = \sum_{m=0}^{P} h_i(m)w_m(n), \qquad i = 0, 1, \ldots, L-1$$

After computing the outputs of the L subfilters, the internal state $\mathbf{w}$ may be updated to the next time instant by a call to the routine `delay`, which shifts the contents:

$$w_m(n+1) = w_{m-1}(n), \qquad m = 1, 2, \ldots, P$$

This leads to the following *sample processing algorithm* for the polyphase form: Initialize the internal state vector $\mathbf{w}(n) = [w_0(n), w_1(n), \ldots, w_P(n)]$ by filling it with the first M low-rate input samples, $x(0), x(1), \ldots, x(M-1)$, that is, at time $n = 0$ start with

$$\mathbf{w}(0) = [0, x_{M-1}, x_{M-2}, \ldots, x_0, \underbrace{0, 0, \ldots, 0}_{M-1 \text{ zeros}}]$$

The value $w_0(0)$ need not be initialized—it is read as the current input sample. If the low-rate samples are being read sequentially from a file or an input port, then this initialization can be implemented by the following algorithm:

$$\boxed{\begin{array}{l} \textit{for } m = M \textit{ down to } m = 1 \textit{ do:} \\ \quad \textit{read low-rate input sample } x \\ \quad w_m = x \end{array}} \qquad (12.2.19)$$

Then, proceed by reading each successive low-rate sample, $x(M + n)$, $n = 0, 1, \ldots$, and processing it by the algorithm:

$$\boxed{\begin{array}{l} \textit{for each low-rate input sample } x \textit{ do:} \\ \quad w_0 = x \\ \quad \textit{for } i = 0, 1, \ldots, L - 1 \textit{ compute:} \\ \quad\quad y_i = \text{dot}(P, \mathbf{h}_i, \mathbf{w}) \\ \quad \text{delay}(P, \mathbf{w}) \end{array}} \qquad (12.2.20)$$

12.2.3 Frequency Domain Characteristics

Finally, we look in more detail at the passband and stopband characteristics of the ideal lowpass interpolation filter. Let $T = 1/f_s$ and $T' = 1/f_s' = T/L$ be the sampling time periods with respect to the low and high rates f_s and f_s'. With reference to Fig. 12.1.2, let $x_a(t)$ be the output of the prefilter and let $X_a(f)$ be its bandlimited spectrum. The spectrum of the sampled low-rate signal $x(n) = x_a(nT)$, shown in Fig. 12.1.3, will be related to $X_a(f)$ by the Poisson summation formula:

$$X(f) = \sum_n x(n) e^{-2\pi jfnT} = \frac{1}{T} \sum_{m=-\infty}^{\infty} X_a(f - mf_s)$$

The upsampled signal $x_{\text{up}}(n')$ at the output of the L-fold rate expander of Fig. 12.2.1 has exactly the *same spectrum* as $x(n)$, as indicated in Fig. 12.1.8. Indeed, using Eqs. (12.2.1) and (12.2.2) and $T = LT'$, we have

$$X_{\text{up}}(f) = \sum_{n'} x_{\text{up}}(n') e^{-2\pi jfn'T'} = \sum_n x_{\text{up}}(nL) e^{-2\pi jfnLT'}$$

$$= \sum_n x_{\text{up}}(nL) e^{-2\pi jfnT} = \sum_n x(n) e^{-2\pi jfnT} = X(f)$$

Thus,

$$\boxed{X_{\text{up}}(f) = X(f) = \frac{1}{T} \sum_{m=-\infty}^{\infty} X_a(f - mf_s)} \qquad (12.2.21)$$

The same relationship can be expressed in terms of the digital frequencies as:

$$X_{\text{up}}(\omega') = X(\omega) = X(\omega'L)$$

where

$$\omega = \frac{2\pi f}{f_s} = 2\pi f T, \qquad \omega' = \frac{2\pi f}{f_s'} = 2\pi f T', \qquad \omega = \omega' L$$

and

$$X_{\text{up}}(\omega') = \sum_{n'} x_{\text{up}}(n') e^{-j\omega' n'}, \qquad X(\omega) = \sum_{n} x(n) e^{-j\omega n}$$

Similarly, using Eq. (12.2.13), their z-transforms will be related by

$$X_{\text{up}}(\zeta) = X(z) = X(\zeta^L)$$

where the slow and fast z variables are related to the corresponding digital frequencies by

$$z = e^{j\omega} = e^{2\pi jf/f_s}, \qquad \zeta = e^{j\omega'} = e^{2\pi jf/f_s'} = e^{2\pi jf/Lf_s}$$

If the analog signal $x_a(t)$ had actually been *resampled* at the higher rate f_s' giving rise to the sampled signal $x'(n') = x_a(n'T')$, then the corresponding spectrum, depicted in Fig. 12.1.6, would be:

$$X'(f) = \sum_{n'} x_a(n'T') e^{-2\pi jf n'T'} = \frac{1}{T'} \sum_{m'} X_a(f - m'f_s') \qquad (12.2.22)$$

The difference between $x'(n')$ and $x_{\text{up}}(n')$ is that $x'(n')$ contains the correct interpolated values between low-rate samples, whereas the $x_{\text{up}}(n')$ is zero there.

In the time domain, the job of an ideal interpolation filter is to reproduce the interpolated samples correctly, that is, its output is required to be $y_{\text{up}}(n') = x'(n')$ for all n'. In the frequency domain, its job is to *reshape* the low-rate sampled spectrum $X(f)$, shown in Fig. 12.1.8, into the high-rate spectrum $X'(f)$ shown in Fig. 12.1.6. Denoting the ideal interpolation filter by $D(f)$, we have for the spectrum of the output $y_{\text{up}}(n')$:

$$Y_{\text{up}}(f) = D(f) X_{\text{up}}(f) = D(f) X(f)$$

The filter output is required to be $Y_{\text{up}}(f) = X'(f)$, thus,

$$\boxed{X'(f) = D(f) X(f)} \qquad \text{(ideal interpolation)} \qquad (12.2.23)$$

for all f. This condition determines the ideal passband and stopband specifications for $D(f)$. Using Eqs. (12.2.21) and (12.2.22) and separating out the central replica of $X'(f)$ and the first L replicas of $X(f)$, we rewrite the above condition as

$$\frac{1}{T'} X_a(f) + \text{replicas} = \underbrace{\frac{1}{T} D(f) X_a(f)}_{\text{passband}} + \underbrace{\frac{1}{T} D(f) \sum_{m=1}^{L-1} X_a(f - mf_s)}_{\text{stopband}} + \text{replicas}$$

Because $X_a(f)$ is bandlimited to within $[-f_s/2, f_s/2]$, it follows that the $L-1$ intermediate replicas $\sum_{m=1}^{L-1} X_a(f - mf_s)$ will be bandlimited to within $[f_s/2, Lf_s - f_s/2]$. The filter $D(f)$ is required to remove these replicas, that is,

$$D(f) = 0, \qquad \frac{f_s}{2} \le |f| \le Lf_s - \frac{f_s}{2}$$

as shown in Fig. 12.2.3. Similarly, within the low-rate Nyquist interval $-f_s/2 \le f \le f_s/2$, the filter must satisfy:

$$\frac{1}{T'} X_a(f) = \frac{1}{T} D(f) X_a(f) \qquad \Rightarrow \qquad D(f) = \frac{T}{T'} = L$$

This justifies the choice L for the passband gain. In summary, the ideal digital interpolation filter $D(f)$ is defined as follows over the high-rate Nyquist interval $[-f_s'/2, f_s'/2]$:

$$\text{(ideal interpolator)} \qquad \boxed{D(f) = \begin{cases} L, & \text{if } |f| \le \dfrac{f_s}{2} \\[2mm] 0, & \text{if } \dfrac{f_s}{2} < |f| \le \dfrac{f_s'}{2} \end{cases}} \qquad (12.2.24)$$

and is periodically extended outside that interval. It is depicted in Figs. 12.1.9 and 12.2.3. Its impulse response is given by Eq. (12.2.6).

The operation of the ideal interpolation filter, expressed by Eq. (12.2.23), can also be understood in the time domain in terms of the sampling theorem. The sampled analog signal $x(n) = x_a(nT)$ can be reconstructed to analog form by the analog reconstructor :

$$x_a(t) = \sum_n x_a(nT) h(t - nT)$$

where $h(t)$ is the ideal reconstructor for the rate f_s:

$$h(t) = \frac{\sin(\pi t/T)}{\pi t/T} \qquad (12.2.25)$$

Resampling at the higher rate f_s' gives the sampled signal:

$$x'(n') = x_a(n'T') = \sum_n x_a(nT) h(n'T' - nT) = \sum_n x_a(nLT') h(n'T' - nLT')$$

Denoting

$$d(k') = h(k'T') = \frac{\sin(\pi k'T'/T)}{\pi k'T'/T} = \frac{\sin(\pi k'/L)}{\pi k'/L} \qquad (12.2.26)$$

and using $x_a(nLT') = x_a(nT) = x(n)$, we obtain $h(n'T' - nLT') = d(n' - nL)$ and the filtering equation:

$$x'(n') = \sum_n d(n' - nL) x(n) = \sum_{m'} d(n' - m') x_{\text{up}}(m')$$

which is recognized as the time-domain version of Eq. (12.2.23).

In summary, the effect of the ideal interpolator in the frequency domain is shown in Fig. 12.2.6. The input spectrum consists of replicas at multiples of the *input* sampling rate f_s. The filter removes all of these replicas, *except* those that are multiples of the *output* rate Lf_s. The output spectrum consists only of replicas at multiples of Lf_s. (The scaling by the gain L is not shown.)

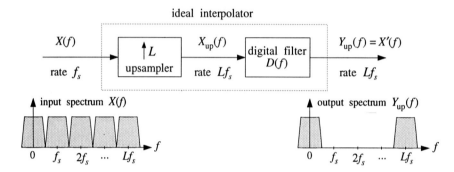

Fig. 12.2.6   Ideal interpolator in frequency domain.

12.2.4 Kaiser Window Designs

Digital interpolation filters can be designed by a variety of filter design methods, such as the Fourier series method with windowing, Parks-McClellan, or IIR designs. Here we summarize FIR designs based on the Kaiser window method.

We follow the design steps of Section 10.2, but use f_s' in place of f_s, because the interpolation filter is operating at the fast rate f_s'. For any length-N window $w(n')$, the interpolator's impulse response is computed by

$$h(n') = w(n')d(n' - LM), \qquad n' = 0, 1, \ldots, N - 1 = 2LM \qquad (12.2.27)$$

The L length-$(2M)$ *polyphase subfilters* are defined in terms of $h(n')$ as follows. For $i = 0, 1, \ldots, L - 1$:

$$h_i(n) = h(nL + i), \qquad n = 0, 1, \ldots, 2M - 1 \qquad (12.2.28)$$

For a Kaiser window design, we start by specifying the desired stopband attenuation A in dB, and desired transition width Δf about the ideal cutoff frequency:

$$f_c = \frac{f_s'}{2L} = \frac{f_s}{2}$$

so that the passband and stopband frequencies are:

$$f_{\text{pass}} = f_c - \frac{1}{2}\Delta f, \quad f_{\text{stop}} = f_c + \frac{1}{2}\Delta f$$

The Kaiser window parameters are calculated by:

$$\delta = 10^{-A/20}$$

$$D = \frac{A - 7.95}{14.36}$$

$$\alpha = 0.1102\,(A - 8.7) \qquad \text{(because, typically, } A > 50 \text{ dB)}$$

$$N - 1 \geq \frac{Df_s'}{\Delta f} = \frac{DLf_s}{\Delta f} = \frac{DL}{\Delta F}$$

(12.2.29)

where we used f_s' in the formula for N and set $\Delta F = \Delta f/f_s$. Then, N must be rounded up to the smallest odd integer of the form $N = 2LM + 1$ satisfying the above inequality.

The design specifications are shown in Fig. 12.2.7. The designed length-N impulse response is given by Eq. (12.2.27), with $w(n')$ given for $n' = 0, 1, \ldots, N - 1$:

$$w(n') = \frac{I_0\left(\alpha\sqrt{1 - (n' - LM)^2/(LM)^2}\right)}{I_0(\alpha)} = \frac{I_0\left(\alpha\sqrt{n'\,(2LM - n')}/LM\right)}{I_0(\alpha)}$$

The frequency response of the designed filter may be computed by:

$$H(f) = \sum_{n'=0}^{N-1} h(n')e^{-2\pi j f n'/f_s'} = \sum_{n'=0}^{N-1} h(n')e^{-2\pi j f n'/(Lf_s)}$$

The designed filter $h(n')$ can be implemented in its direct or polyphase forms.

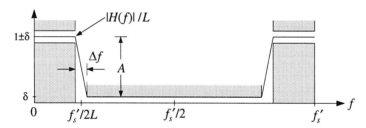

Fig. 12.2.7 Kaiser design specifications for L-fold interpolation filter.

12.2.5 Multistage Designs

Interpolation filters can also be implemented in a multistage form, whereby the sampling rate is gradually increased in stages until the final rate is reached. This is shown in Fig. 12.2.8. The first filter increases the sampling rate by a factor of L_0, the second by a factor of L_1, and the third by a factor of L_2, so that the overall interpolation factor is $L = L_0L_1L_2$. Such multistage realizations allow additional savings in the overall computational rate of the interpolator.

The first filter $H_0(f)$ must have the most *stringent* specifications in the sense that it has the desired transition width Δf, which is typically very narrow. The

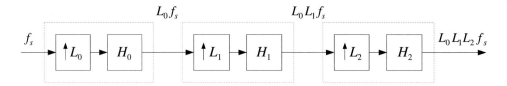

Fig. 12.2.8 Three-stage interpolation filter.

remaining stages have much *wider* transition widths and therefore smaller filter lengths.

To see this, consider the design of a 4-fold interpolator realized as the cascade of two 2-fold interpolators, $L = L_0 L_1$, with $L_0 = L_1 = 2$. The desired ideal frequency characteristics of the two interpolation filters are depicted in Fig. 12.2.9.

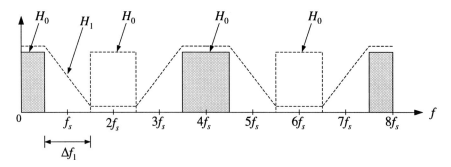

Fig. 12.2.9 Two-stage 2×2 interpolation filter.

The first interpolator H_0 is operating at the intermediate rate $f_s' = L_0 f_s = 2f_s$ and is designed to act as an ideal lowpass filter with cutoff $f_c = f_s/2 = f_s'/4$. It removes all the replicas at multiples of its input rate f_s, except those that are multiples of its output rate $2f_s$.

It can be designed using a Kaiser window. For example, assuming a narrow transition width Δf about f_c, and a stopband attenuation A, we obtain from Eqs. (12.2.29):

$$N_0 - 1 = \frac{D f_s'}{\Delta f} = \frac{D(2f_s)}{\Delta f} = \frac{2D}{\Delta F}$$

where again $\Delta F = \Delta f / f_s$. The second interpolator H_1 is operating at the rate $2f_s' = 4f_s$, and must remove all replicas at multiples of its input rate $2f_s$, except those that are multiples of its output rate $4f_s$. Therefore, it has a wide transition width given by

$$\Delta f_1 = f_s' - f_s = 2f_s - f_s = f_s$$

Its Kaiser length will be:

$$N_1 - 1 = \frac{D(4f_s)}{\Delta f_1} = \frac{D(4f_s)}{f_s} = 4D$$

The combined effect of the two interpolation filters is to remove every three intervening replicas leaving only the replicas at multiples of $4f_s$. Because H_0 is operating at rate $2f_s$ and H_1 at rate $4f_s$, the corresponding frequency responses will be:

$$H_0(f) = \sum_{n'=0}^{N_0-1} h_0(n') e^{-2\pi j f n'/(2f_s)}, \qquad H_1(f) = \sum_{n'=0}^{N_1-1} h_1(n') e^{-2\pi j f n'/(4f_s)}$$

Assuming that both filters $h_0(n')$ and $h_1(n')$ are realized in their polyphase forms, the total computational rate of the multistage case will be, in MACs per second:

$$R_{\text{multi}} = N_0 f_s + N_1(2f_s) \simeq \left(\frac{2D}{\Delta F} + 8D\right) f_s = \frac{2D}{\Delta F}(1 + 4\Delta F) f_s$$

By contrast, a single stage design would have filter length:

$$N - 1 = \frac{D(Lf_s)}{\Delta f} = \frac{4D}{\Delta F}$$

and polyphase computational rate:

$$R_{\text{single}} = N f_s \simeq \frac{4D}{\Delta F} f_s$$

The relative performance of the multistage versus the single stage designs will be

$$\frac{R_{\text{multi}}}{R_{\text{single}}} = \frac{1 + 4\Delta F}{2} = \frac{1}{2} + 2\Delta F \qquad\qquad (12.2.30)$$

We note that this ratio is *independent* of the filter lengths and stopband attenuations; it depends only on the transition width. Computational savings will take place whenever:

$$\frac{1}{2} + 2\Delta F < 1 \qquad \Leftrightarrow \qquad \Delta F < \frac{1}{4}$$

which is usually satisfied because typical values of ΔF are of the order of 0.1. As another example, consider an 8-fold interpolator which can be realized in three different multistage ways:

$$8 = 2 \times 2 \times 2 = 2 \times 4 = 4 \times 2$$

The frequency characteristics of the different stages are shown in Fig. 12.2.10. The interpolator at each stage removes all replicas at multiples of *its input* rate, except those that are multiples of *its output* rate. In all three cases, the combined effect is to remove every seven intervening replicas leaving only the replicas at the multiples of $8f_s$. For the $2\times2\times2$ case, the transition widths of the three stages are taken to be:

$$\Delta f_0 = \Delta f, \quad \Delta f_1 = 2f_s - f_s = f_s, \quad \Delta f_2 = 4f_s - f_s = 3f_s$$

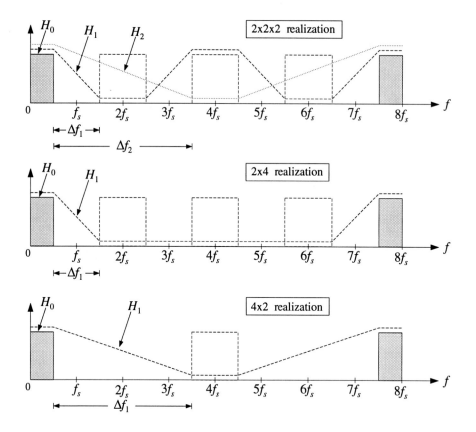

Fig. 12.2.10   Frequency characteristics of multistage 8-fold interpolators.

resulting in Kaiser filter lengths:

$$N_0 - 1 = \frac{D(2f_s)}{\Delta f_0} = \frac{2D}{\Delta F}, \quad N_1 - 1 = \frac{D(4f_s)}{\Delta f_1} = 4D, \quad N_2 - 1 = \frac{D(8f_s)}{\Delta f_2} = \frac{8D}{3}$$

and total polyphase computational rate:

$$R_{\text{multi}} = N_0 f_s + N_1(2f_s) + N_2(4f_s) \simeq \left(\frac{2D}{\Delta F} + 8D + \frac{32D}{3}\right) f_s$$

By contrast, the single stage design would have filter length:

$$N - 1 = \frac{D(8f_s)}{\Delta f} = \frac{8D}{\Delta F}$$

and polyphase computational cost:

$$R_{\text{single}} = N f_s = \frac{8D}{\Delta F} f_s$$

The relative performance of the multistage versus the single stage design is then

$$\frac{R_{\text{multi}}}{R_{\text{single}}} = \frac{1}{4} + \frac{7}{3}\Delta F$$

with savings whenever $\Delta F < 9/28 = 0.321$. In a similar fashion, we find the multistage versus single stage polyphase computational costs of the 2×4 and 4×2 cases:

$$\frac{R_{\text{multi}}}{R_{\text{single}}} = \frac{1}{4} + 2\Delta F \qquad (2\times4 \text{ case})$$

$$\frac{R_{\text{multi}}}{R_{\text{single}}} = \frac{1}{2} + \frac{4}{3}\Delta F \qquad (4\times2 \text{ case})$$

Comparing the three multistage cases, it appears that the 2×4 case is more efficient than the 2×2×2 case, which is more efficient than the 4×2 case. Indeed,

$$\frac{1}{4} + 2\Delta F < \frac{1}{4} + \frac{7}{3}\Delta F < \frac{1}{2} + \frac{4}{3}\Delta F$$

the second inequality being valid for $\Delta F < 1/4$. Some specific design examples will be presented later on.

The *general multistage design* procedure is as follows [273]. Assume there are K interpolation stages $H_0, H_1, \ldots, H_{K-1}$ that increase the sampling rate successively by the factors $L_0, L_1, \ldots, L_{K-1}$. The sampling rate at the input of the ith interpolation filter H_i will be $F_{i-1}f_s$ and at its output it will be increased by a factor L_i, that is, $F_i f_s = L_i F_{i-1} f_s$, where:

$$F_i = L_i F_{i-1} = L_0 L_1 \cdots L_i, \qquad i = 0, 1, \ldots, K - 1$$

We set $F_{-1} = 1$, so that $F_0 = L_0$. The total interpolation factor will be:

$$L = F_{K-1} = L_0 L_1 \cdots L_{K-1}$$

For a Kaiser design, we assume a given transition width Δf about the ideal cutoff frequency of the L-fold interpolator $f_c = f_s/2$ and given *stopband* attenuations in dB for each stage $A_0, A_1, \ldots, A_{K-1}$. Typically, these attenuations will be the same.[†] Next, compute the Kaiser D factors, α parameters, and passband/stopband ripples δ. For $i = 0, 1, \ldots, K - 1$

$$\delta_i = 10^{-A_i/20}$$

$$D_i = \frac{A_i - 7.95}{14.36}$$

$$\alpha_i = 0.1102\,(A_i - 8.7) \qquad (\text{assuming } A_i > 50 \text{ dB})$$

Then, compute the effective transition widths for the interpolation filters:

[†]If the overall multistage output is reconstructed by a DAC and analog postfilter, then the second and later stages can have less attenuation than A_0 because their suppression of the replicas will be aided by the postfilter. See Section 12.4.5.

$$\Delta f_0 = \Delta f$$

$$\Delta f_i = F_{i-1}f_s - f_s = (F_{i-1} - 1)f_s, \quad i = 1, 2, \ldots, K - 1 \tag{12.2.31}$$

The theoretical cutoff frequencies of these filters will be:

$$f_{ci} = \frac{F_i f_s}{2L_i} = \frac{1}{2}F_{i-1}f_s \quad \Rightarrow \quad \omega'_{ci} = \frac{2\pi f_{ci}}{F_i f_s} = \frac{\pi}{L_i}$$

for $i = 0, 1, \ldots, K - 1$. In particular, $f_{c0} = f_s/2$. For Kaiser designs, the above choices of widths imply the following passband and stopband frequencies for the filters. For $i = 0$, we have

$$f_{\text{pass},0} = f_{c0} - \frac{1}{2}\Delta f_0 = \frac{f_s}{2} - \frac{1}{2}\Delta f, \qquad f_{\text{stop},0} = f_{c0} + \frac{1}{2}\Delta f_0 = \frac{f_s}{2} + \frac{1}{2}\Delta f$$

and for $i = 1, 2, \ldots, K - 1$

$$f_{\text{pass},i} = f_{ci} - \frac{1}{2}\Delta f_i = \frac{f_s}{2}, \qquad f_{\text{stop},i} = f_{ci} + \frac{1}{2}\Delta f_i = F_{i-1}f_s - \frac{f_s}{2}$$

Alternatively, we can demand that all filters have exactly the same passband frequency, namely, $f_s/2 - \Delta f/2$. This changes all the stopband frequencies by shifting them down by $\Delta f/2$, that is, we can define for $i = 1, 2, \ldots, K - 1$

$$f_{\text{pass},i} = \frac{f_s}{2} - \frac{1}{2}\Delta f, \qquad f_{\text{stop},i} = F_{i-1}f_s - \frac{f_s}{2} - \frac{1}{2}\Delta f$$

Note that the transition widths remain the same, but the ideal cutoff frequencies also shift down by $\Delta f/2$, that is, for $i = 1, 2, \ldots, K - 1$

$$f_{ci} = \frac{1}{2}(f_{\text{pass},i} + f_{\text{stop},i}) = \frac{1}{2}F_{i-1}f_s - \frac{1}{2}\Delta f \quad \Rightarrow \quad \omega'_{ci} = \frac{2\pi f_{ci}}{F_i f_s} = \frac{\pi}{L_i} - \pi\frac{\Delta F}{F_i}$$

where $\Delta F = \Delta f/f_s$. The frequency characteristics of the first filter H_0 are as shown in Fig. 12.2.7, with $L = L_0$. The specifications of the ith stage are shown in Fig. 12.2.11.

The filter H_i removes all replicas at multiples of its input rate $F_{i-1}f_s$ except those that are multiples of its output rate $F_i f_s$. The replicas between the first replica and that at $F_{i-1}f_s$ have already been removed by the previous stages; hence, the wide transition width Δf_i. The corresponding Kaiser filter lengths are:

$$N_0 - 1 = \frac{D_0 (F_0 f_s)}{\Delta f_0} = \frac{D_0 L_0}{\Delta F}$$

$$N_i - 1 = \frac{D_i (F_i f_s)}{\Delta f_i} = \frac{D_i F_i}{F_{i-1} - 1}, \qquad i = 1, 2, \ldots, K - 1 \tag{12.2.32}$$

where N_i must be rounded to the next smallest integer of the form:

$$N_i = 2L_i M_i + 1, \qquad i = 0, 1, \ldots, K - 1$$

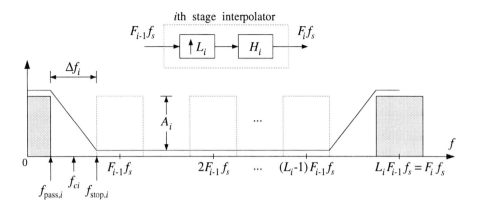

Fig. 12.2.11 H_i removes replicas at multiples of $F_{i-1}f_s$, but not at $F_i f_s$.

The windowed impulse responses of the filters will be, for $i = 0, 1, \ldots, K - 1$

$$h_i(n') = d(L_i, n' - L_iM_i) w(\alpha_i, N_i, n'), \qquad n' = 0, 1, \ldots, N_i - 1$$

where $d(L_i, k')$ is the ideal interpolation filter with cutoff frequency[†] $\omega'_{ci} = \pi/L_i$

$$d(L_i, k') = \frac{\sin(\omega'_{ci} k')}{\omega'_{ci} k'} = \frac{\sin(\pi k'/L_i)}{\pi k'/L_i}$$

and $w(\alpha_i, N_i, n')$ is the corresponding Kaiser window:

$$w(\alpha_i, N_i, n') = \frac{I_0(\alpha_i\sqrt{1 - (n' - L_iM_i)^2/(L_iM_i)^2})}{I_0(\alpha_i)}$$

The polyphase subfilters of each H_i are defined by

$$h_{ij}(n) = h(nL_i + j), \qquad j = 0, 1, \ldots, L_i - 1, \qquad n = 0, 1, \ldots, P_i$$

where $P_i = 2M_i - 1$. Finally, compute the frequency responses of the individual stages:

$$H_i(f) = \sum_{n'=0}^{N_i-1} h_i(n') e^{-2\pi j f n'/(F_i f_s)}, \qquad i = 0, 1, \ldots, K - 1$$

and the total frequency response:

$$H_{\text{tot}}(f) = H_0(f) H_1(f) \cdots H_{K-1}(f)$$

Note that the *effective passband ripple* of the combined filter $H_{\text{tot}}(f)$ is worse than the ripples of the individual factors. This can be seen as follows. In the passband frequency range, the individual filters satisfy

---

[†]For the alternative case, replace ω'_{ci} by its shifted version $\omega'_{ci} = \pi/L_i - \pi\Delta F/F_i$.

$$1 - \delta_i \le \left| \frac{H_i(f)}{L_i} \right| \le 1 + \delta_i$$

where the DC gain L_i has been factored out. Multiplying these inequalities together, we obtain the bounds:

$$\prod_{i=0}^{K-1} (1 - \delta_i) \le \left| \frac{H_0(f) H_1(f) \cdots H_{K-1}(f)}{L_0 L_1 \cdots L_{K-1}} \right| \le \prod_{i=0}^{K-1} (1 + \delta_i)$$

For small δ_i we may use the approximation

$$\prod_{i=0}^{K-1} (1 \pm \delta_i) \simeq 1 \pm \sum_{i=0}^{K-1} \delta_i$$

to get the total passband ripple of the cascaded filter

$$1 - \delta_{\text{tot}} \le \left| \frac{H_{\text{tot}}(f)}{L} \right| \le 1 + \delta_{\text{tot}}, \quad \text{where} \quad \delta_{\text{tot}} = \sum_{i=0}^{K-1} \delta_i$$

For equal ripples $\delta_i = \delta$ and K stages, we have $\delta_{\text{tot}} = K\delta$. If so desired, this effect can be compensated by starting the design of the $H_i(f)$ using δ_i's that are smaller by a factor of K. It should be noted, however, that this may not always be necessary because the individual filters may not reach their extremum values simultaneously over the passband, and therefore the bounds $(1 \pm \delta_{\text{tot}})$ may be too conservative. This is illustrated in the design examples later.

Also, in Kaiser window designs, it is the stopband attenuation A that essentially determines the passband ripples. In order to achieve reasonably high stopband attenuations, for example, in the range 70-100 dB, the corresponding passband ripples will be so small that even if they are multiplied by any K of the order of 10, they will still give rise to excellent passbands.

The total polyphase computational rate in MACs per second is obtained by adding up the computational rates of all the stages, that is,

$$R_{\text{multi}} = \sum_{i=0}^{K-1} R_i, \quad \text{where} \quad R_i = N_i F_{i-1} f_s, \quad i = 0, 1, \ldots, K - 1$$

These follow from the observation that the ith filter operates at rate $F_i f_s$ and would have computational rate $N_i F_i f_s$ in its direct form. But in its polyphase form we save a factor of L_i, resulting in the rate $N_i F_i f_s / L_i = N_i F_{i-1} f_s$. By comparison, the single-stage design will have Kaiser length and polyphase computational rate:

$$N - 1 = \frac{D(L f_s)}{\Delta f} = \frac{DL}{\Delta F}, \quad R_{\text{single}} = N f_s$$

where we may take $A = A_0$ for the stopband attenuation and $D = D_0$. It follows that the relative performance of multistage versus single stage designs will be:

$$\boxed{\frac{R_{\text{multi}}}{R_{\text{single}}} = \frac{N_0 + \sum_{i=1}^{K-1} N_i F_{i-1}}{N}} \tag{12.2.33}$$

Assuming that all the attenuations are the same, $A_i = A$, and therefore all the D_i are the same, we may use the approximations:

$$N_0 \simeq \frac{DL_0}{\Delta F}, \qquad N_i \simeq \frac{DF_i}{F_{i-1} - 1}, \qquad N \simeq \frac{DL}{\Delta F}$$

to obtain the simplified expression:

$$\boxed{\frac{R_{\text{multi}}}{R_{\text{single}}} = \frac{L_0}{L} + \Delta F \sum_{i=1}^{K-1} \frac{F_i F_{i-1}}{(F_{i-1} - 1) L}} \tag{12.2.34}$$

For a two-stage design with $L = L_0 L_1$, $F_1 = L$, and $F_0 = L_0$, we have

$$\frac{R_{\text{multi}}}{R_{\text{single}}} = \frac{L_0}{L} + \frac{F_1 F_0}{(F_0 - 1) L} \Delta F = \frac{1}{L_1} + \frac{L_0}{L_0 - 1} \Delta F$$

Setting $L_0 = L_1 = 2$, or $L_0 = 2$, $L_1 = 4$, or $L_0 = 4$, $L_1 = 2$, we recover the results obtained previously for the 2×2, 2×4, and 4×2 cases. The condition that the multistage form be *more efficient* than the single-stage one is:

$$\frac{1}{L_1} + \frac{L_0}{L_0 - 1} \Delta F < 1 \qquad \Leftrightarrow \qquad \Delta F < \left(1 - \frac{1}{L_0}\right)\left(1 - \frac{1}{L_1}\right)$$

Given this condition, then the most efficient *ordering* of the two filters is to place first the filter with the *smaller* oversampling ratio. For example, assuming $L_0 < L_1$ then the ordering $H_0 H_1$ is more efficient than $H_1 H_0$ because

$$\frac{1}{L_1} + \frac{L_0}{L_0 - 1} \Delta F < \frac{1}{L_0} + \frac{L_1}{L_1 - 1} \Delta F < 1$$

For a three-stage design with $F_0 = L_0$, $F_1 = L_0 L_1$, and $F_2 = L_0 L_1 L_2 = L$, we find

$$\frac{R_{\text{multi}}}{R_{\text{single}}} = \frac{L_0}{L} + \Delta F \left[\frac{F_1 F_0}{(F_0 - 1) L} + \frac{F_2 F_1}{(F_1 - 1) L}\right] = \frac{1}{L_1 L_2} + \Delta F \left[\frac{L_0}{L_2 (L_0 - 1)} + \frac{L_0 L_1}{L_0 L_1 - 1}\right]$$

Setting $L_0 = L_1 = L_2 = 2$, we recover the results of the 2×2×2 case. Because the designed filter lengths N_i are slightly larger than those given by the Kaiser formulas, the correct relative computational rate should be computed using Eq. (12.2.33), whereas Eq. (12.2.34) gives only an approximation.

12.3 *Linear and Hold Interpolators**

We saw in Section 12.2.3 that the ideal interpolator may be thought of as the *sampled* version of the ideal *analog* reconstructor, sampled at the high rate f_s', that is,

$$d(k') = h(k'T') \tag{12.3.1}$$

This relationship can be applied to other analog reconstructors, resulting in simpler interpolators. For any analog reconstructor $h(t)$ that reconstructs the low-rate samples by

$$y_a(t) = \sum_m h(t - mT)x(m)$$

we can obtain the interpolated samples by resampling $y_a(t)$ at the high rate f_s':

$$y_a(n'T') = \sum_m h(n'T' - mT)x(m)$$

which can be written in the form

$$y_{\text{up}}(n') = \sum_m d(n' - mL)x(m) \tag{12.3.2}$$

where $d(k')$ is obtained from $h(t)$ via Eq. (12.3.1). The interpolation equation can be written also in terms of the *upsampled* version of $x(n)$

$$y_{\text{up}}(n') = \sum_{m'} d(n' - m')x_{\text{up}}(m') = \sum_{k'} d(k')x_{\text{up}}(n' - k') \tag{12.3.3}$$

Two of the most common interpolators are the *hold* and *linear* interpolators resulting from the sample/hold and linear analog reconstructors having impulse responses:

$$h(t) = \begin{cases} 1, & \text{if } 0 \le t < T \\ 0, & \text{otherwise} \end{cases} \quad \text{and} \quad h(t) = \begin{cases} 1 - \dfrac{|t|}{T}, & \text{if } |t| \le T \\ 0, & \text{otherwise} \end{cases}$$

They are shown in Fig. 12.3.1. Setting $t = k'T'$ in these definitions, and using the relationship $T = LT'$, we find the following discrete-time versions:

$$\text{(hold)} \quad d(k') = \begin{cases} 1, & \text{if } 0 \le k' \le L - 1 \\ 0, & \text{otherwise} \end{cases} = u(k') - u(k' - L)$$

$$\text{(linear)} \quad d(k') = \begin{cases} 1 - \dfrac{|k'|}{L}, & \text{if } |k'| \le L - 1 \\ 0, & \text{otherwise} \end{cases}$$

Note that in the linear case, the endpoints $k' = \pm L$ are not considered because $d(k')$ vanishes there. Figure 12.3.1 shows the sampled impulse responses for $L = 8$.

The filtering operations of these interpolators are very simple. The hold interpolator holds each low-rate sample constant for L high-rate sampling times. In other words, each low-rate sample is *repeated* L times at the high rate. The linear interpolator interpolates *linearly* between a given low-rate sample and the next one.

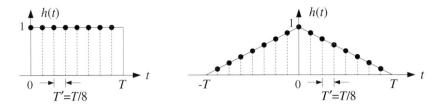

Fig. 12.3.1 Hold and linear interpolator impulse responses for $L = 8$.

To see this, we rewrite the filtering equation (12.3.3) in its polyphase form. As we argued for Eq. (12.2.12), we set $n' = nL + i$ and $k' = kL + j$ and use the fact that $x_{\text{up}}(nL + i - kL - j) = 0$, if $i \neq j$, to get

$$y_{\text{up}}(nL + i) = \sum_k d_i(k)x(n - k) \tag{12.3.4}$$

where $d_i(k)$ are the corresponding polyphase subfilters:

$$d_i(k) = d(kL + i), \qquad i = 0, 1, \ldots, L - 1$$

The summation over k must be determined for each case. In the hold case, we have the restriction on k:

$$0 \leq k' \leq L - 1 \quad \Rightarrow \quad 0 \leq kL + i \leq L - 1 \quad \Rightarrow \quad 0 \leq k \leq 1 - \frac{1 + i}{L}$$

Because $0 \leq i \leq L - 1$, the only allowed value of k in that range is $k = 0$. Similarly, in the linear case, we have:

$$|k'| \leq L - 1 \quad \Rightarrow \quad |kL + i| \leq L - 1 \quad \Rightarrow \quad -1 + \frac{1 - i}{L} \leq k \leq 1 - \frac{1 + i}{L}$$

The only possible integer value in the left-hand side is $k = -1$, which is realized when $i = 1$, and the only integer value of the right-hand side is $k = 0$. Therefore, the polyphase subfilters are in the two cases:

$$\text{(hold)} \qquad d_i(k) = d_i(0)\,\delta(k)$$

$$\text{(linear)} \qquad d_i(k) = d_i(0)\,\delta(k) + d_i(-1)\,\delta(k + 1)$$

where for the hold case, we have:

$$d_i(0) = d(i) = u(i) - u(i - L) = 1 - 0 = 1$$

and for the linear case:

$$d_i(0) = d(i) = 1 - \frac{|i|}{L} = 1 - \frac{i}{L}$$

$$d_i(-1) = d(-L + i) = 1 - \frac{|i - L|}{L} = 1 - \frac{L - i}{L} = \frac{i}{L}$$

Thus, the polyphase subfilters are:

$$\text{(hold)} \quad d_i(k) = \delta(k)$$

$$\text{(linear)} \quad d_i(k) = \left(1 - \frac{i}{L}\right)\delta(k) + \frac{i}{L}\delta(k+1) \tag{12.3.5}$$

for $i = 0, 1, \ldots, L - 1$. Inserting these impulse responses in Eq. (12.3.4), we find the interpolation equations in the two cases. For the hold case:

$$\boxed{y_{up}(nL + i) = x(n)}, \qquad i = 0, 1, \ldots, L - 1 \tag{12.3.6}$$

Thus, each low-rate sample is repeated L times. For the linear case we have:

$$\boxed{y_{up}(nL + i) = \left(1 - \frac{i}{L}\right)x(n) + \frac{i}{L}x(n+1)}, \qquad i = 0, 1, \ldots, L - 1 \tag{12.3.7}$$

They correspond to linearly weighting the two successive low-rate samples $x(n)$ and $x(n + 1)$. For example, when $L = 8$ the eight interpolated samples between $x(n)$ and $x(n + 1)$ are calculated by:

$$y_{up}(8n) = x(n)$$

$$y_{up}(8n + 1) = 0.875\,x(n) + 0.125\,x(n+1)$$

$$y_{up}(8n + 2) = 0.750\,x(n) + 0.250\,x(n+1)$$

$$y_{up}(8n + 3) = 0.625\,x(n) + 0.375\,x(n+1)$$

$$y_{up}(8n + 4) = 0.500\,x(n) + 0.500\,x(n+1)$$

$$y_{up}(8n + 5) = 0.375\,x(n) + 0.625\,x(n+1)$$

$$y_{up}(8n + 6) = 0.250\,x(n) + 0.750\,x(n+1)$$

$$y_{up}(8n + 7) = 0.125\,x(n) + 0.875\,x(n+1)$$

Figure 12.3.2 shows the interpolated signal using 8-fold hold and linear interpolators. To understand the frequency domain properties of the hold and linear interpolators and the extent to which they differ from the ideal interpolator, we compute their high-rate ζ-transforms using Eq. (12.2.15). For the hold case, taking the low-rate z-transform of $d_i(k)$ given in Eq. (12.3.5), we find

$$D_i(z) = 1 \qquad \Rightarrow \qquad D_i(\zeta^L) = 1$$

Then, it follows from Eq. (12.2.15)

$$D(\zeta) = \sum_{i=0}^{L-1} \zeta^{-i} D_i(\zeta^L) = \sum_{i=0}^{L-1} \zeta^{-i} = \frac{1 - \zeta^{-L}}{1 - \zeta^{-1}} \tag{12.3.8}$$

Setting $\zeta = e^{j\omega'} = e^{2\pi jf/f_s'} = e^{2\pi jf/Lf_s}$ we obtain the frequency response of the hold interpolator:

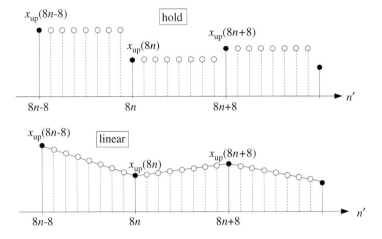

Fig. 12.3.2 8-fold hold and linear interpolators.

$$D(f) = \frac{1 - e^{-jL\omega'}}{1 - e^{-j\omega'}} = \frac{\sin(L\omega'/2)}{\sin(\omega'/2)}e^{-j(L-1)\omega'/2}$$

$$= \frac{\sin(\pi f/f_s)}{\sin(\pi f/Lf_s)}e^{-j\pi(L-1)f/Lf_s}$$

(12.3.9)

Similarly, the low-rate z-transform of $d_i(k)$ for the linear case is:

$$D_i(z) = 1 - \frac{i}{L} + \frac{i}{L}z \quad \Rightarrow \quad D_i(\zeta^L) = 1 + \frac{i}{L}(\zeta^L - 1)$$

From Eq. (12.2.15) we find:

$$D(\zeta) = \sum_{i=0}^{L-1} \zeta^{-i}D_i(\zeta^L) = \sum_{i=0}^{L-1} \zeta^{-i} + \frac{\zeta^L - 1}{L}\sum_{i=0}^{L-1} i\zeta^{-i}$$

and using the identity:

$$\sum_{i=0}^{L-1} i\zeta^{-i} = \frac{(1 - \zeta^{-L})\zeta^{-1}}{(1 - \zeta^{-1})^2} - \frac{L\zeta^{-L}}{1 - \zeta^{-1}}$$

we obtain (see also Problem 5.11):

$$D(\zeta) = \frac{1}{L}\frac{(1 - \zeta^{-L})(1 - \zeta^L)}{(1 - \zeta^{-1})(1 - \zeta)} = \frac{1}{L}\left(\frac{1 - \zeta^{-L}}{1 - \zeta^{-1}}\right)^2 \zeta^{L-1}$$

which leads to the frequency response:

$$D(f) = \frac{1}{L}\left|\frac{\sin(L\omega'/2)}{\sin(\omega'/2)}\right|^2 = \frac{1}{L}\left|\frac{\sin(\pi f/f_s)}{\sin(\pi f/Lf_s)}\right|^2$$

(12.3.10)

Both responses (12.3.9) and (12.3.10) are periodic in f with period $f_s' = Lf_s$ and vanish at all multiples of f_s which are *not* multiples of Lf_s. Therefore, they partially remove the spectral replicas that are between multiples of f_s'. They are shown in Fig. 12.3.3 for the case $L = 8$, together with the ideal response.

Because of their simple structure, linear and hold interpolators are used in *multistage* implementations of interpolators, especially in the latter stages that have higher sampling rates. Some example designs are discussed in Section 12.4.5.

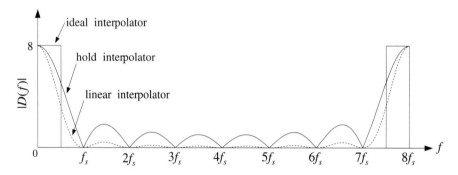

Fig. 12.3.3 Hold and linear interpolator frequency responses for $L = 8$.

12.4 Design Examples*

12.4.1 4-fold Interpolators

Consider the case of a 4-fold interpolator having $L = 4$ and polyphase filter length $2M = 4$ or $M = 2$. This corresponds to a filter length $N = 2LM + 1 = 17$. The ideal impulse response will be:

$$d(k') = \frac{\sin(\pi k'/4)}{\pi k'/4}, \qquad -8 \le k' \le 8$$

or, numerically,

$$\mathbf{h} = \mathbf{d} = [0, -0.13, -0.21, -0.18, 0, 0.30, 0.64, 0.90, 1, 0.90, 0.64,$$

$$0.30, 0, -0.18, -0.21, -0.13, 0]$$

(12.4.1)

where $\mathbf{h}$ is the causal version, with time origin shifted to the *left* of the vector, and $\mathbf{d}$ is the symmetric one with time origin at the *middle* of the vector. This truncated ideal impulse response is shown in Fig. 12.4.1. The four polyphase subfilters are defined by Eq. (12.2.10), that is, for $i = 0, 1, 2, 3$,

$$d_i(k) = d(4k + i), \qquad -2 \le k \le 1$$

They are extracted from $\mathbf{h}$ by taking every fourth entry, starting with the ith entry:

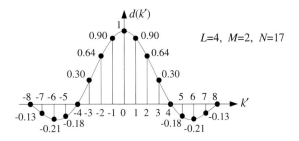

Fig. 12.4.1 Length-17 symmetric impulse response of 4-fold FIR interpolator.

$$\mathbf{h}_0 = \mathbf{d}_0 = [0, 0, 1, 0]$$

$$\mathbf{h}_1 = \mathbf{d}_1 = [-0.13, 0.30, 0.90, -0.18]$$

$$\mathbf{h}_2 = \mathbf{d}_2 = [-0.21, 0.64, 0.64, -0.21] \tag{12.4.2}$$

$$\mathbf{h}_3 = \mathbf{d}_3 = [-0.18, 0.90, 0.30, -0.13]$$

The interpolated samples between $x(n) = x_{up}(4n)$ and $x(n+1) = x_{up}(4n+4)$ are calculated from Eqs. (12.2.18). All four subfilters act on the time-advanced low-rate input samples $\{x(n+2), x(n+1), x(n), x(n-1)\}$, or, $\{x_{up}(4n+8), x_{up}(4n+4), x_{up}(4n), x_{up}(4n-4)\}$. Equations (12.2.12) can be cast in a compact matrix form:

$$\begin{bmatrix} y_{up}(4n) \\ y_{up}(4n+1) \\ y_{up}(4n+2) \\ y_{up}(4n+3) \end{bmatrix} = \begin{bmatrix} 0 & 0 & 1 & 0 \\ -0.13 & 0.30 & 0.90 & -0.18 \\ -0.21 & 0.64 & 0.64 & -0.21 \\ -0.18 & 0.90 & 0.30 & -0.13 \end{bmatrix} \begin{bmatrix} x_{up}(4n+8) \\ x_{up}(4n+4) \\ x_{up}(4n) \\ x_{up}(4n-4) \end{bmatrix} \tag{12.4.3}$$

These results can be understood more intuitively using the LTI form of convolution, that is, superimposing the full length-17 symmetric impulse response $\mathbf{d}$ at the four contributing low-rate samples and summing up their contributions at the four desired time instants, that is, at $n' = 4n + i, i = 0, 1, 2, 3$. This is illustrated in Fig. 12.4.2.

For example, referring to the impulse response values indicated on Fig. 12.4.1, we find that at time instant $4n + 1$, the input sample $x_{up}(4n + 8)$ will contribute an amount $-0.13x_{up}(4n + 8)$, the sample $x_{up}(4n + 4)$ will contribute an amount $0.30x_{up}(4n + 4)$, the sample $x_{up}(4n)$ will contribute $0.90x_{up}(4n)$, and the sample $x_{up}(4n - 4)$ an amount $-0.18x_{up}(4n - 4)$. The interpolated value is built up from these four contributions:

$$y_{up}(4n + 1) = -0.13x_{up}(4n + 8) + 0.30x_{up}(4n + 4) + 0.90x_{up}(4n) - 0.18x_{up}(4n - 4)$$

Similarly, it should be evident from Fig. 12.4.2 that $y_{up}(4n) = x_{up}(4n)$, with the contributions of the other low-rate inputs vanishing at time instant $4n$. We may also use the flip-and-slide form of convolution, in which the impulse response $d(k')$ is

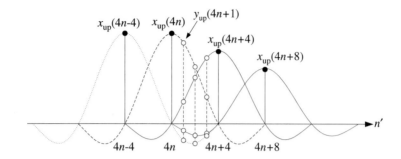

Fig. 12.4.2 LTI form of convolution by superposition of FIR impulse responses.

flipped, delayed, and positioned at the sampling instant n' to be computed. For example, at $n' = 4n + 1$, we have:

$$y_{up}(4n + 1) = \sum_{k'} d(4n + 1 - k')x_{up}(k')$$

Figure 12.4.3 shows this operation. Because of symmetry, the flipped impulse response is the same as that in Fig. 12.4.1. It is then translated to $n' = 4n + 1$ and the above linear combination is performed.

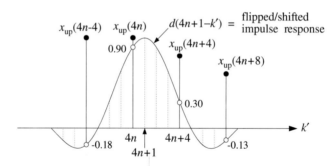

Fig. 12.4.3 Flip-and-slide form of convolution.

The only contributions come from the low-rate samples that fall within the finite extent of the impulse response. Thus, only the terms $k' = 4n - 4, 4n, 4n + 8, 4n + 8$ contribute, and each is weighted by the appropriate impulse response values that are read off from the figure, that is, $\{-0.18, 0.90, 0.30, -0.13\}$, so that again:

$$y_{up}(4n + 1) = -0.18x_{up}(4n - 4) + 0.90x_{up}(4n) + 0.30x_{up}(4n + 4) - 0.13x_{up}(4n + 8)$$

The Hamming windowed version of the filter is obtained by multiplying the full length-17 filter response $\mathbf{h}$ by a length-17 Hamming window. The resulting impulse response becomes:

$$\mathbf{h} = [0, -0.02, -0.05, -0.07, 0, 0.22, 0.55, 0.87, 1, 0.87, 0.55, 0.22, 0,$$

$$-0.07, -0.05, -0.02, 0]$$

The polyphase interpolation equations become in this case:

$$
\begin{bmatrix}
y_{up}(4n) \\
y_{up}(4n+1) \\
y_{up}(4n+2) \\
y_{up}(4n+3)
\end{bmatrix}
=
\begin{bmatrix}
0 & 0 & 1 & 0 \\
-0.02 & 0.22 & 0.87 & -0.07 \\
-0.05 & 0.55 & 0.55 & -0.05 \\
-0.07 & 0.87 & 0.22 & -0.02
\end{bmatrix}
\begin{bmatrix}
x_{up}(4n+8) \\
x_{up}(4n+4) \\
x_{up}(4n) \\
x_{up}(4n-4)
\end{bmatrix}
$$

The graphs in Fig. 12.4.4 compare the *magnitude responses* of the rectangularly and Hamming windowed interpolation filters. A *block diagram* realization of the polyphase form for this example is shown in Fig. 12.4.5. It is based on Eqs. (12.2.14) and (12.2.15), that is,

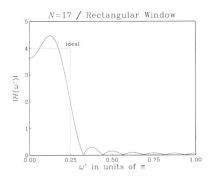
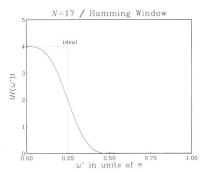

Fig. 12.4.4 Magnitude response $|H(\omega')|$ versus $\omega' = 2\pi f/f_s'$.

$$
H(\zeta) = H_0(\zeta^4) + \zeta^{-1}H_1(\zeta^4) + \zeta^{-2}H_2(\zeta^4) + \zeta^{-3}H_3(\zeta^4)
$$

$$
= H_0(z) + z^{-1/4}H_1(z) + z^{-2/4}H_2(z) + z^{-3/4}H_3(z)
$$

with all the subfilters using the *same tapped delay line* holding the incoming low-rate samples. The block diagram is equivalent to the commutator model of Fig. 12.2.5. The polyphase subfilters are defined by:

$$
\mathbf{h}_i = [h_{i0},\, h_{i1},\, h_{i2},\, h_{i3}], \qquad H_i(z) = h_{i0} + h_{i1}z^{-1} + h_{i2}z^{-2} + h_{i3}z^{-3}
$$

for $i = 0, 1, 2, 3$, where $\mathbf{h}_i$ are given by Eq. (12.4.2).

The possibility of a parallel multiprocessor implementation is evident from this diagram. The four outputs of the filters $H_i(z)$ are produced simultaneously in a parallel implementation, but they are not sent to the overall output simultaneously. During each low-rate sampling period T, the sample $y_0(n)$ is sent out first, then $T/4$ seconds later (represented by the delay $z^{-1/4}$) the second computed interpolated sample $y_1(n)$ is sent out, another $T/4$ seconds later the third sample $y_2(n)$ is sent out, and $T/4$ seconds after that, the fourth interpolated sample $y_3(n)$ is sent out.

As a concrete filtering example, consider the following low-rate input signal $x(n)$ consisting of 25 DC samples, and depicted in Fig. 12.4.6 with respect to the fast time scale:

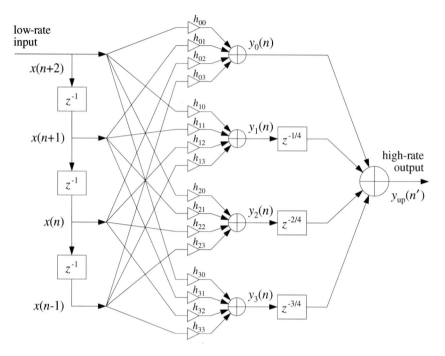

Fig. 12.4.5 Polyphase realization of 4-fold interpolator.

$$x(n) = \{1, 1, 1, 1, 1, 1, 1, 1, 1, 1, 1, 1, 1, 1, 1, 1, 1, 1, 1, 1, 1, 1, 1, 1\}$$

The interpolated values between these low-rate samples are shown in Fig. 12.4.7 for the cases of the rectangularly and Hamming windowed interpolating filters. They were computed by the polyphase sample processing algorithm of Eq. (12.2.20). The input-on and input-off transients are evident.

As another example, consider the case of $L = 4$ and $M = 12$, that is, interpolation filter length $N = 2LM + 1 = 97$. This is a more realistic length for typical 4-fold oversampling digital filters used in CD players. The corresponding rectangularly and Hamming windowed magnitude responses are shown in Fig. 12.4.8. The interpolated output signals from these two filters are shown in Fig. 12.4.9 for the same low-rate input signal $x(n)$. Note the longer input-on and input-off transients.

12.4.2 Multistage 4-fold Interpolators

Here, we follow the discussion of Section 12.2.5 and design multistage and single stage digital interpolation filters using the Kaiser window method. Such filters may be used as oversampling digital filters in CD players.[†]

[†]See Ref. [282] for actual DSP chips with comparable design characteristics.

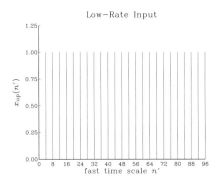

Fig. 12.4.6 Low-rate input samples $x_{up}(n')$.

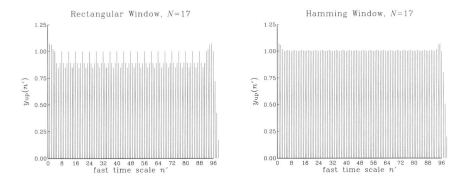

Fig. 12.4.7 High-rate interpolated output samples $y_{up}(n')$.

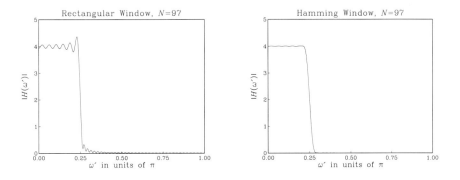

Fig. 12.4.8 Magnitude responses of length-97 interpolation filters.

We take $L = 4$ for the oversampling ratio, and assume a nominal digital audio sampling rate of $f_s = 40$ kHz, so that the fast rate will be $f_s' = L f_s = 4 \times 40 = 160$ kHz. We take the transition width to be $\Delta f = 5$ kHz about the ideal cutoff frequency

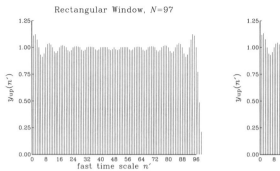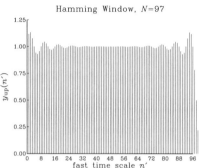

Fig. 12.4.9 High-rate interpolated outputs $y_{up}(n')$.

$f_c = f_s'/(2L) = f_s/2 = 20$ kHz. Therefore, the passband and stopband frequencies of the filter will be

$$f_{pass} = 20 - \frac{5}{2} = 17.5 \text{ kHz}, \quad f_{stop} = 20 + \frac{5}{2} = 22.5 \text{ kHz}$$

and the normalized transition width

$$\Delta F = \frac{\Delta f}{f_s} = \frac{5}{40} = 0.125$$

The stopband attenuation is taken to be $A = 80$ dB, which gives rise to a passband/stopband ripple $\delta = 10^{-4}$, passband attenuation $A_{pass} = 0.0017$ dB, and the following values for the Kaiser window parameters D and α:

$$D = \frac{A - 7.95}{14.36} = 5.017, \quad \alpha = 0.1102\,(A - 8.7) = 7.857$$

For a 2×2 multistage design, shown in Fig. 12.4.10, we may use the general design equations (12.2.31) and (12.2.32) to find the Kaiser lengths of the two filters H_0 and H_1:

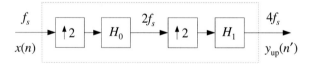

Fig. 12.4.10 2×2 = 4-fold oversampling filter.

$$N_0 - 1 = \frac{DL_0}{\Delta F} = \frac{2D}{\Delta F} = 80.27, \quad N_1 - 1 = \frac{DF_1}{F_0 - 1} = 4D = 20.07$$

which get rounded up to the values:

$$N_0 = 85 = 2L_0M_0 + 1 = 4M_0 + 1 \qquad \Rightarrow \qquad M_0 = 21$$

$$N_1 = 25 = 2L_1M_1 + 1 = 4M_1 + 1 \qquad \Rightarrow \qquad M_1 = 6$$

The magnitude response of the filter $H_0(f)$ in dB and a typical low-rate sinusoidal input to be interpolated are shown in Fig. 12.4.11. The 2-fold interpolated output of $H_0(f)$ is shown in Fig. 12.4.12. It serves as the input to the next filter $H_1(f)$ whose magnitude response is also shown in Fig. 12.4.12.

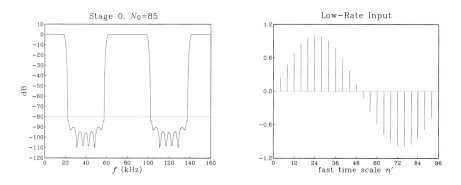

Fig. 12.4.11 Filter $H_0(f)$ and its low-rate input signal.

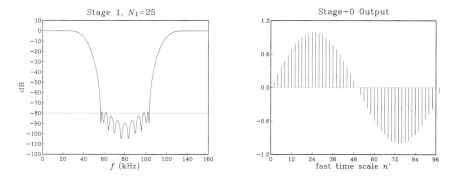

Fig. 12.4.12 Filter $H_1(f)$ and its input which is the output of $H_0(f)$.

The 2-fold interpolated output of $H_1(f)$ will be the final 4-fold interpolated output. It is shown in Fig. 12.4.13 together with the superimposed plots of the filters $H_0(f)$ and $H_1(f)$. In these figures, the frequency responses have been normalized by their DC values, that is, $H_0(f)/L_0$, $H_1(f)/L_1$.

Finally, we compare the multistage design to an equivalent single stage Kaiser design. In this case the Kaiser filter length will be

$$N - 1 = \frac{4D}{\Delta F} = 160.56$$

which is rounded up to the value

$$N = 169 = 2LM + 1 = 8M + 1 \qquad \Rightarrow \qquad M = 21$$

Its magnitude response $H(f)$ is shown in Fig. 12.4.14 together with the magnitude response of the combined multistage filter $H_{\text{tot}}(f) = H_0(f)H_1(f)$. Again, we have normalized them to their DC values, namely, $H(f)/L$, and $H_{\text{tot}}(f)/L$.

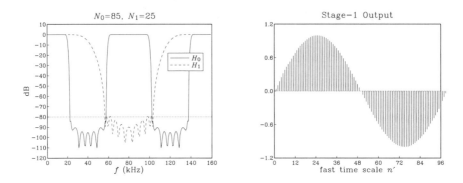

Fig. 12.4.13 Filters $H_0(f)$, $H_1(f)$ and the overall interpolated output.

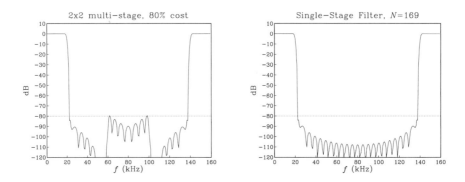

Fig. 12.4.14 Multistage filter $H_0(f)H_1(f)$ and single-stage filter $H(f)$.

The multistage realization requires only 80 percent of the computational rate of the single-stage design. Indeed, the relative computational rate of the multistage versus the single stage designs is given according to Eq. (12.2.33) by:

$$\frac{R_{\text{multi}}}{R_{\text{single}}} = \frac{N_0 + N_1 F_0}{N} = \frac{85 + 25 \cdot 2}{169} = 0.80$$

which compares well with the approximate result of Eq. (12.2.30).

$$\frac{R_{\text{multi}}}{R_{\text{single}}} = \frac{1}{2} + 2\Delta F = 0.5 + 2 \cdot 0.125 = 0.75$$

Finally, as we saw earlier, the passband of the total filter $H_{\text{tot}}(f) = H_0(f)H_1(f)$ tends to be worse than the passbands of the individual factors. Let δ be the common passband ripple, as calculated from Eq. (12.2.29). Then, the two individual filters will satisfy within their passbands:

$$1 - \delta \leq \left| \frac{H_0(f)}{L_0} \right| \leq 1 + \delta, \qquad 1 - \delta \leq \left| \frac{H_1(f)}{L_1} \right| \leq 1 + \delta$$

Multiplying the two inequalities, we find for the total filter

$$(1 - \delta)^2 \leq \left| \frac{H_0(f)H_1(f)}{L_0 L_1} \right| \leq (1 + \delta)^2$$

or, approximately if δ is small,

$$1 - 2\delta \leq \left| \frac{H_{\text{tot}}(f)}{L} \right| \leq 1 + 2\delta$$

Thus, the passband ripple is effectively doubled, $\delta_{\text{tot}} = 2\delta$. Taking logs of both sides, we obtain the following bounds for the passband attenuations in dB:

$$-8.7\delta \leq 20 \log_{10} \left| \frac{H_0(f)}{L_0} \right| \leq 8.7\delta, \qquad -8.7\delta \leq 20 \log_{10} \left| \frac{H_1(f)}{L_1} \right| \leq 8.7\delta$$

$$-8.7(2\delta) \leq 20 \log_{10} \left| \frac{H_{\text{tot}}(f)}{L} \right| \leq 8.7(2\delta)$$

where we used the small-δ approximation:

$$20 \log_{10}(1 + \delta) \simeq 8.7\delta$$

Figure 12.4.15 shows a magnified plot of the passband region of the individual and total filters for the above designs, with the passband bounds placed on the figure. It is evident that the actual passband ripple of $H_{\text{tot}}(f)$ is less than the worst-case ripple $\delta_{\text{tot}} = 2\delta$.

12.4.3 DAC Equalization

In an oversampling DSP system, the interpolator output samples are reconstructed by a staircase D/A converter operating at the high rate $f_s' = L f_s$. Its frequency response (normalized to unity gain at DC) is

$$H_{\text{dac}}(f) = \frac{\sin(\pi f / f_s')}{\pi f / f_s'} e^{-j\pi f / f_s'}$$

It causes some attenuation within the Nyquist interval, with maximum of about 4 dB at the Nyquist frequency $f_s'/2$. For an L-fold interpolation filter which has cutoff at $f_c = f_s/2 = f_s'/2L$, the maximum attenuation within the filter's passband will be:

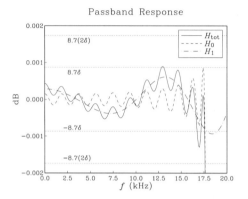

Fig. 12.4.15   Magnified passband of 2×2 interpolator.

$$|H_{\text{dac}}(f_c)| = \left| \frac{\sin(\pi f_c/f_s')}{\pi f_c/f_s'} \right| = \frac{\sin(\pi/2L)}{\pi/2L} \tag{12.4.4}$$

For large values of the oversampling ratio L, this attenuation is insignificant, approaching 0 dB. Thus, one of the benefits of oversampling is that the aperture effect of the DAC can be neglected.

However, for smaller values of L (for example, $L \leq 8$) it may be desirable to compensate this attenuation by designing the interpolation filter to have an *inverse* shape to the $\sin x/x$ DAC response over the relevant passband range. The desired *equalized* ideal interpolation filter can then be defined by the following equation, replacing Eq. (12.2.24):

$$D(f) = \begin{cases} LD_{\text{eq}}(f), & \text{if } |f| \leq \dfrac{f_s}{2} \\ 0, & \text{if } \dfrac{f_s}{2} < |f| \leq \dfrac{f_s'}{2} \end{cases} \tag{12.4.5}$$

where $D_{\text{eq}}(f)$ is essentially the inverse response $1/H_{\text{dac}}(f)$ with the phase removed in order to keep $D(f)$ real and even in f:

$$D_{\text{eq}}(f) = \frac{\pi f/f_s'}{\sin(\pi f/f_s')}, \qquad |f| \leq \frac{f_s}{2} \tag{12.4.6}$$

In units of the high-rate digital frequency $\omega' = 2\pi f/f_s'$, Eq. (12.4.5) becomes:

$$D(\omega') = \begin{cases} L\dfrac{\omega'/2}{\sin(\omega'/2)}, & \text{if } |\omega'| \leq \dfrac{\pi}{L} \\ 0, & \text{if } \dfrac{\pi}{L} < |\omega'| \leq \pi \end{cases} \tag{12.4.7}$$

Such a filter can be designed by the *frequency sampling* design method of Section 10.3. If the filter order is known, say $N = 2LM + 1$, then we can compute the desired filter weights by the inverse N-point DFT:

$$\tilde{d}(k') = \frac{1}{N} \sum_{i=-LM}^{LM} D(\omega_i') e^{j\omega_i' k'}, \qquad -LM \le k' \le LM \qquad (12.4.8)$$

where ω_i' are the N DFT frequencies spanning the symmetric Nyquist interval $[-\pi, \pi]$:

$$\boxed{\omega_i' = \frac{2\pi i}{N}}, \qquad -LM \le i \le LM$$

The designed causal windowed filter will be

$$h(n') = \tilde{d}(n' - LM) w(n'), \qquad 0 \le n' \le N - 1 \qquad (12.4.9)$$

In the Hamming window case, we must assume a desired value for N. In the Kaiser case, we may start with a desired stopband attenuation A and transition width Δf, and then determine the filter length N and the window parameter α. Because the filter is sloping upwards in the passband, to achieve a true attenuation A in the stopband, we may have to carry out the design with a slightly larger value of A. This is illustrated in the examples below.

Note also that because $\tilde{d}(k')$ and $D(\omega')$ are real-valued, we may replace the right-hand side of Eq. (12.4.8) by its real part and write it in the cosine form:

$$\tilde{d}(k') = \frac{1}{N} \sum_{i=-LM}^{LM} D(\omega_i') \cos(\omega_i' k'), \qquad -LM \le k' \le LM$$

and because $D(\omega')$ is even in ω'

$$\tilde{d}(k') = \frac{1}{N} \left[D(\omega_0') + 2 \sum_{i=1}^{LM} D(\omega_i') \cos(\omega_i' k') \right], \qquad -LM \le k' \le LM$$

where $D(\omega_0') = D(0) = L$. This expression can be simplified even further by noting that $D(\omega_i')$ is non-zero only for

$$0 < \omega_i' < \frac{\pi}{L} \quad \Rightarrow \quad 0 < \frac{2\pi i}{N} < \frac{\pi}{L} \quad \Rightarrow \quad 0 < i < \frac{N}{2L} = \frac{2LM + 1}{2L} = M + \frac{1}{2L}$$

Thus, the summation can be restricted over $1 \le i \le M$, giving

$$\tilde{d}(k') = \frac{L}{N} \left[1 + 2 \sum_{i=1}^{M} \frac{\omega_i'/2}{\sin(\omega_i'/2)} \cos(\omega_i' k') \right], \qquad -LM \le k' \le LM$$

As a first example, consider a 2-times oversampling filter for a CD player. Assume a nominal audio rate of $f_s = 40$ kHz, transition width $\Delta f = 5$ kHz, and stopband attenuation $A = 80$ dB. The normalized width is $\Delta F = \Delta f/f_s = 0.125$. The Kaiser D parameter and filter length N will be

$$D = \frac{A - 7.95}{14.36} = 5.017, \qquad N - 1 \ge = \frac{DL}{\Delta F} = 80.3$$

which rounds up to $N = 85$. Fig. 12.4.16 shows the designed filter together with the inverse DAC response $1/|H_{dac}(f)|$ over the high-rate Nyquist interval. It also shows the passband in a magnified scale. Notice how the inverse DAC response reaches 4 dB at the Nyquist frequency of $f_s'/2 = 40$ kHz.

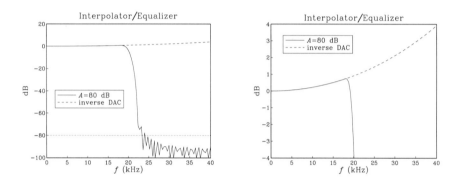

Fig. 12.4.16   2-fold interpolator/equalizer designed with $A = 80$ dB.

The actual stopband attenuation is somewhat less than the prescribed 80 dB, namely, about 72 dB at $f = f_s/2 + \Delta f/2 = 22.5$ kHz. Thus, we may wish to redesign the filter starting out with a somewhat larger attenuation. For example, assuming $A = 90$ dB, we obtain filter length $N = 93$. See [282] for a similar design. The redesigned filter is shown in Fig. 12.4.17. It achieves 80 dB attenuation at 22.5 kHz.

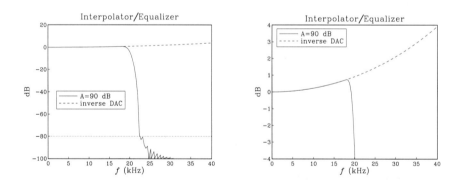

Fig. 12.4.17   2-fold interpolator/equalizer redesigned with $A = 90$ dB.

As another example, we design an equalized 4-times oversampling filter for digital audio, assuming 40 kHz audio rate, 5 kHz transition width, and a stopband attenuation of 60 dB. The Kaiser parameters are:

$$D = \frac{A - 7.95}{14.36} = 3.625, \qquad N - 1 \geq = \frac{DL}{\Delta F} = 115.98 \quad \Rightarrow \quad N = 121$$

The designed filter and its passband are shown in Fig. 12.4.18.

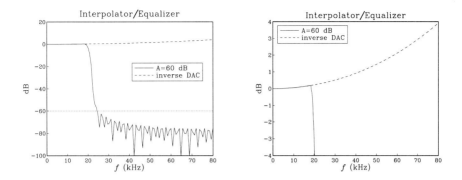

Fig. 12.4.18 4-fold interpolator/equalizer filter and its magnified passband.

12.4.4 Postfilter Design and Equalization

In addition to compensating for the attenuation of the DAC, one may wish to compensate for other effects. For example, the staircase output of the DAC will be fed into an analog anti-image postfilter which introduces its own slight attenuation within the desired passband. This attenuation can be equalized digitally by the interpolation filter. Figure 12.4.19 shows this arrangement.

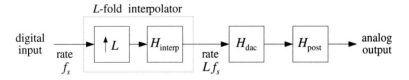

Fig. 12.4.19 Interpolation filter equalizes DAC and postfilter responses.

As we saw in Section 12.1, because of oversampling, the postfilter will have a wide transition region resulting in low filter order, such as 2 or 3. The postfilter must provide enough attenuation in its stopband to remove the spectral images of the interpolated signal at multiples of f_s', as shown in Fig. 12.4.20. Its passband extends up to the low-rate Nyquist frequency[†] and its stopband begins at the left edge of the first spectral image, that is,

$$\boxed{f_{\text{pass}} = \frac{f_s}{2}, \qquad f_{\text{stop}} = f_s' - f_{\text{pass}} = Lf_s - \frac{f_s}{2}}$$

At f_{stop}, the DAC already provides a certain amount of attenuation given by:

$$|H_{\text{dac}}(f_{\text{stop}})| = \left| \frac{\sin(\pi f_{\text{stop}}/f_s')}{\pi f_{\text{stop}}/f_s'} \right| = \frac{\sin(\pi - \pi/2L)}{\pi - \pi/2L} = \frac{\sin(\pi/2L)}{\pi - \pi/2L}$$

[†]If the interpolator is not ideal but has transition width Δf about $f_s/2$, we may use the more accurate expressions $f_{\text{pass}} = f_s/2 - \Delta f/2$ and $f_{\text{stop}} = Lf_s - f_s/2 - \Delta f/2$.

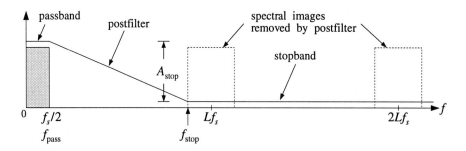

Fig. 12.4.20 Analog anti-image postfilter specifications.

which, for large L, becomes approximately:

$$|H_{dac}(f_{stop})| = \frac{\sin(\pi/2L)}{\pi - \pi/2L} \simeq \frac{1}{2L}$$

or, in dB

$$A_{dac} \simeq 20 \log_{10}(2L) \tag{12.4.10}$$

The analog postfilter must supply an additional amount of attenuation A_{stop}, raising the total attenuation at f_{stop} to a desired level, say A_{tot} dB:

$$A_{tot} = A_{dac} + A_{stop}$$

For example, suppose $f_s = 40$ kHz and $L = 4$, and we require the total suppression of the replicas to be more than $A_{tot} = 60$ dB. The stopband frequency will be $f_{stop} = Lf_s - f_s/2 = 160 - 20 = 140$ kHz. At that frequency the DAC will provide an attenuation $A_{dac} = 20 \log_{10}(8) = 18$ dB, and therefore the postfilter must provide the rest:

$$A_{stop} = A_{tot} - A_{dac} = 60 - 18 = 42 \text{ dB}$$

Suppose we use a third-order Butterworth filter with magnitude response [269–271]:

$$|H_{post}(f)|^2 = \frac{1}{1 + \left(\dfrac{f}{f_0}\right)^6}$$

and attenuation in dB:

$$A_{post}(f) = -10 \log_{10}|H_{post}(f)|^2 = 10 \log_{10}\left[1 + \left(\frac{f}{f_0}\right)^6\right] \tag{12.4.11}$$

where f_0 is the 3-dB normalization frequency to be determined. Then, the requirement that at f_{stop} the attenuation be equal to A_{stop} gives:

$$A_{\text{stop}} = 10 \log_{10} \left[1 + \left(\frac{f_{\text{stop}}}{f_0} \right)^6 \right]$$

which can be solved for f_0:

$$f_0 = f_{\text{stop}} \left[10^{A_{\text{stop}}/10} - 1 \right]^{-1/6} = 140 \cdot \left[10^{42/10} - 1 \right]^{-1/6} = 28 \text{ kHz}$$

The third-order Butterworth analog transfer function of the postfilter will be:

$$H_{\text{post}}(s) = \frac{1}{1 + 2\left(\dfrac{s}{\Omega_0} \right) + 2\left(\dfrac{s}{\Omega_0} \right)^2 + \left(\dfrac{s}{\Omega_0} \right)^3} \qquad (12.4.12)$$

where $\Omega_0 = 2\pi f_0$. This postfilter will adequately remove the spectral images at multiples of f_s', but it will also cause a small amount of attenuation within the desired passband. The maximum passband attenuations caused by the postfilter and the DAC at $f_{\text{pass}} = f_s/2$ can be computed from Eqs. (12.4.11) and (12.4.4):

$$A_{\text{post}}(f_{\text{pass}}) = 10 \log_{10} \left[1 + \left(\frac{f_{\text{pass}}}{f_0} \right)^6 \right] = 10 \log_{10} \left[1 + \left(\frac{20}{28} \right)^6 \right] = 0.54 \text{ dB}$$

$$A_{\text{dac}}(f_{\text{pass}}) = -20 \log_{10} \left[\frac{\sin(\pi/2L)}{\pi/2L} \right] = 0.22 \text{ dB}$$

resulting in a total passband attenuation of $0.54 + 0.22 = 0.76$ dB. This combined attenuation of the DAC and postfilter can be equalized by the interpolator filter. Using the frequency sampling design, we replace the interpolator's defining equation (12.4.7) by the equalized version:

$$D(\omega') = \begin{cases} L \left[\dfrac{\omega'/2}{\sin(\omega'/2)} \right] \left[1 + \left(\dfrac{\omega'}{\omega_0} \right)^6 \right]^{1/2}, & \text{if } |\omega'| \leq \dfrac{\pi}{L} \\ 0, & \text{if } \dfrac{\pi}{L} < |\omega'| \leq \pi \end{cases} \qquad (12.4.13)$$

where $\omega_0 = 2\pi f_0/f_s'$. The impulse response coefficients will be calculated by:

$$\tilde{d}(k') = \frac{L}{N} \left[1 + 2 \sum_{i=1}^{M} \left[\frac{\omega_i'/2}{\sin(\omega_i'/2)} \right] \left[1 + \left(\frac{\omega_i'}{\omega_0} \right)^6 \right]^{1/2} \cos(\omega_i' k') \right]$$

for $-LM \leq k' \leq LM$. These coefficients must be weighted by an appropriate window, as in Eq. (12.4.9).

Figure 12.4.21 shows a Kaiser design corresponding to interpolator stopband attenuation of $A = 60$ dB and a transition width of $\Delta f = 5$ kHz. As before, the resulting filter length is $N = 121$.

For reference, the DAC response $H_{\text{dac}}(f)$, postfilter response $H_{\text{post}}(f)$, and total response $H_{\text{dac}}(f) H_{\text{post}}(f)$ are superimposed on the figure. Notice how they meet their respective specifications at $f_{stop} = 140$ kHz. The DAC response vanishes (i.e., it has infinite attenuation) at $f_s' = 160$ kHz and all its multiples. The figure also shows the filter's passband in a magnified scale, together with the plots of the total filter $H_{dac}(f)H_{post}(f)$ and total inverse filter $1/(H_{dac}(f)H_{post}(f))$.

The effective overall analog reconstructor $H_{interp}(f)H_{dac}(f)H_{post}(f)$, consisting of the equalized interpolator, DAC, and postfilter, is shown in Fig. 12.4.22. The spectral images at multiples of $f_s' = 160$ kHz are suppressed by more than 60 dB and the 20 kHz passband is essentially flat. The figure also shows the passband in a magnified scale.

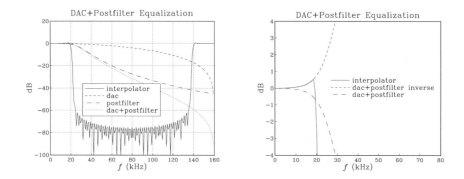

Fig. 12.4.21   4-fold interpolator with equalization of DAC and Butterworth postfilter.

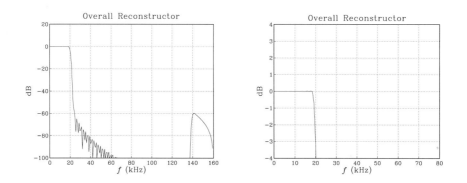

Fig. 12.4.22   Effective reconstructor has flat passband.

In digital audio applications, Bessel postfilters may also be used instead of Butterworth filters; see [279] for an example. Bessel filters provide the additional benefit that they have approximately *linear phase* over their passband. In particular, the transfer function and magnitude response of a third-order Bessel filter are given by [269-272]:

$$H_{\text{post}}(s) = \frac{15}{15 + 15\left(\dfrac{s}{\Omega_0}\right) + 6\left(\dfrac{s}{\Omega_0}\right)^2 + \left(\dfrac{s}{\Omega_0}\right)^3} \tag{12.4.14}$$

$$|H_{\text{post}}(f)|^2 = \frac{225}{225 + 45\left(\dfrac{f}{f_0}\right)^2 + 6\left(\dfrac{f}{f_0}\right)^4 + \left(\dfrac{f}{f_0}\right)^6} \tag{12.4.15}$$

where $\Omega_0 = 2\pi f_0$ and f_0 is related to the 3-dB frequency of the filter by

$$f_{3\text{dB}} = 1.75 f_0$$

The passband attenuation of this filter can be equalized digitally in a similar fashion. For equal 3-dB frequencies, Bessel filters fall off somewhat less sharply than Butterworth ones, thus, suppressing the spectral images by a lesser amount.

For example, for the previous 3-dB frequency $f_{3\text{dB}} = 28$ kHz, we find the normalization frequency $f_0 = f_{3\text{dB}}/1.75 = 16$ kHz. The corresponding postfilter attenuations at the passband and stopband frequencies, $f_{\text{pass}} = 20$ kHz and $f_{\text{stop}} = 140$ kHz, calculated from Eq. (12.4.15) are:

$$A_{\text{post}}(f_{\text{pass}}) = 1.44 \text{ dB}, \qquad A_{\text{post}}(f_{\text{stop}}) = 33 \text{ dB}$$

Thus, the DAC/postfilter combination will only achieve a total stopband attenuation of $33 + 18 = 51$ dB for the removal of the spectral images. Similarly, the total passband attenuation to be compensated by the interpolator will be $0.22 + 1.44 = 1.66$ dB.

If 51 dB suppression of the spectral images is acceptable[†] then we may redesign the interpolator so that it suppresses its stopband also by 51 dB. With $A = 51$ and $L = 4$, the redesigned interpolator will have Kaiser parameters:

$$D = \frac{A - 7.95}{14.36} = 2.998$$

$$N - 1 \geq= \frac{DL}{\Delta F} = 95.93 \quad \Rightarrow \quad N = 97$$

The redesigned equalized interpolation filter and the effective overall reconstruction filter are shown in Fig. 12.4.23. The overall reconstructor has a flat passband and suppresses all spectral images by at least 51 dB.

12.4.5 Multistage Equalization

Another application where equalization may be desirable is in *multistage* implementations of interpolators. The first stage is usually a high-quality lowpass digital filter, whereas the subsequent stages may be interpolators with *simplified* structure, such as linear or hold interpolators which do not have a flat passband. In

---

[†]The first Philips CD player had a similar 4-times oversampling interpolator of order $N = 97$ and stopband attenuation of $A = 50$ dB; see [279] for details.

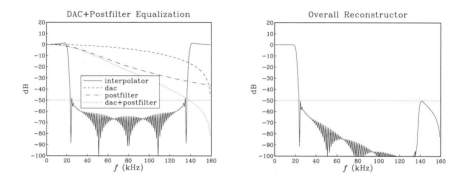

Fig. 12.4.23 4-fold interpolator with equalization of DAC and Bessel postfilter.

such cases, the filter in the first stage can be designed to equalize the attenuation
of all the subsequent stages *within* the passband. In addition, the analog reconstructing DAC and postfilter may also be equalized by the first stage. The overall
system is shown in Fig. 12.4.24.

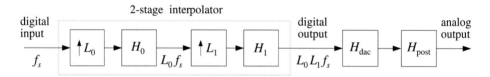

Fig. 12.4.24 First stage equalizes the passband responses of all remaining stages.

As an example, consider a 2-stage interpolator with oversampling factors $L = L_0 L_1$. The first interpolation filter is a lowpass digital filter with sharp cutoff at $f_s/2$,
and the second is a *linear* interpolator. The first filter H_0 increases the sampling
rate from f_s to $L_0 f_s$. The second filter H_1 increases it by a factor L_1, from $L_0 f_s$ to
$L_1 L_0 f_s$. It has magnitude response (normalized to unity at DC):

$$|H_1(f)| = \left| \frac{\sin(\pi f / L_0 f_s)}{L_1 \sin(\pi f / L_1 L_0 f_s)} \right|^2 = \left| \frac{\sin(\omega'/2)}{L_1 \sin(\omega'/2L_1)} \right|^2 \equiv D_{\text{lin}}(\omega')$$

where $\omega' = 2\pi f / (L_0 f_s)$ is the digital frequency with respect to the rate $L_0 f_s$ of the
filter H_0.

The job of the filter H_0 is to remove the $(L_0 - 1)$ spectral replicas that lie between
replicas at multiples of $L_0 f_s$. Because it is periodic with period $L_0 f_s$, the filter H_0
cannot remove the replicas at multiples of $L_0 f_s$. Those are *partially* removed by
the second filter H_1, which vanishes at multiples of $L_0 f_s$ that are *not* multiples
of $L_1 L_0 f_s$. Even though these replicas vanish at their centers, their edges are only
suppressed by about 33 dB. Further suppression requires the aid of a postfilter. The
combined effect of the H_0 and H_1 filters is to leave only the replicas at multiples
of the final sampling rate $L_1 L_0 f_s$. Those are also removed by the postfilter. Figures
(12.4.25) and (12.4.27) illustrate these remarks for $L_0 = L_1 = 4$.

The output of H_1 is applied to a staircase DAC operating at the high rate $L_1 L_0 f_s$. Its frequency response (normalized to unity at DC) is:

$$|H_{\text{dac}}(f)| = \left| \frac{\sin(\pi f/L_1 L_0 f_s)}{\pi f/L_1 L_0 f_s} \right| = \left| \frac{\sin(\omega'/2L_1)}{\omega'/2L_1} \right| \equiv D_{\text{dac}}(\omega')$$

The DAC's $\sin x/x$ response vanishes at multiples of the final rate $L_1 L_0 f_s$. Finally, the staircase output of the DAC will be smoothed by an analog postfilter, say, a third-order Butterworth filter with 3-dB cutoff frequency f_0, having magnitude response:

$$|H_{\text{post}}(f)| = \frac{1}{\left[1 + \left(\dfrac{f}{f_0}\right)^6\right]^{1/2}} = \frac{1}{\left[1 + \left(\dfrac{\omega'}{\omega_0}\right)^6\right]^{1/2}} \equiv D_{\text{post}}(\omega')$$

where $\omega_0 = 2\pi f_0/L_0 f_s$. The three responses $D_{\text{lin}}(\omega')$, $D_{\text{dac}}(\omega')$, and $D_{\text{post}}(\omega')$ can be equalized simultaneously by the interpolation filter H_0 by defining its passband specifications as:

$$D(\omega') = \begin{cases} \dfrac{L_0}{D_{\text{lin}}(\omega') D_{\text{dac}}(\omega') D_{\text{post}}(\omega')} & \text{if } |\omega'| \le \dfrac{\pi}{L_0} \\ 0, & \text{if } \dfrac{\pi}{L_0} < |\omega'| \le \pi \end{cases} \qquad (12.4.16)$$

Assuming a filter length $N_0 = 2L_0 M_0 + 1$, we obtain the coefficients of the filter H_0 by the following frequency sampling design expression:

$$\tilde{d}(k') = \frac{L_0}{N_0} \left[1 + 2 \sum_{i=1}^{M_0} \frac{1}{D_{\text{lin}}(\omega_i') D_{\text{dac}}(\omega_i') D_{\text{post}}(\omega_i')} \cos(\omega_i' k') \right]$$

for $-L_0 M_0 \le k' \le L_0 M_0$. Note that if L_0 and L_1 are large, two simplifications may be introduced in the definition (12.4.16). First, we may omit the DAC equalization factor $D_{\text{dac}}(\omega')$ because *within* the passband $0 \le \omega' \le \pi/L_0$ it reaches a maximum attenuation of:

$$\frac{\sin(\pi/2L_1 L_0)}{\pi/2L_1 L_0}$$

which will be extremely small. Second, again within the passband $0 \le \omega' \le \pi/L_0$, we can approximate the linear interpolator response by its analog equivalent $(\sin x/x)^2$ response, which is *independent* of L_1:

$$D_{\text{lin}}(\omega') = \left| \frac{\sin(\omega'/2)}{L_1 \sin(\omega'/2L_1)} \right|^2 \simeq \left| \frac{\sin(\omega'/2)}{\omega'/2} \right|^2$$

In deciding the specifications of the postfilter, we must consider the total attenuation of the replica at $L_0 f_s$, which will have the *worst* attenuation. Because of the downward sloping of the postfilter, the remaining replicas will be attenuated more. At the left edge of that replica, that is, at frequency

$$f_{\text{stop}} = L_0 f_s - \frac{f_s}{2} \quad \Rightarrow \quad \omega'_{\text{stop}} = \frac{2\pi f_{\text{stop}}}{L_0 f_s} = \frac{\pi(2L_0 - 1)}{L_0}$$

the attenuations of the linear interpolator, DAC, and postfilter can be calculated by

$$A_{\text{lin}}(f_{\text{stop}}) = -20 \log_{10} \left| \frac{\sin(\omega'_{\text{stop}}/2)}{L_1 \sin(\omega'_{\text{stop}}/2L_1)} \right|^2$$

$$A_{\text{dac}}(f_{\text{stop}}) = -20 \log_{10} \left| \frac{\sin(\omega'_{\text{stop}}/2L_1)}{\omega'_{\text{stop}}/2L_1} \right| \tag{12.4.17}$$

$$A_{\text{post}}(f_{\text{stop}}) = 10 \log_{10} \left[1 + \left(\frac{f_{\text{stop}}}{f_0} \right)^6 \right]$$

From these equations, one may determine the 3-dB frequency f_0 of the postfilter in order that the total attenuation at f_{stop} be equal to a desired level, say A_{tot}

$$A_{\text{tot}} = A_{\text{lin}}(f_{\text{stop}}) + A_{\text{dac}}(f_{\text{stop}}) + A_{\text{post}}(f_{\text{stop}})$$

Similarly, one can calculate the attenuations of the linear interpolator, DAC, and postfilter at the edge of the passband, that is, at $f_{\text{pass}} = f_s/2$ or $\omega'_{\text{pass}} = \pi/L_0$

$$A_{\text{lin}}(f_{\text{pass}}) = -20 \log_{10} \left| \frac{\sin(\omega'_{\text{pass}}/2)}{L_1 \sin(\omega'_{\text{pass}}/2L_1)} \right|^2$$

$$A_{\text{dac}}(f_{\text{pass}}) = -20 \log_{10} \left| \frac{\sin(\omega'_{\text{pass}}/2L_1)}{\omega'_{\text{pass}}/2L_1} \right| \tag{12.4.18}$$

$$A_{\text{post}}(f_{\text{pass}}) = 10 \log_{10} \left[1 + \left(\frac{f_{\text{pass}}}{f_0} \right)^6 \right]$$

Their sum is the total passband attenuation that must be equalized by the interpolator H_0. Finally, one must verify that the last replica at $L_1 L_0 f_s$, which survives the combined interpolation filters H_0 and H_1, is attenuated sufficiently by the DAC/postfilter combination. At the left edge of that replica, that is, at frequency

$$f_{\text{last}} = L_1 L_0 f_s - \frac{f_s}{2} \quad \Rightarrow \quad \omega'_{\text{last}} = \frac{2\pi f_{\text{last}}}{L_0 f_s} = \frac{\pi(2L_1 L_0 - 1)}{L_0}$$

the attenuations of the linear interpolator, DAC, and postfilter are

$$A_{\text{lin}}(f_{\text{last}}) = -20 \log_{10} \left| \frac{\sin(\omega'_{\text{last}}/2)}{L_1 \sin(\omega'_{\text{last}}/2L_1)} \right|^2$$

$$A_{\text{dac}}(f_{\text{last}}) = -20 \log_{10} \left| \frac{\sin(\omega'_{\text{last}}/2L_1)}{\omega'_{\text{last}}/2L_1} \right| \qquad (12.4.19)$$

$$A_{\text{post}}(f_{\text{last}}) = 10 \log_{10} \left[1 + \left(\frac{f_{\text{last}}}{f_0} \right)^6 \right]$$

As a concrete design example,[†] suppose $f_s = 40$ kHz and $L_0 = L_1 = 4$, so that the total interpolation factor will be $L = 16$. For the first stage H_0, we use a Kaiser design with stopband attenuation $A = 60$ dB and transition width $\Delta f = 5$ kHz. The Kaiser parameters are:

$$D = \frac{A - 7.95}{14.36} = 3.625$$

$$N_0 - 1 \geq \frac{DL_0}{\Delta F} = 115.98 \quad \Rightarrow \quad N_0 = 121$$

The linear interpolator and DAC will have frequency responses:

$$|H_1(f)| = \left| \frac{\sin(\pi f/4f_s)}{4 \sin(\pi f/16f_s)} \right|^2, \qquad |H_{\text{dac}}(f)| = \left| \frac{\sin(\pi f/16f_s)}{\pi f/16f_s} \right|$$

It is evident that $|H_1(f)|$ vanishes at all multiples of $4f_s$ which are not multiples of $16f_s$, whereas $|H_{\text{dac}}(f)|$ vanishes at all non-zero multiples of $16f_s$.

Using Eqs. (12.4.17), we find that the requirement that at $f_{\text{stop}} = L_0 f_s - f_s/2 = 140$ kHz the total attenuation be more than 60 dB gives the value $f_0 = 50$ kHz for the 3-dB frequency of the postfilter.

Table 12.4.1 shows the attenuations calculated by Eqs. (12.4.17–12.4.19) at the three frequencies $f_{\text{pass}} = 20$ kHz, $f_{\text{stop}} = 140$ kHz, and $f_{\text{last}} = 620$ kHz, or equivalently, the normalized ones $\omega'_{\text{pass}} = \pi/4$, $\omega'_{\text{stop}} = 7\pi/4$, and $\omega'_{\text{last}} = 31\pi/4$.

It is evident from this table that most of the attenuation in the passband arises from the linear interpolator—one could have equalized only the linear interpolator, ignoring the DAC and postfilter, without much degradation.

At 140 kHz, the linear interpolator and postfilter contribute almost equally towards the suppression of the 160 kHz replica—the linear interpolator providing an attenuation of about 33 dB and the postfilter supplementing it with another 27 dB for a total of 60 dB. The DAC's contribution is minimal there.

At 620 kHz, the main contributors towards the suppression of the 640 kHz replica are the postfilter and the DAC providing a total of 95 dB suppression. The linear interpolator's contribution is minimal (by symmetry, its attenuation is the same at 20 kHz and 620 kHz).

[†]See [281] for a similar example having three stages $L_0 = 4$, $L_1 = 32$ linear, and $L_2 = 2$ hold interpolator, with overall $L = 256$.

	f_{pass}	f_{stop}	f_{last}
A_{lin}	0.421	32.863	0.421
A_{dac}	0.014	0.695	29.841
A_{post}	0.018	26.838	65.605
A_{tot}	0.453	60.396	95.867

Table 12.4.1 Attenuations in dB.

Figure 12.4.25 shows the magnitude responses $H_0(f), H_1(f), H_{\text{dac}}(f), H_{\text{post}}(f)$. It also shows the total interpolator response $H_0(f)H_1(f)$, which removes the replicas at multiples of $L_0 f_s$ only at the 33 dB level.

Figure 12.4.26 shows the magnified passband of the filter $H_0(f)$, together with the inverse filter $\left[H_1(f)H_{\text{dac}}(f)H_{\text{post}}(f)\right]^{-1}$. It also shows the magnified passband of the total interpolator filter $H_0(f)H_1(f)$, which is essentially flat since most of the equalizing action of $H_0(f)$ goes to equalize $H_1(f)$. Figure 12.4.27 shows the effective total reconstruction filter

$$H_{\text{rec}}(f) = H_0(f)H_1(f)H_{\text{dac}}(f)H_{\text{post}}(f)$$

which has a flat passband and suppresses all spectral images by at least 60 dB.

In summary, the spectral images of the original sampled signal at multiples $mf_s = m40$ kHz, are removed in several stages: First, the interpolator H_0 removes the replicas at $m = (1,2,3)$, $(5,6,7)$, $(9,10,11)$, $(13,14,15)$, and so on. Then, the second interpolator H_1, with the help of the postfilter, removes the replicas at $m = 4, 8, 12$, and so on. Finally, the postfilter removes the replica $m = 16$ (and all others at multiples of 16 beyond that).

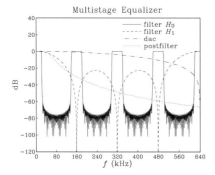
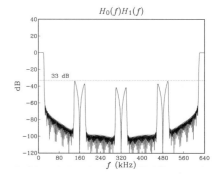

Fig. 12.4.25   16-times interpolator with DAC, postfilter, and multistage equalization.

This design example raises some additional questions: Could we have used a second-order Butterworth postfilter? A first-order one? Given a desired level, say

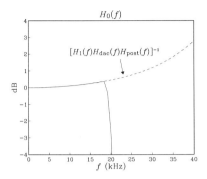 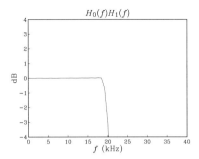

Fig. 12.4.26 Magnified passbands of $H_0(f)$ and $H_0(f)H_1(f)$.

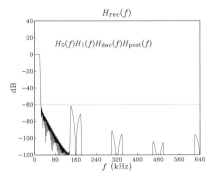 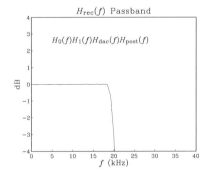

Fig. 12.4.27 Effective reconstructor has flat passband and 60 dB stopband.

A dB, of suppression of the images in the overall equalized reconstructor, can we predict what the lowest order of the postfilter would be? To answer these questions and to give a more accurate design technique, let us define the total filter being equalized by $H_0(f)$, namely,

$$H_{\text{eq}}(f) = H_1(f)H_{\text{dac}}(f)H_{\text{post}}(f)$$

so that the overall reconstructor is:

$$H_{\text{rec}}(f) = H_0(f)H_{\text{eq}}(f)$$

Assuming that H_0 and H_{rec} are normalized to unity at DC by dividing out a factor L_0, we write the corresponding attenuations in dB:

$$A_{\text{rec}}(f) = -20\log_{10}|H_{\text{rec}}(f)|$$

$$A_0(f) = -20\log_{10}|H_0(f)|$$

$$A_{\text{eq}}(f) = -20\log_{10}|H_{\text{eq}}(f)|$$

and therefore, we have:

$$A_{\text{rec}}(f) = A_0(f) + A_{\text{eq}}(f)$$

The attenuation achieved by $H_{\text{rec}}(f)$ at frequency f_{stop} should be at least A dB, that is, $A_{\text{rec}}(f_{\text{stop}}) \geq A$, and therefore

$$A_0(f_{\text{stop}}) + A_{\text{eq}}(f_{\text{stop}}) \geq A \qquad (12.4.20)$$

Because the filter $H_0(f)$ is periodic with period $L_0 f_s$ and the passband and stopband frequencies satisfy $f_{\text{stop}} = L_0 f_s - f_{\text{pass}}$, it follows that

$$|H_0(f_{\text{stop}})| = |H_0(L_0 f_s - f_{\text{pass}})| = |H_0(-f_{\text{pass}})| = |H_0(f_{\text{pass}})|$$

or, in dB:

$$A_0(f_{\text{stop}}) = A_0(f_{\text{pass}})$$

But at frequency f_{pass}, the filter $H_0(f)$ is designed to equalize the total filter $H_{\text{eq}}(f)$, that is,

$$|H_0(f_{\text{pass}})| = \frac{1}{|H_{\text{eq}}(f_{\text{pass}})|} \qquad \Rightarrow \qquad A_0(f_{\text{stop}}) = A_0(f_{\text{pass}}) = -A_{\text{eq}}(f_{\text{pass}})$$

Therefore, we rewrite the design condition (12.4.20) as:

$$A_{\text{eq}}(f_{\text{stop}}) - A_{\text{eq}}(f_{\text{pass}}) \geq A \qquad (12.4.21)$$

For the designed example above, we can subtract the two entries in the last row of Table 12.4.1 to get the value $60.396 - 0.453 = 59.943$, which is almost the desired 60 dB—it would exceed 60 dB had we chosen a slightly smaller 3-dB normalization frequency for the postfilter, for example, $f_0 = 49.8$ kHz.

Writing A_{eq} as the sum of the individual attenuations of the linear interpolator, the DAC, and the postfilter, and solving Eq. (12.4.21) for the difference of attenuations of the postfilter, we find

$$A_{\text{post}}(f_{\text{stop}}) - A_{\text{post}}(f_{\text{pass}}) \geq A - A_d \qquad (12.4.22)$$

where

$$A_d = A_{\text{lin}}(f_{\text{stop}}) - A_{\text{lin}}(f_{\text{pass}}) + A_{\text{dac}}(f_{\text{stop}}) - A_{\text{dac}}(f_{\text{pass}})$$

Given the value of A, the right-hand side of Eq. (12.4.22) can be calculated using Eqs. (12.4.17) and (12.4.18). Then, Eq. (12.4.22) imposes a certain restriction on the order of the postfilter. For a Butterworth filter of order N_b, we have

$$A_{\text{post}}(f) = 10 \log_{10}\left[1 + \left(\frac{f}{f_0}\right)^{2N_b}\right]$$

and therefore, the design condition (12.4.22) becomes

$$10\log_{10}\left[\frac{1 + (f_{stop}/f_0)^{2N_b}}{1 + (f_{pass}/f_0)^{2N_b}}\right] \geq A - A_d \qquad (12.4.23)$$

Given N_b, Eq. (12.4.23) can be solved for the 3-dB frequency f_0. For large values of N_b, the passband term $A_{post}(f_{pass})$ can be ignored because it is much smaller than the stopband term $A_{post}(f_{stop})$, as was the case in Table 12.4.1. For small values of N_b, f_0 must also get smaller in order to provide sufficient attenuation at f_{stop}, but this also causes more attenuation within the passband, so that the difference $A_{post}(f_{stop}) - A_{post}(f_{pass})$ may never become large enough to satisfy Eq. (12.4.22).

In fact, thinking of the left-hand side of Eq. (12.4.23) as a function of f_0, one can easily verify that it is a decreasing function of f_0 and its maximum value, reached at $f_0 = 0$, is $20N_b \log_{10}(f_{stop}/f_{pass})$. Therefore, Eq. (12.4.23) will have a solution for f_0 only if N_b is large enough to satisfy

$$20N_b\log_{10}\left(\frac{f_{stop}}{f_{pass}}\right) \geq A - A_d \qquad \Rightarrow \qquad N_b \geq \frac{A - A_d}{20\log_{10}\left(\frac{f_{stop}}{f_{pass}}\right)}$$

Because $f_{stop}/f_{pass} = 2L_0 - 1$, we can rewrite this condition as

$$N_b \geq \frac{A - A_d}{20\log_{10}(2L_0 - 1)} \qquad (12.4.24)$$

If Eq. (12.4.24) is satisfied, then the solution of Eq. (12.4.23) for f_0 is

$$f_0 = f_{pass}\left[\frac{(2L_0 - 1)^{2N_b} - 1}{10^{(A - A_d)/10} - 1} - 1\right]^{1/2N_b} \qquad (12.4.25)$$

Figure 12.4.28 shows the minimum Butterworth order given in Eq. (12.4.24) as a function of the desired attenuation A, for various values of the interpolation factor L_0. One should pick, of course, the next integer above each curve. As L_0 increases, separating the $L_0 f_s$ replicas more, the allowed filter order becomes less. In the figure, $L_1 = 4$. For a given value of L_0, the dependence of the curves on L_1 is very minimal. Only the intercept A_d, not the slope, of the straight lines is slightly changed if L_1 is changed.

Inspecting this figure, we find for the above example that any filter order $N_b \geq 2$ can be used, but $N_b = 1$ cannot. Alternatively, we can calculate Eq. (12.4.24) directly. Using Table 12.4.1, we find $A_d = 33.123$ dB and therefore Eq. (12.4.24) yields $N_b \geq (60 - 33.123)/20\log_{10}(8 - 1) = 1.59$.

12.5 *Decimation and Oversampling**

Decimation by an integer factor L is the reverse of interpolation, that is, *decreasing* the sampling rate from the high rate f_s' to the lower rate $f_s = f_s'/L$.

An ideal interpolator replaces a low-rate signal $x(n)$ by the high-rate interpolated signal $x'(n')$, which would ideally correspond to the resampling of the analog

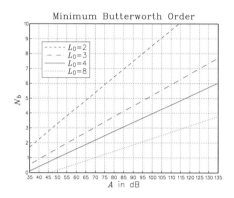

Fig. 12.4.28 Minimum Butterworth postfilter order for given stopband attenuation.

signal at the higher rate. As shown in Fig. 12.2.6, the spectrum $X'(f)$ of $x'(n')$ is the spectrum of $x(n)$ with $L-1$ spectral images removed between multiples of f_s'. This ideal interpolation process can be reversed by keeping from $x'(n')$ every Lth sample and discarding the $L-1$ samples that were interpolated between the low-rate ones.

This process of *downsampling* and its effect in the time and frequency domains is depicted in Fig. 12.5.1. Formally, the downsampled signal is defined in terms of the slow time scale as follows:

$$\boxed{x_{\text{down}}(n) = x'(n')\big|_{n'=nL} = x'(nL)} \tag{12.5.1}$$

For the ideal situation depicted in Fig. 12.5.1, the downsampled signal $x_{\text{down}}(n)$ *coincides* with the low-rate signal $x(n)$ that would have been obtained had the analog signal been resampled at the lower rate f_s, that is,

$$x(n) = x_{\text{down}}(n) = x'(nL) \tag{12.5.2}$$

The gaps in the input spectrum $X'(f)$ are necessary to guarantee this equality. Dropping the sampling rate by a factor of L, shrinks the Nyquist interval $[-f_s'/2, f_s'2]$ by a factor of L to the new interval $[-f_s/2, f_s/2]$. Thus, if the signal had frequency components outside the new Nyquist interval, aliasing would occur and $x_{\text{down}}(n) \neq x(n)$.

In Fig. 12.5.1, the input spectrum was already restricted to the f_s Nyquist interval, and therefore, aliasing did not occur. The rate decrease causes the spectral images of $X'(f)$ at multiples of f_s' to be *down shifted* and become images of $X(f)$ at multiples of f_s without overlapping. The mathematical justification of this downshifting property is derived by expressing Eq. (12.5.2) in the frequency domain. It can be shown (see Problem 12.12) that:

$$\boxed{X(f) = X_{\text{down}}(f) = \frac{1}{L}\sum_{m=0}^{L-1} X'(f - mf_s)} \tag{12.5.3}$$

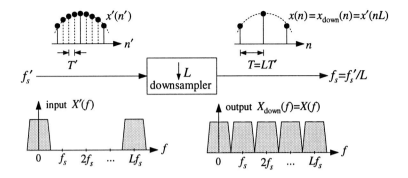

Fig. 12.5.1 Downsampler keeps one out of every L high-rate samples.

Therefore, the downsampling process causes the periodic replication of the original spectrum $X'(f)$ at multiples of the low rate f_s. This operation is depicted in Fig. 12.5.2 for $L = 4$.

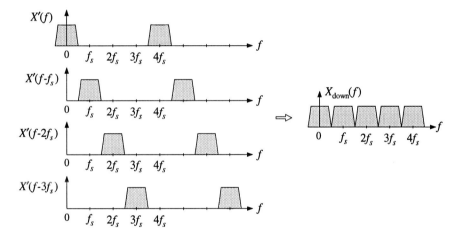

Fig. 12.5.2 Downsampled spectrum is sum of shifted high-rate spectra.

In general, if the high-rate signal $x'(n')$ has frequency components outside the low-rate Nyquist interval $[-f_s/2, f_s/2]$, then downsampling alone is not sufficient to perform decimation. For example, noise in the signal, such as quantization noise arising from the A/D conversion process, will have such frequency components.

To avoid the aliasing that will arise by the spectrum replication property (12.5.3), the high-rate input $x'(n')$ must be *prefiltered* by a digital lowpass filter, called the *decimation filter*. The combined filter/downsampler system is called a *decimator* and is depicted in Fig. 12.5.3.

The filter operates at the high rate f_s' and has cutoff frequency $f_c = f_s/2 = f_s'/2L$. It is similar to the ideal interpolation filter, except its DC gain is unity instead of L. The high-rate output of the filter is downsampled to obtain the desired low-rate decimated signal, with non-overlapping down-shifted replicas:

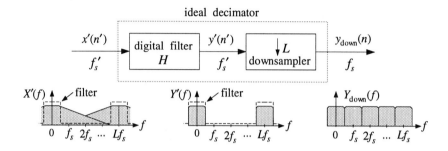

Fig. 12.5.3 Ideal digital decimator in frequency domain.

$$y_{\text{down}}(n) = y'(nL), \qquad Y_{\text{down}}(f) = \frac{1}{L}\sum_{m=0}^{L-1} Y'(f - mf_s) \qquad (12.5.4)$$

The design of the decimation filter is identical to that of the interpolation filter. For example, a length-N FIR decimator can be obtained by windowing the (causal) ideal impulse response:

$$h(n') = w(n')d(n' - LM), \qquad \text{where} \quad d(k') = \frac{\sin(\pi k'/L)}{\pi k'}$$

where $n' = 0, 1, \ldots, N-1$, and $N = 2LM + 1$. A Kaiser window $w(n')$ may be used. The downsampled output is obtained by:

$$\boxed{y_{\text{down}}(n) = y'(nL) = \sum_{m'=0}^{N-1} h(m')x'(nL - m')} \qquad (12.5.5)$$

Because only every Lth output of the filter is needed, the overall computational rate is reduced by a factor of L, that is,

$$R = \frac{1}{L}Nf_s' = Nf_s \qquad (12.5.6)$$

This is similar to the savings of the polyphase form of interpolation. A simple implementation uses a length-N tapped delay line into which the high-rate input samples are shifted at the high rate f_s'. Every L inputs, its contents are used to perform the filter's dot product output computation. A circular buffer implementation of the delay-line would, of course, avoid the time it takes to perform the shifting. Denoting by $\mathbf{w} = [w_0, w_1, \ldots, w_{N-1}]$ the N-dimensional internal state vector of the filter, we may state this filtering/downsampling algorithm as follows:

$$\boxed{\begin{array}{l} \textit{for each high-rate input sample } x' \textit{ do:} \\ \quad w_0 = x' \\ \textit{for every } L\textit{th input compute:} \\ \quad y_{\text{down}} = \text{dot}(N-1, \mathbf{h}, \mathbf{w}) \\ \quad \text{delay}(N-1, \mathbf{w}) \end{array}} \qquad (12.5.7)$$

Multistage implementations of decimators are also possible [273–276]. The proper ordering of the decimation stages is the *reverse* of the interpolation case, that is, the decimator with the most *stringent* specifications is placed *last*.

Often, the earlier decimators, which also have the highest rates, are chosen to have *simplified* structures, such as simple averaging filters [277]. For example, the decimation version of the hold interpolator of Section 12.3 is obtained by dividing Eq. (12.3.8) by L to restore its DC gain to unity:

$$H(\zeta) = \frac{1}{L} \frac{1 - \zeta^{-L}}{1 - \zeta^{-1}} = \frac{1}{L} \left[1 + \zeta^{-1} + \zeta^{-2} + \cdots + \zeta^{-(L-1)} \right] \qquad (12.5.8)$$

where ζ^{-1} is one high-rate delay. Thus, the decimator is a simple FIR *averaging* filter that averages L successive high-rate samples:

$$y_{\text{down}}(n) = \frac{x'(nL) + x'(nL-1) + x'(nL-2) + \cdots + x'(nL-L+1)}{L} \qquad (12.5.9)$$

If so desired, the cruder passbands of the earlier decimators can be equalized by the last decimator, which can also equalize any imperfect passband of the analog antialiasing prefilter used prior to sampling.

Indeed, one of the main uses of decimators is to alleviate the need for high-quality analog prefilters, much as the interpolators ease the specifications of the anti-image postfilters. This idea is used in many current applications, such as the sampling systems of DAT machines, PC sound cards, speech CODECs, and various types of delta-sigma A/D converter chips.

Sampling an analog signal, such as audio, at its nominal Nyquist rate f_s would require a high-quality analog prefilter to bandlimit the input to the Nyquist frequency $f_{\max} = f_s/2$. In a sampling system that uses oversampling and decimation, the analog input is first prefiltered by a simple prefilter and then sampled at the higher rate $f_s' = L f_s$. The decimation filter then reduces the bandwidth of the sampled signal to $f_s/2$. The sharp cutoffs at the Nyquist frequency $f_s/2$ are provided by the digital decimation filter instead of the prefilter.

The specifications of the prefilter are shown in Fig. 12.5.4. The decimator removes all frequencies from the range $[f_s/2, Lf_s - f_s/2]$. But because of periodicity, it cannot remove any frequencies in the range $Lf_s \pm f_s/2$. Such frequencies, if present in the analog input, must be removed by the prefilter prior to sampling; otherwise they will be aliased back into the desired Nyquist interval $[-f_s/2, f_s/2]$. Therefore, the prefilter's passband and stopband frequencies are:

$$\boxed{f_{\text{pass}} = \frac{f_s}{2}, \qquad f_{\text{stop}} = Lf_s - \frac{f_s}{2}} \qquad (12.5.10)$$

The transition width of the prefilter is $\Delta f = f_{\text{stop}} - f_{\text{pass}} = (L-1)f_s$ and gets wider with the oversampling ratio L. Hence, the filter's complexity reduces with increasing L. (See Problem 12.16 for a quantitative relationship between L and filter order N.)

In summary, oversampling in conjunction with decimation and interpolation alleviates the need for high-quality analog prefilters and postfilters by assigning

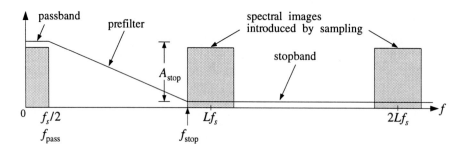

Fig. 12.5.4 Analog prefilter specifications for L-fold decimation.

the burden of achieving sharp transition characteristics to the *digital* filters. Figure 12.5.5 shows an oversampling DSP system in which sampling and reconstruction are carried out at the fast rate f_s', and any intermediate digital processing at the low rate f_s.

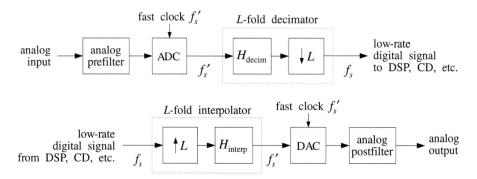

Fig. 12.5.5 Oversampling DSP system.

A second major benefit of oversampling is that it also simplifies the structure of the A/D and D/A converters shown in the figure, so that they require fewer bits without sacrificing quality. This is accomplished by the principle of *feedback quantization*, which we discuss in Section 12.7. The changes in Fig. 12.5.5 are to replace the conventional ADC block by a delta-sigma ADC operating at fewer bits (even 1 bit), and insert between the output interpolator and the DAC a noise shaping quantizer that requantizes the output to fewer bits.

12.6 Sampling Rate Converters*

Interpolators and decimators are examples of sampling rate converters that change the rate by *integer* factors. A more general sampling rate converter [273–276] can change the rate by an arbitrary *rational* factor, say L/M, so that the output rate will be related to the input rate by:

$$f_s' = \frac{L}{M} f_s \qquad (12.6.1)$$

Such rate changes are necessary in practice for interfacing DSP systems operating at different rates. For example, to convert digital audio for broadcasting, sampled at 32 kHz, to digital audio for a DAT machine, sampled at 48 kHz, one must use a conversion factor of $48/32 = 3/2$. Similarly, to convert DAT audio to CD audio at 44.1 kHz, one must use the factor $44.1/48 = 147/160$.

The rate conversion can be accomplished by first increasing the rate by a factor of L to the high rate $f_s'' = Lf_s$ using an L-fold interpolator, and then decreasing the rate by a factor of M down to $f_s' = f_s''/M = Lf_s/M$ using an M-fold decimator.

Note that f_s'' is an integer multiple of both the input and output rates, and the corresponding sampling time interval $T'' = 1/f_s''$ is an integer fraction of both the input and output sampling times T and T':

$$f_s'' = Lf_s = Mf_s', \qquad T'' = \frac{T}{L} = \frac{T'}{M} \qquad (12.6.2)$$

Because both the interpolation and decimation filters are operating at the same high rate f_s'' and both are lowpass filters, they may be combined into a single lowpass filter preceded by an upsampler and followed by a downsampler, as shown in Fig. 12.6.1.

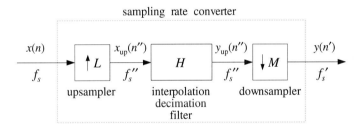

Fig. 12.6.1 Sampling rate conversion by a factor of L/M.

The interpolation filter must have cutoff frequency $f_s''/2L = f_s/2$ and the decimation filter $f_s''/2M = f_s'/2$. Thus, the cutoff frequency of the common filter must be chosen to be the *minimum* of the two:

$$f_c = \frac{1}{2} \min(f_s, f_s') \qquad (12.6.3)$$

which can be written also in the alternative forms:

$$f_c = \min\left(1, \frac{L}{M}\right) \frac{f_s}{2} = \min\left(\frac{M}{L}, 1\right) \frac{f_s'}{2} = \min\left(\frac{1}{L}, \frac{1}{M}\right) \frac{f_s''}{2}$$

In units of the high-rate digital frequency $\omega'' = 2\pi f/f_s''$, we have:

$$\boxed{\omega_c'' = \frac{2\pi f_c}{f_s''} = \min\left(\frac{\pi}{L},\frac{\pi}{M}\right) = \frac{\pi}{\max(L,M)}}\qquad(12.6.4)$$

When $f_s' > f_s$, the common filter acts as an *anti-image postfilter* for the upsampler, removing the spectral replicas at multiples of f_s but not at multiples of Lf_s. When $f_s' < f_s$, it acts as an *antialiasing prefilter* for the downsampler, making sure that the down-shifted replicas at multiples of f_s' do not overlap.

The design of the filter is straightforward. Assuming a filter length N of the form[†] $N = 2LK + 1$ and passband gain of L, we define the windowed impulse response, with respect to the high-rate time index $n'' = 0, 1, \ldots, N - 1$:

$$h(n'') = w(n'')d(n'' - LK),\quad\text{where}\quad d(k'') = L\,\frac{\sin(\omega_c'' k'')}{\pi k''}\qquad(12.6.5)$$

where $w(n'')$ is any desired length-N window. Its L polyphase subfilters of length $2K$ are defined for $i = 0, 1, \ldots, L - 1$:

$$h_i(n) = h(Ln + i),\qquad n = 0, 1, \ldots, 2K - 1\qquad(12.6.6)$$

Next, we discuss the time-domain operation and implementation of the converter. The input signal $x(n)$ is upsampled to the high rate f_s''. Then, the upsampled input $x_{\text{up}}(n'')$ is filtered, generating the interpolated output $y_{\text{up}}(n'')$, which is then downsampled by keeping one out of every M samples, that is, setting $n'' = Mn'$ to obtain the desired signal $y(n')$ resampled at rate f_s'. Thus, we have:

$$y_{\text{up}}(n'') = \sum_{m''=0}^{N-1} h(m'')x_{\text{up}}(n'' - m'')\quad\text{and}\quad y(n') = y_{\text{up}}(Mn')$$

The interpolation operation can be implemented efficiently in its polyphase realization. Setting $n'' = Ln + i$, with $i = 0, 1, \ldots, L - 1$, we obtain the ith sample interpolated between the input samples $x(n)$ and $x(n + 1)$, from Eq. (12.2.18):

$$y_i(n) = y_{\text{up}}(Ln + i) = \sum_{m=0}^{P} h_i(m)x(n - m) = \text{dot}(P, \mathbf{h}_i, \mathbf{w}(n))\qquad(12.6.7)$$

where we set $P = 2K - 1$ for the *order* of the polyphase subfilters (the time-advance required for causal operation is not shown here). As we saw in Eq. (12.2.20), its implementation requires a low-rate tapped delay line $\mathbf{w} = [w_0, w_1, \ldots, w_P]$, which is used by all polyphase subfilters before it is updated.

Because the downsampler keeps only every Mth filter output, it is not necessary to compute all L interpolated outputs between input samples. Only those interpolated values that correspond to the output time grid need be computed. Given an output sample time $n'' = Mn'$, we can write it uniquely in the form $Mn' = Ln + i$,

---

[†]Here, we use K instead of M to avoid confusion with the downsampling factor M.

where $0 \le i \le L - 1$. It follows that the downsampled output will be the ith inter-polated value arising from the current input $x(n)$ and computed as the output of the ith polyphase subfilter $\mathbf{h}_i$:

$$y(n') = y_{\mathrm{up}}(Mn') = y_{\mathrm{up}}(Ln + i) = y_i(n)$$

The pattern of polyphase indices i that correspond to successive output times n' *repeats* with period L, and depends only on the relative values of L and M. Therefore, for the purpose of deriving a sample processing implementation of the converter, it proves convenient to think in terms of *blocks* of output samples of length-L. The total time duration of such an output block is LT'. Using Eq. (12.6.2), we have:

$$\boxed{T_{\mathrm{block}} = LT' = LMT'' = MT} \tag{12.6.8}$$

Thus, within each output time block there are M input samples, LM high-rate interpolated samples, and L output samples. The M input samples get interpolated into the LM high-rate ones, from which the L output samples are selected.

The computational rate is M times smaller than the polyphase rate Nf_s required for full interpolation. Indeed, we have $2K$ MACs per polyphase filter output and L polyphase outputs in each period T_{block}, that is, $R = 2KL/T_{\mathrm{block}} = N/T_{\mathrm{block}} = N/MT = Nf_s/M$. Equivalently, we have *one* polyphase output in each output period T', $R = 2K/T' = 2Kf_s'$. Thus,

$$\boxed{R = \frac{Nf_s}{M} = \frac{Nf_s'}{L} = 2Kf_s'} \tag{12.6.9}$$

Figure 12.6.2 shows an example with $L = 5$ and $M = 3$, so that $f_s' = 5f_s/3$. The interpolating high rate is $f_s'' = 5f_s = 3f_s'$. The top and bottom figures show the input and output signals and their spectra. The two middle figures show the high-rate interpolated signal, viewed both with respect to the input and output time scales.

Because $f_s' > f_s$, the interpolation filter has cutoff $f_s/2$, and acts as an antialias-ing prefilter removing the four input replicas up to $f_s'' = 5f_s$. The downsampling operation then downshifts the replicas at multiples of f_s'.

In the time domain, each block period $T_{\mathrm{block}} = 15T'' = 3T = 5T'$ contains three input samples, say $\{x_0, x_1, x_2\}$, five output samples, say $\{y_0, y_1, y_2, y_3, y_4\}$, and 15 interpolated high-rate samples.

As can be seen in the figure, the first input period from x_0 to x_1 contains two outputs: y_0, y_1. We have time-aligned the samples so that $y_0 = x_0$. The output y_1 is the third ($i = 3$) interpolated value, and therefore, it is obtained as the output of the polyphase filter $\mathbf{h}_3$ with current input x_0. After this operation, the input sample x_0 is no longer needed and the delay-line $\mathbf{w}$ holding the input samples may be shifted and the next input x_1 read into it.

During the next input period from x_1 to x_2, there are two more outputs: y_2, y_3. The output y_2 is the first ($i = 1$) interpolated value, and therefore, it is the output of the filter $\mathbf{h}_1$, whereas the output y_3 is the fourth ($i = 4$) interpolated value, or

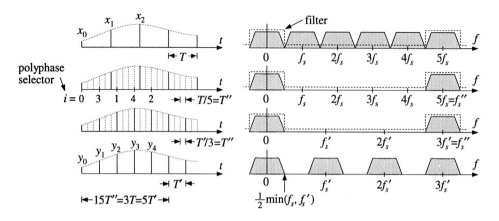

Fig. 12.6.2 Sampling rate conversion by a factor of 5/3.

the output of $\mathbf{h}_4$. After this operation, the delay-line $\mathbf{w}$ may be updated and x_2 read into it.

Finally, the third input period starting at x_2 contains only one output, namely, y_4, which is the second ($i = 2$) interpolated value, or the output of $\mathbf{h}_2$ with input x_2. After this operation, the delay-line may be shifted and the same computational cycle involving the next three inputs repeated. The above steps may be summarized in the following sample processing algorithm:

$$
\begin{aligned}
&\textit{for each input block } \{x_0, x_1, x_2\} \textit{ do:} \\
&\quad w_0 = x_0 \\
&\qquad y_0 = \mathrm{dot}\,(P, \mathbf{h}_0, \mathbf{w}) = x_0 \\
&\qquad y_1 = \mathrm{dot}\,(P, \mathbf{h}_3, \mathbf{w}) \\
&\quad \mathrm{delay}\,(P, \mathbf{w}) \\
&\quad w_0 = x_1 \\
&\qquad y_2 = \mathrm{dot}\,(P, \mathbf{h}_1, \mathbf{w}) \\
&\qquad y_3 = \mathrm{dot}\,(P, \mathbf{h}_4, \mathbf{w}) \\
&\quad \mathrm{delay}\,(P, \mathbf{w}) \\
&\quad w_0 = x_2 \\
&\qquad y_4 = \mathrm{dot}\,(P, \mathbf{h}_2, \mathbf{w}) \\
&\quad \mathrm{delay}\,(P, \mathbf{w})
\end{aligned}
$$

(12.6.10)

The outputs $\{y_0, y_1, y_2, y_3, y_4\}$ were computed by the five polyphase filters $\{\mathbf{h}_0, \mathbf{h}_3, \mathbf{h}_1, \mathbf{h}_4, \mathbf{h}_2\}$ corresponding to the sequence of polyphase indices $i = \{0, 3, 1, 4, 2\}$. The input samples that were used in the computations were $\{x_0, x_0, x_1, x_1, x_2\}$, so that the corresponding index of x_n was $n = \{0, 0, 1, 1, 2\}$. When the index was repeated, the delay line was not updated.

It is easily seen from Fig. 12.6.2 that the patterns of i's and n's get repeated for every group of five outputs. These patterns can be *predetermined* as the solutions of the equations $5n + i = 3m$ for $m = 0, 1, \ldots, 4$. In general, we can calculate the patterns by solving the L equations:

$$Ln_m + i_m = Mm, \qquad m = 0, 1, \ldots, L - 1 \qquad (12.6.11)$$

with solution (where % denotes the modulo operation):

$$
\boxed{
\begin{aligned}
&for \ m = 0, 1, \ldots, L - 1 \ compute: \\
&\quad i_m = (Mm)\%L \\
&\quad n_m = (Mm - i_m)/L
\end{aligned}
}
\qquad \text{(polyphase selectors)} \qquad (12.6.12)
$$

Assuming that the sequences $\{i_m, n_m\}$, $m = 0, 1, \ldots, L - 1$, have been *precomputed*, the general sample rate conversion algorithm that transforms each length-M input block $\{x_0, x_1, \ldots, x_{M-1}\}$ into a length-L output block $\{y_0, y_1, \ldots, y_{L-1}\}$, can be stated as follows:

$$
\boxed{
\begin{aligned}
&for \ each \ input \ block \ \{x_0, x_1, \ldots, x_{M-1}\} \ do: \\
&\quad for \ n = 0, 1, \ldots, M - 1 \ do: \\
&\qquad w_0 = x_n \\
&\qquad for \ Ln/M \le m < L(n+1)/M \ do: \\
&\qquad\quad y_m = \mathrm{dot}(P, \mathbf{h}_{i_m}, \mathbf{w}) \\
&\qquad \mathrm{delay}(P, \mathbf{w})
\end{aligned}
}
\qquad (12.6.13)
$$

The inner loop ensures that the output time index m lies between the two input times $Ln \le Mm < L(n+1)$, with respect to the T'' time scale. Because $Mm = Ln_m + i_m$, it follows that such m's will have $n_m = n$. The index i_m serves as a *polyphase filter selector*.

In the special cases of interpolation ($M = 1$), or decimation ($L = 1$), the algorithm reduces to the corresponding sample processing algorithms given in Eqs. (12.2.20) and (12.5.7). For causal processing, the initialization of the algorithm must be as in Eq. (12.2.19) (with K replacing M).

Another example is shown in Fig. 12.6.3 that has $L = 3$, $M = 5$ and decreases the sampling rate by a factor of $3/5$ so that $f_s' = 3f_s/5$. The interpolating high rate is now $f_s'' = 3f_s = 5f_s'$. Because $f_s' < f_s$, the filter's cutoff frequency must be $f_c = f_s'/2$, and therefore, the filter acts as an antialiasing filter for the downsampler. The filter necessarily chops off those high frequencies from the input that would otherwise be aliased by the downsampling operation, that is, the frequencies in the range $f_s'/2 \le f \le f_s/2$.

In the time domain, each block of five input samples $\{x_0, x_1, x_2, x_3, x_4\}$ generates a block of three output samples $\{y_0, y_1, y_2\}$. The solution of Eq. (12.6.12) gives the polyphase selector sequences, for $m = 0, 1, 2$:

$$n_m = \{0, 1, 3\}, \qquad i_m = \{0, 2, 1\}$$

which means that only the inputs $\{x_0, x_1, x_3\}$ will generate interpolated outputs, with the polyphase subfilters $\{\mathbf{h}_0, \mathbf{h}_2, \mathbf{h}_1\}$. The inputs $\{x_2, x_4\}$ will not generate outputs, but still must be shifted into the delay-line buffer. The same conclusions can also be derived by inspecting Fig. 12.6.3. The corresponding sample processing algorithm, which is a special case of Eq. (12.6.13), is:

$$
\boxed{\begin{aligned}
&\textit{for each input block } \{x_0, x_1, x_2, x_3, x_4\} \textit{ do:}\\
&\quad w_0 = x_0\\
&\qquad y_0 = \mathrm{dot}(P, \mathbf{h}_0, \mathbf{w}) = x_0\\
&\quad \mathrm{delay}(P, \mathbf{w})\\
&\quad w_0 = x_1\\
&\qquad y_1 = \mathrm{dot}(P, \mathbf{h}_2, \mathbf{w})\\
&\quad \mathrm{delay}(P, \mathbf{w})\\
&\quad w_0 = x_2\\
&\quad \mathrm{delay}(P, \mathbf{w})\\
&\quad w_0 = x_3\\
&\qquad y_2 = \mathrm{dot}(P, \mathbf{h}_1, \mathbf{w})\\
&\quad \mathrm{delay}(P, \mathbf{w})\\
&\quad w_0 = x_4\\
&\quad \mathrm{delay}(P, \mathbf{w})
\end{aligned}}
\qquad (12.6.14)
$$

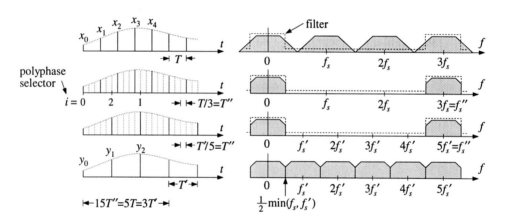

Fig. 12.6.3 Sampling rate conversion by a factor of $3/5$.

The type of converter discussed above is useful when the sampling rates remain *fixed* and synchronous. In some applications, it may be desirable to have *asynchronous* rate changes that can accommodate slowly changing input and output sampling clocks. Such converters must be able to change the rate by *arbitrary* factors, not just rational ones.

Theoretically, the sampling rate can be changed by an arbitrary factor by reconstructing the sampled signal to analog form and then resampling it at the output rate. Digitally, one can use an *extremely large* interpolation factor L to effectively obtain an analog signal and then resample it at the new rate.

The large value of L creates a very dense time grid for the interpolated signal. Every output time instant falls between two such time grid points and the output sample can be chosen to be the nearest of the two interpolated samples, or, it can be formed by linear or higher-order interpolations. References [283–288] discuss practical implementations of this idea.

For example, a recent sample rate conversion chip, designed by Analog Devices for use with digital audio [286–288], uses an interpolation ratio of $L = 2^{16} = 65536$. It can convert sampling rates from 8 to 56 kHz.

The built-in interpolation filter has length $N = 64 \times 2^{16} = 2^{22} \simeq 4 \times 10^6$ and is realized in its polyphase form. Thus, there are $L = 2^{16}$ polyphase filters of length $2K = 64$. The input delay-line buffer also has length 64. The computational rate of the chip is only $R = (2K)f_s{}' = 64f_s{}'$ MAC/sec, where $f_s{}'$ is the output rate.

The chip has a polyphase filter selector (like the quantity i_m) that selects the appropriate polyphase filter to use for each output sampling time. To minimize coefficient storage for the 2^{22} filter coefficients, only 1 out of every 128 impulse response coefficients are saved in ROM; the intermediate values are computed when needed by linear interpolation. Thus, the ROM storage is $2^{22}/128 = 32768$ words.

The interpolation filter has a variable cutoff frequency $f_c = \min(f_s/2, f_s{}'/2)$. To avoid having to redesign the filter every time the cutoff changes, the filter is designed once based on a nominal input frequency f_s, such as 44.1 kHz, and then it is "time-stretched" to accommodate the variable cutoff [285,286]. To understand this, we define the scale factor $\rho = \min(1, f_s{}'/f_s)$, such that $\rho < 1$ whenever the output rate is less than the input rate. Then, we may write f_c in the form:

$$f_c = \rho \frac{f_s}{2} \quad \Rightarrow \quad \omega_c'' = \frac{2\pi f_c}{f_s''} = \rho \frac{\pi}{L}$$

If $\rho < 1$, the corresponding ideal impulse response is:

$$d_\rho(k'') = L \frac{\sin(\omega_c'' k'')}{\pi k''} = \rho \frac{\sin(\pi \rho k''/L)}{\pi \rho k''/L}$$

The fixed filter has response corresponding to $\rho = 1$:

$$d(k'') = \frac{\sin(\pi k''/L)}{\pi k''/L}, \quad -LK \le k'' \le LK$$

It follows that $d_\rho(k'')$ will be the "stretched" version of $d(k'')$:

$$d_\rho(k'') = \rho\, d(\rho k'') \tag{12.6.15}$$

The effective length of this filter must also stretch commensurately. Indeed, because the argument $\rho k''$ must lie in the designed index range of the original filter, that is, $-LK \le \rho k'' \le LK$, we must have in Eq. (12.6.15):

$$-\frac{1}{\rho}LK \le k'' \le \frac{1}{\rho}LK$$

Thus, K increases to $K_\rho = K/\rho$, and the effective length of the filter becomes $N_\rho = 2LK_\rho = 2LK/\rho = N/\rho$. The length of the tapped delay line also becomes longer, $2K_\rho = 2K/\rho$. Because the coefficients $d(k'')$ are stored in ROM only for integer values of k'', the argument of $d(\rho k'')$ must be rounded to the nearest integer. Because of the highly oversampled nature of $d(k'')$, this rounding causes only a small distortion [286].

12.7 Noise Shaping Quantizers*

The main purpose of noise shaping is to reshape the spectrum of quantization noise so that most of the noise is filtered out of the relevant frequency band, such as the audio band. Noise shaping is used in four major applications:

- Oversampled *delta-sigma* A/D converters.
- Oversampled *requantizers* for D/A conversion.
- Non-oversampled dithered noise shaping for requantization.
- Non-oversampled roundoff noise shaping in digital filters.

In the oversampled cases, the main objective is to trade off bits for samples, that is, increasing the sampling rate but reducing the number of bits per sample. The resulting increase in quantization noise is compensated by a noise shaping quantizer that pushes the added noise out of the relevant frequency band in such a way as to preserve a desired level of signal quality. The reduction in the number of bits simplifies the structure of the A/D and D/A converters. See [276,277] for a review and earlier references.

In the non-oversampled cases, one objective is to minimize the accumulation of roundoff noise in digital filter structures [70–76]. Another objective is to reduce the number of bits without reducing quality. For example, in a digital audio recording and mixing system where all the digital processing is done with 20 bits, the resulting audio signal must be rounded eventually to 16 bits in order to place it on a CD. The rounding operation can cause unwanted granulation distortions. Adding a dither signal helps remove such distortions and makes the quantization noise sound like steady background white noise. However, further noise shaping can make this white noise even more inaudible by concentrating it onto spectral bands where the ear is least sensitive [59–67].

A related application in digital audio is to actually keep the bits saved from noise shaping and use them to carry extra data on a conventional CD, such as compressed images, speech, or text, and other information [68,69]. This "buried" data channel is encoded to look like pseudorandom dither which is then added (subtractively) to the CD data and subjected to noise shaping. As many as 4 bits from each 16-bit CD word may be dedicated to such hidden data without sacrificing the quality of the CD material. The resulting data rates are $4\times44.1 = 176.4$ kbits/sec or double that for two stereo channels.

In Section 2.2, we introduced noise shaping quantizers and discussed some of their implications, such as the tradeoff between oversampling ratio and number of bits, but did not discuss how they are constructed.

Figure 12.7.1 shows a typical oversampled first-order *delta-sigma* A/D converter system.[†] The analog input is assumed to have been prefiltered by an antialiasing prefilter whose structure is simplified because of oversampling. The relevant frequency range of the input is the low-rate Nyquist interval $f_s/2$. Such converters are commonly used in oversampling DSP systems, shown in Figs. 2.2.5 and 12.5.5.

The analog part of the converter contains an ordinary A/D converter operating at the fast rate $f_s' = Lf_s$ and having a small number of bits, say B' bits. The most

---

[†] Also called a sigma-delta converter or a feedback quantizer.

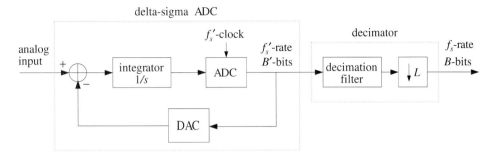

Fig. 12.7.1   Oversampled first-order delta-sigma A/D converter.

useful practical choice is $B' = 1$, that is, a two-level ADC. The output of the ADC is reconstructed back into analog form by the DAC (i.e., a two-level analog signal, if $B' = 1$) and subtracted from the input.

The difference signal (the "delta" part) is accumulated into the integrator (the "sigma" part) and provides a local average of the input. The feedback loop causes the quantization noise generated by the ADC to be *highpass* filtered, pushing its energy towards the higher frequencies (i.e., $f_s'/2$) and away from the signal band.

The digital part of the converter contains an L-fold decimator that reduces the sampling rate down to f_s and increases the number of bits up to a desired resolution, say B bits, where $B > B'$. In practice, the analog and digital parts reside usually on board the same chip.

The lowpass decimation filter does three jobs: (1) It removes the high-frequency quantization noise that was introduced by the feedback loop, (2) it removes any undesired frequency components beyond $f_s/2$ that were not removed by the simple analog prefilter, and (3) through its filtering operation, it increases the number of bits by linearly combining the coarsely quantized input samples with its coefficients, which are taken to have enough bits.

To see the filtering action of the feedback loop on the input and quantization noise, we consider a sampled-data equivalent model of the delta-sigma quantizer, shown in Fig. 12.7.2. The time samples, at rate f_s', are denoted by $x'(n')$ in accordance with our notation in this chapter.

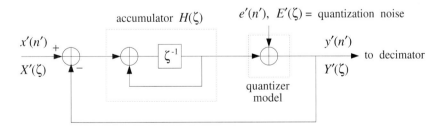

Fig. 12.7.2   Discrete-time model of first-order delta-sigma quantizer.

The ADC is replaced by its equivalent additive-noise model of Fig. 2.1.3 and the integrator by a discrete-time accumulator $H(\zeta)$ with transfer function:

$$H(\zeta) = \frac{\zeta^{-1}}{1 - \zeta^{-1}} \tag{12.7.1}$$

where ζ^{-1} denotes a high-rate unit delay. The numerator delay ζ^{-1} is necessary to make the feedback loop computable.

Working with ζ-transforms, we note that the input to $H(\zeta)$ is the difference signal $X'(\zeta) - Y'(\zeta)$. Its output is added to $E'(\zeta)$ to generate $Y'(\zeta)$. Thus,

$$H(\zeta)\,(X'(\zeta) - Y'(\zeta)) + E'(\zeta) = Y'(\zeta)$$

which may be solved for $Y'(\zeta)$ in terms of the two inputs $X'(\zeta)$ and $E'(\zeta)$:

$$Y'(\zeta) = \frac{H(\zeta)}{1 + H(\zeta)}\,X'(\zeta) + \frac{1}{1 + H(\zeta)}\,E'(\zeta) \tag{12.7.2}$$

It can be written in the form:

$$Y'(\zeta) = H_x(\zeta)\,X'(\zeta) + H_{NS}(\zeta)\,E'(\zeta) \tag{12.7.3}$$

where the noise shaping transfer function $H_{NS}(\zeta)$ and the transfer function for the input $H_x(\zeta)$ are defined as:

$$H_x(\zeta) = \frac{H(\zeta)}{1 + H(\zeta)}, \qquad H_{NS}(\zeta) = \frac{1}{1 + H(\zeta)} \tag{12.7.4}$$

Inserting $H(\zeta)$ from Eq. (12.7.1), we find for the first-order case:

$$\boxed{H_x(\zeta) = \zeta^{-1}, \qquad H_{NS}(\zeta) = 1 - \zeta^{-1}} \tag{12.7.5}$$

Thus, $H_{NS}(\zeta)$ is a simple highpass filter, and $H_x(\zeta)$ an allpass plain delay. The I/O equation (12.7.3) becomes:

$$Y'(\zeta) = \zeta^{-1}X'(\zeta) + (1 - \zeta^{-1})\,E'(\zeta) \tag{12.7.6}$$

or, in the time domain:

$$\boxed{y'(n') = x'(n' - 1) + \varepsilon(n')} \tag{12.7.7}$$

where we defined the filtered quantization noise:

$$\varepsilon(n') = e'(n') - e'(n' - 1) \quad \Leftrightarrow \quad \mathcal{E}(\zeta) = (1 - \zeta^{-1})\,E'(\zeta) \tag{12.7.8}$$

Thus, the quantized output $y'(n')$ is the (delayed) input plus the filtered quantization noise. Because the noise is highpass filtered, further processing of $y'(n')$ by the lowpass decimation filter will tend to average out the noise to zero and also replace the input by its locally averaged, decimated, value. A typical example of a decimator is the hold decimator of Eq. (12.5.9), which averages L successive high-rate samples.

By comparison, had we used a conventional B-bit ADC and sampled the input at the low rate f_s, the corresponding quantized output would be:

$$y(n) = x(n) + e(n)$$ (12.7.9)

where $e(n)$ is modeled as white noise over $[-f_s/2, f_s/2]$.

The "design" condition that renders the quality of the two quantizing systems equivalent and determines the tradeoff between oversampling ratio L and savings in bits, is to require that the rms quantization errors of Eqs. (12.7.7) and (12.7.9) be the *same* over the desired frequency band $[-f_s/2, f_s/2]$. As we saw in Section 2.2, the mean-square errors are obtained by integrating the power spectral densities of the noise signals over that frequency interval, yielding the condition:

$$\sigma_e^2 = \sigma_{e'}^2 \frac{1}{f_s'} \int_{-f_s/2}^{f_s/2} |H_{NS}(f)|^2 \, df$$ (12.7.10)

Setting $f_s' = Lf_s$ and $\sigma_e/\sigma_{e'} = 2^{-B}/2^{-B'} = 2^{-\Delta B}$, where $\Delta B = B - B'$, we obtain the desired relationship between L and ΔB given by Eq. (2.2.10).

Higher-order delta-sigma quantizers have highpass noise shaping transfer functions of the form:

$$H_{NS}(\zeta) = (1 - \zeta^{-1})^p$$ (12.7.11)

where p is the order. The input/output equations for such quantizers are still of the form of Eq. (12.7.3), where $H_x(\zeta)$ is typically a multiple delay. The frequency and magnitude responses of $H_{NS}(\zeta)$ are obtained by setting $\zeta = e^{2\pi j f/f_s'}$:

$$H_{NS}(f) = \left(1 - e^{-2\pi j f/f_s'}\right)^p, \qquad |H_{NS}(f)|^2 = \left|2\sin\left(\frac{\pi f}{f_s'}\right)\right|^{2p}$$ (12.7.12)

resulting in the expressions used in Eq. (2.2.8).

There exist many architectures for higher-order delta-sigma quantizers that address various circuit limitations and limit-cycle instability problems [277,290–294]. Some examples of such architectures are given in the problems.

Example 12.7.1: To illustrate the time-domain operation of a delta-sigma quantizer, consider the common 1-bit case that has a two-level ADC. Let $Q(x)$ denote the two-level quantization function defined by:

$$Q(x) = \text{sign}(x) = \begin{cases} +1, & \text{if } x \geq 0 \\ -1, & \text{if } x < 0 \end{cases}$$ (12.7.13)

The corresponding block diagram of the quantizer is shown below, together with the computational sample processing algorithm. The quantity w_1 is the content of the accumulator's delay:

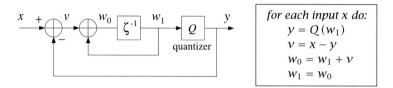

The following table shows the computed outputs for the two constant inputs, $x = 0.4$ and $x = -0.2$, with the algorithm iterated ten times:

x	w_1	y	v	w_0	x	w_1	y	v	w_0
0.4	0.0	1.0	−0.6	−0.6	−0.2	0.0	1.0	−1.2	−1.2
0.4	−0.6	−1.0	1.4	0.8	−0.2	−1.2	−1.0	0.8	−0.4
0.4	0.8	1.0	−0.6	0.2	−0.2	−0.4	−1.0	0.8	0.4
0.4	0.2	1.0	−0.6	−0.4	−0.2	0.4	1.0	−1.2	−0.8
0.4	−0.4	−1.0	1.4	1.0	−0.2	−0.8	−1.0	0.8	0.0
0.4	1.0	1.0	−0.6	0.4	−0.2	0.0	1.0	−1.2	−1.2
0.4	0.4	1.0	−0.6	−0.2	−0.2	−1.2	−1.0	0.8	−0.4
0.4	−0.2	−1.0	1.4	1.2	−0.2	−0.4	−1.0	0.8	0.4
0.4	1.2	1.0	−0.6	0.6	−0.2	0.4	1.0	−1.2	−0.8
0.4	0.6	1.0	−0.6	0.0	−0.2	−0.8	−1.0	0.8	0.0

The average of the ten successive values of y are in the two cases, $\bar{y} = 0.4$ and $\bar{y} = -0.2$. Such averaging would take place in the decimator, for example, using a 10-fold hold decimator of the form of Eq. (12.5.9).   □

Example 12.7.2: To illustrate the capability of a delta-sigma quantizer/decimator system to accurately sample an analog signal, consider the first-order quantizer of the previous example, but with a time-varying input defined with respect to the fast time scale as:

$$x'(n') = 0.5 \sin(2\pi f_0 n' / f_s'), \qquad n' = 0, 1, \ldots, N_{\text{tot}} - 1$$

We choose the values $f_0 = 8.82$ kHz, $f_s = 44.1$ kHz, $L = 10$, and $N_{\text{tot}} = 200$ samples. The fast rate is $f_s' = 10 \times 44.1 = 441$ kHz, and the normalized frequency $f_0/f_s' = 0.02$.

We want to see how the two-level quantized output $y'(n')$ of the delta-sigma quantizer is filtered by the decimation filter to effectively recover the input (and resample it at the lower rate). We compare three different decimation filters, whose frequency responses are shown in Fig. 12.7.3, with magnified passbands on the right.

The first one is an L-fold averaging decimator with transfer function given by Eq. (12.5.8). The other two are designed by the window method, and have impulse responses:

$$h(n') = w(n') \frac{\sin(\pi(n' - LM)/L)}{\pi(n' - LM)}, \qquad n' = 0, 1, \ldots, N - 1$$

where $N = 2LM + 1$. One has the minimum possible length, that is, $N = 2LM + 1$, with $M = 1$, giving $N = 21$, and uses a rectangular window, $w(n') = 1$. The other one is designed by the Kaiser method using a stopband attenuation of $A = 35$ dB

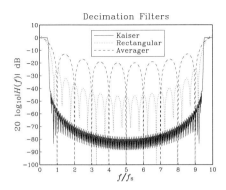 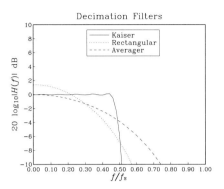

Fig. 12.7.3 Magnitude responses of decimation filters.

and transition width $\Delta f = 4.41$ kHz, or $\Delta f / f_s = 0.1$ (about the cutoff frequency $f_c = f_s/2 = 22.05$ kHz). It has length $N = 201$, $M = 10$, and Kaiser parameters $D = 1.88$ and $\alpha = 2.78$.

The output of the quantizer $y'(n')$, which is the input to the three decimators, is shown on the left of Fig. 12.7.4; the output of the averaging decimator is on the right. The outputs of the rectangular and Kaiser decimators are shown in Fig. 12.7.5.

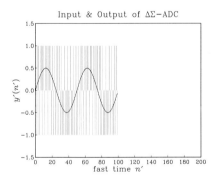 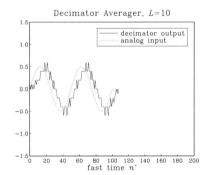

Fig. 12.7.4 Delta-sigma quantizer output and averaging decimator's output.

The averager recovers the input sinusoid only approximately and with a delay of $(L-1)/2 = 4.5$. Some of the high frequencies in $y'(n')$ get through, because they cannot be completely removed by the filter. This can be seen from the decimator's frequency response, shown in Fig. 12.7.3,

$$|H(f)| = \left| \frac{\sin(\pi f / f_s)}{L \sin(\pi f / 10 f_s)} \right|$$

which does not vanish everywhere between $[f_s/2, 10f_s - f_s/2]$, although it does vanish at the multiples $m f_s$, $m = 1, 2, \ldots, 9$.

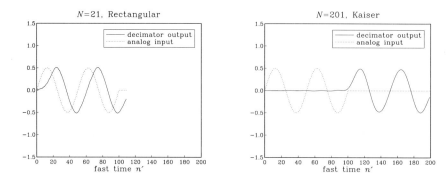

Fig. 12.7.5 Decimator filter output for rectangular and Kaiser designs.

The outputs of the window designs are faithful representations of the input sinusoid, up to the filter delay of LM samples, that is, $LM = 10$ and $LM = 100$, respectively. The Kaiser decimator gives the best output because it acts as a better lowpass filter.

What is being plotted in these graphs is the output of the decimation filter *before* it is downsampled by a factor of $L = 10$. The downsampled signal is extracted by taking every tenth output. The nine intermediate samples which are to be discarded need not be computed. However, we did compute them here for plotting purposes.

We chose simple specifications for our designs in order to get small values for the filter delays LM. In practice, stricter specifications can result in long filter lengths, for example, for a third-order noise shaper to give CD quality audio, we need $L = 64$ (see Table 2.2.1) which would require $N = DLf_s/\Delta f \simeq 4100$ for $A = 100$ dB and $\Delta f = 0.1f_s$. In such cases, a practical approach is to use *multistage* decimators. □

Next, we discuss oversampled noise shaping *requantizers* for D/A conversion. A typical requantizer system is shown in Fig. 12.7.6. The digital input is incoming at rate f_s and B-bits per sample. It is upsampled and interpolated by an L-fold interpolator, which increases the rate to f_s'. The noise shaping requantizer reduces the number of bits to $B' < B$. This output is, then, fed into an ordinary B'-bit DAC, followed by an anti-image postfilter (whose structure is greatly simplified because of oversampling).

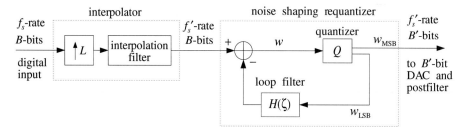

Fig. 12.7.6 Oversampled noise shaping requantizer for D/A conversion.

The quantizer Q rounds the incoming B-bit word w by keeping the B' most significant bits, say w_{MSB}, which become the output, $y = w_{\mathrm{MSB}}$. The requantization error, that is, the $B - B'$ least significant bits of w, $w_{\mathrm{LSB}} = w - w_{\mathrm{MSB}}$, are fed back through a loop filter and subtracted from the input.

The feedback loop causes the quantization noise to be highpass filtered, reducing its power within the input's baseband by just the right amount to counteract the increase in noise caused by the reduction in bits.

Figure 12.7.7 shows a model of the requantizer in which the quantizer Q is replaced by its equivalent noise model and the difference of the signals around the quantizer generates the LSB signal and feeds it back.

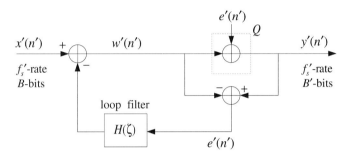

Fig. 12.7.7 Noise shaping requantizer model.

The quantized output is $y'(n') = w'(n') + e'(n')$, so that $y'(n') - w'(n') = e'(n')$. Therefore, the input to the loop filter is $e'(n')$ itself. In the ζ-domain, we have:

$$Y'(\zeta) = W'(\zeta) + E'(\zeta) \quad \text{and} \quad W'(\zeta) = X'(\zeta) - H(\zeta)E'(\zeta)$$

which gives the I/O equation:

$$Y'(\zeta) = X'(\zeta) + (1 - H(\zeta))E'(\zeta) = X'(\zeta) + H_{\mathrm{NS}}(\zeta)E'(\zeta) \qquad (12.7.14)$$

Thus, the effective noise shaping filter is

$$H_{\mathrm{NS}}(\zeta) = 1 - H(\zeta) \qquad (12.7.15)$$

First-, second-, or higher-order filters $H_{\mathrm{NS}}(\zeta)$ can be constructed easily by choosing the loop filter as $H(\zeta) = 1 - H_{\mathrm{NS}}(\zeta)$, for example:

$$
\begin{aligned}
H(\zeta) &= \zeta^{-1} & H_{\mathrm{NS}}(\zeta) &= (1 - \zeta^{-1}) \\
& & \Rightarrow & \\
H(\zeta) &= 2\zeta^{-1} - \zeta^{-2} & H_{\mathrm{NS}}(\zeta) &= (1 - \zeta^{-1})^2
\end{aligned}
$$

Noise shaping requantizers are based on the same principle of feedback quantization as delta-sigma A/D converters. Therefore, the tradeoff between L and ΔB remains the same. They are used routinely in the playback systems of CD players, DATs, and speech CODECs. For example, the first CD player built by Philips employed a first-order requantizer with $H(\zeta) = \zeta^{-1}$ and a 4-times oversampling interpolator [279].

12.8 Problems

12.1 Consider the 4-fold, length-17 interpolator defined in Eq. (12.4.1). Write down the low-rate transfer functions $D_i(z)$, $i = 0, 1, 2, 3$ and their causal versions $H_i(z)$, corresponding to the polyphase subfilters of Eq. (12.4.2).

Then, replace $z = \zeta^4$ and verify explicitly that the high-rate overall transfer function of the sequence **d** of Eq. (12.4.1) is given by the polyphase decomposition Eq. (12.2.15):

$$D(\zeta) = D_0(\zeta^4) + \zeta^{-1}D_1(\zeta^4) + \zeta^{-2}D_2(\zeta^4) + \zeta^{-3}D_3(\zeta^4)$$

12.2 Design a 2-fold interpolator of length $N = 9$, using a rectangular window. Show that the polyphase form of the interpolator is:

$$\begin{bmatrix} y_{\mathrm{up}}(2n) \\ y_{\mathrm{up}}(2n+1) \end{bmatrix} = \begin{bmatrix} 0 & 0 & 1 & 0 \\ -0.21 & 0.64 & 0.64 & -0.21 \end{bmatrix} \begin{bmatrix} x_{\mathrm{up}}(2n+4) \\ x_{\mathrm{up}}(2n+2) \\ x_{\mathrm{up}}(2n) \\ x_{\mathrm{up}}(2n-2) \end{bmatrix}$$

By superimposing impulse responses, give a graphical interpretation of the above result using the LTI form of convolution, as was done in Fig. 12.4.2.

12.3 Design a 3-fold interpolator of length $N = 13$, using a rectangular and a Hamming window. Show that the polyphase form of the interpolator in the rectangular case is:

$$\begin{bmatrix} y_{\mathrm{up}}(3n) \\ y_{\mathrm{up}}(3n+1) \\ y_{\mathrm{up}}(3n+2) \end{bmatrix} = \begin{bmatrix} 0 & 0 & 1 & 0 \\ -0.17 & 0.41 & 0.83 & -0.21 \\ -0.21 & 0.83 & 0.41 & -0.17 \end{bmatrix} \begin{bmatrix} x_{\mathrm{up}}(3n+6) \\ x_{\mathrm{up}}(3n+3) \\ x_{\mathrm{up}}(3n) \\ x_{\mathrm{up}}(3n-3) \end{bmatrix}$$

Determine a similar expression for the Hamming case. For the rectangular case, give a graphical interpretation of the above result using the LTI form of convolution, as in Fig. 12.4.2.

12.4 Using the LTI form of convolution, that is, superimposing impulse responses, justify the interpolation equations (12.1.3) of a length-25 rectangularly windowed ideal interpolator. Then, rewrite them in the form of Eq. (12.4.3) using the appropriate 4×6 coefficient matrix on the right.

12.5 Design a 3-fold FIR interpolation filter that uses *at most* four low-rate samples to compute the interpolated values between $x(n)$ and $x(n + 1)$, that is,

$$y_{\mathrm{up}}(3n+i) = a_i x(n+2) + b_i x(n+1) + c_i x(n) + d_i x(n-1)$$

for $i = 0, 1, 2$. Determine the values of the coefficients $\{a_i, b_i, c_i, d_i\}$, $i = 0, 1, 2, 3$, for the two cases:

   a. When the filter is an *ideal* interpolator.

   b. When the filter is a *linear* interpolator.

12.6 *Computer Experiment: Interpolation Filter Design.* Consider the following triangular and sinusoidal low-rate signals:

$$x(n) = \{0, 1, 2, 3, 4, 5, 6, 7, 8, 9, 10, 11, 12, 11, 10, 9, 8, 7, 6, 5, 4, 3, 2, 1, 0\}$$

$$x(n) = \sin(2\pi F_0 n), \qquad n = 0, 1, \dots, 24$$

where $F_0 = 0.04$ cycles per sample. Design a length-17 4-fold interpolation filter using a rectangular window, as in Section 12.4.1. Using the polyphase form implemented by the circular buffer version of the sample processing algorithm (12.2.20) and initialized by Eq. (12.2.19), process the above signals to get the interpolated signals $y_{up}(n')$, $n' = 0, 1, \ldots, 99$, and plot them versus the fast time n'.

Repeat by designing the corresponding Hamming windowed interpolation filter and filtering the two signals $x(n)$ through it.

12.7 *Computer Experiment: Multistage* $8\times$ *Interpolation Filter Design.* Design a multistage 8-times oversampling interpolation filter for digital audio applications (see [282] for a comparable design). The sampling rate, transition width, and stopband attenuation for all stages are taken to be $f_s = 40$ kHz, $\Delta f = 5$ kHz, $A = 80$ dB. There are three possible multistage designs, as shown in Fig. 12.2.10:

$$2 \times 4 = 4 \times 2 = 2 \times 2 \times 2 = 8$$

a. For each possibility, use the Kaiser method to determine the filter lengths N_0, N_1, (and N_2 for the 3-stage case). Determine also the length N of a single-stage design with the same specifications.

b. Compute the frequency responses of each stage $H_0(f)$, $H_1(f)$, (and $H_2(f)$ in the 3-stage case) and plot their magnitudes in dB and on the same graph over the range $0 \le f \le 320$ kHz. Normalize them to 0 dB at DC. Plot also the total response of the stages, that is, $H_{tot}(f) = H_0(f)H_1(f)$, (or, $H_0(f)H_1(f)H_2(f)$ in the 3-stage case), and compare it with the response $H(f)$ of the single-stage design.

Note that in order to keep the overall stopband attenuation in $H_{tot}(f)$ below 80 dB, you may have to increase slightly the value of A that you put into the design equations for some of the stages, for example, $A = 84$ dB.

c. Assuming all filters are realized in their polyphase form, calculate the relative computational cost R_{multi}/R_{single} and its approximation using Eq. (12.2.34). Which of the three possibilities is the most efficient?

12.8 It is desired to design a $4\times$ oversampling digital FIR interpolation filter for a CD player. Assume the following specifications: audio sampling rate of 44.1 kHz, passband range $[0, 20]$ kHz, stopband range $[24.1, 88.2]$ kHz, and stopband attenuation of 80 dB.

Using the Kaiser window design method, determine the *filter length* and the total *computational rate* in MAC/sec for the following cases:

a. Single-stage design implemented in its polyphase form.

b. Two-stage (2×2) design implemented in its polyphase form. What are the design specifications of the two stages?

Draw a sketch of the magnitude responses of the designed filters versus frequency in the range $0 \le f \le 176.4$ kHz, and of the two individual filter responses in the two-stage design case. What are the computational savings of design (b) versus design (a)? Can a 20 MIPS DSP chip handle the computational rates?

12.9 *Computer Experiment: Bessel Postfilters.* Bessel analog filters have almost linear phase response *within* their passband. Consider the Butterworth and Bessel filters designed in Section 12.4.4, and given by Eqs. (12.4.12) and (12.4.14). Compute and on the same graph plot their phase response over the passband interval $0 \le f \le 20$ kHz. On a separate graph, plot their phase response over the range $0 \le f \le 160$ kHz. Moreover, plot their magnitude response in dB over the same range.

12.10 Consider a three-stage interpolator H_0, H_1, H_2 with oversampling factors L_0, L_1, L_2 respectively, so that the total interpolation factor is $L = L_0L_1L_2$. The filter H_0 is a very sharp lowpass filter designed by some method, such as Kaiser's. The filter H_1 is a *linear* interpolator, and H_2 a *hold* interpolator. The output of H_2 is fed into a noise shaping requantizer to reduce the number of bits and then fed at rate Lf_s into a staircase DAC, H_{dac}, and then into a final analog postfilter H_{post}. Such a system is used, for example, in the Philips Bitstream 1-bit DAC system for CD players [281], with $L_0 = 4$, $L_1 = 32$, $L_2 = 2$.

    a. Write expressions for the magnitude responses $|H_1(f)|$, $|H_2(f)|$, $|H_{dac}(f)|$, in terms of f and L_0, L_1, L_2.

    b. Using part (a), show that the combined effect of the hold interpolator H_2 followed by the DAC is *equivalent* to a staircase DAC operating at the *reduced* sampling rate $L_0L_1f_s$.

       Why, then, do we need the hold interpolator at all? Why not use only a two-stage interpolator and an oversampling factor of L_0L_1?

    c. Consider the special case $L_0 = 4$, $L_1 = 2$, $L_2 = 2$. On the same graph, sketch roughly over the frequency range $0 \leq f \leq 16f_s$, the spectra at the input and output of H_0, at the output of H_1, at the output of H_2, at the output H_{dac}, at the output of H_{post}. What transition width did you choose for H_{post}?

    d. Sketch the *time-domain* signals at the input and output of H_2 and the output of H_{dac}. Does that explain part (b) in the time domain?

12.11 Show that the ideal L-fold interpolation filter $D(f)$, defined in Eq. (12.2.24) over the high-rate Nyquist interval $[-f_s'/2, f_s'/2]$ and shown in Fig. 12.2.3, satisfies the replication property:

$$\frac{1}{L} \sum_{m=0}^{L-1} D(f - mf_s) = 1$$

for all f, where f_s is the low rate $f_s = f_s'/L$.

12.12 Consider the sampling of an analog signal $x_a(t)$ at the two sampling rates f_s and $f_s' = Lf_s$. The corresponding signal samples are $x(n) = x_a(nT)$ and $x'(n') = x_a(n'T')$. Because $T = LT'$, it follows that $x(n)$ will be the *downsampled* version of $x'(n')$ in the sense of Eq. (12.5.1), that is, $x(n) = x_a(nT) = x_a(nLT') = x'(nL)$. The spectra of $x(n)$ and $x'(n')$ are given by the Poisson summation formulas:

$$X(f) = \frac{1}{T} \sum_{k=-\infty}^{\infty} X_a(f - kf_s), \qquad X'(f) = \frac{1}{T'} \sum_{k'=-\infty}^{\infty} X_a(f - k'f_s')$$

Using the change of variables $k = k'L + m$, where $m = 0, 1, \ldots, L - 1$, show that the spectrum of the downsampled signal is given by the *discrete-time* version of the Poisson summation formula:

$$X(f) = \frac{1}{L} \sum_{m=0}^{L-1} X'(f - mf_s) \tag{12.8.1}$$

Why is the factor L needed? Show that the same equation can be expressed in terms of the normalized digital frequencies $\omega = 2\pi f/f_s$ and $\omega' = 2\pi f/f_s'$ as

$$X(\omega) = \frac{1}{L} \sum_{m=0}^{L-1} X'(\omega' - \frac{2\pi m}{L}) \tag{12.8.2}$$

12.13 The downsampled signal $x(n)$, defined in Eq. (12.5.1), can be thought of as re-sampling of $x'(n')$. More precisely, the upsampled version of the downsampled signal $x(n)$, that is, the samples $x(n)$ with $L-1$ zeros inserted between them, can be thought of as the multiplication of $x'(n')$ by a discrete-time sampling function:

$$x_{\text{up}}(n') = \sum_{n=-\infty}^{\infty} x'(nL) \delta(n' - nL) = s'(n') x'(n'), \quad \text{where} \quad s'(n') = \sum_{n=-\infty}^{\infty} \delta(n' - nL)$$

First, show that $s'(n')$, being periodic in n' with period L, can be written in terms of the following discrete Fourier series, which is essentially an L-point inverse DFT:

$$s'(n') = \sum_{n=-\infty}^{\infty} \delta(n' - nL) = \frac{1}{L} \sum_{m=0}^{L-1} e^{2\pi jmn'/L} \tag{12.8.3}$$

Then, prove the downsampling property Eq. (12.8.2) using the representation Eq. (12.8.3).

12.14 Prove the downsampling equation (12.8.1) by using the property $X'(f) = D(f)X(f)$ where $D(f)$ is the ideal interpolator defined by Eq. (12.2.24), and using the results of Problem 12.11. Why can't we write $X(f) = X'(f)/D(f)$?

12.15 Consider a third-order analog Butterworth antialiasing prefilter that precedes an L-fold decimator. The passband attenuation is required to be less than 0.1 dB. Show that the *minimum* oversampling ratio L that must be used in order for the prefilter to suppress the spectral images by at least A_{stop} dB is given approximately by:

$$L = 0.94 \cdot 10^{A_{\text{stop}}/60} + 0.5$$

Make a plot of the above formula versus A_{stop} in the range $20 < A_{\text{stop}} < 100$ dB.

12.16 Show that the order N of an analog Butterworth antialiasing prefilter to be used in conjunction with an L-fold decimator and designed with specifications $\{A_{\text{pass}}, A_{\text{stop}}, f_{\text{pass}}, f_{\text{stop}}\}$, as shown in Fig. 12.5.4, is given by:

$$N = \frac{\ln\left(\frac{10^{A_{\text{stop}}/10} - 1}{10^{A_{\text{pass}}/10} - 1}\right)}{2\ln(2L - 1)}$$

Determine N for the values $A_{\text{pass}} = 0.1$ dB, $A_{\text{stop}} = 60$ dB, $L = 16$. Round N up to the next integer, say N_0. For what range of Ls does the filter order remain fixed at N_0?

12.17 Using the Kaiser window method, design a sample rate converter for up-converting CD audio at 44.1 kHz to DAT audio at 48 kHz. The required ratio is $L/M = 160/147$. Assume a transition region of $[20, 24.41]$ kHz and stopband attenuation of 95 dB.

What is the filter length N? What is the computational cost in MAC/sec assuming a polyphase realization? Can a modern DSP chip handle this cost? What are the memory requirements for such a converter?

12.18 A DAT-recorded digital audio signal is to be broadcast digitally. Using the Kaiser method, design a sampling rate converter filter for down-converting the 48 kHz DAT rate to a 32 kHz broadcast rate.

What is the filter's cutoff frequency? Assume reasonable values for the filter's transition width and stopband attenuation. What is the filter length N? What is the computational cost in MAC/sec assuming a polyphase realization? Write explicitly (i.e., in the form of Eq. (12.6.10)) the sample processing algorithm implementing the conversion algorithm.

12.19 Consider two sample rate converters for converting by the ratios 7/4 and 4/7. For each case, sketch figures similar to Figs. 12.6.2 and 12.6.3 showing the conversion stages in the time and frequency domains. For both cases, determine the polyphase filter selection indices i_m, n_m, and write explicitly (i.e., in the form of Eq. (12.6.10)) the corresponding sample processing algorithms implementing the conversion process.

12.20 Show that the time-stretching property given in Eq. (12.6.15) is preserved if the impulse response is windowed by a Hamming or Kaiser window (or, any other window).

12.21 *Computer Experiment: Sample Rate Conversion.* Write a general C or MATLAB program that implements sample rate conversion (SRC) by a factor L/M.

The SRC filter may be designed by the Kaiser method. The program must have as inputs the parameters L, M, stopband attenuation A, and normalized transition width $\Delta F = \Delta f/f_s$, where f_s is the input rate. Then, it must process an arbitrary file or array of input-rate data and output a file or array of output-rate data.

The program must initialize the $(P+1)$-dimensional state vector $\mathbf{w}$ correctly by reading in the first K input samples, as in Eq. (12.2.19). Then, it must continue processing input samples via the sample processing algorithm of Eq. (12.6.13), until the last input sample. Finally, it must calculate an additional K input-off transients to compensate for the initial delay. [*Hint:* You need to call Eq. (12.6.13) approximately K/M more times with zero input.]

As a filtering example, consider a 10 msec portion of a 100 Hz sinusoid, that is, $x(t) = \sin(2\pi t/10)$, where $0 \le t \le 10$ msec. Show that if this signal is sampled at 3 kHz, at 5 kHz, or at 15 kHz, its samples will be given respectively by:

$$x(n) = \sin(2\pi n/30), \qquad n = 0,1,\ldots,29$$
$$x'(n') = \sin(2\pi n'/50), \qquad n' = 0,1,\ldots,49$$
$$x''(n'') = \sin(2\pi n''/150), \qquad n'' = 0,1,\ldots,149$$

Design a 5/3 converter that has $A = 30$ dB and $\Delta F = 0.1$. Filter the signal $x(n)$ through the SRC filter to generate the output $y(n')$. Compare the digitally resampled signal $y(n')$ with the analog resampled signal $x'(n')$.

To compare $x(n)$ with $y(n')$, you must work with respect to the same time scale. That is, upsample the input $x(n)$ by a factor of $L = 5$ and the output $y(n')$ by a factor of $M = 3$. Then, plot the upsampled signals versus the fast time n'' and compare them.

A typical output is shown in Fig. 12.8.1. Within each 15-sample period, there are 3 input-rate samples and 5 output-rate samples. Had we not downsampled the output of the SRC filter by a factor of 3, it would be equal (for a perfect filter) to the signal $x''(n'')$.

Next, design a reverse sample rate converter to convert back by a factor of 3/5. The SRC filter has the same A and ΔF. Then, process $y(n')$ through it to see how well you may recover the original signal $x(n)$. Finally, plot the magnitude responses of the 5/3 and

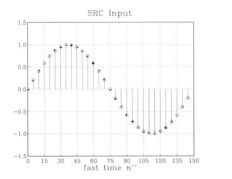 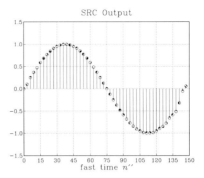

Fig. 12.8.1 Sample rate conversion by 5/3.

3/5 filters versus frequency in the range $0 \leq f \leq 15$ kHz, both in absolute and decibel scales.

12.22 An alternative discrete-time model for a first-order delta-sigma quantizer is shown in Fig. 12.8.2. It uses a conventional accumulator, but puts a delay in the feedback loop to make it computable. Replace the quantizer Q by its equivalent noise model and work out the I/O relationship in the form of Eq. (12.7.3). What are the transfer functions $H_x(\zeta)$ and $H_{NS}(\zeta)$? Write Eq. (12.7.3) in the n' time domain.

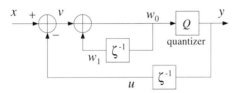

Fig. 12.8.2 Alternative model of first-order delta-sigma quantizer.

12.23 For the delta-sigma quantizer model shown in Fig. 12.8.2, define the action of the quantizer Q by the two-level function $Q(x)$ of Eq. (12.7.13). Using the indicated intermediate variables on the figure, write the corresponding sample processing algorithm. For the two constant inputs, $x = 0.4$ and $x = -0.2$, iterate the algorithm ten times and make a table of the values of all the variables, as in Example 12.7.1. Compute the average of the quantized outputs y.

12.24 A discrete-time model for a second-order delta-sigma quantizer is shown in Fig. 12.8.3. Write the I/O equation in the form of Eq. (12.7.3) and determine the signal and noise transfer functions $H_x(\zeta)$ and $H_{NS}(\zeta)$ in terms of the loop filters $H_1(\zeta)$ and $H_2(\zeta)$. Then, determine $H_1(\zeta)$ and $H_2(\zeta)$ such that

$$H_x(\zeta) = 1, \qquad H_{NS}(\zeta) = (1 - \zeta^{-1})^2$$

Redraw the full block diagram by replacing each $H_i(\zeta)$ by its realization, and write the sample processing algorithm assuming a quantizer function of the form $Q(x)$.

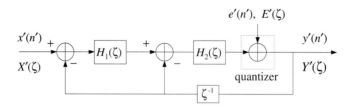

Fig. 12.8.3 Discrete-time model of second-order delta-sigma quantizer.

12.25 An alternative discrete-time model for the second-order $\Delta\Sigma$ quantizer is obtained by removing the delay ζ^{-1} from the feedback loop in Fig. 12.8.3. Determine $H_1(\zeta)$ and $H_2(\zeta)$ in order that the signal and noise transfer functions be:

$$H_x(\zeta) = \zeta^{-1}, \qquad H_{NS}(\zeta) = (1 - \zeta^{-1})^2$$

12.26 *Computer Experiment: First-Order Delta-Sigma ADC.* Write C or MATLAB programs to reproduce all the results and graphs of Example 12.7.2. Implement the decimator filtering operations in their sample-by-sample processing form using the routines `fir` or `cfir` of Chapter 4. In computing the outputs of the filters, you must also compute the input-off transients. In particular, for the window designs, you need to compute an extra LM input-off transients to compensate for the filter's delay.

Better decimators are obtained by raising the simple averaging decimator to some power. For example, a second-order L-fold "comb" decimator is defined by:

$$H(\zeta) = \left[\frac{1}{L}\frac{1 - \zeta^{-L}}{1 - \zeta^{-1}}\right]^2$$

It is similar to a linear interpolator normalized by L. For $L = 10$, determine its impulse response **h**. Then, compute its output for the same quantized input as above, and compare it with the outputs of the averaging and length-21 decimators. Also, plot the magnitude response of this decimator on the same graph with the other three.

12.27 *Computer Experiment: Second-Order Delta-Sigma ADC.* Using the second-order delta-sigma quantizer and its sample processing algorithm defined in Problem 12.24 and using the same quantizer function $Q(x)$ of Eq. (12.7.13), repeat all the questions and graphs of the above computer experiment.

12.28 A second-order multistage delta-sigma quantizer architecture (known as MASH [277,289–291]) is shown in Fig. 12.8.4. It employs two identical first-order quantizers of the type of Fig. 12.8.2, with $H(\zeta) = 1/(1 - \zeta^{-1})$, and $D(\zeta) = 1 - \zeta^{-1}$.

The negative of the quantization error e_1, obtained by subtracting the two signals around the first quantizer, becomes the input to the second stage and its output is postfiltered by the differencing filter $D(\zeta)$ and added to the output of the first stage.

Using the I/O equation derived in Problem 12.22, show that the overall I/O equation of Fig. 12.8.4 involves only the second quantization error e_2, and is given by

$$Y'(\zeta) = X'(\zeta) + (1 - \zeta^{-1})^2 E_2'(\zeta)$$

12.29 A third-order delta-sigma MASH quantizer, can be obtained by adding a third first-order quantizer to the diagram of Fig. 12.8.4 [277,289-291]. The signal $-e_2$ can be generated

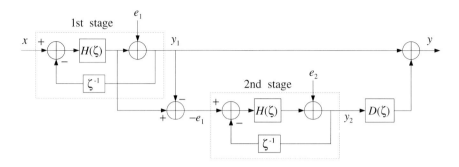

Fig. 12.8.4 MASH architecture of second-order delta-sigma quantizer.

from the second stage just like $-e_1$ is generated by the first stage. Then, the signal $-e_2$ is fed into the third stage, which has its own quantization noise e_3.

Draw the 3-stage block diagram and add an additional differentiator $D(\zeta)$ so that when you sum the outputs of the three stages, you get a third-order noise shaping I/O relationship, that is, combine the I/O equations of the three stages so that:

$$Y'_1 = X' + DE'_1$$
$$Y'_2 = -E'_1 + DE'_2 \quad \Rightarrow \quad Y' = X' + D^3 E'_3$$
$$Y'_3 = -E'_2 + DE'_3$$

12.30 A third-order delta-sigma MASH quantizer, can also be obtained by using a cascade combination of first- and second-order quantizers [277,289–291]. In the block diagram of Fig. 12.8.4, replace the first-order quantizer of the second stage by the second-order quantizer of Fig. 12.8.3. The differencer $D(\zeta)$ remains unchanged. Show that the overall I/O equation is now:

$$Y'(\zeta) = X'(\zeta) + (1 - \zeta^{-1})^3 E'_2(\zeta)$$

12.31 Delta-sigma A/D converters are not always appropriate and must be used with caution, especially in multiplexing the sampling of several input channels or in feedback control systems. Why would you say this is so?

13

Appendices

A Random Signals*

A.1 Autocorrelation Functions and Power Spectra

One of the most important applications of DSP is removing noise from noisy signals. The design of noise reduction and signal enhancement filters is discussed in Section 8.3. Here, we review briefly some basic facts regarding random signals, autocorrelation functions, and power spectra; see [2,25,26] for more details.

The *autocorrelation function* of a zero-mean[†] random signal is defined as the correlation between two samples $x(n)$ and $x(n + k)$ separated by a time lag k. It is a measure of the dependence of successive samples on the previous ones:

$$\boxed{R_{xx}(k) = E[x(n + k)x(n)]} \qquad \text{(autocorrelation function)} \qquad \text{(A.1)}$$

For stationary signals, $R_{xx}(k)$ depends only on the relative time lag k, and not on the absolute time n. Note that $R_{xx}(k)$ is a double-sided sequence and, as a consequence of stationarity, it is symmetric in k, that is, $R_{xx}(-k) = R_{xx}(k)$.

The *power spectrum* of the random signal $x(n)$ is defined as the discrete-time Fourier transform of its autocorrelation function $R_{xx}(k)$. It represents the frequency content of the random signal $x(n)$ in an average sense:

$$\boxed{S_{xx}(\omega) = \sum_{k=-\infty}^{\infty} R_{xx}(k)e^{-j\omega k}} \qquad \text{(power spectrum)} \qquad \text{(A.2)}$$

where $\omega = 2\pi f / f_s$ is the digital frequency in radians per sample. The inverse DTFT relationship expresses $R_{xx}(k)$ in terms of $S_{xx}(\omega)$:

$$R_{xx}(k) = E[x(n + k)x(n)] = \int_{-\pi}^{\pi} S_{xx}(\omega)e^{j\omega k}\frac{d\omega}{2\pi} \qquad \text{(A.3)}$$

In particular, setting $k = 0$, we obtain the *average power*, or variance, of the signal $x(n)$:

[†]If the mean m is not zero, we replace $x(n)$ by its zero-mean version $x(n) - m$.

$$\sigma_x^2 = R_{xx}(0) = E[x(n)^2] = \int_{-\pi}^{\pi} S_{xx}(\omega)\,\frac{d\omega}{2\pi} = \int_{-f_s/2}^{f_s/2} S_{xx}(f)\,\frac{df}{f_s} \qquad \text{(A.4)}$$

where

$$S_{xx}(f) = \sum_{k=-\infty}^{\infty} R_{xx}(k)\,e^{-2\pi j f k/f_s} \qquad \text{(A.5)}$$

The quantity $S_{xx}(f)/f_s$ represents the *power per unit frequency* interval. Hence, the name "power spectrum" or "power spectral density" (psd). It describes how the signal's power is *distributed* among different frequencies. Its integral over the Nyquist interval, Eq. (A.4), gives the *total* power in the signal.

Often it is more convenient to work with the z-transform of the autocorrelation and replace $z = e^{j\omega} = e^{2\pi j f/f_s}$ to obtain the power spectrum $S_{xx}(\omega)$ or $S_{xx}(f)$:

$$S_{xx}(z) = \sum_{k=-\infty}^{\infty} R_{xx}(k)\,z^{-k} \qquad \text{(A.6)}$$

The above results can be applied to the important special case of a zero-mean *white noise* signal $x(n)$. White noise has a delta-function autocorrelation and a *flat* spectrum, as shown in Fig. A.1.

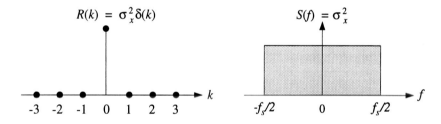

Fig. A.1 Autocorrelation and power spectrum of white noise.

Because (by definition) successive signal samples are independent of each other, the autocorrelation function (A.1) will factor for $k \neq 0$ into the product of the means which are zero:

$$R_{xx}(k) = E[x(n+k)x(n)] = E[x(n+k)]\cdot E[x(n)] = 0$$

whereas for $k = 0$, we get the variance

$$R_{xx}(0) = E[x(n)^2] = \sigma_x^2$$

Combining them into a single equation, we have:

$$R_{xx}(k) = \sigma_x^2 \delta(k) \qquad \text{(white noise autocorrelation)} \qquad \text{(A.7)}$$

Inserting Eq. (A.7) into Eq. (A.5), only the $k = 0$ term will survive the sum giving the flat spectral density (over the Nyquist interval):

$$S_{xx}(f) = \sigma_x^2, \qquad \text{for } -\frac{f_s}{2} \le f \le \frac{f_s}{2} \qquad \text{(white noise spectrum)} \qquad \text{(A.8)}$$

Given a length-N block of signal samples $x(n)$, $n = 0, 1, \ldots, N - 1$, one can compute an *estimate* of the statistical quantity $R_{xx}(k)$ by the so-called *sample autocorrelation* obtained by replacing the statistical average in Eq. (A.1) by the time average:

$$\hat{R}_{xx}(k) = \frac{1}{N} \sum_{n=0}^{N-1-k} x(n+k)x(n) \qquad \text{(sample autocorrelation)} \qquad \text{(A.9)}$$

for $k = 0, 1, \ldots, N-1$. The negative tail can be defined using the symmetry property $\hat{R}_{xx}(-k) = \hat{R}_{xx}(k)$.

The rule of thumb is that only about the first 5–10% of the lags are statistically reliable, that is, $0 \le k \le 0.1N$. The following routine `corr.c` computes Eq. (A.9) for $0 \le k \le M$, with any $M \le N - 1$.

```
/* corr.c - sample cross correlation of two length-N signals */

void corr(N, x, y, M, R)                 computes R[k], k = 0, 1, ... , M
double *x, *y, *R;                       x, y are N-dimensional
int N, M;                                R is (M + 1)-dimensional
{
        int k, n;

        for (k=0; k<=M; k++)
            for (R[k]=0, n=0; n<N-k; n++)
                R[k] += x[n+k] * y[n] / N;
}
```

Actually, the routine computes the more general sample *cross correlation* between two length-N signal blocks $x(n)$, $y(n)$, $n = 0, 1, \ldots, N - 1$, defined as:

$$\hat{R}_{xy}(k) = \frac{1}{N} \sum_{n=0}^{N-1-k} x(n+k)y(n), \qquad k = 0, 1, \ldots, M \qquad \text{(A.10)}$$

It can be shown that for wide-sense stationary signals, $\hat{R}_{xx}(k)$ is a good estimate of $R_{xx}(k)$, converging to the latter for large N (in the mean-square sense):

$$\hat{R}_{xx}(k) \to R_{xx}(k) \quad \text{as} \quad N \to \infty$$

The DTFT of $\hat{R}_{xx}(k)$ is called the *periodogram spectrum* and can be thought of as an *estimate* of the power spectrum $S_{xx}(\omega)$:

$$\hat{S}_{xx}(\omega) = \sum_{k=-(N-1)}^{N-1} \hat{R}_{xx}(k)e^{-j\omega k} \qquad \text{(A.11)}$$

Using the definition (A.9) in (A.11) and rearranging summations, we can express the periodogram in the alternative way:

$$\boxed{\hat{S}_{xx}(\omega) = \frac{1}{N}|X_N(\omega)|^2} \qquad \text{(periodogram spectrum)} \qquad \text{(A.12)}$$

where $X_N(\omega)$ is the DTFT of the length-N data block $x(n)$, which can be computed efficiently using FFTs:

$$X_N(\omega) = \sum_{n=0}^{N-1} x(n) e^{-j\omega n}$$

It can be shown [3,25,26], that for wide-sense stationary random signals the mean of the periodogram (A.12) converges to the true power spectrum $S_{xx}(\omega)$ in the limit of large N, that is,

$$\boxed{S_{xx}(\omega) = \lim_{N\to\infty} E[\hat{S}_{xx}(\omega)] = \lim_{N\to\infty} E\left[\frac{1}{N}|X_N(\omega)|^2\right]} \qquad \text{(A.13)}$$

Unfortunately, the periodogram is not a good estimator of the power spectrum. It does not approximate $S_{xx}(\omega)$ well, even in the limit of large N. That is, even though the mean of the periodogram tends to $S_{xx}(\omega)$, the periodogram itself $\hat{S}_{xx}(\omega)$ does not. The subject of *classical spectral analysis* is essentially the subject of fixing the periodogram to provide a good estimate of the power spectrum.

There are two basic techniques that improve the periodogram: periodogram *averaging* and periodogram *smoothing*. The averaging method tries to emulate the ensemble averaging operation $E[\,]$ of Eq. (A.13). In its simplest form, it consists of dividing the signal into contiguous blocks, computing the ordinary periodogram of each block using Eq. (A.12), and then averaging the computed periodograms.

The method is depicted in Fig. A.2, where there are K blocks, each of length N, so that the total length of the data record is $L = KN$. The signal is required to remain stationary at least over the length L. The block size N must be chosen to provide sufficient *frequency resolution*. This point is discussed further in Chapter 9. Denoting the ith block by $x_i(n)$, $n = 0, 1, \ldots, N-1$, we compute its ordinary periodogram:

$$\hat{S}_i(\omega) = \frac{1}{N}|X_i(\omega)|^2, \qquad i = 1, 2, \ldots, K$$

where $X_i(\omega)$ is its DTFT:

$$X_i(\omega) = \sum_{n=0}^{N-1} x_i(n) e^{-j\omega n}$$

and then average the K periodograms:

$$\begin{aligned}
\hat{S}(\omega) &= \frac{1}{K}\left[\hat{S}_1(\omega) + \hat{S}_2(\omega) + \cdots + \hat{S}_K(\omega)\right] \\
&= \frac{1}{KN}\left[|X_1(\omega)|^2 + |X_2(\omega)|^2 + \cdots + |X_K(\omega)|^2\right]
\end{aligned} \qquad \text{(A.14)}$$

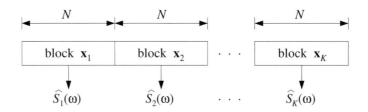

Fig. A.2   Periodogram averaging improves power spectrum estimate.

It can be shown that $\hat{S}(\omega)$ is a good estimator of $S_{xx}(\omega)$, with the (mean-square) error between the two decreasing like $1/K$, for large K. The periodogram smoothing method has similar performance.

There are two basic shortcomings with such classical spectral analysis methods: One is that to achieve high statistical reliability, a large value of K must be used, which implies a long total signal length $L = KN$. Such long blocks may not be possible to obtain in certain applications. The second is that even if a long data record could be measured, it may not be usable because the signal may not remain stationary over such long periods of time, as for example, in speech.

These shortcomings have led to the development of *modern spectral analysis* methods, which are based on *parametric models* of the signal [25,26].

A.2 Filtering of Random Signals

In designing filters to remove noise, it is necessary to know the effect of filtering on the autocorrelation function and on the power spectrum of a random signal.

Suppose the input to a *strictly stable* filter $H(z)$ with impulse response $h(n)$ is a wide-sense stationary signal $x(n)$. Then, the corresponding output $y(n)$ will also be a wide-sense stationary random signal:

$$y(n) = \sum_m h(m)x(n-m)$$

It can be shown [2,28] that the power spectrum of the output is related to that of the input by:

$$\boxed{S_{yy}(\omega) = |H(\omega)|^2 S_{xx}(\omega)} \qquad (A.15)$$

Thus, the input spectrum is reshaped by the filter spectrum. A simple way to justify this result is in terms of periodograms. The filtering equation in the z-domain is $Y(z) = H(z)X(z)$, and in the frequency domain $Y(\omega) = H(\omega)X(\omega)$. It follows that the output periodogram will be related to the input periodogram by a similar equation as (A.15):

$$\frac{1}{N}|Y(\omega)|^2 = |H(\omega)|^2 \cdot \frac{1}{N}|X(\omega)|^2$$

Applying this result to the special case of a *white noise input* with a flat spectral density $S_{xx}(\omega) = \sigma_x^2$ gives

$$\boxed{S_{yy}(\omega) = |H(\omega)|^2 \sigma_x^2} \tag{A.16}$$

Similarly, in the z-transform notation of Eq. (A.6):

$$\boxed{S_{yy}(z) = H(z)H(z^{-1})\sigma_x^2} \tag{A.17}$$

where we replaced $H(\omega) = H(z)$ and $H(\omega)^* = H(z^{-1})$, the latter following from the fact that $h(n)$ is real-valued. Indeed, with $z = e^{j\omega}$ and $z^{-1} = z^* = e^{-j\omega}$, we have:

$$H(\omega)^* = \left(\sum_n h(n)e^{-j\omega n}\right)^* = \sum_n h(n)e^{j\omega n} = H(z^{-1})$$

Equation (A.16) implies that the filtered noise $y(n)$ is no longer white. Its power spectrum acquires the shape of the filter's spectrum. Its autocorrelation function is no longer a delta function. It can be computed by taking the (stable) inverse z-transform of Eq. (A.17).

A measure of whether the filter attenuates or magnifies the input noise is given by the variance of the output σ_y^2. Using Eq. (A.4) applied to $y(n)$, we have:

$$\sigma_y^2 = \int_{-\pi}^{\pi} S_{yy}(\omega)\frac{d\omega}{2\pi} = \sigma_x^2 \int_{-\pi}^{\pi} |H(\omega)|^2 \frac{d\omega}{2\pi}$$

which can be written in the form:

$$\boxed{NRR = \frac{\sigma_y^2}{\sigma_x^2} = \int_{-\pi}^{\pi} |H(\omega)|^2 \frac{d\omega}{2\pi} = \sum_n h(n)^2} \quad \text{(NRR)} \tag{A.18}$$

where we used Parseval's equation discussed in Chapter 5.

This ratio will be referred to as the *noise reduction ratio* (NRR). If it is less than one, the input noise will be attenuated by the filter. It can be used as a useful criterion for designing noise-reducing filters—the objective being to design $H(z)$ such that (A.18) is minimized as much as possible.

A necessary assumption for the derivation of the results (A.15) or (A.18) is that the filter $h(n)$ be strictly stable. The stability of $h(n)$ is required to ensure that the stationary input signal $x(n)$ will generate, after the filter transients die out, a stationary output signal $y(n)$.

Thus, even marginally stable filters with poles on the unit circle are not allowed. To illustrate the problems that may arise, consider the simplest marginally stable filter, namely, an accumulator/integrator:

$$H(z) = \frac{1}{1 - z^{-1}}, \qquad h(n) = u(n)$$

It has I/O difference equation:

$$y(n) = y(n-1) + x(n)$$

Assuming zero initial conditions, we can write it in the convolutional form:

$$y(n) = x(n) + x(n-1) + \cdots + x(1) + x(0) \tag{A.19}$$

If $x(n)$ is a zero-mean, white noise signal with variance σ_x^2, the resulting accumulated output signal $y(n)$ is a version of the *random walk* process [321].

The signal $y(n)$ is not stationary and becomes unstable as n increases, in the sense that its mean-square value (i.e., its variance) $\sigma_y^2(n) = E[y(n)^2]$ diverges. Indeed, using the property that the variance of a sum of independent random variables is equal to the sum of the individual variances, we obtain from Eq. (A.19):

$$\sigma_y^2(n) = E[y(n)^2] = E[x(n)^2] + E[x(n-1)^2] + \cdots + E[x(0)^2] = \sigma_x^2 + \sigma_x^2 + \cdots + \sigma_x^2$$

where all the terms have a common variance σ_x^2, by assumption. It follows that:

$$\sigma_y^2(n) = E[y(n)^2] = (n+1)\sigma_x^2 \tag{A.20}$$

Thus, the mean-square value of $y(n)$ grows linearly in n. In a digital implementation, the growing amplitude of $y(n)$ will quickly saturate the hardware registers.

Analog integrators also behave in a similar fashion, growing unstable when their input is random noise. Therefore, one should never accumulate or integrate white noise. A standard remedy is to use a so-called *leaky integrator*, which effectively stabilizes the filter by pushing its pole slightly into the unit circle. This is accomplished by the replacement, with $0 < \rho \lesssim 1$

$$H(z) = \frac{1}{1 - z^{-1}} \quad \longrightarrow \quad H(z) = \frac{1}{1 - \rho z^{-1}}$$

More generally, a marginally stable filter can be stabilized by the substitution $z \to \rho^{-1}z$, which pushes all the marginal poles into the inside of the unit circle. The substitution amounts to replacing the transfer function and impulse response by

$$\boxed{H(z) \longrightarrow H(\rho^{-1}z), \qquad h(n) \longrightarrow \rho^n h(n)} \tag{A.21}$$

B Random Number Generators

B.1 Uniform and Gaussian Generators

Random number generators are useful in DSP for performing *simulations* of various algorithms, for example, in simulating noisy data. They are also useful in real-time applications, such as adding dither noise to eliminate quantization distortions as we saw in Chapter 2, or in computer music synthesis and in the implementation of digital audio effects, such as chorusing.

Most computer systems and languages have built-in routines for the generation of random numbers. Typically, these routines generate random numbers that are distributed *uniformly* over the standardized interval $[0, 1)$, although Gaussian-distributed random numbers can be generated just as easily. Figure B.1 shows the probability density functions in the uniform and Gaussian cases.

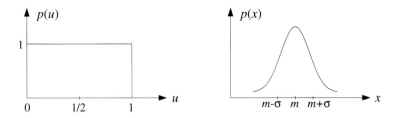

Fig. B.1 Uniform and Gaussian probability distributions.

There is a large literature on random number generators; see [301–319] and references therein. As reviewed by Park and Miller [304], it is hard to find good random number generators, that is, generators that pass all or most criteria of randomness.

By far the most common generators are the so-called *linear congruential generators* (LCG). They can generate fairly long sequences of independent random numbers, typically, of the order of two billion numbers before repeating. For longer sequences, one may use *shift-register* and *lagged-Fibonacci* generators [308–312], which can generate astronomically long sequences of order of 2^{250} or 2^{931}.

In C, a typical call to a random number generator routine takes the form:

```
u = ran(&iseed);
```

where the output is a real number in the interval $0 \le u < 1$.

The input is an integer seed, `iseed`, which is passed by *address* because it is modified by the routine internally and its new value serves as the *next* seed.[†] Thus, the routine has one input, namely `iseed`, and two outputs, u and the new value of `iseed`. Figure B.2 shows the effect of a single call to such a routine, as well as successive calls which generate a sequence of independent random numbers, starting from an arbitrary initial seed.

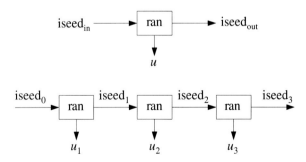

Fig. B.2 Single and successive calls to routine ran.

The LCG algorithm for generating u and updating the seed is defined by three integer parameters, $\{a, c, m\}$, called the multiplier, the displacement, and the mod-

---

[†]In some implementations, the seed is hidden from the user.

ulus. Given an initial integer seed I_0 in the interval[‡] $0 \le I_0 \le m - 1$, the LCG algorithm is the recursion:

$$\boxed{\begin{aligned} I_n &= (aI_{n-1} + c) \bmod (m) \\ u_n &= \frac{I_n}{m} \end{aligned}}$$ (LCG algorithm) (B.1)

Because of the modulo-m operation, all the seeds I_n are restricted to the interval:

$$0 \le I_n \le m - 1$$

This implies that u_n will be in the interval $0 \le u_n < 1$, and that the length of such a sequence can be *at most* $m - 1$. The parameters $\{a, c, m\}$ must be chosen carefully, such that every initial seed must result in a maximal-length sequence [301]. We will use the following generator which has maximal length $m - 1$. It was originally proposed in [313] and has withstood the test of time [304]. It has $c = 0$ and:

$$\boxed{a = 7^5 = 16807, \quad m = 2^{31} - 1 = 2147483647}$$ (B.2)

With these parameters, Eq. (B.1) cannot be implemented in a straightforward fashion because the product aI can take on extremely large values that exceed the integer range of many computers. For example, if $I = m - 1 = 2147483646$, then $aI \simeq 3.6 \times 10^{13} \simeq 2^{45}$, which exceeds the typical signed "long" (4-byte) integer range of most micros:

$$-2^{31} \le I \le 2^{31} - 1$$ (B.3)

A *portable* implementation suggested by Schrage [302,314] rearranges the computation of $(aI) \bmod (m)$ in such a way that *all* intermediate results remain bounded by the range (B.3). The technique is based on using the quotient q and remainder r of the division of m by a, that is,

$$\boxed{m = aq + r}$$ (B.4)

where r is in the range $0 \le r \le a - 1$. The key requirement for the method to work is that r satisfy the additional constraint:

$$r < q$$ (B.5)

For the choice (B.2), we have the values for q and r satisfying (B.4) and (B.5):

$$\boxed{q = 127773, \quad r = 2836}$$ (B.6)

Given an integer seed I in the range $0 \le I \le m-1$, the quantity $J = (aI) \bmod (m)$ is the remainder of the division of aI by m, that is,

---

[‡]If $c = 0$, one must pick $I_0 \ne 0$.

$$aI = mK + J, \qquad \text{where} \quad 0 \le J \le m - 1 \qquad (B.7)$$

Schrage's method calculates J without directly performing the multiplication aI. As a preliminary step, the seed I is divided by q, giving the quotient and remainder:

$$I = qk + j, \qquad \text{where} \quad 0 \le j \le q - 1 \qquad (B.8)$$

Then, the quantity aI can be expressed as follows:

$$aI = a(qk + j) = aqk + aj = (m - r)k + aj = mk + (aj - rk) \qquad (B.9)$$

where we used $aq = m - r$ from Eq. (B.4).

Comparing Eqs. (B.9) and (B.7), it appears that $K = k$ and $J = aj - rk$. This would be true by the uniqueness of Eq. (B.7) if $aj - rk$ were in the range $0 \le aj - rk \le m - 1$. Note that both quantities aj and rk lie in this range:

$$0 \le aj \le m - 1, \qquad 0 \le rk \le m - 1 \qquad (B.10)$$

Indeed, the first follows from the fact that $j < q$ from Eq. (B.8), so that

$$0 \le aj < aq = m - r < m$$

The second one follows from Eqs. (B.5) and (B.8):

$$0 \le rk < qk = I - j \le I \le m - 1$$

Combining the inequalities (B.10) we find the range of the quantity $aj - rk$:

$$-(m - 1) \le aj - rk \le m - 1$$

If $aj - rk$ lies in the positive half, $0 \le aj - rk \le m - 1$, then we must necessarily have $J = aj - rk$ and $K = k$. But, if it lies in the negative half, $-m + 1 \le aj - rk \le -1$, we must *shift* it by m so that $1 \le aj - rk + m \le m - 1$. In this case, we have $aI = mk + aj - rk = m(k - 1) + (m + aj - rk)$; therefore, $J = m + aj - rk$ and $K = k - 1$.

Denoting $k = \lfloor I/q \rfloor$ and $j = I\%q$, we can summarize the computation of the new seed $J = (aI) \bmod (m)$ as follows:

$$
\boxed{
\begin{array}{l}
\textit{given a seed } I \textit{ in the range } 0 \le I \le m - 1 \textit{ do:} \\
\quad \textit{compute } \ J = a(I\%q) - r\lfloor I/q \rfloor \\
\quad \textit{if } \ J < 0, \textit{ then shift} \\
\qquad J = J + m
\end{array}
}
$$

The following routine `ran.c` is a C implementation based on Schrage's Fortran version [302]. Note that `iseed` is declared `long` and passed by reference:

```
/* ran.c - uniform random number generator in [0, 1) */

#define  a    16807              that is, a = 7^5
#define  m    2147483647         that is, m = 2^31 - 1
#define  q    127773             note, q = m/a = quotient
#define  r    2836               note, r = m%a = remainder

double ran(iseed)               usage: u = ran(&iseed);
long *iseed;                    iseed passed by address
{
    *iseed = a * (*iseed % q) - r * (*iseed / q);    update seed

    if (*iseed < 0)             wrap to positive values
            *iseed += m;

    return (double) *iseed / (double) m;
}
```

The following program segment illustrates the usage of the routine. It generates an array of N uniform random numbers. The initial value of the seed is arbitrary:

```
long iseed;                 seed must be long int

iseed = 123456;             initial seed is arbitrary

for (n=0; n<N; n++)
    u[n] = ran(&iseed);
```

There exist methods that improve the quality of random number generators by making them "more random" than they already are [301–319].

The generated random numbers u are uniformly distributed over the interval $0 \le u < 1$, as shown in Fig. B.1. Over this interval, the probability density is flat, $p(u) = 1$. Therefore, the mean and variance of u will be:

$$E[u] = \int_0^1 up(u)\, du = \int_0^1 u\, du = \frac{1}{2}$$

$$\sigma_u^2 = E[u^2] - E[u]^2 = \int_0^1 u^2\, du - \frac{1}{4} = \frac{1}{3} - \frac{1}{4} = \frac{1}{12}$$

(B.11)

To generate a random number which is uniformly distributed over a different interval, say $a \le v < b$, we generate a uniform u over $[0, 1)$ and then shift and scale it to obtain:

$$v = a + (b - a)u$$

(B.12)

The mean and variance of v will be:

$$E[v] = a + (b - a)\frac{1}{2} = \frac{a+b}{2}, \qquad \sigma_v^2 = \frac{(b-a)^2}{12}$$

In particular, the transformation $v = u - 0.5$ will generate *zero-mean* random numbers over the unit interval $[-0.5, 0.5)$, and the transformation $v = 2u - 1$ will generate *zero-mean* random numbers over the length-2 interval $[-1, 1)$.

More complicated transformations and combinations of uniform random numbers can be used to generate random numbers that are distributed according to other probability distributions, such as Gaussian, exponential, Poisson, binomial, etc. [301-319].

A method of generating *Gaussian-distributed* random numbers is based on the *central limit theorem*, which states that the sum of a large number of independent random variables is Gaussian. In particular, summing only 12 independent *uniform* random numbers gives a very good approximation to a Gaussian:

$$v = u_1 + u_2 + \cdots + u_{12} \tag{B.13}$$

The mean of v is the sum of the individual means, and because u_i are independent, the variance of v will be the sum of the variances:

$$E[v] = E[u_1] + \cdots + E[u_{12}] = \frac{1}{2} + \cdots + \frac{1}{2} = 12 \times \frac{1}{2} = 6$$

$$\sigma_v^2 = \sigma_{u_1}^2 + \cdots + \sigma_{u_{12}}^2 = \frac{1}{12} + \cdots + \frac{1}{12} = 12 \times \frac{1}{12} = 1$$

Because each u_i has finite range $0 \leq u_i < 1$, the range of v will also be finite: $0 \leq v < 12$, with mean at 6. Even though v has finite range, it represents an adequate approximation to a Gaussian because there are $\pm 6\sigma_v$ on either side of the mean, and we know that for a Gaussian distribution more than 99.99% of the values fall within $\pm 4\sigma_v$.

To generate a Gaussian random number x with a given mean $E[x] = m$ and variance $\sigma_x^2 = s^2$, we may shift and scale v:

$$x = m + s(v - 6)$$

The following routine `gran.c` implements this method using the uniform routine `ran`. Its inputs are $\{m, s\}$ and a seed. Its outputs are the random number x and the updated seed:

```
/* gran.c - gaussian random number generator */

double ran();                          uniform generator

double gran(m, s, iseed)               usage: x = gran(m, s, &iseed);
double m, s;                           m = mean, s² = variance
long *iseed;                           iseed passed by address
{
        double v = 0;
        int i;

        for (i = 0; i < 12; i++)       sum 12 uniform random numbers
                v += ran(iseed);

        return s * (v - 6) + m;        adjust mean and variance
}
```

Its usage is demonstrated by the following program segment. As in the case of `ran`, the seed must be declared to be `long`:

```
iseed = 123456;                                    initial seed is arbitrary

for (n=0; n<N; n++)
        x[n] = gran(m, s, &iseed);
```

B.2 Low-Frequency Noise Generators*

A sequence of zero-mean uniform random numbers generated by successive calls to ran, such as,

$$u_n = \text{ran}(\&\text{iseed}) - 0.5, \qquad n = 0, 1, 2, \ldots$$

corresponds to a *white noise* signal because the generated numbers are mutually independent. The autocorrelation function and power spectral density of such signal are given by Eqs. (A.7) and (A.8) of Section A.1 and shown in Fig. A.1:

$$R_{uu}(k) = \sigma_u^2 \delta(k), \qquad S_{uu}(f) = \sigma_u^2 \tag{B.14}$$

with variance $\sigma_u^2 = 1/12$.

Such a sequence is *purely random* in the sense that each sample has no memory or dependence on the previous samples. Because $S_{uu}(f)$ is flat, the sequence will contain all frequencies in equal proportions and will exhibit equally slow and rapid variations in time.

The rate at which this sequence is produced is equal to the sampling rate f_s, that is, one random number per sampling instant. In some applications, such as computer music [94-98,100], or for generating $1/f$ noise, it is desired to generate random numbers at a *slower* rate, for example, one random number every D sampling instants. This corresponds to a generation frequency of f_s/D random numbers per second.

If a new random number is generated every D sampling instants, that is, at times $n = 0, D, 2D, 3D, \ldots$, then the signal values filling the gaps between these random numbers must be calculated by *interpolation*. Two simple ways of interpolating are to use hold or linear interpolators [94]. They are shown in Fig. B.3 for $D = 5$.

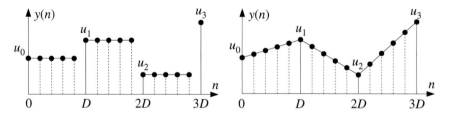

Fig. B.3 Low-frequency noise generation using hold and linear interpolators.

In the hold interpolator, each random number is held constant for D sampling instants. In the linear interpolator, two successive random numbers, separated by D time units, are connected by a straight line and the intermediate samples lie on that line.

The following routine `ranh.c` implements the hold generator. Its inputs are the desired period D and a seed. Its outputs are a *zero-mean* random number and an updated seed.

```
/* ranh.c - hold random number generator of period D */

double ran();                          uniform generator
void cdelay2();                        circular delay

double ranh(D, u, q, iseed)            usage: y = ranh(D, u, &q, &iseed);
int D, *q;                             q is cycled modulo-D
double *u;                             u = 1-dimensional array
long *iseed;                           q, iseed are passed by address
{
        double y;

        y = u[0];                              hold sample for D calls

        cdelay2(D-1, q);                       decrement q and wrap mod-D

        if (*q == 0)                           every D calls,
                u[0] = ran(iseed) - 0.5;       get new u[0] (zero mean)

        return y;
}
```

The temporary variable u is a 1-dimensional array that holds the current value of the random number for D calls. The index q is cycled modulo-D with the help of the circular delay routine `cdelay2`, which decrements it circularly during each call. Every D calls, the index q cycles through zero, and a new zero-mean random number is obtained by a call to `ran`, which overwrites the value of $u[0]$ and also updates the seed. Before the first call, the array $u[0]$ must be filled with an initial (zero-mean) random number. The initialization and usage of the routine are illustrated by the following program segment:

```
double *u;                          u is a 1-dimensional array
int D, q;
long iseed = 654321;                initial seed is arbitrary

u[0] = ran(&iseed) - 0.5;           initialize u (zero mean)
q = 0;                              initialize q

for (n=0; n<N; n++)
        y[n] = ranh(D, u, &q, &iseed);   q, iseed are passed by address
```

For the linear interpolation case, we need to keep track of *two* successive random values u, say $u[0]$ and $u[1]$, and connect them linearly. Because the slope of the straight line between $u[0]$ and $u[1]$ is $(u[1]-u[0])/D$, the linearly interpolated samples will be

$$y = u[0] + (u[1]-u[0])\frac{i}{D}, \qquad i = 0, 1, \ldots, D-1$$

Because we use the routine `cdelay2`, the circular index q takes periodically the successive values: $q = 0, D-1, D-2, \ldots, 1$. These may be mapped to the interpolation index i by:

$$i = (D - q)\%D = 0, 1, \ldots, D - 1$$

where the modulo-D operation is effective only when $q = 0$ giving in that case $i = D\%D = 0$. The following routine `ran1.c` implements the linearly interpolated periodic generator, where now the temporary variable u is a *two-dimensional* array:

```
/* ran1.c - linearly interpolated random generator of period D */

double ran();                          uniform generator
void cdelay2();                        circular delay

double ran1(D, u, q, iseed)            usage: y = ran1(D, u, &q, &iseed);
int D, *q;                             q is cycled modulo-D
double *u;                             u = 2-dimensional array
long *iseed;                           q, iseed are passed by address
{
        double y;
        int i;

        i = (D - *q) % D;              interpolation index

        y = u[0] + (u[1] - u[0]) * i / D;    linear interpolation

        cdelay2(D-1, q);               decrement q and wrap mod-D

        if (*q == 0) {                 every D calls,
                u[0] = u[1];           set new u[0] and
                u[1] = ran(iseed) - 0.5;   get new u[1] (zero mean)
                }

        return y;
}
```

Every D calls, as q cycles through zero, the value of $u[1]$ is shifted into $u[0]$ and $u[1]$ is replaced by a new zero-mean random number by a call to `ran`. The initialization and usage of the routine are illustrated by the program segment:

```
double u[2];                           u is a 2-dimensional array
int D, q;
long iseed = 654321;                   initial seed is arbitrary

u[0] = ran(&iseed) - 0.5;              initialize u[0] and u[1]
u[1] = ran(&iseed) - 0.5;              zero-mean initial values
q = 0;                                 initialize q

for (n=0; n<N; n++)
        y[n] = ran1(D, u, &q, &iseed);    q, iseed are passed by address
```

Figure B.4 shows typical sequences $y[n]$ both for the hold and linear interpolator generators, for the cases $D = 5$ and $D = 10$. The sequence length was $N = 100$.

The routines `ranh` and `ran1` generate random numbers y in the range $-0.5 \le y < 0.5$, with mean $E[y] = 0$. In the hold case, the variance is $\sigma_y^2 = \sigma_u^2 = 1/12$, and in the linear case $\sigma_y^2 = (2D^2 + 1)\sigma_u^2/3D^2$.

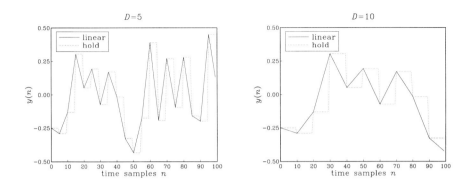

Fig. B.4 Hold and linearly interpolated low-frequency random sequences.

The hold and linear interpolator generators can be given a convenient *filtering* interpretation, as shown in Fig. B.5. The interpolated random sequence $y(n)$ can be thought of as the *output* of an *interpolation filter* whose input is the *low-rate* sequence of random numbers u_m occurring at rate f_s/D and being separated from each other by $D - 1$ zeros. Each input random number causes the filter to produce its impulse response filling the gap till the next random number D samples later. The impulse responses for the hold and linear cases are shown in Fig. 12.3.1.

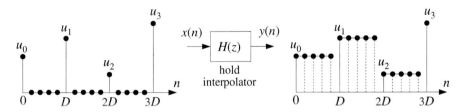

Fig. B.5 Filtering interpretation of hold interpolator.

To turn the input and output sequences $x(n)$ and $y(n)$ into stationary random sequences, we must make a slight modification to the generation model [320]. Instead of assuming that the random numbers u_m, $m = 0, 1, \ldots$ are generated exactly at multiples of D, that is, at times $n = mD$, we introduce a random delay shift in the entire sequence and assume that the u_m's are generated at times:

$$n = mD + d$$

where the delay d is a *discrete-valued random variable* taking on the possible values $\{0, 1, \ldots, D - 1\}$ with *uniform probability*, that is, $p(d) = 1/D$. In this case, the sequences $x(n)$ and $y(n)$ defining the generation model are obtained by:

$$x(n) = \sum_{m=-\infty}^{\infty} \delta(n - mD - d)u_m$$

$$y(n) = \sum_{m=-\infty}^{\infty} h(n - mD - d)u_m$$

(B.15)

Each realization of $x(n)$ and $y(n)$ is defined by a random value of d from the set $\{0, 1, \ldots, D - 1\}$ and the random numbers u_m. The particular case $D = 5$ and $d = 3$ is shown in Fig. B.6 for the hold interpolator.

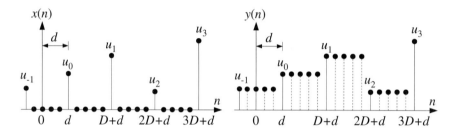

Fig. B.6 Randomly delayed input and output sequences.

If we assume that the random numbers u_m are zero-mean, mutually independent, and uniformly distributed over the interval $[-0.5, 0.5)$, then the sequence $x(n)$ becomes a zero-mean *white noise* sequence with variance $\sigma_x^2 = \sigma_u^2/D$, and therefore, its filtered version $y(n)$ will be stationary (in the steady state) and have power spectrum as given by Eq. (A.16):

$$S_{yy}(f) = |H(f)|^2 \frac{\sigma_u^2}{D}$$

(B.16)

where for the hold and linear interpolators, we have from Section 12.3:

$$H(f) = \frac{\sin(\pi f D/f_s)}{\sin(\pi f/f_s)} e^{-j\pi f(D-1)/f_s}, \qquad H(f) = \frac{1}{D}\left[\frac{\sin(\pi f D/f_s)}{\sin(\pi f/f_s)}\right]^2$$

(B.17)

The above technique of introducing a random delay to guarantee the stationarity of the output process is standard in digital communication applications [321]. For the purpose of generating low-frequency random sequences, such a delay is a welcome feature that improves the flexibility of the model.

The implementation of an overall random delay shift in the generated sequence can be accomplished by initializing the routines ranh or ran1 not at $q = 0$, but at $q = d$, where the random integer $d = 0, 1, \ldots, D - 1$ can be generated by an initial call to ran:

$$d = \lfloor D \cdot \text{ran}(\&\text{iseed}) \rfloor$$

The interpolation between the low-rate random numbers u_m can also be accomplished using more sophisticated interpolation filters, whose frequency response $H(f)$ closely approximates an ideal lowpass filter with cutoff $f_s/2D$. The subject of interpolation filter design was discussed in Chapter 12. Any of those designs and their efficient, so-called polyphase, realizations can be used in place of the hold and linear interpolators. For example, using the polyphase sample processing algorithm of Eq. (12.2.20), we may write the low-frequency random number generation algorithm:

$$
\boxed{
\begin{aligned}
&\textit{repeat forever:}\\
&\quad \textit{if } (q = 0)\\
&\qquad w[0] = \mathrm{ran}\,(\&\mathrm{iseed}) - 0.5\\
&\quad i = (D - q)\%D\\
&\quad y = \mathrm{dot}\,(P, \mathbf{h}_i, \mathbf{w}) = \text{output}\\
&\quad \mathrm{cdelay2}\,(D - 1, \&q)\\
&\quad \textit{if } (q = 0)\\
&\qquad \mathrm{delay}\,(P, \mathbf{w})
\end{aligned}
}
\tag{B.18}
$$

where $\mathbf{h}_i$ is the ith polyphase subfilter of order P, $\mathbf{w}$ is the low-rate delay line holding the random numbers u_m. The length of the interpolation filter is $N = 2DM + 1$, and $P = 2M - 1$. As in the routines ranh and ran1, q is passed by address and gets decremented circularly with the help of cdelay2. As q cycles through zero every D calls, the low-rate delay line is shifted and a new (zero-mean) random number is entered into $\mathbf{w}$. See Problem B.5 for a simulation.

B.3 1/f Noise Generators*

$1/f$-noise is also known as *flicker* or *pink* noise, depending on the context. It is characterized by a power spectrum that falls in frequency like $1/f$:

$$
S(f) = \frac{A}{f}
\tag{B.19}
$$

To avoid the infinity at $f = 0$, this behavior is assumed valid for $f \geq f_{\min}$, where $f_{\min}$ is a desired minimum frequency. The spectrum (B.19) is characterized by a 3-dB *per octave* drop, that is, whenever f doubles, $S(f)$ drops by a factor of $1/2$. Indeed, we have:

$$
S(2f) = \frac{A}{2f} = \frac{1}{2}S(f) \quad \Rightarrow \quad 10\log_{10}\left[\frac{S(f)}{S(2f)}\right] = 10\log_{10}(2) = 3 \text{ dB}
$$

The amount of power contained within a frequency interval $[f_1, f_2]$ is

$$
\int_{f_1}^{f_2} S(f)\, df = A \ln\left(\frac{f_2}{f_1}\right)
$$

This implies that the amount of power contained in *any* octave interval is the same. That is, if $f_2 = 2f_1$, then $A\ln(f_2/f_1) = A\ln(2f_1/f_1) = A\ln(2)$, which is *independent* of the octave interval $[f_1, 2f_1]$.

$1/f$ noise is ubiquitous in nature. It is observed in solid-state circuits, astrophysics, oceanography, geophysics, fractals, and music; see [322–331] and references therein. In audio engineering, it is known as *pink noise* and is used to test the frequency response of audio equipment such as loudspeakers. It represents the *psychoacoustic equivalent of white noise* because our auditory system is better matched to the logarithmic octave frequency scale than the linear scale.

In this section, we present a $1/f$ noise generator that uses the low-frequency generator `ranh`. It is based on an algorithm by Voss, mentioned in [323]. The algorithm is a variant of the so-called "spreading of time constants" models that have been proposed to explain the physics of $1/f$ noise [322,327,331].

Such models assume that the noise consists of the sum of several white noise processes that are filtered through first-order lowpass filters having time constants that are successively larger and larger, forming a geometric progression. In Voss's algorithm, the role of the lowpass filters is played by the hold interpolation filters.

In our notation, Voss's $1/f$-noise generator is defined by taking the *average* of several periodically held random numbers with periods that form a geometric progression, $D_b = 2^b$, with $b = 0, 1, 2, \ldots, B - 1$. That is, we define the random number sequence:

$$y(n) = \frac{1}{B} \sum_{b=0}^{B-1} y_b(n) = \frac{1}{B} \left[y_0(n) + y_1(n) + \cdots + y_{B-1}(n) \right] \qquad \text{(B.20)}$$

where the term $y_b(n)$ is a periodically held random number with period $D_b = 2^b$, produced by calling the routine `ranh`:

$$y_b(n) = \text{ranh}(D_b, \&u[b], \&q[b], \&iseed) \qquad \text{(B.21)}$$

Each term $y_b(n)$ must have its own 1-dimensional array $u[b]$ and circular index $q[b]$. The same seed variable `iseed` is used by all terms, but because it is updated at different periods, the terms $y_b(n)$ will be mutually independent.

The following routine `ran1f.c` implements this algorithm. Its inputs are the number of "bits" B, the B-dimensional arrays u and q, and a seed. Its outputs are a $1/f$-noise random number and an updated seed:

```
/* ran1f.c - 1/f random number generator */

double ranh();                          random hold periodic generator

double ran1f(B, u, q, iseed)            usage: y = ran1f(B, u, q, &iseed);
int B, *q;                              q, u are B-dimensional
double *u;
long *iseed;                            passed by address
{
    double y;
    int b;

    for(y=0, b=0; b<B; b++)
        y += ranh(1<<b, u+b, q+b, iseed);      period = (1<<b) = 2^b

    return y / B;
```

```
        }
```

Because the component signals $y_b(n)$ are mutually independent with mean $E[y_b] = 0$ and variance $\sigma_{y_b}^2 = \sigma_u^2 = 1/12$, it follows that the mean and variance of the $1/f$-noise sequence $y(n)$ will be:

$$E[y] = 0, \qquad \sigma_y^2 = \frac{\sigma_u^2}{B} = \frac{1}{12B} \tag{B.22}$$

The initialization and usage of the routine are illustrated by the following program segment, which generates N random numbers $y(n)$:

```
double *u;
int *q;
long iseed=123456;                              initial seed is arbitrary

u = (double *) calloc(B, sizeof(double));       B-dimensional
q = (int *) calloc(B, sizeof(int));             B-dimensional

for (b=0; b<B; b++) {
        u[b] = ran(&iseed) - 0.5;               initialize u's
        q[b] = (1<<b) * ran(&iseed);            random initial q's
        }

for (n=0; n<N; n++)                             N is arbitrary
        y[n] = ranlf(B, u, q, &iseed);
```

As discussed in the previous section, to guarantee stationarity the initial values of q must be selected randomly from the set $\{0, 1, \ldots, D - 1\}$. To see how the various terms $y_b(n)$ combine to generate $y(n)$, consider the case $B = 4$

$$y(n) = \frac{1}{4}\left[y_0(n) + y_1(n) + y_2(n) + y_3(n)\right]$$

Figures B.7 and B.8 show the generated signal $y(n)$ and its four component signals $y_b(n)$, for $n = 0, 1, \ldots, 199$. The periods of the four component signals are $1, 2, 2^2, 2^3$. For convenience, all initial random delays were set to zero, that is, $q[b] = 0, b = 0, 1, 2, 3$.

The power spectrum of the model (B.20) does not have an exact $1/f$ shape, but is close to it. Therefore, it can be used in practice to simulate $1/f$ noise. The $1/f$ shape is approximated for frequencies $f \geq f_{min}$, where:

$$\boxed{f_{min} = \frac{f_s}{2^B}} \tag{B.23}$$

This expression can be used to pick the proper value of B for a particular simulation. That is, given a desired minimum frequency we calculate $B = \log_2(f_s/f_{min})$.

Because the $y_b(n)$ components are mutually independent, the power spectrum of $y(n)$ will be equal to the *sum* of the power spectra of the individual parts. Using Eqs. (B.16) and (B.17), we have:

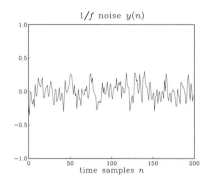

Fig. B.7 $1/f$-noise with $B = 4$.

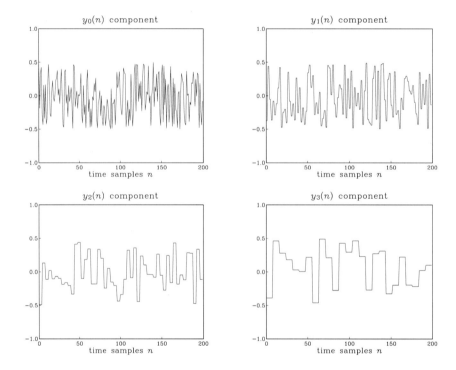

Fig. B.8   Components of $1/f$ noise.

$$S_{yy}(f) = \frac{1}{B^2} \sum_{b=0}^{B-1} S_{y_b y_b}(f) = \frac{1}{B^2} \sum_{b=0}^{B-1} \frac{1}{2^b} \frac{\sin^2(\pi f 2^b / f_s)}{\sin^2(\pi f / f_s)} \sigma_u^2 \qquad \text{(B.24)}$$

The DC value of the spectrum at $f = 0$ is not infinite as suggested by Eq. (B.19); it is finite, but large. Taking the limit $f \to 0$, we find: $S_{yy}(0) = (2^B - 1)\sigma_u^2 / B^2$. Figure

B.9 shows the theoretical spectrum computed via Eq. (B.24) for $B = 8$, together with the exact $1/f$ curve, and the estimated spectra obtained by the periodogram averaging method, for the two cases of averaging $K = 2$ and $K = 200$ zero-mean blocks of length $N = 256$, generated by calls to ran1f.

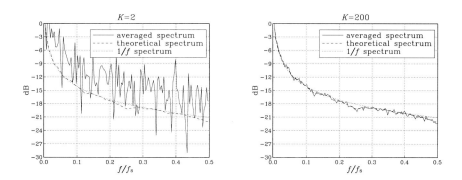

Fig. B.9 Theoretical and estimated $1/f$ power spectra, for $B = 8$.

All spectra have been normalized to unity at $f = f_{\min} = f_s 2^{-B} = f_s/256$, and are plotted in dB, that is, $10 \log_{10}(S(f)/S(f_{\min}))$. Basically, they attenuate by 3 dB per octave. The calculation of the averaged periodograms was done by the following program segment, which implements the periodogram averaging method mentioned in Section A.1:

```
for (k=0; k<K; k++) {                                 average K periodograms
    for (b=0; b<B; b++) {                             initialize kth block
        u[b] = ran(&iseed) - 0.5;
        q[b] = (1<<b) * ran(&iseed);                  randomized initial q's
        }

    for (n=0; n<N; n++)                               generate kth block
        Y[n] = cmplx(ran1f(B,u,q,&iseed), 0.0);       complexify for FFT

    fft(N, Y);                                        FFT of kth block

    for(i=0; i<N; i++)
        S[i] += cabs(Y[i]) * cabs(Y[i]);              accumulate kth periodogram
    }
```

B.4 Problems

B.1 Consider the random signal model of Eq. (B.15). Show that $x(n)$ is a zero-mean white noise sequence with variance $\sigma_x^2 = \sigma_u^2/D$. Show that the output signal $y(n)$ has zero mean and variance is in the two cases:

$$\sigma_y^2 = \sigma_u^2 \quad \text{(hold)}, \qquad \sigma_y^2 = \frac{2D^2 + 1}{3D^2}\sigma_u^2 \quad \text{(linear)}$$

B.2 For the hold interpolator case of Eq. (B.15), show that the autocorrelation function of the output signal is:

$$R_{yy}(k) = E[y(n+k)y(n)] = \left(1 - \frac{|k|}{D}\right)\sigma_u^2, \quad -D \le k \le D$$

B.3　*Computer Experiment: Autocorrelation Function of Held Noise.* Generate a length-100 block of randomly held zero-mean random numbers with period $D = 5$. Using the routine `corr`, compute the sample autocorrelation $\hat{R}_{yy}(k)$ of the block for $k = 0, 1, \ldots, 99$, and plot it together with the theoretical autocorrelation of Problem B.2.

B.4　*Computer Experiment: Power Spectrum of Held Noise.* For the cases $D = 2$, 5, and 10, plot the theoretical power spectrum of the hold interpolation noise given by Eq. (B.16), over the interval $0 \le f \le f_s$. Then, for the case $D = 5$, generate $K = 200$ blocks of held numbers $y(n)$ of length $N = 256$, compute the periodogram of each block using a 256-point FFT, and average the K periodograms to get an estimate of the power spectrum. Plot that estimate together with the theoretical power spectrum. Use absolute scales (not dB) and normalize all spectra to unity at DC. (The steps for such a computation were illustrated at the end of Section B.3.)

B.5　*Computer Experiment: Interpolated Random Number Generators.* The algorithm of Eq. (B.18) generates low-frequency random numbers using a general interpolation filter. The input random numbers are generated at a rate f_s/L and are interpolated by an L-fold interpolator resulting in a random sequence at rate f_s. Write a general routine, say `rani.c`, that implements this algorithm. Its inputs should be the $L \times (P+1)$ polyphase filter matrix $h[i][n]$ (designed independently), the low-rate delay line vector **w**, an input/output seed variable declared as in `ranh` or `ranl`, and the circular index q that cycles modulo-L.

Using this routine, write a test program that generates $N_{tot} = 150$ random numbers of frequency $f_s/10$, i.e., $L = 10$. The delay line **w** must be initialized as in Eq. (12.2.19). Use three interpolator designs—all implemented by your routine `rani`: a hold, a linear, and a Kaiser interpolator with given stopband attenuation A and transition width Δf (you may choose values such that $P = 5$). Plot and compare the three length-150 random number sequences.

B.6　Using the result of Problem B.2, show that the autocorrelation of the $1/f$ noise model of Eq. (B.20) is:

$$R_{yy}(k) = \left[1 - \frac{b(k)}{B} - \frac{2|k|}{B}\left(2^{-b(k)} - 2^{-B}\right)\right]\frac{\sigma_u^2}{B} \tag{B.25}$$

where $b(k)$ is the ceiling quantity $b(k) = \lceil \log_2(|k|+1)\rceil$. Draw a sketch of $R_{yy}(k)$ for $B = 4$. Show that the maximum correlation length is $k_{max} = 2^{B-1} - 1$.

B.7　*Computer Experiment: Autocorrelation Function of $1/f$ Noise.* Generate a block of $1/f$ noise samples $y(n)$, $n = 0, 1, \ldots, N-1$, where $N = 2000$ assuming $B = 8$. Using the correlation routine `corr.c` of Appendix A.1, compute and plot the sample autocorrelation of the sequence for lags $0 \le k \le 150$. Compare it Eq. (B.25).

B.8　*Computer Experiment: Alternative $1/f$ Noise Generator.* An alternative $1/f$ noise generator is based on the "spreading of time constants" model [322,327,331] in which white noise signals are filtered through first-order lowpass filters with time constants in geometric progression, $\tau_b = \tau_0 c^b$, $b = 0, 1, \ldots$, where $c > 1$. Discrete-time versions of such filters are of the form $H(z) = G/(1 - az^{-1})$, where a is related to the sampling interval and time constant by $a = e^{-T/\tau}$ and the gain is chosen to be $G = \sqrt{1 - a^2}$ in order for the NRR of the filter to be unity. For small T/τ, we may use the first-order approximation $a = 1 - T/\tau$. The generation model is based on summing the outputs of B such filters:

$$y(n) = \frac{1}{\sqrt{B}} \sum_{b=0}^{B-1} y_b(n)$$

(B.26)

$$y_b(n) = a_b y_b(n-1) + G_b x_b(n), \quad b = 0, 1, \ldots, B-1$$

where $G_b = \sqrt{1 - a_b^2}$ and $x_b(n)$ are mutually independent, zero-mean, unit-variance, white Gaussian signals that can be generated by calls to `gran`. The $1/\sqrt{B}$ factor normalizes $y(n)$ to unit variance. The power spectrum of the signal $y(n)$ will be:

$$S(\omega) = \frac{1}{B} \sum_{b=0}^{B-1} |H_b(\omega)|^2 = \frac{1}{B} \sum_{b=0}^{B-1} \frac{1 - a_b^2}{1 - 2a_b \cos \omega + a_b^2}$$

(B.27)

The filter parameters can be expressed as $a_b = e^{-T/\tau_b} \simeq 1 - T/\tau_b = 1 - c^{-b}T/\tau_0$. As a practical matter, we would like the model to approximate $1/f$ noise over a given interval $\omega_{\min} \leq \omega \leq \omega_{\max}$. These limits are inversely proportional to the longest and shortest time constants τ_b [331]. Thus, we may set $c^{B-1} = \omega_{\max}/\omega_{\min}$. Using the approximation $a = 1 - \omega_c$ of Example 8.3.1, we obtain the "design" equations for the filter parameters:

$$c = \left(\frac{\omega_{\max}}{\omega_{\min}} \right)^{1/(B-1)}, \qquad a_b = 1 - \omega_{\min} c^{B-1-b} = 1 - \omega_{\max} c^{-b},$$

(B.28)

for $b = 0, 1, \ldots, B-1$. The model works well over a wide range of frequencies, especially when $\omega_{\min}$ is very small [327,332]. The positivity of a_b requires $\omega_{\max} < 1$ in rads/sample. However, the model also works if we allow negative a_b's as long as they have $|a_b| < 1$. This requires that $\omega_{\max} < 2$ or in terms of the sampling frequency $f_{\max} < f_s/\pi$. To get a feeling for the range of applicability of this model consider the values:

$$\omega_{\min} = 0.01\pi, \; 0.001\pi$$

$$\omega_{\max} = 0.1\pi, \; 0.2\pi, \; 0.3\pi, \; 0.4\pi, 0.5\pi, \; 0.6\pi$$

For each pair $\{\omega_{\min}, \omega_{\max}\}$, compute the model spectrum (B.27) over the interval $\omega_{\min} \leq \omega \leq \omega_{\max}$ and plot it together with the desired $1/\omega$ spectrum. Use dB scales and normalize each spectrum to 0 dB at $\omega_{\min}$. In each case, you need to experiment to find the best value for B, but typically, $B = 2\text{-}6$.

Next, for each frequency pair, generate K sequences $y(n)$ each of length L by the Eqs. (B.26). For each sequence compute its L-point FFT periodogram and average the K periodograms; (such a computation was outlined at the end of Section B.3). Use $K = 200$ and $L = 256$. Plot the averaged periodogram together with the model spectrum. Use dB scales and normalize all spectra to the same DFT frequency (for this problem, it is better to normalize them at the second DFT frequency; that is, $\omega = 2\pi/L$.)

B.9 *Computer Experiment: Yet Another $1/f$ Noise Generator.* A simple and effective $1/f$ noise generator that covers almost the entire Nyquist interval is obtained by sending a zero-mean white noise sequence $x(n)$ of variance σ_x^2 through the following third-order filter [333]:

$$H(z) = G \frac{(1 - 0.98444z^{-1})(1 - 0.83392z^{-1})(1 - 0.07568z^{-1})}{(1 - 0.99574z^{-1})(1 - 0.94791z^{-1})(1 - 0.53568z^{-1})}$$

(B.29)

The resulting output sequence $y(n)$ has power spectrum $S_{yy}(\omega) = |H(\omega)|^2 \sigma_x^2$, according to Eq. (A.15). The filter is lowpass with finite gain at DC; its 3-dB frequency is approximately $\omega_c = 0.0015\pi$ (e.g., 30 Hz at the audio rate $f_s = 40$ kHz). Beyond ω_c the filter's magnitude response squared behaves approximately like $1/\omega$ over the rest of the Nyquist interval, that is,

$$|H(\omega)|^2 \simeq \frac{\text{const.}}{\omega}, \qquad \text{for } \omega_c \lesssim \omega \leq \pi$$

Thus, the output sequence $y(n)$ will imitate $1/f$ noise. To generate $y(n)$, one needs to filter a white noise input $x(n)$ through $H(z)$. To avoid the transients introduced by the filter, the first n_{eff} outputs must be discarded, where $n_{\text{eff}} = \log\epsilon/\log a$ is the ϵ-level time constant of the filter (for example, with $\epsilon = 0.05$, $a = \max_i |p_i| = 0.99574$, we have $n_{\text{eff}} = 702$).

This model is similar to the spreading of time constants model discussed in Problem B.8, except it is the cascade instead of the sum of factors with time constants increasing in geometric progression. Indeed, a more general such model would be of the form:

$$H(z) = G\frac{(1 - bz^{-1})(1 - b^c z^{-1})(1 - b^{c^2} z^{-1})}{(1 - az^{-1})(1 - a^c z^{-1})(1 - a^{c^2} z^{-1})} \qquad (\text{B.30})$$

The time constants n_{eff} of the three poles are in geometric proportions $1 : c : c^2$.

   a. Determine the parameters $\{a, b, c\}$ of the model (B.30) by matching them to those of the model (B.29), that is, set $a = 0.99574$, $a^c = 0.94791$, and $b = 0.98444$. Then solve for c and determine the remaining pole and zeros: a^{c^2}, b^c, b^{c^2}.

   b. Evaluate $|H(\omega)|^2$ of the filters (B.29) and (B.30) at 500 equally spaced frequencies over the interval $0.002\pi \leq \omega \leq \pi$ and plot them on the same graph together with the curve $1/\omega$ evaluated over the same frequencies. For plotting convenience, use dB scales for all responses and normalize them to 0 dB at $\omega = 0.01\pi$. Note the characteristic 3 dB/octave drop.

   c. Define the overall gain factor G such that the *NRR* of the filter (B.29) is unity and therefore the generated $1/f$ noise sequence $y(n)$ has variance $\sigma_y^2 = \sigma_x^2$. Verify that $G = 0.57534$. Then, write the sample processing algorithm for generating the output samples $y(n)$ using the cascade realization of the three sections.

C Complex Arithmetic in C

The following set of C functions implements complex arithmetic in C. These routines are used by the DFT/FFT routines. They are based on the routines given in Ref. [28].

We use the definition of a complex number defined in Microsoft C, Turbo C, and Borland C as a *structure* containing the real and imaginary part of the number. This definition as well as the function `cabs` to compute the absolute value of a complex number are declared in `math.h` in the above C compilers.

The C functions are named after their Fortran counterparts and are included in the following file `complex.c`. This file must be compiled and linked with any other routine that invokes these functions, such as `fft`.

```
/* complex.c - complex arithmetic functions */

#include <math.h>                                    for MSC and TC/BC, it declares:
                                                     struct complex and cabs(z)
/* struct complex {double x, y;}; */        uncomment if not MSC or TC/BC

                                             uncomment if not MS or TC/BC
/*   double cabs(z)
 *   complex z;
 *   {
 *       return sqrt(z.x * z.x + z.y * z.y);
 *   }
 */


typedef struct complex complex;

complex cmplx(x, y)                                  z = cmplx(x,y) = x+jy
double x, y;
{
        complex z;

        z.x = x;   z.y = y;

        return z;
}

complex conjg(z)                                     complex conjugate of z=x+jy
complex z;
{
        return cmplx(z.x, -z.y);                     returns z* = x-jy
}

complex cadd(a, b)                                   complex addition
complex a, b;
{
        return cmplx(a.x + b.x, a.y + b.y);
}

complex csub(a, b)                                   complex subtraction
complex a, b;
{
        return cmplx(a.x - b.x, a.y - b.y);
}

complex cmul(a, b)                                   complex multiplication
complex a, b;
{
        return cmplx(a.x * b.x - a.y * b.y, a.x * b.y + a.y * b.x);
}

complex rmul(a, z)                                   multiplication by real
double a;
complex z;
{
        return cmplx(a * z.x, a * z.y);
}
```

```
complex cdiv(a, b)                                      complex division
complex a, b;
{
    double D = b.x * b.x + b.y * b.y;

    return cmplx((a.x * b.x + a.y * b.y) / D, (a.y * b.x - a.x * b.y) / D);
}

complex rdiv(z, a)                                      division by real
complex z;
double a;
{
        return cmplx(z.x / a, z.y / a);
}

double real(z)                                          real part Re(z)
complex z;
{
        return z.x;
}

double aimag(z)                                         imaginary part Im(z)
complex z;
{
        return z.y;
}

complex cexp(z)                                         complex exponential
complex z;
{
        double R = exp(z.x);

        return cmplx(R * cos(z.y), R * sin(z.y));
}
```

All the necessary declarations are contained in the following header file cmplx.h that must be included in every routine that uses complex arithmetic. This header file must not be confused with the file complex.h that is part of the above compilers and is used in the C++ implementation of complex numbers.

```
/* cmplx.h - complex arithmetic declarations */

#include <math.h>                         in MSC and TC/BC, it declarares:
                                          struct complex and cabs(z)

/* struct complex{double x, y;}; */       uncomment if neccessary
/* double cabs(struct complex); */        uncomment if neccesary

typedef struct complex complex;

complex cmplx(double, double);            define complex number
complex conjg(complex);                   complex conjugate

complex cadd(complex, complex);           complex addition
complex csub(complex, complex);           complex subtraction
complex cmul(complex, complex);           complex multiplication
complex cdiv(complex, complex);           complex division
```

```
complex rmul(double, complex);        multiplication by real
complex rdiv(complex, double);        division by real

double real(complex);                 real part
double aimag(complex);                imaginary part

complex cexp(complex);                complex exponential
```

D MATLAB Functions

The emphasis of the included functions is on FIR and IIR filter design. The inputs to the functions follow closely the discussion and notation of the text. The FIR and IIR filtering functions `fir.m` and `cas.m` are not as fast as MATLAB's filtering function `filter`, but they are included here to illustrate sample-by-sample processing. The routines, listed by function, are as follows:

Routine	Function
fir, delay, cas, sos, cas2can	FIR and IIR filtering
cfir2, cdelay2, wrap2	circular FIR filtering
dtft	DTFT computation
sigav, sg, sgfilt, ecg	Signal averaging, SG smoothing
kwind, I0, kparm, kparm2	Kaiser window
klh, dlh, kbp, dbp, kdiff, ddiff, khilb, dhilb	FIR filter design
parmeq, combeq	Parametric equalizer design
lhbutt, bpsbutt, lhcheb1, lhcheb2, bpcheb2, bscheb2	IIR filter design

1. *FIR Sample Processing Algorithm.*

```
% fir.m - sample processing algorithm for FIR filter.
%
% [y, w] = fir(M, h, w, x)
%
% h = order-M filter (row vector)
% w = filter state (row vector)
% x = scalar input
% y = scalar output
% based on fir2.c

function [y, w] = fir(M, h, w, x)

w(1) = x;                 % read input
y = h * w';               % compute output
w = delay(M, w);          % update delay
```

where the delay is implemented by the function:

```
% delay.m - M-fold delay
%
% w = delay(M, w)
%
```

```
% w is the (M + 1)-dimensional linear delay-line buffer
% based on delay.c

function w = delay(M, w)

w(M+1:-1:2) = w((M+1:-1:2)-1);
```

The state vector **w** must be initialized to zero before using `fir`; for example, `w = zeros(1, M+1)`. The function `firfilt` of Problem 4.10 may be used to filter a long input vector. Alternatively, one may use MATLAB's built-in filtering function `filter`. The filter vector **h** may be obtained from the output of the Kaiser window FIR filter design functions, such as `klh.m`, given below.

2. *FIR Filtering Using Circular Delays*

```
% cfir2.m - FIR filter using circular delay-line buffer
%
% [y, w, q] = cfir2(M, h, w, q, x)
%
% h = order-M filter (row vector)
% w = circular filter state (row vector)
% q = circular index into w
% x = scalar input
% y = scalar output
% based on cfir2.c

function [y, w, q] = cfir2(M, h, w, q, x)

w(q+1) = x;                               % read input
y = h * w(rem(q+(0:M), M+1)+1)';          % compute output
q = cdelay2(M, q);                        % update delay
```

where the circular delay-line buffer index q is updated by the function:

```
% cdelay2.m - circular delay
%
% q = cdelay2(D, q)
%
% q = 0, 1, ..., D = circular index pointing to w(q + 1)
% based on cdelay2.c

function q = cdelay2(D, q)

q = q - 1;                      % decrement index and wrap mod-(D + 1)
q = wrap2(D, q);                % when q = -1, it wraps around to q = D
```

where the modulo wrapping of the index is done by:

```
% wrap2.m - circular wrapping of delay line buffer
%
% q = wrap2(D, q)
%
% q = 0, 1, ..., D = circular index pointing to w(q + 1)
% based on wrap2.c
```

```
function q = wrap2(D, q)

if q > D,
        q = q - (D+1);                    % if q = D + 1, it wraps to q = 0
end

if q < 0,
        q = q + (D+1);                    % if q = −1, it wraps to q = D
end
```

As discussed in Chapter 4, $\mathbf{w}$ is $(M+1)$-dimensional and must be initialized to zero. The index q must also be initialized to $q = 0$. A MATLAB version of the circular tap routine `tap2` can easily be written to return the i-th content of a circular tapped delay line, e.g., with a MATLAB statement of the form: `si = w(rem(q+i, D+1)+1)`.

3. *IIR Cascade Sample Processing Algorithm.*

```
% cas.m - filtering by cascade of 2nd order sections
%
% [y, W] = cas(K, B, A, W, x)
%
% B  =  K×3 numerator matrix
% A  =  K×3 denominator matrix
% W  =  K×3 state matrix
% x = scalar input
% y = scalar output
% based on cas.c

function [y, W] = cas(K, B, A, W, x)

y = x;

for i = 1:K,
        [y, W(i,:)] = sos(B(i,:), A(i,:), W(i,:), y);
end
```

where the basic second-order section is implemented by the function:

```
% sos.m - second order section
%
% [y, w] = sos(b, a, w, x)
%
% b  =  [b₀, b₁, b₂] = 3-dim numerator
% a  =  [1, a₁, a₂] = 3-dim denominator
% w = 3-dim filter state (row vector)
% x = scalar input
% y = scalar output
% based on sos.c

function [y, w] = sos(b, a, w, x)

w(1) = x - a(2) * w(2) - a(3) * w(3);
y = b(1) * w(1) + b(2) * w(2) + b(3) * w(3);
```

```
w(3) = w(2);
w(2) = w(1);
```

The state matrix must be initialized to zero; for example, W=zeros(K,3). The function casfilt of Problem 7.15 may be used to filter a long input vector. Alternatively, one may use cas2can to get the direct form coefficient vectors and apply MATLAB's built-in filtering function filter. The cascade matrices *A* and *B* are obtained from the outputs of the IIR filter design functions, such as lhbutt.m, given below. (For the bandpass/bandstop designs, one must rewrite cas.m and sos.m so that they are cascades of fourth-order sections.)

4. *Cascade to Canonical.*

```
% cas2can.m - cascade to canonical
%
% a = cas2can(A)
%
% convolves the rows of A
% A is K×3 coefficient matrix (or, K×5 for bandpass filters)
% based on cas2can.c

function a = cas2can(A)

[K, L] = size(A);

a = [1];
for i=1:K,
        a = conv(a, A(i,:));
end
```

The input matrices *A* or *B* of cas2can are obtained from the outputs of the IIR design functions, such as lhbutt.m.

5. *DTFT Computation.*

```
% dtft.m - DTFT of a signal at a frequency vector w
%
% X = dtft(x, w);
%
% x = row vector of time samples
% w = row vector of frequencies in rads/sample
% X = row vector of DTFT values
%
% based on and replaces both dtft.c and dtftr.c

function X = dtft(x, w)

[L1, L] = size(x);

z = exp(-j*w);

X = 0;
for n = L-1:-1:0,
        X = x(n+1) + z .* X;
end
```

It uses Hörner's rule to evaluate the DTFT of a finite-length signal at any desired frequency or frequencies, not just DFT frequencies. It is convenient for evaluating the frequency response of FIR filters, for example, by H=dtft(h,w), and of IIR filters, by H = dtft(b,w)./dtft(a,w), where b, a are the numerator and denominator coefficient vectors.

6. *Signal Averaging.*

```
% sigav.m - signal averaging
%
% y = sigav(D, N, x)
%
% D = length of each period
% N = number of periods
% x = row vector of length at least ND (doesn't check it)
% y = length-D row vector containing the averaged period
% It averages the first N blocks in x

function y = sigav(D, N, x)

y = 0;

for i=0:N-1,
        y = y + x((i*D+1) : (i+1)*D);         % accumulate ith period
end

y = y / N;
```

7. *Savitzky-Golay Smoother Design.*

```
% sg.m - Savitzky-Golay length-N order-d smoother design.
%
% [B, S] = sg(d, N);
%
% N = 2M + 1 = filter length, d = polynomial order
% S = [s_0, s_1, ..., s_d], F = S^T S
% G = SF^{-1} = derivative filters
% B = SF^{-1}S^T = smoothing filters
% indexing: B(M + 1 + m, M + 1 + k) = B_mk, m, k = -M : M
% m-th SG filter = B(:, M + 1 + m) = b_m, m = -M : M
% NRRs = diagonal entries of B.

function [B, S] = sg(d, N)

M = (N-1)/2;

for m=-M:M,
        for i=0:d,
                S(m+M+1, i+1) = m^i;
        end
end

F = S' * S;
B = S * F^(-1) * S';
```

8. *Filtering with Savitzky-Golay Smoother.*

```
% sgfilt.m - filtering with length-N order-d SG smoother.
%
% y = sgfilt(d, N, x);
%
% x and y are L×1 column vectors; and N = 2M + 1. Must have L > N + 1.
% B(:, i) = b_{M+1-i} = input-on transient filters, i = 1 : M + 1
% B(:, M + 1) = b_0  = steady-state filter
% B(:, M + 1 + i) = b_i = input-off transient filters, i = 0 : M

function  y = sgfilt(d, N, x)

M = (N-1)/2;
[L, L1] = size(x);

B = sg(d, N);                                    % design filter

for i = 1:M+1,                                   % input-on transients
        y(i,1) = B(:,i)' * x(1:N);
end

for n = M+2:L-M-1,                               % steady-state
        y(n,1) = B(:,M+1)' * x(n-M:n+M);
end

for i = 0:M,                                      % input-off transients
        y(L-M+i,1) = B(:,M+1+i)' * x(L-N+1:L);
end
```

9. *Simulated ECG.*

```
% ecg.m - ECG generator.
%
% x = ecg(L) = column vector
%
% generates piecewise linear ECG signal of length L
% must post-smooth it with an N-point smoother:
% y = sgfilt(d, N, x), usually with d = 0, and N = 3, 5, 9, etc.

function x = ecg(L)

a0 = [0,1,40,1,0,-34,118,-99,0,2,21,2,0,0,0];                  % template
d0 = [0,27,59,91,131,141,163,185,195,275,307,339,357,390,440];
a = a0 / max(a0);
d = round(d0 * L / d0(15));              % scale them to fit in length L
d(15)=L;

for i=1:14,
        m = d(i) : d(i+1) - 1;
        slope = (a(i+1) - a(i)) / (d(i+1) - d(i));
        x(m+1) = a(i) + slope * (m - d(i));
end
```

The output of `ecg` must be post-smoothed with a Savitzky-Golay smoother, for example, of order 0, using the function `sgfilt.m` and then renormalized to unity maximum.

10. *Kaiser Window.*

```
% kwind.m - Kaiser window.
%
% w = kwind(alpha, N) = row vector
%
% alpha = Kaiser window shape parameter
% N   = 2M + 1 = window length (must be odd)

function w = kwind(alpha, N)

M = (N-1) / 2;
den = I0(alpha);

for n = 0:N-1,
        w(n+1) = I0(alpha * sqrt(n * (N - 1 - n)) / M) / den;
end
```

The modified Bessel function is given by:

```
% I0.m - modified Bessel function of 1st kind and 0th order.
%
% S = I0(x)
%
% defined only for scalar x ≥ 0
% based on I0.c

function S = I0(x)

eps = 10^(-9);
n = 1; S = 1; D = 1;

while D > (eps * S),
        T = x / (2*n);
        n = n+1;
        D = D * T^2;
        S = S + D;
end
```

For FIR filter design, the Kaiser window shape parameter α and window length N can be calculated from the filter specifications by:

```
% kparm.m - Kaiser window parameters for filter design.
%
% [alpha, N] = kparm(DF, A)
%
% alpha = window shape parameter α
% N = window length (odd)
% DF = Δf/fs = transition width in units of fs
% A = ripple attenuation in dB; ripple δ = 10^(-A/20)

function [alpha, N] = kparm(DF, A)

if A > 21,                              % compute D factor
        D = (A - 7.95) / 14.36;
```

```
else
      D = 0.922;
end

if A <= 21,                                      % compute shape parameter α
      alpha = 0;
elseif A < 50
      alpha = 0.5842 * (A - 21)^0.4 + 0.07886 * (A - 21);
else
      alpha = 0.1102 * (A - 8.7);
end

N = 1 + ceil(D / DF);                            % compute window length
N = N + 1 - rem(N, 2);                           % next odd integer
```

For spectral analysis, the window parameters can be calculated from a given mainlobe width Δf_w (recall our definition is half the mainlobe base width), and relative sidelobe level R in dB:

```
% kparm2.m - Kaiser window parameters for spectral analysis.
%
% [alpha, L] = kparm2(DF, R)
%
% alpha = window shape parameter
% L = window length (odd)
% DF  =  Δ f /f_s = mainlobe width in units of f_s
% R = relative sidelobe level in dB
% R must be less than 120 dB.

function [alpha, L] = kparm2(DF, R)

c = 6 * (R + 12) / 155;

if R < 13.26
      alpha = 0;
elseif R < 60
      alpha = 0.76609 * (R - 13.26)^0.4 + 0.09834 * (R - 13.26);
else
      alpha = 0.12438 * (R + 6.3);
end

L = 1 + ceil(c / DF);
L = L + 1 - rem(L, 2);                           % next odd integer
```

11. *Lowpass/Highpass FIR Filter Design Using Kaiser Window.*

```
% klh.m - lowpass/highpass FIR filter design using Kaiser window.
%
% h = klh(s, fs, fpass, fstop, Apass, Astop)
%
% s  =  1, −1 = lowpass, highpass
% dlh(s, wc, N) = ideal lowpass/highpass FIR filter

function h = klh(s, fs, fpass, fstop, Apass, Astop)

fc = (fpass + fstop) / 2;   wc = 2 * pi * fc / fs;
```

```
Df = s * (fstop - fpass);   DF = Df / fs;

dpass = (10^(Apass/20) - 1) / (10^(Apass/20) + 1);
dstop = 10^(-Astop/20);
d = min(dpass, dstop);
A = -20 * log10(d);

[alpha, N] = kparm(DF, A);
h = dlh(s, wc, N) .* kwind(alpha, N);
```

where the ideal lowpass/highpass filter impulse response is calculated by:

```
% dlh.m - ideal lowpass/highpass FIR filter
%
% h = dlh(s, wc, N) = row vector
%
% s  =  1, −1 = lowpass, highpass
% N  =  2M + 1 = filter length (odd)
% wc = cutoff frequency in [rads/sample]

function h = dlh(s, wc, N)

M = (N-1)/2;

for k = -M:M,
    if k == 0,
        h(k+M+1) = (1-s) / 2 + s * wc / pi;
    else
        h(k+M+1) = s * sin(wc * k) / (pi * k);
    end
end
```

The output filter vector **h** of this and the following FIR filter design functions can be passed to the function `firfilt` of Problem 4.10 for filtering a long input vector, or to MATLAB's `filter` function.

12. *Bandpass FIR Filter Design Using Kaiser Window.*

```
% kbp.m - bandpass FIR filter design using Kaiser window.
%
% h = kbp(fs, fpa, fpb, fsa, fsb, Apass, Astop, s)
%
% s  =  1, −1 = standard, alternative design
% dbp(wa, wb, N) = ideal bandpass FIR filter

function h = kbp(fs, fpa, fpb, fsa, fsb, Apass, Astop, s)

Df = min(fpa-fsa, fsb-fpb);   DF = Df / fs;
fa = ((1+s) * fpa + (1-s) * fsa - s * Df) / 2;   wa = 2 * pi * fa / fs;
fb = ((1+s) * fpb + (1-s) * fsb + s * Df) / 2;   wb = 2 * pi * fb / fs;

dpass = (10^(Apass/20) - 1) / (10^(Apass/20) + 1);
dstop = 10^(-Astop/20);
d = min(dpass, dstop);
A = -20 * log10(d);
```

```
[alpha, N] = kparm(DF, A);
h = dbp(wa, wb, N) .* kwind(alpha, N);
```

where the ideal bandpass filter impulse response is calculated by:

```
% dbp.m - ideal bandpass FIR filter
%
% h = dbp(wa, wb, N) = row vector
%
% N  =  2M + 1 = filter length (odd)
% wa, wb = cutoff frequencies in [rads/sample]

function h = dbp(wa, wb, N)

M = (N-1)/2;

for k = -M:M,
   if k == 0,
       h(k+M+1) = (wb - wa) / pi;
   else
       h(k+M+1) = sin(wb * k) / (pi * k) - sin(wa * k) / (pi * k);
   end
end
```

13. *Lowpass FIR Differentiator Filter Design Using Kaiser Window.*

```
% kdiff.m - lowpass FIR differentiator design using Kaiser window.
%
% h = kdiff(fs, fc, Df, A)
%
% fc = cutoff frequency in [Hz]
% Df = transition width in [Hz]
% A = stopband ripple attenuation in [dB]
% ddiff(wc, N) = ideal FIR differentiator

function h = kdiff(fs, fc, Df, A)

wc = 2 * pi * fc / fs;
DF = Df / fs;

[alpha, N] = kparm(DF, A);
h = ddiff(wc, N) .* kwind(alpha, N);
```

where the ideal differentiator impulse response is calculated by:

```
% ddiff.m - ideal lowpass differentiator FIR filter
%
% h = ddiff(wc, N) = row vector
%
% N  =  2M + 1 = filter length (odd)
% wc in rads/sample

function h = ddiff(wc, N)
```

```
M = (N-1)/2;

for k = -M:M,
  if k == 0,
    h(k+M+1) = 0;
  else
    h(k+M+1) = wc * cos(wc * k) / (pi * k) - sin(wc * k) / (pi * k^2);
  end
end
```

See Problem 10.3 for details. A full-band differentiator is obtained by setting $f_c = f_s/2$ or, $\omega_c = \pi$.

14. *Lowpass FIR Hilbert Transformer Filter Design Using Kaiser Window.*

```
% khilb.m - lowpass FIR Hilbert trasformer design using Kaiser window.
%
% h = khilb(fs, fc, Df, A)
%
% fc = cutoff frequency in [Hz]
% Df = transition width in [Hz]
% A = stopband ripple attenuation in [dB]
% dhilb(wc, N) = ideal FIR Hilbert transformer

function h = khilb(fs, fc, Df, A)

wc = 2 * pi * fc / fs;
DF = Df / fs;

[alpha, N] = kparm(DF, A);
h = dhilb(wc, N) .* kwind(alpha, N);
```

where the ideal Hilbert transformer impulse response is calculated by:

```
% dhilb.m - ideal lowpass Hilbert transformer FIR filter
%
% h = dhilb(wc, N) = row vector
%
% N = 2M + 1 = filter length (odd)
% wc = cutoff frequency in [rads/sample]

function h = dhilb(wc, N)

M = (N-1)/2;

for k = -M:M,
    if k == 0,
        h(k+M+1) = 0;
    else
        h(k+M+1) = (1 - cos(wc * k)) / (pi * k);
    end
end
```

15. *Parametric Equalizer Filter Design.*

```
% parmeq.m - second-order parametric EQ filter design
%
% [b, a, beta] = parmeq(G0, G, GB, w0, Dw)
%
% b  =  [b0, b1, b2] = numerator coefficients
% a  =  [1,  a1, a2] = denominator coefficients
% G0, G, GB = reference, boost/cut, and bandwidth gains
% w0, Dw = center frequency and bandwidth in [rads/sample]
% beta = design parameter β
%
% for plain PEAK use:  G0 = 0, G = 1, GB = 1/√2
% for plain NOTCH use: G0 = 1, G = 0, GB = 1/√2

function [b, a, beta] = parmeq(G0, G, GB, w0, Dw)

beta = tan(Dw/2) * sqrt(abs(GB^2 - G0^2)) / sqrt(abs(G^2 - GB^2));
b = [(G0 + G*beta), -2*G0*cos(w0), (G0 - G*beta)] / (1+beta);
a = [1, -2*cos(w0)/(1+beta), (1-beta)/(1+beta)];
```

16. *Periodic Comb/Notch Equalizer Filter Design.*

```
% combeq.m - periodic comb/notch EQ filter design
%
% [a, b, c, beta] = combeq(G0, G, GB, D, Dw, s)
%
% s  =  1, −1 for peaks at 2kπ/D, or, (2k + 1)π/D
% G0, G, GB = reference, boost/cut, and bandwidth gains
% D = period, Dw = width in [rads/sample]
% note Dw < π/D, beta = design parameter β
%
% for plain COMB use:  G0 = 0, G = 1, GB = 1/√2
% for plain NOTCH use: G0 = 1, G = 0, GB = 1/√2
%
% H(z) = (b − c z^−D)/(1 − az^−D)    (caution: note minus signs)

function [a, b, c, beta] = combeq(G0, G, GB, D, Dw, s)

beta = tan(D * Dw / 4) * sqrt(abs((GB^2 - G0^2) / (G^2 - GB^2)));
a = s * (1 - beta) / (1 + beta);
b = (G0 + G * beta) / (1 + beta);
c = s * (G0 - G * beta) / (1 + beta);
```

17. *Lowpass/Highpass Butterworth Filter Design.*

```
% lhbutt.m - lowpass/highpass Butterworth digital filter design
%
% [A, B, P] = lhbutt(s, fs, fpass, fstop, Apass, Astop)
%
% s  =  1, −1 = lowpass, highpass
% design parameters:
% P = [Wpass, Wstop, epass, estop, Nex, N, Astop, W0, f0];
% A, B are K×3 matrices of cascade second-order sections
```

```
function [A, B, P] = lhbutt(s, fs, fpass, fstop, Apass, Astop)

Wpass = tan(pi * fpass / fs);   Wpass = Wpass^s;        % cot() for HP
Wstop = tan(pi * fstop / fs);   Wstop = Wstop^s;

epass = sqrt(10^(Apass/10) - 1);
estop = sqrt(10^(Astop/10) - 1);

Nex = log(estop/epass) / log(Wstop/Wpass);
N = ceil(Nex);   r = rem(N,2);   K = (N - r) / 2;        % K = no. sections

W0 = Wpass * (epass^(-1/N));
Astop = 10 * log10(1 + (Wstop/W0)^(2*N));                % actual Astop
f0 = (fs/pi) * atan(W0^s);                               % 3-dB freq. in Hz
P = [Wpass, Wstop, epass, estop, Nex, N, Astop, W0, f0];

if r==1,                                                 % N = odd
    G = W0 / (1 + W0);                                    % 1st order section
    B(1,:) = G * [1, s, 0];
    A(1,:) = [1, s*(2*G-1), 0];
else                                                     % N = even
    B(1,:) = [1, 0, 0];
    A(1,:) = [1, 0, 0];
end

for i=1:K,
    th = pi * (N - 1 + 2 * i) / (2 * N);
    D = 1 - 2 * W0 * cos(th) + W0^2;
    G = W0^2 / D;
    a1 = 2 * (W0^2 - 1) / D;
    a2 = (1 + 2 * W0 * cos(th) + W0^2) / D;
    B(i+1,:) = G * [1, 2*s, 1];
    A(i+1,:) = [1, s*a1, a2];
end
```

The output matrices A and B of this and the following IIR filter design functions can be passed to the filtering function `cas.m`, or `casfilt` of Problem 7.15 to filter a long input vector. (For the bandpass/bandstop cases, you need to write versions of `cas.m` for cascading fourth-order sections.)

18. *Bandpass/Bandstop Butterworth Filter Design.*

```
% bpsbutt.m - bandpass/bandstop Butterworth digital filter design
%
% [A, B, P] = bpsbutt(s, fs, fpa, fpb, fsa, fsb, Apass, Astop)
%
% s = 1, -1 = bandpass, bandstop
% design parameters:
% P = [Wpass,Wstop,Wsa,Wsb,c,fc,epass,estop,Nex,N,Astop,W0,f0a,f0b];
% A,B are K×5 matrices of cascade of fourth-order sections

function [A, B, P] = bpsbutt(s, fs, fpa, fpb, fsa, fsb, Apass, Astop)

c = sin(2*pi*(fpa + fpb)/fs) / (sin(2*pi*fpa/fs) + sin(2*pi*fpb/fs));
fc = 0.5 * (fs/pi) * acos(c);
Wpass = (abs((c - cos(2*pi*fpb/fs)) / sin(2*pi*fpb/fs)))^s;
```

```
Wsa = (c - cos(2*pi*fsa/fs)) / sin(2*pi*fsa/fs);  Wsa = Wsa^s;
Wsb = (c - cos(2*pi*fsb/fs)) / sin(2*pi*fsb/fs);  Wsb = Wsb^s;
Wstop = min(abs(Wsa), abs(Wsb));

epass = sqrt(10^(Apass/10) - 1);
estop = sqrt(10^(Astop/10) - 1);

Nex = log(estop/epass) / log(Wstop/Wpass);
N = ceil(Nex);   r = rem(N,2);   K = (N - r) / 2;

W0 = Wpass * (epass^(-1/N));   W0s = W0^s;
Astop = 10 * log10(1 + (Wstop/W0)^(2*N));
f0a = (fs/pi) * atan((sqrt(W0s^2 - c^2 + 1) - W0s)/(c+1));
f0b = (fs/pi) * atan((sqrt(W0s^2 - c^2 + 1) + W0s)/(c+1));
P = [Wpass,Wstop,Wsa,Wsb,c,fc,epass,estop,Nex,N,Astop,W0,f0a,f0b];

if r==1,
    G = W0 / (1 + W0);
    a1 = -2 * c / (1 + W0s);
    a2 = (1 - W0s) / (1 + W0s);
    A(1,:) = [1, a1, a2, 0, 0];
    B(1,:) = G * [1, (s-1)*c, -s, 0, 0];
else
    A(1,:) = [1, 0, 0, 0, 0];
    B(1,:) = [1, 0, 0, 0, 0];
end

for i=1:K,
    th = pi * (N - 1 + 2 * i) / (2 * N);
    D = 1 - 2 * W0s * cos(th) + W0s^2;
    G = W0^2 / (1 - 2 * W0 * cos(th) + W0^2);
    a1 = 4 * c * (W0s * cos(th) - 1) / D;
    a2 = 2 * (2*c^2 + 1 - W0s^2) / D;
    a3 = - 4 * c * (W0s * cos(th) + 1) / D;
    a4 = (1 + 2 * W0s * cos(th) + W0s^2) / D;
    A(i+1,:) = [1, a1, a2, a3, a4];
    B(i+1,:) = G * conv([1, (s-1)*c, -s], [1, (s-1)*c, -s]);
end
```

19. *Lowpass/Highpass Chebyshev Type 1 Filter Design.*

```
% lhcheb1.m - lowpass/highpass Chebyshev type 1 filter design
%
% [A, B, P] = lhcheb1(s, fs, fpass, fstop, Apass, Astop)
%
% s = 1, −1 = lowpass, highpass
% design parameters:
% P = [Wpass, Wstop, epass, estop, Nex, N, f3, a];
% A, B are K×3 matrices of cascaded second-order sections

function [A, B, P] = lhcheb1(s, fs, fpass, fstop, Apass, Astop)

Wpass = tan(pi * fpass / fs);   Wpass = Wpass^s;
Wstop = tan(pi * fstop / fs);   Wstop = Wstop^s;

epass = sqrt(10^(Apass/10) - 1);
estop = sqrt(10^(Astop/10) - 1);
```

```
Nex = acosh(estop/epass) / acosh(Wstop/Wpass);
N = ceil(Nex);   r = rem(N,2);   K = (N - r) / 2;

a = asinh(1/epass) / N;
W3 = Wpass * cosh(acosh(1/epass)/N);
f3 = (fs/pi) * atan(W3^s);                              % 3dB frequency
P = [Wpass, Wstop, epass, estop, Nex, N, f3, a];
W0 = sinh(a) * Wpass;

if r==1,
    G = W0 / (1 + W0);
    A(1,:) = [1, s*(2*G-1), 0];
    B(1,:) = G * [1, s, 0];
else
    G = 1 / sqrt(1 + epass^2);
    A(1,:) = [1, 0, 0];
    B(1,:) = G * [1, 0, 0];
end

for i=1:K,
    th = pi * (N - 1 + 2 * i) / (2 * N);
    Wi = Wpass * sin(th);
    D = 1 - 2 * W0 * cos(th) + W0^2 + Wi^2;
    G = (W0^2 + Wi^2) / D;
    a1 = 2 * (W0^2 + Wi^2 - 1) / D;
    a2 = (1 + 2 * W0 * cos(th) + W0^2 + Wi^2) / D;
    A(i+1,:) = [1, s*a1, a2];
    B(i+1,:) = G * [1, s*2, 1];
end
```

20. *Lowpass/Highpass Chebyshev Type 2 Filter Design.*

```
% lhcheb2.m - lowpass/highpass Chebyshev type 2 filter design
%
% [A, B, P] = lhcheb2(s, fs, fpass, fstop, Apass, Astop)
%
% s  = 1, -1 = lowpass, highpass
% design parameters:
% P = [Wpass, Wstop, epass, estop, Nex, N, f3dB, a];
% A, B are Kx3 matrices of cascaded second-order sections

function [A, B, P] = lhcheb2(s, fs, fpass, fstop, Apass, Astop)

Wpass = tan(pi * fpass / fs);   Wpass = Wpass^s;
Wstop = tan(pi * fstop / fs);   Wstop = Wstop^s;

epass = sqrt(10^(Apass/10) - 1);
estop = sqrt(10^(Astop/10) - 1);

Nex = acosh(estop/epass) / acosh(Wstop/Wpass);
N = ceil(Nex);   r = rem(N,2);   K = (N - r) / 2;

a = asinh(estop) / N;
W3 = Wstop / cosh(acosh(estop)/N);
f3 = (fs/pi) * atan(W3^s);
P = [Wpass, Wstop, epass, estop, Nex, N, f3, a];
```

```
WO = sinh(a) / Wstop;                                    % reciprocal of text

if r==1,
    G = 1 / (1 + WO);
    A(1,:) = [1, s*(2*G-1), 0];
    B(1,:) = G * [1, s, 0];
else
    A(1,:) = [1, 0, 0];
    B(1,:) = [1, 0, 0];
end

for i=1:K,
    th = pi * (N - 1 + 2 * i) / (2 * N);
    Wi = sin(th) / Wstop;                                % reciprocal of text
    D = 1 - 2 * WO * cos(th) + WO^2 + Wi^2;
    G = (1 + Wi^2) / D;
    b1 = 2 * (1 - Wi^2) / (1 + Wi^2);
    a1 = 2 * (1 - WO^2 - Wi^2) / D;
    a2 = (1 + 2 * WO * cos(th) + WO^2 + Wi^2) / D;
    A(i+1,:) = [1, s*a1, a2];
    B(i+1,:) = G * [1, s*b1, 1];
end
```

21. *Bandpass Chebyshev Type 2 Filter Design.*

```
% bpcheb2.m - bandpass Chebyshev type 2 digital filter design
%
% [A, B, P] = bpcheb2(fs, fpa, fpb, fsa, fsb, Apass, Astop)
%
% design parameters:
% P = [Wpass, Wstop, Wsa, Wsb, c, epass, estop, Nex, N, f3a, f3b, a];
% A, B are K×5 matrices of cascaded fourth-order sections

function [A, B, P] = bpcheb2(fs, fpa, fpb, fsa, fsb, Apass, Astop)

c = sin(2*pi*(fpa + fpb)/fs)/(sin(2*pi*fpa/fs) + sin(2*pi*fpb/fs));
Wpass = abs((c - cos(2*pi*fpb/fs)) / sin(2*pi*fpb/fs));
Wsa = (c - cos(2*pi*fsa/fs)) / sin(2*pi*fsa/fs);
Wsb = (c - cos(2*pi*fsb/fs)) / sin(2*pi*fsb/fs);
Wstop = min(abs(Wsa), abs(Wsb));

epass = sqrt(10^(Apass/10) - 1);
estop = sqrt(10^(Astop/10) - 1);

Nex = acosh(estop/epass) / acosh(Wstop/Wpass);
N = ceil(Nex);   r = rem(N,2);   K = (N - r) / 2;

a = asinh(estop) / N;
W3  = Wstop / cosh(acosh(estop)/N);
f3a = (fs/pi) * atan((sqrt(W3^2 - c^2 + 1) - W3)/(c+1));
f3b = (fs/pi) * atan((sqrt(W3^2 - c^2 + 1) + W3)/(c+1));
P = [Wpass,Wstop,Wsa,Wsb,c,epass,estop,Nex,N,f3a,f3b,a];
WO = sinh(a) / Wstop;                                    % reciprocal of text

if r==1,
    G = 1 / (1 + WO);
    a1 = -2 * c * WO / (1 + WO);
```

```
        a2 = -(1 - WO) / (1 + WO);
        A(1,:) = [1, a1, a2, 0, 0];
        B(1,:) = G * [1, 0, -1, 0, 0];
    else
        A(1,:) = [1, 0, 0, 0, 0];
        B(1,:) = [1, 0, 0, 0, 0];
    end

    for i=1:K,
        th = pi * (N - 1 + 2 * i) / (2 * N);
        Wi = sin(th) / Wstop;                          % reciprocal of text
        D = 1 - 2 * WO * cos(th) + WO^2 + Wi^2;
        G = (1 + Wi^2) / D;
        b1 = - 4 * c * Wi^2 / (1 + Wi^2);
        b2 = 2 * (Wi^2 * (2*c*c+1) - 1) / (1 + Wi^2);
        a1 = 4 * c * (WO * cos(th) - WO^2 - Wi^2) / D;
        a2 = 2 * ((2*c*c + 1)*(WO^2 + Wi^2) - 1) / D;
        a3 = - 4 * c * (WO * cos(th) + WO^2 + Wi^2) / D;
        a4 = (1 + 2 * WO * cos(th) + WO^2 + Wi^2) / D;
        A(i+1,:) = [1, a1, a2, a3, a4];
        B(i+1,:) = G * [1, b1, b2, b1, 1];
    end
```

22. *Bandstop Chebyshev Type 2 Filter Design.*

```
% bscheb2.m - bandstop Chebyshev type 2 digital filter design
%
% [A, B, P] = bscheb2(fs, fpa, fpb, fsa, fsb, Apass, Astop)
%
% design parameters:
% P = [Wpass, Wstop, Wsa, Wsb, c, epass, estop, Nex, N, f3a, f3b, a];
% A, B are K×5 matrices of cascaded fourth-order sections

function [A, B, P] = bscheb2(fs, fpa, fpb, fsa, fsb, Apass, Astop)

c = sin(2*pi*(fpa + fpb)/fs)/(sin(2*pi*fpa/fs) + sin(2*pi*fpb/fs));
Wpass = abs(sin(2*pi*fpb/fs) / (cos(2*pi*fpb/fs) - c));
Wsa = sin(2*pi*fsa/fs) / (cos(2*pi*fsa/fs) - c);
Wsb = sin(2*pi*fsb/fs) / (cos(2*pi*fsb/fs) - c);
Wstop = min(abs(Wsa), abs(Wsb));

epass = sqrt(10^(Apass/10) - 1);
estop = sqrt(10^(Astop/10) - 1);

Nex = acosh(estop/epass) / acosh(Wstop/Wpass);
N = ceil(Nex);  r = rem(N,2);  K = (N - r) / 2;

a = asinh(estop) / N;
W3 = Wstop / cosh(acosh(estop)/N);
f3a = (fs/pi) * atan((sqrt(1 + W3^2 *(1 - c^2)) - 1)/(W3*(c+1)));
f3b = (fs/pi) * atan((sqrt(1 + W3^2 *(1 - c^2)) + 1)/(W3*(c+1)));
P = [Wpass, Wstop, Wsa, Wsb, c, epass, estop, Nex, N, f3a, f3b, a];
WO = sinh(a) / Wstop;                                 % reciprocal of text

if r==1,
    G = 1 / (1 + WO);
    a1 = -2 * c / (1 + WO);
```

```
    a2 = (1 - W0) / (1 + W0);
    A(1,:) = [1, a1, a2, 0, 0];
    B(1,:) = G * [1, -2*c, 1, 0, 0];
else
    A(1,:) = [1, 0, 0, 0, 0];
    B(1,:) = [1, 0, 0, 0, 0];
end

for i=1:K,
    th = pi * (N - 1 + 2 * i) / (2 * N);
    Wi = sin(th) / Wstop;                            % reciprocal of text
    D = 1 - 2 * W0 * cos(th) + W0^2 + Wi^2;
    G = (1 + Wi^2) / D;
    b1 = -4 * c / (1 + Wi^2);
    b2 = 2 * ((2*c^2 + 1) - Wi^2) / (1 + Wi^2);
    a1 = -4 * c * (1 - W0 * cos(th)) / D;
    a2 = 2 * (2*c^2 + 1 - W0^2 - Wi^2) / D;
    a3 = -4 * c * (1 + W0 * cos(th)) / D;
    a4 = (1 + 2 * W0 * cos(th) + W0^2 + Wi^2) / D;
    A(i+1,:) = [1, a1, a2, a3, a4];
    B(i+1,:) = G * [1, b1, b2, b1, 1];
end
```

References

Texts

[1] B. Gold and C. M. Rader, *Digital Processing of Signals*, McGraw-Hill, New York, 1969.

[2] A. V. Oppenheim and R. W. Schafer, *Discrete-Time Signal Processing*, Prentice Hall, Englewood Cliffs, NJ, 1989.

[3] A. V. Oppenheim and R. W. Schafer, *Digital Signal Processing*, Prentice Hall, Englewood Cliffs, NJ, 1975.

[4] L. R. Rabiner and B. Gold, *Theory and Application of Digital Signal Processing*, Prentice Hall, Englewood Cliffs, NJ, 1975.

[5] S. K. Mitra and J. F. Kaiser, eds., *Handbook of Digital Signal Processing*, Wiley, New York, 1993.

[6] T. W. Parks and C. S. Burrus, *Digital Filter Design*, Wiley, New York, 1987.

[7] A. Antoniou, *Digital Filters: Analysis and Design*, 2nd ed., McGraw-Hill, New York, 1993.

[8] D. F. Elliott, ed., *Handbook of Digital Signal Processing*, Academic Press, New York, 1987.

[9] L. R. Rabiner and C. M. Rader, eds., *Digital Signal Processing*, IEEE Press, New York, 1972.

[10] *Selected Papers in Digital Signal Processing, II*, edited by the Digital Signal Processing Committee and IEEE ASSP, IEEE Press, New York, 1976.

[11] *Programs for Digital Signal Processing*, edited by the Digital Signal Processing Committee, IEEE ASSP Society, IEEE Press, New York, 1979.

[12] A. V. Oppenheim, ed., *Applications of Digital Signal Processing*, Prentice Hall, Englewood Cliffs, NJ, 1978.

[13] J. S. Lim and A. V. Oppenheim, eds., *Advanced Topics in Signal Processing*, Prentice Hall, Englewood Cliffs, NJ, 1988.

[14] R. A. Roberts and C. T. Mullis, *Digital Signal Processing*, Addison-Wesley, Reading, MA, 1987.

[15] P. A. Lynn and W. Fuerst, *Introductory Digital Signal Processing with Computer Applications*, Wiley, New York, 1989.

[16] J. G. Proakis and D. G. Manolakis, *Introduction to Digital Signal Processing*, 2nd ed., Macmillan, New York, 1988.

[17] E. C. Ifeachor and B. W. Jervis, *Digital Signal Processing: A Practical Approach*, Addison-Wesley, Reading, MA, 1993.

773

[18] R. A. Haddad and T. W. Parsons, *Digital Signal Processing: Theory, Applications, and Hardware*, Computer Science Press, W. H. Freeman, New York, 1991.

[19] L. B. Jackson, *Digital Filters and Signal Processing*, Kluwer Academic Publishers, , Norwell, MA 1989.

[20] A. Bateman and W. Yates, *Digital Signal Processing Design*, Computer Science Press, W. H. Freeman, New York, 1991.

[21] S. D. Stearns and D. R. Hush, *Digital Signal Analysis*, 2nd ed., Prentice Hall, Englewood Cliffs, NJ, 1990.

[22] S. D. Stearns and R. A. David, *Signal Processing Algorithms*, Prentice Hall, Englewood Cliffs, NJ, 1988.

[23] D. J. DeFatta, J. G. Lucas, and W. S. Hodgkiss, *Digital Signal Processing: A System Design Approach*, Wiley, New York, 1988.

[24] E. Robinson and S. Treitel, *Geophysical Signal Analysis*, Prentice Hall, Englewood Cliffs, NJ, 1980.

[25] S. M. Kay, *Modern Spectral Estimation: Theory and Application*, Prentice Hall, Englewood Cliffs, NJ, 1988.

[26] S. L. Marple, *Digital Spectral Analysis with Applications*, Prentice Hall, Englewood Cliffs, NJ, 1987.

[27] B. Widrow and S. D. Stearns, *Adaptive Signal Processing*, Prentice Hall, Englewood Cliffs, NJ, 1985.

[28] S. J. Orfanidis, *Optimum Signal Processing*, 2nd ed., McGraw-Hill, New York, 1988.

[29] S. J. Orfanidis, *Digital Signal Processing Laboratory Manual*, ECE Department, Rutgers University, Piscataway, NJ, 1989–94.

[30] I. S. Gradshteyn and I. M. Ryzhik, *Table of Integrals, Series, and Products*, Academic Press, New York, 1980.

[31] H. R. Chillingworth, *Complex Variables*, Pergamon, Oxford, 1973.

[32] P. H. Scholfield, *The Theory of Proportion in Architecture*, Cambridge Univ. Press, London, 1958.

[33] J. Kappraff, *Connections: The Geometric Bridge Between Art and Science*, McGraw-Hill, New York, 1990.

[34] Internet Resources: (1) Signal processing information base at `spib.rice.edu`, (2) `comp.dsp` newsgroup, (3) `comp.dsp.faq` frequently asked questions from `evans.ee.adfa.oz.au` in /pub/dsp.

Sampling

[35] D. A. Linden, "A Discussion of Sampling Theorems," *Proc. IRE*, **47**, 1219 (1959).

[36] A. J. Jerri, "The Shannon Sampling Theorem—Its Various Extensions and Applications: A Tutorial Review," *Proc. IEEE*, **65**, 1565 (1977).

[37] P. L. Butzer and R. L. Strauss, "Sampling Theory for not Necessarily Band-Limited Functions: A Historical Overview," *SIAM Review*, **34**, 40 (1992).

[38] R. J. Marks II, *Introduction to Shannon Sampling and Interpolation Theory*, Springer-Verlag, New York, 1991.

[39] R. J. Marks II, ed., *Advanced Topics in Shannon Sampling and Interpolation Theory*, Springer-Verlag, New York, 1993.

A/D & D/A Conversion, Quantization, Dithering, and Noise Shaping

[40] G. B. Clayton, *Data Converters*, Halsted Press, Wiley, New York, 1982.

[41] M. J. Demler, *High-Speed Analog-to-Digital Conversion*, Academic Press, New York, 1991.

[42] G. F. Miner and D. J. Comer, *Physical Data Acquisition for Digital Processing*, Prentice Hall, Englewood Cliffs, NJ, 1992.

[43] D. Seitzer, G. Pretzl, and N. A. Hamdy, *Electronic Analog-to-Digital Converters*, Wiley, New York, 1983.

[44] D. H. Sheingold, ed., *Analog-Digital Conversion Handbook*, 3d ed., Prentice Hall, Englewood Cliffs, NJ, 1986.

[45] A. VanDoren, *Data Acquisition Systems*, Reston Publishing, Reston, VA, 1982.

[46] W. R. Bennett, "Spectra of Quantized Signals," *Bell Syst. Tech. J.*, **27**, 446 (1948).

[47] B. Widrow, "Statistical Analysis of Amplitude-Quantized Sampled-Data Systems," *AIEE Trans. Appl. Ind.*, pt.2, **79**, 555 (1961).

[48] P. F. Swaszek, ed., *Quantization*, Van Nostrand Reinhold, New York, 1985.

[49] A. B. Sripad and D. L. Snyder, "A Necessary and Sufficient Condition for Quantization Errors to Be Uniform and White," *IEEE Trans. Acoust., Speech, Signal Process.*, **ASSP-25**, 442 (1977).

[50] C. W. Barnes, et al., "On the Statistics of Fixed-Point Roundoff Error," *IEEE Trans. Acoust., Speech, Signal Process.*, **ASSP-33**, 595 (1985).

[51] R. M. Gray, "Quantization Noise Spectra," *IEEE Trans. Inform. Theory*, **IT-36**, 1220 (1990), and earlier references therein. Reprinted in Ref. [276], p. 81.

[52] L. G. Roberts, "Picture Coding Using Pseudo-Random Noise," *IRE Trans. Inform. Th.*, **IT-8**, 145 (1962).

[53] L. Schuchman, "Dither Signals and Their Effect on Quantization Noise," *IEEE Trans. Commun.*, **COM-12**, 162 (1964).

[54] N. S. Jayant and P. Noll, *Digital Coding of Waveforms*, Prentice Hall, Englewood Cliffs, NJ 1984.

[55] J. F. Blinn, "Quantization Error and Dithering," *IEEE Comput. Graphics & Appl. Mag.*, (July 1994), p. 78.

[56] S. P. Lipshitz, R. A. Wannamaker, and J. Vanderkooy, "Quantization and Dither: A Theoretical Survey," *J. Audio Eng. Soc.*, **40**, 355 (1992).

[57] J. Vanderkooy and S. P. Lipshitz, "Resolution Below the Least Significant Bit in Digital Systems with Dither," *J. Audio Eng. Soc.*, **32**, 106 (1984).

[58] J. Vanderkooy and S. P. Lipshitz, "Dither in Digital Audio," *J. Audio Eng. Soc.*, **35**, 966 (1987).

[59] J. Vanderkooy and S. P. Lipshitz, "Digital Dither: Signal Processing with Resolution Far Below the Least Significant Bit," *Proc. 7th Int. Conf.: Audio in Digital Times*, Toronto, May 1989, p. 87.

[60] M. A. Gerzon and P. G. Graven, "Optimal Noise Shaping and Dither of Digital Signals," presented at 87th Convention of the AES, New York, October 1989, *AES Preprint 2822, J. Audio Eng. Soc.*, (Abstracts) **37**, 1072 (1989).

[61] S. P. Lipshitz, J. Vanderkooy, and R. A. Wannamaker, "Minimally Audible Noise Shaping," *J. Audio Eng. Soc.*, **39**, 836 (1991).

[62] R. A. Wannamaker, "Psychoacoustically Optimal Noise Shaping," *J. Audio Eng. Soc.*, **40**, 611 (1992).

[63] M. A. Gerzon, P. G. Graven, J. R. Stuart, and R. J. Wilson, "Psychoacoustic Noise Shaped Improvements in CD and Other Linear Digital Media," presented at 94th Convention of the AES, Berlin, May 1993, *AES Preprint no. 3501.*

[64] R. van der Waal, A. Oomen, and F. Griffiths, "Performance Comparison of CD, Noise-Shaped CD and DCC," presented at 96th Convention of the AES, Amsterdam, February 1994, *AES Preprint no. 3845.*

[65] J. A. Moorer and J. C. Wen, "Whither Dither: Experience with High-Order Dithering Algorithms in the Studio," presented at 95th Convention of the AES, New York, October 1993, *AES Preprint no. 3747.*

[66] R. A. Wannamaker, "Subtractive and Nonsubtractive Dithering: A Comparative Analysis," presented at 97th Convention of the AES, San Francisco, November 1994, *AES Preprint no. 3920.*

[67] D. Ranada, "Super CD's: Do They Deliver The Goods?," *Stereo Review*, July 1994, p. 61.

[68] M. A. Gerzon and P. G. Craven, "A High-Rate Buried-Data Channel for Audio CD," *J. Audio Eng. Soc.*, **43**, 3 (1995).

[69] A. Oomen, M. Groenewegen, R. van der Waal, and R. Veldhuis, "A Variable-Bit-Rate Buried-Data Channel for Compact Disc," *J. Audio Eng. Soc.*, **43**, 25 (1995).

[70] Trân-Thông and B. Liu, "Error Spectrum Shaping in Narrow Band Recursive Filters," *IEEE Trans. Acoust., Speech, Signal Process.*, **ASSP-25**, 200 (1977).

[71] W. E. Higgins and D. C. Munson, "Noise Reduction Strategies for Digital Filters: Error Spectrum Shaping Versus the Optimal Linear State-Space Formulation," *IEEE Trans. Acoust., Speech, Signal Process.*, **ASSP-30**, 963 (1982).

[72] C. T. Mullis and R. A. Roberts, "An Interpretation of Error Spectrum Shaping in Digital Filters," *IEEE Trans. Acoust., Speech, Signal Process.*, **ASSP-30**, 1013 (1982).

[73] J. Dattoro, "The Implementation of Digital Filters for High-Fidelity Audio," *Proc. AES 7th Int. Conf., Audio in Digital Times*, Toronto, 1989, p. 165.

[74] R. Wilson, et al., "Filter Topologies," *J. Audio Eng. Soc.*, **41**, 455 (1993).

[75] U. Zölser, "Roundoff Error Analysis of Digital Filters," *J. Audio Eng. Soc.*, **42**, 232 (1994).

[76] W. Chen, "Performance of the Cascade and Parallel IIR Filters," presented at 97th Convention of the AES, San Francisco, November 1994, *AES Preprint no. 3901.*

[77] D. W. Horning and R. Chassaing, "IIR Filter Scaling for Real-Time Signal Processing," *IEEE Trans. Educ.*, **34**, 108 (1991).

[78] K. Baudendistel, "Am Improved Method of Scaling for Real-Time Signal Processing Applications," *IEEE Trans. Educ.*, **37**, 281 (1994).

DSP Hardware

[79] E. A. Lee, "Programmable DSP Architectures: Part I," *IEEE ASSP Mag.*, **5**, no. 4, 4 (1988), and "Programmable DSP Architectures: Part II," ibid., **6**, no. 1, 4 (1989).

[80] K-S. Lin, ed., *Digital Signal Processing Applications with the TMS320 Family*, vol. 1, Prentice Hall, Englewood Cliffs, NJ, 1987.

[81] P. E. Papamichalis, ed., *Digital Signal Processing Applications with the TMS320 Family*, vols. 2 and 3, Prentice Hall, Englewood Cliffs, NJ, 1991.

[82] *TMS320C2x, TMS320C3x, TMS320C4x, TMS320C5x, User Guides*, Texas Instruments, Dallas, TX, 1989–93.

[83] R. Chassaing, *Digital Signal Processing with C and the TMS320C30*, Wiley, New York, 1992.

[84] M. El-Sharkawy, *Real Time Digital Signal Processing Applications with Motorola's DSP56000 Family*, Prentice Hall, Englewood Cliffs, NJ, 1990.

[85] M. El-Sharkawy, *Signal Processing, Image Processing and Graphics Applications with Motorola's DSP96002 Processor*, vol. I, Prentice Hall, Englewood Cliffs, NJ, 1994.

[86] V. K. Ingle and J. G. Proakis, *Digital Signal Processing Laboratory Using the ADSP-2101 Microcomputer*, Prentice Hall, Englewood Cliffs, NJ, 1991.

[87] *WE® DSP32 Digital Signal Processor, Application Guide*, AT&T Document Management Organization, 1988.

[88] J. Tow, "Implementation of Digital Filters with the WE®DSP32 Digital Signal Processor," AT&T Application Note, 1988.

[89] S. L. Freeny, J. F. Kaiser, and H. S. McDonald, "Some Applications of Digital Signal Processing in Telecommunications," in Ref. [12], p. 1.

[90] J. R. Boddie, N. Sachs, and J. Tow, "Receiver for TOUCH-TONE Service," *Bell Syst. Tech. J.*, **60**, 1573 (1981).

[91] J. Hartung, S. L. Gay, and G. L. Smith, "Dual-Tone Multifrequency Receiver Using the WE®DSP32 Digital Signal Processor," AT&T Application Note, 1988.

[92] P. Mock, "Add DTMF Generation and Decoding to DSP-μP Designs," *EDN*, March 21, (1985). Reprinted in Ref. [80], p. 543.

[93] A. Mar, ed., *Digital Signal Processing Applications Using the ADSP-2100 Family*, Prentice Hall, Englewood Cliffs, NJ, 1990.

Computer Music

[94] M. V. Mathews, et al., *The Technology of Computer Music*, MIT Press, Cambridge, MA, 1969.

[95] F. R. Moore, *Elements of Computer Music*, Prentice Hall, Englewood Cliffs, NJ, 1990.

[96] C. Roads and J. Strawn, eds., *Foundations of Computer Music*, MIT Press, Cambridge, MA, 1988.

[97] C. Roads, ed., *The Music Machine*, MIT Press, Cambridge, MA, 1989.

[98] C. Dodge and T. A. Jerse, *Computer Music*, Schirmer/Macmillan, New York, 1985.

[99] G. Haus, ed., *Music Processing*, A-R Editions, Inc., Madison, WI, 1993.

[100] B. Vercoe, "Csound: A Manual for the Audio Processing System and Support-
ing Programs with Tutorials," *MIT Media Lab*, 1993. Csound is available via
ftp from cecelia.media.mit.edu in pub/Csound and MSDOS versions from
ftp.maths.bath.ac.uk in pub/dream.

[101] J. M. Chowning, "The Synthesis of Complex Audio Spectra by Means of Frequency
Modulation," *J. Audio Eng. Soc.*, **21**, 526, 1973. Reprinted in Ref. [96].

[102] F. R. Moore, "Table Lookup Noise for Sinusoidal Digital Oscillators," *Computer Music
J.*, **1**, 26, 1977. Reprinted in Ref. [96].

[103] S. Mehrgardt, "Noise Spectra of Digital Sine-Generators Using the Table-Lookup
Method," *IEEE Trans. Acoust., Speech, Signal Process.*, **ASSP-31**, 1037 (1983).

[104] W. M. Hartmann, "Digital Waveform Generation by Fractional Addressing," *J. Acoust.
Soc. Am.*, **82**, 1883 (1987).

[105] P. D. Lehrman, "The Computer as a Musical Instrument," *Electronic Musician*, **7**, no. 3,
30 (1991).

[106] M. N. McNabb, "Dreamsong: The Composition," *Computer Music J.*, **5**, no. 4, 1981.
Reprinted in Ref. [97].

[107] G. De Poli, "A Tutorial on Digital Sound Synthesis Techniques," *Computer Music J.*, **7**,
no. 4, (1983). Reprinted in Ref. [97].

[108] R. Karplus and A. Strong, "Digital Synthesis of Plucked String and Drum Timbres,"
Computer Music J., **7**, 43 (1983). Reprinted in Ref. [97].

[109] D. A. Jaffe and J. O. Smith, "Extensions of the Karplus-Strong Plucked-String Algorithm," *Computer Music J.*, **7**, 56 (1983). Reprinted in Ref. [97].

[110] C. R. Sullivan, "Extending the Karplus-Strong Algorithm to Synthesize Electric Guitar
Timbres with Distortion and Feedback," *Computer Music J.*, **14**, 26 (1990).

[111] J. O. Smith, "Physical Modeling Using Digital Waveguides," *Computer Music J.*, **16**, 74
(1992).

[112] G. Mayer-Kress, et al., "Musical Signals from Chua's Circuit," *IEEE Trans. Circuits
Syst.—II: Analog and Digital Signal Process.*, **40**, 688 (1993).

[113] X. Rodet, "Models of Musical Instruments from Chua's Circuit with Time Delay," *IEEE
Trans. Circuits Syst.—II: Analog and Digital Signal Process.*, **40**, 696 (1993).

[114] S. Wilkinson, "Model Music," *Electronic Musician*, **10**, no. 2, 42 (1994).

[115] J. A. Moorer, "Signal Processing Aspects of Computer Music: A Survey," *Proc. IEEE*,
65, 1108, (1977). Reprinted in Ref. [149].

Digital Audio Effects

[116] *IEEE ASSP Mag.*, **2**, no. 4, October 1985, Special Issue on Digital Audio.

[117] Y. Ando, *Concert Hall Acoustics*, Springer-Verlag, New York, 1985.

[118] D. Begault, "The Evolution of 3-D Audio," *MIX*, October 1993, p. 42.

[119] P. J. Bloom, "High-Quality Digital Audio in the Entertainment Industry: An Overview
of Achievements and Challenges," in Ref. [116], p. 2.

[120] B. Blesser and J. M. Kates, "Digital Processing of Audio Signals," in Ref. [12], p. 29.

[121] N. Brighton and M. Molenda, "Mixing with Delay," *Electronic Musician,* **9**, no. 7, 88 (1993).

[122] N. Brighton and M. Molenda, "EQ Workshop" *Electronic Musician,* October 1993, p. 105.

[123] D. Cronin, "Examining Audio DSP Algorithms," *Dr. Dobbs Journal,* **19**, no. 7, 78, (1994).

[124] G. Davis and R. Jones, *Sound Reinforcement Handbook,* 2nd ed., Yamaha Corp., Hal Leonard Publishing, Milwaukee, WI., 1989.

[125] J. M. Eargle, *Handbook of Recording Engineering,* 2nd ed., Van Nostrand Reinhold, New York, 1992.

[126] P. Freudenberg, "All About Dynamics Processors, Parts 1 & 2," *Home & Studio Recording,* March 1994, p. 18, and April 1994, p. 44.

[127] D. Griesinger, "Practical Processors and Programs for Digital Reverberation," *Proc. AES 7th Int. Conf., Audio in Digital Times,* Toronto, 1989, p. 187.

[128] G. Hall, "Effects, The Essential Musical Spice," *Electronic Musician,* **7**, no. 8, 62 (1991).

[129] B. Hurtig, "Pumping Gain: Understanding Dynamics Processors," *Electronic Musician,* **7**, no. 3, 56 (1991).

[130] B. Hurtig, "The Engineer's Notebook: Twelve Ways to Use Dynamics Processors," *Electronic Musician,* **7**, no. 3, 66 (1991).

[131] B. Blesser, "Audio Dynamic Range Compression for Minimum Perceived Distortion," *IEEE Trans. Audio Electroacoust.,* **AU-17**, 22 (1969).

[132] G. W. McNally, "Dynamic Range Control of Digital Audio Signals," *J. Audio Eng. Soc.,* **32**, 316 (1984).

[133] G. Hall, "Solving the Mysteries of Reverb," *Electronic Musician,* **7**, no. 6, 80 (1991).

[134] M. Kleiner, B. Dalebäck, and P. Svensson, "Auralization—An Overview," *J. Audio Eng. Soc.,* **41**, 861 (1993).

[135] P. D. Lehrman, "The Electronic Orchestra, Parts I and II," *Electronic Musician,* September 1993, p. 41, and October 1993, p. 46.

[136] J. Meyer, "The Sound of the Orchestra," *J. Audio Eng. Soc.,* **41**, 203 (1993).

[137] J. A. Moorer, "About this Reverberation Business," *Computer Music J.,* **3**, 13 (1979). Reprinted in Ref. [96].

[138] D. Moulton, "Spectral Management, Parts 1 & 2," *Home & Studio Recording,* July 1993, p. 22, and August 1993, p. 50.

[139] E. Persoon and C. Vanderbulcke, "Digital Audio: Examples of the Application of the ASP Integrated Signal Processor," *Philips Tech. Rev.,* **42**, 201 (1986).

[140] J. R. Pierce, *The Science of Musical Sound,* W. H. Freeman and Company, New York, 1992.

[141] K. Pohlmann, *The Principles of Digital Audio,* 2nd ed., H. W. Sams, Carmel, IN, 1989.

[142] K. Pohlmann, ed., *Advanced Digital Audio,* H. W. Sams, Carmel, IN, 1991.

[143] M. R. Schroeder, "Natural Sounding Artificial Reverberation," *J. Audio Eng, Soc.,* **10**, 219, (1962).

[144] M. R. Schroeder, "Digital Simulation of Sound Transmission in Reverberant Spaces", *J. Acoust. Soc. Am.,* **47**, 424 (1970).

[145] M. R. Schroeder, D. Gottlob, and K. F. Siebrasse, "Comparative Study of European Concert Halls: Correlation of Subjective Preference with Geometric and Acoustic Parameters," *J. Acoust. Soc. Am.*, **56**, 1195 (1974).

[146] W. L. Sinclair and L. I. Haworth, "Digital Recording in the Professional Industry, parts I and II," *Electronics & Communications Eng. J.*, June and August 1991.

[147] *SE-50 Stereo Effects Processor*, User Manual, Roland/Boss Corp., 1990.

[148] J. O. Smith, "An Allpass Approach to Digital Phasing and Flanging," *Proc. Int. Computer Music Conf. (ICMC)*, IRCAM, Paris, Oct. 1984, p. 236.

[149] J. Strawn, ed., *Digital Audio Signal Processing: An Anthology*, W. Kaufmann, Los Altos, CA, 1985.

[150] J. Watkinson, *The Art of Digital Audio*, Focal Press, London, 1988.

[151] P. White, *Creative Recording: Effects and Processors*, Music Maker Books, Cambridgeshire, UK, 1989.

Biomedical Signal Processing

[152] M. L. Ahlstrom and W. J. Tompkins, "Digital Filters for Real-Time ECG Signal Processing Using Microprocessors," IEEE Trans. Biomed. Eng., **BME-32**, 708 (1985).

[153] I. I. Christov and I. A. Dotsinsky, "New Approach to the Digital Elimination of 50 Hz Interference from the Electrocardiogram," *Med. & Biol. Eng. & Comput.*, **26** 431 (1988).

[154] "The Design of Digital Filters for Biomedical Signal Processing," Part-1, *J. Biomed. Eng.*, **4**, 267 (1982), Part-2, ibid., **5**, 19 (1983), Part-3, ibid., **5**, 91 (1983).

[155] M. Della Corte, O. Cerofolini, and S. Dubini, "Application of Digital Filter to Biomedical Signals," *Med. & Biol. Eng.*, **12** 374 (1974).

[156] G. M. Friesen, et al., "A Comparison of the Noise Sensitivity of Nine QRS Detection Algorithms," *IEEE Trans. Biomed. Eng.*, **BME-37**, 85 (1990).

[157] C. Levkov, et al., "Subtraction of 50 Hz Interference from the Electrocardiogram," *Med. & Biol. Eng. & Comput.*, **22** 371 (1984).

[158] C. L. Levkov, "Fast Integer Coefficient FIR Filters to Remove the AC Interference and the High-Frequency Noise Components in Biological Signals," *Med. & Biol. Eng. & Comput.*, **27** 330 (1989).

[159] P. A. Lynn, "Online Digital Filters for Biological Signals: Some Fast designs for a Small Computer," *Med. & Biol. Eng. & Comput.*, **15** 534 (1977).

[160] P. A. Lynn, "Transversal Resonator Digital Filters: Fast and Flexible Online Processors for Biological Signals," *Med. & Biol. Eng. & Comput.*, **21** 718 (1983).

[161] R. M. Lu and B. M. Steinhaus, "A Simple Digital Filter to Remove Line-Frequency Noise in Implantable Pulse Generators," *Biomed. Instr. & Technol.*, **27**, 64 (1993).

[162] N. R. Malik, "Microcomputer Realisations of Lynn's Fast Digital Filtering Designs," *Med. & Biol. Eng. & Comput.*, **18** 638 (1980). (This reference uses circular addressing to implement delays and refers to its earlier use by J. D. Schoeffler (1971) as a wraparound queue.)

[163] C. J. Marvell and D. L. Kirk, "Use of a Microprocessor to Simulate Precise Electrocardiograms," *J. Biomed. Eng.*, **2**, 61 (1980).

[164] V. T. Rhyne, "A Digital System for Enhancing the Fetal Electrocardiogram," IEEE Trans. Biomed. Eng., **BME-16**, 80 (1969).

[165] J. E. Sheild and D. L. Kirk, "The Use of Digital Filters in Enhancing the Fetal Electrocardiogram," *J. Biomed. Eng.*, **3**, 44 (1981).

[166] T. P. Taylor and P. W. Macfarlane, "Digital Filtering of the ECG–A Comparison of Lowpass Digital Filters on a Small Computer," *Med. & Biol. Eng.*, **12** 493 (1974).

[167] N. V. Thakor and D. Moreau, "Design and Analysis of Quantised Coefficient Digital Filters: Application to Biomedical Signal Processing With Microprocessors," *Med. & Biol. Eng. & Comput.*, **25** 18 (1987).

[168] W. J. Tompkins and J. G. Webster, eds., *Design of Microcomputer-Based Medical Instrumentation*, Prentice Hall, Englewood Cliffs, NJ, 1981.

[169] W. J. Tompkins, ed., *Biomedical Digital Signal Processing*, Prentice Hall, Englewood Cliffs, NJ, 1993.

[170] R. Wariar and C. Eswaran, "Integer Coefficient Bandpass Filter for the Simultaneous Removal of Baseline Wander, 50 and 100 Hz Interference from the ECG," *Med. & Biol. Eng. & Comput.*, **29** 333 (1991).

Digital TV

[171] A. Acampora, "Wideband Picture Detail Restoration in a Digital NTSC Comb-Filter System," *RCA Engineer*, **28-5**, 44, Sept./Oct. (1983).

[172] A. A. Acampora, R. M. Bunting, and R. J. Petri, "Noise Reduction in Video Signals by a Digital Fine-Structure Process," *RCA Engineer*, **28-2**, 48, March/April (1983).

[173] A. A. Acampora, R. M. Bunting, and R. J. Petri, "Noise Reduction in Video Signals Using Pre/Post Signal Processing in a Time Division Multiplexed Component System," *RCA Review*, **47**, 303 (1986).

[174] M. Annegarn, A. Nillesen, and J. Raven, "Digital Signal Processing in Television Receivers," *Philips Tech. Rev.*, **42**, 183 (1986).

[175] J. F. Blinn, "NTSC: Nice Technology, Super Color," *IEEE Comput. Graphics & Appl. Mag.*, (March 1993), p. 17.

[176] J. Isailovic, *Videodisc Systems: Theory and Applications*, Prentice Hall, Englewood Cliffs, NJ, 1987.

[177] H. E. Kallmann, "Transversal Filters," *Proc. IRE*, **28**, 302 (1940). Perhaps, the earliest reference on FIR and comb filters.

[178] P. Mertz and F. Gray, "A Theory of Scanning and Its Relation to the Characteristics of the Transmitted Signal in Telephotography and Television," *Bell Syst. Tech. J.*, **13**, 464 (1934).

[179] D. E. Pearson, *Transmission and Display of Pictorial Information*, Wiley, New York, 1975.

[180] J. O. Limb, C. B. Rubinstein, and J. E. Thompson, "Digital Coding of Color Video Signals—A Review," *IEEE Trans. Commun.*, **COM-25**, 1349 (1977).

[181] C. H. Lu, "Subcarrier Phase Coherence in Noise Reduction of Composite NTSC Signals—Three Approaches and Their Comparison," *RCA Review*, **47**, 287 (1986).

[182] D. H. Pritchard and J. J. Gibson, "Worldwide Color TV Standards—Similarities and Differences," *RCA Engineer*, **25-5**, 64, Feb./Mar. (1980).

[183] D. H. Pritchard, "A CCD Comb Filter for Color TV Receiver Picture Enhancement," *RCA Review*, **41**, 3 (1980).

[184] J. P. Rossi, "Color Decoding a PCM NTSC Television Signal," *SMPTE J.*, **83**, 489 (1974).

[185] J. P. Rossi, "Digital Television Image Enhancement," *SMPTE J.*, **84**, 546 (1975).

[186] C. P. Sandbank, ed., *Digital Television*, Wiley, Chichester, England, 1990.

[187] K. B. Benson and D. G. Fink, *HDTV, Advanced Television for the 1990s*, Intertext/Multiscience, McGraw-Hill, New York, 1991.

[188] M. I. Krivocheev and S. N. Baron, "The First Twenty Years of HDTV: 1972-1992," *SMPTE J.*, **102**, 913 (1993).

[189] Special Issue on All Digital HDTV, *Signal Processing: Image Commun.*, **4**, no. 4-5 (Aug. 1992).

[190] Special Issue on Digital High Definition Television, *Signal Processing: Image Commun.*, **5**, no. 5-6 (Dec. 1993).

[191] R. Hopkins, "Digital Terrestrial HDTV for North America: The Grand Alliance HDTV System," *IEEE Trans. Consum. Electr.*, **40**, 185 (1994).

Signal Averaging

[192] D. G. Childers, "Biomedical Signal Processing," in *Selected Topics in Signal Processing*, S. Haykin, ed., Prentice Hall, Englewood Cliffs, NJ, 1989.

[193] A. Cohen, *Biomedical Signal Processing*, vols. 1 and 2, CRC Press, Boca Raton, FL, 1986.

[194] H. G. Goovaerts and O. Rompelman, "Coherent Average Technique: A Tutorial Review," *J. Biomed. Eng.*, **13**, 275 (1991).

[195] P. Horowitz and W. Hill, *The Art of Electronics*, 2nd ed., Cambridge University Press, Cambridge, 1989.

[196] O. Rompelman and H. H. Ros, "Coherent Averaging Technique: A Tutorial Review, Part 1: Noise Reduction and the Equivalent Filter," *J. Biomed. Eng.*, **8**, 24 (1986); and "Part 2: Trigger Jitter, Overlapping Responses, and Non-Periodic Stimulation," ibid., p. 30.

[197] V. Shvartsman, G. Barnes, L. Shvartsman, and N. Flowers, "Multichannel Signal Processing Based on Logic Averaging," *IEEE Trans. Biomed. Eng.*, **BME-29**, 531 (1982).

[198] C. W. Thomas, M. S. Rzeszotarski, and B. S. Isenstein, "Signal Averaging by Parallel Digital Filters," *IEEE Trans. Acoust., Speech, Signal Process.*, **ASSP-30**, 338 (1982).

[199] T. H. Wilmshurst, *Signal Recovery from Noise in Electronic Instrumentation*, 2nd ed., Adam Hilger and IOP Publishing, Bristol, England, 1990.

Smoothing Filters

[200] M. Bromba and H. Ziegler, "Efficient Computation of Polynomial Smoothing Digital Filters," *Anal. Chem.*, **51**, 1760 (1979).

[201] M. Bromba and H. Ziegler, "Explicit Formula for Filter Function of Maximally Flat Nonrecursive Digital Filters," *Electron. Lett.*, **16**, 905 (1980).

[202] M. Bromba and H. Ziegler, "Application Hints for Savitzky-Golay Digital Smoothing Filters," *Anal. Chem.*, **53**, 1583 (1981).

[203] T. H. Edwards and P. D. Wilson, "Digital Least Squares Smoothing of Spectra," *Appl. Spectrosc.*, **28**, 541 (1974).

[204] T. H. Edwards and P. D. Wilson, "Sampling and Smoothing of Spectra," *Appl. Spectrosc. Rev.*, **12**, 1 (1976).

[205] C. G. Enke and T. A. Nieman, "Signal-to-Noise Ratio Enhancement by Least-Squares Polynomial Smoothing," *Anal. Chem.*, **48**, 705A (1976).

[206] R. R. Ernst, "Sensitivity Enhancement in Magnetic Resonance," in *Advances in Magnetic Resonance*, vol. 2, J. S. Waugh, ed., Academic Press, 1966.

[207] R. W. Hamming, *Digital Filters*, 2nd ed., Prentice Hall, Englewood Cliffs, NJ, 1983.

[208] M. Kendall, *Time-Series*, 2nd ed., Hafner Press, Macmillan, New York, 1976.

[209] M. Kendall and A. Stuart, *Advanced Theory of Statistics*, vol. 3, 2nd ed., Charles Griffin & Co., London, 1968.

[210] H. H. Madden, "Comments on the Savitzky-Golay Convolution Method for Least-Squares Fit Smoothing and Differentiation of Digital Data," *Anal. Chem.*, **50**, 1383 (1978).

[211] A. Savitzky and M Golay, "Smoothing and Differentiation of Data by Simplified Least Squares Procedures," *Anal. Chem..* **36**, 1627 (1964).

[212] H. W. Schüssler and P. Steffen, "Some Advanced Topics in Filter Design," in Ref. [13].

[213] P. Steffen, "On Digital Smoothing Filters: A Brief Review of Closed Form Solutions and Two New Filter Approaches," *Circ., Syst., and Signal Process.*, fb5, 187 (1986).

[214] J. Steinier, Y. Termonia, and J. Deltour, "Comments on Smoothing and Differentiation of Data by Simplified Least Squares Procedures," *Anal. Chem..* **44**, 1627 (1972).

[215] C. S. Williams, *Designing Digital Filters*, Prentice Hall, Englewood Cliffs, NJ, 1986.

[216] H. Ziegler, "Properties of Digital Smoothing Polynomial (DISPO) Filters," *Appl. Spectrosc.*, **35**, 88 (1981).

[217] J. F. Kaiser and W. A. Reed, "Data Smoothing Using Lowpass Digital Filters," *Rev. Sci. Instrum.*, **48**, 1447 (1977).

[218] J. F. Kaiser and R. W. Hamming, "Sharpening the Response of a Symmetric Nonrecursive Filter by Multiple Use of the Same Filter," *IEEE Trans. Acoust., Speech, Signal Process.*, **ASSP-25**, 415 (1975).

Spectral Analysis and DFT/FFT Algorithms

[219] F. J. Harris, "On the Use of Windows for Harmonic Analysis with the Discrete Fourier Transform," *Proc. IEEE*, **66**, 51 (1978).

[220] N. C. Geçkinli and D. Yavuz, "Some Novel Windows and a Concise Tutorial Comparison of Window Families," *IEEE Trans. Acoust., Speech, Signal Process.*, **ASSP-26**, 501 (1978).

[221] J. F. Kaiser and R. W. Schafer, "On the Use of the I_0-Sinh Window for Spectrum Analysis," *IEEE Trans. Acoust., Speech, Signal Process.*, **ASSP-28**, 105 (1980).

[222] A. H. Nuttal, "Some Windows with Very Good Sidelobe Behavior," *IEEE Trans. Acoust., Speech, Signal Process.*, **ASSP-29**, 84 (1981).

[223] E. O. Brigham, *The Fast Fourier Transform*, Prentice Hall, Englewood Cliffs, NJ, 1988.

[224] R. W. Ramirez, *The FFT, Fundamentals and Concepts*, Prentice Hall, Englewood Cliffs, NJ, 1985.

[225] C. S. Burrus and T. W. Parks, *DFT/FFT and Convolution Algorithms*, Wiley, New York, 1985.

[226] C. Van Loan, *Computational Frameworks for the Fast Fourier Transform*, SIAM, Philadelphia, 1992.

[227] W. L. Briggs and V. E. Henson, *The DFT: An Owner's Manual for the Discrete Fourier Transform*, SIAM, Philadelphia, 1995.

[228] J. W. Cooley, P. A. W. Lewis, and P. D. Welch, "The Fast Fourier Transform and Its Applications," *IEEE Trans. Educ.*, **12**, 27 (1969).

[229] G. D. Bergland, "A Guided Tour of the Fast Fourier Transform," IEEE Spectrum, **6**, 41, July 1969.

[230] J. W. Cooley, P. A. W. Lewis, and P. D. Welch, "The Fast Fourier Transform Algorithm: Programming Considerations in the Calculation of Sine, Cosine, and Laplace Transforms," *J. Sound Vib.*, **12**, 315 (1970). Reprinted in Ref. [9].

[231] F. J. Harris, "The Discrete Fourier Transform Applied to Time Domain Signal Processing," *IEEE Commun. Mag.*, May 1982, p. 13.

[232] J. W. Cooley, P. A. W. Lewis, and P. D. Welch, "Historical Notes on the Fast Fourier Transform," *IEEE Trans. Audio Electroacoust.*, **AU-15**, 76 (1967). Reprinted in Ref. [9].

[233] M. T. Heideman, D. H. Johnson, and C. S. Burrus, "Gauss and the History of the Fast Fourier Transform," *IEEE ASSP Mag.*, **4**, no. 4, 14 (1984).

[234] J. W. Cooley, "How the FFT Gained Acceptance," *IEEE Signal Proc. Mag.*, **9**, no. 1, 10 (1992).

[235] P. Kraniauskas, "A Plain Man's Guide to the FFT," *IEEE Signal Proc. Mag.*, **11**, no. 2, 36 (1994).

[236] J. R. Deller, "Tom, Dick, and Mary Discover the DFT," *IEEE Signal Proc. Mag.*, **11**, no. 2, 36 (1994).

[237] P. Duhamel and M. Vetterli, "Fast Fourier Transforms: A Tutorial Review and a State of the Art," *Signal Processing*, **19**, 259 (1990).

[238] N. Ahmed, T. Natarajan, and K. R. Rao, "Discrete Cosine Transform," *IEEE Trans. Comput.*, **C-23**, 90 (1974).

[239] H. S. Hou, "A Fast Recursive Algorithm for the Discrete Cosine Transform," *IEEE Trans. Acoust., Speech, Signal Process.*, **ASSP-35**, 1455 (1987).

[240] J. F. Blinn, "What's the Deal with the DCT?," *IEEE Comput. Graphics & Appl. Mag.*, (July 1993), p. 78.

[241] R. C. Singleton, "A Method for Computing the Fast Fourier Transform with Auxiliary Memory and Limited High-Speed Storage," *IEEE Trans. Audio Electroacoust.*, **AU-15**, 91 (1967). Reprinted in Ref. [9].

[242] N. M. Brenner, "Fast Fourier Transform of Externally Stored Data," *IEEE Trans. Audio Electroacoust.*, **AU-17**, 128 (1969).

[243] W. K. Hocking, "Performing Fourier Transforms on Extremely Long Data Streams," *Comput. Phys.*, **3**, 59 (1989).

[244] H. V. Sorensen and C. S. Burrus, "Efficient Computation of the DFT with Only a Subset of Input or Output Points," *IEEE Trans. Acoust., Speech, Signal Process.*, **ASSP-41**, 1184 (1993).

FIR Filter Design

[245] J. F. Kaiser, "Design Methods for Sampled Data Filters," *Proc. 1st Allerton Conf. Circuit System Theory*, p. 221, (1963), and reprinted in Ref. [9], p. 20.

[246] J. F. Kaiser, "Digital Filters," in F. F. Kuo and J. F. Kaiser, eds., *System Analysis by Digital Computer*, Wiley, New York, 1966, p. 228.

[247] J. F. Kaiser, "Nonrecursive Digital Filter Design Using the I_0-Sinh Window Function," *Proc. 1974 IEEE Int. Symp. on Circuits and Systems*, p. 20, (1974), and reprinted in [10], p. 123.

[248] H. D. Helms, "Nonrecursive Digital Filters: Design Methods for Achieving Specifications on Frequency Response," *IEEE Trans. Audio Electroacoust.*, **AU-16**, 336 (1968).

[249] H. D. Helms, "Digital Filters with Equiripple or Minimax Response," *IEEE Trans. Audio Electroacoust.*, **AU-19**, 87 (1971), and reprinted in Ref. [9], p. 131.

[250] C. L. Dolph, "A Current Distribution Which Optimizes the Relationship Between Beamwidth and Side-lobe Level," *Proc. I.R.E*, **34**, 335 (1946).

[251] D. Barbiere, "A Method for Calculating the Current Distribution of Tschebyscheff Arrays," *Proc. I.R.E*, **40**, 78 (1952).

[252] R. J. Stegen, "Excitation Coefficients and Beamwidths of Tschebyscheff Arrays," *Proc. I.R.E*, **41**, 1671 (1953).

[253] R. C. Hansen, "Linear Arrays," in A. W. Rudge, et al., eds., *Handbook of Antenna Design*, vol. 2, 2nd ed., P. Peregrinus and IEE, London, 1986.

[254] T. Saramäki, "Finite Impulse Response Filter Design," in Ref. [5], p. 155.

[255] K. B. Benson and J. Whitaker, *Television and Audio Handbook*, McGraw-Hill, New York, 1990.

[256] P. L. Schuck, "Digital FIR Filters for Loudspeaker Crossover Networks II: Implementation Example," *Proc. AES 7th Int. Conf., Audio in Digital Times*, Toronto, 1989, p. 181.

[257] R. Wilson, et al., "Application of Digital Filters to Loudspeaker Crossover Networks," *J. Audio Eng. Soc.*, **37**, 455 (1989).

[258] K. Steiglitz, T. W. Parks, and J. F. Kaiser, "METEOR: A Constraint-Based FIR Filter Design Program," *IEEE Trans. Acoust., Speech, Signal Process.*, **ASSP-40**, 1901 (1992). This program is available via anonymous ftp from princeton.edu.

[259] K. Steiglitz and T. W. Parks, "What is the Filter Design Problem?," *Proc. 1986 Princeton Conf. Inform. Sci. Syst.*, p. 604 (1986).

Second-Order IIR Filter Design

[260] K. Hirano, S. Nishimura, and S. Mitra, "Design of Digital Notch Filters," *IEEE Trans. Commun.*, **COM-22**, 964 (1974).

[261] M. N. S. Swami and K. S. Thyagarajan, "Digital Bandpass and Bandstop Filters with Variable Center Frequency and Bandwidth," *Proc. IEEE*, **64**, 1632 (1976).

[262] J. A. Moorer, "The Manifold Joys of Conformal Mapping: Applications to Digital Filtering in the Studio," *J. Audio Eng. Soc.*, **31**, 826 (1983).

[263] S. A. White, "Design of a Digital Biquadratic Peaking or Notch Filter for Digital Audio Equalization," *J. Audio Eng. Soc.*, **34**, 479 (1986).

[264] P. A. Regalia and S. K. Mitra, "Tunable Digital Frequency Response Equalization Filters," *IEEE Trans. Acoust., Speech, Signal Process.*, **ASSP-35**, 118 (1987).

[265] D. J. Shpak, "Analytical Design of Biquadratic Filter Sections for Parametric Filters," *J. Audio Eng. Soc.*, **40**, 479 (1992).

[266] D. C. Massie, "An Engineering Study of the Four-Multiply Normalized Ladder Filter," *J. Audio Eng. Soc.*, **41**, 564 (1993).

[267] F. Harris and E. Brooking, "A Versatile Parametric Filter Using Imbedded All-Pass Sub-Filter to Independently Adjust Bandwidth, Center Frequency, and Boost or Cut," presented at the 95th Convention of the AES, New York, October 1993, *AES Preprint 3757*.

[268] R. Bristow-Johnson, "The Equivalence of Various Methods of Computing Biquad Coefficients for Audio Parametric Equalizers," presented at the 97th Convention of the AES, San Francisco, November 1994, *AES Preprint 3906*.

Analog Filter Design

[269] M. E. Van Valkenburg, *Analog Filter Design*, Holt, Rinehart and Winston, New York, 1982.

[270] A. B. Williams and F. J. Taylor, *Electronic Filter Design Handbook*, 2nd ed., McGraw-Hill, New York, 1988.

[271] L. P. Huelsman, *Active and Passive Analog Filter Design: An Introduction*, McGraw-Hill, New York, 1993.

[272] W. J. Thompson, "Chebyshev Polynomials: After the Spelling the Rest Is Easy," *Comput. Phys.*, **8**, 161 (1994).

Interpolation, Decimation, Oversampling, and Noise Shaping

[273] R. E. Crochiere and L. R. Rabiner, *Multirate Digital Signal Processing*, Prentice Hall, Englewood Cliffs, NJ, 1983.

[274] R. E. Crochiere and L. R. Rabiner, "Multirate Processing of Digital Signals" in Ref. [13].

[275] P. P. Vaidyanathan, *Multirate Systems and Filter Banks*, Prentice Hall, Englewood-Cliffs, NJ, 1993.

[276] J. C. Candy and G. C. Temes, eds., *Oversampling Delta-Sigma Data Converters*, IEEE Press, Piscataway, NJ, 1992.

[277] J. C. Candy and G. C. Temes, "Oversampling Methods for A/D and D/A Conversion," in Ref. [276].

[278] R. M. Gray, "Oversampled Sigma-Delta Modulation," *IEEE Trans. Commun.*, **COM-35**, 481 (1987). Reprinted in Ref. [276], p. 73.

[279] D. Goedhart, et al., "Digital-to-Analog Conversion in Playing a Compact Disc," *Philips Tech. Rev.*, **40**, 174-179, (1982).

[280] M. W. Hauser, "Principles of Oversampling A/D Conversion," *J. Audio Eng. Soc.*, **39**, 3-26, (1991).

[281] P. J. A. Naus, et al., "A CMOS Stereo 16-bit D/A Converter for Digital Audio," *IEEE J. Solid-State Circuits*, **SC-22**, 390-395, (1987). Reprinted in Ref. [276].

[282] *SONY Semiconductor IC Data Book, A/D, D/A Converters*, 1989 and 1990.

[283] R. Legadec and H. O. Kunz, "A Universal, Digital Sampling Frequency Converter for Digital Audio," *Proc. 1981 IEEE Int. Conf. Acoust., Speech, Signal Process.*, ICASSP-81, Atlanta, GA, p. 595.

[284] T. A. Ramstad, "Digital Methods for Conversion Between Arbitrary Sampling Frequencies," *IEEE Trans. Acoust., Speech, Signal Process.*, **ASSP-32**, 577 (1984).

[285] J. O. Smith and P. Gossett, "A Flexible Sampling-Rate Conversion Method," *Proc. 1984 IEEE Int. Conf. Acoust., Speech, Signal Process.*, ICASSP-84, San Diego, CA, p. 19.4.1. C code is available via `ftp` from `ftp.netcom.com` in directory `pub/thinman/resample.*`.

[286] R. Adams and T. Kwan, "Theory and VLSI Architectures for Asynchronous Sample-Rate Converters," *J. Audio Eng. Soc.*, **41**, 539 (1993).

[287] "SamplePort Stereo Asynchronous Sample Rate Converters, AD1890/AD1891," Data Sheet, Analog Devices, Norwood, MA, 1994.

[288] R. Adams, "Asynchronous Sample-Rate Converters," *Analog Dialogue*, **28**, no. 1, 9 (1994), Analog Devices, Inc., Norwood, MA.

[289] K. Uchimura, et al., "Oversampling A-to-D and D-to-A Converters with Multistage Noise Shaping Modulators," *IEEE Trans. Acoust., Speech, Signal Process.*, **ASSP-36**, 259 (1988). Reprinted in Ref. [276].

[290] D. B. Ribner, "A Comparison of Modulator Networks for High-Order Oversampled ΣΔ Analog-to-Digital Converters," *IEEE Trans. Circuits Syst.*, **CAS-38**, 145 (1991).

[291] L. A. Williams, III and B. A. Wooley, "Third-Order Cascaded Sigma-Delta Modulators," *IEEE Trans. Circuits Syst.*, **CAS-38**, 489 (1991).

[292] R. W. Adams, et al., "Theory and Practical Implementation of a Fifth-Order Sigma-Delta A/D Converter," *J. Audio Eng. Soc.*, **39**, 515 (1991).

[293] S. Harris, "How to Achieve Optimum Performance from Delta-Sigma A/D and D/A Converters," *J. Audio Eng. Soc.*, **41**, 782 (1993).

[294] O. Josefsson, "Using Sigma-Delta Converters—Part 1," *Analog Dialogue*, **28**, no. 1, 26 (1994), Analog Devices, Inc., Norwood, MA, and "Part 2," ibid., no. 2, 24 (1994).

[295] R. N. J. Veldhuis, M. Breeuwer, and R. G. Van Der Waal, "Subband Coding of Digital Audio Signals," *Philips J. Res.*, **44**, 329 (1989).

[296] R. N. J. Veldhuis, "Bit Rates in Audio Source Coding," *IEEE J. Select. Areas Commun.*, **10**, 86 (1992).

[297] G. C. P. Lockoff, "DCC—Digital Compact Cassette," *IEEE Trans. Consum. Electr.*, **37**, 702 (1991).

[298] G. C. P. Lockoff, "Precision Adaptive Subband Coding (PASC) for the Digital Compact Cassette (DCC)," *IEEE Trans. Consum. Electr.*, **38**, 784 (1992).

[299] A. Hoogendoorn, "Digital Compact Cassette," *Proc. IEEE*, **82**, 1479 (1994).

[300] T. Yoshida, "The Rewritable MiniDisc System," *Proc. IEEE*, **82**, 1492 (1994).

Random Number Generators

[301] D. E. Knuth, *The Art of Computer Programming*, vol. 2, 2nd ed., Addison-Wesley, Reading, MA, 1981.

[302] P. Bratley, B. L. Fox, and L. Schrage, *A Guide to Simulation*, Springer-Verlag, New York, 1983.

[303] W. H. Press, B. P. Flannery, S. A. Teukolsky, and W. T. Vetterling, *Numerical Recipes in C*, 2nd ed., Cambridge Univ. Press, New York, 1992.

[304] S. K. Park and K. W. Miller, "Random Number Generators: Good Ones Are Hard to Find," *Comm. ACM*, **31**, 1192 (1988).

[305] B. D. Ripley, "Computer Generation of Random Variables: A Tutorial," *Int. Stat. Rev.*, **51**, 301 (1983).

[306] D. J. Best, "Some Easily Programmed Pseudo-Random Normal Generators," *Austr. Comput. J.*, **11**, 60 (1979).

[307] C. A. Whitney, "Generating and Testing Pseudorandom Numbers," *BYTE*, (October 1984), p. 128.

[308] S. Kirkpatrick and E. Stoll, "A Very Fast Shift-Register Sequence Random Number Generator," *J. Computational Phys.*, **40**, 517 (1981).

[309] G. Marsaglia, "A Current View of Random Number Generators," in *Proceedings, Computer Science and Statistics, 16th Symposium on the Interface*, L. Billard, ed., Elsevier, North-Holland, 1985.

[310] G. Marsaglia, "Toward a Universal Random Number Generator," *Statist. & Prob. Lett.*, **8**, 35 (1990).

[311] G. Marsaglia and A. Zaman, "A New Class of Random Number Generators," *Annals Appl. Prob.*, **1**, 462 (1991).

[312] G. Marsaglia and A. Zaman, "Some Portable Very-Long Random Number Generators," *Comput. Phys.*, **8**, 117 (1994).

[313] P. Lewis, A. Goodman, and J. Miller, "A Pseudo-Random Number Generator for the System/360," *IBM Syst. J.*, **8**, 136 (1969).

[314] L. Schrage, "A More Portable Fortran Random Number Generator," *ACM Trans. Math. Soft.*, **5**, 132 (1979).

[315] C. Bays and S. D. Durham, "Improving a Poor Random Number Generator," *ACM Trans. Math. Soft.*, **2**, 59 (1976).

[316] G. Marsaglia and T. A. Bray, "On-line Random Number Generators and their Use in Combinations," *Comm. ACM*, **11**, 757 (1968).

[317] B. A. Wichmann and I. D. Hill, "An Efficient and Portable Pseudo-Random Number Generator," *Appl. Stat.*, **31**, 188 (1982).

[318] B. A. Wichmann and D. Hill, "Building a Random Number Generator," *BYTE*, March 1987, p. 127.

[319] D. Lorrain, "A Panoply of Stochastic Cannons," in Ref. [97], p. 351.

[320] Suggested by Dr. N. Moayeri, private communication, 1993.

[321] W. A. Gardner, *Introduction to Random Processes*, 2nd ed., McGraw-Hill, New York, 1989.

1/f **Noise**

[322] D. A. Bell, *Noise and the Solid State*, Halsted Press, (Wiley), New York, 1985.

[323] M. Gardner, "White and Brown Music, Fractal Curves and One-Over-f Fluctuations," *Sci. Amer.*, **288**, 16 (1978).

[324] M. S. Keshner, "1/*f* Noise," *Proc. IEEE*, **70**, 212 (1982).

[325] B. B. Mandelbrot, *The Fractal Geometry of Nature*, W. H. Freeman, New York, 1983.

[326] W. H. Press, "Flicker Noise in Astronomy and Elsewhere," *Comments Astrophys.*, **7**, no. 4, 103 (1978).

[327] M. Schroeder, *Fractals, Chaos, Power Laws*, W. H. Freeman, New York, 1991.

[328] R. F. Voss and J. Clark, "1/*f* Noise in Music and Speech," *Nature*, **258**, no. 5533, 317 (1975).

[329] R. F. Voss and J. Clark, "1/*f* Noise in Music: Music from 1/*f* Noise," *J. Acoust. Soc. Amer.*, **63**, 258 (1978).

[330] B. J. West and M. Shlesinger, "The Noise in Natural Phenomena," *Amer. Scientist*, **78**, 40, Jan-Feb, (1990).

[331] A. van der Ziel, *Noise in Solid State Devices and Circuits*, Wiley, New York, 1986.

[332] M. Reichbach, "Modeling 1/*f* Noise by DSP," Graduate Special Problems Report, ECE Department, Rutgers University, Fall 1993.

[333] R. Bristow-Johnson, reported in `comp.dsp` newsgroup, Jan. 1995, and private communication.

Index